Achieving sustainable cultivation of bananas

Volume 3: Diseases and pests

It is widely recognised that agriculture is a significant contributor to global warming and climate change. Agriculture needs to reduce its environmental impact and adapt to current climate change whilst still feeding a growing population, i.e. become more 'climate-smart'. Burleigh Dodds Science Publishing is playing its part in achieving this by bringing together key research on making the production of the world's most important crops and livestock products more sustainable.

Based on extensive research, our publications specifically target the challenge of climate-smart agriculture. In this way we are using 'smart publishing' to help achieve climate-smart agriculture.

Burleigh Dodds Science Publishing is an independent and innovative publisher delivering high quality customer-focused agricultural science content in both print and online formats for the academic and research communities. Our aim is to build a foundation of knowledge on which researchers can build to meet the challenge of climate-smart agriculture.

For more information about Burleigh Dodds Science Publishing simply call us on +44 (0) 1223 839365, email info@bdspublishing.com or alternatively please visit our website at www.bdspublishing.com.

Related titles:

Achieving sustainable cultivation of bananas Volume 1: Cultivation techniques
Print (ISBN 978-1-78676-156-9); Online (ISBN 978-1-78676-158-3, 978-1-78676-159-0)

Achieving sustainable cultivation of bananas Volume 2: Germplasm and genetic improvement
Print (ISBN 978-1-78676-344-0); Online (ISBN 978-1-78676-346-4, 978-1-78676-347-1)

Achieving sustainable cultivation of tropical fruits
Print (ISBN 978-1-78676-284-9); Online (ISBN 978-1-78676-286-3, 978-1-78676-287-0)

Achieving sustainable cultivation of mangoes
Print (ISBN 978-1-78676-132-3); Online (ISBN 978-1-78676-134-7, 978-1-78676-135-4)

Chapters are available individually from our online bookshop:
https://shop.bdspublishing.com

climate•SMART•publishing
IN AGRICULTURAL SCIENCE

BURLEIGH DODDS SERIES IN AGRICULTURAL SCIENCE
NUMBER 123

Achieving sustainable cultivation of bananas

Volume 3: Diseases and pests

Edited by Professor André Drenth, University of Queensland, Australia and Professor Gert H. J. Kema, Wageningen University, The Netherlands

burleigh dodds
SCIENCE PUBLISHING

Published by Burleigh Dodds Science Publishing Limited
82 High Street, Sawston, Cambridge CB22 3HJ, UK
www.bdspublishing.com

Burleigh Dodds Science Publishing, 1518 Walnut Street, Suite 900, Philadelphia, PA 19102-3406, USA

First published 2024 by Burleigh Dodds Science Publishing Limited

Library of Congress Control Number: 2023940912

British Library Cataloguing in Publication Data
A catalogue record for this book is available from the British Library

ISBN 978-1-78676-981-7 (print)
ISBN 978-1-78676-984-8 (PDF)
ISBN 978-1-78676-983-1 (ePub)
ISSN 2059-6936 (print)
ISSN 2059-6944 (online)

DOI: 10.19103/AS.2022.0108

Typeset by Deanta Global Publishing Services, Dublin, Ireland

Contents

Series list

Title	Series number
Achieving sustainable cultivation of maize - Vol 1 From improved varieties to local applications Edited by: Dr Dave Watson, CGIAR Maize Research Program Manager, CIMMYT, Mexico	001
Achieving sustainable cultivation of maize - Vol 2 Cultivation techniques, pest and disease control Edited by: Dr Dave Watson, CGIAR Maize Research Program Manager, CIMMYT, Mexico	002
Achieving sustainable cultivation of rice - Vol 1 Breeding for higher yield and quality Edited by: Prof. Takuji Sasaki, Tokyo University of Agriculture, Japan	003
Achieving sustainable cultivation of rice - Vol 2 Cultivation, pest and disease management Edited by: Prof. Takuji Sasaki, Tokyo University of Agriculture, Japan	004
Achieving sustainable cultivation of wheat - Vol 1 Breeding, quality traits, pests and diseases Edited by: Prof. Peter Langridge, The University of Adelaide, Australia	005
Achieving sustainable cultivation of wheat - Vol 2 Cultivation techniques Edited by: Prof. Peter Langridge, The University of Adelaide, Australia	006
Achieving sustainable cultivation of tomatoes Edited by: Dr Autar Mattoo, USDA-ARS, USA and Prof. Avtar Handa, Purdue University, USA	007
Achieving sustainable production of milk - Vol 1 Milk composition, genetics and breeding Edited by: Dr Nico van Belzen, International Dairy Federation (IDF), Belgium	008
Achieving sustainable production of milk - Vol 2 Safety, quality and sustainability Edited by: Dr Nico van Belzen, International Dairy Federation (IDF), Belgium	009
Achieving sustainable production of milk - Vol 3 Dairy herd management and welfare Edited by: Prof. John Webster, University of Bristol, UK	010
Ensuring safety and quality in the production of beef - Vol 1 Safety Edited by: Prof. Gary Acuff, Texas A&M University, USA and Prof. James Dickson, Iowa State University, USA	011
Ensuring safety and quality in the production of beef - Vol 2 Quality Edited by: Prof. Michael Dikeman, Kansas State University, USA	012
Achieving sustainable production of poultry meat - Vol 1 Safety, quality and sustainability Edited by: Prof. Steven C. Ricke, University of Arkansas, USA	013
Achieving sustainable production of poultry meat - Vol 2 Breeding and nutrition Edited by: Prof. Todd Applegate, University of Georgia, USA	014

Introduction

Bananas are a key food source for millions of people and their production supports the livelihood of farmers across the tropics. However, recent research has identified the vulnerability of commonly grown banana varieties to major pests and diseases, such as the threat of Fusarium wilt of banana (FWB, previously known as Panama disease) caused by Tropical Race 4 and black leaf streak disease (BLSD or Black Sigatoka). With the global spread of these and a range of other major diseases, the production of bananas has come under serious strain. It is evident that more research is required to understand the biology of the diseases and pests that impact the production of bananas and that more effective and sustainable management strategies are required to reduce the impact.

This volume provides comprehensive and novel insights into the major diseases and pests affecting global banana production. Chapters are organised into six parts: Part 1 chapters focus on fungal diseases of bananas and chapters in Part 2 cover bacterial and phytoplasma diseases. In Part 3, viral diseases are discussed followed by nematode pests in Part 4. Part 5 chapters focus on insect pests affecting bananas and finally, in Part 6 fruit diseases are covered, focusing on in-field and postharvest management.

Part 1 Fungal diseases

Part 1 of the book opens with a chapter that reviews the Sigatoka disease complex. Chapter 1 first highlights the three major leaf spot diseases of banana that form the Sigatoka disease complex: Sigatoka disease (or Yellow Sigatoka) caused by *Pseudocercospora musae*, black leaf streak disease caused by *P. fijiensis* and *Eumusae* leaf spot disease caused by P. eumusae. The chapter gives an overview of the symptoms of the diseases, their taxonomy, diagnosis, population history, and geographical distribution. The chapter also provides some new insights into the disease cycle, epidemiology, and the host-pathogen interactions which are required to underpin the development of more sustainable methods to control these diseases through field management and the identification and deployment of effective resistance in banana germplasm.

The next chapter draws attention to controlling BLSD, using Costa Rica as a case study. It provides an overview of the fungicides currently being used for BLSD management and their application programs. The chapter outlines the issues related to reduced sensitivity to fungicides in the pathogen population and provides an overview of the trends in reducing this risk. This chapter provides a compelling case that alternative and sustainable disease control methods are urgently required in large-scale global banana production systems.

Chapter 3 discusses freckle disease of banana, which is of economic importance in some parts of the world. The chapter first reviews the disease symptoms and highlights how the three different pathogens species causing freckle disease can infect the leaf at any stage of plant growth while they also can infect the fruit. It then moves on to review pathogen characteristics and diagnosis which is followed by an overview of their disease cycles, taxonomy, infection biology and epidemiology. The chapter also discusses the disease's impact on bananas once it is infected as well as the various options currently available to manage the disease.

Chapter 4 provides an overview of the impact, epidemiology and management of FWB caused by Race 1 and Race 2 in banana. The chapter begins by providing an overview of FWB and its symptoms, then moves on to focus on the pathogen characteristics, specifically drawing attention to Race 1 and Race 2 whilst also briefly mentioning how Race 4 affects banana cultivars. The chapter also looks at the disease cycle and epidemiology of FWB, highlighting the factors that can affect FWB development in the plant. A section on resistance to FWB in different banana germplasm and varieties released from breeding programs is included. The chapter concludes with a review of the management and control of FWB.

Expanding on topics previously discussed in Chapter 4, Chapter 5 draws attention to FWB caused by Tropical Race 4. The chapter examines the origin, spread, and biology of Tropical Race 4 to provide an overview of the current prevalence and impact of the disease in bananas around the world. It then goes on to review the present situation regarding Tropical Race 4, focusing on ways to limit the spread of the disease through improving detection methods, understanding the disease's dissemination, phylogeography, and epidemiology to improve control and containment methods. The chapter also critically reviews how disease control and breeding can be improved to better manage Tropical Race 4 in the future.

To summarise topics previously discussed in Chapters 4 and 5, Chapter 6 focuses on sustainable management of FWB. The chapter outlines the development of the disease, the mode of infection of the pathogens, and their epidemiology. An account is given of the various disease control methods that were tried and practiced over the last century and the underlying biological factors are outlined to explain why most of these management options have been ineffective to control FWB in a monoculture plantation setting. At the same time, some of the successful options to manage FWB are explored, together with the challenges around implementing long-term effective disease control measures to combat this disease.

Chapter 7 closes Part 1 of the book by examining several minor fungal diseases of banana. Although they are not considered serious constraints in the banana industry, these diseases can still cause significant economic loss in countries with particular climatic conditions. The chapter focuses on the following minor diseases: Cordana (*Cordana musae*), Deightoniella leaf spot (*Deightoniella torulosa*), Speckle (*Mycosphaerella musae*), Anthracnose (*Colletotrichum musae*), Cigar-end rot (*Verticillium theobromae*), Banana blast disease (*Pyricularia angulata*), Freckle disease (*Guignardia musae*) and Black cross leaf spot (*Phyllachora musicola*). For each disease, the chapter provides an overview of the occurrence, causal agent, key disease symptoms, host susceptibility and discusses ways in which each disease can be controlled.

Part 2 Bacterial and phytoplasma diseases

Chapter 8 is the first chapter of Part 2 and discusses Moko disease, a serious bacterial wilt disease affecting bananas and plantains. The chapter starts by describing the main external and internal symptoms of the disease, and then reviews advances in understanding the taxonomy and classification of disease-causing strains, before looking at host range, infection mechanisms, and impact. Finally, the chapter reviews management strategies such as exclusion, disease identification and tracking, eradication techniques, the role of cultural practices in disease transmission, and prospects for developing disease-resistant cultivars.

Chapter 9 focuses on Blood disease, a bacterial wilt caused by *Ralstonia syzygii* subsp. *celebesensis*. The chapter begins by highlighting the disease's symptoms and how their expression can differ depending on the time and mode of infection. It goes on to describe pathogen characteristics and diagnosis and provides details to help distinguish the difference between Moko and Blood disease, which are both major bacterial wilt diseases of banana. A section on the disease cycle and epidemiology of Blood disease is also included, focusing specifically on inoculum sources, transmission, survival outside the host and host range. The occurrence and impact of Blood disease are also discussed, followed by an overview of how Blood disease can be managed.

Chapter 10 draws attention to Xanthomonas wilt (BXW) of banana, a major banana disease affecting small-holder farms in East and Central Africa. The chapter presents an overview of the epidemiology of the disease followed by a discussion on BXW control and prevention measures. It addresses management practices that have shown variable adoption successes and presents innovations recently introduced (including Single Diseased Stem Removal and improved tool sterilisation options) or underway (including the development of resistant cultivars, and risk mitigation through early warning systems). The chapter then discusses lessons learned from problems that have arisen during phases of implementation of management, the ongoing constraints that local farmers face, and the shift towards sustainable management through integrated participatory extension services.

Chapter 11 focuses on bacterial soft rots of bananas, and although considered minor diseases, they cause losses of individual plants, yield reduction, and raise the cost of production. The chapter first describes the different forms of soft rots and their symptoms, such as head rot or rhizome soft rot, pseudostem wet rot, and rhizome rot. It then highlights the different pathogens causing the diseases and how they can be diagnosed, their epidemiology and the disease cycles of various soft rots, and how they can occur and impact the infected banana plant. The chapter also reviews how the different bacterial soft rot diseases can be managed.

Chapter 12 is the final chapter of Part 2 and looks at phytoplasma diseases of banana plants. Phytoplasmas have been recorded from banana plants in different parts of the world, but they have presently a limited distribution and are therefore still considered pathogens of minor importance in most banana-producing countries. The chapter describes the symptoms, main characteristics, and taxonomy of phytoplasmas, the disease cycle, epidemiology, and available diagnostic methods. The latest information on diversity, occurrence, impact, and management is provided for banana phytoplasmas occurring in Papua New Guinea and the Solomon Islands, known collectively as banana wilt-associated phytoplasmas. The chapter also highlights hurdles encountered in disease management and suggests possible strategies to prevent further losses.

Part 3 Viral diseases

Part 3 opens with Chapter 13 that describes banana bunchy top disease, globally the most important virus disease of banana. It reviews the current knowledge on banana bunchy top disease, provides an overview of the historical aspects of the disease and the early studies, outlines the symptoms of the disease, and biological characteristics of the banana bunchy top virus (BBTV) related to host range, transmission and spread. This is followed by an overview of control strategies which involve detection of BBTV, its epidemiology and resistance in the host, and how this can be deployed in management strategies.

Chapter 14 discusses banana streak disease, which is considered the most widely distributed viral disease of bananas and plantains in the world. The chapter first describes common characteristic symptoms of banana streak disease, and then highlights the pathogen characteristics and the diagnostic methods that can be used to identify its presence in banana plants. The chapter then provides an overview of the disease cycle and the pathogens' epidemiology, as well as how the various viruses causing banana streak disease can occur and impact the banana industry. A section on current disease management strategies for banana streak disease is also included.

Part 3 closes with Chapter 15 which focuses on other viral pathogens of banana. The chapter draws attention to several minor viral pathogens, specifically cucumber mosaic virus, banana bract mosaic virus, sugarcane mosaic virus, and banana mild mosaic virus. For these viral pathogens, the next sections focus on their occurrence and impact, taxonomy, disease symptoms, diagnostic methods, epidemiology, and control. The chapter also describes why banana virus X and banana die-back virus are considered minor viral pathogens and why some viral pathogens - such as banana bract mosaic virus and the abacá strain of sugarcane mosaic virus - have the potential to cause much more serious problems if they are allowed to spread unchecked around the world.

Part 4 Nematode pests

Part 4 opens with Chapter 16 addressing some of the key nematodes that affect banana production. The chapter focuses first on some of the most damaging plant-parasitic nematodes in banana, such as the burrowing nematode, root lesion nematodes, and spiral nematodes. Each type has a dedicated section that discusses the life cycle, symptoms, spread and distribution, and current control strategies. The chapter then provides a more in-depth overview of some of the control and management strategies that can be used, focusing specifically on alternate cropping systems, agro-ecosystems biodiversity, chemical control, resistance breeding, and genetic modification.

Expanding on topics previously touched upon in Chapter 16, Chapter 17 discusses ways to sustainably manage nematodes in banana production. The chapter begins by providing an overview of plant-parasitic nematode management, and then focuses on the importance of cultural control for nematode pest management. Sections on biological and chemical control are provided, and the chapter then moves on to review the potential of genetic resistance as an effective control method to limit plant-parasitic nematodes. Integrated pest management systems for plant-parasitic nematodes are described, followed by examples of integrated nematode control management programs in different countries such as Australia, Costa Rica, the French West Indies, and Ivory Coast. The challenges of managing nematodes in subtropical banana production systems are also discussed.

Part 5 Insect pests

Part 5 opens with Chapter 18 which discusses the impact of the banana weevil borer (Banana black weevil), *Cosmopolites sordidus*. The chapter provides an overview of the spread of the pest, its life cycle, and its impact on plant health and yields and then discusses control measures. These include chemical control, cultural/agronomic practices, the use of pheromones in monitoring and mass trapping, genetic resistance, and the use of predator species, such as nematodes and entomopathogens. The chapter also looks at the link between the occurrence of this pest and diseases such as FWB.

Chapter 19 focuses on banana Red Rust thrips. It begins by discussing pest nomenclature and classification of the Red Rust thrip, then describes the distribution and the economic impact it has on banana production. The chapter provides an overview of Red Rust thrip morphology and the stages of the insect's life cycle. Ways to detect pest infestations are discussed, followed by an overview of how symptoms of Red Rust thrips can be confused with 'Maturity Stain' and 'Maturity Bronzing', both of which are disorders that affect the cosmetic appearance of the banana fruit. Integrated pest management strategies are discussed, focusing specifically on biological and cultural control as well as combined physical and chemical control using bunch bags. A section on integrated pest management in organic banana production finalises the chapter.

Chapter 20 closes Part 5 of the book and focuses on the potential of more ecological-based approaches in banana bunch pest management. The chapter reviews management trends of banana bunch pests for export which cause cosmetic damage or trigger quarantine prohibitions in the context of changing regulatory and market requirements. The authors review six major insect pests of the banana bunch with a particular focus on understanding their life cycle as a basis for designing more ecologically-based pest management practices for red rust and flower thrips, aphids, scale insects, mealybugs and fruit scarring beetles. Sections include monitoring pests and beneficial organisms, creating environments that reduce pests and favour beneficial organisms, and direct control with insecticides. The authors propose five cornerstone practices including; early bagging with non-toxic repellents containing bags, field sanitation, measures to promote conservation biocontrol, minimal pesticide use, and increased hedge and cover plant diversity to ensure blemish-free bananas with a low level of rejects and reduced impact on human health and the environment.

Part 6 Fruit diseases

The concluding Part 6 of this book opens with Chapter 21 which outlines the management of bacterial and fungal diseases on banana fruit from a practical and real-life banana farmers' perspective and can be used as a guide to identifying and manage diseases occurring on banana fruits. Different options for control and prevention are outlined and discussed from a grower's perspective, experienced in dealing with the day-to-day diseases in bananas. The chapter focuses on diseases affecting fruit quality, market acceptance, and overall farm profitability, while other fruit diseases controlled by normal crop husbandry and of low significance are briefly discussed. Banana growers, from large plantation owners and managers to smallholder farmers throughout the world may benefit from the practical information in this chapter. Starting with the selection of soil type, water availability and use affecting disease pressure and hence, the ability to increase sustainability and adaptation to climate change, the chapter also discusses changes in plantation management to increase productivity and profitability. Insights into crop hygiene, plant and bunch care are discussed and aim to reduce the impact of diseases on the fruit. This chapter enables growers to reduce their reliance on chemical solutions.

Chapter 22 is the final chapter of this book and provides an overview of postharvest diseases of banana and their sustainable management. The chapter discusses crown rot and anthracnose in detail, with a more concise coverage of the minor diseases such as mokillo, fruit speckle, squirter, and ceratocystis fruit stem-end rot. For each disease, the chapter focuses on causal agents, the specific symptoms associated with infection, the length of the disease cycle and epidemiology. A range of options for control of postharvest diseases is discussed with an emphasis on sustainable control involving non-chemical treatments to secure food safety and environmental sustainability of banana production and consumption.

Acknowledgement

Cover image copyright:

- Fusarium wilt/Tropical Race 4 affecting Cavendish bananas in China, 2016, © Andre Drenth.

Part 1

Fungal diseases

Chapter 1

The Sigatoka leaf disease complex in banana

Jean Carlier, Luc De Lapeyre De Bellaire, Catherine Abadie, Marc Chillet and Dominique Carval, CIRAD, France; Robert Neil Gerard Miller, Universidade de Brasília, Brazil; Raman Thangavelu, National Centre for Banana Research – ICAR, India; and Josué Essoh Ngando, CARBAP, Cameroon

1 Introduction

There are three major leaf diseases, now called *Pseudocercospora* leaf spot diseases of banana, which form the Sigatoka leaf spot complex (review in Crous et al., 2021):

- Sigatoka disease (SD or yellow Sigatoka);
- Black leaf streak disease (BLSD or black Sigatoka); and
- Eumusae leaf spot disease (ELSD).

Sigatoka disease was first described from Indonesia (Zimmerman, 1902) but gained attention when it first caused an epidemic in banana plantations in the Sigatoka Valley of Fiji in 1912 and spread globally in the 1930s. This disease is caused by the fungus *Pseudocercospora musae* (*Pm*) (Zimm.) Deighton (previously *Mycosphaerella musicola* R. Leach ex J.L. Mulder). BLSD was first described in 1963 in the Sigatoka District of Fiji and is caused by the fungus *Pseudocercospora fijiensis* (*Pf*) (M. Morelet) Deighton (previously *Mycosphaerella fijiensis* M.Morelet) and spread globally in the 1960s. Eumusae leaf spot disease was described in 2000 and is caused by the fungus *Pseudocercospora*

http://dx.doi.org/10.19103/AS.2022.0108.01

eumusae (*Pe*) P. Crous & X. Mourichon (previously *Mycosphaerella eumusae* P. Crous & X. Mourichon). This disease is mostly prevalent in Asia and was initially confused with SD and BLSD.

These diseases, particularly BLSD, are considered a major constraint to banana cultivation and they occur across most of the global production zones of banana. They are mainly controlled by regular application of fungicides (see Chapter 2 by Chong and Kema in this volume). Due to their environmental, health and economic impacts, there is a need to develop sustainable disease management strategies.

2 Symptoms and impact

The three pathogens initially cause chlorosis and eventually severe necrosis as a result of intercellular colonization by ramifying hyphae in the foliar mesophyll tissue. Under conducive conditions, the first symptoms appear approximately 10–14 days after infection as barely visible small specks, usually on the top of the younger leaves. These specks develop into streaks parallel to leaf venation and eventually coalesce into necrotic grey-to-dark spots, with a yellow halo. These spots further coalesce resulting in collapse of the leaf where the grey-to-dark spots remain clearly visible with light-coloured centres and dark borders depending on the banana variety.

Focussing on the symptomatology in Cavendish varieties the main differences between the three diseases comprise the colour and location of the first streaks, the form of the necrotic spots and the colour of the halo in the spot stages (Fig. 1). Brun (1963) described five stages for SD from minute yellowish-green specks to large grey and sunken spots with well-defined yellow halos (Fig. 1). Fouré (1987) described six stages for BLSD from the initial specks on the lower leaf surface to coalesced necrotic spots with dry grey centres, usually surrounded by well-defined black borders and water-soaked or yellow halos (Fig. 1). The stages for ELSD start with faint, minute, yellowish-to-reddish brown specks on the lower surface of mostly older leaves, to depressed coalesced spots with greyish-white centres that remain clearly visible with light-coloured centres and dark borders after the leaves collapse and wither (Fig. 1).

Figure 1 illustrates that it is not easy to differentiate the individual diseases. In most cases, BLSD has overtaken SD in areas where both diseases occur. Trained staff in large-scale, commercial Cavendish plantations for export markets have helped in tracking the spread of BLSD in the Americas and across many countries in Africa (Fouré and Ganry, 2008; Ganry et al., 2008). The development of molecular diagnostics has helped to identify the three causal agents (Henderson et al., 2006; Arzanlou et al., 2007; Devi and Thangavelu, 2018). BLSD has a broader host range and also affects plantains in Latin America

Figure 1 Description of symptoms of BLSD, SD and ELSD. First streaks (stage 2) are respectively reddish/brown and visible on the abaxial (lower) leaf surface for BLSD and ELSD and yellow on the adaxial leaf surface for SD. The spot symptoms (stage 4) are elliptical and respectively black for BLSD and ELSD and brown for SD. The necrotic symptoms (stage 6 for BLSD and stage 5 for SD and ELSD) are depressed with a greyish white centre coalescing for BLSD and ELSD or isolated for SD, surrounded by a yellow halo except for ELSD.

(Stover, 1983) and Africa (Fouré, 1982) as well as East African cooking bananas (Tushemereirwe and Waller, 1993; Kimunye et al., 2020).

SD has had a major direct (yield and fruit size) and indirect (early ripening) impact on banana cropping systems (Wardlaw, 1939; Brun, 1948; Calpouzos, 1955; Wardlaw, 1961; Meredith, 1970; Stover, 1972; Ramsey et al., 1990). Control of SD requires regular fungicide applications (Stover, 1980). Since defoliation due to BLSD is faster than for SD, more frequent fungicide applications are required which have quickly resulted in the development of fungicide resistance (see Chapter 2 by Chong and Kema in this volume; Marin et al., 2003; de Lapeyre de Bellaire et al., 2010; Chong et al., 2021). For BLSD in plantains, yield losses of up to 30% have been reported in Africa (Mobambo et al., 1993). Since ELSD does not occur in the main banana exporting countries, where disease surveillance is strongest, there are limited data on its impact on banana yield and fruit quality.

Leaf necrosis caused by the three foliar banana diseases reduces photosynthetic capacity (Hidalgo et al., 2016), which affects fruit filling and fruit size (Ramsey et al, 1990). There are two critical periods for fruit size and for fruit filling determined, respectively, by (i) the time from flowering to +350 degree days (dd; which is the physiological age calculated as the sum of daily average temperatures above 14°C) when fruit size is determined by cell divisions; and (ii) from +350 dd to harvest when the rate of fruit filling is dependent on the source-sink ratio (foliar area/bunch size) (Jullien et al., 2001). Typically, adequate fruit filling requires five to seven effective leaves from flowering to harvest (Ramsey et al., 1990; Robinson et al., 1992; Daniells et al., 1994; Vargas et al., 2009; Lassois et al., 2010). However, banana is a semi-perennial plant with successor suckers mobilizing assimilates during the fruit-filling phase (Eckstein et al., 1995). Hence, a reduced canopy may also compromise the yield of the following cycle (Dens et al., 2008), underscoring that disease severity and yield evaluations should be monitored across several crop cycles.

Post-harvest damage is the most significant effect of the three foliar diseases. The time between harvest and the onset of ripening is called green life, which is a key quality trait of bananas because they cannot be stored for long periods (Thompson, 1998). Therefore, fruits destined for export need to be harvested green, shipped and then artificially ripened in dedicated ripening houses (Popma, 2018). Green life is strongly related to the fruit maturity at harvest, which is determined by the sum of daily average temperatures above 14°C (in degree.days) from flowering to harvest (Jullien et al., 2008). Recent research provides a comprehensive relationship between foliar diseases and yield and green life in Cavendish concerning SD under tropical conditions in Guadeloupe, and BLSD under tropical conditions in Cameroon as well as under subtropical conditions in Brazil, particularly since harvest was always at 900 dd, unlike former studies, and different disease severities were always taken into account.

The SD (Guadeloupe) and BLSD (Cameroon) severity at flowering was similar and varied between <5% and >40%, whereas the average severity from flowering to harvest could reach 57% in Guadeloupe or even 80% in Cameroon. This significantly reduced fruit weight in proportion to disease severity reaching 20% yield loss in the most severely affected plants. In Guadeloupe, due to SD, green life was significantly lower (11 days at 20°C, which is ~40%) in severely affected farms (Chillet et al., 2008). The relationship between the rate of disease development – measured as area under the disease progress curve (AUDPC) – and green life was linear, showing a proportional effect of the necrotic surface area per plant from flowering to harvest (Fig. 2, Castelan et al., 2013). In the case of BLSD in Cameroon, a similar linear relationship between disease severity and green life was shown and green life was also significantly lower (40% reduction), but outliers were also observed for the highest disease severity levels (Fig. 3). This is very different from SD and likely linked with severe defoliation at or even prior to harvest due to BLSD. The effect of SD and BLSD on green life can be even greater under subtropical conditions due to the longer timespan between flowering and harvest (Castelan et al., 2012). All these studies clearly show that

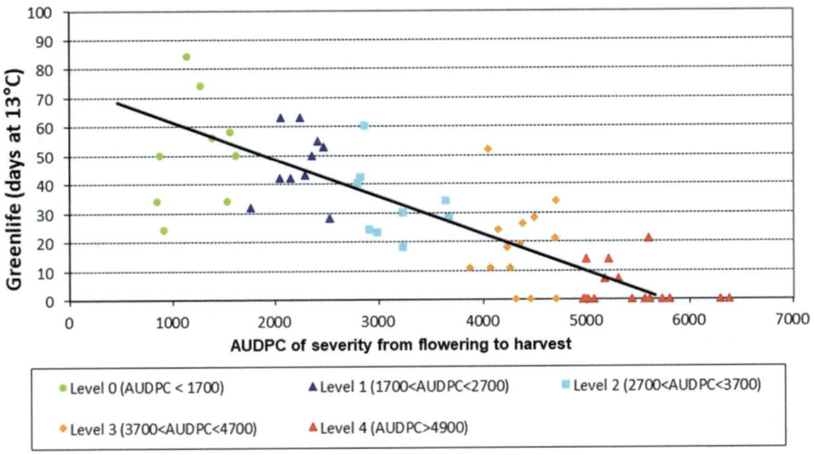

Figure 2 The relationship between green life measured at bananas harvest at 900 dd and the area under the disease progress curve of SD severity from flowering to harvest (from Castelan et al., 2013).

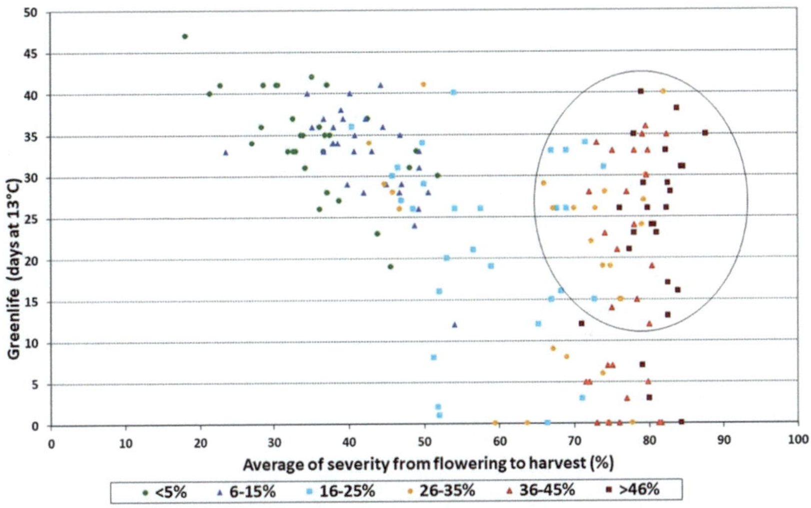

Figure 3 The relationship between average BLSD severity from flowering to harvest and green life measured at harvest at 900 dd. Presented data include BLSD severity <5->46% at harvest.

foliar necrosis directly reduces green life. There is some evidence that a foliar signal is involved. The intensity of this signal depends not only on the quantity of necrotic stages on banana plants as shown by the linear relationship between disease severity from flowering to harvest and green life but also on its duration as seen in subtropical conditions in Brazil. The foliar signal is also alleviated through mechanical or natural elimination of necrotic stages (as observed for high disease levels for BLSD). Regular mechanical removal of affected leaves between flowering and harvest reduces the inoculum production on affected leaves but is also an efficient tool to alleviate effects on green life (Chillet et al., 2013; Guillermet et al., 2018; Wielemaker, 2018). These results raise questions about the physiological mechanisms involved in leaf necrosis and the relationship between premature ripening and green life. *Pf* secretes secondary metabolites (such as toxins) that induce necrotic lesions (Hoss et al., 2000; El Hadrami et al., 2005) and are (in)directly involved in pathogenesis (Molina and Krausz, 1989; Harelimana et al., 1997). We hypothesize that these compounds might initiate a cascade of biochemical events that affect green life, while plant metabolites (including hormones) might also be involved in this relationship.

3 The pathogens

3.1 Taxonomy and diagnosis

Taxonomic studies resulted in more than 30 *Mycosphaerellaceae* species associated with banana, often coexisting on the same leaf or even in the same lesion (Crous et al., 2020). This interspecific diversity must be considered in the development of molecular diagnostics as recently demonstrated by van Westerhoven et al. (2023) who showed that existing diagnostics for *Pf* lacked sufficient specificity (Henderson et al., 2006; Arzanlou et al., 2007). The three *Pseudocercospora* spp. have an asexual reproduction cycle involving the production of splash-dispersed conidiospores and a sexual cycle producing airborne ascospores. The three species can be distinguished using classical mycology and guidelines for *in situ* observation, spore isolation and culturing were recently updated (Pérez-Vicente et al., 2021). Sequencing and identification of the mating-type alleles of the three species confirmed that all are heterothallic (Conde-Ferráez et al., 2007; Arango Isaza et al., 2016).

3.2 Genomes and phylogeny

Pf was the first species whose genome was sequenced (Arango Isaza et al., 2016). Pulsed-field gel electrophoresis and synteny studies suggested a core genome containing 11 or 12 chromosomes representing 87% of the genome and including more than 96% of the predicted genes. Several smaller scaffolds with lower GC content and gene density could be accessory chromosomes whose exact number is not yet known. The total genome size of *Pf* is 74 Mb, and among the largest genome across all related Dothideomycetes (Ohm et al., 2012; Haridas et al., 2020). The extension of the genome is largely due to long terminal repeats (LTR) retrotransposons which represent >50% of the genome (Arango Isaza et al., 2016). Transposons may play an important role in genome evolution and gene expression (Moller and Stukenbrock, 2017) and thus in the adaptive potential of *Pf*. Both *Pm* and *Pe* were subsequently sequenced and compared to the *Pf* reference genome (Chang et al., 2016), but the assemblies remain very fragmented (56 vs. 2600 scaffolds) with estimated genome sizes to be 54 Mb and 83 Mb, with significant repetitive genome content of 36% and 62% for *Pe* and *Pm*, respectively. The total number of genes in the three species is similar (10 548 for *Pm*, 11 064 for *Pe* and 13 107 for *Pf*), and approximately 6300 genes are shared, and about 26% (*Pf*), 16% (*Pm*) and 18% (*Pe*) of the proteome is species specific. Recently, the mitochondrial genomes of the species and other fungal species in the *Mycosphaerellaceae* have been compared (Arcila-Galvis et al., 2021). These are highly variable in size - 74.089 bp for *Pf*, 59.236 bp for *Pe* and 51.645 for *Pm* - and in gene order. Phylogenetic analyses suggest that the three species had a common ancestor, with a first split between *Pf* and an ancestor of the other two species and a subsequent divergence between *Pm* and *Pe* (Arzanlou et al., 2008; Chang et al., 2016; Arcila-Galvis et al., 2021).

3.3 Population history and geographical distribution

Studies of *Pf* and *Pm* populations using molecular markers have shown a high level of genetic diversity (Rivas et al., 2004; Halkett et al., 2010; Rieux et al., 2011, 2013; Robert et al., 2012; Carlier et al., 2021; Hayden et al., 2005; Brito et al., 2020). The geographical distribution of this diversity, combined with historical data on invasions, has enabled us to trace the population history and current distribution of the species. Compared to other continents, a higher level of genetic diversity has been identified in *Pm* and *Pf* in Southeast Asia (Carlier et al., 1996; Robert et al., 2012; Hayden et al., 2003, Table 1), which identifies it as the centre of origin of these species. The lower level of diversity detected in *Pf* and *Pm* outside Southeast Asia shows founder effects (Hayden et al., 2003; Carlier et al., 2021, Table 1) The species *Pe* is hypothesized as of Asian origin since it is mainly found in India and parts of Southeast Asia. Furthermore, phylogenetic studies suggest that the species of the Sigatoka leaf spot complex appeared after the genus *Musa* but before the colonization of Southeast Asia by banana

Table 1 Genetic diversity[a] using microsatellite markers[b] and restriction fragment length polymorphisms for *P. fijiensis*[c] and *P. musae*[d]

Origin	*P. fijiensis*[b]	*P. fijiensis*[c]	*P. musae*[d]
SEA	0.65	0.57	0.44
Oceania	0.33	0.28	0.35
Africa	0.22	0.25	0.29
America	0.35	0.40	0.22

[a]Nei (1978).
[b]Robert et al. (2012).
[c]Carlier et al. (1996).
[d]Hayden et al. (2003).

ancestors, and before the divergence of *Musa balbisiana* and *Musa acuminata*, the ancestors of most edible bananas (Rouard et al., 2018; Arcila-Galvis et al., 2021). This suggests that speciation in the complex may have occurred via host-tracking coevolution (Roy, 2001).

Pm was first described in 1902 in Indonesia (Zimmerman, 1902) and subsequent reports were confined to Asia and Oceania (Simmonds, 1962; Stover, 1962). Between 1939 and 1962, it also was widely reported in Africa (Stover, 1962, 1972; Meredith, 1970) as well as in different countries in Latin America and in the Caribbean between the mid-1930s and mid-1940s (Stover, 1962; Meredith, 1970). In the Middle East *Pm* was reported in Yemen in 1971 and Iran in 2014. In the Americas, it was reported in Ecuador in the 1950s (Wardlaw, 1961), Florida and Montserrat in 1955 and in Bolivia in 1968 (Meredith, 1970). Later, it was reported in Antigua and Barbuda in 1997 and finally in French Guyana in 2018. It has not been reported in the Caribbean islands Anguilla, Aruba, St Martin, Turks and Caicos, and the Virgin Islands. The population structure of *Pm* suggests an introduction from Southeast Asia to Australia, and a linked but independent introduction into Africa and Latin America (Hayden et al., 2003), the latter likely derived from infected material from Africa during the establishment of commercial plantations.

Pf was first reported in the Sigatoka Valley in Fiji (where SD was also first reported, Rhodes, 1964). Within 10 years, BLSD was reported on the majority of islands in Oceania and the majority of humid tropical Asian countries. Examination of old specimens and reassessment of older descriptions of leaf spot disease suggest an early presence of *Pf* in countries such as in Taiwan (~1927) (Stover, 1978). This suggests the presence of *Pf* prior to initial reports, which is consistent with its origin in Southeast Asia, ranging from Malaysia to Papua New Guinea where its genetic diversity is high (Carlier et al., 1996; Robert et al., 2012). It has been suggested that *Pf* may have originated in Papua New Guinea where the other two species do not occur (Stover, 1978). The colonization of Oceania probably occurred via infected germplasm during human migration (Kagy et al., 2016; Sardos et al., 2022).

In Latin America, *Pf* was first detected in Honduras in 1972. In less than 14 years, *Pf* was reported in all the main banana exporting countries of Central and South America (Mourichon and Fullerton, 1990). Further spread into the Caribbean occurred from 1990 to 2012 (Carlier et al., 2021). Since its first occurrence, the spread of *Pf* has been closely monitored (Mourichon and Fullerton, 1990), including its dissemination to Brazil (Da Silveira et al., 2016) and the Caribbean (Carlier et al., 2021). There are also reports from French Guyana (Anonymous, 2018) and Suriname (de Lapeyre de Bellaire et al., 2022, pers. comm.).

In Africa, *Pf* was first reported in Zambia in 1973 but was first confirmed in Gabon in 1978. In less than 12 years, *Pf* was reported in most countries of the tropical humid belt of Africa, including the key Cavendish exporting countries of Central and West Africa (such as Cameroon, Cote d'ivoire and Ghana) and in the countries of Central and Eastern Africa where edible bananas, such as AAB cooking bananas of the plantain subgroup and East African AAA cooking bananas of the Lujujira group, are important subsistence crops (Mourichon and Fullerton, 1990; Blomme et al., 2013). More recently, *Pf* was identified on Réunion Island (Rieux et al., 2019) and Mauritius (Vally et al, 2023), Senegal, and in Guinea where official reports are being prepared. The global population structure of *Pf* shows that there have been independent introductions from Southeast Asia into Oceania, Africa and the Americas (Robert et al., 2012). In Africa, studies support the existence of a single introduction event with a strong bottleneck effect (Gilabert et al, 2023). In contrast, an admixture likely originating from Southeast Asia and Oceania was detected in the Americas (Robert et al., 2015; Carlier et al., 2021).

The distribution of *Pe* is probably underestimated because symptoms are confused with other pathogens, but it has been reported across Asia in India, Sri Lanka, Thailand and Vietnam (Carlier et al., 2000). In Africa, it was first reported in Nigeria in 2000 from samples collected in 1989 (Carlier et al., 2000) as well as Mauritius and Réunion, but has not yet been reported in the Americas.

3.4 Disease cycle and epidemiology

Since SD was the first banana leaf spot disease affecting the banana industry, important studies were conducted from the 1930s to the end of the 1970s, mainly in the Caribbean, South America and Africa (Stahel, 1937; Calpouzos, 1955; Brun, 1963; Meredith and Lawrence, 1970; Meredith, 1970). The introduction of BLSD in the Americas and in Africa initiated additional research into its epidemiology (Fouré and Moreau, 1992; Gauhl, 1994). Although its epidemiology is often studied at plot level, a focus on the individual plant is also important. Bananas have a constant leaf emission rate, and a new leaf emerges every 5–9 days at the top of the canopy prior to flowering (Stahel, 1937; Champion, 1963). Every new leaf appears as a tightly rolled cylinder (also called heartleaf or cigar leaf) which unfurls in six stages (Brun, 1963). Once unfolding starts the heartleaf is exposed to airborne inoculum which starts the foliar infection that gradually evolves in secondary spread. Therefore, one can observe a disease gradient for *Pm*, *Pf* and *Pe* towards the lower leaves of the plant.

3.4.1 Spore deposition

Deposition of airborne inoculum is more important on the heartleaf due to its vertical position that intercepts spores during lateral movement. Leach (1946) and Brun (1963), counted the number of *Pm* ascospores on the heartleaf at different stages of the unfurling, showing that ascospore deposition started on the left side of the leaf and continued on the right side when the heartleaf unfurled. Such studies have not been conducted for *Pf* nor for *Pe*, but weekly disease monitoring for *Pf* in commercial plantations have revealed that symptoms are always more numerous on the left side of the leaves assuming it would be similar to *Pm*. Finally, spore deposition on fully developed leaves has been poorly documented and cannot be excluded for secondary foliar invasions, especially through the run-off of conidia in dew or rainwater (Meredith, 1970).

3.4.2 Germination and penetration

The presence of free water at the leaf surface is essential for germination of conidia for *Pm* (Calpouzos, 1955), but not for its ascospores that can germinate under 94% relative humidity (RH) (Brun, 1963). Neither *Pf* conidia nor ascospores require free water and they germinate at 92% RH and 85% RH, respectively (Brun, 1963; Jacome et al., 1991). It is unknown why *Pf* conidia germination does not depend on free water but it could be linked to their liberation process and their role in secondary infections. Ascospore germination is faster than conidial germination (Jacome et al., 1991), and starts for *Pm* (Brun, 1963) and *Pf* (Fouré and Pefoura, 1988) 2–3 h after deposition. The optimum temperature for germination of ascospore and conidia is for both species between 25°C and 28°C. Below 20°C, *Pm* germinates better and develops faster than *Pf* (MouliomPefouraet al., 1996; Porras and Perez-Vicente, 1998), providing a competitive advantage for *Pm* in cooler conditions. Spore germination is followed by epiphytic growth, which is longer for *Pm* than for *Pf* and can last for 6 days, exposing the germ tubes and hyphae to (a)biotic stress on the foliar surface before stomatal penetration. The three species exclusively penetrate the leaves through the stomates and primarily at the abaxial side (Stahel, 1937; Brun, 1963; Fouré and Pefoura, 1988; Balint-Kurti et al., 2001). The penetration efficiency is low, e.g. 1.6% for plantain in the Dominican Republic (Landry et al., 2017), independant of the host (Beveraggi et al., 1993) and likely decreases with foliar age (Stahel, 1937; Leach, 1946).

3.4.3 Incubation period

The time between penetration and the appearance of the first symptoms is the incubation period (IP) which is an important epidemiological parameter that can last 13–14 days for *Pm* and *Pf* under very conducive conditions in a susceptible host like Cavendish (Brun, 1963; Fouré and Moreau, 1992). Under tropical conditions, longer IPs have been observed for *Pm* (50–80 days: Klein, 1963; Brun, 1963) than for *Pf* (30–36 days: Fouré et al., 1984; Poeydebat et al., 2018). The IP may also differ for individual *Pf* strains (Romero and Sutton, 1997) but also varies due to host resistance, fungicide treatments (de Lapeyre de Bellaire et al., 2010) and inoculum concentration (Brun, 1963; Guzman et al., 2018). During regular weekly disease monitoring in commercial plantations, the youngest leaf-bearing streaks is a proxy of the IP.

3.4.4 Transition period

The transition period (TP) is the time lapse between the first symptom and the last necrotic stage and is determined by hygrometric conditions and host resistance. Some authors (Fouré,1985) have postulated that TP was a better descriptor of disease resistance than IP. For *Pm* and *Pf*, it has been shown that the TP is shorter when the leaf infection level is high since symptoms coalesce rapidly to form large necrotic blotches (Brun, 1964; Fouré, 1985). This is more frequently observed for BLSD and inoculum density has therefore been included as a parameter for the adaptation of disease forecasting systems to BLSD (Ganry et al., 2008; Fouré and Ganry, 2008). This parameter was not used initially for SD in the French West Indies where the forecasting system was first established (Ganry et Laville, 1983; Ganry et al., 2018). During regular weekly disease monitoring in commercial plantations, the youngest leaf-bearing spots (YLS) are proxy of the entire epidemiological cycle (IP+TP) as influenced both by climate and by fungicide treatments (de Lapeyre de Bellaire et al., 2010).

3.4.5 Asexual reproduction

The morphology of conidiophores and conidia are characteristic of each of the three species and asexual reproduction plays a very different role in their epidemiology. *Pf* conidia are produced on single conidiophores emerging from stomata in the first visible stages (small brown dots/streaks) of the disease (Meredith and Lawrence, 1969; Stover, 1980; Zapater et al., 2008). Conidial formation depends on climatic conditions and the number of conidiophores. mm^{-2} is higher during the rainy season than in the dry season (Fouré and Moreau, 1992). Stover (1980) estimated that a single lesion might produce up to 1200 conidia, while Ngando et al. (2015) counted up to 1110 conidia per lesion per day. For *Pm* and *Pe*, conidia are formed on sporodochia located at the adaxial side of the leaves during more advanced stages of disease development (Fig. 4; Stover, 1970; Zapater et al., 2008) then, unlike *Pf*, the conidial latency is longer than the incubation period for both species. A single lesion might produce 75–125 sporodochia, resulting in 30 000 conidia (Stover, 1980). Conidia may play a major role in *Pm* epidemics, particularly during rain-free periods when

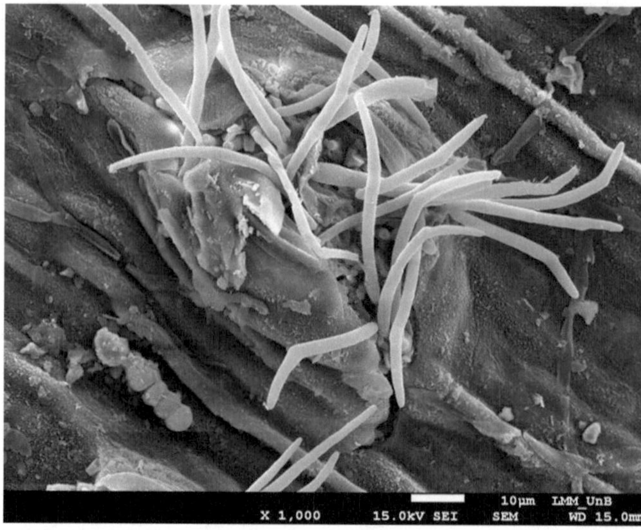

Figure 4 Scanning electron microscopy of adaxial leaf surface of field-infected *Musa acuminata* 'Grande Naine' (AAA, Cavendish subgroup) showing *Pm* conidiophores emerging through stomata (Bar = 10 µm). Image courtesy of Erica de Castro Costa.

dew is formed, but the role of *Pf* conidia in the epidemiology of BLSD may be underestimated.

3.4.6 Sexual reproduction

For the three leaf spot diseases, sexual reproduction occurs during the late necrotic stage when asci are formed in globose subepidermal and slightly erumpent dark ostiolate pseudothecia (Stover, 1964; Brun, 1963; Meredith and Lawrence, 1969; Crous and Mourichon, 2002), which are very similar for the three *Pseudocercospora* spp. (Zapater et al., 2008). Since all three species are heterothallic (Arzanlou et al., 2010), ascospore formation is highly dependent on lesion density and might not occur when lesion density is low (Brun, 1963; Fouré and Moreau, 1992). Therefore, ascospore formation is higher where the density of lesions is high, especially at the tip of banana leaves (Brun, 1963). The number of pseudothecia depends on necrotic leaf tissue and has been used to evaluate the influence of climatic conditions on ascospore production. Various studies have shown a higher production during wet periods (Stover, 1968; Fouré and Moreau, 1992). A relationship between necrotic surface area and available ascospores has also been proposed for *Pf* (Burt et al., 1999). Rieux et al. (2013) used this relationship, along with conidia counting, to estimate the total inoculum production on untreated banana plants and found that ascospore production was 2000-fold higher than conidial production, underscoring the importance of sexual reproduction in the epidemiology of BLSD in unmanaged fields where necrotic spots are abundant. Therefore, the removal of necrotic stages in banana plantations is an important prophylactic practice since necrotic leaves hanging on the pseudostem can continue to produce ascospores for 12–21 weeks (Gauhl, 1994).

3.4.7 Spore liberation and dispersal

Ascospores and conidia have very different liberation and dispersion patterns. For the three species, these processes are very similar for ascospores, but differ for conidia. Ascospores are violently discharged and become airborne when necrotic tissues are wetted, which has been widely documented for *Pm* and *Pf* (Meredith, 1970; Meredith et al., 1973; Burt et al., 1999) and is probably very similar for *Pe*. Ascospore discharge requires rain or abundant dew (Brun, 1963; Meredith et al, 1973) but heavy rains are not required since at least 0.5 mm of rain can induce ascospore release (Brun, 1963; Gauhl, 1994) within 20–60 min (Brun, 1963). In the absence of rain, ascospore release has a diurnal pattern (Meredith et al., 1973). Various ascospore trapping studies – with or without inoculum sources – showed evidence of wind dispersal (Meredith, 1962; Meredith et al, 1973; Burt and Rutter, 1997; Burt et al., 1998; Gauhl, 1994; Rutter et al, 1998). Rieux et al. (2014) showed that ascospores are dispersed on average over a distance of 200 m and could be recovered 1000 m from an inoculum source. The pathogen populations structure in the Caribbean suggests potential aerial dispersion over longer distances between islands. Conidia of *Pm* are mainly released in dew or rainwater (Calpouzos, 1954; Stover, 1970) and are not airborne (Meredith, 1962). Their dispersion occurs at short distances through the run-off of rain or dew (Stover and Fulton, 1966; Meredith, 1970). Conidia might also be splashed from sporulating lesions to heartleaves of neighbouring plants (Fulton, 1963). *Pf* conidia can also be released by rain or dew, but, unlike *Pm* conidia, they are also released by wind (Burt et al., 1997; Gauhl, 1994). Conidia are trapped inside highly BLSD-infested banana fields (Burt et al., 1997; Gauhl, 1994), and on average infect between 2.5 m and 12 m from a source (Rieux et al., 2013).

3.4.8 Comparative epidemiology

The primary inoculum for *Pe* are ascospores but other data are largely lacking and hence the epidemiological comparison is largely between *Pm* and *Pf* (Table 2). The main difference between *Pm* and *Pf* is the developmental rate. All epidemic steps from germination and epiphytic growth to the transition period are longer for *Pm* (Table 2), independent of weather conditions. Moreover, the density of infection is higher for *Pf*, which is probably the main competitive

Table 2 Differences between *P. musae* and *P. fijiensis* during the various epidemiological phases in Cavendish bananas under tropical conditions

Phase	*P. musae*	*P. fijiensis*
Epiphytic growth	3-6 days	2-3 days
Relative humidity (RH) requirements for germination and epiphytic growth	Conidia: FW Ascospores: FW - RH 94%	Conidia: FW - RH 92% Ascospores: FW - RH 98%
Incubation period (IP)	19-76 days	13-34 days
Transition period (TP)	8-60 days	10-32 days
Conidial latency	26-127 days	13-34 days
Ascospore latency	27-136 days	23-66 days

FW, free water.

reason why *Pf* progressively replaces *Pm* in the field. However, the differences in ascospore quantities cannot explain infection densities in commercial plots because necrotic foliage is eliminated. The conidia of *Pf* are formed sooner and are liberated and disseminated by wind. Their role in BLSD epidemics is probably more important than previously thought, especially in commercial plantations where their role might be also more important than the conidia of *Pm*.

4 Host resistance and host-pathogen interactions

4.1 Field evaluations and host responses

Plants have developed complex defense systems to resist pests and pathogens. Methods for determining the degree of resistance to *Pseudocercospora* species under controlled and field conditions have recently been published (Pérez-Vicente L. et al., 2021). During field evaluations, key parameters are measured monthly during the vegetative development of banana plants (from 3 to 5 months after planting), at flowering and harvest stages during two crop cycles on 3-10 plants per accession, depending on plot size and compared with reference varieties.

Up to eight parameters can be measured (see Section 2). Two parameters relate to plant development: leaf emission rate per week and the total number of leaves per plant. Two parameters relate to disease severity: the total necrotic surface per plant and the proportion of non-spotted leaves. Four parameters relate to symptom development: the youngest leaf with streaks (YLS Str), the youngest spotted leaf (YLS), the time for complete symptom development and the total symptom development rate. These parameters allow approximation of the incubation time (YLS Str) and the sexual latency (YLS). Because monitoring the rate of symptom development is time-consuming, field evaluations can be based on the YLS and YLS Str.

As a complement to measure the infection efficacy (number of lesions produced by spores) and exact incubation time, phenotyping can also be conducted under controlled conditions by inoculating 2-month-old plants or detached leaves from 2- to 4-month-old plants with mycelium or conidial suspensions (Pérez-Vicente L. et al., 2021), followed by quantifying lesion numbers and surface area. Usually, three responses against BLSD (Guzman et al., 2018) and SD (Jones, 2018) are recognized: qualitative hypersensitive resistance (HR) such as in some wild germplasm, partial resistance (PR) characterized by slow disease progress with a large variation from nearly HR to almost complete susceptibility, and susceptibility (S) characterized by rapid disease progress and abundant (a)sexual sporulation, such as in Cavendish cultivars. Resistance to ELSD is less well studied, but the various phases of disease development are very similar to SD and BLSD (Saravanakumar et al., 2016).

The International Musa Testing Program (IMTP) is an important and comprehensive attempt to evaluate banana germplasm under different conditions to SD, BLSD and ELSD (Orjeda et al., 1999). In total, 300 banana accessions have been evaluated for BLSD in Cameroon (Fouré et al., 2000) and were recently updated for eight genome groups and 37 species/subspecies and subgroups/clones (Guzman et al., 2018). SD and ELSD data in the IMTP are limited but are regularly updated (Daniells et al., 2016; Vakili, 1968; Jones, 2018; Sulliman et al., 2012; Saravanakumar et al., 2016; Uma et al., 2016). The various phenotypes of BLSD, SD and ELSD on the same banana species/subspecies and groups can differ with varying environmental conditions, hence comparisons require the use of a similar reference variety between different field evaluations as proposed by IMTP. Based on these data, a classification of the phenotypic resistance to BLSD, SD, and ELSD of five banana genomic groups (AA, AAA, AB, AAB, ABB groups) and various sub-groups is shown in Fig. 5. In most cases, the resistance or susceptibility of a genomic group against BLSD occurs also against SD and ELSD. Nevertheless, the response of this germplasm is more diverse for ELSD than for BLSD. It also shows that HR phenotypes are more frequent for ELSD, compared to SD and BLSD, but that PR and S phenotypes occur for all three diseases.

Phenotypic diversity has been observed among individual strains of *Pf* and *Pe* (Zandjanakou-Tachin et al., 2013), as well as the breakdown of resistance to *Pf* in genotypes that were previously considered resistant (Fullerton and Olsen, 1995). The cultivar Paka was considered highly resistant (high level of PR) but was severely infected after 8 years of exposure to *Pf*. Resistance can be expressed during the different phases of the disease cycle (Abadie et al, 2009), and erosion of resistance – such as in FHIA18 and FHIA21 hybrids after 5 years of large-scale cultivation in Cuba and the Dominican Republic (Guzman et al., 2018) has been observed (Dumartinet et al., 2020). Changes in pathogenicity

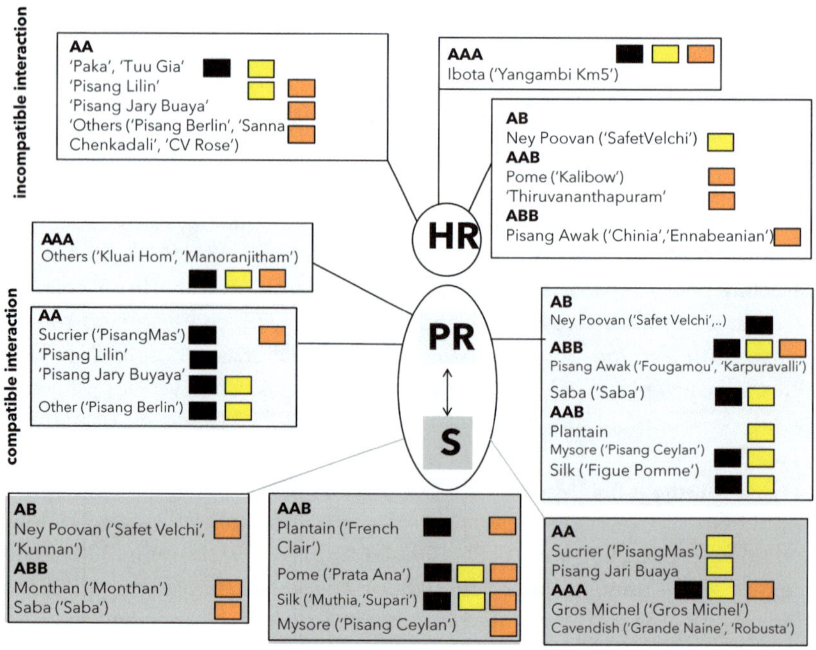

Figure 5 Classification of *Musa* species/subspecies, banana groups and banana clones (genomic group and ploidy indicated by AA, AAA, AB, AAB, ABB; names of popular cultivars in groups in parentheses; adapted by Abadie C., CIRAD, France from Figure published by Guzman et al. (2018) according to their response (HR = highly resistant, PR = partially resistant and S = susceptible) against BLSD (black), SD (yellow) and ELSD (orange) under field conditions. Evaluations for BLSD were mostly recorded in Cameroon, for SD at various locations in Asia and Guadeloupe, and for ELSD in India and Mauritius.

have also confirmed under controlled conditions, and some isolates caused disease in Yamganbi Km 5 (AAA, Ibota subgroup) which was previously considered as highly resistant. Interestingly, this breakdown was also observed in the field in different countries. Understanding of banana-*Pseudocercospora* spp. interactions initially relied on histopathological examination after inoculations with individual strains and provided fundamental information about pathogen penetration, disease progress and host responses. Recent advances in next-generation sequencing and other genomics-based technologies have increased our understanding of the genes and cellular pathways involved in plant immunity and fungal pathogenicity (Arango et al., 2016; van Westerhoven et al., 2023). Recently, Carlier et al. (2021) and Dumartinet et al. (2022) identified candidate genes in more than 30 genomic regions that are involved in the genetic basis of pathogen adaptation, using a population genomics approach.

4.2 Compatible and incompatible interactions

Examination of compatible interactions suggests that *Pseudocercospora* pathogens are hemi-biotrophic, with hyphae appearing between living host cells ahead of necrotic regions during early phases of infection, indicative of an initial biotrophic phase. Beveraggi et al. (1995) have compared both healthy and infected tissues in partially resistant 'Fougamou' (ABB Pisang Awak subgroup) and susceptible ´Grande Naine` (AAA Cavendish subgroup) during interaction with *Pf*. Clear differences were apparent between the two cultivars, with non-infected healthy tissues in the partially resistant 'Fougamou' harbouring an abundance of parenchyma cells in palisade and spongy mesophyll tissues with polyphenol-rich vacuoles, in contrast to ´Grande Naine'. Evidence also suggests the release of polyphenols occurring in the partial resistance response upon cell necrosis following infection, potentially with an affinity for the fungal pathogen cell wall and reduction in mycelial advance in host parenchyma.

Incompatible response to *Pf* in 'Yangambi Km 5' (AAA, Ibota subgroup) (Beveraggi et al., 1995) and in *Musa acuminata* ssp. *burmannica* (AAw, accession 'Calcutta 4') (Cavalcante et al., 2011; Passos et al., 2012) showed a clear rapid death of guard cells or sub-stomatal cells, providing evidence of HR during the incompatible response, together with the release of electron-dense polyphenolic compounds into the intercellular space. These compounds show an affinity for both host cell walls and pathogen hyphae cell walls, with the subsequent degeneration of hyphae indicative of fungitoxic activity.

Further cell death in leaf tissues surrounding the point of infection can also occur, with Cavalcante et al. (2011) reporting in *Musa acuminata* ssp. *burmannica* a later resistance response that involved cell wall thickening/ lignification, H_2O_2 accumulation and peroxidase enzyme activity to limit the development of the pathogen, associated with the hypersensitive response. At 3 weeks after inoculation, significantly fewer stomata appeared infected in the resistant 'Calcutta 4', in contrast to the partially resistant ´Pisang Madu´ (AA Musa subgroup) and the susceptible ´Grande Naine`. Rodriguez et al. (2020) also reported early detection of reactive oxygen in infected tissues and mesophyllic leaf tissues in *Musa acuminata* ssp. *burmannica* (AAw, accession 'Calcutta 4') at 72 h and 144 h after inoculation, while this was only detected after 360 h in the susceptible 'Williams' (AAA, Cavendish subgroup). Apoptotic events were also observed 19 days after inoculation in 'Calcutta 4' by Passos et al. (2012), with no subsequent cell death progression after this time.

4.3 Biochemical host responses during pathogenesis

Investigations of biochemical reactions during infection by *Pseudocercospora* pathogens have reported numerous responses in the host, ranging from lignification of plant cell walls, to the hypersensitive response, peroxidase (POX), hydrogen peroxide, phenylalanine ammonia-lyase (PAL) and β-1, 3-glucanase (GLU) (Alakonya et al., 2018 and references therein). Antimicrobial phytoprotectants, classified as either low molecular weight phytoalexins or phytoanticipins, are typically broad spectrum in action against phytopathogens. Differences exist between the two, with phytoalexins defined as antimicrobial compounds synthesized exclusively after plant exposure to microorganisms (Paxton, 1981), whilst phytoanticipins represent antimicrobial compounds present in plants prior to pathogen challenge or produced after infection solely from pre-existing constituents (vanEtten et al., 1994).

A phytoanticipin with a saponin structure was isolated from 'Grande Naine', with antimicrobial action confirmed against *Pf* (Cruz-Cruz et al., 2010). A number of phytoalexins have been characterized in banana, with phenalenones (perinaphthenone-type) compounds and derivatives documented as native defense compounds that are induced in leaves or roots during biotic and abiotic stresses (Luis et al., 1993; Hirai et al., 1994; Binks et al., 1997). Perinaphthenones isolated from rhizome material in Yangambi Km 5, a cultivar resistant to *Pf*, were reported to display significant antifungal activity against this pathogen, based on inhibition of mycelial growth (Otálvaro et al., 2007). Hidalgo et al. (2016) reported differences in phenalenone-type compounds between the susceptible 'Williams' (AAA, Cavendish subgroup) and the resistant 'Khai Thong Ruang' (AAA, Ibota subgroup), with a total of 15 such metabolites in the defense response of the latter compared to the former.

Finally, analysis of enzyme activities potentially involved in plant defense responses have also been investigated in banana genotypes contrasting in resistance to *Pf*. Hoss et al. (2000) observed higher levels of PAL at 24 h after inoculation in inoculated Yangambi Km 5 (resistant to *Pf*) in comparison with ´Cachaco` (ABB, Bluggoe subgroup, PR) and ´Dominico Hartón` (AAB, Plantain subgroup, S) and several defense-related enzymes – such as PAL, POX and GLU – have also been observed across different time points in 'Calcutta 4' (HR), within 6–18 h after inoculation, followed by H_2O_2 at 72 h after inoculation compared to the later induction susceptible 'Williams' (AAA, Cavendish subgroup) (Torres et al., 2012). Such data provides evidence for a rapid host enzymatic response to limit pathogen advance in diverse incompatible interactions.

4.4 Genetics of host resistance

Genetic mapping in banana is complicated due to sterility, poor fertility, heterozygosity and polyploidy in many *Musa* accessions (Jones, 2019), although BLSD-resistant Calcutta 4 has been successfully crossed with susceptible plantain cultivars (AAB) (Ortiz and Vuylsteke, 1994; Craenen and Ortiz, 1997). In these studies, segregation ratios amongst the hybrids revealed a single major locus with a recessive allele with dosage effects and two independent minor loci with additive effects. As an alternative to the development of segregating populations and genetic mapping, the Genome-Wide Association Study (GWAS) approach is now also possible using high-quality genetic markers for application across existing diversity in sets of accessions from natural populations. Applied to date for marker and candidate gene discovery associated with domestication traits of parthenocarpy and seed sterility in *Musa* (Sardos et al., 2016), as well as bunch weight and component traits (Nyine et al., 2019).

Plant immune system mechanisms. In *Musa*, nucleotide-binding and leucine-rich repeat immune receptors (NLRs) were initially isolated using degenerate primer approaches targeting conserved gene motifs. Sequence characterization has been conducted across both wild and cultivated bananas, including diploid wild species within *Musa acuminata* and *Musa balbisiana*, cultivated species Zhongshandajiao (ABB), Fenjiao (AAB), 'Williams' (AAA, Cavendish subgroup) and *Musa balbisiana* 'Pisang Klutuk Wulung' (BB), together with accessions of *Musa schizocarpa*, *Musa textilis*, *Musa velutina* and *Musa ornata* (Pei et al., 2007; Azhar and Heslop-Harrison, 2008; Miller et al., 2008; Lu et al., 2011; Baurens et al., 2010). Later whole genome sequencing of *Musa* species has enabled more complete analysis of NLR gene members. From such data, Wu et al. (2016) identified a total of 117, 93, and 62 NLR genes in *Musa acuminata*, *Musa balbisiana*, and *M. itinerans*, respectively. Improved long read-based whole genome sequencing of the reference *Musa acuminata* ssp. *malaccensis* (AAw, accession DH-Pahang) genome revealed a total of 128 NLRs in the genome assembly version 4 (Belser et al., 2021). Rijzaani et al. (2021), identified a total of 136 NLRs from three *Musa* species, from a diverse group of cultivated banana clones and from one *Ensete* species. However, both functional Pattern Recognition Receptors (PRRs) and NLR resistance genes for the *Pseudocercospora* pathogens have yet to be identified. Nevertheless, genomic resources, which in the case of NLRs include many candidate genes from genotypes resistant to both *Pf* and *Pm*, such as *Musa acuminata* ssp. *burmannica* (AAw, accession 'Calcutta 4'), offer considerable potential for characterisation of resistance and genetic improvement of susceptible commercial cultivars through genetic modification or gene editing approaches.

4.5 Candidate genes involved in plant defense

Transcriptome analysis of banana–*Pseudocercospora* interactions also enabled the identification of candidate genes potentially involved in plant defense responses. Portal et al. (2011) investigated the compatible *Musa acuminata* 'Grande Naine'–*Pf* interaction, identifying evidence, at stages 3 and 4 of disease progress, of upregulation of genes involved in the flavonoid pathway (CHR-I, CHS and F3′ 5 ′H), detoxification (Glutathione S-transferase, Betaine-aldehyde dehydrogenase) and production of pathogenesis-related (PR) proteins 1, 4, 5, 10 and 14. Upregulated genes also included a number involved in ethylene (ET) and jasmonic acid (JA) phytohormone pathway signalling, likely induced as a response in the susceptible cultivar to the necrotrophic phase of the pathogen.

Passos et al. (2012) also investigated the *Musa-Pf* interaction, through a detached leaf-bioassay approach, and identified differential early-stage expression of germin/oxalate oxidase (OXO), metallothionein MT2 and peroxidase in 'Calcutta 4' compared with ´Grande Naine`. D'Hont et al. (2012) also used detached leaf bioassays to study compatible, partially resistant and incompatible responses. In the compatible and partially resistant interactions, RLK receptor genes, chitinases, germin-like proteins, hevein-like proteins, glutathione-S-transferases, WRKY transcription factors and PR 1 proteins were all upregulated, but not in the incompatible interaction, likely due to late time point sampling. However, employing detached leaf tissues in bioassays with *Pseudocercospora* pathogens may not enable typical disease development (Arzanlou et al., 2007; Donzelli and Churchill, 2007; Churchill, 2011).

Overall, various authors have reported differential gene expression during pathogenesis in hosts with HR, PR and S responses, comprising genes such as PAL, cinnamate-4-hydroxylase (C4H), chalcone synthase (CHS) and isoflavone reductase (IRL) in 'Calcutta 4' (Alvarez et al., 2013; Mendoza-Rodrigues et al., 2018) as well as metallothioneins, proteins containing zinc finger domains, and a putative NLR disease resistance protein RGA1 (Timm et al., 2016) and activation of JA and ET biosynthetic pathways in the incompatible response of 'Calcutta 4' to *Pf*, in contrast to the susceptible cultivar 'Williams' (Rodriguez et al., 2020).

Regarding the banana-*Pm* interaction, Passos et al. (2013) conducted a large-scale transcriptome analysis during the interaction of contrasting genotypes with the pathogen and identified, e.g. transcripts involved in ROS detoxification associated with HR, as well as superoxide dismutases (SODs) and glutathione-S-transferases. Such gene expression associated with ROS detoxification and HR provide evidence for conserved host incompatible reactions against *Pseudocercospora* pathogens, with accumulation of H_2O_2 and peroxidase activity in 'Calcutta 4' also previously reported to peak up to 10 DAI with *Pf* (Cavalcante et al., 2011). In addition, Passos et al. (2013) also observed upregulation of transcripts for PAL in 'Calcutta 4' during the early incompatible interaction with *Pm*. A total of 14 expressed NLR genes were also identified from *Pm*-inoculated and non-inoculated leaf tissues. Later examination of early-stage gene expression in 'Calcutta 4' during interaction with *Pm* through Illumina RNAseq also revealed upregulation of numerous genes potentially involved in PTI responses. Finally, investigation of the banana-*Pe* incompatible interaction through subtractive suppression hybridization (SSH) (cv Manoranjitham-AAA) revealed responses of host genes encoding metallothioneins, ethylene responsive factors, flavin-containing monoxygenase and serine glyoxylate lipoxygenase (Saravanakumar et al., 2016).

4.6 Comparative pathogen genomics and evolution

Despite the differences in genome sizes of *Pm*, *Pf* and *Pe*, the numbers of protein-coding genes are largely similar and vary between 10 000 and 13 000 genes (Chang et al., 2016). Of these, a total of 6307 represent a core set of protein coding genes across the three species, with 234 exclusive to these fungal pathogens and potentially involved in adaptation and virulence to banana. Evidence for species-specific adaptation is considerable, despite the recent evolution of the species from a common ancestor, with over 1000 orphan genes in each species potentially indicative of species-specific evolution of virulence.

Pairwise comparisons of gene content indicates that *Pf* and *Pe* share a greater number of gene families, which may indicate either a shorter evolutionary distance between these two or a similar functional bias in gene duplication or loss. The similar patterns of expansion and contraction in the core gene families in these two pathogens are related mostly to cell metabolism and plant cell wall enzymatic degradation, potentially associated with virulence in these two species through uptake of nutrients from the infected host (Chang et al., 2016). Previously, numerous specific gene families associated with pathogenicity were identified in the genome sequences of *Pf* isolates CIRAD86 and CIRAD139a (Arango et al., 2016). Reduction in plant cell wall degrading enzymes in *Zymoseptoria tritici* pathogenic on wheat has been suggested to represent a mechanism for host defense evasion during stealth pathogenesis (Goodwin et al., 2011). Indeed, as hemi-biotrophs, the *Pseudocercospora* species of banana are expected to suppress host defense during initial biotrophy, whilst later able to kill host cells during necrotrophy through secretion of secondary metabolites and proteins (Noar and Daub, 2016a; Giraldo and Valent, 2013; Stergiopoulos et al., 2013). Despite the recent evolution from a common ancestor and shared host specificity, a diverse array of 105, 110 and 112 putative secreted effector proteins theoretically able to modulate resistance responses in banana have been predicted in the genomes of *Pf*, *Pm* and *Pe*, respectively, with approximately 50% specific to each species, implying the evolution of different strategies in each species for host immune system manipulation. A total of 22 core effector families have also been identified, with seven exclusive to *Pm*, *Pf* and *Pe*, potentially essential for adaptation to banana. Of the core effectors, homologs and paralogs of the *Cladosporium fulvum Ecp2*, *Ecp6* and

Avr4 chitin-binding effectors were also identified (Arango et al., 2016; Chang et al., 2016).

4.7 Candidate effectors

Analysis of *Pf* gene expression has been more limited than observed for the host. Initial studies focused on gene expression in culture media, with homologs of the *C. fulvum* tomato pathogen effector genes *Avr4* and *Ecp2* identified amongst the expressed genes in Fries liquid medium with and without banana leaf extract (Cho et al., 2008). *Pf MfAvr4* and *MfEcp2* effector proteins were subsequently reported to induce *Cf-Ecp4*- and *Cf-Ecp2*-mediated HR in tomato, respectively (Stergiopoulos et al., 2010) and allelic variation was assessed in global populations revealing polymorphisms in effector alleles that can break down resistance (Stergiopoulos et al., 2014). Noar and Daub (2016a) performed a *Pf* transcriptome analysis during interaction with banana leaves on solid media which revealed the upregulation of genes associated with fumonisins and solanapyrone clusters, in contrast to melanin biosynthesis-associated clusters that did not show differential expression. In addition, upregulated pathogenicity-related genes in infected leaf tissues included genes for secondary metabolism, such as cytochrome P450s, short-chain dehydrogenases and oxidoreductases as well as genes encoding secreted cysteine-rich proteins that are characteristic of pathogen effectors. Analysis of mitogen-activated protein kinases (MAPK) has also revealed involvement of these signalling pathways in virulence in *Pf*, including roles for *Fus3* and *Slt2* genes in pathogen advance in East African Highland Banana (Onyilo et al., 2018). Recent proteomics-based analysis of the cell wall in *Pf* strains contrasting in virulence reported the isolation of 24 pathogenicity factors, including the effector proteins *PfAvr4* and *PfEcp6*, suggesting their binding to the pathogen cell wall (Burgos et al., 2019). Gene expression studies of *Pm* during interaction with banana are limited but have reported transcripts for the *C. fulvum Ecp6* homolog, which were more abundant on resistant 'Calcutta 4' than in ´Grande Naine` (Passos et al., 2013).

4.8 Pathogen secondary metabolites

Molina and Krauz (1987) speculated that *Pf* may cause leaf tissue necrosis ahead of fungal growth by secreting toxins. Churchill (2011) and references therein reported the identification of secondary metabolites from culture filtrates, such as fijiensin, juglone and the melanin shunt metabolite 2,4,8-trihydroxytetralone, subsequent disruption of the latter revealed no negative effect on pathogenicity. Juglone has been reported to be highly phytotoxic and light dependent (Busogoro et al., 2004a,b). Indeed, investigations have revealed chloroplasts as a potential target site, with greater damage observed in susceptible 'Grande Naine' than in partially resistant 'Fougamou', which was confirmed by El Hadrami et al. (2005). Neither 2,4,8-trihydroxytetralone nor juglone, however, has been shown to play primary roles in pathogenicity, with no evidence of early involvement during the biotrophic phase (Harelimana et al., 1997). Therefore, the employment of secondary metabolites for the selection of resistant banana genotypes is not recommended until proof of action of a genetically characterized toxin is obtained (Churchill, 2011).

5 Conclusion and future trends

With the regular aerial application of fungicides under increasing pressure to cease (see Chapter 2 by Chong and Kema in this volume), new strategies for disease control are urgently needed. The use of disease-resistant varieties may play a major role. However, we know that pathogen populations continuously evolve and can adapt to different biotic conditions. Therefore, it is necessary to develop and sustainably deploy varietal resistance. However, the current knowledge regarding the biology of the pathogens and host-pathogen interactions – even after almost a century of research – is not well advanced. The coevolutionary centre of origin of banana and its pathogens is Southeast Asia. Furthermore, a thorough understanding of the distribution and population structure of the three species in the centre of origin is necessary. This will help to identify areas of host-pathogen coevolution that may contain valuable sources of resistance.

Many population genetic studies, in parallel with epidemiological studies, have resulted in an understanding of pathogen dispersal at the plantation as well as continental level. Evidence suggests that the movement of infected plant material contributed to intercontinental movement. Hence, strengthening quarantine measures is necessary to prevent new introductions and to limit gene flow between established populations. From the agricultural landscape perspective, spore dispersal has been estimated by direct and indirect approaches, but detailed epidemiological studies are still required to estimate inoculum coming from leaves of the same or different plants as well as the ratio between asexual and sexual reproduction at the plant and plantation scales.

Population genomic methods may unveil demographic and genetic parameters to decipher the mechanisms underlying genetic diversity in the pathogen population and adaptation to the host. The identification of resistance genes is necessary to enable plant improvement, irrespective of the approach. Conventional breeding and biotechnological methods would be greatly supported by well characterized and mapped genes. Furthermore, an understanding of genes involved in pathogenicity is important which requires genetic studies and functional analyses. The tools for such analyses are currently being developed and hence, within the coming years we expect to gain substantially more insight in the mechanism underlying host resistance and pathogenicity. Ideally, such studies should encompass the three pathogens.

Detailed analysis on the plant side have resulted in considerable advances in our understanding of banana–*Pseudocercospora* interactions at the molecular level which revealed numerous candidate genes at the receptor or downstream pathway level that represent resources for validation and subsequent engineering of resistance to the *Pseudocercospora* pathogens. Stacking of NLR genes, e.g. has recently shown considerable promise in conferring both specific and broad-spectrum resistance to fungal pathogens in other host plants (e.g. Witek et al., 2021; Luo et al., 2021). Recent genomics, transcriptomics and proteomics-based approaches for the pathogens enabled the identification of effector molecules conserved across the three *Pseudocercospora* species, as well as species-specific candidates. Conserved effectors may be appropriate for the development of broad-spectrum resistance to the Sigatoka complex. Pathogenicity genes other than effectors can be also identified through populations genomics studies. These studies will not only provide a better understanding of the pathogenicity of these fungi but may also facilitate the discovery of resistance genes. Banana proteins that recognize or are targeted by pathogen effectors could be genes that are relevant for banana breeding (van Esse et al., 2020). Finally, host diversification is required for overall sustainability. Since most bananas are produced in smallholder settings and agroforestry landscapes, novel disease management strategies should first and foremost focus on these agricultural systems.

6 Where to look for further information

The best single site for further information is Promusa: https://www.promusa.org/Black+leaf+streak.

7 References

Abadie, C., Chilin-Charles, Y., Huat, J., Salmon, F., Pignolet, L., Carlier, J., Lescot, T., Côte, F. and Jenny, C. 2009. New approaches to select cultivars of banana with durable resistance to Mycosphaerella leaf spot diseases. *Acta Horticulturae* 828, 171–178.

Alakonya, A. E., Kimunye, J., Mahuku, G., Amah, D., Uwimana, B., Brown, A. and Swennen, R. 2018. Progress in understanding *Pseudocercospora* banana pathogens and the development of resistant germplasm. *Plant Pathology* 67(4), 759–770.

Alvarez, J. C., Rodriguez, H. A., Rodriguez-Arango, E., Monsalve, Z. I., Morales, O. J. G. and Arango, I. R. E. 2013. Characterization of a differentially expressed phenylalanine ammonia-lyase gene from banana induced during *Mycosphaerella fijiensis* infection. *Journal of Plant Studies* 2(2), 35–46.

Anonymous. 2018. Principaux organismes nuisibles par plantes hôtes en Guyane, DAAF de Guyana, 68 p. Available at: https://daaf.guyane.agriculture.gouv.fr/IMG/pdf/Organismes_nuisibles_par_plantes-20180824_cle42ced4.pdf.

Arango Isaza, R. E., Diaz-Trujillo, C., Dhillon, B., Aerts, A., Carlier, J., Crane, C. F., V de Jong, T., de Vries, I., Dietrich, R., Farmer, A. D., Fortes Fereira, C., Garcia, S., Guzman, M., Hamelin, R. C., Lindquist, E. A., Mehrabi, R., Quiros, O., Schmutz, J., Shapiro, H., Reynolds, E., Scalliet, G., Souza, M., Stergiopoulos, I., Van der Lee, T. A., De Wit, P. J., Zapater, M. F., Zwiers, L. H., Grigoriev, I. V., Goodwin, S. B. and Kema, G. H. 2016. Combating a global threat to a clonal crop: banana *Black sigatoka* pathogen *Pseudocercospora fijiensis* (synonym *Mycosphaerella fijiensis*) genomes reveal clues for disease control. *PLoS Genet* 12(8), e1005876.

Arcila-Galvis, J. E., Arango, R. E., Torres-Bonilla, J. M. and Arias, T. 2021. The mitochondrial genome of a plant fungal pathogen *Pseudocercospora fijiensis* (Mycosphaerellaceae), comparative analysis and diversification times of the sigatoka disease complex using fossil calibrated phylogenies. *Life* 11(3), 215. DOI: 10.3390/life11030215.

Arzanlou, M., Abeln, E. C. A., Kema, G. H. J., Waalwijk, C., Carlier, J., de Vries, I., Guzman, M. and Crous, P. W. 2007. Molecular diagnostics for the sigatoka disease complex of banana. *Phytopathology* 97(9), 1112–1118.

Arzanlou, M., Groenewald, J. Z., Fullerton, R. A., Abeln, E. C. A., Carlier, J., Zapater, M. F., Buddenhagen, I. W., Viljoen, A. and Crous, P. W. 2008. Multiple gene genealogies and phenotypic characters differentiate several novel species of Mycosphaerella and related anamorphs on banana. *Persoonia* 20, 19–37.

Azhar, M. and Heslop-Harrison, J. S. 2008. Genomes, diversity and resistance gene analogues in *Musa* species. *Cytogenetic and Genome Research* 121(1), 59–66.

Balint-Kurti, P. J., May, G. D. and Churchill, A. C. L.2001. Development of a transformation system for *Mycosphaerella* pathogens of banana: a tool for the study of host/pathogen interactions. *FEMS Microbiology Letters* 195(1), 9–15.

Baurens, F. C., Bocs, S., Rouard, M., Matsumoto, T., Miller, R. N., Rodier-Goud, M., MBéguié-A-MBéguié, D. and Yahiaoui, N. 2010. Mechanisms of haplotype divergence at the RGA08 nucleotide-binding leucine-rich repeat gene locus in wild banana (*Musa balbisiana*). *BMC Plant Biology* 10, 149.

Beveraggi, A., Mourichon, X. and Salle, G. 1993. Study of host-parasite interactions in susceptible and resistant bananas inoculated with *Cercospora fijiensis*, pathogen of black leaf streak disease. In: Ganry, J. (Ed.) *Breeding Banana and Plantain for Resistance to Disease and Pests*. Proceedings of the International Symposium on Genetic Improvement of Bananas to Diseases and Pests, pp. 171–192. Organized by CIRAD-FLHOR, Montpellier, France, 7–9 September 1992. CIRAD, Montpellier.

Beveraggi, A., Mourichon, X. and Salle, G. 1995. Etude des interactions hôte-parasite chez des bananiers sensibles et résistants inoculés par *Cercospora fijiensis* (*Mycosphaerella fijiensis*) responsable de la maladie des raies noires. *Canadian Journal of Botany* 73(9), 1328–1337.

Binks, R. H., Greenham, J. R., Luis, J. G. and Gowen, S. R. 1997. A phytoalexin from roots of *Musa acuminata* var. Pisang sipulu. *Phytochemistry* 45(1), 47–49.

Blomme, G., Ploetz, R., Jones, D., De Langhe, E., Price, N., Gold, C., Geering, A., Viljoen, A., Karamura, D., Pillay, M., Tinzaara, W., Teycheney, P.-Y., Lepoint, P., Karamura, E. and Buddenhagen, I. 2013. A historical overview of the appearance and spread of *Musa* pests and pathogens on the African continent: highlighting the importance of clean *Musa* planting materials and quarantine measures. *Annals of Applied Biology* 162(1), 4–26.

Brito, F. S. D., Santos, J. R. P., Azevedo, V. C. R., Peixouto, Y. S., de Oliveira, S. A., Ferreira, C. F., Haddad, F., Amorim, E. P., Fraaije, B. and Miller, R. N. G. 2020. Genetic diversity and azole fungicide sensitivity in *Pseudocercospora musae* field populations in Brazil. *Frontiers in Microbiology* 11, 99.

Brun, J. 1948. La maladie de Sigatoka du bananier causée par *Cercospora musae* Zimm. *Fruits* 3(1), 13–20.

Brun, J. 1963. La cercosporiose du bananier en Guinée. Etude de la phase ascoporée du *Mycosphaerella musicola* Leach. Thèse de Docteur de Science, Orsay, Paris, F 190 pp.

Burgos, Y. Y., Canto-Canche, B., Berezovski, M. V., Mironov, G., Loyola-Vargas, V. M. and Barba de Rose, A. P. 2019. The cell wall proteome from two strains of *Pseudocercospora fijiensis* with differences in virulence. *World Journal of Microbiology and Biotechnology* 35(7), 105.

Busogoro, J. P., Etamé, J. J., Harelimana, G., Lognay, G., Messiaen, J., Lepoivre, P. and van Cutsem, P. 2004a. Experimental evidence for the action of *M. fijiensis* toxins on banana photosynthetic apparatus. In: Jain, S. M. and Swennen, R. (Eds) *Banana Improvement: Cellular, Molecular Biology, and Induced Mutations*, pp. 171–181. Science Publishers Inc, Enfield, NH.

Busogoro, J. P., Etamé, J. J., Lognay, G., Messiaen, J., van Cutsem, P. and Lepoivre, P. 2004b. Analysis of the mechanisms of action of *Mycosphaerella fijiensis* toxins during the development of black leaf streak disease. In: Jain, S. M. and Swennen, R. (Eds) *Banana Improvement: Cellular, Molecular Biology, and Induced Mutations*, pp. 171–181. Science Publishers Inc, Enfield, NH.

CABI. 1997. CABI/EPPO distribution maps of plant diseases. Map N°7. *Mycosphaerella musicola* Leach. Wallingford, UK. Available at: https://www.cabi.org/isc/abstract/20066500007.

Calpouzos, L. 1955. Studies on the SD of bananas and its fungus pathogen. *Atkins Garden and Research Laboratory, Cuba*, 70 p.

Carlier, J., Lebrun, M. H., Zapater, M. F., Dubois, C. and Mourichon, X. 1996. Genetic structure of the global population of banana black leaf streak fungus, *Mycosphaerella fijiensis*. *Molecular Ecology* 5(4), 499–510.

Carlier, J., Zapater, M. F., Lapeyre, F., Jones, D. R. and Mourichon, X. 2000. Septoria leaf spot of banana: a newly discovered disease caused by *Mycosphaerella eumusae* (anamorph *Septoria eumusae*). *Phytopathology* 90(8), 884–890.

Carlier, J., Bonnot, F., Roussel, V., Ravel, S., Martinez, R. T., Perez-Vicente, L., Abadie, C., Wright, S. and Turgeon, B. G. 2021. Convergent adaptation to quantitative host resistance in a major plant pathogen. *mBio* 12(1), e03129-20.

Castelan, F. P., Saraiva, L. A., Lange, F., de Lapeyre de Bellaire, L., Cordenunsi, B. R. and Chillet, M. 2012. Effects of black leaf streak disease and Sigatoka disease on fruit quality and maturation process of bananas produced in the subtropical conditions of Brazil. *Crop Protection* 35, 127–131.

Castelan, F. P., Abadie, C., Hubert, O., Chilin-Charles, Y., de Lapeyre de Bellaire, L. and Chillet, M. 2013. Relation between the severity of Sigatoka disease and banana quality characterized by pomological traits and fruit green life. *Crop Protection* 50, 61–65.

Cavalcante, M. J. B., Escoute, J., Madeira, J. P., Romero, R. E., Nicole, M. R., Oliveira, L. C., Hamelin, C., Lartaud, M. and Verdeil, J. L. 2011. Reactive oxygen species and cellular interactions between *Mycosphaerella fijiensis* and banana. *Tropical Plant Biology* 4(2), 134–143.

Chang, T. C., Salvucci, A., Crous, P. W. and Stergiopoulos, I. 2016. Comparative genomics of the Sigatoka disease complex on banana suggests a link between parallel evolutionary changes in *Pseudocercospora fijiensis* and *Pseudocercospora eumusae* and increased virulence on the banana host. *PLoS Genetics* 12(8), e1005904.

Chillet, M., Olivier, H., Yolande, C.-C. and de Bellaire Luc, D. L. (2008). Sigatoka disease reduces the green life of bananas. *Crop Protection* 28(1), 41–45.

Chillet, M., Castelan, F. P., Abadie, C., Hubert, O. and de Lapeyre de Bellaire, L. 2013. Necrotic leaf removal: a key component of integrated management of *Mycospaerella* leaf spot diseases to improve the quality of banana: the case of Sigatoka disease. *Fruits* 68(4), 271–277.

Chong, P., Essoh, J. N., Arango Isaza, R. E., Keizer, P., Stergiopoulos, I., Seidl, M. F., Guzman, M., Sandoval, J., Verweij, P. E., Scalliet, G., Sierotzski, H., de Lapeyre de Bellaire, L., Crous, P. W., Carlier, J., Cros, S., Meijer, H. J. G., Peralta, E. L. and Kema, G. H. J. 2021. A world-wide analysis of reduced sensitivity to DMI fungicides in the banana pathogen *Pseudocercospora fijiensis*. *Pest Management Science*. DOI: 10.1002/ps.6372.

Churchill, A. C. L. 2011. *Mycosphaerella fijiensis*, the black leaf streak pathogen of banana: progress towards understanding pathogen biology and detection, disease development, and the challenges of control. *Molecular Plant Pathology* 12(4), 307–328.

Conde-Ferráez, L., Waalwijk, C., Canto-Canché, B. B., Kema, G. H., Crous, P. W., James, A. C. and Abeln, E. C. 2007. Isolation and characterization of the mating type locus of *Mycospaerella fijiensis*, the causal agent of black leaf streak disease of banana. *Molecular Plant Pathology* 8(1), 111–120.

Craenen, K. and Ortiz, R. 1997. Effect of the *bs1* gene in plantain-banana hybrids on response to black Sigatoka. *Theoretical and Applied Genetics* 95(3), 497–505.

Crous, P. W. and Mourichon, X. 2002. Mycosphaerella eumusae and its anamorph *Pseudocercospora eumusae* spp. nov.: causal agent of eumusae leaf spot disease of banana. *Sydowia* 44(1), 35–43.

Crous, P. W., Groenewald, J. Z., Aptroot, A., Braun, U., Mourichon, X. and Carlier, J. 2020. Integrating morphological and molecular data sets on *Mycosphaerella*, with specific reference to species occuring on *Musa*. In: IPGRI and INIBAP (Eds) International workshop on *Mycosphaerella* leaf spot diseases of bananas, San José, Costa Rica. INIBAP, pp. 43–57.

Crous, P. W., Carlier, J., Roussel, V. and Groenewald, J. Z. 2021. *Pseudocercospora* and allied genera associated with leaf spots of banana (*Musa* spp.). *Fungal Systematics and Evolution* 7, 1–19.

Cruz-Cruz, C. A., Ramírez-Tec, G., García-Sosa, K., Escalante-Erosa, F., Hill, L., Osbourn, A. E. and Peña-Rodríguez, L. M. 2010. Phytoanticipins from banana (*Musa acuminata* cv. Grande Naine) plants, with antifungal activity against *Mycosphaerella fijiensis*, the causal agent of black Sigatoka. *European Journal of Plant Pathology* 126, 459–463.

Da Silveira, O. R., de Jesus Bagolin, D., Neto, D. C., Martins, M. B., da Silva, R. M. and de Araújo, J. A. N. (2016). Distribution of black leaf steak disease (*Mycosphaerella fijiensis* Morelet) in Matto Grosso, Brazil. *Bioscience Journal* 32, 384–388.

Daniells, J. W., Lisle, A. T. and Bryde, N. J. 1994. Effect of bunch trimming and leaf removal at flowering on maturity bronzing, yield, and other aspects of fruit-quality of bananas in North Queensland. *Australian Journal of Experimental Agriculture* 34(2), 259–265.

de Lapeyre de Bellaire, L., Foure, E., Abadie, C. and Carlier, J. 2010. Black leaf streak disease is challenging the banana industry. *Fruits* 65(6), 327–342.

de Lapeyre de Bellaire, L., Guillermet, C. and le Guen, R. 2016. La Sigatoka en las Antillas francesas y en la Republica Dominicana: desafios y nuevas perspectivas para el control. In: *Unidos por el Desarrollo Bananero*. VI Congreso Internacional sobre banano (Corbana) y XXI Reunion Internacional de Acorbat, Miami, 19–22 avril, p. 5.

Dens, K. R., Romero, R. A., Swennen, R. and Turner, D. W. 2008. Removal of bunch, leaves, or pseudostem alone, or in combination, influences growth and bunch weight of ratoon crops in two banana cultivars. *Journal of Horticultural Science and Biotechnology* 83(1), 113–119.

Devi, P. G. and Thangavelu, R. 2018. Development of species specific SCAR based molecular marker for the detection of *Pseudocercospora eumusae*, the causal agent of eumusae leaf spot disease in banana. *Journal of Plant Pathology* 101, 295–305.

D'Hont, A., Denoeud, F., Aury, J. M., Baurens, F. C., Carreel, F., Garsmeur, O., Noel, B., Bocs, S., Droc, G., Rouard, M., Da Silva, C., Jabbari, K., Cardi, C., Poulain, J., Souquet, M., Labadie, K., Jourda, C., Lengellé, J., Rodier-Goud, M., Alberti, A., Bernard, M., Correa, M., Ayyampalayam, S., Mckain, M. R., Leebens-Mack, J., Burgess, D., Freeling, M., Mbéguié-A-Mbéguié, D., Chabannes, M., Wicker, T., Panaud, O., Barbosa, J., Hribova, E., Heslop-Harrison, P., Habas, R., Rivallan, R., Francois, P., Poiron, C., Kilian, A., Burthia, D., Jenny, C., Bakry, F., Brown, S., Guignon, V., Kema, G., Dita, M., Waalwijk, C., Joseph, S., Dievart, A., Jaillon, O., Leclercq, J., Argout, X., Lyons, E., Almeida, A., Jeridi, M., Dolezel, J., Roux, N., Risterucci, A. M., Weissenbach, J., Ruiz, M., Glaszmann, J. C., Quétier, F., Yahiaoui, N. and Wincker, P. 2012. The banana (*Musa acuminata*) genome and the evolution of monocotyledonous plants. *Nature* 488(7410), 213–217.

Donzelli, B. G. G. and Churchill, A. C. L. 2007. A quantitative assay using mycelial fragments to assess virulence of *Mycosphaerella fijiensis*. *Phytopathology* 97(8), 916–929.

Dumartinet, T., Abadie, C., Bonnot, F., Carreel, F., Roussel, V., Habas, R., Martinez, R. T., Perez-Vicente, L. and Carlier, J. 2020 Pattern of local adaptation to quantitative host resistance in a major pathogen of a perennial crop. *Evolutionary Applications* 13(4), 824–836.

Dumartinet, T., Ravel, S., Roussel, V., Perez-Vicente, L., Aguayo, J., Abadie, C. and Carlier, J. 2022. Complex adaptive architecture underlies adaptation to quantitative host resistance in a fungal plant pathogen. *Molecular Ecology* 31(4), 1160–1179.

Eckstein, K., Robinson, J. C. and Davie, S. J. 1995. Physiological-responses of banana (Musa-AAA, cavendish subgroup) in the subtropics. 3. Gas-exchange, growth analysis and source-sink interaction over a complete crop cycle. *Journal of Horticultural Science* 70(1), 169–180.

El Hadrami, A., Kone, D. and Lepoivre, P. 2005. Effect of juglone on active oxygen species and antioxidant enzymes in susceptible and partially resistant banana cultivars to Black Leaf Streak Disease. *European Journal of Plant Pathology* 113(3), 241–254.

Fouré, E. 1982. Les cercosporioses du bananier et leurs traitements. Comportement des variétés. Etude de la sensibilité variétale des bananiers et des plantains à *Mycosphaerella fijiensis* Morelet au Gabon (I) – I Incubation et Évolution de la Maladie – II Etude de quelques paramètres. *Fruits* 37, 749–771.

Fouré, E. 1987 In: Persley, G. J. and De Langhe, E. A. (Eds) Banana and plantain breeding strategies. Proceedings of the International Workshop Held in Cairns, Australia, 13–17 October 1986. *ACIAR Proceedings* No. 21, pp. 110–113. Australian Center for International Agricultural Research, Canberra, Australia.

Fouré, E. and Moreau, A. 1992. Contribution à l'étude épidemiologique de la cercosporiose noire dans la zone bananière du Mungo de 1987 à 1989. *Fruits* 47, 3–16.

Fouré, E., Mourichon, X. and Jones, D. R. 2000. Host reaction-evaluating germplasm for reaction to black leaf streak. In: Jones, D. R. (Ed.) *Diseases of Banana, Abaca and Enset*, pp. 62–67. CABI Publishing, Wallingford.

Fouré, É. and Ganry, J. 2008. A biological forecasting system to control black leaf streak disease of bananas and plantains. *Fruits* 63(5), 311–317.

Fullerton, R. A. and Olsen, T. L. 1995. Pathogenic variability in *Mycosphaerella fijiensis* Morelet, cause of black Sigatoka in banana and plantain. *New Zealand Journal of Crop and Horticultural Science* 23(1), 39–48.

Ganry, J. and Meyer, J. P. 1975. Recherche d'une loi d'action de la température sur la croissance des fruits du bananier. *Fruits* 30(6), 375–392.

Ganry, J., de Lapeyre de Bellaire, L. and Mourichon, X. 2008. A biological forecasting system to control Sigatoka disease of bananas and plantains. *Fruits* 63(6), 381–387.

Gauhl, F. 1994. Epidemiology and ecology of Black Sigatoka (*Mycosphaerella fijiensis*) on plantain and banana (*Musa sp.*) in Costa Rica, Central America. PhD Thesis, Göttingen University, Germany.

Gilabert, A., Rieux, A., Robert, S., Vitalis, R., Zapater, M. F., Abadie, C., Carlier, J. and Ravigné, V. 2023. Revisiting the historical scenario of a disease dissemination using genetic data and approximate Bayesian computation methodology: the case of *Pseudocercospora fijiensis* invasion in Africa. *Ecology and Evolution*. DOI: 10.1002/ece3.10013.

Goodwin, S. B., Ben M'Barek, S. B., Dhillon, B., Wittenberg, A. H. J., Crane, C. F., Hane, J. K., Foster, A. J., Van der Lee, T. A., Grimwood, J., Aerts, A., Antoniw, J., Bailey, A., Bluhm, B., Bowler, J., Bristow, J., van der Burgt, A., Canto-Canché, B., Churchill, A. C., Conde-Ferràez, L., Cools, H. J., Coutinho, P. M., Csukai, M., Dehal, P., De Wit, P., Donzelli, B., van de Geest, H. C., van Ham, R. C., Hammond-Kosack, K. E., Henrissat, B., Kilian, A., Kobayashi, A. K., Koopmann, E., Kourmpetis, Y., Kuzniar, A., Lindquist, E., Lombard, V., Maliepaard, C., Martins, N., Mehrabi, R., Nap, J. P., Ponomarenko, A., Rudd, J. J., Salamov, A., Schmutz, J., Schouten, H. J., Shapiro, H., Stergiopoulos, I., Torriani, S. F., Tu, H., de Vries, R. P., Waalwijk, C., Ware, S. B., Wiebenga, A., Zwiers, L. H., Oliver, R. P., Grigoriev, I. V. and Kema, G. H. 2011. Finished genome of the fungal wheat pathogen *Mycosphaerella graminicola* reveals dispensome structure, chromosome plasticity, and stealth pathogenesis. *PLoS Genetics* 7(6), e1002070.

Guillermet, C., Le Guen, R., Dorel, M., Lescot, T. and de Lapeyre de Bellaire, L. 2018. Experimental approaches for agro-ecological management of BLSD in dry and humid tropical conditions. In: Proceedings of the X International Symposium on Banana: ISHS-Promusa. Symposiium on Agroecological approaches to promote innovative banana production systems, Van den Bergh I, Risede JM, Johnson V, Eds. *Acta Horticulturae* 1196, 113–120.

Guzman, M., Pérez-Vicente, L., Carlier, J., Abadie, C., de Lapeyre de Bellaire, L., Carreel, F., Marin, D. H., Romero, R. A., Gauhl, F., Pasberg-Gauhl, C. and Jones, R. 2018. Black Leaf streak. In: *Handbook of Diseases of Banana, Abacá and Enset*, pp. 41-115. CABI Publishing, Wallingford.

Halkett, F., Coste, D., Platero, G. G., Zapater, M. F., Abadie, C. and Carlier, J. 2010. Genetic discontinuities and disequilibria in recently established populations of the plant pathogenic fungus *Mycosphaerella fijiensis*. *Molecular Ecology* 19(18), 3909-3923.

Harelimana, G., Lepoivre, P., Jijakli, H. and Mourichon, X. 1997. Use of *Mycosphaerella fijiensis* toxins for the selection of banana cultivars resistant to black leaf streak. *Euphytica* 96(1), 125-128.

Haridas, S., Albert, R., Binder, M., Bloem, J., Labutti, K., Salamov, A., Andreopoulos, B., Baker, S. E., Barry, K., Bills, G., Bluhm, B. H., Cannon, C., Castanera, R., Culley, D. E., Daum, C., Ezra, D., Gonzalez, J. B., Henrissat, B., Kuo, A., Liang, C., Lipzen, A., Lutzoni, F., Magnuson, J., Mondo, S. J., Nolan, M., Ohm, R. A., Pangilinan, J., Park, H. J., Ramirez, L., Alfaro, M., Sun, H., Tritt, A., Yoshinaga, Y., Zwiers, L. H., Turgeon, B. G., Goodwin, S. B., Spatafora, J. W., Crous, P. W. and Grigoriev, I. V. 2020. 101 Dothideomycetes genomes: a test case for predicting lifestyles and emergence of pathogens. *Studies in Mycology* 96, 141-153.

Hayden, H. L., Carlier, J. and Aitken, E. A. B. 2003. Population differentiation in the banana leaf spot pathogen *Mycosphaerella musicola*, examined at a global scale. *Plant Pathology* 52(6), 713-719.

Hayden, H. L., Carlier, J. and Aitken, E. A. B. 2005. The genetic structure of Australian populations of *Mycosphaerella musicola* suggests restricted gene flow at the continental scale. *Phytopathology* 95(5), 489-498.

Henderson, J., Pattemore, J. A., Porchun, S. C., Hayden, H. L., Van Brunschot, S., Grice, K. R. E., Peterson, R. A., Thomas-Hall, S. R. and Aitken, E. A. B. 2006. Black Sigatoka disease: new technologies to strengthen eradication strategies in Australia. *Australasian Plant Pathology* 35, .

Hidalgo, W., Chandran, J. N., Menezes, R. C., Otálvaro, F. and Schneider, B. 2016. Phenylphenalenones protect banana plants from infection by *Mycosphaerella fijiensis* and are deactivated by metabolic conversion. *Plant, Cell and Environment* 39(3), 492-513.

Hirai, N., Ishida, H. and Koshimizu, K. 1994. A phenalenone-type phytoalexin from *Musa acuminata*. *Phytochemistry* 37(2), 383-385.

Hoss, R., Helbig, J. and Bochow, H. 2000. Function of host and fungal metabolites in resistance response of banana and plantain in the BLSD disease pathosystem (*Musa* spp. *Mycosphaerella fijiensis*). *Journal of Phytopathology* 148(7-8), 387-394.

Jacome, L. H., Schuh, W. and Stevenson, R. E. 1991. Effect of temperature and relative humidity on germination and germ tube development of *Mycosphaerella fijiensis* var. *difformis*. *Phytopathology* 81(12), 1480-1485.

Jones, D. R. (Ed.) 2018. Sigatoka leaf spot. In: *Handbook of Diseases of Banana, Abacá and Enset*, pp. 115-126. CABI Publishing, Wallingford.

Jones, D. R. (Ed.) 2019. In: *Handbook of Diseases of Banana, Abacá and Enset*, pp. 74-88. CABI Publishing, Wallingford.

Jullien, A., Munier-Jolain, N., Malézieux, E., Chillet, M. and Ney, B. 2001. Effect of pulp cell number and assimilate availability on dry matter accumulation rate in a banana fruit [*Musa* sp. AAA group 'Grande Naine' (*Cavendish* subgroup)]. *Annals of Botany* 88(2), 321-330.

Jullien, A., Chillet, M. and Malezieux, E. 2008. Pre-harvest growth and development, measured as accumulated degree days, determine the post-harvest green life of banana fruit. *Journal of Horticultural Science and Biotechnology* 83(4), 506–512.

Kagy, V., Wong, M., Vandenbroucke, H., Jenny, C., Dubois, C., Ollivier, A., Cardi, C., Mournet, P., Tuia, V., Roux, N., Doležel, J. and Perrier, X. 2016. Traditional banana diversity in Oceania: an endangered heritage. *PLoS ONE* 11(3). DOI: 10.1371/journal.pone.0151208.

Kimunye, J. N., Were, E., Mussa, F., Tazuba, A., Jomanga, K., Viljoen, A., Swennen, R., Muthoni, F. K. and Mahuku, G. 2020. Distribution of *Pseudocercospora* species causing Sigatoka leaf diseases of banana in Uganda and Tanzania. *Plant Pathology* 69(1), 50–59.

Klein, H. H. 1963. Control of Cercospora leaf spot of bananas with application of oil sprays based on the disease cycle. *Phytopathology* 50, 488–489.

Lassois, L., Bastiaanse, H., Chillet, M., Jullien, A., Jijakli, M. H. and de Lapeyre de Bellaire, L. 2010. Hand position on the bunch and source-sink ratio influence the banana fruit susceptibility to crown rot disease. *Annals of Applied Biology* 156(2), 221–229.

Leach, R. 1946. *Banana Leaf Spot (Mycosphaerella musicola) on the Gros Michel Variety in Jamaica. Investigations into the Aetiology of the Disease and the Principles of Control by Spraying*. Government Printer, Kingston, Jamaica.

Lu, Y., Xu, W. H., Xie, Y. X., Zhang, X., Pu, J. J., Qi, Y. X. and Li, H. P. 2011. Isolation and characterization of nucleotide binding site and C-terminal leucine-rich repeat-resistance gene candidates in bananas. *Genetics and Molecular Research* 10(4), 3098–3108.

Luis, J. G., Echeverri, F., Quiñones, W., Brito, I., López, M., Torres, F., Cardona, G., Aguiar, Z., Pelaez, C. and Rojas, M. 1993. Irenolone and emenolone: two new types of phytoalexins from *Musa paradisiaca*. *Journal of Organic Chemistry* 58(16), 4306–4308.

Luo, M., Xie, L., Chakraborty, S., Wang, A., Matny, O., Jugovich, M., Kolmer, J. A., Richardson, T., Bhatt, D., Hoque, M., Patpour, M., Sørensen, C., Ortiz, D., Dodds, P., Steuernagel, B., Wulff, B. B. H., Upadhyaya, N. M., Mago, R., Periyannan, S., Lagudah, E., Freedman, R., Lynne Reuber, T., Steffenson, B. J. and Ayliffe, M. 2021. A five-transgene cassette confers broad-spectrum resistance to a fungal rust pathogen in wheat. *Nature Biotechnology* 39(5), 561–566.

Marin, D. H., Romero, R. A., Guzmán, M. and Sutton, T. B. 2003. Black Sigatoka: an increasing threat to banana cultivation. *Plant Disease* 87(3), 208–222.

Mendoza-Rodríguez, M. F., Portal, O., Oloriz, M. I., Ocaña, B., Rojas, L. E., Acosta-Suárez, M., Roque, B., Canales, E., Borrás-Hidalgo, O. and Jiménez, E. 2018. Early regulation of primary metabolism, antioxidant, methyl cycle and phenylpropanoid pathways during the *Mycosphaerella fijiensis-Musa* spp. interaction. *Tropical Plant Pathology* 43(1), 1–9.

Meredith, M. A. 1970. *Banana Leaf Spot Disease (Sigatoka) Caused by Mycosphaerella musicola Leach*. Commonwealth Mycol. Inst, Kew. Phytopathological papers, paper 11, 147 p.

Meredith, D. S. and Lawrence, J. S. 1970. Black leaf streak disease (*Pseudocercospora fijiensis*): susceptibility of cultivars. *Tropical Agriculture* 47, 275–287.

Miller, R. N., Bertioli, D. J., Baurens, F. C., Santos, C. M., Alves, P. C., Martins, N. F., Togawa, R. C., Souza, M. T. and Pappas, G. J. 2008. Analysis of non-TIR NBS-LRR resistance gene analogs in *Musa acuminata* Colla: isolation, RFLP marker development, and physical mapping. *BMC Plant Biology* 8, 15.

Mobambo, K. N., Gauhl, F., Vuylsteke, D., Ortiz, R., Pasberggauhl, C. and Swennen, R. 1993. Yield loss in plantain from BLSD Leaf Spot and field performance of resistant hybrids. *Field Crops Research* 35(1), 35–42.

Molina, G. and Krausz, J. P. 1987. Production of toxin by *Mycosphaerella fijiensis* var. difformis: its relevance in the breeding for resistance to BLSD. *Informe Mensual UPEB* 11(82), 60–61.

Molina, G. C. and Krausz, J. P. 1989. A phytotoxic activity in extracts of broth cultures of *Mycosphaerella fijiensis* var. *difformis* and its use to evaluate host resistance to black Sigatoka. *Plant Disease* 73(2), 142–143.

Moller, M. and Stukenbrock, E. H. 2017. Evolution and genome architecture in fungal plant pathogens. *Nature Reviews Microbiology* 15(12), 756–771.

MouliomPefoura, A., Lassoudiere, A., Foko, J. and Fontem, D. A. 1996. Comparison of development of *Mycosphaerella fijiensis* and *Mycosphaerella musicola* on banana and plantain in the various ecological zones in Cameroon. *Plant Disease* 80(8), 950–954.

Mourichon, X. and Fullerton, R. A. 1990. Geographical distribution of the two species *Mycosphaerella musicola* Leach (*Cercospora musae*) and *M. fijiensis* Morelet (*C. fijiensis*), respectively agents of Sigatoka disease and black leaf streak disease in bananas and plantains. *Fruits* 45, 213–218.

Ngando, J. E., Rieux, A., Nguidjo, O., Pignolet, L., Dubois, C., Mehl, A., Zapater, M. F., Carlier, J. and de Lapeyre de Bellaire, L. 2015. A novel bioassay to monitor fungicide sensitivity in *Mycosphaerella fijiensis*. *Pest Management Science* 71(3), 441–451.

Noar, R. D. and Daub, M. E. 2016a. Bioinformatics prediction of polyketide synthase gene clusters from *Mycosphaerella fijiensis*. *PLoS ONE* 11(7), e0158471.

Noar, R. D. and Daub, M. E. 2016b. Transcriptome sequencing of *Mycosphaerella fijiensis* during association with *Musa acuminata* reveals candidate pathogenicity genes. *BMC Genomics* 17(1), 690.

Nyine, M., Uwimana, B., Akech, V., Brown, A., Ortiz, R., Doležel, J., Lorenzen, J. and Swennen, R. 2019. Association genetics of bunch weight and its component traits in East African highland banana (*Musa* spp. AAA group). *Theoretical and Applied Genetics* 132(12), 3295–3308.

Ohm, R. A., Feau, N., Henrissat, B., Schoch, C. L., Horwitz, B. A., Barry, K. W., Condon, B. J., Copeland, A. C., Dhillon, B., Glaser, F., Hesse, C. N., Kosti, I., Labutti, K., Lindquist, E. A., Lucas, S., Salamov, A. A., Bradshaw, R. E., Ciuffetti, L., Hamelin, R. C., Kema, G. H., Lawrence, C., Scott, J. A., Spatafora, J. W., Turgeon, B. G., de Wit, P. J., Zhong, S., Goodwin, S. B. and Grigoriev, I. V. 2012. Diverse lifestyles and strategies of plant pathogenesis encoded in the genomes of eighteen Dothideomycetes fungi. *PLoS Pathogens* 8(12), e1003037.

Onyilo, F., Tusiime, G., Tripathi, J. N., Chen, L. H., Falk, B., Stergiopoulos, I., Tushemereirwe, W., Kubiriba, J. and Tripathi, L. 2018. Silencing of the mitogen-activated protein kinases (MAPK) Fus3 and Slt2 in *Pseudocercospora fijiensis* reduces growth and virulence on host plants. *Frontiers in Plant Science* 9, 291.

Orjeda, G., Escalant, J. V. and Moore, N. 1999. The International Musa Testing Programme (IMTP) phase II overview of final report and summary of results. *InfoMusa* 8(1), 3–10. ISNN 1023-0068.

Ortiz, R. and Vuylsteke, D. 1994. Inheritance of black sigatoka disease resistance in plantain banana (*Musa* spp.) hybrids. *Theoretical and Applied Genetics* 89(2-3), 146–152.

Otálvaro, F., Nanclares, J., Vaáasquez, L. E., Quiñones, W., Echeverri, F., Arango, R. and Schneider, B. 2007. Phenalenone-type compounds from *Musa acuminata* var. "Yangambi km 5" (AAA) and their activity against *Mycosphaerella fijiensis*. *Journal of Natural Products* 70(5), 887-890.

Poeydebat, C., Carval, D., Tixier, P., Daribo, M. O. and De Bellaire, L. L. 2018. Ecological regulation of black leaf streak disease driven by plant richness in banana ecosystems. *Phytopathology* 108, 1185–1195.

Passos, M. A., de Oliveira Cruz, V., Emediato, F. L., de Camargo Teixeira, C., Souza, M. T., Matsumoto, T., Rennó Azevedo, V. C., Ferreira, C. F., Amorim, E. P., de Alencar Figueiredo, L. F., Martins, N. F., de Jesus Barbosa Cavalcante, M., Baurens, F. C., da Silva, O. B., Pappas, G. J., Pignolet, L., Abadie, C., Ciampi, A. Y., Piffanelli, P. and Miller, R. N. 2012. Development of expressed sequence tag and expressed sequence tag-simple sequence repeat marker resources for *Musa acuminata*. *AoB Plants* 2012, pls030.

Passos, M. A. N., de Cruz, V. O., Emediato, F. L., de Teixeira, C. C., Azevedo, V. C. R., Brasileiro, A. C. M., Amorim, E. P., Ferreira, C. F., Martins, N. F., Togawa, R. C., Júnior, G. J., da Silva, O. B. and Miller, R. N. 2013. Analysis of the leaf transcriptome of *Musa acuminata* during interaction with *Mycosphaerella musicola*: gene assembly, annotation and marker development. *BMC Genomics* 14, 78.

Paxton, J. D. 1981. Phytoalexins: a working redefinition. *Journal of Phytopathology* 101(2), 106–109.

Pei, X., Li, S., Jiang, Y., Zhang, Y., Wang, Z. and Jia, S. 2007. Isolation, characterization and phylogenetic analysis of the resistance gene analogues (RGAs) in banana (*Musa* spp.). *Plant Science* 172(6), 1166–1174.

Pérez-Vicente, L., Carreel, F., Roussel, V., Carlier, J. and Abadie, C. 2021. In Dita, M., et al. Practical guidelines for early screening and field evaluation of banana against Fusarium wilt, *Pseudocercospora leaf spots and drought*. Bioversity International, Montpellier, France.

Philpot, J. and Knowles, C. H. 1913. *Report on a Visit to Sigatoka*. Pamphlet of the Department of Agriculture, Suva, Fiji.

Popma, F. 2018. Ripening systems for bananas. In: Kema, G. and Drenth, A. (Eds) *Achieving Sustainable Cultivation of Bananas Volume 1: Cultivation Techniques*. Burleigh Dodds Science Publishing, Cambridge.

Portal, O., Izquierdo, Y., Vleesschauwer, D., Sánchez-Rodríguez, A., Mendoza-Rodríguez, M., Acosta-Suárez, M., Ocaña, B., Jiménez, E. and Höfte, M. 2011. Analysis of expressed sequence tags derived from a compatible *Mycosphaerella fijiensis*-banana interaction. *Plant Cell Reports* 30(5), 913–928.

Ramsey, M. D., Daniells, J. W. and Anderson, V. J. 1990. Effects of Sigatoka leaf spot (*Mycosphaerella musicola* Leach) on fruit yields, field ripening and green life of bananas in North Queensland. *Scientia Horticulturae* 41(4), 305–313.

Rieux, A., Halkett, F., De Lapeyre De Bellaire, L., Zapater, M. F., Rousset, F., Ravigne, V. and Carlier, J. 2011. Inferences on pathogenic fungus population structures from microsatellite data: new insights from spatial genetics approaches. *Molecular Ecology* 20(8), 1661–1674.

Rieux, A., De Lapeyre De Bellaire, L., Zapater, M. F., Ravigne, V. and Carlier, J. 2013. Recent range expansion and agricultural landscape heterogeneity have only minimal effect on the spatial genetic structure of the plant pathogenic fungus *Mycosphaerella fijiensis*. *Heredity (Edinb)* 110(1), 29–38.

Rieux, A., Hostachy, B., de Lapeyre de Bellaire, L., Martin, Y., Maratchia, G., Dupuis, A. S., Ioos, R., Jeandel, C. and Hubert, J. 2019. First report of black Sigatoka disease in banana caused by *Mycosphaerella fijiensis* on Reunion Island. *New Disease Reports* 39(1), 12.

Rijzaani, H., Bayer, P. E., Rouard, M., Doležel, J., Batley, J. and Edwards, D. 2021. The pangenome of banana highlights differences between genera and genomes. *Plant Genome*, e20100.

Rivas, G. G., Zapater, M. F., Abadie, C. and Carlier, J. 2004. Founder effects and stochastic dispersal at the continental scale of the fungal pathogen of bananas *Mycosphaerella fijiensis*. *Molecular Ecology* 13(2), 471–482.

Robert, S., Ravigne, V., Zapater, M. F., Abadie, C. and Carlier, J. 2012. Contrasting introduction scenarios among continents in the worldwide invasion of the banana fungal pathogen *Mycosphaerella fijiensis*. *Molecular Ecology* 21(5), 1098–1114.

Robert, S., Zapater, M. F., Carlier, J., Abadie, C. and Ravigne, V. 2015. Multiple introductions and admixture at the origin of The Continental spread of the fungal banana pathogen Mycosphaerella fijiensis in Central America: a statistical test using approximate bayesian computation. *Revue d'Ecologie (La Terre et la Vie)* 70, 127–138.

Robinson, J. C., Anderson, T. and Eckstein, K. 1992. The influence of functional leaf removal at flower emergence on components of yield and photosynthetic compensation in banana. *Journal of Horticultural Science* 67(3), 403–410.

Rodriguez, H. A., Rodriguez-Arango, E., Morales, J. G., Kema, G. and Arango, R. E. 2016. Defense gene expression associated with biotrophic phase of *Mycosphaerella fijiensis* M. Morelet infection in banana. *Plant Disease* 100(6), 1170–1175.

Rodriguez, H. A., Hidalgo, W. F., Sanchez, J. D., Menezes, R. C., Schneider, B., Arango, R. E. and Morales, J. G. 2020. Differential regulation of jasmonic acid pathways in resistant (Calcutta 4) and susceptible (Williams) banana genotypes during the interaction with *Pseudocercospora fijiensis*. *Plant Pathology* 69(5), 872–882.

Romero, R. A. and Sutton, T. B. 1997. Sensitivity of *Pseudocercospora fijiensis*, causal agent of black Sigatoka of banana, to propiconazole. *Phytopathology* 87(1), 96–100.

Rouard, M., Droc, G., Martin, G., Sardos, J., Hueber, Y., Guignon, V., Cenci, A., Geigle, B., Hibbins, M. S., Yahiaoui, N., Baurens, F. C., Berry, V., Hahn, M. W., D'Hont, A. and Roux, N. 2018. Three new genome assemblies support a rapid radiation in *Musa acuminata* (wild banana). *Genome Biology and Evolution* 10(12), 3129–3140.

Roy, B. A. 2001. Patterns of association between crucifers and their flower-mimic pathogens: host jumps are more common than coevolution or cospeciation. *Evolution* 55(1), 41–53.

Saravanakumar, A. S., Uma, S., Thangavelu, R., Backiyarani, S., Saraswathi, M. S. and Sriram, V. 2016. Preliminary analysis on the transcripts involved in resistance responses to eumusae leaf spot disease of banana caused by *Mycosphaerella eumusae*, a recent add-on of the sigatoka disease complex. *Turkish Journal of Botany* 40, 461–471.

Sardos, J., Rouard, M., Hueber, Y., Cenci, A., Hyma, K. E., van den Houwe, I., Hribova, E., Courtois, B. and Roux, N. 2016. A genome-wide association study on the seedless phenotype in banana (*Musa* spp.) reveals the potential of a selected panel to detect candidate genes in a vegetatively propagated crop. *PLoS ONE* 11(5), e0154448.

Sardos, J., Breton, C., Perrier, X., van den Houwe, I., Carpentier, S., Paofa, J., Rouard, M. and Roux, N. 2022. Hybridization, missing wild ancestors and the domestication of cultivated diploid bananas. *Frontiers in Plant Science* 13, 969220. DOI: 10.3389/fpls.2022.969220.

Simmonds, N. W. 1962. *The Evolution of the Bananas*. Tropical Sciences Publishers, Longman Publishers, London.

Stergiopoulos, I., Cordovez, V., Okmen, B., Beenen, H. G., Kema, G. H. J. and De Wit, P. J. G. M. 2014. Positive selection and intragenic recombination contribute to high allelic diversity in effector genes of *Mycosphaerella fijiensis*, causal agent of the black leaf streak disease of banana. *Molecular Plant Pathology* 15(5), 447–460.

Stover, R. H. 1962. Intercontinental spread of banana leaf spot (*Mycosphaerella musicola* Leach). *Tropical Agriculture* 39(4), 327–338.

Stover, R. H. 1972. *Banana, Plantain and Abaca Diseases*, 316 p. Commonwealth Mycological Institute, Kew, Surrey, England.

Stover, R. H. 1978. Distribution and probable origin of *Mycosphaerella fijiensis* in Southeast Asia. *Tropical Agriculture, Trinidad* 55, 65–68.

Stover, R. H. 1980. Sigatoka leaf spots of bananas and plantains. *Plant Disease* 64(8), 750–756.

Stover, R. H. 1983. Effet du Cercospora noir sur les plantains en Amérique centrale. *Fruits* 38(4), 326–329.

Sulliman, K. S.-U. S., Pillay, M. and Yasmina, J.-F. 2012. Polymorphism at selected defence gene analogs (DGAs) of *Musa* accessions in Mauritius. *African Journal of Biotechnology* 11(51), 11207–11220. DOI: 10.5897/AJB11.2211.

Thompson, A. K. 1998. *Controlled Atmosphere Storage of Fruits and Vegetables*. CAB International, Wallingford, UK, 278 p.

Timm, E. S., Pardo, H. L., Coello, R. P., Navarrete, T. C., Villegas, O. N. and Ordóñez, E. S. 2016. Identification of differentially-expressed genes in response to *Mycosphaerella fijiensis* in the resistant *Musa* accession 'Calcutta-4' using suppression subtractive hybridization. *PLoS ONE* 11(8), e0160083.

Torres, J. M., Calderón, H., Rodríguez-Arango, E., Morales, J. G. and Arango, R. 2012. Differential induction of pathogenesis-related proteins in banana in response to *Mycosphaerella fijiensis* infection. *European Journal of Plant Pathology* 133(4), 887–898.

Tushemereirwe, W. K. and Waller, J. M. 1993. Black leaf streak (*Mycosphaerella fijiensis*) in Uganda. *Plant Pathology* 42(3), 471–472.

Uma, S., Backiyarani, S., Saravanakumar, A. S., Chandrasekar, A., Thangavelu, R. and Saraswathi, M. S. 2016. Identification of *Mycosphaerella eumusae* responsive unique genes/transcripts from a resistant banana cultivar. *Acta Horticulturae*, 1114, 111–118.

Vakili, N. G. 1968. Responses of *Musa acuminata* species and edible cultivars to infection by *Mycosphaerella musicola*. *Tropical Agriculture* 45(1), 13–22.

Vally, V., Jouen, E., Pecheur, B., Carval, D., De Lapeyre, L., Teycheney, P. Y., Roussel, V., Fabre, S., Pages, C., Adreit, H., Carlier, J. and Rieux, A. 2023. First report of black leaf streak disease in bananas caused by *Pseudocercospora fijiensis* on Mauritius island. *Plant Disease*. DOI: 10.1094/PDIS-09-23-1787-PDN.

van Esse, H. P. ,Reuber, T. L. and van der Does, D. 2020. Genetic modification to improve disease resistance in crops. *New Phytologist* 225(1), 70–86.

VanEtten, H. D., Mansfield, J. W., Bailey, J. A. and Farmer, E. E. 1994. Two classes of plant antibiotics: phytoalexins versus "Phytoanticipins". *Plant Cell* 6(9), 1191–1192.

Van Westerhoven, A. C. 2023. A chromosome-level genome assembly of *Zasmidium syzgii* isolated from banana leaves. *bioRxiv*. DOI: 10.1101/2023.08.24.554682.

Vargas, A., Araya, M., Guzman, M. and Murillo, G. 2009. Effect of leaf pruning at flower emergence of banana plants (Musa AAA) on fruit yield and BLSD (*Mycosphaerella fijiensis*) disease. *International Journal of Pest Management* 55(1), 19-25.

Wardlaw, C. W. 1939. Cercospora leaf spot disease of bananas. *Nature* 144(3635), 11-14.

Wardlaw, C. W. 1961. *Banana Diseases Including Plantains and Abaca* (1st edn), 648 p. Longmans.

Wielemaker, F. 2018. Organic banana cultivation and sustainability. In: Kema, G. and Drenth, A. (Eds) *Achieving Sustainable Cultivation of Bananas Volume 1: Cultivation Techniques*. Burleigh Dodds Science Publishing, Cambridge.

Witek, K., Lin, X., Karki, H. S., Jupe, F., Witek, A. I., Steuernagel, B., Stam, R., van Oosterhout, C., Fairhead, S., Heal, R., Cocker, J. M., Bhanvadia, S., Barrett, W., Wu, C. H., Adachi, H., Song, T., Kamoun, S., Vleeshouwers, V. G. A. A., Tomlinson, L., Wulff, B. B. H. and Jones, J. D. G. 2021. A complex resistance locus in *Solanum americanum* recognizes a conserved *Phytophthora* effector. *Nature Plants* 7(2), 198-208.

Wu, W., Yang, Y. L., He, W. M., Rouard, M., Li, W. M., Xu, M., Roux, N. and Ge, X. J. 2016. Whole genome sequencing of a banana wild relative *Musa itinerans* provides insights into lineage-specific diversification of the *Musa* genus. *Scientific Reports* 6, 31586.

Zandjanakou-Tachin, M., Ojiambo, P. S., Vroh-Bi, I., Tenkouano, A., Gumedzoe, Y. M. and Bandyopadhyay, R. 2013. Pathogenic variation of *Mycosphaerella* species infecting banana and plantain in Nigeria. *Plant Pathology* 62(2), 298-308.

Zapater, M. F., Abadie, C., Pignolet, L., Carlier, J. and Mourichon, X. 2008. Diagnosis of *Mycosphaerella* spp., responsible for Mycosphaerella leaf spot diseases of bananas and plantains, through morphotaxonomic observations. *Fruits* 63(6), 389-393.

Zimmerman, A. 1902. Uber einige tropisher Kulturpflanzen beobachtete Pilze. *Zentralblatt für Nakteriologie, Parasitenkunde, Infektionskrankheiten und Hygiene* 8, 219.

Chapter 2

Controlling black leaf streak disease (BLSD) in banana: the case of Costa Rica

Pablo Chong, ESPOL Polytechnic University, Ecuador; Claudiana Carr and Gilberth Murillo, National Banana Corporation (CORBANA) and Cropland Biosciences, Costa Rica; Mauricio Guzmán, Cropland Biosciences, Costa Rica; Randy Villalobos and Jorge Sandoval, National Banana Corporation (CORBANA) and Cropland Biosciences, Costa Rica; and Gert H. J. Kema, Wageningen University, The Netherlands

1 Introduction

Bananas are among the most important global crops (Ploetz et al., 2015). According to the Food and Agriculture Organization of the United Nations (FAO), global production reached 125 million tons of which 20.2 million tons were exported in 2019 (FAO, 2020). The majority of bananas are a key staple food for millions of people in tropical countries (Singh et al., 2016).

The majority of bananas grown either for export or local consumption are susceptible to black leaf streak disease (BLSD), an important foliar disease caused by the Dothideomycete fungus *Pseudocercospora fijiensis* (Ploetz et al., 2015). Airborne *P. fijiensis* ascospores can travel hundreds of meters, promoting rapid local spread (de Lapeyre de Bellaire et al., 2010; Rieux et al., 2014). The disease causes extensive defoliation (Fig. 1) that, in turn, induces premature

http://dx.doi.org/10.19103/AS.2022.0108.02

Figure 1 A banana leaf with the late stages of BLSD.

fruit ripening making fruit unsuitable for transport to distant markets. BLSD thus both directly reduces yields and also indirectly reduces export volumes through inferior fruit quality (Ploetz, 2000; Ploetz et al., 2015). Since the 1950s, the export banana trade has been dominated by the 'Cavendish' banana varieties which are highly vulnerable to BLSD (Drenth and Kema, 2021). BLSD also has a major effect on subsistence production of banana and plantain since most of the smallholders cannot afford the costs of the fungicides typically used to control the disease. The average cost for the chemical control of BLSD is 1000–1500 US$.ha^{-1}.year^{-1} equivalent to 25% of the production cost (Ploetz, 2000; Ploetz et al., 2015; Becker et al., 2021), and smallholders will lack the equipment necessary to effectively apply it to the canopy.

The main control measure in export-oriented banana plantations involves the frequent application of fungicides resulting in a very high economic and environmental burden (Ploetz et al., 2015; Risède et al., 2010; de Lapeyre de Bellaire et al., 2010). This excessive use of fungicides for BLSD control has caused substantial changes in fungicide sensitivity in the pathogen populations and resistance has reached such levels that effective control is becoming very challenging (Chong et al., 2021).

The development of resistance to fungicides in the pathogen populations is exemplified by BLSD management in Costa Rica, a major exporter of bananas, which had to deal with the disease soon after its incursion in Central America in 1972 (Marín et al., 2003). BLSD was first reported in Fiji in the Pacific in 1963 (Rhodes, 1964) and arrived in Costa Rica in 1979 (de Lapeyre de Bellaire et al., 2010; Rivas et al., 2004), As in other countries, BLSD has displaced yellow Sigatoka as the principal leaf spot disease affecting banana production worldwide (Guzmán et al., 2013).

Well-documented records of historical disease management at the San Pablo Farm of the National Banana Corporation (CORBANA) enables

Table 1 Historical timeline of the evolution and important events in the chemical control of BLSD

Year	BLSD strategies, evolution and important events	Reference
1934	Arrival of *Pseudocercospora musae* (*Mycosphaerella musicola*), causal agent of yellow Sigatoka, to the Americas and the beginning of the use of Bordeaux mixture.	Stover (1962)
1941	Leach identifies the unfurling cigar (heart) leaf as the major target for the *P. musae* spores. He also proves that the heart leaf cannot be protected by protectant fungicides.	Leach (1946)
1946	Leach publishes his discoveries in 'Banana leaf spot (*M. musicola*)' on the Gros Michel variety in Jamaica: 'Investigations on the etiology of the disease and the principles of control by spraying'.	Leach (1946)
1950	The first dithiocarbamate, Maneb, introduced for yellow Sigatoka control.	Stover (1990)
1956	Guyot and Guillé discovered oil sprays application.	Guyot and Guillé (1954)
1958	Introduction of the dithiocarbamate mancozeb (Dithane) for the control of yellow Sigatoka. Maneb successfully controls leaf spot diseases together with copper applications by aircraft.	Stover (1990)
1960	Klein publishes the first forecasting method of oil spray in Honduras. He shows that oil controls the streak disease stage and that the stages of the disease can be used as a decision moment for fungicide application.	Klein (1960)
1962	Discovery that oil-in-water mixtures are more effective than oil alone.	Stover (1990)
1963	First report of BLSD in Fiji.	Rhodes (1964)
1967	DuPont's first systemic fungicide (the first benzimidazole), compound number 1991 or Benlate (benomyl), is evaluated in Honduras. In the same year, Calixin (tridemorph) was also tested in Honduras.	Stover (1990)
1972	• The second forecasting method was developed in Guadeloupe by Ganry and Meyer, who divided the disease symptoms in five stages. They correlated disease development with temperature and evaporation rates. This was the first system introducing oil and systemic fungicides (Benlate). • BLSD is discovered in Honduras and from there it dispersed throughout Latin America and the Caribbean.	Ganry and Meyer (1972) and Guzmán et al. (2013)
1973	First epidemic of BLSD in Honduras. Benlate was effectively and extensively used for disease control. First use of in-oil mixtures and later in 'cocktails' (different mode of action fungicide mixtures) with Mancozeb.	Stover (1990)
1976	First fungicide resistance appears against Benlate in Honduras, followed by a serious outbreak of BLSD.	Stover (1990)

(*Continued*)

Table 1 (*Continued*)

Year	BLSD strategies, evolution and important events	Reference
1979	Chlorothalonil was introduced but did not provide adequate control under conducive conditions for the disease (heavy rain promoting abundant ascospore production).	Stover (1990)
1980	The Fungicide Resistance Action Committee (FRAC) was founded as an organization designed to discuss resistance problems and to formulate plans for cooperative efforts to avoid or manage fungicide resistance.	http://www.frac.info/about-frac/why-frac-
1981	Dithane, Benlate and Calixin were reintroduced to control BLSD. At that time protectant fungicides were used combined with cocktails during rainy periods and almost exclusively in dry periods.	Stover (1990)
1983	First report of BLSD in Cameroon. The symptom development method was modified to six stages by Fouré.	Fouré (1985) and Stover (1990)
1984	The first triazole fungicide propiconazole belonging to the DMI group was introduced together with a forecasting method to control BLSD.	Stover (1990)
1986	Flusilazole fungicide introduction for BLSD control. Propiconazole and flusilazole were extremely effective having curative effect when applied 14 days after the cigar leaf unfurled. Both compounds were extremely persistent in the leaf (up to 60 days after application) and were applied as oil-in-water emulsions. Decision support systems (forecasting) and the use of triazoles reduced the number of application cycles for BLSD control from 35–45 to 20.	Stover (1990)
1987	Guidelines were established to reduce the development of resistance in *Pseudocercospora fijiensis* to azoles and included no year-round applications (maximum 4–8 cycles over 6 months).	Stover (1990)
1988	Fungicide resistance monitoring methods were discussed/described during a FRAC meeting on the use of DMI fungicides.	Stover (1990)
1997	The first quinol oxidation inhibitor (QoI) or strobilurin was introduced.	Amil et al. (2007)
2000	Resistance to the QoIs already common in some farms in the production zones of Costa Rica.	Amil et al. (2007)
2003	• Reduced efficacy of DMIs toward *P. fijiensis* is prevalent in many countries. • FRAC Banana Working Group is established.	Marín et al. (2003) and http://www.frac.info/about-frac/why-frac-
2004	Anilinopyrimidines introduced.	Guzmán (2007)
2006	DMI sensitivity is not 'restored' after refraining from application over a period of 4 years in Costa Rica.	Guzmán (2007)
2008	*P. fijiensis* populations in Costa Rica are nearly fixed for QoI resistance (92–100%).	Arango Isaza et al. (2016)

(*Continued*)

Table 1 (*Continued*)

Year	BLSD strategies, evolution and important events	Reference
2009	Correlation between reduced efficacy of triazoles and point mutations in the *Pfcyp51* gene discovered.	Cañas-Gutiérrez et al. (2009)
2010	• Irrespective of the country of origin in Latin America, the resistance baseline for boscalid, fluopyram and isopyrazam is high. • BLSD incursion in Martinique.	FRAC (2012) and Ioos et al. (2011)
2012	BLSD incursion in Guadeloupe.	Guzmán et al. (2013)
2016	Genome analyses discover high frequencies of QoI resistant genotypes in fungicide-treated plantations. Resistance is fixed in the population.	Arango Isaza et al. (2016)
2018	Overexpression of *Pfcyp51* gene is found. Costa Rican isolates from 2009 have insertions in the promoter region.	Diaz-Trujillo et al. (2018)
2019	First genetic mapping of DMI resistance in *P. fijiensis*. DMI resistance is mainly driven by changes in the *Pfcyp51* gene.	Chong et al. (2019)
2021	Global survey and molecular analyses of DMI resistance in *P. fijiensis*. Samples taken from eight different countries from Southeast Asia, Africa, Latin America and the Caribbean. Martinique and Guadeloupe strains are sensitive to DMI. Strains from non-treated zones in many production countries are sensitive to DMIs.	Chong et al. (2021)

a critical review of trends in BLSD management (Table 1). Since Cavendish varieties dominate exports (98%) and global production (45%) (Ploetz, 2015), the pro blem of excessive fungicide applications and resistance in *P. fijiensis* populations to these fungicides is not restricted to Costa Rica. Similar trends are observed in the Philippines and to a lesser extent in other Latin American countries such as the Dominican Republic, Colombia, and Ecuador (Chong et al., 2021). In this chapter, Costa Rica is used as an example to highlight the challenges and the urgent need for alternative, and more sustainable disease management approaches.

2 Fungicides used for black leaf streak disease management

Although cultural practices affect BLSD development (Pérez-Vicente, 2012), fungicides remain the cornerstone of contemporary disease control, particularly in banana plantations in wet tropical areas producing fruit for export (Table 2). Overall, the international banana industry deploys two groups of fungicides for BLSD management that are categorized according to their phytomobility

Table 2 Fungicides used in BLSD control ordered by their chemical classifications (FRAC, 2018)

Chemical class		Active ingredient	Fungicide trade name[a]	Description
Protectant fungicides	Benzene derivatives	Chlorothalonil	Bravo, Bronco, Daconil	This compound interferes with the glutathione pathway, a coenzyme and 2-mercaptoethanol reducing thiol-group-based metabolism in the cell. Under prohibitive EU legislation.
	Carbamates	Mancozeb	Dithane	Mancozeb is placed in the subclass of carbamate pesticides called dithiocarbamates. As a cholinesterase inhibitor, it affects the nerves. Under prohibitive EU legislation.
Systemic fungicides	Demethylation inhibitors (DMI)	Bitertanol	Baycor	These compounds inhibit the lanosterol 14α-demethylase, an essential enzyme in the ergosterol biosynthesis pathway.
		Difenoconazole	Score, Sico	
		Epoxiconazole	Opus	
		Fenbuconazole	Indar	
		Myclobutanil	Rally, Sisthane	
		Propiconazole	Tilt, Bumper	
		Tebuconazole	Silvacur	
		Tetraconazole	Eminente	
		Triadimenol	Bulldock, Caporal,	
	Amine fungicides	Spiroxamine	Impulse	Amines are ergosterol synthesis inhibitions. Tridemorph inhibits the Δ8–Δ7 isomerase and the C14 reductase in the ergosterol metabolic pathway.
		Fenpropimorph	Volley	
		Tenpropidin	Seeker 750	
		Tridemorph	Calixin, Musaclean	

Anilinopyrimidines (AP)	Pyrimethanil	Siganex	AP's inhibit the methionine biosynthesis. It should only be used in mixtures and in full alternation. To reduce selection pressure, the total number of applications is limited to eight per year and these should not represent more than 50% of total number of sprays.
Benzimidazoles (BCM)	Benomyl	Benlate	Fungicides with high systemic and curative activity that allow long intervals between applications. Resistant *P. fijiensis* strains were detected 2 years after first application.
	Carbendazim	Curacarb	
	Thiophanate	Cycosin	
	Thiabendazole	Mertect	
	Thiophanate-methyl	Nucilate, Thiophol, Topsin	
N-Phenylcarbamates	Dietofencarb,	Powmyl	No resistance data are yet available. A maximum of 33% of the total number of sprays can be applied with *N*-phenylcarbamates.
SDHI fungicides	Boscalid	Cumora	Irrespective of the country of origin in Latin America, the overall resistance baseline for boscalid, fluopyram, and isopyrazam is high.
	Fluopyram	Luna	
	Bsopyrazam	Reflect	
Guanidines	Dodine	Syllit	Irrespective of the country of origin in Latin America, the resistance baseline for dodine is variable. However, in Ecuador, the resistance baseline did not significantly change after years of applications.

[a] Some trade names also include mixes with other active ingredients.

into protectant (contact) or penetrant (systemic) fungicides. Prior to the 1970s, almost all fungicides were protectants (Ma and Michailides, 2005). They remain on the surface of the leaf and are only redistributed by precipitation, irrigation, and dew.

Penetrant or systemic fungicides, on the other hand, are absorbed by plant tissue and can be subdivided into acropetal penetrants, local penetrants, or systemic penetrants. Acropetal penetration move between cells along with the water potential gradient. They are xylem mobile and therefore translocated toward the leaf tips and margins (Latin, 2011). Local penetrants diffuse into the leaf cuticle and remain there due to their chemical affinity with the cuticle wax. However, movement of some compounds throughout the cuticle wax and between the cells toward the cuticle of the opposite side of the leave is also possible (Latin, 2011). Finally, systemic penetrants are compounds that move through the cells (live protoplast) following a sugar gradient of high to low concentrations, e.g. from expanded leaves to the roots and newly emerging leaves (Latin, 2011).

The main compound currently used for BLSD management is the protectant fungicide mancozeb. Although mancozeb has been banned in the European Union (European Commission, 2020), it is still being used for BLSD control, and revised mancozeb´s minimum residual levels are in place (Anastassiadou, et al., 2020b). This may change in the future and for that reason the industry is desperately searching for replacements for this compound. Other compounds used for BLSD control are acropetal-penetrant fungicides such as the sterol DMIs: propiconazole, epoxiconazole, difenoconazole, and tebuconazole. Currently, DMIs are the major systemic fungicides used for BLSD management in banana. They affect the lanosterol 14α-demethylase (CYP51) pathway (Cañas-Gutiérrez et al., 2009), which is essential for ergosterol biosynthesis. This enzyme catalyzes the removal of a methyl group from lanosterol in fungi, allowing sterol metabolism to produce ergosterol. Triazoles inhibit the CYP51 function by binding to the heme cofactor in the active site of the enzyme. Therefore, the sterol metabolism is hampered due to the accumulation of 14-α-methyl-3,6-diol, a toxic sterol produced by the Δ-5,6-desaturase encoded by the ERG3gene, which affects membrane integrity and leads to growth inhibition and cell death (Lupetti et al., 2002; Pietila et al., 2006; Shapiro et al., 2011).

Propiconazole was the first DMI introduced in 1984 (Stover, 1990) and has been widely used ever since. However, its use was banned in the EU in 2021 (European Commission, 2021), but in Latin America, it is still used to control BLSD.

In 1997, strobilurins or QoIs were introduced (Amil et al., 2007; Knight et al., 2003). These new fungicides block the respiration pathway by interfering with the cytochrome bc1 (*Pfcytb*) complex in the mitochondria (Sierotzki et al.,

2000). Due to their entirely new chemistry and efficacy, strobilurins (such as azoxystrobin, pyraclostrobin, and trifloxystrobin) were quickly widely used accounting for >20% of the global fungicide market within the first ten years after their introduction (Fernández et al., 2010).

Amines represent another important group of systemic fungicides that inhibit ergosterol synthesis. The first amine compound, tridemorph, was introduced in 1967 in Honduras to control BLSD (Stover, 1990) and is still used in bananas to control BLSD, along with spiroxamine, fenpropimorph, and fenpropidin. Tridemorph inhibits the Δ8–Δ7 isomerase and the C14 reductase in the ergosterol metabolic pathway. Thus far, no resistance to these compounds has been found in *P. fijiensis*, most probably due to their dual site mode of action (FRAC, 2022).

The most recent fungicides for BLSD management are the 'second and third generation' carboxamides or succinate dehydrogenase inhibitors (SDHIs). First-generation carboxamides were discovered in the mid-1960s for the control of basidiomycete fungi. Their fungicidal activity is based on the disruption of the mitochondrial tricarboxylic acid cycle (TCA) through inhibition of the succinate dehydrogenase (SDH) enzyme. At the molecular level, carboxamides inhibit ubiquinone reduction by binding to the ubiquinone binding site (Qp site) of the SDH enzyme (Scalliet et al., 2012). The SDHIs in the banana industry comprise boscalid, fluopyram, fluxapyroxad, and isopyrazam.

3 Fungicide application programs

In most cases, several fungicides are simultaneously applied as 'cocktails' comprising two or three fungicides with different modes of action, mostly one or two systemic fungicides and a protectant. An example would be a DMI, an amine, and mancozeb. Depending on the proportion of resistant strains in the pathogen population, QoIs or SDHIs (both acting on the fungal mitochondria) are sometimes used instead of DMIs. The QoIs are mostly mixed with amines or pyrimidines.

Due to the epidemiology of BLSD, with almost continuous ascospore production (de Lapeyre de Bellaire et al., 2010; Marín et al., 2003), protectants remain very important components of tank mixes, sometimes combined with new or specialty products such as mineral oils, vegetable wax, and various biologicals which also have a protective function. However, the use of protectants is becoming restricted, since the ban on chlorothalonil and mancozeb compounds in the European Union in 2020 (Anastassiadou, et al., 2020a,b; Arena et al., 2018; European Commission, 2020). These restrictions also affect the use of these fungicides on plant products destined for the European market.

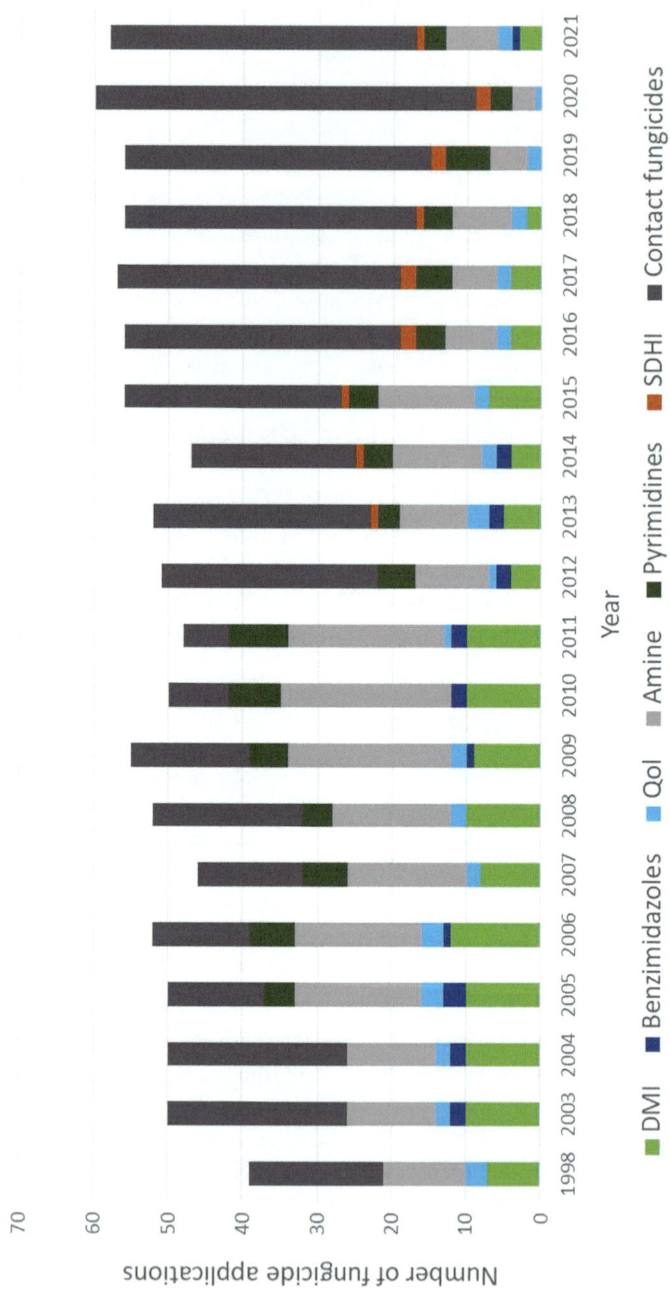

Figure 2 An overview of different chemicals used to control BLSD at the San Pablo farm in Limón, Costa Rica, in 1998 and during the period 2003–21. One application can include several fungicides in mixtures.

At present, weekly spray schedules are required with up to five applications per month (depending on rainfall). A summary of the number of applications and the specific systemic fungicide groups used on the San Pablo Farm since 1998 is shown in Fig. 2. In that year, the number of applications comprised two applications per month from May to August and three per month in the rest of the year and most fungicides were applied separately. Systemic fungicides were applied over a 6-month period, alternating weekly between products with a different mode of action or by using a protectant fungicide.

By 2003, applications had increased to four per month, with five per month in some periods. Although some protectants were used in isolation, most fungicides were mixed in 'cocktails': either systemic/protectant or protectant/protectant tank mixtures. From 2008 onwards, spray schedules were consistently at least four applications per month using systemic fungicides applied in different mixtures with alternating mode of actions over a 6-month period. The number of applications increased to five per month in 2009, a trend that continued up to the present (Fig. 2).

4 Assessing levels of fungicide reaching pathogens

The actual doses that reach the fungus on the leaf is the key element related to the selection pressure that, e.g. DMI fungicides exert on *P. fijiensis* populations. The translation from laboratory efficacy trials to efficacy under field conditions is difficult and little tested.

In general, most farmers follow the manufacturer's technical instructions as printed on the label of commercial fungicide products. However, the effective dose also depends on the 'carrier', mineral oil or oil/water emulsions with different emulsifiers. Fisher and Didio (1991) reported that a mix of propiconazole in emulsion penetrated the leaf better (90%) than a mix with mineral oil (43%) although the latter had a significantly better leaf coverage than emulsions, particularly under moderate dew conditions (Ciba-Geigy, 1991).

Details on the final propiconazole dose under field conditions are available in the 1991 Ciba-Geigy Tilt technical dossier which provides a comprehensive description of propiconazole application, leaf penetration, and degradation inside the leaf. This showed that an oil-in-water emulsion is the best combination for leaf penetration from 13 $mg.L^{-1}$ of fungicide inside the leaf at the first hour down to a final dose of 2 $mg.L^{-1}$ at 96 h after application. However, the most stable application consisted of an oil mixture that, despite a relatively low leaf penetration (5 $mg.L^{-1}$ at the first hour after application), maintained a concentration of 6 $mg.L^{-1}$ for the next 11 h and finally decreased to 4 $mg.L^{-1}$ between 24 h and 96 h. A water formulation showed a penetration of 6 $mg.L^{-1}$ during the first hour but this dropped to only 0.5 mg.L^{-1} after 96 h. In general, most of the fungicide penetrating the leaf lamina is no longer detectable on the leaf surface after 12 h (due to plant absorption, biological degradation, run-off, or dilution), but a residual low level of the active ingredient remains in the tissue for more than 4 days (Ciba-Geigy, 1991). Doses on the leaf are usually 1–1.5 times higher on the leaf surface compared to the inside of the leaf during the first 12 h (depending largely on the type of carrier) (Ciba-Geigy, 1991). These results suggest that the actual propiconazole doses that most pathogens face will be 1–5 mg.L^{-1} for at least several days (Ciba-Geigy, 1991). This provides a good basis for experimental studies on resistance.

5 Trends in fungicide resistance

The two main procedures to measure the level of resistance to fungicides in fungal populations are to measure ascospores germ tube lengths or to determine the half-maximal effective concentration (EC_{50}) after exposing the fungus to different concentrations of the fungicide. Ascospore germ tube length evaluations are measured with two or three doses of the fungicide 48 h after application with a considerable number of samples (on average 50–100 spores per treatment with more than three repetitions) (Calvo et al., 1997). However, such data at most approximate changes in the level of resistance in the pathogen population. For a more accurate assessment of the level of resistance in the pathogen population, the EC_{50} is used to determine how much of a particular compound is required to exert a given effect. Typically, mono-ascosporic colonies can be evaluated with eight different fungicide concentrations after 5–10 days of incubation in 96-well microtiter tests (Diaz-Trujillo et al., 2018). This method is precise and provides an overall level of resistance of an individual strain. Apart from sufficient samples, these tests require a range of preliminary experiments for standardization and to be able to determine the resistance baseline of a population with a certain level of statistical accuracy. Once these baselines are known, this technique provides a compelling tool to measure the development of resistance to any class of fungicides.

The sublime efficacy of the QoIs rapidly deteriorated after the first reports of resistant isolates of *P. fijiensis*, due to a point mutation in *Pfcytb* that leads to a change of glycine to alanine at position 143 (G143A) of the protein (Sierotzki et al., 2000). Such resistant strains were already common in some farms in Costa Rica by 2000 (Amil et al., 2007) and in 2008, *P. fijiensis* populations in three commercial plantations in Costa Rica were almost fixed for resistance to strobilurins (Arango Isaza et al., 2016). Despite their overall resistance, QoIs are still used, possibly due to their 'greening effect' in delaying senescence and thus extending the harvesting season (Bennett and Arneson, 2003; Pérez-Vicente, 2006).

Contrary to resistance to strobilurins, which is based on a single gene mutation and thus qualitative, several studies have shown a correlation between the gradual or quantitative resistance to triazoles and point mutations in the *cyp51* gene (Lepesheva and Waterman, 2004; Cañas-Gutiérrez et al., 2009), which belongs to the cytochrome P450 monooxygenase (CYP) superfamily and mediates a crucial step of the synthesis of ergosterol. In *P. fijiensis,* such mutations are abundant, and resistance is also due to overexpression of *Pfcyp51*, as a result of promoter insertions (Diaz-Trujillo et al., 2018; Chong et al., 2019, 2021).

Between 2012 and 2018, isolates with *in vitro* resistance toward SDHI fungicides were identified in Costa Rica, Ecuador, Colombia and Guatemala (FRAC, 2018), although the resistance mechanism is still unknown.

The resistance baseline in *P. fijiensis* populations toward DMI fungicides in Costa Rica is continuously rising. In 2003, the average EC_{50} for propiconazole in Costa Rica was 0.15 $mg.L^{-1}$ with a maximum value of 0.5 $mg.L^{-1}$ (Marín et al., 2003). Diaz-Trujillo et al. (2018) reported a significant increase to an average of 1.10 $mg.L^{-1}$ among four Costa Rican isolates sampled in 2009 with a maximum value of 1.53 $mg.L^{-1}$. Recently Chong et al. (2021) determined an EC_{50} average of 5.8 $mg.L^{-1}$ and a maximum value of 18.4 $mg.L^{-1}$ among 107 *P. fijiensis* isolates sampled in 2014. This suggests that the EC_{50} between 2003 and 2014 gradually increased by a factor of 1.4 per year. Hence, DMI sensitivity is currently 38 times lower than in 2003.

Due to loss of effectiveness, the number of DMI applications reduced over time from ten in 2003 to three applications per year in 2021, but with a temporary increase in applications in 2014 and 2015 (Fig. 3), suggesting that the selection pressure during these years pushed the population toward resistance. The changes in DMI cycles at this farm between 2011 and 2021 were driven by weather forecasts and information systems predicting disease development and the associated fungicide application costs.

General EC_{50} values among *P. fijiensis* isolates across Costa Rica also indicate the selective pressure exerted by DMI fungicides on the pathogen population (Chong et al., 2021). Records from 2001 to 2021 of DMI resistance monitoring from different Costa Rican banana regions (Fig. 4) show fluctuations in sensitivity but an overall increase demonstrating the reduced efficacy of DMIs for Costa Rican *P. fijiensis* populations over time.

It is noticeable that ceasing DMI applications during field experiments for a period of 4 years had little effect on the frequency of resistance strains in the population (Guzmán, 2007), suggesting no or a low fitness penalty for attaining fungicide resistance in this population. This is supported by previous comparative studies between sensitive *P. fijiensis* isolates and isolates with resistance to DMIs. These did not show significant differences in the incubation period, latency period, conidial sporulation, and resulting disease severity

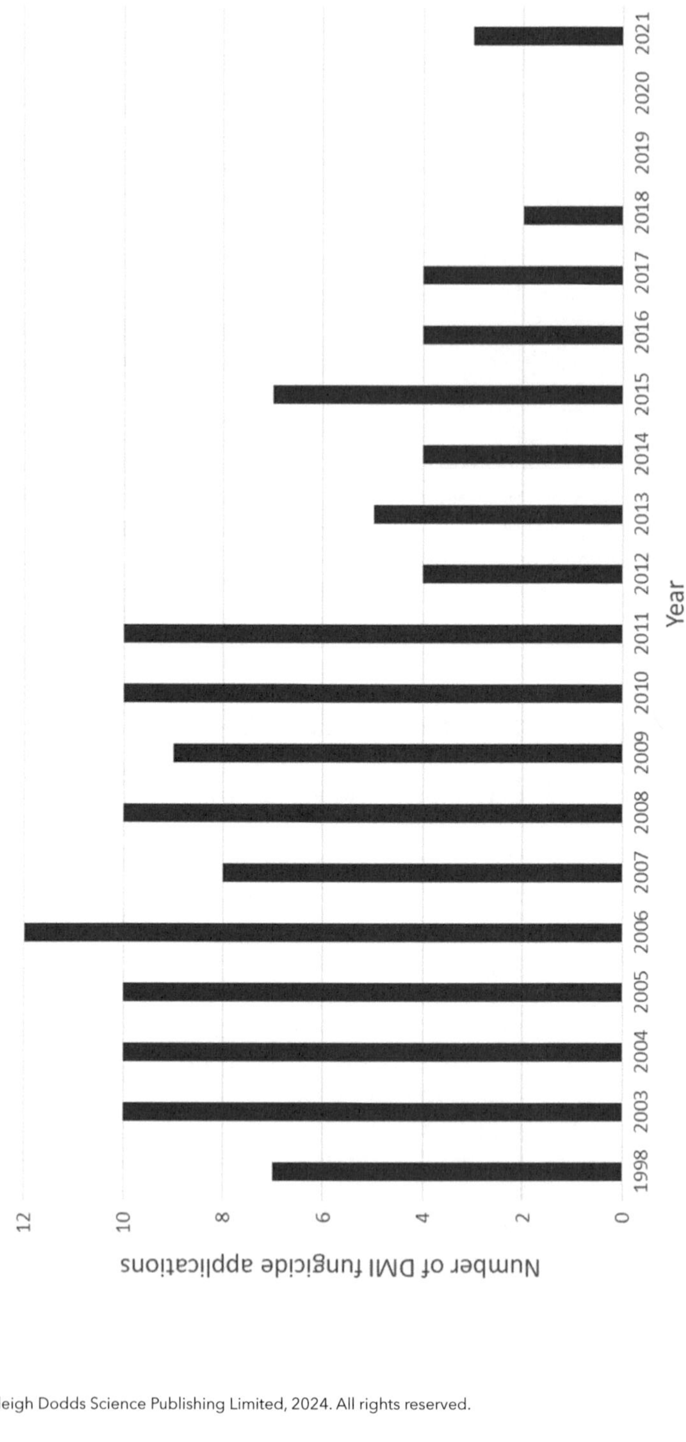

Figure 3 An overview of the number of DMI applications for the control of BLSD at the San Pablo farm in Limon, Costa Rica, in 1998 and during the period 2003–21.

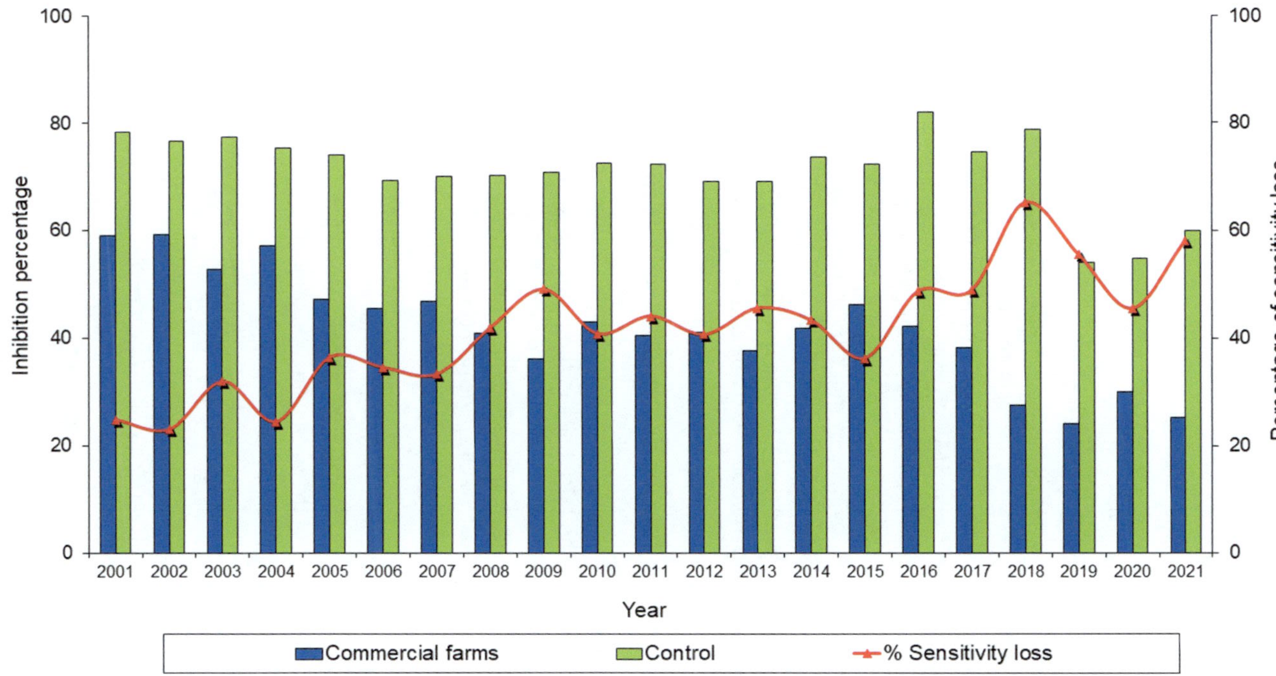

Figure 4 The inhibition of *P. fijiensis* ascospore germ-tube lengths at 0.1 mg.L^{-1} as a measure of sensitivity to the combined effect of the DMI fungicides bitertanol, propiconazole, tebuconazole, difenoconazole, and epoxyconazole in San Carlos, Costa Rica, and commercial Cavendish plantations in the main producing area in Limón in 1998 and during the period 2003–21. The red line represents the trend of the sensitivity reduction.

(Romero and Sutton, 1997, 1998; Guzmán, 2007). This suggests that the use of DMI-free periods does not seem to affect population shifts in sensitivity. An experiment showed that by the end of the 4-year period, the frequency of strains with resistance also increased in control plots that were not treated with DMI fungicides, although this could also be due to gene flow, hence the movement of airborne ascospores across experimental plots.

The increased use of mancozeb, azoxystrobin, and tridemorph, to compensate for reduced cycles with DMIs, has also exerted significant selective pressure for azoxystrobin and tridemorph, and has resulted in *P. fijiensis* populations that are now nearly fixed for strobilurin resistance (Arango Isaza et al., 2016). Whilst isolates can vary in the degree of sensitivity for some fungicides (Latin, 2011; Vincelli, 2014), such variation in sensitivity to strobilurin resistance has not been observed (Arango Isaza et al., 2016), probably due to the different mode of action and genetically different mechanisms that contribute to resistance to fungicides. As mentioned earlier, strobilurin resistance is qualitative and is caused by a point mutation in the *Pfcytb* gene (Amil et al., 2007), whereas resistance to DMIs is quantitative and caused by a range of point mutations in the *cyp51* gene that contribute to conformational changes of the protein, thereby affecting the docking of the active ingredients (Chong, 2016).

6 Understanding fungicide resistance in *P. fijiensis*

One of the main adaptations of exposure to fungicides is underpinned by the genetic variation of the target site in the fungus. Site-specific compounds often select for complete resistance based on point mutations - often referred as 'major gene' resistance or 'monogenic' resistance - that render these compounds ineffective (Latin, 2011), including the mechanisms for resistance to benzimidazoles and strobilurins (Sierotzki et al., 2000; Cañas-Gutiérrez et al., 2006). Under selection pressure, due to the continuous use of such fungicides, the frequency of mutant alleles will increase in the population, conferring reduced sensitivity, or total resistance to such fungicides (Grünwald et al., 2003). The epidemiology of the organism often amplifies the effect and rate of dissemination (Kema et al., 2018).

The target of DMIs is lanosterol 14α-demethylase that is encoded by *Pfcyp51* (Cañas-Gutiérrez et al., 2009). Many studies show a correlation between resistance to propiconazole and point mutations in the *Pfcyp51* gene (Lepesheva and Waterman, 2004; Cañas-Gutiérrez et al., 2009; Diaz-Trujillo et al., 2018; Chong et al., 2021), which has caused cross resistance to epoxiconazole and difenoconazole (Chong et al., 2021). The plethora of *Pfcyp51* mutations has resulted in a total of 28 amino acid substitutions (Chong et al., 2021), of which substitutions at positions 136, 313, 380, 381, and 460–463 are associated with resistance to DMIs (Cañas-Gutiérrez et al., 2009; Diaz-Trujillo et al., 2018;

Chong et al., 2021). These amino acid changes are near central positions of lanosterol 14α-demethylase, surrounding the Substrate Recognition Site (SRS) (e.g. positions Y136, A313, 381) and inside a loop close to the L α helix (e.g. Y460 to Y463) (Lepesheva and Waterman, 2004; Cañas-Gutiérrez et al., 2009; Chong et al., 2021).

Pfcyp51 promoter insertions were recently discovered as a driving mechanism for *Pfcyp51* expression contributing to quantitative variation for DMI resistance (Diaz-Trujillo et al., 2018). Similar mechanisms were identified in *Aspergillus fumigatus* isolates with resistance to medical azole fungicides (Mellado et al., 2007; Snelders et al., 2012; Verweij et al., 2013). Interestingly, none of the DMI-sensitive strains found in a global survey contained promoter insertions, while they were very common in strains with resistance and with the number of insertions correlating with levels of resistance to DMIs (Diaz-Trujillo et al., 2018; Chong et al., 2021). Overexpression of *cyp51* also correlated with promoter insertions in *Venturia inaequalis* and *Blumeriella jaapii* (Schnabel and Jones, 2001; Ma et al., 2006), but their frequency in *P. fijiensis* is higher. In *B. jaapii*, the overexpression results from upstream insertions of various truncated derivatives of LINE-like retrotransposons (Ma et al., 2006). However, the underlying mechanism and function of these repeated elements remains to be deciphered (Schnabel and Jones, 2001; Ma et al., 2006; Diaz-Trujillo et al., 2018) and might involve the generation of binding sites for positive regulators that enhance the promoter.

7 Improving black leaf streak disease management: integrating molecular parameters

Biological parameters/indicators such as the 'Stage of Evolution of Disease' (SED), the 'Youngest Leaf bearing Streaks' (YLSt), The 'Youngest Leaf Spotted' (YLS), and the 'Number of Functional Leaves at Harvest' (NLH) are very important for decision-making in the use of fungicides and the evaluation of the efficacy of BLSD control strategies (de Lapeyre de Bellaire et al., 2010; Fouré and Ganry, 2008).

Current resistance monitoring procedures typically use the ascospores germ-tube length as a parameter for fungicide efficacy. However, the information retrieved using this method makes it hard to identify the underlying mechanism involved. More information is needed to understand the origin and distribution of genotypes with resistance. For example, the currently available genetic information on modulation of the *Pfcyp51* gene, which seems the main driver for resistance (Chong et al., 2019), is a potential add-on for optimizing forecasting programs and disease control.

Quantitative information on *P. fijiensis* population characteristics such as aerial spore quantity, mycelium biomass, target site mutations, and promoter insertions can be easily quantified by PCR technologies such as ddPCR, qT-PCR, or LAMP, in conjunction with well-positioned spore trap devices. This information can potentially be used to assess inoculum pressure, cross-resistance, and multidrug resistance. Molecular monitoring of target site mutations in neighboring plantations can show population profiles, immigration, and gene flow patterns that can be used to predict the success of spray applications. Understanding the dynamics and quantification of the spore load in aerial samples (with simultaneous amplification of the target sites) can improve early warning systems by identifying infection conditions weeks before symptoms are visible on the leaf (Rogers et al., 2009; Villari et al., 2017; Newlands, 2018).

Current costs of generating such data can be substantial due to sample/material preparation, but such analyses provide a much broader view of the evolution of *P. fijiensis* populations and enables site-specific applications rather than aerial blanket fungicide applications over entire plantations. With the continuously reducing costs for genome-based technologies, molecular monitoring of target site mutations is becoming more feasible in optimizing BLSD management similar to monitoring fungicide resistance in other crops.

As an example, the lack of cross resistance between pyraclostrobin and azoxystrobin in *P. fijiensis* can be shown by a simple PCR test with specific molecular markers for the G143A substitution in the *Pfcytb* gene (Sepúlveda and Torres, 2016). A similar analysis for DMIs shows that amino acid changes at positions 313, 136, and 463 (or even combinations of these substitutions) are the most important substitutions causing reduced efficacy. Promoter insertions can be visualized by a simple PCR and are important indicators for reduced efficacy of DMIs as recently demonstrated for Costa Rican isolates (Diaz-Trujillo et al., 2018). Ideally, qPCR or ddPCR tests should be run directly on DNA preps from infected leaves and air samples to rapidly quantify DMI resistance in pathogen populations (Singh and Mustapha, 2013).

8 Conclusion: the long road to sound black leaf streak disease management

This chapter clearly shows that the ceiling for chemical control of BLSD is approaching or has already been reached. The continuously reducing sensitivity to systemic fungicides in the field is a clear wake-up call for an industry that is becoming increasingly reliant on protectants which are now being banned because they are both environmentally unfriendly and threaten the health of farm workers and local communities (Van Wendel De Joode et al., 2016).

Alternative compounds with different modes of action and broader, multisite targets are needed for continued disease control. New biopesticide technologies based on RNA inhibition need to be explored as they may be safer and more environmentally friendly but their effectiveness in banana plantations in the wet tropics remains untested to date (Mumbanza et al., 2013; Mamta and Rajam, 2017; McLoughlin et al., 2018; Islam and Sherif, 2020; Nerva et al., 2020).

An alternative and far more sustainable approach is to deploy genetically based resistance in the host plant. Host resistance is the cornerstone for effective management of fungal diseases in many crops. Resistance to black leaf streak exists in wild as well as cultivated bananas and several breeding programs have developed varieties with adequate resistance to BLSD. Albeit these varieties are important for local markets, no BLSD-resistant varieties have been developed that meet international market demands, leading to an industry that remains bound to fungicides for BLSD control. Recent research into the genetics of resistance and *P. fijiensis* biology provides a foundation for improvements in breeding (Ahmad et al., 2020; Chong et al., 2019, 2021; Diaz-Trujillo et al., 2018). However, developing diversified resistant banana germplasm will require time and resources.

Governments and industry need to invest in modern data-driven resistance monitoring strategies rather than maintaining old-fashioned low-throughput methodologies. Continuous and improved monitoring of fungicide resistance is key to timely and an accurate disease forecasting and control. Improved monitoring techniques can improve decision support systems in forecasting disease, tracking resistance, and informing fungicide application programs (de Lapeyre de Bellaire et al., 2010; Risède et al., 2010). Current disease forecasting programs can work effectively under suboptimal conditions for BLSD development such as the dry tropics (de Lapeyre de Bellaire et al., 2010). An important challenge is to translate these to the wet tropics where - since *P. fijiensis* thrives and undergoes continuous sexual reproduction - it is more challenging to monitor and control. More broadly, there is a need for all actors in the banana value chain, including consumers, to accept that a low price for a susceptible banana variety in Western supermarkets grown with the aid of frequent fungicide applications at the expense of producers and workers in low-wage countries is both unethical and unsustainable in the long term.

9 Acknowledgements

This work was financially supported by the Dutch Dioraphte Foundation, a scholarship to PC from the Ecuadorian government through the Superior Education, Science, Technology, and Innovation Secretary (SENESCYT, grant number 20100202) as well as a fellowship from the ESPOL Polytechnic University, 'Escuela Superior Politécnica del Litoral' (ESPOL, grant number CAc-2011-008) and the Ecuadorian Biotechnology Research Centre 'Centro de Investigaciones Biotecnológicas del Ecuador' (CIBE). We gratefully acknowledge Syngenta AG, Switzerland, for financial support.

10 Where to look for further information

A standard introduction on fungicide-resistant subject is presented by Richard Latin, *A Practical Guide to Turfgrass Fungicides* (see Latin 2011 in the References and further reading section for full details).

The best single source of information on worldwide DMI fungicide resistant in BLSD is found in the Fungicide Resistance Action Committee, FRAC (https://www.frac.info/). The site includes information resources and key contacts.

Centres of expertise included:

- Banana Research at Wageningen University and Research, WUR (https://bananaresearch.org/).
- The French agricultural research and cooperation organization working for the sustainable development of tropical and Mediterranean regions, CIRAD (https://www.cirad.fr/en).
- 'La Corporación Bananera Nacional', CORBANA (https://www.corbana.co.cr/).
- 'El Centro de Investigaciones Biotecnológicas del Ecuador', CIBE (http://www.cibe.espol.edu.ec/).

There are a few current research projects to improve control and management of BLSD:

- Molecular biology of banana Black Sigatoka disease project by the WUR, with the aim of using *P. fijiensis* effectors to select BLSD-resistant banana germplasms (https://www.wur.nl/en/project/molecular-biology-of-banana-black-sigatoka-disease.htm).
- Achieving sustainable resistance to black Sigatoka disease in new dessert banana varieties – DuRéBan by CIRAD, the project aims to generate BLSD-resistant banana germplasms (https://www.cirad.fr/en/worldwide/cirad-worldwide/projects/dureban-project).

11 References

Ahmad, F., Martawi, N. M., Poerba, Y. S., de Jong, H., Schouten, H. J. and Kema, G. H. J. (2020). Genetic mapping of Fusarium wilt resistance in a wild banana *Musa acuminata* ssp. *malaccensis* accession, *Theoretical and Applied Genetics* 133(12), 3409–3418. https://doi.org/10.1007/s00122-020-03677-y.

Amil, A. F., Heaney, S. P., Stanger, C. and Shaw, M. W. (2007). Dynamics of QoI sensitivity in *Mycosphaerella fijiensis* in Costa Rica during 2000 to 2003, *Phytopathology* 97(11), 1451–1457. https://doi.org/10.1094/PHYTO-97-11-1451.

Arango Isaza, R. E., Diaz-Trujillo, C., Dhillon, B., Aerts, A., Carlier, J., Crane, C. F., V de Jong, T., de Vries, I., Dietrich, R., Farmer, A. D., Fortes Fereira, C., Garcia, S., Guzman, M., Hamelin, R. C., Lindquist, E. A., Mehrabi, R., Quiros, O., Schmutz, J., Shapiro, H., Reynolds, E., Scalliet, G., Souza, M., Stergiopoulos, I., Van der Lee, T. A., De Wit, P. J., Zapater, M.

F., Zwiers, L. H., Grigoriev, I. V., Goodwin, S. B. and Kema, G. H. (2016). Combating a global threat to a clonal crop: banana black sigatoka pathogen *Pseudocercospora fijiensis* (Synonym *Mycosphaerella fijiensis*) genomes reveal clues for disease control, *PLoS Genetics* 12(8), 1–36. https://doi.org/10.1371/journal.pgen.1005876.

Becker, P., Esker, P. and Umaña, G. (2021). Incorporation of microorganisms to reduce chemical fungicide usage in black sigatoka control programs in Costa Rica by use of biological fungicides, *Crop Protection* 146(December 2020). https://doi.org/10.1016/j.cropro.2021.105657.

Bennett, R. S. and Arneson, P. A. (2003). Black sigatoka, *The Plant Health Instructor* [Preprint]. https://doi.org/10.1094/phi-i-2003-0905-01.

Cañas-Gutiérrez, G. P., Patiño, L. F., Rodríguez-Arango, E. and Arango, R. (2006). Molecular characterization of benomyl-resistant isolates of *Mycosphaerella fijiensis*, collected in Colombia, *Journal of Phytopathology* 154(7–8), 403–409. https://doi.org/10.1111/j.1439-0434.2006.01113.x.

Cañas-Gutiérrez, G. P., Angarita-Velásquez, M. J., Restrepo-Flórez, J. M., Rodríguez, P., Moreno, C. X. and Arango, R. (2009). Analysis of the CYP51 gene and encoded protein in propiconazole-resistant isolates of *Mycosphaerella fijiensis*, *Pest Management Science* 65(8), 892–899. https://doi.org/10.1002/ps.1770.

Chong, P. (2016). *The Origin, Versatility and Distribution of Azole Fungicide Resistance in the Banana Black Sigatoka Pathogen Pseudocercospora fijiensis*. Wageningen University. https://doi.org/10.18174/387237.

Chong, P., Vichou, A. E., Schouten, H. J., Meijer, H. J. G., Arango Isaza, R. E. and Kema, G. H. J. (2019). Pfcyp51 exclusively determines reduced sensitivity to 14α-demethylase inhibitor fungicides in the banana black sigatoka pathogen *Pseudocercospora fijiensis*, *PLoS ONE* 14(10), pp. 1–17. https://doi.org/10.1371/journal.pone.0223858.

Chong, P., Essoh, J. N., Arango Isaza, R. E., Keizer, P., Stergiopoulos, I., Seidl, M. F., Guzman, M., Sandoval, J., Verweij, P. E., Scalliet, G., Sierotzski, H., de Lapeyre de Bellaire, L., Crous, P. W., Carlier, J., Cros, S., Meijer, H. J. G., Peralta, E. L. and Kema, G. H. J. (2021). A world-wide analysis of reduced sensitivity to DMI fungicides in the banana pathogen *Pseudocercospora fijiensis*, *Pest Management Science* 77(7), 3273–3288. https://doi.org/10.1002/ps.6372.

Ciba-Geigy (1991). *Tilt Thecnical Dossier*. Basle: Ltd. C-G.

de Lapeyre de Bellaire, L., Fouré, E., Abadie, C. and Carlier, J. (2010). Black leaf streak disease is challenging the banana industry, *Fruits* 65(6), 327–342. https://doi.org/10.1051/fruits/2010034.

de Lapeyre de Bellaire, L., Abadie, C., Carlier, J., Ngando, J. and Kema, G. H. J. (2010). Mycosphaerella foliar diseases of bananas : towards an integrated protection. In: Food Quality and Safety (Ed.) *From Science to Field Banana Case Study*. Endure. Paris. 1–6.

Diaz-Trujillo, C., Chong, P., Stergiopoulos, I., Cordovez, V., Guzman, M., De Wit, P. J. G. M., Meijer, H. J. G., Scalliet, G., Sierotzki, H., Lilia Peralta, E., Arango Isaza, R. E. and Kema, G. H. J. (2018). A new mechanism for reduced sensitivity to demethylation-inhibitor fungicides in the fungal banana black sigatoka pathogen *Pseudocercospora fijiensis*, *Molecular Plant Pathology* 19(6), 1491–1503. https://doi.org/10.1111/mpp.12637.

Drenth, A. and Kema, G. (2021). The vulnerability of bananas to globally emerging disease threats, *Phytopathology®*, 111(12), 2146–2161. https://doi.org/10.1094/PHYTO-07-20-0311-RVW.

European Commission (2020). Commission implementing regulation (EU) 2020/2087, *Official Journal of the European Union* 68, 48–119. https://doi.org/10.2903/j.efsa.2019.5755.

European Commission (2021). Commission implementing regulation (EU) 2021/2081 of 26 November 2021 concerning the non-renewal of approval of the active substance indoxacarb, in accordance with regulation (EC) no 1107/2009 of the European Parliament and of the council concerning the p, *Official Journal of the European Union* 64(L 426), 28–31.

European Food Safety Authority (EFSA), Anastassiadou, M., Bernasconi, G., Brancato, A., Carrasco Cabrera, L., Ferreira, L., Greco, L., Jarrah, S., Kazocina, A., Leuschner, R., Magrans, J. O., Miron, I., Nave, S., Pedersen, R., Reich, H., Rojas, A., Sacchi, A., Santos, M., Theobald, A., Vagenende, B. and Verani, A. (2020). Evaluation of confirmatory data following the Article 12 MRL review for chlorothalonil, including assessments for import tolerances for banana, papaya and peanuts, *EFSA Journal* 18(9), e06239. https://doi.org/10.2903/j.efsa.2020.6239.

European Food Safety Authority (EFSA), Anastassiadou, M., Bernasconi, G., Brancato, A., Carrasco Cabrera, L., Greco, L., Jarrah, S., Kazocina, A., Leuschner, R., Magrans, J. O., Miron, I., Nave, S., Pedersen, R., Reich, H., Rojas, A., Sacchi, A., Santos, M., Stanek, A., Theobald, A., Vagenende, B. and Verani, A. (2020). Modification of the existing maximum residue levels for mancozeb in various crops, *EFSA Journal* 18(8), e06108. https://doi.org/10.2903/j.efsa.2020.6108.

European Food Safety Authority (EFSA), Arena, M., Auteri, D., Barmaz, S., Bellisai, G., Brancato, A., Brocca, D., Bura, L., Byers, H., Chiusolo, A., Court Marques, D., Crivellente, F., De Lentdecker, C., Egsmose, M., Erdos, Z., Fait, G., Ferreira, L., Goumenou, M., Greco, L., Ippolito, A., Istace, F., Jarrah, S., Kardassi, D., Leuschner, R., Lythgo, C., Magrans, J. O., Medina, P., Miron, I., Molnar, T., Nougadere, A., Padovani, L., Parra Morte, J. M., Pedersen, R., Reich, H., Sacchi, A., Santos, M., Serafimova, R., Sharp, R., Stanek, A., Streissl, F., Sturma, J., Szentes, C., Tarazona, J., Terron, A., Theobald, A., Vagenende, B., Verani, A. and Villamar-Bouza, L. (2018). Peer review of the pesticide risk assessment of the active substance chlorothalonil, *EFSA Journal* 16(1), 1-40. https://doi.org/10.2903/j.efsa.2018.5126.

FAO (2020). *Banana Market Review: Preliminary Results 2019, Food and Agriculture Organization of the United Nations*. Rome: FAO.

Fernández, D., Tores, J., Vicente, A. and Pérez-García, A. (2010). Fungicides: the QoI fungicides, the rise and fall of a successful class of agricultural fungicides. In: Carisse, O. (Ed.). *Fungicides*. Spain: InTech, pp. 203–220. https://doi.org/10.5772/13205.

Fisher, J. and Didio, R. (1991). Influencia de los vehículos en el depósito, distribución y dispersión del propiconazole en el banano. In: Contreras, M., Guzmán, J. and Carrasco, L. (Eds) *Reunión ACORBAT 91*. San José, Costa Rica: CORBANA, pp. 329–334. http://www.sidalc.net/cgi-bin/wxis.exe/?IsisScript=INIBA.xis&method=post&formato=2&cantidad=1&expresion=mfn=002584.

Fouré, É. (1985). Black leaf streak disease of banana and plantains (*Mycosphaerella fijiensis Morelet*). *Study of the Symptoms and Stage of the Disease in Gabon.* IRFA-CIRAD. París. 20.

Fouré, É. and Ganry, J. (2008). A biological forecasting system to control black leaf streak disease of bananas and plantains, *Fruits* 63(5), 311-317. https://doi.org/10.1051/fruits:2008029.

FRAC (2012) *Banana Working Group Meeting Minutes 2012*. https://www.frac.info/docs/default-source/working-groups/banana-group/group/banana-wg-meeting-minutes-2022---english.pdf?sfvrsn=fd1d4e9a_2.

FRAC (2018) *Banana Working Group Meeting Minutes 2018*. https://www.frac.info/docs/default-source/working-groups/banana-group/group/banana-wg-meeting-minutes-2018---english.pdf?sfvrsn=af68499a_2.

FRAC (2022) *Banana Working Group Meeting Minutes 2022*. https://www.frac.info/docs/default-source/working-groups/banana-group/group/banana-wg-meeting-minutes-2022---english.pdf?sfvrsn=fd1d4e9a_2.

Ganry, J. and Meyer, C. (1972). La lutte contrôlée la Cercosporiose aux Antilles. Bases climatiques de l'avertissement technique d'observation et de numération de la maladie, *Fruits* 27(10), 665–676.

Grünwald, N. J., Goodwin, S. B., Milgroom, M. G. and Fry, W. E. (2003). Analysis of genotypic diversity data for populations of microorganisms, *Phytopathology* 93(6), 738–746. https://doi.org/10.1094/PHYTO.2003.93.6.738.

Guyot, H. and Guillé, J. (1954). Les formules fongicides huileuses pour le traitement de bananerais, *Fruits*, 9 289–292.

Guzmán, M. (2007). *Estrategias para el manejo de la resistencia de Mycosphaerella fijiensis Morelet, agente causal de la Sigatoka negra en banano, a fungicidas triazoles*. Universidad de Costa Rica. San José.

Guzmán, M., Orozco, M. and Pérez-Vicente, L. (2013). Sigatoka leaf spot diseases of bananas: dispersion, impact and evolution of management strategies in Latin America-Caribbean region. In: *XX Reunião Internacional da Associação para a Cooperação em Pesquisa e Desenvolvimento Integral das Musáceas (Bananas e Plátanos) 9 a 13 de setembro de 2013*. Vol 9. Fortaleza, Brazil.

Ioos, R., Hubert, J., Abadie, C., Duféal, D., Opdebeeck, G. and Iotti, J. (2011). First report of black sigatoka disease in banana caused by *Mycosphaerella fijiensis* on Martinique Island, *Plant Disease* 95(3), 359. https://doi.org/10.1094/PDIS-11-10-0850.

Islam, M. T. and Sherif, S. M. (2020). Rnai-based biofungicides as a promising next-generation strategy for controlling devastating gray mold diseases, *International Journal of Molecular Sciences* 21(6), 1–10. https://doi.org/10.3390/ijms21062072.

Kema, G. H. J., Mirzadi Gohari, A., Aouini, L., Gibriel, H. A. Y., Ware, S. B., van den Bosch, F., Manning-Smith, R., Alonso-Chavez, V., Helps, J., Ben M'Barek, S., Mehrabi, R., Diaz-Trujillo, C., Zamani, E., Schouten, H. J., van der Lee, T. A. J., Waalwijk, C., de Waard, M. A., de Wit, P. J. G. M., Verstappen, E. C. P., Thomma, B. P. H. J., Meijer, H. J. G. and Seidl, M. F. (2018). Stress and sexual reproduction affect the dynamics of the wheat pathogen effector AvrStb6 and strobilurin resistance, *Nature Genetics* 50(3), 375–380. https://doi.org/10.1038/s41588-018-0052-9.

Klein, H. (1960). Control of Cercospora leaf spot of bananas with applications of oil spray based on the disease cycle, *Phytopathology* 50, 488–490.

Knight, S., et al. (2003). The role of managing resistance to fungicides in maintaining the effectiveness of integrated strategies to control black leaf streak disease. *Mycosphaerella leaf spot diseases of bananas: present status and outlook 2nd International workshop on Mycosphaerella leaf spot diseases*. INIBAP, San José, Costa Rica. Vol. 303, pp. 303-307.

Latin, R. (2011). *A Practical Guide to Turfgrass Fungicides* (1st edn.) Edited by R. Latin. Minnesota: The American Phytopathological Society. https://doi.org/10.1094/9780890544976.

Leach, R. (1946). Banana leaf spot (*Mycosphaerella musicola*) on the Gros Michel variety in Jamaica: investigations on the aetiology of the disease and the principles of control by spraying. Kingston, Jamiaca: The Government Printer. https://search.ebscohost.com/login.aspx?direct=true&db=ldh&AN=19471100634&site=ehost-live&scope=site.

Lepesheva, G. I. and Waterman, M. R. (2004). CYP51 - the omnipotent P450, *Molecular and Cellular Endocrinology* 215(1-2), 165–170. https://doi.org/10.1016/j.mce.2003.11.016.

Lupetti, A., Danesi, R., Campa, M., Del Tacca, M. and Kelly, S. (2002). Molecular basis of resistance to azole antifungals, *Trends in Molecular Medicine* 8(2), 76–81. https://doi.org/10.1016/S1471-4914(02)02280-3.

Ma, Z., Proffer, T. J., Jacobs, J. L. and Sundin, G. W. (2006). Overexpression of the 14α-demethylase target gene (CYP51) mediates fungicide resistance in *Blumeriella jaapii*, *Applied and Environmental Microbiology* 72(4), 2581–2585. https://doi.org/10.1128/AEM.72.4.2581-2585.2006.

Ma, Z. and Michailides, T. J. (2005). Advances in understanding molecular mechanisms of fungicide resistance and molecular detection of resistant genotypes in phytopathogenic fungi, *Crop Protection* 24(10), 853–863. https://doi.org/10.1016/j.cropro.2005.01.011.

Mamta, B. and Rajam, M. V. (2017). RNAi technology: a new platform for crop pest control, *Physiology and Molecular Biology of Plants: an International Journal of Functional Plant Biology* 23(3), 487–501. https://doi.org/10.1007/s12298-017-0443-x.

Marín, D. H., Romero, R. A., Guzmán, M. and Sutton, T. B. (2003). Black Sigatoka: an increasing threat to banana cultivation, *Plant Disease* 87(3), 208–222.

McLoughlin, A. G., Wytinck, N., Walker, P. L., Girard, I. J., Rashid, K. Y., de Kievit, T., Fernando, W. G. D., Whyard, S. and Belmonte, M. F. (2018). Identification and application of exogenous dsRNA confers plant protection against *Sclerotinia sclerotiorum* and *Botrytis cinerea*, *Scientific Reports* 8(1), pp. 1–14. https://doi.org/10.1038/s41598-018-25434-4.

Mellado, E., Garcia-Effron, G., Alcázar-Fuoli, L., Melchers, W. J., Verweij, P. E., Cuenca-Estrella, M. and Rodríguez-Tudela, J. L. (2007). A new *Aspergillus fumigatus* resistance mechanism conferring in vitro cross-resistance to azole antifungals involves a combination of *cyp51A* alterations, *Antimicrobial Agents and Chemotherapy* 51(6), 1897–1904. https://doi.org/10.1128/AAC.01092-06.

Mumbanza, F. M., Kiggundu, A., Tusiime, G., Tushemereirwe, W. K., Niblett, C. and Bailey, A. (2013). In vitro antifungal activity of synthetic dsRNA molecules against two pathogens of banana, *Fusarium oxysporum* f. sp. *cubense* and *Mycosphaerella fijiensis*, *Pest Management Science* 69(10), 1155–1162. https://doi.org/10.1002/ps.3480.

Nerva, L., Sandrini, M., Gambino, G. and Chitarra, W. (2020). Double-stranded RNAS (Dsrnas) as a sustainable tool against gray mold (botrytis cinerea) in grapevine: effectiveness of different application methods in an open-air environment, *Biomolecules* 10(2). https://doi.org/10.3390/biom10020200.

Newlands, N. K. (2018). Model-based forecasting of agricultural crop disease risk at the regional scale, integrating airborne inoculum, environmental, and satellite-based monitoring data, *Frontiers in Environmental Science* 6. https://doi.org/10.3389/fenvs.2018.00063.

Pérez-Vicente, L. (2006). Manejo convencional y alternativo de la sigatoka negra en bananos: Estado actual y perspectivas, *Fitosanidad* 10(1), 55–72.

Pérez-Vicente, L. (2012). *A Holistic Integrated Management Approach to Control Black Sigatoka Disease of Banana Caused by Mycosphaerella fijiensis, TCP/SLC/3402 Technical Manual*. https://www.fao.org/fileadmin/templates/agphome/documents/caribbeanspci/IntApproachBSDManagementLVP.pdf.

Pietila, M. P., Vohra, P. K., Sanyal, B., Wengenack, N. L., Raghavakaimal, S. and Thomas, C. F. (2006). Cloning and characterization of CYP51 from Mycobacterium avium, *American Journal of Respiratory Cell and Molecular Biology* 35(2), 236–242. https://doi.org/10.1165/rcmb.2005-0398OC.

Ploetz, R. (2000). Black sigatoka, *Pesticide Outlook* 11(1), 19–23. https://doi.org/10.1039/b006308h.

Ploetz, R. C. (2015). Fusarium wilt of banana, *Phytopathology* 105(12), 1512–1521. https://doi.org/10.1094/PHYTO-04-15-0101-RVW.

Ploetz, R. C., Kema, G. H. J. and Ma, L. J. (2015). Impact of diseases on export and smallholder production of banana, *Annual Review of Phytopathology* 53(May), 269–288. https://doi.org/10.1146/annurev-phyto-080614-120305.

Rhodes, P. L. (1964). A new banana disease in Fiji, *Commonwealth Phytopathological News* 10, 38–41.

Rieux, A., Soubeyrand, S., Bonnot, F., Klein, E. K., Ngando, J. E., Mehl, A., Ravigne, V., Carlier, J. and de Lapeyre de Bellaire, L. (2014). Long-distance wind-dispersal of spores in a fungal plant pathogen: estimation of anisotropic dispersal kernels from an extensive field experiment, *PLoS ONE* 9(8), e103225. https://doi.org/10.1371/journal.pone.0103225.

Risède, J., Lescot, T., Cabrera Cabrera, J., Guillon, M., Tomekpé, K., Kema, G. H. J. and Côte, F. (2010). *Challenging Short and Mid-Term Strategies to Reduce the use of Pesticides in Banana Production, From Science to Field Banana Case Study – Guide Number 1*. Endure. https://agritrop.cirad.fr/553854/1/document_553854.pdf.

Rivas, G. G., Zapater, M. F., Abadie, C. and Carlier, J. (2004). Founder effects and stochastic dispersal at the continental scale of the fungal pathogen of bananas *Mycosphaerella fijiensis*, *Molecular Ecology* 13(2), 471–482. https://doi.org/10.1046/j.1365-294X.2003.02043.x.

Rogers, S. L., Atkins, S. D. and West, J. S. (2009). Detection and quantification of airborne inoculum of *Sclerotinia sclerotiorum* using quantitative PCR, *Plant Pathology* 58(2), 324–331. https://doi.org/10.1111/j.1365-3059.2008.01945.x.

Romero, R. A. and Sutton, T. B. (1997). Sensitivity of *Mycosphaerella fijiensis*, causal agent of black Sigatoka of banana, to propiconazole, *Phytopathology* 87(1), 96–100. https://doi.org/10.1094/PHYTO.1997.87.1.96.

Romero, R. A. and Sutton, T. B. (1998). Characterization of benomyl resistance in *Mycosphaerella fijiensis*, cause of black sigatoka of banana, in Costa Rica, *Plant Disease* 82(8), 931–934. https://doi.org/10.1094/PDIS.1998.82.8.931.

Scalliet, G., Bowler, J., Luksch, T., Kirchhofer-Allan, L., Steinhauer, D., Ward, K., Niklaus, M., Verras, A., Csukai, M., Daina, A. and Fonné-Pfister, R. (2012). Mutagenesis and functional studies with succinate dehydrogenase inhibitors in the wheat pathogen *Mycosphaerella graminicola*, *PLoS ONE* 7(4), e35429. https://doi.org/10.1371/journal.pone.0035429.

Schnabel, G. and Jones, A. L. (2001). The 14α-Demethylase (CYP51A1) gene is overexpressed in Venturia inaequalis strains resistant to myclobutanil, *Phytopathology* 91(1), 102–110. https://doi.org/10.1094/PHYTO.2001.91.1.102.

Sepúlveda, L. and Torres, J. (2016). Lack of cross resistance between pyraclostrobin and azoxystrobin in *Mycosphaerella fijiensis*. In: *VI Congreso Internacional sobre banano CORBANA y XXI Reunión internacional ACORBAT*. Miami, Florida: ACORBAT, 6.

Shapiro, R. S., Robbins, N. and Cowen, L. E. (2011). Regulatory circuitry governing fungal development, drug resistance, and disease, *Microbiology and Molecular Biology Reviews: MMBR* 75(2), 213–267. https://doi.org/10.1128/mmbr.00045-10.

Sierotzki, H., Parisi, S., Steinfeld, U., Tenzer, I., Poirey, S. and Gisi, U. (2000). Mode of resistance to respiration inhibitors at the cytochrome bc1 enzyme complex of *Mycosphaerella fijiensis* field isolates, *Pest Management Science* 56(10), 833–841. https://doi.org/10.1002/1526-4998(200010)56:10<833::AID-PS200>3.0.CO;2-Q.

Singh, B., Singh, J. P., Kaur, A. and Singh, N. (2016). Bioactive compounds in banana and their associated health benefits - a review, *Food Chemistry* 206, 1–11. https://doi.org/10.1016/j.foodchem.2016.03.033.

Singh, P. and Mustapha, A. (2013). Multiplex TaqMan® detection of pathogenic and multi-drug resistant Salmonella, *International Journal of Food Microbiology* 166(2), 213-218. https://doi.org/10.1016/j.ijfoodmicro.2013.07.023.

Snelders, E., Camps, S. M., Karawajczyk, A., Schaftenaar, G., Kema, G. H., van der Lee, H. A., Klaassen, C. H., Melchers, W. J. and Verweij, P. E. (2012). Triazole fungicides can induce cross-resistance to medical triazoles in *Aspergillus fumigatus*, *PLoS ONE* 7(3), e31801. https://doi.org/10.1371/journal.pone.0031801.

Stover, R. H. (1962). Intercontinental spread of banana leaf spot (*Mycosphaerella musicola* Leach), *Tropical Agriculture* 39, 327–338.

Stover, R. H. (1990). Sigatoka leaf spots: thirty years of changing control strategies 1959-1989. In: Fullerton, R. A. (Ed.) *INIBAP Poceedings International Worshop*. San José, Costa Rica: INIBAP, pp. 66–74.

Van Wendel De Joode, B., Barbeau, B., Bouchard, M. F., Mora, A. M., Skytt, Å., Córdoba, L., Quesada, R., Lundh, T., Lindh, C. H. and Mergler, D. (2016). Manganese concentrations in drinking water from villages near banana plantations with aerial mancozeb spraying in Costa Rica: results from the Infants' Environmental Health Study (ISA), *Environmental Pollution* 215, 247–257. https://doi.org/10.1016/j.envpol.2016.04.015.

Verweij, P. E., Kema, G. H., Zwaan, B. and Melchers, W. J. (2013). Triazole fungicides and the selection of resistance to medical triazoles in the opportunistic mould *Aspergillus fumigatus*, *Pest Management Science* 69(2), 165–170. https://doi.org/10.1002/ps.3390.

Villari, C., Mahaffee, W. F., Mitchell, T. K., Pedley, K. F., Pieck, M. L. and Hand, F. P. (2017). Early detection of airborne inoculum of *Magnaporthe oryzae* in turfgrass fields using a quantitative LAMP assay, *Plant Disease* 101(1), 170–177. https://doi.org/10.1094/PDIS-06-16-0834-RE.

Vincelli, P. (2014). *Some Principles of Fungicide Resistance*. Kentucky. https://mail-attachment.googleusercontent.com/attachment/u/0/?ui=2&ik=701cc69c3b&view=att&th=15f77a6a11b24056&attid=0.1&disp=inline&safe=1&zw&saddbat=ANGjdJ9WK5SX74mjRfanU-vi_bOYqucuP7UhfW5a5sUBtRybGEcFPMzSL1DExBFuJkMLY6aCnoAZwevb1XbDJ_jVfnBx2cLsOlKlzIbRN.

Chapter 3

Freckle disease of banana

André Drenth, The University of Queensland, Australia; and Mee Hua Wong, Agriculture Research Centre Semongok, Malaysia

1 Introduction

Freckle is a disease of the foliage, peduncle and fruit with unsightly freckle-like spots on infected plant parts. On infected leaves, discrete brown to black spots, circular or streaking lesions, appear on leaf surfaces and/or along leaf veins. Infection of fruits is expressed as reddish-brown to black spots surrounded by dark green, water-soaked halos which may completely cover the fruit at harvest (Meredith, 1968). Freckle disease has become a limiting factor in the production of banana, as it not only affects the yield and quality but also marketability as blemished fruit lacks acceptance in the banana export industry (Corcolon and Raymundo, 2008). Freckle disease is commonly found on a wide range of banana cultivars in Southeast Asia, South Asia (India, Pakistan and China), Oceania and East Africa (La Reunion) (Jones, 2019). There are three different species of *Phyllosticta*, namely, *P. musarum*, *P. maculata* and *P. cavendishii*, associated with the disease (Wong et al., 2012). The distinction of these three *Phyllosticta* species is important in view of their differences in host specificity and geographic distribution.

http://dx.doi.org/10.19103/AS.2022.0108.03

2 Disease symptoms

The pathogens can infect the leaf during any stage of plant growth but, in general, older plants are more susceptible and more seriously affected by the disease (Chuang, 1984). Disease symptoms on the leaf are expressed as numerous small discrete reddish-brown spots (Fig. 1a). Over time, these spots turn black and may coalesce to form streaks or blotches of lesions on the leaf lamina or veins (Fig. 1b and c) or aggregate on the midrib of the leaf (Fig. 1d). Disease symptoms are more abundantly expressed on the adaxial leaf surface (Wardlaw, 1935; Meredith, 1968). The peduncle of the fruit bunch is similarly affected with dense aggregations of black spots on the surface (Fig. 1e).

The spots or lesions of freckle are made up of either individual or clusters of fruiting bodies consisting of conidiomata, ascomata or both, and frequently spermagonia are also present (Van der Aa, 1973). A diagnostic feature of freckle disease is the rough, sandpaper-like texture of the leaf surface due to the presence of these protruding fruiting bodies from the lesions (Wardlaw, 1935).

Freckle spots on the fruit are expressed as reddish-brown to black spots with a dark green water-soaked halo (Fig. 1f), and mature fruit may be covered completely with dense aggregations of spots. The maturity of the fruit may influence disease expression. Infection may commence a few weeks after emergence of the inflorescence and the disease continues to progress as the

Figure 1 Freckle disease symptoms on leaf and fruit of banana. (a) Initial symptoms showing numerous reddish-brown spots on the leaf surface. (b) Necrotic streaks of spots on leaf veins. (c) Advanced stage of infection with blotches on the leaf laminar. (d) Spots coalesced and aggregated on the midrib. (e) Dense aggregation of black spots along the peduncle. (f) Reddish-brown to black spots with associated water-soaked lesions on fruit surfaces. Photos: Courtesy Department of Agriculture, Sarawak.

fruit matures and becomes most severe at the time of harvest (Meredith, 1968). In severely infected plants, leaves senesce prematurely (Chuang, 1984) and yields are significantly reduced due to early defoliation (Tsai et al., 1993). Fruits produced from these plants are discolored, undersized and ripen unevenly (Carpenter, 1919). Though spots on the fruit surface do not go deeper than the exocarp (Wardlaw, 1961), their presence adversely affects the aesthetic and commercial value of the fruit.

3 Pathogen characteristics and diagnosis

Prior to the detailed study by Wong et al. (2012), freckle disease in banana was considered to be caused by a single fungal pathogen, *Guignardia musae*, with its anamorph *P. musarum* based on morphological descriptions of the conidial, ascigerous and spermatial states provided by Van der Aa (1973) and Punithalingam (1974), respectively. Since its first description, the pathogen has undergone several name changes. *P. musarum* was first described as *Sphaeropsis musarum* Cooke 1880 and has several synonyms: *Phoma musae* (Cooke) Sacc. 1884, *Macrophoma musae* (Cooke) Berl. & Volg. 1886, *Phyllostictina musarum* (Cooke) Petr. 1931 and *Phyllosticta musarum* (Cooke) van der Aa (Van der Aa, 1973). *Guignardia musae* Racib. 1909 was the earliest name given to the sexual stage of the fungus. Subsequently, several synonyms, including *Guignardia musae* Syd. 1912, *Guignardia musae* Stevens 1925, *Guignardia sydowiana* Trotter 1928 and *Phomatospora musae* Yen 1972, were also introduced. However, both *G. musae* Syd. 1912 and *G. musae* Stevens 1925 were later excluded due to their relatively small ascospores (Van der Aa, 1973; Punithalingam and Holliday, 1975; Van der Aa and Vanev, 2002). The abolishment of dual nomenclature for fungi (Hawksworth et al., 2011; Wingfield et al., 2012) gives priority to the older name *Phyllosticta* (1818) over that of *Guignardia* (1892) (Glienke et al., 2011; Wikee et al., 2011).

A detailed polyphasic approach involving both morphological and molecular studies revealed that freckle disease in banana was caused by three different *Phyllosticta* species: *P. cavendishii, P. maculata* and *P. musarum* (Wong et al., 2012). Apart from certain morphological variations, the three species also differ in host specificity and geographic distribution, as summarized in Table 1.

Phyllosticta cavendishii is a novel species that is pathogenic toward both Cavendish and non-Cavendish host varieties of the AAA, AAB and ABB genomic groups. Having the ability to infect varieties within the AAA genomic group, *P. cavendishii* is potentially damaging the export-oriented banana plantations that focus on Cavendish cultivars. *P. maculata*, the new name introduced for *G. musae*, was found to infect only non-Cavendish cultivars of the AAB and ABB genomic groups over a wide geographic range. *P. musarum* can infect varieties of both AAB and ABB genomic groups and occurs in a

Table 1 Distribution and host range of the three different *Phyllosticta* species

	P. cavendishii	*P. musarum*	*P. maculata*
Known distribution	East Timor, Sri Lanka, India, Hawaii, Indonesia, the Philippines, Malaysia, Taiwan, Vietnam, South Pacific Islands, Federated States of Micronesia Australia (eradicated)	Limited distribution around the Bay of Bengal region in India and Thailand	Wide geographic distribution in Southeast Asia and Oceania, Australia, American Samoa, Malaysia, Indonesia, Papua New Guinea and the Philippines
Host range	AAA, AAB and ABB genomic groups	AAB and ABB genomic groups	AAB and ABB genomic groups

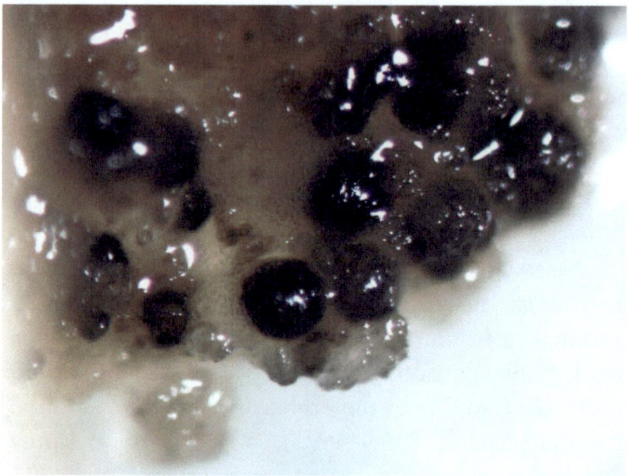

Figure 2 Sporulation of *P. maculata* with formation of black pynidia and hyaline conidia grown on oatmeal agar. Source: Wong et al. (2012).

restricted geographic range as it has only been identified around the Bay of Bengal region. Morphologically, both the ascospore and conidial mean size for *P. maculata* is comparatively bigger than those for *P. cavendishii* and *P. musarum*.

Over time, the difficulty in isolating the causal organism of freckle disease has been reported several times (Carpenter, 1919; Lee, 1922; Wardlaw, 1961; Meredith, 1968; Stover, 1972). Only Chuang (1981) reported successful isolation and sporulation of *Phyllosticta* species from Cavendish banana in Taiwan, but without detailed information on the morphological structures apart from pycnidial color and conidial size. This has since changed when sporulation under artificial growth conditions for *P. maculata* was achieved from a pure culture isolated from banana (Fig. 2), which enabled detailed examination and description of cultural characteristics and morphological structures of *P. maculata* (Wong et al., 2012).

The growth of *P. maculata* and *P. cavendishii* cultures *in vitro* is very slow and it takes several months of incubation for *P. cavendishii* to reach a sizable colony diameter. Both species produce cultures that lack aerial mycelium, with *P. maculata* displaying a rosy-buff colony interspersed with hazel-brown at the edge and *P. cavendishii* displaying hazel-brown to black cultures interspersed with a rosy-buff sector. The ability to culture the pathogen has solved the problem that *Phyllosticta capitalensis*, a closely related common endophyte of many plant species, was often incorrectly reported as the causal agent of freckle disease (Yen, 1972; Punithalingam, 1974; Wulandari et al., 2010). Given the difficulty to isolate and cultivate *Phyllosticta* species associated with freckle disease and the ease to obtain cultures of the commonly cooccurring endophytic *P. capitalensis* (Okane et al., 2003; Pandey et al., 2003; Carlier et al., 2021; Glienke et al., 2011), it is reasonable to assume that the causal agent identified in several previous reports was *P. capitalensis* (Baayen et al., 2002; Wulandari et al., 2010).

In recent years, DNA-based molecular techniques have been widely used to detect and identify plant pathogenic fungi, including a closely related species, *Guignardia citricarpa*, causing black spot of citrus (Bonants et al., 2003; Meyer et al., 2006; Peres et al., 2007; Van Gent-Pelzer et al., 2007). These commonly used PCR-based assays have limitations, such as lack of quantitative data and risk of contamination, and are time-consuming due to post-amplification processing steps (Van Gent-Pelzer et al., 2007). A real-time PCR assay using TaqMan probe chemistry, subsequently developed for detecting *G. citricarpa* on citrus fruit, was shown to be a robust assay (Van Gent-Pelzer et al., 2007). Despite the success of real-time PCR as a diagnostic tool in many plant pathological applications (Schena et al., 2004; Arzanlou et al., 2007; Tan et al., 2009), its requirement for fluorescent-labeled probes has rendered the assay relatively expensive.

A High Resolution Melt (HRM) analysis was developed for the three closely related *Phyllosticta* species of freckle disease (Wong et al., 2013b). This assay is able to identify and differentiate each of the three species as well as a mixture of all three species or any combination of these species. Furthermore, both cultures and herbarium specimens can be genotyped using the assay, and the results of validation tests demonstrated concordance with the sequence data (Wong et al., 2013b). The HRM assay has the advantages of allowing its amplicons to be sequenced directly for further confirmation of results, precluding the need for further sample processing such as electrophoresis or restriction digestion, thereby reducing the potential for contamination and increasing sample throughput (Taylor, 2009; Vossen et al., 2009). However, a drawback to the technique is the lack of a database system for storing melt curves generated for use as reference genotypes to facilitate subsequent determination or comparison of unknown samples (Mackay et al., 2008; Goldschmidt et al., 2012). Another viable alternative for freckle disease diagnosis is the direct sequencing of selected target genes which has increased in popularity due to reductions in cost and rapid turnaround time. An example is the freckle disease eradication campaign in the Northern Territory, Australia, where such a direct sequencing approach was used. The two genes, ITS and Actin, with amplicon sizes of 444 bp and 238 bp, respectively, were amplified and sequenced, and the resulting sequences were compared to the holotype using phylogenetic analysis (McMaster et al., 2020).

4 Disease cycle and epidemiology

This overview contains some earlier studies where the causal agent of freckle disease was known as a single species, *G. musae* (anamorph: *P. musarum*), prior to the reclassification of the disease being associated with three different *Phyllosticta* species (Wong et al., 2012). Based on their morphological similarity, we presume the epidemiology and infection biology are similar among these three closely related species.

The freckle pathogens produce two different fruiting bodies, asexual conidiomata/pycnidia and sexual ascomata (Van der Aa, 1973; Punithalingam, 1974). Although it has not been experimentally verified, one can assume that the spermagonia play a role as male gametes but serve no direct role as infective propagules in the epidemiology of the disease. In mature pycnidia, conidia are produced by and borne at the apex of the conidiogenous cells (Ocfemia, 1927) that line the inner wall of the pycnidial chamber during conidiogenesis (Van der Aa, 1973; Janex-Favre et al., 1993; Watanabe et al., 1997). In the presence of moisture, white tendrils of conidial masses are extruded from the pycnidia through the ostiole which are then disseminated by rain splash to neighboring banana plants initiating primary infection (Meredith, 1968; Tsai et al., 1993) or by rainwater dripping down from leaves to plant tissues beneath causing secondary infection, as depicted by Wong (2013) in the disease cycle of *P. musarum* (Fig. 3). This highlighted the biological significance of pycnidia and conidia under natural conditions in the field and suggests that conidia, as a source of inoculum, have great potential to disperse and to start primary or secondary infections (Wong et al., 2013a). However, successful establishment of the disease is also strongly dependent on both biotic and abiotic factors further influenced by environmental conditions.

Infection is initiated within a few hours after conidia are deposited on banana tissues under favorable conditions (Jones, 2019). Upon germination of these spores, germ tubes elongate and swell and each of them forms an appressorium. An infection hypha develops from the appressorium and

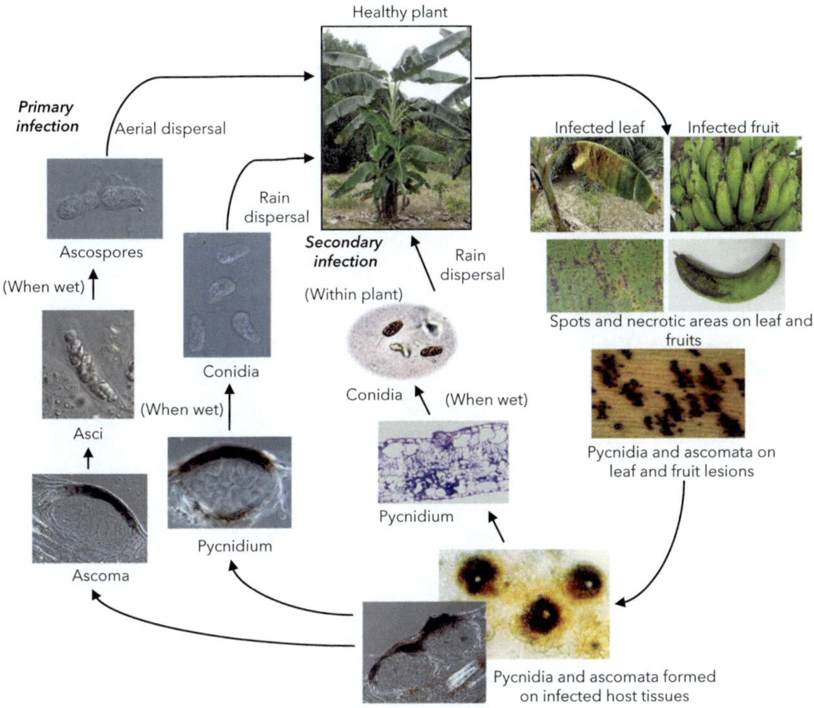

Figure 3 Disease cycle of *P. musarum* causing freckle disease on banana. Source: Wong (2013).

subsequently invades young leaves initiating an infection. The infected leaf produces abundant pycnidia which provide conidia for subsequent secondary infection of host tissues.

In studies on inoculated banana leaves and fruits in the field, disease symptoms were expressed 4–6 days after inoculation with minute reddish-brown spots plus a water-soaked halo on the fruits and green flecks on the leaves. Two to three weeks after infection, spots on the fruits enlarged while green flecks on the leaves turned reddish-brown and later black, with pycnidial formation producing large numbers of conidia four to six weeks after infection (Meredith, 1968). These spots subsequently coalesced to form patches of black lesions on the infected surface.

Studies on detached banana leaf and fruit tissues showed that conidial germination occurs within 2–3 hours after inoculation (Meredith, 1968; Pu et al., 2008; Wong et al., 2013a). On the leaf surface, appressoria are formed at the tip of long, slender germ tubes within 18 hours after inoculation (Wong et al., 2013a) as compared to shorter times of 6–12 hours reported in earlier studies (Meredith, 1968; Pu et al., 2008). On the fruit surface, most of the appressoria, which are sessile and variable in shape, are formed directly from conidia 3 hours after inoculation (Wong et al., 2013a). Although the time required for conidial germination on leaf and fruit is similar, the formation of appressoria has been reported to occur earlier on the fruit surface. This temporal variation may be due to factors inherent to the host plant such as differences in the amount of surface wax between leaf and fruit.

Direct non-stomatal penetration of host tissues has been reported. On the leaf tissue, the appressorium of *P. musarum* penetrated the cuticle directly causing a reddish-brown necrosis in the leaf epidermal cell under the appressorium (Pu et al., 2008). Similarly, on fruit, penetration of the appressorium through the cuticle occurred within 24–72 hours after its formation with the fruit epidermal cells turning reddish-brown (Meredith, 1968). Detailed studies based on scanning electron microscopy (SEM) on *P. maculata* did not find evidence of stomatal penetration as the pathogen penetrates the host cuticle directly, with or without appressorium formation (Wong et al., 2013a). The freckle pathogen subsequently colonizes host tissues through cell to cell penetration and proliferation of intercellular hyphae. After colonization, pycnidia may develop with thick intercellular hyphae beneath the epidermis where the fruiting bodies enlarge and break through the epidermis to release the conidia (Meredith, 1968). As the pathogen does not gain ingress through stomata, the density of stomata on the host surface presumably will have little or no effect on its infectivity and neither any bearing on its disease severity nor host plant susceptibility (Wong et al., 2013a). This also implies that banana fruit and leaf are equally susceptible to pathogen infection despite the fruit having 30 times fewer stomata than the leaf (Johnson and Brun, 1966).

SEM studies further revealed circular, darkened areas around the appressorium of *P. maculata*, suggesting dissolution of cuticular wax as part of the infection process. In addition, an extracellular matrix found at the host-pathogen interface surrounding and underlying conidia and appressoria (Wong et al., 2013a) may act as an adhesive to enable attachment of fungal propagules and offer protection to the pathogen from desiccation and host metabolites (Nicholson and Epstein, 1991; Braun and Howard, 1994; Kuo and Hoch, 1995). Adhesion of conidia to host tissues through the extracellular matrix may have significant biological implications as it enables conidia to attach to the waxy surfaces of leaf and fruit under wet conditions, facilitating the infection process of conidial germination, appressoria formation and pathogen penetration (Wong et al., 2013a).

The infection process as well as disease spread and development is greatly influenced by climatic and environmental conditions, especially availability of free water and high relative humidity. During periods of high rainfall and relative humidity, more conidia are produced and dispersed which aid infection (Meredith, 1968), and it has been shown that disease severity is positively correlated with rainfall (Abayasekera et al., 1993). In Taiwan, infection and disease severity is increased toward the end of wet season but reduced during the cool, dry season (Chuang, 1984).

Apart from rain dispersal of conidia to disseminate fungal propagules, aerial dispersal of ascospores also plays an important role in transmission of freckle disease (Chuang, 1983b; Jones, 2019). The sexual ascoma houses asci, each containing eight ascospores. Upon wetting, the mature ascoma utilizes a special mechanism to violently eject ascospores, aiding aerial dispersal of these spores over long distances (Chuang, 1983b). In Taiwan, though ascospores are present throughout the year in infected leaf tissues (Chuang, 1983b), their production is reduced during the hot, wet season and increased during the cold, dry season (Chuang and Chen, 1982). It has been postulated that freckle disease in the Torres Strait and the northern Cape York Peninsula in Australia was due to wind dispersal of ascospores from Papua New Guinea (Jones and Alcorn, 1982). Long-distance dispersal through movement of infected plant material or fruit was reported to be another possible pathway of spread for freckle disease from the Philippines to Hawaii (Lee, 1922).

5 Occurrence and impact

According to historical records, *P. musarum* was found on leaves of *Musa paradisiaca* collected from India as early as 1879 (based on a disease specimen from Kew Herbarium, UK) (Wong, 2013). Subsequently, an infected specimen with the fungus described as *G. musae* in 1909 was found on *Musa acuminata* from Indonesia (Van der Aa, 1973). Freckle disease, the name of which was coined by Carpenter when he found it in Hawaii in 1917 (Carpenter, 1919), has since been reported from several banana-growing regions in the world, including the Philippines (Lee, 1922), Taiwan (Chuang, 1981), China (Liang, 1990), India (Chandra, 1991), Sri Lanka (Abayasekera et al., 1993), Thailand, Malaysia, Vietnam, the Pacific and Oceania Islands (Meredith, 1968; Punithalingam and Holliday, 1975; Jones, 2019), Papua New Guinea and Australia (Jones and Alcorn, 1982) (Table 1).

Outside the Asia-Pacific region, Congo in Africa and the Dominican Republic in the Caribbean have reported the presence of freckle disease (Wardlaw, 1961). It is also found in Central America as shown in the distribution map of *Guignardia musae* Racib. (CABI, 1990), although earlier reports stated otherwise (Stover, 1972; Punithalingam and Holliday, 1975). In Africa and Latin America, the occurrence, distribution, importance and the identities of *Phyllosticta* species are unknown beyond the information provided in Table 1, and detailed surveys are therefore required to determine the status of freckle disease in these two regions.

The occurrence of freckle disease can be a serious constraint to the production of high-quality export bananas. Countries such as the Philippines and Taiwan that produce bananas for the export market experienced problems with the disease since blemished fruit is not acceptable for export (Corcolon and Raymundo, 2008; Jones, 2019). In the Philippines, the disease is becoming a serious problem in Cavendish plantations, causing more than three quarters of the fruits being downgraded and more than half being rejected (Corcolon and Raymundo, 2008). Apart from quality, yields are significantly reduced in severely infected plants. In Taiwan, freckle disease has replaced black Sigatoka (Arzanlou et al., 2007; Arzanlou et al., 2010) as the major foliar disease (Tsai et al., 1993).

6 Disease management

6.1 Disease resistance

In South and Southeast Asia, freckle disease affects a wide range of banana varieties in the AAA, AAB and ABB genomic groups (Jones, 2019). In Hawaii, the majority of cultivars in the Cavendish subgroup of the AAA genomic group are infected, but the disease has not been reported from Gros Michel. The disease can be severe on several cultivars of the AAB but not the ABB genomic group, including Bluggoe (Meredith, 1968). A similar situation was reported in Taiwan (Hwang et al., 1984; Tsai et al., 1993). However, the situation is reversed in Australia and the South Pacific regions, where cultivars in the Cavendish subgroup have not shown disease symptoms but Bluggoe was severely infected (Jones, 2000). These observations have been made prior to identification of the three different *Phyllosticta* species causing freckle disease (Table 1) (Wong et al., 2012). The presence of *P. cavendishii* is relevant, especially in regions that grow bananas for the export trade, where large-scale monoculture systems with cultivars in the Cavendish subgroup dominate plantations. The planting of a single susceptible host cultivar under a high-fertilizer regime in a high-rainfall environment is ideal for the establishment and spread of the pathogen.

In view of the fact that freckle disease only occurs on certain host genomic groups and cultivars and in certain geographic locations, planting those cultivars that have not shown infection in any particular location (Meredith, 1968; Hwang et al., 1984; Tsai et al., 1993) may provide an effective control strategy. Thus, in areas where *P. maculata* and *P. musarum*, which both have not shown pathogenicity on cultivars of the AAA genomic group, are prevalent, planting of non-susceptible cultivars in such genomic groups, including Cavendish, may be considered. To ensure the viability of implementing this option in different geographic locations, detailed knowledge on the distribution of different *Phyllosticta* species and the level of susceptibility of any cultivar within different host genomic groups is required. At present, limited information is available due to lack of cultivar evaluation among the rich genetic diversity in bananas.

6.2 Disease avoidance

Pests and diseases of banana have spread around the world in the last century making large-scale monoculture of cultivars in the Cavendish subgroup increasingly difficult (Drenth and Kema, 2021). Preventing the introduction and stopping the invasions of pathogens through biosecurity measures, eradicating or containing outbreaks can be very effective ways of controlling diseases. Banana bunchy top virus has been contained in a small production area in Australia for almost a century which allows growers to enjoy significant exclusion benefits, with an estimated saving of 27 million AUD per annum (Cook et al., 2012). Australia also has succeeded in excluding black leaf streak (black Sigatoka) and managed to eradicate the incursion of this pathogen in the main production area in North Queensland (Henderson et al., 2006; Cook et al., 2013).

In 2013, an outbreak of *P. cavendishii* capable of infecting Cavendish bananas was identified near Darwin in the Northern Territory (NT) of Australia. Since *P. cavendishii* was classified as an emergency plant pest in Australia, an eradication response was initiated. The eradication campaign involved visiting a total of 676 properties, of which 479 showed visible symptoms of freckle disease. Molecular analysis revealed that 463 of these properties were infected with *P. cavendishii*, while the remaining 17 were of *P. maculata*. These surveys resulted in an eradication program involving all banana plants within the affected zones, instigation of a host fee period, followed by sentinel planting and regular inspections of susceptible checks. The freckle disease eradication program was completed in 2017 resulting in a declaration of Proof of Freedom on 1 February 2019 (McMaster et al., 2020). Although these eradication campaigns are expensive and offer no guarantee of success at the start, they do confer significant benefits to banana growers in the long run. The absence of the causative species, *P. cavendishii*, averts infection of Cavendish bananas and protects Australian banana growers from increased management costs, yield losses, production and fruit quality constraints which are being experienced in countries where *P. cavendishii* is present.

6.3 Cultural control

As diseased leaf and fruit are sources of inoculum, cultural practices such as field sanitation can reduce disease severity (Hwang, 1991). Removing sources of conidial inoculum harboring on infected leaves and fruits as well as diseased plant parts and debris reduces disease severity (Meredith, 1968; Hwang, 1991). Trimming infected leaves before senescence has been recommended to reduce in-field inoculum in plantations (Corcolon and Raymundo, 2008).

Bagging the inflorescences and emerging fruit bunches protects plants from infection (Meredith, 1968; Abayasekera et al., 1993). Since conidia are dislodged

Figure 4 Aggregate of spots running in lines is evidence of spore-laden water droplets running down the leaf. Photo: Courtesy Department of Agriculture, Sarawak.

during rainfall and run down the leaves and the peduncle leaving a typical trail of infection (Fig. 4), it is important to cover the inflorescence with a bag and secure it tightly with rope onto the peduncle to prevent rainwater laden with spores from seeping in from the top and infecting the young fruit within (Fig. 5).

6.4 Chemical control

Chemical control is achieved through application of fungicides such as dithiocarbamate or chlorothalonil (Chuang, 1983a), triadimefon, tebuconazole or propiconazole (Tsai et al., 1993), mancozeb or mancozeb plus copper oxychloride (Corcolon and Raymundo, 2008; Thammaiah et al., 2009). However, to improve the efficacy of these fungicides, regular and timely applications just prior to the start of the rainy season is critical as the infection process requires the presence of free water. Other issues such as chemical and labor costs, pesticide residues, emergence of fungicide-resistant strains and environmental effects have to be taken into consideration when deploying chemical control measures.

6.5 Integrated pest and disease management

To mitigate the impact of freckle disease in a sustainable manner, an integrated management strategy is needed. These options as outlined earlier include

Figure 5 Tightening of the bunch bag to prevent freckle spores from running along the peduncle infecting the fruit inside. Photo: A. Drenth.

planting resistant host cultivars, adopting cultural practices such as de-leafing and bagging as well as applying recommended fungicides. Options for biological control are very limited if a narrow definition of biological control, defined as the one-on-one use of an introduced microorganism, is used to control a disease (Cook, 2000). At present, there is no known effective biological control option for the management of freckle disease in banana.

7 Conclusion and future trends

Current knowledge on the geographic distribution of the three different *Phyllosticta* species causing freckle disease is based on limited surveys (Wong et al., 2012). More surveys in the center of origin of bananas in Asia may reveal additional species such as the recently described *Phyllosticta musaechinensis* associated with freckle disease on bananas in China (Wu et al., 2014). Similar to the association of the common endophytic species, *Phyllosticta capitalensis*, with freckle disease in China (Sun et al., 2016), these authors stated that *P. capitalensis* and *P. musaechinensis* are both weakly pathogenic to banana, although Wu et al. (2014) state that leaves could not be infected by artificial inoculation of *P. musaechinensis*; thus, it remains unclear if these species are the actual causal agents of disease or are only able to infect older senescing leaves. Due to the challenges of isolating and culturing of certain *Phyllosticta* species but with the ease of cultivating the cooccurring endophytic species, *P. capitalensis*, on the same lesion, care must be taken not to confuse the pathogens with the endophyte. Clearly, more extensive surveys and experimental work are required to get a better insight into the species involved and their respective distributions and ability to cause disease.

The molecular basis of *Phyllosticta* species recognition of the banana host surface to commence germination and, subsequently, relay information to activate the developmental and penetration response is yet to be understood. This information is integral to identify the various genes involved and to establish their respective roles as such genes may form potential targets for novel control methods and new fungicides against these pathogens.

More research to improve management of the disease is required. In areas where black leaf streak disease is a problem and fungicides are applied on a regular basis, the impact of freckle disease may be masked as it is to a large extent controlled by the same broad-spectrum fungicides. Accurate assessment on the true impact of freckle disease will only be feasible when black leaf streak resistant cultivars are introduced, and growers reduce application of these fungicides.

Breeding for disease resistance is another aspect that needs to be considered. Plant breeders and pathologists must work in concert to identify freckle disease resistance traits in *Musa* germplasms from different geographic locations for future breeding programs to mitigate the problem so that more effective and sustainable control of the disease can be achieved.

8 Where to look for further information

- Ploetz, R. C., Thomas, J. E. and Slabaugh, W. R. (2003). Diseases of banana and plantain. In: Ploetz, R. C., Ed., *Diseases of Tropical Fruit Crops*, CABI Publishing, CABI International, Wallingford, Oxon, 73–134. http://dx.doi.org/10.1079/9780851993904.0073.
- Anderson, J., O'Dwyer, C., Phanthavong, S., Sengsoulichan, K., Drenth, A., Vilavong, S. and Burgess, L. W. (2021). First report of *Phyllosticta* spp. associated with banana freckle disease in southern Lao PDR. *Australasian Plant Disease Notes* 16(29): 1–4.
- Pécheur, M. J. B., Vally, V., Maudarbaccus, F., Henderson, J. and O'Dwyer, C. (2022). First report of banana freckle disease caused by *Phyllosticta cavendishii* in Mauritius. *New Disease Reports* 45: e12060.
- https://www.business.qld.gov.au/industries/farms-fishing-forestry/agriculture/biosecurity/plants/priority-pest-disease/banana-freckle.
- https://www.planthealthaustralia.com.au/wp-content/uploads/2013/01/Banana-freckle-FS.pdf.

9 References

Abayasekera, C., Karunaratne, A. and Adikaram, N. (1993). Freckle disease of banana. In: Champ, B., Highlay, C. and Johnson, G. (Eds.), *Proceedings of an International Conference on Postharvest Handling of Tropical Fruits*, Chiang Mai. Proceedings of the No. 50, Australian Center for International Agricultural Research, Thailand, 458–459.

Arzanlou, M., Abeln, E. C., Kema, G. H., Waalwijk, C., Carlier, J., Vries, Id, Guzmán, M. and Crous, P. W. (2007). Molecular diagnostics for the Sigatoka disease complex of banana. *Phytopathology* 97(9): 1112–1118.

Arzanlou, M., Crous, P. W. and Zwiers, L. H. (2010). Evolutionary dynamics of mating-type loci of *Mycosphaerella spp.* occurring on banana. *Eukaryotic Cell* 9(1): 164–172.

Baayen, R. P., Bonants, P. J., Verkley, G., Carroll, G. C., van der Aa, H. A., de Weerdt, M., van Brouwershaven, I. R., Schutte, G. C., Maccheroni, W., de Blanco, C. G. and Azevedo, J. L. (2002). Nonpathogenic isolates of the citrus black spot fungus, *Guignardia citricarpa*, identified as a cosmopolitan endophyte of woody plants, *G. mangiferae* (*Phyllosticta capitalensis*). *Phytopathology* 92(5): 464–477.

Bonants, P. J. M., Carroll, G. C., De Weerdt, M., van Brouwershaven, I. R. and Baayen, R. P. (2003). Development and validation of a fast PCR-based detection method for pathogenic isolates of the citrus black spot fungus, *Guignardia citricarpa*. *European Journal of Plant Pathology* 109(5): 503–513.

Braun, E. J. and Howard, R. J. (1994). Adhesion of *Cochliobolus heterostrophus* conidia and germlings to leaves and artificial surfaces. *Experimental Mycology* 18(3): 211–220.

CABI. (1990). *Guignardia musae Racib: Distribution Maps of Plant Diseases* (Vol. 623). Wallingford, UK: CAB International.

Carlier, J., Robert, S., Roussel, V., Chilin-Charles, Y., Lubin-Adjanoh, N., Gilabert, A. and Abadie, C. (2021). Central American and Caribbean population history of the *Pseudocercospora fijiensis* fungus responsible for the latest worldwide pandemics on banana. *Fungal Genetics and Biology* 148: 103528.

Carpenter, C. (1919). Banana freckle or black spot disease. *Report of the Hawaii Agricultural Experimental Station* 36–40.

Chandra, K. (1991). Status of banana diseases in India. In: Valmayor, R., Umali, B. and Bejosano, C. (Eds.), *Proceedings of a Regional Technical Meeting on Disease Affecting Banana and Plantain in Asia and the Pacific*. Brisbane, Australia: INIBAP-ASPNET, 73–83.

Chuang, T. (1983a). Chemical control of banana freckle caused by *Phyllostica musarum*. *Plant Protection Bulletin* 25: 15–22.

Chuang, T. (1983b). *Ecological and Physiological Study on Freckle of Banana. Research Abstract No. 8 (1981/1982)*. R.O.C, Taiwan: Taiwan Banana Research Institute.

Chuang, T. (1984). Ecological study of banana freckle caused by *Phyllosticta musarum*. *Plant Protection Bulletin* 26(4): 335–345.

Chuang, T. and Chen, S. (1982). Perfect stage of the pathogen of banana freckle. *Plant Protection Bulletin* 24: 272.

Chuang, T. Y. (1981). Isolation of *Phyllosticta musarum*, causal organism of banana freckle. *Transactions of the British Mycological Society* 77(3): 670–671.

Cook, D. C., Liu, S., Edwards, J., Villalta, O. N., Aurambout, J. P., Kriticos, D. J., Drenth, A. and De Barro, P. J. (2012). Predicting the benefits of banana bunchy top virus exclusion from commercial plantations in Australia. *PLoS ONE* 7(8): e42391.

Cook, D. C., Liu, S., Edwards, J., Villalta, O. N., Aurambout, J., Kriticos, D. J., Drenth, A. and De Barro, P. J. (2013). Predicted economic impact of black Sigatoka on the Australian banana industry. *Crop Protection* 51: 48–56.

Cook, R. J. (2000). Advances in plant health management in the twentieth century. *Annual Review of Phytopathology* 38(1): 95–116.

Corcolon, B. M. and Raymundo, A. (2008). Estimating yield losses in banana due to freckle disease caused by *Phyllosticta musarum* (Cke.) Van der Aa. *Philippine Journal of Crop Science* 33(2): 75–85.

Drenth, A. and Kema, G. H. J. (2021). The vulnerability of bananas to globally emerging disease threats. *Phytopathology* 111(12): 2146–2161.

Glienke, C., Pereira, O. L., Stringari, D., Fabris, J., Kava-Cordeiro, V., Galli-Terasawa, L., Cunnington, J., Shivas, R. G., Groenewald, J. Z. and Crous, P. W. (2011). Endophytic and pathogenic *Phyllosticta* species, with reference to those associated with Citrus Black Spot. *Persoonia* 26: 47–56.

Goldschmidt, P., Degorge, S., Benallaoua, D., Semoun, O., Borsali, E., Le Bouter, A., Batellier, L., Borderie, V., Laroche, L. and Chaumeil, C. (2012). New strategy for rapid diagnosis and characterization of keratomycosis. *Ophthalmology* 119(5): 945–950.

Hawksworth, D. L., Crous, P. W., Redhead, S. A., Reynolds, D. R., Samson, R. A., Seifert, K. A., Taylor, J. W., Wingfield, M. J., Abaci, O., Aime, C., Asan, A., Bai, F. Y., de Beer, Z. W., Begerow, D., Berikten, D., Boekhout, T., Buchanan, P. K., Burgess, T., Buzina, W., Cai, L., Cannon, P. F., Crane, J. L., Damm, U., Daniel, H. M., van Diepeningen, A. D., Druzhinina, I., Dyer, P. S., Eberhardt, U., Fell, J. W., Frisvad, J. C., Geiser, D. M., Geml, J., Glienke, C., Gräfenhan, T., Groenewald, J. Z., Groenewald, M., de Gruyter, J., Guého-Kellermann, E., Guo, L. D., Hibbett, D. S., Hong, S. B., de Hoog, G. S., Houbraken, J., Huhndorf, S. M., Hyde, K. D., Ismail, A., Johnston, P. R., Kadaifciler, D. G., Kirk, P. M., Kõljalg, U., Kurtzman, C. P., Lagneau, P. E., Lévesque, C. A., Liu, X., Lombard, L., Meyer, W., Miller, A., Minter, D. W., Najafzadeh, M. J., Norvell, L., Ozerskaya, S. M., Oziç, R., Pennycook, S. R., Peterson, S. W., Pettersson, O. V., Quaedvlieg, W., Robert, V. A., Ruibal, C., Schnürer, J., Schroers, H. J., Shivas, R., Slippers, B., Spierenburg, H., Takashima, M., Taşkın, E., Thines, M., Thrane, U., Uztan, A. H., van Raak, M., Varga, J., Vasco, A., Verkley, G., Videira, S. I., de Vries, R. P., Weir, B. S., Yilmaz, N., Yurkov, A. and Zhang, N. (2011). The Amsterdam Declaration on fungal nomenclature. *IMA Fungus* 2(1): 105–112.

Henderson, J., Pattemore, J. A., Porchun, S. C., Hayden, H. L., Van Brunschot, S., Grice, K. R. E., Peterson, R. A., Thomas-Hall, S. R. and Aitken, E. A. B. (2006). Black Sigatoka disease: New technologies to strengthen eradication strategies in Australia. *Australasian Plant Pathology* 35(2): 181–193.

Hwang, S. (1991). Status of banana diseases in Taiwan. In: Valmayor, R., Umali, B. and Bejosano, C. (Eds.), *Proceedings of a Regional Technical Meeting on Disease Affecting Banana and Plantain in Asia and the Pacific*. INIBAP-ASPNET, Brisbane, Australia, 73–83.

Hwang, S., Chen, C. and Wu, F. (1984). An investigation on susceptibility of banana clones to fusarial wilt, freckle and marginal scorch disease in Taiwan. *Plant Protection Bulletin, Taiwan* 26(2): 155–161.

Janex-Favre, M. C., Parguey-Leduc, A. and Jailloux, F. (1993). The ontogeny of pycnidia of *Guignardia bidwellii* in culture. *Mycological Research* 97(11): 1333–1339.

Johnson, B. E. and Brun, W. A. (1966). Stomatal density and responsiveness of banana fruit stomates. *Plant Physiology* 41(1): 99–101.

Jones, D. (2000). Speckles, freckle and rust. In: Jones, D. R. (Ed.), *Diseases of Banana, Abaca and Enset*. Wallingford, UK: CABI Publishing, 120–125.

Jones, D. (2019). Freckle. In: Jones, D. R. (Ed.), *Handbook of Diseases of Banana, Abacá and Enset*. Wallingford, Oxfordshire: CABI Publishing International, 166–171.

Jones, D. and Alcorn, J. (1982). Freckle and black Sigatoka diseases of banana in far north Queensland. *Australasian Plant Pathology* 11(1): 7–9.

Kuo, K. and Hoch, H. C. (1995). Visualization of the extracellular matrix surrounding pycnidiospores, germlings, and appressoria of *Phyllosticta ampelicida*. *Mycologia* 87(6): 759–771.

Lee, H. (1922). Banana freckle in the Philippine Islands. *Phytopathology* 12: 101–102.

Liang, L. (1990). Banana production and research in Guangdong (China). In: Valmayor, R. (Ed.), *Proceedings of a Regional Consultation on Banana and Plantain R&D Networking*. Manila and Davao, Philippines, INIBAP-ASPNET, 41–48.

Mackay, J. F., Wright, C. D. and Bonfiglioli, R. G. (2008). A new approach to varietal identification in plants by microsatellite high resolution melting analysis: Application to the verification of grapevine and olive cultivars. *Plant Methods* 4(1): 8.

McMaster, C. A., Tran-Nguyen, L. T. T., Voutsinos, M. Y., Cook, S. E., Condé, B. D., West, S. J., Nguyen, T. V. and Liberato, J. R. (2020). Outbreak of freckle disease (*Phyllosticta cavendishii*) on Cavendish bananas in Australia and the initial biosecurity response. *Plant Pathology* 69(1): 38–49.

Meredith, D. S. (1968). Freckle disease of banana in Hawaii caused by *Phyllostictina musarum* (Cke.) Petr. *Annals of Applied Biology* 62(2): 329–340.

Meyer, L., Sanders, G. M., Jacobs, R. and Korsten, L. (2006). A one-day sensitive method to detect and distinguish between the citrus black spot pathogen *Guignardia citricarpa* and the endophyte *Guignardia mangiferae*. *Plant Disease* 90(1): 97–101.

Nicholson, R. L. and Epstein, L. (1991). Adhesion of fungi to the plant surface: Prerequisite for pathogenesis. In: Cole, G. and Hoch, H. (Eds.), *The Fungal Spore and Disease Initiation in Plants and Animals*. New York: Plenum Press, 3–23.

Ocfemia, G. (1927). *Macrophoma musae* (Cke.) Berl. and Vogl. and *Phoma musae* Carpenter. *The Philippine Agriculturist* 15: 467–469.

Okane, I., Nakagiri, A., Ito, T. and Lumyong, S. (2003). Extensive host range of an endophytic fungus, *Guignardia endophyllicola* (anamorph: *Phyllosticta capitalensis*). *Mycoscience* 44(5): 353–363.

Pandey, A. K., Reddy, M. S. and Suryanarayanan, T. S. (2003). ITS-RFLP and ITS sequence analysis of a foliar endophytic *Phyllosticta* from different tropical trees. *Mycological Research* 107(4): 439–444.

Peres, N. A., Harakava, R., Carroll, G. C., Adaskaveg, J. E. and Timmer, L. W. (2007). Comparison of molecular procedures for detection and identification of *Guignardia citricarpa* and *G. mangiferae*. *Plant Disease* 91(5): 525–531.

Pu, J., Xie, Y., Zhang, X., Qi, Y., Zhang, C. and Liu, X. (2008). Preinfection behaviour of *Phyllosticta musarum* on banana leaves. *Australasian Plant Pathology* 37(1): 60–64.

Punithalingam, E. (1974). Studies on *Sphaeropsidales* in culture II. *Mycological Papers* 136: 1–63.

Punithalingam, E. and Holliday, P. (1975). *Guignardia musae*. CMI Descriptions of Pathogenic Fungi and Bacteria No 467, Commonwealth Mycological Institute, Kew, UK.

Schena, L., Nigro, F., Ippolito, A. and Gallitelli, D. (2004). Real-time quantitative PCR: A new technology to detect and study phytopathogenic and antagonistic fungi. *European Journal of Plant Pathology* 110(9): 893–908.

Stover, R. H. (1972). *Banana, Plantain and Abaca Diseases*. Kew, UK: Commonwealth Mycological Institute, 102–103.

Sun, J., Zhang, Y., Zhang, J., Lan, X. and Lu, J. (2016). First report of freckle disease of banana caused by *Phyllosticta capitalensis* in Guangxi, Southwest China. *Journal of Plant Pathology* 98(1): 175.

Tan, M.-K., Ghalayini, A., Sharma, I., Yi, J., Shivas, R., Priest, M. and Wright, D. (2009). A one-tube fluorescent assay for the quarantine detection and identification of *Tilletia indica* and other grass bunts in wheat. *Australasian Plant Pathology* 38(2): 101–109.

Taylor, C. F. (2009). Mutation scanning using high-resolution melting. *Biochemical Society Transactions* 37(2): 433–437.

Thammaiah, N., Kanamadi, V., Shirol, A., Kulkarni, M. S. and Reddy, B. S. (2009). Management of freckle leaf spot disease of banana. *International Journal of Plant Protection* 2(1): 42–44.

Tsai, Y., Chen, H. and Liu, S. (1993). Freckle disease of banana in Taiwan. In: Valmayor, R. (Ed.), *Proceedings of the International Symposium on Recent Developments in Banana Cultivation Technology*, INIBAP-ASPNET, Pingtung, Taiwan, 298–307.

Van der Aa, H. A. (1973). Studies in *Phyllosticta* I. *Studies in Mycology* 5: 1–110.

Van der Aa, H. A. and Vanev, S. (2002). *A Revision of the Species Described in Phyllosticta*. Utrecht, The Netherlands: Centraalbureau voor Schimmelcultures (CBS).

Van Gent-Pelzer, M. P. E., Van Brouwershaven, I. R., Kox, L. F. F. and Bonants, P. J. M. (2007). A TaqMan PCR method for routine diagnosis of the quarantine fungus *Guignardia citricarpa* on citrus fruit. *Journal of Phytopathology* 155(6): 357–363.

Vossen, R. H., Aten, E., Roos, A. and den Dunnen, J. T. (2009). High-resolution melting analysis (HRMA)–More than just sequence variant screening. *Human Mutation* 30(6): 860–866.

Wardlaw, C. W. (1935). *Diseases of the Banana and of the Manila Hemp Plant*. London, UK: Macmillan and Co. Ltd.

Wardlaw, C. W. (1961). *Banana Diseases, Including Plantains and Abaca*. London, UK: Macmillan and Co. Ltd.

Watanabe, K., Doi, Y. and Kobayashi, T. (1997). Pycnidial development of *Phyllosticta harai* and *Sphaeropsis sp*. *Mycoscience* 38(3): 259–265.

Wikee, S., Udayanga, D., Crous, P. W., Chukeatirote, E., McKenzie, E. H. C., Bahkali, A. H., Dai, D. and Hyde, K. D. (2011). *Phyllosticta*–An overview of current status of species recognition. *Fungal Diversity* 51(1): 43–61.

Wingfield, M. J., De Beer, Z. W., Slippers, B., Wingfield, B. D., Groenewald, J. Z., Lombard, L. and Crous, P. W. (2012). One fungus, one name promotes progressive plant pathology. *Molecular Plant Pathology* 13(6): 604–613.

Wong, M.-H. (2013). *Phyllosticta species causing freckle disease of banana*. PhD Thesis, School of Agriculture and Food Sciences, The University of Queensland.

Wong, M.-H., Crous, P. W., Henderson, J., Groenewald, J. Z. and Drenth, A. (2012). *Phyllosticta* species associated with freckle disease of banana. *Fungal Diversity* 56(1): 173–187.

Wong, M.-H., Henderson, J., Aitken, E. A. B. and Drenth, A. (2013a). Mode of infection of *Phyllosticta maculata* on banana as revealed by scanning electron microscopy. *Journal of Phytopathology* 161(3): 135–141.

Wong, M. -H., Henderson, J. and Drenth, A. (2013b). Identification and differentiation of *Phyllosticta* species causing freckle disease of banana using high resolution melting (HRM) analysis. *Plant Pathology* 62(6): 1285–1293.

Wu, S.-P., Liu, Y.-X., Yuan, J., Wang, Y., Hyde, K. D. and Liu, Z. (2014). *Phyllosticta* species from banana (*Musa* sp.) in Chongqing and Guizhou Provinces, China. *Phytotaxa* 188(3): 135–144.

Wulandari, N. F., To-Anun, C., Cai, L., Abdelsalam, K. A. and Hyde, K. D. (2010). *Guignardia/Phyllosticta* species on banana. *Cryptogamie-Mycologie* 31(4): 403.

Yen, J. (1972). Étude sur les champignons parasites du sud-est asiatique. XVIII. *Maladies des taches foliaires des bananiers provoquées, á formose, par deux nouveaux champignons* III(88): 221–225.

Chapter 4

Fusarium wilt of banana: impact, epidemiology and management of Fusarium wilt Race 1 and 2

Einar Martínez de la Parte, Wageningen University and Research, The Netherlands and Instituto de Investigaciones de Sanidad Vegetal (INISAV), Cuba; and Luis Pérez-Vicente, Instituto de Investigaciones de Sanidad Vegetal (INISAV), Cuba

1 Introduction

Banana production is affected by numerous diseases and pests around the world (Jones, 2019). Fusarium wilt of banana (FWB), also known as Panama disease, is caused by *Fusarium oxysporum* f. sp. *cubense* (E.F. Smith), W.C. Snyder, and H.N. Hansen (Foc). Phylogenetic studies reveal that Foc is an asexual polyphyletic fungus with various strains due to convergent evolution (Bentley et al., 1998; O'Donnell et al., 1998; Ploetz, 2006a; Fourie et al., 2011). More recently, analyses of the sequences of the translation elongation factor-1alpha (*tef1*), the RNA polymerase II largest subunit (*rpb1*), and the RNA polymerase II second largest subunit (*rpb2*) of a large collection of isolates revealed nine different lineages within Foc, which were described as new species (Maryani et al., 2019). This awaits further confirmation and hence in this chapter, we will refer to the causal agents of FWB as Foc.

http://dx.doi.org/10.19103/AS.2022.0108.04

FWB is considered to be one of the most destructive diseases affecting banana (Stover, 1962; Ploetz and Pegg, 1999; Ploetz et al., 2015) with a significant impact on productivity, local and regional economies, and society. The disease was first reported in Australia but gained notoriety from the Foc R1 epidemic and its impact on Gros Michel (AAA) in Latin American export plantations during the first half of the twentieth century (Wardlaw, 1961; Stover, 1962; Ploetz, 2005). Contrary to R1, Foc Race 2 (R2), which affects cooking bananas, such as Bluggoe (ABB) and some bred tetraploids (Ploetz, 2006a), did not affect banana-exporting industry. However, R2 threatens the livelihoods of smallholder farmers growing diverse susceptible cultivars, affecting local markets and food security in those countries where cooking bananas are an important staple food. There have been many causes that contributed to the ongoing proliferation of the disease, including the fact that infested areas were abandoned and cultivation moved to new areas without preventive measures; the long and variable latent period – depending on varieties or agroecological settings – between infection and symptom development; the lack of healthy planting material of susceptible cultivars; the long survival of the causal agent in infested soils; and growers' lack of information on disease cycles and pathogen biology (Pérez-Vicente, 2004).

Virtually all banana-producing areas for export in the Americas and Africa were affected by FWB during the past century (Ploetz, 2019). Currently, R1 and R2 are present in most banana-producing regions of the world, while Tropical Race 4 (TR4), which is pathogenic to popular Cavendish banana varieties, continues to spread (van Westerhoven et al., 2022). The distribution of R1 includes the Americas, Africa, Asia, Australia, Pacific and Melanesia islands, the Middle East, and countries surrounding the Mediterranean Sea (Stover and Simmonds, 1987; Pérez-Vicente, 2004; Ploetz, 2015, 2019; Ploetz et al., 2015; Dita et al., 2018).

In Latin America and the Caribbean Islands, R1 and R2 are widely distributed, affecting Gros Michel (AAA), Silk (AAB), Prata and Prata Ana (AAB), Mai'a Maoli' (Maqueño and Mai'a Maoli', AAB), Bluggoe (ABB), and Pisang Awak (ABB) subgroups of cultivars (Stover and Waite, 1960; Pocasangre et al. (2011). The cultivars Mysore (Mysore subgroups, AAB), Yangambi (Ibota subgroup, AAA), Isla (Iholena subgroup, AAB, in Peru), and FHIA 03 (AABB hybrid in Cuba) are sometimes affected. Similarly, Bluggoe, Gros Michel, Silk, and Pisang Awak are the banana subgroups more affected in Africa and India, together with Pome (AAB), Ney Poovan (AB), and Monthan (ABB) (Mustaffa and Thangavelu, 2011; Blomme et al., 2013).

FWB occurs on *Musa acuminata*, *Musa balbisiana*, *Musa schizocarpa*, *Musa textilis*, and their hybrids (Ploetz and Pegg, 1999). *Heliconia caribaea*, *Heliconia crassa*, *Heliconia collinsiana*, *Heliconia latispatha*, *Heliconia mariae*, *Heliconia rostrata*, and *Heliconia vellerigera* were reported as hosts of Foc in

Latin America (Stover, 1962), but these isolates belong to Race 3, which are no longer considered causal agents of FWB, as they were either weakly pathogenic or avirulent on Gros Michel and Bluggoe. Little is known as to whether other *Musa* spp. used as ornamentals (e.g. *M. velutina* and *M. coccinea*) can act as hosts to Foc.

2 Disease symptoms

FWB is a typical vascular wilt disease. Infection takes place in secondary or tertiary roots, while larger roots are rarely infected directly. Following infection of the roots, the fungus progresses into the rhizome, followed by a systemic invasion of the pseudostem (Fig. 1). The time between root infection and the expression of symptoms is usually from 2 up to 5 months under field conditions and varies among varieties (Stover, 1962). The first external symptom is the appearance of pale green streaks on the base of the petiole and fading of the normal green color at the base and borders of mature leaves. The most common symptom is chlorosis of the leaf ('yellow leaf syndrome') (Pérez-Vicente, 2004).

Yellow leaf syndrome, characterized by a chlorotic border on older leaves, can at times be confused with phosphorus or potassium deficiency, or be

Figure 1 External symptoms of Fusarium wilt of banana: (a) yellow leaf wilt syndrome; (b) green leaf wilt syndrome; (c and d) shortening of the emerging leaf; (e) pseudostem splitting; (f) asymptomatic sucker from an infected mother plant; (g) brown spots in the inner and outer sides of leaf sheaths.

associated with dry and cold environmental conditions. Affected leaves have a characteristic bright yellow tint. The chlorosis progresses from older to younger leaves (Fig. 1a). The leaves collapse gradually, bending at the petiole, commonly close to the midrib and hang down, forming a 'skirt' of dead leaves around the pseudostem. These symptoms can develop over a couple of weeks. In contrast to yellow leaf syndrome, in some cultivars the leaves of affected plants remain predominantly green until the petioles bend down (green leaf syndrome) and the plant collapses (Fig. 1b). Pseudostems can remain erect for 1 or 2 months after the death of the plant.

In actively growing plants, pseudostem splitting frequently occurs just above soil level (Fig. 1e). In susceptible cultivars, the pseudostem is eventually colonized, and faint brown streaks and flecks are evident on the inner and outer sides of leaf sheaths (Fig. 1g). New emerging leaves can be shorter than normal (Fig. 1c, d). If a bunch has emerged, it is usually stunted and produces small fruits. There are no known external or internal fruit symptoms.

The first internal symptoms of the disease occur in the root hairs at the initial point of infection and during initial progress into the rhizome. Accentuated and conspicuous necrosis of vascular tissues becomes visible in the limits of the cortex and the central cylinder of the pseudostem. The pathogen has the ability to pass through the vascular system of the plant to new emerging shoots (Fig. 1f). A characteristic symptom of the disease is a reddish-brown discoloration of the internal vessels of the pseudostem (Fig. 2a, b, c). In highly susceptible cultivars, the extension of vascular discoloration to leaf petioles (Fig. 2d) and the rachis can be observed. There are no obviously visible differences in symptoms by the races causing FWB, including TR4, in a wide spectrum of susceptible cultivars (Fig. 3).

Figure 2 Internal symptoms in Fusarium wilt-infected banana plants: (a) reddish colored vessels of the pseudostem; (b and c) transversal and longitudinal view, respectively, of affected vessels of pseudostem of infected Burro CEMSA (Bluggoe ABB) plants; (d) discolored vessels of leaf petioles of highly susceptible FWB-affected Silk plant; (e) necrosis of vessels of rhizome xylem; (f) longitudinal section view of infected rhizome and roots of Pisang Awak ABB; (g) FWB infected xylem showing the development of necrotic vessels to the sucker in Pisang Awak.

Figure 3 *Fusarium oxysporum* f. sp. *cubense* symptoms in different banana cultivars: no difference has been observed among varieties with regard to symptom expression caused by different races. R1 symptoms on (a) Apple, (b) Gros Michel, (c) Cocos, (d) Pisang Awak, (e) Prata-Anã, and (f) Lakatan. R2 symptoms on (g) Dwarf Bluggoe and (h) Burro CEMSA.

FWB can be readily distinguished from bacterial wilt, caused by *Ralstonia solanacearum*, by the presence of internal symptoms in the fruits and the absence of (a) symptoms in young shoots of less than 4–5 months of age and (b) of bacterial exudates from the male bell or when the pseudostem of the affected plant is cut across.

Spores of Foc remain dormant until stimulated to germinate by host or non-host root exudates or by direct contact with susceptible root tissues (Stover, 1962; Ploetz and Pegg, 1999; Guo et al., 2015). Spores produce germ tubes that attach to the roots, 2 days post-inoculation (DPI) (Guo et al., 2015). Hyphae grow along the grooves at junctions of epidermal cells at 4 DPI and appear in xylem vessels of banana roots at 5-6 DPI. At 15 DPI, the xylem vessels are heavily populated by fungal hyphae and chlamydospores in susceptible cultivars (Guo et al., 2015). However, in partially resistant cultivars, colonization of tissues is delayed and a lower level of chlamydospores formation is observed, indicating the presence of defense response processes, inhibiting fungal colonization and reproduction in the host (Pérez-Vicente et al., 2014).

Once the fungus reaches the xylem vessels inside the lateral roots, it is well established and will eventually colonize the rhizome of susceptible banana varieties (Rishbeth, 1955). Once it has reached the rhizome, the pathogen can become distributed throughout the pseudostem in under 2 weeks (MacHardy and Beckman, 1981). Colonization of the vascular bundles in the rhizome is a crucial step in disease development as it blocks the vessels and thus interferes with nutrient uptake and upward water transport to the pseudostem and leaves (Li et al., 2017). Despite slow initial movement through the roots, requiring some 4 weeks to advance 75 cm, in mature aboveground xylem vessels of the pseudostem, the pathogen can advance in surges of 30 cm in 2–3 days, generating new macro-, microconidia, and chlamydospores (Li et al., 2011; Pegg et al., 2019).

Nevertheless, even when the pathogen successfully attaches its structures to the roots, most infection attempts appear to be blocked by the host. In response to infection, the cell wall of banana roots trigger a series of coordinated responses and change in the spatial distribution and abundance of hydroxyproline-rich glycoproteins, pectin methylesterases, and pectins (Fan et al., 2017), which have a direct impact on the ability of the pathogen to infect. In addition, the living xylem constitutes a barrier against the advance of the pathogen and the probability that the infection may reach the rhizome is low, estimated by Rishbeth (1955) to be about one root in 20. In the xylem vessels, host defense responses may involve gels, tyloses, and lignification, leading to vascular occlusion and blocking the pathogen's advance. This response is present even in susceptible cultivars (Ploetz and Pegg, 1999). These gels will not persist unless there are other host reactions in the xylem elements, and therefore only impede colonization for a short time. If the gel persists long enough to allow tyloses to form and then become impregnated by phenolic materials, the pathogen can be successfully contained (Pegg et al., 2019).

After the pathogen has systemically invaded the xylem vessel elements with appreciable colonization of the rhizome, a severe water shortage develops in susceptible banana plants due to vascular plugging. This limitation in water movement leads to a reduction in transpiration and the expression of external symptoms (Pegg et al., 2019). In resistant cultivars, infection is restricted by the rapid deployment of the host defenses in the rootlets, roots, or at the root base of resistant varieties (MacHardy and Beckman, 1981).

3 Pathogen characteristics

F. oxysporum comprises anamorphic, filamentous, and morphologically undifferentiated fungal species featuring saprophytes, antagonists, and pathogens (O'Donnell et al., 1998). In the case of plant pathogens, they mostly cause wilting, damping off, and rotting of roots and organs. From an agricultural and economic point of view, *F. oxysporum* is the most important taxon of *Fusarium* (Ploetz, 2006b).

Many *Fusarium* spp. have the ability to cause disease in certain plant genera and families and hence are classified in formae speciales that specialize on a particular hosts. The addition of *cubense* is unusual because the name is not taken from the host but from the geographic origin of the first characterized strains (Cuba). The first person to isolate the pathogen from banana tissues was E.F. Smith in 1910, who named it *Fusarium cubense* due to its Cuban source. It was Ashby in 1913 who gave the first detailed description of the causal agent in culture. Nevertheless, a decade passed before Brandes confirmed Koch's postulates in 1919 for *F. cubense* on the cultivars Gros Michel (AAA), Silk (AAB), and Bluggoe (ABB). By 1935, Wollenweber and Reinking had recognized *F. cubense* as a variant of *F. oxysporum* and then the species was renamed Foc (E.F. Sm.), W.C. Snyder, and H.N. Hansen (Ploetz, 2005).

Foc cannot be morphologically distinguished from other formae speciales that cause wilting in other hosts and other non-pathogenic *F. oxysporum*. Foc is an anamorphic fungus without a known sexual stage (teleomorph). The fungus produces macroconidia, microconidia, and chlamydospores for asexual reproduction and dispersal (Fig. 4). Macroconidia and microconidia can be produced in sporodochia. The sexual stage (teleomorph) has not been found, although the mating type idiomorphs *MAT-1* and *MAT-2* have been detected in isolates from different populations of Foc (Fourie et al., 2009; Ordoñez, 2018). See McTaggart et al. (2021) for a more comprehensive analysis.

Foc populations have been classified in races, vegetative compatible groups (VCG), and by the production of volatiles. To date, four different races of the pathogen have been described based on the pathogenicity to reference host cultivars (Pérez-Vicente, 2004; Ploetz, 2015a; Dita et al., 2018):

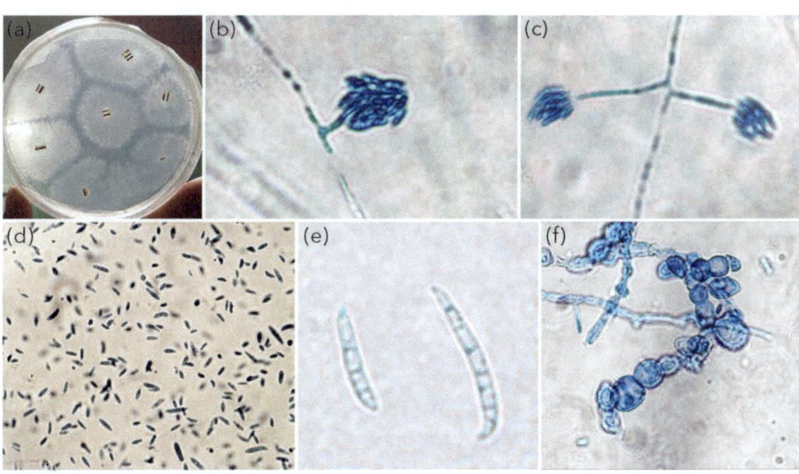

Figure 4 *Fusarium oxysporum* f. sp. *cubense*: (a) colony growths from pieces of infected vascular tissues; (b and c) false heads of microconidia developed in monophialides; (d) microconidia; (e) macroconidia; (f) chlamydospores in chains.

- Race 1 (R1) affects Gros Michel (AAA), Manzano/Apple/Latundan (Silk subgroup, AAB), Pisang Awak (Pisang Awak subgroup, ABB) and Maqueño (Mai'a Maoli–Popoulu subgroup, AAB).
- Race 2 (R2) affects cooking bananas of the Bluggoe (ABB) subgroup.
- Race 4 (R4) affects all cultivars in the Cavendish (AAA) subgroup in addition to those susceptible to R1 and R2 and has been split into two subgroups (Ploetz, 2015):
  - Subtropical race 4 (STR4), which affects 'Cavendish' in the subtropics under abiotic stress conditions; and
  - Tropical race 4 (TR4), which affects many of the same cultivars as STR4, independent of abiotic conditions in the tropics as well as the subtropics.

Race 3 causes disease in *Heliconia* spp. and is not considered to be pathogenic on banana (Ploetz et al., 2015).

Due to the lack of a genetic basis for the discrimination of Foc into races, heterocompatibility or heterokaryon development between isolates has been used to characterize populations (Pérez-Vicente et al., 2014). At present, 24 VCGs or VCG complexes have been identified in Foc (Ordoñez et al., 2015; Mostert et al., 2017; Dita et al., 2018). In addition, there are numerous Foc isolates that do not fit into any known VCG, despite being self-compatible. These isolates might be putative novel Foc VCGs not previously detected. The distribution of and relationships among the various VCGs have been used as phenotypic markers to provide insight into the dissemination and evolution of Foc.

Strains belonging to diverse VCGs can belong to the same race, suggesting that pathogenicity toward a specific cultivar evolved convergently (O'Donnell et al., 1998; Ploetz, 2006; Fourie et al., 2009) is the product of past sexual recombination (Drenth et al., 2019) or results from horizontal transfer of mini chromosomes among members of the *F. oxysporum* complex (Ma et al., 2010). However, from a genetic point of view, race designations in Foc are influenced by environmental conditions and their use as phenotypic markers is not an ideal system to understand the spread and the evolution of this pathogen (Dita et al., 2020).

While vegetative compatibility gives a measure of phenotypic diversity, molecular genetic markers and specific DNA sequences provide more detailed information into the genetic relationships among isolates within each VCG and among different VCGs. Different genetic markers have been used, such as translation elongation factor-1α (*tef1*), the mitochondrial small subunit (MtSSU), the Intergenic Spacer (IGS), and Internal Transcribed Spacer of the ribosomal DNA, the RNA polymerase II largest subunit (*rpb1*), the RNA polymerase II second largest subunit (*rpb2*), and a repeat region encoded in the mitochondrial genome (MtR). These have shown that Foc is polyphyletic, consisting of several clonal lineages (Bentley et al., 1998; Fourie et al., 2009, 2011; Mostert et al., 2017; Maryani et al., 2019). These studies show that TR4 isolates are less diverse than Foc R1 (and probably R2), which can be found in almost all lineages (Maryani et al., 2019).

The presence or absence of a strong volatile odor when isolates are grown on a plain rice medium has been used to classify pathogen strains that are respectively classified as 'odoratum' or 'inodoratum' (Moore et al., 1991; Pérez-Vicente, 2004; Battle and Pérez-Vicente, 2009). Stover (1962) stated that R1 isolates, from different Latin America and Caribbean countries, were either 'odoratum' or 'inodoratum' and classified all R2 isolates as 'inodoratum'. Reinforcing Stover's reports, Cuban Foc R1 and R2 isolates belonging to VCGs 01210, 0124, 0124/0125, and 0128 indiscriminately produce volatiles (Battle and Pérez-Vicente, 2009). However, according to Moore et al. (1991), R1 (VCGs 0124, 0125) and R2 (VCG 0128) isolates from Australia were classified as 'inodoratum', while Race 4 (VCGs 0120, 0129) isolates were 'odoratum' (Moore et al., 1991).

4 Disease cycle and epidemiology

A survey carried out in Cuba involving 51 farms and plots in different provinces of the country revealed factors related to the presence and development of FWB. Related factors, in order of importance, included

1) cultivar susceptibility;
2) use of apparently healthy planting material;
3) lack of preventive biosecurity practices regarding movement of equipment and people between fields;
4) age of plantation;
5) potassium content in soil and plants; and
6) detection and destruction of infected plants and treatment of soil.

Infected plants as well as the suckers produced from infected plants have a long incubation period before symptoms appear. The historical practice of establishing plantations with asymptomatic but infected suckers has been a major factor in pathogen dispersal, resulting in new outbreaks of the disease. Once soils are contaminated with the pathogen, they cannot be successfully replanted with susceptible cultivars due to the long-term survival of the pathogen in soil.

Foc can survive for decades in the dead host material and as a non-pathogenic parasite of weeds in banana plantations (cortical tissues of roots are infected, but symptoms are not produced) (Ploetz, 2015; Dita et al., 2018;

Pegg et al., 2019). Its persistence and the multiple ways in which it can be moved increase the risk of spread. It is disseminated by infected plant material ('suckers'), surface water flow, and irrigation water from contaminated rivers or ponds, contaminated tools and equipment, and spores attached to machinery and footwear (Fig. 5) (Stover, 1962; Ploetz, 1990; Dita et al., 2013a, 2018; Pérez-Vicente et al., 2014; Pegg et al., 2019). FWB is a polycyclic disease as studies carried out with published data from the R1 epidemic in Gros Michel in Honduras and TR4 in Malaysia show that the disease progress curves fit the Gompertz and Logistic mathematical models (Pérez-Vicente et al., 2017).

Temperature is a critical factor in FWB development (Pegg et al., 2019). Usually, diseased plants express more symptoms during the warmer and wetter months of the year and studies have shown that isolates of different VCGs grow similarly between 9°C and 38°C, with an optimum growth rate between 23°C and 28°C (Battle and Pérez Vicente, 2009), indicating that temperature is not a limiting factor in the development of the disease in most banana-growing areas in tropical countries.

Different cropping and edaphic factors have been related to FWB incidence in the field. Soil factors that favor or suppress the disease include surface and internal soil drainage, as well as physical, chemical, and microbiological

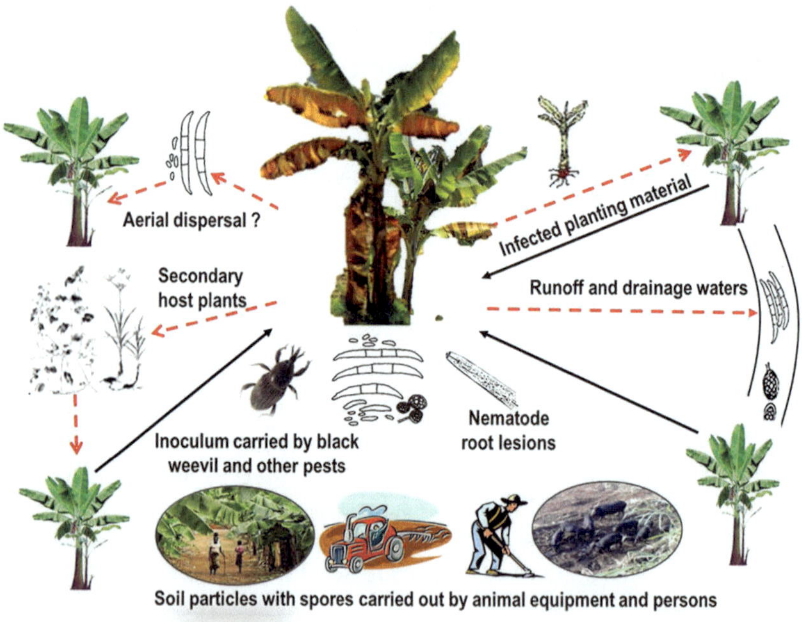

Figure 5 Disease cycle of Fusarium wilt of banana caused by *Fusarium oxysporum* f. sp. *cubense*.

properties. Mineral nutrients can influence wilt development. An increase of NO_3^- in soil reduced disease development, while an increase of NH_4^+ favored the development of FWB (Dita et al., 2018; Pegg et al., 2019). A high level of Ca^{2+} increases soil suppressiveness and reduces chlamydospore germination (Peng et al., 1999). A reduction in Fe^{2+} levels increases soil suppressiveness (Scher and Baker, 1982) and a correlation between FWB incidence and soil potassium level has been described, with suppressive soils having three times more potassium than conducive soils (Pegg et al., 2019). Soil pH, which can be altered by different factors, can influence the incidence of FWB. A pH lower than 5.6 favors disease development (Peng et al., 1999; Segura-Mena et al., 2021).

The composition of the microbial community in the rhizosphere, where microbes interact with pathogens, may also influence infection near the root surface (Raaijmakers et al., 2009). Disease-suppressive soils showed more diverse and complex microbial communities (Shen et al., 2015; Köberl et al., 2017).

Foc R1 has been found in weeds such as *Paspalum fasciculatum*, *Panicum purpurescens*, *Ixophorus unisetus* (Poaceae), and *Commelina diffusa* (Commelinaceae) in Central America (Waite and Dunlap, 1953). In all cases, weeds did not show external symptoms resembling FWB. These findings suggest that weed hosts aid the survival of Foc in the soil.

Foc can be spread from plant to plant by root proximity. The fungus is present in some of the major roots of an infected plant, and as the plant dies, these roots decay and spores are released into the soil. The short-lived secondary and tertiary roots are more likely to be infected than the major roots and will release a small but constant supply of inoculum into the soil while the plant is still growing (Stover, 1962). Roots of adjacent healthy banana plants will grow into the root zone of the diseased plants, thus accounting for mat-to-mat spread (Pegg et al., 2019).

The pathogen is dispersed in contaminated soil, by natural or artificial means. Natural means include soil drift due to wind or rain (erosion). Artificial means are related to soil adhering (the movement being greater in sticky clay soils) to agricultural implements, containers, tools, animals, footwear, and clothes and the use of soil as a substrate in nurseries (Dita et al., 2013a; Pérez-Vicente et al., 2014). Foc can be dispersed to disease-free areas through irrigation water, rainfall, surface flow after rainfall, and in rivers and streams. If a Foc-contaminated water reservoir is used to irrigate a disease-free area, the disease may disperse rapidly (Pérez-Vicente et al., 2014).

Chlamydospores survive better in a running river than in dam water due to aeration. Stover (1962) found that a large percentage of chlamydospores die after 40 days of submersion in stagnant water, such as in dams, but survive for much longer in oxygenated running water, such as in rivers. He suggested that the lack of free oxygen in dams and saturated soils are major factors causing the loss in viability of chlamydospores.

In smallholder farming systems and plantations close to human settlements, livestock and domestic animals are frequently found within banana fields. These animals can transport soil particles from infested areas to Foc-free areas, contributing to farm-to-farm dissemination of FWB (Dita et al., 2018).

As a soil-borne pathogen, Foc can also penetrate the host through wounds caused by, for example, plant parasitic nematodes or other external factors promoting root damage (Dita et al., 2018). Under greenhouse conditions, the combined inoculation of Foc and different plant parasitic nematodes species, such as *Radopholus similis*, *Pratylenchus coffeae*, and *Meloidogyne incognita*, resulted in the early onset and increased severity of FWB symptoms compared to plants solely inoculated with Foc, suggesting a positive interaction among the pathogens (Dinesh et al., 2014; Sanjeevkumar et al., 2018; Rocha et al., 2020). Although studies to assess the interaction between nematodes and Foc at the field level are limited, it can be deduced that reducing plant parasitic nematode populations in the soil will help reduce FWB infections.

Insects, especially the banana weevil *Cosmopolites sordidus* (Coleoptera: Curculionidae), are other vectors of Foc (Meldrum et al., 2013; Guillen Sanchez et al., 2021). Banana weevils are found wherever bananas are grown, moving through the soil, feeding on the roots and corm of the plants (Guillén Sanchez et al., 2021, Chapter 18 of this book). Guillen Sanchez et al. (2021) proved that inoculum of Foc R1 can be found on external teguments in the digestive tract and feces of *C. sordidus* following contact with an infected source. They found that external and internal inoculum decreased exponentially within 50 hours, but weevils remained infested for 2–3 days, enabling inoculum dispersal. Another insect, *Metamasius hemipterus* L. (Coleoptera: Curculionidae), a weevil frequently found in banana fields, as well as other subterranean vertebrate pests found in banana field soils, may also need to be considered as vectors of Foc.

Local (on-farm) or long-distance (other farms, regions, or countries) dispersal occurs mainly through the movement and planting of asymptomatic plants but already-infected suckers (Pérez-Vicente et al., 2014). To avoid this, tissue-cultured plantlets, grown under aseptic conditions in accredited nurseries with high standards of hygiene, should be used as a source of disease-free planting material (Pegg et al., 2019). The movement of pseudostem tissues, leaves, and bunch stalks from infected plants can be another method of dispersal of Foc. Banana leaves are frequently used for wrapping or packing bananas intended for local consumption and other handicraft uses (Pérez-Vicente et al., 2014).

5 Disease occurrence and impact

A comprehensive review of the occurrence of FWB in different countries was published by Stover (1962), including approximate dates of the first reported appearance, and the varieties attacked. In many cases, the movement of the causal fungus from country to country is related to the introduction of new cultivars to the growing areas (Stover, 1962). Subsequent spread within an area then occurs via infected planting material, contaminated water, and infested soil transported by people, tools, and equipment (Blomme et al., 2013).

In the Indio-Malayan area, FWB affects several varieties of edible bananas (Stover, 1962; Ploetz and Pegg, 1997) but in the western tropics, the disease is mainly associated with the two common varieties, Silk and Gros Michel (Stover, 1962). Silk (Apple AAB) was present in the Caribbean before 1750, while Gros Michel (AAA) was not introduced until the early 1800s and was not extensively planted before 1850. During the rapid expansion of Gros Michel in the Caribbean between 1890 and 1910, lands formerly infested with diseased Silk were planted with Gros Michel. Despite the wilt having been present in some areas where Silk was planted before Gros Michel was introduced, it was on the widely planted Gros Michel that most of the first reports of FWB in the Western Hemisphere were made (Stover, 1962). It was reported for the first time in America, Panama, and Costa Rica in the 1890s (Stover, 1962) and subsequently Jamaica in 1903, Suriname in 1906, Trinidad and Tobago in 1907, Cuba in 1908, Guatemala in 1910, Colombia in 1929, and Venezuela in 1930 (Drenth and Kema, 2021). The relevance of the disease was established by its impact on the intensive export production of Gros Michel in Central American plantations, but there is evidence to suggest that the arrival of FWB in the Western Hemisphere may have preceded the production of Gros Michel in large-scale plantations, which increased the spread and impact of the disease (Pérez-Vicente, 2004).

According to data by Stover (1962), the losses caused by R1 to the Gros Michel export industry were calculated at US$2 billion (equal to about US$2.74 billion in 2021) during the first half of the twentieth century (Ploetz, 2005; Drenth and Kema, 2021). These figures only include data registered by exporting companies and not the losses to existing smallholder production systems and cultivars.

Although the economic impact of the disease on the vast Gros Michel monocultures is well known, its impact on subsistence agriculture was not well documented and most likely not recorded or included in the losses reported during the 1950s. Currently, approximately 20 million tons (64% of production) of *Musa* are locally consumed in Latin America and the Caribbean (Dita et al., 2013b).

Although the export industry has moved to Cavendish to deal with R1, the disease still causes serious losses in subsistence agriculture in Central America, mainly in the Gros Michel variety, which is still grown in agroforestry systems and via intercropping with coffee. Other preferred varieties are the highly susceptible Prata-type (AAB) varieties and Apple (Silk, AAB), which account for more than 70% of total production (~500 000 ha) in Brazil. In Northern Paraná (Santa Mariana, Urai, Assaí, Andirá), losses of up to 100% are recorded each year in Silk plantations, and in Corupá and the Ribeira Valley, losses in Prata can exceed 30% (Dita et al., 2020). Pisang Awak and Bluggoe (ABB) types are also grown due to their resistance to stress factors and adaptability to subsistence farming but are heavily affected by R1 and R2, respectively.

A similar scenario as observed in the Americas is also evident in Africa. Most early FWB reports in different African countries between 1924 and 1953 were associated with Gros Michel (AAA) and Bluggoe (ABB) cultivars (Stover, 1962). In West Africa, the disease was first reported during the 1950s, in Zaire's Gros Michel export production areas to Ghana. The disease now occurs sporadically in the remaining small-scale plantations of Gros Michel, the preferred variety for local markets. In East Africa, where insignificant export production occurred, FWB spread was not associated with 'Gros Michel' and 'Silk' cultivars but with the distribution of cultivars Pisang Awak and Bluggoe (ABB) (Pérez-Vicente, 2004). These cultivars, which are highly susceptible to R1 and R2, respectively, were introduced after World War II (Stover, 1962). Although there are reports of FWB affecting East African Highland Bananas (EAHB, AAA) (Tushemereirwe, 1992), as well as other dessert banana cultivars such as 'Gros Michel', 'Sukari Ndizi' ('Pome' AAB), and 'Ney Poovan' (AB) (Blomme et al., 2013), recent studies show that EAHBs are mostly resistant to R1 (Ndayihanzamaso et al., 2020). Currently, few banana-producing areas in East and Central Africa remain free of R1 (Ploetz, 2019) where FWB is a major threat to the sustainability of banana processing and export initiatives (Karangwa et al., 2018).

R1 has been widespread in African plantations of cultivars of the subgroups Gros Michel, Ney Poovan, and Pisang Awak (Blomme et al., 2013). It is one of the main factors responsible for the significant fall in banana production levels over the last few decades in East Africa (Rutherford, 2001). In Kenya, where losses of up to 100% of 'Wang'ae' (Ney Poovan) plantations have been reported, it is estimated that FWB directly affects the livelihood of more than 9.5 million people (Kung'u et al., 2001). In Tanzania, R1 caused the destruction of 75% of the Ney Poovan plantations in an area of the Arusha/Kilimanjaro region, and gradually destroyed the widely planted Pisang Awak and Gros Michel cultivars in the Kagera district (Blomme et al., 2013), to be replaced by more resistant cultivars. In Uganda, FWB caused an estimated yield loss of more than 60% in dessert bananas (Tushemereirwe et al., 2000).

EAHBs dominate East and Central African banana cultivation in the highlands of Uganda and Tanzania, sustaining the livelihoods of millions of small-scale farmers and are resistant to R1. However, the diploids Mchare and Muraru cultivars are susceptible to R1, while the responses of Mchare, NARITA's and Matooke hybrids range from susceptible to resistant (Ndayihanzamaso et al., 2020). The selection of pathogenic variants imposed by the genetic distinctiveness of local cultivars may underpin the diversification of Foc in this region after being introduced from Asia. East and Central Africa harbor six VCGs, including VCGs 01212 and 01214, which have not yet been reported anywhere else in the world (Karangwa et al., 2018).

In Australia, R1 (VCGs 0124 and 0125) is widespread throughout the entire east coast banana production area, affecting mainly Lady Finger (Pome, AAB) but also Ducasse (Pisang Awak, ABB), Gros Michel, and Mysore (AAB) cultivars. Among these, Lady Finger is the only non-Cavendish cultivar of commercial importance in Australia (Pegg et al., 1996; Bentley et al., 1998) and plantations in subtropical eastern Australia currently suffer significant damage from FWB (Pegg et al., 2019).

6 Disease resistance: cultivar susceptibility and host resistance

In Foc-infested soils, planting resistant cultivars is the only consistent and economically sustainable measure of control. Since the early outbreaks of R1 and R2, the cultivars of the subgroups Gros Michel (AAA), Ibota (AAA, Yangambi), Silk and Prata (AAB), Mysore (AAB; with cultivars known as Lady Finger in Australia, and Manzano INIVIT in Cuba), Pisang Awak (ABB), and Bluggoe (ABB) have proven to be highly susceptible to FWB. The R1 problem in the Gros Michel export industry was solved by moving to extensive cultivation of Cavendish varieties (AAA) and plantains (AAB) (Stover, 1962; Ploetz, 2005, 2019; Pérez-Vicente, 2004; Pérez Vicente et al., 2009; Dita et al., 2013a, 2018). After this, the disease was practically forgotten by the export-orientated banana industry and a reduced number of papers were published until the spread of TR4 – first in Asia and Australia and now in Africa and Latin America – reignited research interest in FWB of banana.

The banana export industries in the Caribbean and Central America responded to the devastation in their Gros Michel plantations and the lack of effective disease control measures by establishing breeding programs. This began in Trinidad (Imperial College of Tropical Agriculture) in 1922 and was aimed at developing FWB-resistant replacements for 'Gros Michel' and was continued from Jamaica in 1960 (Smith and Pillay, 2020). In the early 1960s, Cavendish cultivars originating from Southeast Asia replaced Gros Michel in the export trade and the conversion was complete in 1971. Since that time, the most important function of banana breeding has been to develop genetically diverse cultivars for long-term protection against disease epidemics. Today, there are additional objectives such as tolerance to environmental stress and market diversification.

6.1 The Jamaica banana breeding program from 1960 to 1980

The breeding program in Jamaica was based on a single cross, being the female triploid Gros Michel and the male parent Pisang lilin (an edible diploid variety introduced into Jamaica from Malaysia in 1937), which shows resistance to R1 (in most, but not all locations), burrowing nematode, and Sigatoka (Stover and Simmonds, 1987). The breeding program produced the tetraploid 'Bodles Altafort', which was not released commercially due to the height of the plants, short fingers, weak fruit pedicels, and susceptibility to R2 in some locations. Another tetraploid hybrid with long fingers, resistant to R1, and yellow Sigatoka (caused by *Pseudocercospora musae*) showed good eating quality but ripened faster than the Cavendish cultivar Valery. The breeders concluded that tetraploids with good taste and shelf life similar to Cavendish cultivars were difficult to obtain.

6.2 Banana breeding by Federación Hondureña de Investigaciones Agrícolas

The destruction of Gros Michel by FWB prompted the establishment of an exhaustive breeding program by the United Fruit Company (UFC) in 1959 when black leaf streak disease (BLSD, or black Sigatoka caused by *Pseudocercospora fijiensis*) was still unknown. By the 1970s, after the arrival of *P. fijiensis*, the UFC germplasm was evaluated and *M. acuminata* subsp. *burmannica* and *malaccensis* were used to develop male and female dwarf diploid hybrid parents with resistance to BLSD and the burrowing nematodes *R. similis* with ripened fruits that remained attached to the crown (Stover and Simmonds, 1987). The UFC program was donated to FHIA in 1984 and supported by the International Development Research Centre, Food and Agriculture Organization of the United Nations (FAO), United Nations Development Programme (UNDP)/International Network for the Improvement of Banana and Plantain (INIBAP), Windward Islands Banana Growers' Association (WINBAN), U.S. Agency for International Development (USAID), Natural Resources Institute, and the governments of Honduras and Ecuador. Following the full adoption of Cavendish, the program aimed to develop banana hybrids for local and export markets with resistance to BLSD, *R. similis*, and FWB (Rowe and Rosales, 2000). FHIA hybrids have been released and adopted in 50 countries, particularly Cuba. The most commonly known and evaluated hybrids released by the program are shown in Table 1.

Table 1 Characteristics of popular hybrids released by Federación Hondureña de Investigaciones Agrícolas banana breeding program

Cultivar	*Genome*	*Parents*	*Characteristics*[1]
FHIA 01 (Goldfinger)	AAAB	'Santa Catarina Prata' × SH-3142	Resistant to BLSD, R1 and R2; Partial resistance to TR4, *R. similis*, crown rot; exceptional yield of sub-acid dessert fruit.
FHIA 02 Mona Lisa	AAAA	'Williams' × SH-3393	Resistant to BLSD, but susceptible to R1 and subtropical Race 4 in Australia. High bunch weights on relatively short cycle times.
FHIA 03 (Sweetheart)	AABB	SH-3386 ('Gaddatu' × BB) × SH-2471) × SH-3320	Partial resistance to BLSD; susceptible to FWB, resistant to Moko caused by *R. solanacearum* and nematodes (*Radopholus similis*). Adapted to a wide range of climates, including subtropical conditions and high altitudes.
FHIA 17	AAAA	'Highgate' × SH-3362	Partial resistance to R1 and susceptibility to TR4. Partial resistance to banana weevil (*C. sordidus*) and BLSD. Performs well in subtropics and at higher altitudes. Fruit ripens naturally to a golden yellow.
FHIA 18 (Bonanza)	AAAB	'Prata Ana' × SH-3142	Resistant to BLSD, to R1 and R2; Partial resistance to TR4.
FHIA 20	AAAB	'AVP-67' (French plantain) × SH-3437	Resistant to BLSD, FWB R1, and TR4.
FHIA 21	AAAB	'AVP-67' (French plantain) × SH-3142	Resistant to BLSD, FWB R1, and TR4, *R. similis* and banana weevil.
FHIA 23	AAAA	'Highgate' × SH-3362	Partial resistance to BLSD and resistant to R1.
FHIA 25	AAB	SH-3648 × SH-3142	Resistant to BLSD, R1, TR4 and *R. similis*. High yielding. Poor flavor.
SH 3436-9	AAAA	'Highgate' × SH-3142	Susceptible to R1 and TR4.
SH-3640 (High noon)	AAAB	Prata Aña × SH-3393 (AA)	Susceptible to R1 and TR4.

6.3 Banana breeding by Empresa Brasileira de Pesquisa Agropecuária

In Brazil, a banana breeding program was initiated at the end of 1982. The preferred cultivar in Brazil is Maçã (Apple, subgroup Silk, AAB) and, for this reason, the emphasis of the breeding program has been on the AAB group and, more specifically, on the Prata (Prata, Prata ana, Pacovan) and Silk (Maçã) subgroups. Prata cultivars are very tall, with fairly low productivity as they bear relatively small bunches and fruit even in low-density plantings, and do not cycle quickly enough to compensate for these shortcomings. The principal priority of the Brazilian program has always been resistance to FWB, yellow Sigatoka, and BLSD (Amorim et al., 2020). Prata and Prata-Anã are moderately susceptible to R1, and highly susceptible to the yellow Sigatoka and BLSD. The breeding program was first based on hybridization between wild diploids and triploids for the development of Prata-type tetraploid hybrids. After 1993, the program concentrated on the development of tetraploid hybrids similar to Maçã (Silk). Today, the strategy is the development of triploid and tetraploid hybrids following crosses between diploids and tri- and tetraploids and by doubling chromosomes (Amorim et al., 2013, 2020). This program has released several AAAB tetraploids hybrids to growers, such as Caprichosa (susceptible to FWB and resistant to yellow Sigatoka and BLSD); Japira, Vitória, Princesa, Platina, Tropical (resistant to FWB and yellow Sigatoka but susceptible to BLSD), Garantida, Pacovan Ken and Preciosa (resistant to FWB, yellow Sigatoka and BLSD), as well as the triploids Thap Maeo and Conquista (AAB; resistant to FWB, yellow Sigatoka and BLSD) and Caipira (AAA, resistant to yellow Sigatoka, partially resistance to FWB and susceptible to BLSD).

6.4 Banana breeding by International Institute of Tropical Agriculture in Africa

A plantain and banana improvement program started in 1987 at the International Institute of Tropical Agriculture (IITA) in Nigeria. Initially, the program focused on developing plantain hybrids resistant to BLSD but later included a program for EAHBs in Uganda. EAHBs are the primary staple crop for over 13 million Ugandans (representing approximately 35% of the population) and are principally used for cooking (Matooke types) and a small percentage for brewing (Mbidde types) (Batte et al., 2019). EAHBs dominate banana cultivation in East and Central Africa, a region that produces more than half of all bananas produced in Africa (Ndayihanzamaso et al., 2020).

The IITA breeding program developed Mchare (AA), Matooke (AAA), and NARITA (AAA) hybrids for high yield and BLSD resistance (Batte et al., 2019), while including or maintaining resistance to R1.

6.5 Banana breeding programs in India

In India, banana breeding programs exist at the National Research Centre for Banana, Tamil Nadu Agricultural University (TNAU), and Kerala Agricultural University, where numerous intraspecific AA and interspecific AB diploid hybrids have been created (Bakry et al., 2020).

Although banana breeding in India started already in 1949 in Tamil Nadu, it took until 1971 to extensively produce interspecific diploid crosses with the objective of resistance to nematodes, yellow Sigatoka and BLSD, and FWB (Smith and Pillay, 2020). The TNAU breeding program developed the triploid hybrid 'NPH-02-01' with combined resistance to R1 and partial resistance to nematodes (Kumar et al., 2009). However, the main difficulty for smallholders in India, similar to Africa, is the diverse regional and subregional consumer preference which determines the cropping of specific cultivars or landraces irrespective of disease resistance (Smith and Pillay, 2020).

6.6 Banana breeding at Centre de Coopération Internationale en Recherche Agronomique pour le Développement (CIRAD)

Since 1993, CIRAD developed an intensive banana breeding program with a focus on the production of numerous 'dessert' and 'cooking' triploid hybrids (Bakry et al., 2020). By crossing diploid cultivars (AA) with tetraploids (AAAA), the program developed Fhlorban hybrids (AAA). Fhlorban 925 is a Cavendish-like hybrid, which has an smaller size fruit but is resistant to yellow Sigatoka, BLSD, and *R. similis*. Other hybrids such as Fhlorban 916, Fhlorban 918, and Fhlorban 920 are also resistant to yellow Sigatoka and BLSD but are not (completely) resistant to *R. similis* and *P. coffeae*. All these hybrids show a high level of resistance to R1 under greenhouse conditions (García-Bastidas et al., 2019).

7 Management of the disease

Any disease management eventually fails in a large-scale perennial monoculture using a highly susceptible variety. No individual control measure has proved effective in permanent banana production of susceptible cultivars such as Gros Michel in the presence of R1 (Ploetz, 2019). Fortunately, in the case of the R1 epidemic in the first half of the 1900s, the Cavendish subgroup of bananas was identified as a suitable substitute for the Gros Michel-based export industry. The switch to Cavendish and the need to meet high standards required for export bananas, including consumer acceptance, created commercial and operational challenges. Although Cavendish was resistant to R1 it turned out to be highly susceptible to *R. similis* (Stover and Simmonds, 1987) and required different post-harvest handling.

With the extensive cropping of resistant Cavendish (AAA) and plantain clones (AAB) since the 1960s, R1 and R2 remained a problem only in smallholder farms growing susceptible cultivars such as Gros Michel subgroup cultivars, Silk, Prata, Pisang Awak, and Bluggoe, and some others for local consumption and animal feed. Disease management in smallholder plantations growing susceptible varieties can be improved by increasing grower awareness of FWB and its spread, biosecurity principles, and exclusion through the use of healthy planting material (Pérez-Vicente et al., 2009, 2014; Dita et al., 2018).

8 Conclusion and future trends

The R1 epidemics were overcome in the 1960s by replacing Gros Michel with agronomically superior Cavendish (AAA) and plantain (AAB) clones. However, R1 and R2 are still a problem for smallholders who produce locally preferred but susceptible banana cultivars in agroforestry settings, for instance, mixed with coffee, cacao, or forest trees. The current banana export industry is based on an extensive monoculture of Cavendish cultivars, which is under significant threat from BLSD and from FWB caused by TR4 (Drenth and Kema, 2021). Clearly, there is a need to diversify banana cultivation with different productive and resistant hybrids.

The impact of TR4 on Cavendish over the last 20 years has stimulated research interest in FWB and *Fusarium* sp. and has already provided important insights into the pathogen's biology, genetic diversity, epidemiology, cultivar susceptibility, and practical management options, not only for TR4 but also for R1 and R2. There will be a spillover of new technology derived from TR4 research that could be applied to R1 and R2 issues. TR4 will have disastrous consequences for the Cavendish-based export industry, as well as smallholders, and new resistant cultivars are essential to cultivate bananas in a sustainable manner.

First and foremost, FWB management seeks to keep the pathogen out of the plantation but, failing that, prevents inoculum build-up and pathogen dissemination through early detection and destruction of infected plants. An interesting agroecological approach is to study the allelopathic effects produced by different crop and cover plants that eventually could make banana and other cropping plants rotational, as well as different mixed cropping associations, to reduce soil inoculum and infield transmission from plant to plant. However, such approaches may not work when growing highly susceptible Cavendish varieties in the presence of TR4.

The continuous monotonous cultivation of susceptible or partially resistant cultivars in large-scale plantations requires timely identification and systematic elimination of symptomatic plants and surrounding non-symptomatic plants, site disinfestation, replanting with healthy planting material, and improved soil suppressiveness. However, none of these activities allows meaningful long-term permanent production but will enable limited banana production. The adoption of various management methodologies, especially effective resistance, is required to place the export banana industry, as well as smallholder growers, on a sustainable footing.

9 Where to look for further information

In addition to the reference list, the reader can find additional information in scientific papers that have been published in peer-reviewed journals, which number have been steadily increasing in the last years. For details about the infection process of Foc on banana plant, see:

- Cheng, C., Liu, F., Sun, X., Tian, N., Mensah, R. A., Li, D. and Lai, Z. (2019). Identification of *Fusarium oxysporum* f. sp. *cubense* tropical race 4 (Foc TR4) responsive miRNAs in banana root. *Scientific reports* 9(1), 1–16.
- Warman, N. M. and Aitken, E. A. (2018). The movement of *Fusarium oxysporum* f. sp. *cubense* (sub-tropical race 4) in susceptible cultivars of banana. *Frontiers in Plant Science* 9, 1748. DOI: 10.3389/fpls.2018.01748.
- Zhang, L., Yuan, T., Wang, Y., Zhang, D., Bai, T., Xu, S., Wang, Y., Tang, W. and Zheng, S. J. (2018). Identification and evaluation of resistance to *Fusarium oxysporum* f. sp. *cubense* tropical race 4 in *Musa acuminata* Pahang. *Euphytica* 214, 1–12. DOI: 10.1007/s10681-018-2185-4.

For additional information about the geographic distribution and variability of Foc populations, see:

- Chittarath, K., Nguyen, C. H., Bailey, W. C., Zheng, S. J., Mostert, D., Viljoen, A., Tazuba, A. F., Ocimati, W., Kearsley, E., Chi, T. Y., Tho, N. T., Hung, N. T., Dita, M., Shah, T., Karanja, M., Mahuku, G. and Blomme, G. (2022). Geographical distribution and genetic diversity of the banana Fusarium wilt fungus in Laos and Vietnam. *Journal of Fungi* 8(1), 46. DOI: 10.3390/jof8010046.
- Magdama, F., Monserrate-Maggi, L., Serrano, L., García Onofre, J. and Jiménez-Gasco, M. D. M. (2020). Genetic diversity of *Fusarium oxysporum* f. sp. *cubense*, the Fusarium wilt pathogen of banana, in Ecuador. *Plants* 9(9), 1133. DOI: 10.3390/plants9091133.
- Mostert, D., Wicker, E., de Jager, M. M., Al Kaabi, S. M., O'Neill, W. T., Perry, S., Li, C., Ganyun, Y., Pegg, K., Mostert, L. and Viljoen, A. (2022). A polyphasic approach reveals novel genotypes and updates the genetic structure of the banana fusarium wilt pathogen. *Microorganisms* 10(2), 269. DOI: 10.3390/microorganisms10020269.
- Ordóñez, N., Kema, G. H. J., Seidl, M. F. and Meijer, H. J. G. (2018). A global genetic diversity analysis of *Fusarium oxysporum* f. sp. *cubense*: the Panama disease pathogen of banana (Doctoral dissertation, Wageningen University and Research). Available at: https://edepot.wur.nl/453455.
- Reyes-Herrera, P. H., Torres-Bedoya, E., Lopez-Alvarez, D., Burbano-David, D., Carmona, S. L., Bebber, D. P. and Soto-Suarez, M. (2023). Genome

sequence data reveal at least two distinct incursions of the tropical race 4 variant of Fusarium wilt into South America. *Phytopathology* 113(1), 90–97. DOI: 10.1094/PHYTO-01-22-0034-R.

- van Westerhoven, A. C., Meijer, H. J., Houdijk, J., Martínez de la Parte, E., Matabuana, E. L., Seidl, M. F. and Kema, G. H. (2023). Dissemination of Fusarium wilt of banana in Mozambique caused by *Fusarium odoratissimum* Tropical Race 4. *Plant Disease* 107(3), 628–632. DOI: 10.1094/PDIS-07-22-1576-SC.

For more information about the susceptibility of different germplasms to Fusarium wilt, see:

- García-Bastidas, F. A., Kema, G. H. J. and Arango-Isaza, R. (2019). Panama disease in banana: spread, screens and genes (Doctoral dissertation, Wageningen University and Research). Available at: https://edepot.wur.nl/467427.
- Zuo, C., Deng, G., Li, B., Huo, H., Li, C., Hu, C., Kuang, R., Yang, Q., Dong, T., Sheng, O. and Yi, G. (2018). Germplasm screening of *Musa* spp. for resistance to *Fusarium oxysporum* f. sp. *cubense* tropical race 4 (Foc TR4). *European Journal of Plant Pathology* 151, 723–734. DOI: 10.1007/s10658-017-1406-3.
- Zhan, N., Kuang, M., He, W., Deng, G., Liu, S., Li, C., Roux, N., Dita, M., Yi. G. and Sheng, O. (2022). Evaluation of resistance of banana genotypes with AAB genome to Fusarium wilt tropical race 4 in China. *Journal of Fungi* 8(12), 1274. DOI: 10.3390/JOF8121274.

10 References

Amorim, E. P., de Oliveira Amorim, V. B., dos Santos Silva, M., Haddad, F., Ferreira, C. F. and dos Santos Serejo, J. A. (2020). Developing hybrid banana varieties with improved properties. In: Kema, G. H. J., Drenth, A. (Eds.), *Achieving Sustainable Cultivation of Bananas (Vol. 2) - Germplasm and Genetic Improvement.* Cambridge, UK: Burleigh Dodds Science Publishing Limited, pp. 323–338.

Amorim, E. P., Dos Santos-Serejo, J. A., Amorim, V. B. O., Ferreira, C. F. and Silva, S. O. (2013). Banana breeding at EMBRAPA cassava and fruits. *Acta Horticulturae* 986, 171–176. DOI: 10.17660/ActaHortic.2013.986.18.

Bakry, F., Horry, J. P. and Jenny, C. (2020). Making banana breeding more effective. In: Kema, G. H. J., Drenth, A. (Eds.), *Achieving Sustainable Cultivation of Bananas (Vol. 2) - Germplasm and Genetic Improvement.* Cambridge, UK: Burleigh Dodds Science Publishing Limited, pp. 217–256. DOI: 10.19103/AS.2020.0070.12.

Batte, M., Swennen, R., Uwimana, B., Akech, V., Brown, A., Tumuhimbise, R., Hovmalm, H. P., Geleta, M. and Ortiz, R. (2019). Crossbreeding east African highland bananas: Lessons learnt relevant to the botany of the crop after 21 years of genetic enhancement. *Frontiers in Plant Science* 10, 81. DOI: 10.3389/fpls.2019.00081.

Battle, A. and Pérez-Vicente, L. (2009). Variabilidad genética de las poblaciones de *Fusarium oxysporum* f. sp. *cubense* en Bananos y plátanos de Cuba. *Fitosanidad* 13(3), 169–186.

Bentley, S. B. K. G., Pegg, K. G., Moore, N. Y., Davis, R. D. and Buddenhagen, I. W. (1998). Genetic variation among vegetative compatibility groups of *Fusarium oxysporum* f. sp. *cubense* analyzed by DNA fingerprinting. *Phytopathology* 88(12), 1283–1293. DOI: 10.1094/PHYTO.1998.88.12.1283.

Blomme, G., Ploetz, R., Jones, D., De Langhe, E., Price, N., Gold, C., Geering, A., Viljoen, A., Karamura, D., Pillay, M., Tinzaara, W., Teycheney, P.-Y., Lepoint, P., Karamura, E. and Buddenhagen, I. (2013). A historical overview of the appearance and spread of *Musa* pests and pathogens on the African continent: Highlighting the importance of clean *Musa* planting materials and quarantine measures. *Annals of Applied Biology* 162(1), 4–26. DOI: 10.1111/aab.12002.

Dinesh, B. M., Ravichandra, N. G., Somasekhara, Y. M., Reddy, B. M. R. and Kumar, K. H. (2014). Interactions between *Radopholus similis* and *Fusarium oxysporum* f. sp. *cubense* causing wilt complex on banana. *International Journal of Advanced Research (IJAR)* 2(9), 976–985.

Dita, M., Barquero, M., Heck, D., Mizubuti, E. S. G. and Staver, C. P. (2018). Fusarium wilt of banana: Current knowledge on epidemiology and research needs toward sustainable disease management. *Frontiers in Plant Science* 9, 1468. DOI: 10.3389/FPLS.2018.01468.

Dita, M. A., Echegoyen, P. E. and Pérez-Vicente, L. (2013a). *Plan de Contingencia ante un brote de la raza 4 tropical de Fusarium oxysporum f. sp. cubense en un país de la región del OIRSA*. San Salvador, El Salvador: OIRSA-Organismo Internacional Regional de Sanidad Agropecuaria.

Dita, M. A., Garming, H., Van den Bergh, I., Staver, C. and Lescot, T. (2013b). Banana in Latin America and the Caribbean: Current state, challenges and perspectives. *Acta Horticulturae* 986, 365–380. DOI: 10.17660/ActaHortic.2013.986.39.

Dita, M., Teixeira, L. A. J., O'Neill, W., Pattison, A. B., Weinert, M. P., Li, C. Y., Zheng, S. J., Staver, C., Thangavelu, R. and Viljoen, A. (2020). Current state of Fusarium wilt of banana in the subtropics. *Acta Horticulturae* 1272, 45–56. DOI: 10.17660/ActaHortic.2020.1272.7.

Drenth, A. and Kema, G. H. J. (2021). The vulnerability of bananas to globally emerging disease threats. *Phytopathology* 111(12), 2146–2161.

Drenth, A., McTaggart, A. R. and Wingfield, B. D. (2019). Fungal clones win the battle, but recombination wins the war. *IMA Fungus* 10, 18. DOI: 10.1186/s43008-019-0020-8.

Fan, H., Dong, H., Xu, C., Liu, J., Hu, B., Ye, J., Mai, G. and Li, H. (2017). Pectin methylesterases contribute the pathogenic differences between races 1 and 4 of *Fusarium oxysporum* f. sp. *cubense*. *Scientific Reports* 7(1), 13140. DOI: 10.1038/s41598-017-13625-4.

Fourie, G., Steenkamp, E. T., Gordon, T. R. and Viljoen, A. (2009). Evolutionary relationships among the *Fusarium oxysporum* f. sp. *cubense* vegetative compatibility groups. *Applied and Environmental Microbiology* 75(14), 4770–4781. DOI: 10.1128/AEM.00370-09.

Fourie, G., Steenkamp, E. T., Ploetz, R. C., Gordon, T. R. and Viljoen, A. (2011). Current status of the taxonomic position of *Fusarium oxysporum* formae specialis *cubense* within the *Fusarium oxysporum* complex. *Infection, Genetics and Evolution* 11(3), 533–542. DOI: 10.1016/J.MEEGID.2011.01.012.

García-Bastidas, F. A., Quintero-Vargas, J. C., Ayala-Vasquez, M., Schermer, T., Seidl, M. F., Santos-Paiva, M., Noguera, A. M., Aguilera-Galvez, C., Wittenberg, A., Hofstede, R., Sørensen, A. and Kema, G. H. J. (2019). First report of Fusarium wilt tropical race 4 in Cavendish bananas caused by *Fusarium odoratissimum* in Colombia. *Plant Disease* 104(3),1-2.

Guillen Sanchez, C., Tixier, P., Tapia Fernandez, A., Conejo Barboza, A. M., Sandoval Fernández, J. A. and De Lapeyre de Bellaire, L. (2021). Can the banana weevil Cosmopolites sordidus be a vector of *Fusarium oxysporum* f. sp. *cubense* race 1? Unravelling the internal and external acquisition of effective inoculum. *Pest Management Science* 77(6), 3002–3012. DOI: 10.1002/ps.6339.

Guo, L., Yang, L., Liang, C., Wang, G., Dai, Q. and Huang, J. (2015). Differential colonization patterns of bananas (*Musa* spp.) by physiological race 1 and race 4 isolates of *Fusarium oxysporum* f. sp. *cubense*. *Journal of Phytopathology* 163(10), 807–817. DOI: 10.1111/jph.12378.

Jones, D. R. (2019). (Ed.), *Handbook of Diseases of Banana, Abaca and Ensete*. Boston, MA: CAB International.

Karangwa, P., Mostert, D., Ndayihanzamaso, P., Dubois, T., Niere, B., Zum Felde, A., Schouten, A., Blomme, G., Beed, F. and Viljoen, A. (2018). Genetic diversity of *Fusarium oxysporum* f. sp. *cubense* in East and Central Africa. *Plant Disease* 102(3), 552–560. DOI: 10.1094/PDIS-02-17-0282-RE.

Köberl, M., Dita, M., Martinuz, A., Staver, C. and Berg, G. (2017). Members of Gammaproteobacteria as indicator species of healthy banana plants on Fusarium wilt-infested fields in Central America. *Scientific Reports* 7(1), 45318. DOI: 10.1038/srep45318.

Kumar, N., Damodaran, T. and Krishnamoorthy, V. (2009). Breeding banana for combined resistance to Fusarium wilt and nematodes in Tamil Nadu, India. *Acta Horticulturae* 828, 323–332. DOI: 10.17660/ActaHortic.2009.828.33.

Kung'u, J. N., Rutherford, M. A. and Jeffries, P. (2001). Distribution of Fusarium wilt of banana in Kenya and its impact on smallholder farmers. *Infomusa* 10(1), 28–32.

Li, C., Chen, S., Zuo, C., Sun, Q., Ye, Q., Yi, G. and Huang, B. (2011). The use of GFP-transformed isolates to study infection of banana with *Fusarium oxysporum* f. sp. *cubense* race 4. *European Journal of Plant Pathology* 131(2), 327–340. DOI: 10.1007/s10658-011-9811-5.

Li, C., Yang, J., Li, W., Sun, J. and Peng, M. (2017). Direct root penetration and rhizome vascular colonization by *Fusarium oxysporum* f. sp. *cubense* are the key steps in the successful infection of Brazil Cavendish. *Plant Disease* 101(12), 2073-2078. DOI: 10.1094/PDIS-04-17-0467-RE.

Ma, L. J., Van Der Does, H. C., Borkovich, K. A., Coleman, J. J., Daboussi, M. J., Di Pietro, A., Dufresne, M., Freitag, M., Grabherr, M., Henrissat, B., Houterman, P. M., Kang, S., Shim, W. B., Woloshuk, C., Xie, X., Xu, J. R., Antoniw, J., Baker, S. E., Bluhm, B. H., Breakspear, A., Brown, D. W., Butchko, R. A. E., Chapman, S., Coulson, R., Coutinho, P. M., Danchin, E. G. J., Diener, A., Gale, L. R., Gardiner, D. M., Goff, S., Hammond-Kosack, K. E., Hilburn, K., Hua-Van, A., Jonkers, W., Kazan, K., Kodira, C. D., Koehrsen, M., Kumar, L., Lee, Y. H., Li, L., Manners, J. M., Miranda-Saavedra, D., Mukherjee, M., Park, G., Park, J., Park, S. Y., Proctor, R. H., Regev, A., Ruiz-Roldan, M. C., Sain, D., Sakthikumar, S., Sykes, S., Schwartz, D. C., Turgeon, B. G., Wapinski, I., Yoder, O., Young, S., Zeng, Q., Zhou, S., Galagan, J., Cuomo, C. A., Kistler, H. C. and Rep,

M. (2010). Comparative genomics reveals mobile pathogenicity chromosomes in *Fusarium*. *Nature* 464(7287), 367–373. DOI: 10.1038/nature08850.

MacHardy, W. E. and Beckman, C. H. (1981). Vascular wilt Fusaria: Infection and pathogenesis. In: Nelson, P. E., Toussoun, T. A. and Cook, R. J. (Eds.), *Fusarium: Diseases, Biology, and Taxonomy*. University Park, PA: The Pennsylvania State University Press, pp. 365–391.

Maryani, N., Lombard, L., Poerba, Y. S., Subandiyah, S., Crous, P. W. and Kema, G. H. J. (2019). Phylogeny and genetic diversity of the banana Fusarium wilt pathogen *Fusarium oxysporum* f. sp. *cubense* in the Indonesian centre of origin. *Studies in Mycology* 92, 155–194. DOI: 10.1016/j.simyco.2018.06.003.

McTaggart, A. R., James, T. Y., Shivas, R. G., Drenth, A., Wingfield, B. D., Summerell, B. A. and Duong, T. A. (2021). Population genomics reveals historical and ongoing recombination in the *Fusarium oxysporum* species complex. *Studies in Mycology* 99, 100132. DOI: 10.1016/j.simyco.2021.100132.

Meldrum, R. A., Daly, A. M., Tran-Nguyen, L. T. T. and Aitken, E. A. B. (2013). Are banana weevil borers a vector in spreading *Fusarium oxysporum* f. sp. *cubense* tropical race 4 in banana plantations? *Australasian Plant Pathology* 42(5), 543–549. DOI: 10.1007/s13313-013-0214-2.

Moore, N. Y., Hargreaves, P. A., Pegg, K. G. and Irwin, J. A. G. (1991). Characterization of strains of *Fusarium oxysporum* f. sp. *cubense* by production of volatiles. *Australian Journal of Botany* 39(2), 161–166.

Mostert, D., Molina, A. B., Daniells, J., Fourie, G., Hermanto, C., Chao, C. P., Fabregar, E., Sinohin, V. G., Masdek, N., Thangavelu, R., Li, C., Yi, G., Mostert, L. and Viljoen, A. (2017). The distribution and host range of the banana Fusarium wilt fungus, *Fusarium oxysporum* f. sp. *cubense*, in Asia. *PLoS ONE* 12(7), e0181630. DOI: 10.1371/journal.pone.0181630.

Mustaffa, M. M. and Thangavelu, R. (2011). Status of Fusarium wilt in India. *Acta Horticulturae* 897, 323–329.

Ndayihanzamaso, P., Mostert, D., Matthews, M. C., Mahuku, G., Jomanga, K., Mpanda, H. J., Mduma, H., Brown, A., Uwimana, B., Swennen, R., Tumuhimbise, R. and Viljoen, A. (2020). Evaluation of Mchare and Matooke bananas for resistance to *Fusarium oxysporum* f. sp. *cubense* race 1. *Plants* 9(9), 1082. DOI: 10.3390/plants9091082.

O'Donnell, K., Kistler, H. C., Cigelnik, E. and Ploetz, R. C. (1998). Multiple evolutionary origins of the fungus causing Panama disease of banana: Concordant evidence from nuclear and mitochondrial gene genealogies. *Proceedings of the National Academy of Sciences of the United States of America* 95(5), 2044–2049. DOI: 10.1073/pnas.95.5.2044.

Ordoñez, N., Seidl, M. F., Waalwijk, C., Drenth, A., Kilian, A., Thomma, B. P., Ploetz, R. C. and Kema, G. H. J. (2015). Worse comes to worst: Bananas and Panama disease—when plant and pathogen clones meet. *PLOS Pathogens* 11(11), e1005197. DOI: 10.1371/JOURNAL.PPAT.1005197.

Pegg, K. G., Coates, L. M., O'Neill, W. T. and Turner, D. W. (2019). The epidemiology of Fusarium Wilt of banana. *Frontiers in Plant Science* 10, 1395. DOI: 10.3389/fpls.2019.01395.

Pegg, K. G., Moore, N. Y. and Bentley, S. (1996). Fusarium wilt of banana in Australia: A review. *Australian Journal of Agricultural Research* 47(5), 637–650. DOI: 10.1071/AR9960637.

Peng, H. X., Sivasithamparam, K. and Turner, D. W. (1999). Chlamydospore germination and Fusarium wilt of banana plantlets in suppressive and conducive soils are affected by physical and chemical factors. *Soil Biology and Biochemistry* 31(10), 1363–1374. DOI: 10.1016/S0038-0717(99)00045-0.

Pérez-Vicente, L. (2004). Fusarium wilt (Panama disease) of bananas: An updating review of the current knowledge on the disease and its causal agent. *Fitosanidad* 8, 27–38.

Pérez Vicente, L., Batlle Viera, A., Chacón Benazet, J. and Montenegro Moracén, V. (2009). Reacción de clones naturales e híbridos de la FHIA de bananos y plátanos a las poblaciones de Cuba de *Fusarium oxysporum* f. sp. *cubense*, agente causal de la marchitez por Fusarium o mal de Panamá. *Fitosanidad* 13(4), 237–242.

Pérez-Vicente, L., Dita, M. A. and Martinez-de la Parte, E. (2014). Technical manual prevention and diagnostic of Fusarium wilt (Panama disease) of banana caused by *Fusarium oxysporum* f.sp. *cubense* tropical race 4 (TR4). In: *Proceedings of the Regional Workshop on the Diagnosis of Fusarium Wilt (Panama Disease) Caused by Fusarium oxysporum f. sp. cubense Tropical Race 4: Mitigating the Threat and Preventing Its Spread in the Caribbean*. St. Augustine, Trinidad and Tobago: Food and Agriculture Organization of the United Nations (FAO), p. 74.

Pérez-Vicente, L., Martinez-de la Parte, E., Borras-Hidalgo, O. and Canales, E. (2017). *Variabilidad, Diagnóstico, Epidemiología, Resistencia y Manejo de Especies de Fusarium Patógenas en Musáceas, Piña y Soya. Informe Científico Técnico Final del Proyecto P131LH003014*. La Habana, Cuba.

Ploetz, R. C. (1990). Population biology of *Fusarium oxysporum* f. sp. *cubense*. In: Ploetz, R. C. (Ed.), *Fusarium Wilt of Banana*. St. Paul, MN: APS Press, The American Phytopahological Society, pp. 63–76.

Ploetz, R. C. (2005). Panama disease: An old nemesis rears its ugly head part 1: The beginnings of the banana export trades. *Plant Health Progress* 6(1), 18. DOI: 10.1094/PHP-2005-1221-01-RV.

Ploetz, R. C. (2006a). Fusarium wilt of banana is caused by several pathogens referred to as *Fusarium oxysporum* f. sp. *cubense*. *Phytopathology* 96(6), 653–656. DOI: 10.1094/PHYTO-96-0653.

Ploetz, R. C. (2006b). Panama disease: An old nemesis rears its ugly head - Part 2. The Cavendish era and beyond. *Plant Health Progress* 7(1), 36. DOI: 10.1094/PHP-2006-0308-01-RV.

Ploetz, R. C. (2015). Fusarium wilt of banana. *Phytopathology* 105(12), 1512–1521. DOI: 10.1094/PHYTO-04-15-0101-RVW.

Ploetz, R. C. (2019). Fusarium wilt. In: Jones, D. R. (Ed.), *Handbook of Diseases of Banana, Abaca and Ensete*. Boston, MA: CAB International, pp. 207–228.

Ploetz, R. C. and Pegg, K. G. (1997). Fusarium wilt of banana and Wallace's line: Was the disease originally restricted to his Indo-Malayan region? *Australasian Plant Pathology* 26(4), 239. DOI: 10.1071/AP97039.

Ploetz, R. C. and Pegg, K. G. (1999). Fusarium wilt. In: Jones, D. R. (Ed.), *Diseases of Banana, Abaca and Enset*. Wallingford, UK: CABI Publishing, pp. 143–159.

Ploetz, R. C., Kema, G. H. J. and Ma, L. J. (2015). Impact of diseases on export and smallholder production of banana. *Annual Review of Phytopathology* 53(1), 269–288. DOI: 10.1146/annurev-phyto-080614-120305.

Pocasangre, L. E., Ploetz, R. C., Molina, A. B. and Perez Vicente, L. (2011). Raising awareness of the threat of fusarium wilt tropical Race 4 in Latin America and the Caribbean. *Acta Horticulturae* 897, 331–337.

Raaijmakers, J. M., Paulitz, T. C., Steinberg, C., Alabouvette, C. and Moënne-Loccoz, Y. (2009). The rhizosphere: A playground and battlefield for soilborne pathogens and beneficial microorganisms. *Plant and Soil* 321(1-2), 341-361. DOI: 10.1007/s11104-008-9568-6.

Rishbeth, J. (1955). Fusarium wilt of bananas in Jamaica: I. *Annals of Botany* 19(3), 293-328.

Rocha, A. D. J., Ferreira, M. D. S., Rocha, L. D. S., Oliveira, S. A. S., Amorim, E. P., Mizubuti, E. S. G. and Haddad, F. (2020). Interaction between *Fusarium oxysporum* f. sp. *cubense* and *Radopholus similis* can lead to changes in the resistance of banana cultivars to Fusarium wilt. *European Journal of Plant Pathology* 158(2), 403-417.

Rowe, P. R. and Rosales, F. (2000). Conventional banana breeding in Honduras. In: Jones, D. A. (Ed.), *Diseases of Banana, Abaca and Ensette*. Wallingford: CABI Publication, pp. 435-449.

Rutherford, M. A. (2001). Fusarium wilt of banana in East Africa. In: Molina, A. B., Masdek, N. and Liew, K. W. (Eds.), *Banana Fusarium Wilt Management: Towards Sustainable Cultivation*. Proceedings of the International Workshop on the Banana Fusarium Wilt Disease, Genting Highlands Resort (MYS). Los Baños: INIBAP, pp. 86-94.

Sanjeevkumar, K., Balabaskar, P. and Sivakumar, T. (2018). Studies on *F. oxysporum* f. sp. *cubense*-nematode interaction in banana cv. 'Monthan'. *International Journal of Chemical Studies* 6(2), 3708-3710.

Scher, F. M. and Baker, R. (1982). Effect of *Pseudomonas putida* and synthetic iron chelator on induction of soil suppressiveness to Fusarium wilt pathogens. *Phytopathology* 72(12), 1567-1773.

Segura-Mena, R. A., Stoorvogel, J. J., García-Bastidas, F., Salacinas-Niez, M., Kema, G. H. J. and Sandoval, J. A. (2021). Evaluating the potential of soil management to reduce the effect of *Fusarium oxysporum* f. sp. *cubense* in banana (*Musa* AAA). *European Journal of Plant Pathology* 160(2), 441-455. DOI: 10.1007/s10658-021-02255-2.

Shen, Z., Ruan, Y., Xue, C., Zhong, S., Li, R. and Shen, Q. (2015). Soils naturally suppressive to banana Fusarium wilt disease harbor unique bacterial communities. *Plant and Soil* 393(1-2), 21-33. DOI: 10.1007/s11104-015-2474-9.

Smith, M. and Pillay, M. (2020). An overview of genetic improvement in bananas over the last century. In: Kema, G. H. and Drenth, A. (Eds.), *Achieving Sustainable Cultivation of Bananas (Vol. 2): Germplasm and Genetic Improvement*. Cambridge, UK: Burleigh Dodds Science Publishing, pp. 1-28. DOI: 10.19103/AS.2020.0070.20.

Stover, R. H. (1962). *Fusarial Wilt of Bananas and Other Musa species*. Surrey, UK: Commonwealth Mycological Institute.

Stover, R. H. and Simmonds, N. W. (1987). *Bananas*. Essex, UK: Longman Scientific & Technical.

Stover, R. H. and Waite, B. H. (1960). Studies on fusarium wilt of bananas V. Pathogenicity and distribution of *Fusarium oxysporum* f. sp. *cubense* races 1 and 2. *Canadian Journal of Botany* 38(1), 51-61.

Tushemereirwe, W. (1992). Fusarium wilt of banana in Uganda. In: *Proceedings of the International Symposium on Recent Developments in Banana Cultivation Technology*. Chiuju. Pingtung, Taiwan: Taiwan Banana Research Institute, pp. 240-246.

Tushemereirwe, W. K., Kangire, A., Kubiriba, J. and Nowakunda, K. (2000). Fusarium wilt resistant bananas considered appropriate replacements for cultivars susceptible to the disease in Uganda. *Uganda Journal of Agricultural Sciences* 5(1), 62-64.

van Westerhoven, A. C., Meijer, H. J. G., Seidl, M. F. and Kema, G. H. J. (2022). Uncontained spread of Fusarium wilt of banana threatens African food security. *PLoS Pathogens* 18(9), e1010769. DOI: 10.1371/journal.ppat.1010769.

Waite, B. H. and Dunlap, V. C. (1953). Preliminary host range studies with *Fusarium oxysporum* f. sp. *cubense*. *Plant Disease Reporter* 37, 79–80.

Wardlaw, C. W. (1961). *Banana Diseases, Including Plantains and Abaca*. London: Longmans (GBR).

Chapter 5

The past, present and future of Fusarium wilt of banana caused by Tropical Race 4

Fernando García-Bastidas, KeyGene B.V., The Netherlands; André Drenth, The University of Queensland, Australia; and Gert H. J. Kema, Wageningen University, The Netherlands

1 Introduction

Fusarium wilt of banana (FWB) was and is a threat to global banana production. Previous epidemics were caused by Race 1 strains that decimated banana production in Central America, which was dominated by the then-major commercial banana variety Gros Michel (Ploetz, 2005a). Many contemporary varieties are vulnerable to Race 1 strains (Zuo et al., 2018; García-Bastidas, 2019; Martinez de la Parte et al., 2023). Therefore, resistance to Race 1 is a prerequisite for any new banana variety that comes to market, for instance in Brazil or East Africa, from the breeding programs of the Empresa Brasileira de Pesquisa Agropecuária (EMBRAPA) and the International Institute for Tropical Agriculture (IITA), respectively. Beyond resistance, methods of control have not sufficiently reduced the impact of FWB (Kema et al., 2021; Ploetz, 2015b). The replacement of Gros Michel by Cavendish varieties for the export markets provided a solution to the export banana industry which required major investments and changes to the logistic chain to ensure the delivery of marketable bananas in international markets (Ploetz, 2005b). Once these problems were solved through the packaging of the Cavendish bananas in cardboard boxes and changes in refrigeration during transport, followed by

http://dx.doi.org/10.19103/AS.2022.0108.05

on-site ripening, Cavendish production steadily increased. The international banana production continued to grow and gradually developed into a monoculture landscape planted to the Race 1-resistant Cavendish cultivars such as 'Williams', 'Valery', and 'Grand Naine' wherever Race 1 resistance was required, not only in Central America but across the world.

The resistance in the Cavendish varieties to Race 1 turned out to be durable and has held up for over a century. The downside is that with every hectare added to the global dominion of Cavendish – now over 50% of all bananas produced these varieties or to the cavendish group? – the genetic vulnerability of contemporary banana production increases (Drenth and Kema, 2021). Cavendish was already affected by another strain of FWB in Taiwan in the late 1960s, but even after the dissemination of the causal agent that we nowadays know as Tropical Race 4 (TR4), the global industry considered this to be a far-away problem and little, if any, action was undertaken. More recent intercontinental spread of TR4 to the Middle East, Africa, and Latin America has been a wake-up call. The quest for solutions is on now, but after decades of neglect in research and development – demonstrated by little to no investments in plant breeding nor the identification of sources of resistance – it will take an immense effort to develop and deliver sustainable strategies. Meanwhile, Cavendish varieties, once the showstopper of FWB, now rapidly succumb in many areas due to the international and intercontinental dissemination of TR4. The introduction of TR4 in Mozambique is a good example. A 9000 ha farm was established by a group of investors to target the Middle-East market (van Westerhoven et al., 2023). Soon the required labor force was contracted from countries with industrial banana production, such as the Philippines. The exact origin of TR4 incursions in most countries, including Mozambique, is still unknown, but it is very likely anthropogenic and associated with the diverse workforce in the export-driven Cavendish production system (van Westerhoven et al., 2022). After a significant decline in production on infested farms, they are abandoned or at best turned into farms producing other crops. This, however, does not eliminate the causal agents of FWB. On the contrary, since bananas are no longer produced, the attention is reduced and TR4 continues to disseminate from these fields to other regions, threatening other banana varieties, including those that provide the staple food for millions of people in Africa's banana belt. The current expansion of TR4 demonstrates the overall failure of forward planning and quarantine to contain the pathogen and the inability to effectively manage it in the field. These strategies failed for Race 1 100 years ago and are currently failing for TR4. A point of difference is that the industry can now use clean tissue culture plants. The economically driven strategy of unlimited expansions to new un-infested areas failed in the Race 1 era and has already failed for TR4 in Southeast Asia and should not be repeated.

Instead, investments in research and development should turn the tide with science-based solutions and genetic improvement over time as is customary in other crops.

The genetic uniformity of global banana production facilitates the dissemination of diseases (Kema et al., 2021), requires massive fungicide inputs, and hits the global poor despite their diverse cropping systems. TR4 affects many more varieties than just the Cavendish clones as demonstrated in a study focused on Cuba (Martínez de la Parte et al., 2023; see also Chapter 4 in this book). Although smallholders grow a diversity of crops, which reduces the risk of disease outbreaks, the diversity among banana germplasm grown in Africa is very low. Additionally, awareness is also needed that traditional knowledge (Antonelli, 2023), despite its value, typically does not address novel introduced disease problems and does not suffice to feed rapidly growing populations in sub-Saharan Africa. Indeed, food security relies on a range of different crops and diversity within each crop, and the implementation of effective management practices based on solid data and proven strategies.

2 Past: origins, spread and biology

TR4 was first reported in Cavendish bananas in Taiwan in 1968 (Su et al., 1977). Since then, it has spread across Southeast Asia over a period of 50 years with limited documentation. In 2013, TR4 was reported for the first time outside Southeast Asia in Jordan and over the last decade has spread to 15 other countries in all major banana-growing regions of the world.

2.1 Origin of TR4

Whilst the first documented appearance of FWB-affected Cavendish was in 1968 in Taiwan (Su et al., 1977, 1986), the widespread presence in Indonesia suggests it originated there while anthropogenic factors have been involved in its dissemination across the archipelago and beyond (Maryani et al., 2018). Banana originates from Southeast Asia (Heslop-Harrison and Schwarzacher, 2007; Bakry et al., 2009; Donohue and Denham, 2009; Heslop-Harrison, 2011; Volkaert, 2011; Perrier et al., 2011; D'Hont et al., 2012; Li and Ge, 2017; Sardos et al., 2022). Hence, the highest level of inter- and intra-specific diversity of pathogens infecting *Musa* species – referred to in this Volume – originates from this area. Consequently, the highest level of diversity among banana-infecting *Fusarium* species – either pathogens or endophytes – is expected in Southeast Asia (van Westerhoven et al., 2024). Accumulated data confirm this hypothesis as the diversity of vegetative compatibility groups (VCGs); see (Puhalla, 1985; Ploetz, 1990) is larger in Southeast Asia than in any other region (Maryani et al., 2018). High levels of genetic diversity in this region are also confirmed by genomic data (van Westerhoven et al., 2024). Recent unpublished research has shown that TR4 is widely distributed across the Indonesian archipelago which indicates that TR4 is either a good disseminator or that demographic and anthropogenic factors are key for its local and global expansion, which seems to be confirmed by various studies (van Westerhoven et al., 2024; Ordóñez et al., unpublished). Since a substantial number of banana accessions is susceptible to TR4 (García-Bastidas, 2019), the chance for dissemination increases. Its spread throughout Indonesia may be connected to major internal migration in the past. For example, the occurrence of TR4 in the Papua province was confined to settlements of people originating from other Indonesian islands (Maryani et al., 2018). Other occurrences of TR4 in Southeast Asia seem to be associated with large industrial Cavendish plantations. Since susceptible banana varieties are grown worldwide and extensive traveling and exchange of farm workers and technical staff are common in the banana sector, the continuous spread of TR4 is predictable (Zheng et al., 2018).

2.2 History of spread

The large-scale cultivation of the Race 1-resistant Cavendish bananas started when the Standard Fruit Company (now known as Dole) decided to deploy it in Central America in 1956, followed by the United Fruit Company (now known as Chiquita) in 1965. However, the first crack in its reputation already occurred in 1968 in Taiwan. There are no indications that the reemergence of FWB was taken very seriously beyond Taiwan, but gradually it was attributed to the so-called Tropical Race 4 (Ploetz, 1994). Eventually, Cavendish production in Taiwan was no longer profitable. This was largely due to TR4 (Su et al., 1977, 1986), but market developments – such as the production in the Philippines that gradually took over the export to Japan, South Korea, and even China (Qi, 2001) – also contributed to this decline. The export to China, first from the Philippines and later from Latin America, was also driven by the increasing magnitude of TR4 problems in China's national banana production. After initial reports of TR4 from Fujian – the province across Taiwan – banana production gradually intensified in Guangdong, the island of Hainan, and the southern provinces Guangxi and Yunnan, without any measures taken to limit the spread of pathogens thereby facilitating the ongoing dissemination of TR4 (Qi, 2001; Qi et al., 2008; Huang et al., 2016; Wu et al., 2019; Guo et al., 2015), which substantially reduced the national banana production volume. Meanwhile, TR4 had also been reported in the Philippines (Molina et al., 2008), Malaysia (Buddenhagen, 2007; Ong, 1996), Thailand (not official), and again later in Vietnam (Chittarath et al., 2018; Hung et al., 2017; Zheng et al., 2018), Laos, and Myanmar (Chittarath et al., 2018; Zheng et al., 2018). This dissemination took place over a period of 50 years (1968–2018) with relatively poor documentation, except for the last decade, which was largely due to the increasing impact of TR4 on Cavendish production and the switch from VCGs to the use of molecular diagnostics (Dita et al., 2010; Carvalhais et al., 2019) and genome sequencing (Ordóñez et al., 2015) enabling more accurate strain identification. TR4 was for decades mostly an 'Asian story', but since its identification in the Middle East, Africa, and Latin America, it has become a global issue (Kema et al., 2021). Indeed, its international and intercontinental spread have surpassed all quarantine and containment efforts, including the highly structured and legislation-anchored quarantine measures in Australia where it first appeared in the Northern Territory in 1997 but also hit the banana belt in Queensland in 2015. First on a single farm that subsequently was taken out of production, but in March 2023 cases on the seventh farm were reported in the region despite the introduction of intensive on-farm biosecurity measures. The only appropriate conclusion at this stage is that quarantine and FWB management in susceptible varieties such as Cavendish fail and hence lessons need to be learned on how to deal with the TR4 pandemic that threatens fruit and food production (van Westerhoven et al., 2022).

2.3 Biology

The biology of TR4 is not known to be different from other *Fusarium* strains infecting bananas (Martínez de la Parte et al., 2023; Kema and Drenth, 2023; see also Chapters 4 and 6 in this book). It produces micro- and macroconidia as well as chlamydospores (Maryani et al., 2018; Maryani, 2018). The external and internal symptoms caused by TR4 infections in Cavendish are indistinguishable from races causing FWB in other banana varieties (Fig. 1). Hence, FWB in Cavendish should raise an immediate alarm, albeit other genetic lineages can also incur FWB symptoms when predisposed to abiotic stress, such as drought and flooding (Pegg et al., 2019). These strains are referred to as sub-tropical Race 4, but they are genetically different, and belong to different VCGs (Bentley et al., 1998). The different races are defined by their ability to cause FWB in different banana varieties. Race 1 and Race 2 are limited to just a few banana varieties, while TR4 causes disease in the majority of banana accessions that were tested (Zuo et al., 2018; García-Bastidas, 2019; Martinez de la Parte et al., 2023). The occurrence and spread of TR4 have given rise to research into its epidemiology. For instance, its capacity to infect other hosts, such as weeds in banana plantations (Catambacan and Cumagun, 2022; Martínez de la Parte et al., 2023), has raised concerns about the efficacy of control methods. The ability to infect other hosts as an endophyte is likely not confined to TR4, but these aspects of the biology have simply not been thoroughly tested with Race 1 or Race 2 strains (Hennessy et al., 2005). Until now, anthropogenic factors have been considered the major drivers for the international and intercontinental spread of TR4 (Zheng

Figure 1 Fusarium wilt of banana (FWB). External and internal symptoms caused by Race 1 in Gros Michel (a, b) and by TR4 in Grande Naine (c, d). The causal agents of FWB cannot be distinguished by the symptoms they cause in banana plants.

et al., 2018). However, local spread is much more elusive. Once infected plants are present, and consequently, the surrounding soil is infested, and the labor-intensive practices in banana farms where each plant is visited multiple times during the production cycle of 9–12 months contribute to on-farm dissemination. Area-wide irrigation systems and extreme weather events – such as excessive precipitation and floods, as exemplified by typhoons Pablo and Yuka that hit Mindanao in 2012 and Perú in 2023, respectively – further contribute to the spread of TR4 (Nakasato Tagami et al., 2024). Contaminated water streams and rivers exacerbate TR4 dissemination (Salacinas, 2019; Acuña et al., 2022). The aerial spread of TR4 has been considered after sporodochia were observed on aboveground parts of the banana plant under

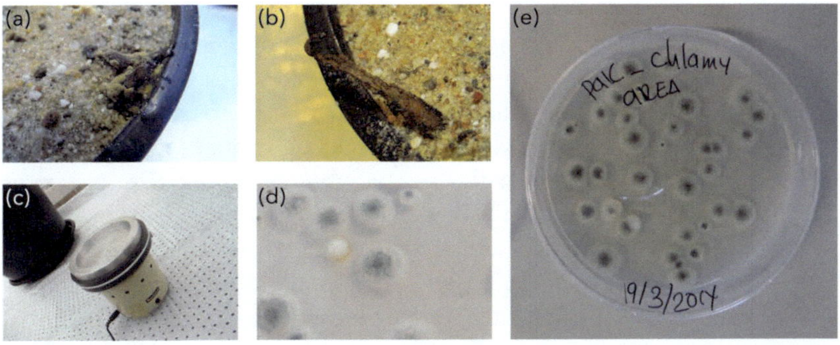

Figure 2 Artificially inoculated Cavendish bananas were subjected to aerial dispersal of Fusarium TR4 under greenhouse conditions. (a) and (b). During a severe infection, fungus covers exposed roots. (c) A sampling device for airborne spores that has a compartment for Petri dishes. (d) Details of a colony that tested positive for TR4 (arrowhead). (e) A Petri dish containing various fungal colonies including mycelium that tested positive for TR4.

greenhouse conditions (Fig. 2) and after the observed seed transmission of TR4 in greenhouse trials using *Setaria* as a model plant (Boggarapu, 2023). However, whether this mode of dissemination also plays a role under in-field conditions has not been elucidated.

3 Present: detection, epidemiology and control

The present situation regarding TR4 requires limiting further spread through improved control, and containment, supported by reliable detection, to distinguish TR4 from other *Fusarium* species occurring in bananas.

3.1 Detection

It is important to recognize that thus far there are no indications that *Fusarium* species in the *F. oxysporum* species complex regularly undergo sexual reproduction, despite the presence of mating-type alleles (Fourie et al., 2009; Leslie et al., 2006) and remnants of historical or occasional sexual reproduction (McTaggart et al., 2021; Fayyaz et al., 2023). The genetic diversity among *Fusarium* strains infecting bananas was elucidated upon after the discovery of VCGs (Puhalla, 1985; Correll et al., 1987; Correll, 1991; Leslie et al., 2006). It involved determining heterokaryon formation between strains after generating various nitrate non-utilizing (*nit*) mutants on a chlorate medium. Once plated in pairs, mutants that complement each other and produce abundant aerial mycelium are considered to belong to the same VCG and are therefore genetically similar. As a result, it was shown that the various *Fusarium* strains infecting bananas are frequently more closely related to species infecting other hosts than to each other. This has been used as an argument for their polyphyletic origin. Based on this, it was concluded that the host specificity within the *Fusarium oxysporum* species complex developed independently and multiple times (Baayen et al., 2000; Koenig et al., 1997; O'Donnell et al., 1998, 2004). Conducting VCG analyses is a time-consuming method to characterize fungal strains, and mutants are not always successfully generated thereby impairing genetic identification. Moreover, the number of VCGs among *Fusarium* strains infecting bananas is larger than in any other *F. oxysporum* species pathogenic on other hosts. For instance, *F. oxysporum* f. sp. *lycopersici* of tomato has only three VCGs (0030–0031 and 0035), and *F. oxysporum* f. sp. *melonis* comprises eight VCGs (0130–0138). Recent VCG analyses indicate that strains belonging to the previously called species *F. oxysporum* f. sp. *cubense* comprised over 26 VCGs, and various isolates belong to unknown VCGs, suggesting that the number of known VCGs is steadily expanding (Bentley et al., 1995; Katan and Di Primo, 1999; Moore et al., 1993; Ploetz, 2015a).

From all accumulated data, it was clear that, contrary to Race 1 and Race 2 strains, TR4 belongs to just one VCG: VCG01213 (Ploetz, 1994; Moore et al., 1993; Bentley et al., 1998). Using genomic analysis, it was shown that TR4 is a unique new lineage belonging to a new species named *F. odoratissimum* (Maryani et al., 2018). The various Race 1 strains belong to multiple VCGs and are also grouped into different species (van Westerhoven et al., 2024; Maryani et al., 2018). The polyphyletic origin of Race 1 and Race 2 complicates the development of molecular diagnostics. However, the genetic similarity of all TR4 strains – which are clones – facilitated the development of molecular diagnostics for TR4 (Aguayo et al., 2017; Ordóñez et al., 2019; Dita et al., 2010; Carvalhais et al., 2019; García-Bastidas et al., 2020a,b; Lin et al., 2013). These have been routinely used to diagnose TR4 and showed that it has spread to 15 countries in just 10 years since its first occurrence outside Southeast Asia in Jordan in 2014 (García-Bastidas et al., 2014; van Westerhoven et al., 2022).

3.2 Dissemination, phylogeography and epidemiology

Although it is known that Race 1 strains are commonly present around the globe (Fig. 3), the details of how they were disseminated in the past are unknown. Southeast Asia is the center of origin of bananas and hence is likely to also be the center of origin of most of its pathogens, due to the co-evolutionary processes taking place in these natural environments. The diversity within and between Race 1 *Fusarium* species (previously referred to as VCGs) is large (Ordóñez et al., 2015; van Westerhoven et al., 2024). It is logical to assume that the dissemination of these species has been facilitated by the transport of infected plant materials. People have taken cuttings or suckers from banana varieties to other areas, including long-distance travel. In addition, weather events such as excessive precipitation, typhoons, and floods may

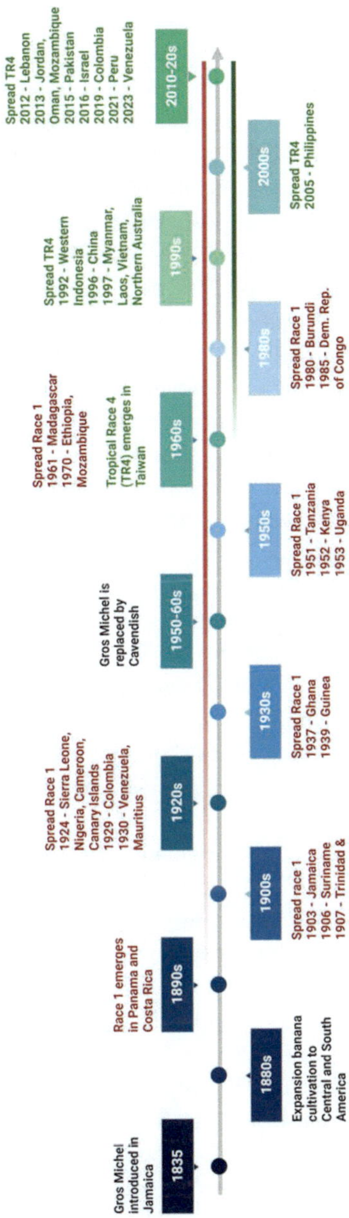

Figure 3 Timeline of the dissemination of Race 1 and TR4 in global banana production.

have contributed to regional dissemination. For TR4, very similar events may have contributed to its dissemination in Southeast Asia. However, its further international expansion has been monitored much more precisely since the development of the first molecular diagnostic (Dita et al., 2010) and upon its first occurrence outside Southeast Asia, as described by van Westerhoven et al. (2022). The most recent outbreak of TR4 has been reported in Venezuela, where it was found in three states (Aragua, Carabobo, and Cojedes), close to Lago Maracay - southwest of the capital Caracas - where many banana plantations are located (Hernández et al., 2023). This is the third reported occurrence in Latin America (Acuña et al., 2022; García-Bastidas et al., 2019) and underscores the failing quarantine strategies for TR4. Clearly, inadequate biosecurity strategies contribute to dissemination (van Westerhoven et al., 2024), such as exemplified in Mozambique, the Mekong delta, and Perú (Zheng et al., 2018; van Westerhoven et al., 2022, 2024), but it also underscores that excluding a soil-borne disease is extremely difficult, maybe even impossible. The first reported occurrence of TR4 outside Indonesia was in Taiwan (Su et al., 1977, 1986). The origin of that incursion seems to be related to a documented banana shipment from Sumatra to Taiwan (Buddenhagen, 2007). These plants may have been infected with TR4 and since then it started its regional, international, and intercontinental spread. Since virtually all further incursions have been closely monitored over the last two decades and many of the causal TR4 strains have been sequenced, it is known that TR4 is a single clone (Ordóñez et al., 2015). Therefore, forensic analyses are difficult as the number of mutations (1271 single nucleotide polymorphisms) is negligible (0.003%) compared to the entire genome size (46 megabases). However, bioinformatic tools enable the tracing of TR4 such as demonstrated in the Mekong Delta, where the occurences in Myanmar, Laos, and Vietnam is clearly linked to the TR4 strain that was described in the Southern Chinese province of Yunnan (Zheng et al., 2018). Chinese growers started banana plantations in the surrounding countries due to the increasing TR4 problems on their farms in China, while moving their equipment and bringing their personnel. It is likely that these processes contributed to the regional dissemination of TR4.

The use of disinfectants is widely recommended to prevent TR4 dissemination (Nel et al., 2007; Meldrum et al., 2013), but the procedures are often inadequate in practice. For example, Salacinas et al. (2022) showed that the efficacy of disinfectants toward TR4 requires an exposure time of 15 s, but recorded time-lapse data of real-world disinfestation stations were 3.42 (± 0.82) and 5.05 (± 3.08) s for footwear and tires, respectively. Hence, despite the good intentions, the practice fails and enables TR4 to survive. Since TR4 is soilborne, measurements of its distribution in soil have shown that propagules can be detected down to a depth of 1 m and can also be present in the irrigation water (Ploetz, 2015b). Apart from its capacity to grow as an endophyte in non-host plants, these epidemiological facts truly complicate the management of TR4 (Martínez de la Parte et al., 2024), and as Race 1 strains continued to spread after the introduction of Cavendish, TR4's dissemination will not stop after deploying resistant varieties (Fig. 3). *Fusarium* strains do not need bananas for survival, they easily survive as endophytes in non-host plant species.

3.3 Control and containment

The distribution of fungal propagules of *Fusarium* in the soil makes disease control unpredictable. Various methods have been used since the first FWB epidemic in Gros Michel, some of which are also being trialed to manage the epidemic in Cavendish. Inundation of whole fields through flood fallowing was a major effort to kill *Fusarium* in the soil (Stover, 1962a). Indeed, generating an anaerobic environment in the soil potentially kills all aerobic organisms. However, many beneficial microbes are also prone to destruction creating a biological vacuum. Inundation failed to control FWB in Gros Michel. Most likely, this was primarily due to the absence of biosecurity measures and the lack of tissue culture and hence, in many cases, asymptomatic infected new suckers were planted in the treated areas thereby restarting the epidemic (Stover, 1962b). However, inundation followed by planting tissue culture plants, which was developed in the 1970s (Ma and Shii, 1972), might be a strategy if followed by enriching the soil with antagonistic microbes after flooding. In practice, this can only be deployed in areas with flat fields and where sufficient water is available. An alternative to inundation is biological soil disinfestation. Initially, this method relied on the decomposition of biomass, ideally of plants from the *Brassicaceae* which produce glucosinolates that kill fungi and in particular *Fusarium* species (Messiha et al., 2007; Blok et al., 2000). Nowadays, various commercial products are available that induce anaerobic conditions and produce volatiles toxic to fungi during their decomposition under plastic. Under greenhouse conditions, these methods have proven efficacy, and field trials are also promising, but the method is expensive and requires substantial soil tilling. Overall, methods that generate anaerobic conditions are not specific and the lasting effects on the soil microbiome are unknown.

Rice hull burning is a method that is practiced in the Philippines and is derived from the control of the soil-borne Moko disease caused by the bacterium *Ralstonia solanacearum* (Molina et al., 2010). After chopping infected plants, a pile of rice hull is used to cover the remains of the plant and the corm and set on fire to smolder for days. Indeed, temperatures in the topsoil layer can go up to >300°C and consequently reduce the TR4 inoculum load, but measurements have also indicated that inoculum in deeper soil layers is unaffected. Variants of rice hull burning using wood or bamboo produce even higher temperatures but do not increase efficacy (Salacinas, 2019). Moreover, the entire process of burning requires substantial manual labor. Several workers are involved and consequently walk around infected plants, thereby potentially spreading the disease with their footwear. Experience has shown that rice hull burning does not stop the further spread of FWB (Fig. 4).

Biocontrol is another option for TR4 management, and many microbes produce compounds with *in vitro* efficacy toward TR4 (Nel et al., 2007; Lakshmanan et al., 1987; Davis et al., 1994; Yuan et al., 2012; Cao et al., 2005;

Figure 4 Rice hull burning is practiced in Philippines to manage Fusarium wilt of banana (FWB) caused by TR4. (a) Aerial photograph of a region on Mindanao that is struck with FWB. (b) Chopping of infected plants. (c) Chopped plant parts are covered with a heap of rice hulls that are set on fire to smolder for up to 3 days. (d) Open spots in a banana planation after practicing rice hull burning. However, FWB progresses through the plantation, see multiple rice hull burning attempts to stop the disease.

Saravanan et al., 2003; Kidane and Laing, 2010; Zhang et al., 2013). However, translation into effective commercial products for FWB management under field conditions has not been successful yet (See Bubici et al. 2019). Like the development of fungicides, it has to take into account aspects such as biological safety and the upscaling potential. The one-size-fits-all strategy that is often advocated with biocontrol, such as the almost mythical status of *Trichoderma* to control TR4 (Thangavelu et al., 2004), has hitherto not resulted in durable FWB management (Bubici et al., 2019). Also, plant volatiles have shown efficacy toward TR4, such as those produced by Chinese leek (Huang et al., 2012), but – again – field trials failed to show control at the plantation level. Mixed cropping is common in smallholder settings and backyard farming, and at a larger scale in agro-forestry landscapes, yet it does not exclude FWB (Fig. 5), but it may slow down its spread. Perennial plantation crops – such as banana, coffee, and cacao – are almost without exception monocultures, incompatible with crop rotation and mixed cropping as practiced in annual crops. However, the potential of cover crops is worth further investigation. For instance, plantations with *Arachis pintoy* as a cover crop showed a 20% lower disease severity (Pattison et al., 2014). Interestingly, this plant species can be infected by TR4 but recovered strains showed reduced pathogenicity on Cavendish bananas (Martínez de la Parte et al., 2024).

Management of foliar diseases in agriculture frequently relies on fungicide applications. Since Cavendish bananas are highly susceptible to black leaf streak disease – also known as Black Sigatoka disease – caused by *Pseudocercospora fijiensis*, the frequent use of fungicides to manage this

Figure 5 Fusarium wilt of banana (FWB) in an agro-forestry setting. (a) External symptoms of infected plant. (b) Internal symptoms, characteristic of FWB.

disease is a common practice in banana cropping (Chong et al., 2024: see also Chapter 4 in this book). However, it is often stated that FWB cannot be controlled with fungicides (Ploetz, 2015b). Indeed, TR4 and Race 1 strains are remarkably insensitive to single-target site fungicides, albeit sensitive to others (Cannon et al., 2022). However, the recently released Tymirium technology – based on succinate dehydrogenase inhibitors – has shown efficacy to TR4, and the commercial Vaniva® fungicide was recently released in the Philippines. This is a new development and time will show whether this will be a useful tool in the management of FWB.

Taken together, the history of the spread and failure of management of Race 1 in the Gros Michel era is currently repeated with TR4. This raises the question as to how TR4 can be managed. Contemporary management options for FWB are limited, and some options have been promoted that are not effective and have resulted in the continuous expansion of TR4 (van Westerhoven et al., 2024).

4 Future: preventative/curative disease control and breeding

If the current trend of TR4 dissemination continues and no effective disease control is implemented, the cultivation of Cavendish varieties will be further threatened (van Westerhoven et al., 2022). Therefore, it is necessary to consider whether FWB can be managed by novel disease control methods.

4.1 Preventative disease control

Presently, the best and cheapest strategy to manage TR4 is to exclude it (Cook et al., 2015). Prevention measures are mostly confined to exclusion, a combination of biosecurity practices aimed at restricting access to the plantation, farm layout with regards to access and water run-off, the use of disinfectants in foot baths, and tire dips combined with cleaning stations. To be efficacious, exclusion strategies should comprise well-coordinated nationwide prevention strategies. People and products entering countries should be assessed for the risk of carrying TR4 and excluded from certain areas to reduce the risk of introducing TR4 into new areas. In some countries, banners are used at airports and harbors to inform travelers about the risks of TR4 (Fig. 6). Ideally, disinfection mats to clean footwear are part of this strategy, but these are hardly used at any airport. Hence, introducing TR4 into a country with contaminated soils under footwear is an imminent risk. Also, the global circulation of containers and equipment to develop new plantations may facilitate the introduction of TR4 in uncontaminated areas as has happened in the Mekong region (Zheng et al., 2018) and Mozambique (van Westerhoven et al., 2022, 2023).

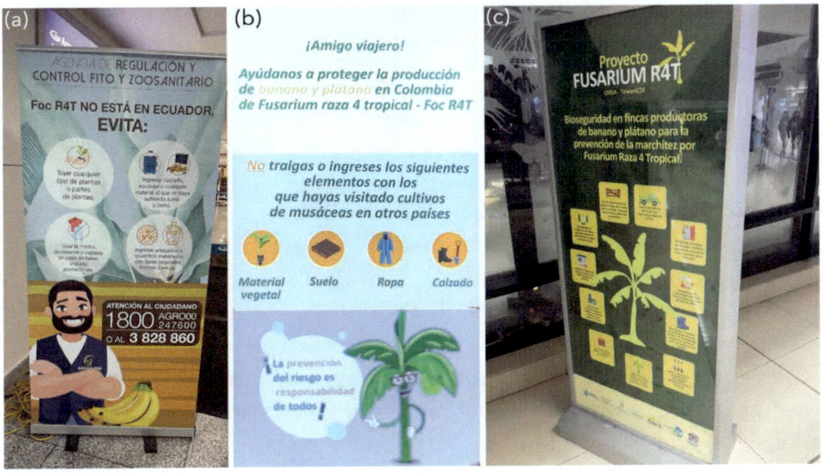

Figure 6 Travelers are informed about the risks of TR4 for banana production in Ecuador (a), Colombia (b), and Guatemala (c) at regional airports.

The next step after country border exclusion is to assure the exclusion at the farm level. This requires the exclusion of all non-essential people and movement on the plantations and the implementation of clean and dirty zones, footwear exchange stations, and disinfection units at each entry of a farm. Traditionally, packing stations are centrally located in banana plantations, and hence, trucks drive through plantations to deliver inputs and pickup packed fruit. A good biosecurity strategy is to use only on-farm equipment for transportation inside the farms. This includes motorcycles for personnel, cars, trucks, and tractors. Clearly, this requires a redesign of farm layouts with sufficient space for parking vehicles, frequent on-farm bus transportation, dedicated on-farm vehicles of every kind, and trucks and drivers having to wait outside the farm gate for deliveries and pickups. The development of zones is far more effective than the costly and time-consuming business trying to thoroughly clean and disinfect any vehicle entering a farm. The same is true for footwear where boot exchange stations are far more effective than footwear and disinfection stations, where people and equipment need to be exposed to disinfectants for a certain amount of time (Salacinas et al., 2022). These activities should supplement fencing off plantations to avoid unauthorized access. In addition, plantations are often not continuous, and different sections are connected, either by roads, rivers, or irrigation canals. A key issue arising here is that these measures involve expensive infrastructure and different working practices. It is also important to realize that these procedures need to be implemented prior to the arrival of TR4 on a plantation. Experience has shown that most plantation owners are reluctant to invest in measures preventing TR4 from arriving, and consequently, once this happens it is too late to implement such measures. Moreover, the efficacy of these measures is higher once implemented in a sector-wide fashion, which further complicates the prevention of TR4.

Once a plantation is infested in a particular area, it is important to delay further spread to other fields. For that purpose, aerial surveys have been explored to proactively identify suspect spots in plantations (Nakasato Tagami et al., 2024). This high-throughput method can map entire farms, regions, or countries and does not require farm entry for personnel to manually inspect farms, which can be a bottleneck. All farm data, geographical and logistical details, and aerial images, from airplanes, drones, or satellites, are loaded into a platform for inspection. The resolution of these images is usually 5–7 cm to identify suspect spots enabling subsequent on-farm inspection to sample and investigate TR4 presence. The frequency of such inspections is key; hence aerial surveys only pay off if they start prior to the arrival of TR4, and surveys are repeated over time at regular intervals and the subsequent inspection activities are streamlined to ensure that suspected areas or plants are identified in a timely manner. Currently, most TR4 incursions have been notified too late and FWB was already widespread and out of control. For instance in Perú, FWB was first observed in Chira Valley Cavendish plantations in April 2021 (Acuña et al., 2022) and by June 8, 2023, a total of 207 cases were reported (Nakasato Tagami et al., 2024). This expansion of TR4 is tightly connected to the production system in the Chira Valley, which is unique as it comprises 8.000 smallholder farm settings of on average one ha and two dozen larger plantations making up a total area of 10.000 ha that are all connected through a generic irrigation system from the Poechos Reservoir in the Sullana province close to the Southern border of Ecuador. The aforementioned aerial survey mapped the entire region, identified suspect open areas that were subsequently sampled, and confirmed the presence of TR4 (Nakasato Tagami et al., 2024). Genomic sequence data suggest that the incursion in Perú is independent of an earlier incursion in Colombia (Reyes-Herrera et al., 2023).

Dissemination by water is a key element of TR4 epidemiology. The tropical storm Pablo in 2012 moved the front of TR4-driven FWB to an unanticipated area in Mindanao in the Philippines, thereby increasing the indirect impact of this storm. Clearly, this impact was only observed after banana plantations were recuperated and many plants simultaneously started suffering from FWB. There is substantial anecdotical information on watershed and FWB dissemination, e.g. in the Pearl River basin of Guangdong (Molina et al., 2005) and the Mekong Delta (Zheng et al., 2018), and the aforementioned rapid dissemination of TR4 in Perú is linked to the shared irrigation system that supplies the entire banana region. Molecular diagnostics have also detected TR4 propagules in water, thereby supporting the important role of water dissemination (Salacinas, 2019). Various methods have been implemented to disinfect water for small-scale use such as in foot baths and vehicle wash-down stations (Nguyen et al., 2018; Nel et al., 2007). However, these systems are not suitable for large-volume irrigation systems where only water sourced from deep wells is safe to use, while decontamination of river water and reservoirs requires the implementation of UV-C systems that are currently being tested. In conclusion, prevention methods can be improved but require substantial investments and coordination at a national and regional scale prior to the arrival of the pathogen. These investments nevertheless pay off over time as unlimited TR4 infestations result in rapidly decreasing production volumes and eventually a non-profitable sector.

4.2 Curative disease control

In many crops, innovative cultivation methods have contributed to the yield and quality of produce. For instance, the increase in tomato production has been equally attributed to breeding efforts and to innovation of cultivation methods (Noordam, 1987; Welleman and Verwer, 1983). Tomatoes were lifted off the soil and are now produced on a substrate of rock wool or other components such as cocoa peat. Similar trends have occurred in berry and kiwi production (Schwab, 2020). Disease control, nutrition, and irrigation have modernized plantation cropping, and innovative post-harvest processes have made bananas available in the most remote places. However, in-field banana cropping has barely changed over the last century (Kema and Drenth, 2018). Rotation schedules, which are difficult to implement in perennial plantation crops, and cultivar diversification have not been practiced, albeit other varieties have been available. Clearly, market forces and economic reasons have dominated the arguments for this century-long status quo.

Fungicides have proven to be a cornerstone for disease management of black leaf streak disease in banana cultivation but for FWB management the first commercial fungicide has only been released recently in the Philippines. As mentioned by Drenth and Kema (2023, Chapter 6), biocontrol has put forward great promises, but few successes in banana production. There are numerous reports of microbes affecting fungal growth in the laboratory (Yadav et al., 2021; Thangavelu et al., 2004; Ploetz, 2015b; Mohandas et al., 2004; Kloepper et al., 2004; Jie et al., 2009; Getha et al., 2005; García-Bastidas et al., 2022), but the efficacy lacks under field conditions (Bubici et al., 2019). Nevertheless, contemporary research into the microbiomes of soils and plants provides intriguing leads for biocontrol (Tao et al., 2023; Martínez de la Parte et al., 2024). Notwithstanding these promising recent developments, the described and proposed options for improved disease control are remarkably similar to those that were unsuccessfully advocated and practiced in stopping the previous FWB epidemic in Gros Michel monocultures (Stover, 1962b). Therefore, new FWB-resistant banana varieties are required to manage TR4 in the current era.

4.3 Breeding

For FWB management, there is one outstanding and almost century-long example of success: the cultivation of Cavendish bananas in areas infested with Race 1 demonstrating outstanding and durable resistance to FWB caused by Race 1 strains. It is therefore remarkable that despite the success of resistance, so little has been invested in understanding the basis of this remarkable resistance in Cavendish to Race 1 strains. Only recently, the genetic basis of this resistance has been explored (Ahmad et al., 2020), and breeding bananas is not any longer seen as a senseless endeavor, prone to failure (Kema and Drenth, 2020). Several breeding efforts have delivered improved varieties with resistance to black leaf streak disease and other constraints (Bakry and Horry, 2016; Bhagwat and Duncan, 1998; Buddenhagen, 1990; Heslop-Harrison, 2011; Menendez and Shepherd, 1975; Ortiz et al., 1995; Persley and De Langhe, 1987; Pillay and Tenkouano, 2011; Shepherd, 1994; Silva et al., 2001; Stover and Buddenhagen, 1986; Vuylsteke, 2000). Notwithstanding these successes, contemporary breeding programs have not released varieties matching the yield and fruit quality of Cavendish bananas (Ploetz, 2015b). However, the development of genomics enables modern breeding techniques, and molecular tools have opened immense opportunities to circumvent hitherto experienced obstacles in banana breeding.

The advent of tissue culture drastically changed banana cultivation. Millions of healthy banana plants are produced every year and distributed globally to start new or recuperate existing plantations. Previously, the multiplication through suckers was the main driver of the FWB Race 1 epidemic in the 1900s. One of the aspects of tissue culture was the observation of off-types caused by somaclonal variation. Most of these variants are discarded, but some had attractive phenotypes, such as reduced tallness resulting in Dwarf Cavendish. Other variants showed a better tolerance to abiotic stress or had better bunch characteristics (Roux, 2004; Roux et al., 2004, 2008; Jain and Maluszynski, 2004; Novak et al., 1990, 1995; Khayat et al., 1998; Khayat and Ortiz, 2011). These observations stimulated a process to select somaclonal variants with improved disease resistance, for instance at the Taiwan Banana Research Institute (TBRI). This resulted in a cultivar called Formosana which was reported to have improved resistance to TR4 (Molina et al., 2011), and through the years other variants – collectively known as Giant Cavendish Tissue Culture Variants, GCTCVs – have been developed and tested in TR4-infested soils. The experience with these clones, however, has been inconclusive. Contrary to agronomical disadvantages such as longer cycling times, tall stature, and poor bunch characteristics, there have been reports of higher sugar content. Cultivation of some of these clones – such as GCTCV219 – revealed that there are many synonyms and hence it is sometimes difficult to ensure that reports are discussing the same germplasm. Nevertheless, some of the GCTCV clones showed a remarkable reduction of FWB compared to controls (Hwang, 1990; Hwang and Ko, 2004; Ploetz, 2015b; Van den Berg et al., 2007). These observations were unfortunately reported and advertised as 'resistance' or 'tolerance' (Molina et al., 2011; Hwang and Ko, 2004; Ploetz, 2015b). Hence, many growers whose farms suffered from FWB caused by TR4 adopted these clones and continued production. Within a short period of time, these clones however started suffering from FWB and increasing the inoculum levels, underscoring that reduced susceptibility or tolerance to TR4 is not a solution. The perennial nature of the crop, combined with the soil-borne nature of FWB, does not enable reduced susceptibility to durably control TR4. The recommendations to grow these variants also resulted in reduced awareness and quarantine measures enabling the further spread of the disease. Thus, in conclusion, the GCTCV clones have done more harm than good and indirectly contributed to the further dissemination of TR4 (van Westerhoven et al., 2022).

It is well known that Cavendish clones can be cultivated in soils that are infested with Race 1 strains and such a level of resistance is required for any new germplasm with resistance to TR4 to solve the current FWB crisis. Mutation breeding may play a role and interesting approaches, such as a novel transposon-based mutation strategy, have been developed which have yielded promising results (Khayat, 2020). Another method to induce mutants is through chemicals or radiation. The International Atomic Energy Agency (IAEA) has invested in mutation programs over the years (Roux, 2004; Roux et al., 2004, 2008; Novak et al., 1995), but the program has not resulted in new germplasm that can withstand FWB.

In conclusion, mutation breeding has made valid contributions to banana improvement and may hold promise for disease resistance. However, for FWB, the complete resistance of Cavendish vs. Race 1 is the ultimate benchmark. Everything that does not meet this level of resistance falls short – such as the aforementioned GCTCV clones – and may contribute to the continued dissemination of the disease.

Plant transformation has enabled genetic engineering in bananas (Wang et al., 2021). Disease resistance has been an important target for banana improvement but The Achilles' heel of banana transformation for disease resistance is the ultimate lack of understanding of the genetic basis of disease resistance. Thus far not a single gene has been functionally tested to prove the genetic control of Race 1 in Cavendish (Ahmad et al., 2020). Once a series of genes conferring resistance to FWB is identified and characterized, genetic engineering attempts can be much more focused and do not have to rely anymore on genes from other plant species. This will avoid at least one aspect of objection to genetic engineering: the use of genes from other plant species. Although resistance to TR4 is currently considered to be vital, any new banana variety should also have resistance to Race 1 strains. Therefore, the improvement of Cavendish is a common strategy in contemporary genetic improvement strategies. Contrary to resistance to Race 1, the only identified gene for resistance to a fungal pathogen in banana is *RGA2*, which provides complete resistance to TR4. Dale et al. (2017) managed to identify and clone the *RGA2* gene and transferred it to the Cavendish cv. Williams. Subsequent greenhouse and field trials have shown the required level of resistance, similar to the donor species *Musa acuminata* ssp. *malaccensis* (D'Hont et al., 2012). Subsequent work should focus on the identification and characterization of more resistance genes in banana, which requires substantial screening for resistance in the widest possible suite of germplasm (García-Bastidas, 2019) to support any improvement strategy, whether genetic engineering or breeding for resistance.

The most recent shift in genetic engineering is undoubtedly revolving around CRISPR/Cas9-based technologies. In banana, Tripathi et al. (2019) were the first to apply it for the inactivation of banana streak virus, which is endogenous in *M. balbisiana* germplasm and is a major hurdle in banana breeding. This technology holds great potential to engineer FWB resistance once resistance genes have been identified. Dale et al. (2017) reported that *RGA2* is also present in Cavendish, but it is not or hardly expressed. CRISPR/Cas9 is a tool that may be able to modulate that expression to render such edited Cavendish bananas resistant to TR4. The rapidly increasing body of genomic data in banana research, both of host and pathogens, may provide leads for FWB resistance. However, CRISPR/Cas9 should not be regarded as a panacea for disease resistance in banana as most likely transfer of genes through hybridization is required to introgress suitable target genes from the wider *Musa* germplasm into suitable varieties.

Breeding bananas is fraught with difficulties (Kema et al., 2021). Currently, there are about 12 banana breeding programs around the world (Smith et al., 2021), which focus on regional markets in Latin America and Africa, for dessert bananas and plantains. Others also focus on dessert bananas for international markets. There are several reasons why banana breeding is challenging such as parthenocarpy, sterility, and the triploid nature of most edible bananas (Amah et al., 2021). A classical – so-called evolutionary breeding – approach was used to try and improve Gros Michel in the first banana breeding program at the Imperial College of Tropical Agriculture in Trinidad that commenced in 1922. The same strategy is also being followed by the breeding program at the Fundación Hondureña de Investigación Agrícola (FHIA) in Honduras. Screening of some lines of this program has shown that FHIA banana and plantain hybrids perform better than many other varieties to TR4, but none were truly resistant (García-Bastidas, 2019; Martínez de la Parte et al., 2023). The EMBRAPA program essentially follows the same strategy and frequently releases new varieties for the domestic market in Brazil (Silva et al., 2001). The IITA and its partners have developed a breeding program for East African Highland Bananas over the last decade, which has adopted a range of innovative genomic tools to accelerate conventional breeding (see https://breedingbetterbananas.org/). The breeding program at the Centre de coopération Internationale en Recherche Agronomique pour le Développement (CIRAD) follows a different strategy through a reconstituting approach. Here, fertile diploids are improved and eventually used in 4x/2x crosses to generate seedless triploids. This approach has the advantage that wild germplasm can be used to enrich the breeding gene pool and that plenty of seeds are being generated. All evolutionary and reconstructive conventional breeding programs listed by Smith et al. (2021) are based in universities or institutions. It is remarkable that thus far commercial breeding companies have not invested in banana breeding, although the FHIA program was initiated by the United Fruit Company in 1959 but is now a governmental endeavor. Recently, a number of commercial start-ups in the United Kingdom (Tropic Biosciences), the United States (Nature Source Improved Plants), and The Netherlands (Yelloway) have emerged that target the development of disease-resistant bananas.

A thorough and reproducible screening method is a key element in breeding for resistance, irrespective of the breeding strategy used. To achieve this, germplasm is first screened under standardized greenhouse conditions prior to evaluation in the field. These analyses are crucial and cannot be conducted on-site in most regions because TR4 is not present in the countries where these breeding programs are based, or it is not allowed to use infested areas as trial sites for quarantine reasons. Another bottleneck is the narrow genetic base of cultivated bananas used to improve varieties. Access to the widest possible genetic diversity in Southeast Asia is essential to achieve genetic gain in banana breeding. Furthermore, genetic studies on the inheritance of resistance to the major diseases and pests are essential to target genes conferring resistance. This information is generic and can be used by any of the aforementioned banana improvement strategies. Therefore, contemporary research programs should focus on discovery research to unveil the genetic basis of required traits of bananas for the future as a foundation for diversification away from Cavendish. Banana is among the last major edible crops where tremendous genetic gain can be potentially achieved, provided we are using a plethora of strategies and increase the critical mass of research that is required for progress.

5 Conclusion

Looking into the future, it is tempting to ask, what is next? Another new race causing FWB? Further dissemination of TR4 within currently affected countries and to other countries is inevitable. But will new resistant varieties, able to grow unimpeded in TR4-infested plantations, come to market in time? We have seen some false dawns with the introduction of so-called 'tolerant' varieties which did not solve the problem and contributed to further dissemination of the pathogen. Hence, authorities may have to play a role in the testing and recommendation of new germplasm. This is common practice in many other crops and countries to transparently inform growers and to prevent the reoccurrence of major plant diseases.

What happens once another new *Fusarium* race or species surfaces? All relevant germplasm must be formally tested with such a variant. At present, it takes years before such information is available and it is often not comprehensive. Is there a danger of new pathogenic variants? Yes, there is, but there is an important difference. Now there are collections of sequenced *Fusarium* strains, hence new seemingly pathogenic variants can be rapidly characterized. Based on that information, existing and new germplasm can be proactively tested for resistance. Moreover, once there are signs of reduced resistance of new germplasm under field conditions, strains should be isolated, sequenced, and tested. Overall, the banana research community has grown, companies are more aware, and new companies are founded with well-trained committed individuals. Hence, the risk of surprises has reduced, and the awareness of global plant disease epidemics is much higher in our more connected world. New incursions of TR4 and other plant diseases are noticed and communicated rapidly. This chapter has provided suggestions on how to act once that happens and above all, what should be done to prevent it. Collective action is required, and thorough, reliable prevention and quarantine measures should be practiced. New promising germplasm is under development and exciting new management options based on disinfection technology, innovation in banana cultivation, biological phenomena, and microbial consortia are being developed. The combination of new insights and technologies will help to better control TR4 and other banana diseases that have spread around the world over the last century (Kema et al., 2021). A well-resourced and connected research community is needed, supported by national and international legislation on plant health, using a comprehensive and rigorous research approach, grounded in genetics and breeding and addressing the various needs of the crop along with rigorously tested and validated disease control options to secure the production of this major food crop and the world's favorite fruit.

6 Where to look for further information

- Global Alliance Against TR4; https://iica.int/en/global-alliance.
- World Banana Forum: https://www.fao.org/world-banana-forum/en/.
- Wageningen University: https://www.wur.nl/en/dossiers/file/banana-cultivation.htm.

7 Acknowledgements

We thank the Wageningen University and Research (WUR) and KeyGene banana research teams for their overall support. The Dutch Dioraphte Foundation is acknowledged for partially funding the WUR banana research program. We thank Job Dirkmaat for preparing Figure 3 (Adapted from 'Timeline (7 Segments, Horizontal) 2,' by BioRender.com (2022). Retrieved from https://app.biorender.com/biorender-templates).

8 References

Acuña, R., Rouard, M., Leiva, A. M., Marques, C., Olortegui, J. A., Ureta, C., Cabrera-Pintado, R. M., Rojas, J. C., Lopez-Alvarez, D., Cenci, A., Cuellar, W. J. and Dita, M. 2022. First report of *Fusarium oxysporum* f. sp. *cubense* tropical race 4 causing fusarium wilt in Cavendish bananas in Peru. *Plant Disease* 106, 2268.

Aguayo, J., Mostert, D., Fourrier-Jeandel, C., Cerf-Wendling, I., Hostachy, B., Viljoen, A. and Ioos, R. 2017. Development of a hydrolysis probe-based real-time assay for the detection of tropical strains of *Fusarium oxysporum* f. sp. cubense race 4. *PLoS ONE* 12(2), e0171767.

Ahmad, F., Martawi, N. M., Poerba, Y. S., De Jong, H., Schouten, H. and Kema, G. H. 2020. Genetic mapping of Fusarium wilt resistance in a wild banana *Musa acuminata* ssp. malaccensis accession. *Theoretical and Applied Genetics*, 133, 3409-3418.

Antonelli, A. 2023. Indigenous knowledge is key to sustainable food systems. *Nature* 613(7943), 239–42.

Baayen, R. P., O'Donnell, K., Bonants, P. J. M., Cigelnik, E., Kroon, L. P., Roebroeck, E. J. and Waalwijk, C. 2000. Gene genealogies and AFLP analyses in the *Fusarium oxysporum* complex identify monophyletic and nonmonophyletic formae speciales causing wilt and rot disease. *Phytopathology* 90(8), 891–900.

Bakry, F., Carreel, F., Jenny, C. and Horry, J.-P. 2009. Genetic improvement of banana. In: Jain, S. M. and Priyadarshan, P. M. *Breeding Plantation Tree Crops: Tropical Species*. Springer, 3–50.

Bakry, F. and Horry, J.-P. 2016. Advances in genomics: applications to banana breeding. *Proceedings of the XXIX International Horticultural Congress on Horticulture: Sustaining Lives, Livelihoods and Landscapes (IHC2014)* IX(1114), 171–80.

Bentley, S., Pegg, K. G. and Dale, J. L. 1995. Genetic variation among a world-wide collection of isolates of *Fusarium oxysporum* f. sp. *cubense* analysed by RAPD-PCR fingerprinting. *Mycological Research* 99(11), 1378–84.

Bentley, S., Pegg, K. G., Moore, N. Y., Davis, R. D. and Buddenhagen, I. W. 1998. Genetic variation among vegetative compatibility groups of *Fusarium oxysporum* f. sp. *cubense* analyzed by DNA fingerprinting. *Phytopathology* 88(12), 1283-93.

Bhagwat, B. and Duncan, E. J. 1998. Mutation breeding of banana cv. Highgate (*Musa* spp., AAA Group) for tolerance to *Fusarium oxysporum* f. sp. *cubense* using chemical mutagens. *Scientia Horticulturae* 73(1), 11–22.

Blok, W. J., Lamers, J. G., Termorshuizen, A. J. and Bollen, G. J. 2000. Control of soilborne plant pathogens by incorporating fresh organic amendments followed by tarping. *Phytopathology* 90(3), 253–9.

Boggarapu, S. S. 2023. Horizontal and Vertical Transmission of *Fusarium odoratissimum*. MSc thesis. Wageningen University, 39.

Bubici, G., Kaushal, M., Prigigallo, M. I., Gómez-Lama Cabanás, C. and Mercado-Blanco, J. 2019. Biological control agents against Fusarium wilt of banana. *Frontiers in microbiology* 10, 616.

Buddenhagen, I. 1990. Banana breeding and Fusarium wilt. *Fusarium Wilt of Banana*, 107–13.

Buddenhagen, I. 2007. Understanding strain diversity in *Fusarium oxysporum* f. sp. *cubense* and history of introduction of 'Tropical Race 4' to better manage banana production. Proceedings of the III International Symposium on Banana: ISHS-ProMusa Symposium on Recent Advances in Banana Crop Protection for Sustainable 828, 193–204.

Cannon, S., Kay, W., Kilaru, S., Schuster, M., Gurr, S. J. and Steinberg, G. 2022. Multi-site fungicides suppress banana Panama disease, caused by *Fusarium oxysporum* f. sp. *cubense* tropical race 4. *PLoS Pathogens* 18(10), e1010860.

Cao, L., Qiu, Z., You, J., Tan, H. and Zhou, S. 2005. Isolation and characterization of endophytic streptomycete antagonists of fusarium wilt pathogen from surface-sterilized banana roots. *FEMS Microbiology Letters* 247(2), 147–52.

Carvalhais, L. C., Henderson, J., Rincon-Florez, V. A., O'Dwyer, C., Czislowski, E., Aitken, E. A. B. and Drenth, A. 2019. Molecular diagnostics of banana Fusarium wilt targeting Secreted-in-Xylem genes. *Frontiers in Plant Science* 10, 547.

Catambacan, D. G. and Cumagun, C. J. R. 2022. The weed *Eleusine indica* as an alternative host of *Fusarium oxysporum* f. sp. *cubense* tropical race 4 causing Fusarium wilt in Cavendish banana. *Journal of Phytopathology* 170(7-8), 437–44.

Chittarath, K., Mostert, D., Crew, K. S., Viljoen, A., Kong, G., Molina, A. B. and Thomas, J. E. 2018. First report of *Fusarium oxysporum* f. sp. *cubense* tropical race 4 (VCG 01213/16) associated with Cavendish bananas in Laos. *Plant Disease* 102(2), 449.

Cook, D. C., Taylor, A. S., Meldrum, R. A. and Drenth, A. 2015. Potential economic impact of Panama disease (tropical race 4) on the Australian banana industry. *Journal of Plant Diseases and Protection* 122(5-6), 229–37.

Correll, J. 1991. The relationship between formae speciales, races, and vegetative compatibility groups in *Fusarium oxysporum*. *Phytopathology* 81, 1061–4.

Correll, J. C., Klittich, C. and Leslie, J. 1987. Nitrate nonutilizing mutants of *Fusarium oxysporum* and their use in vegetative compatibility tests. *Phytopathology* 77(12), 1640–6.

Dale, J., James, A., Paul, J. Y., Khanna, H., Smith, M., Peraza-Echeverria, S., Garcia-Bastidas, F., Kema, G., Waterhouse, P., Mengersen, K. and Harding, R. 2017. Transgenic Cavendish bananas with resistance to Fusarium wilt tropical race 4. *Nature Communications* 8(1), 1496.

Davis, A. J., Say, M., Snow, A. J. and Grant, B. R. 1994. Sensitivity of *Fusarium oxysporum* f. sp. *cubense* to phosphonate. *Plant Pathology* 43(1), 200–5.

D'hont, A., Denoeud, F., Aury, J. M., Baurens, F. C., Carreel, F., Garsmeur, O., Noel, B., Bocs, S., Droc, G., Rouard, M., Da Silva, C., Jabbari, K., Cardi, C., Poulain, J., Souquet, M., Labadie, K., Jourda, C., Lengellé, J., Rodier-Goud, M., Alberti, A., Bernard, M., Correa, M., Ayyampalayam, S., Mckain, M. R., Leebens-Mack, J., Burgess, D., Freeling, M., Mbéguié-A-Mbéguié, D., Chabannes, M., Wicker, T., Panaud, O., Barbosa, J., Hribova, E., Heslop-Harrison, P., Habas, R., Rivallan, R., Francois, P., Poiron, C., Kilian, A., Burthia, D., Jenny, C., Bakry, F., Brown, S., Guignon, V., Kema, G., Dita, M., Waalwijk, C., Joseph, S., Dievart, A., Jaillon, O., Leclercq, J., Argout, X., Lyons, E., Almeida, A., Jeridi, M., Dolezel, J., Roux, N., Risterucci, A. M., Weissenbach, J., Ruiz, M., Glaszmann, J. C., Quétier, F., Yahiaoui, N. and Wincker, P. 2012. The banana (*Musa acuminata*) genome and the evolution of monocotyledonous plants 488(7410), 213–7.

Dita, M. A., Waalwijk, C., Buddenhagen, I. W., Souza Jr, M. T. and Kema, G. H. J. 2010. A molecular diagnostic for tropical race 4 of the banana fusarium wilt pathogen. *Plant Pathology* 59(2), 348–57.

Donohue, M. and Denham, T. 2009. Banana (*Musa* spp.) domestication in the Asia-Pacific region: linguistic and archaeobotanical perspectives. *Ethnobotany Research and Applications* 7, 293–332.

Drenth, A. and Kema, G. H. J. 2021. The vulnerability of bananas to globally emerging disease threats. *Phytopathology* 111, 2146–2161.

Fayyaz, A., Robinson, G., Chang, P. L., Bekele, D., Yimer, S., Carrasquilla-Garcia, N., Negash, K., Surendrarao, A., von Wettberg, E. J. B., Kemal, S. A., Tesfaye, K., Fikre, A., Farmer, A. D. and Cook, D. R. 2023. Hiding in plain sight: genome-wide recombination and a dynamic accessory genome drive diversity in *Fusarium oxysporum* f. sp. *ciceris*. *Proceedings of the National Academy of Sciences of the United States of America* 120(27), e2220570120.

Fourie, G., Steenkamp, E. T., Gordon, T. R. and Viljoen, A. 2009. Evolutionary relationships among the *Fusarium oxysporum* f. sp. *cubense* vegetative compatibility groups. *Applied and Environmental Microbiology* 75(14), 4770–81.

García-Bastidas, F. A. 2019. Panama disease in banana: spread, screens and genes. Doctoral dissertation. Wageningen University.

García-Bastidas, F. A., Arango-Isaza, R., Rodriguez-Cabal, H. A., Seidl, M. F., Cappadona, G., Segura, R., Salacinas, M. and Kema, G. H. J. 2022. Induced resistance to Fusarium wilt of banana caused by Tropical Race 4 in Cavendish cv Grand Naine bananas after challenging with avirulent *Fusarium* spp. *PLoS ONE* 17(9), e0273335.

García-Bastidas, F., Ordóñez, N., Konkol, J., Al-Qasim, M., Naser, Z., Abdelwali, M., Salem, N., Waalwijk, C., Ploetz, R. C. and Kema, G. H. J. 2014. First report of *Fusarium oxysporum* f. sp. *cubense* tropical race 4 associated with Panama disease of banana outside Southeast Asia. *Plant Disease* 98(5), 694.

García-Bastidas, F. A., Pachacama, S. F., Jarrín-Escudero, D. A., León Iza-Arteaga, M. L., Ayala Vásquez, M., Ortiz, H. E., Dix Luna, O. J., Echegaray Buezo, J., Farfán Menéndez, D., Bartolini Martínez, I., Beltrán Montoya, C. and Zeballos Céspedes, G. 2020a. Guía andina para el diagnóstico de Fusarium Raza 4 Tropical (R4T). In: Andina, C. (Ed.). *Comunidad Andina*, 70.

García-Bastidas, F. A., Quintero-Vargas, J. C., Ayala-Vasquez, M., Schermer, T., Seidl, M. F., Santos-Paiva, M., Noguera, A. M., Aguilera-Galvez, C., Wittenberg, A., Hofstede, R., Sørensen, A. and Kema, G. H. J. 2020b. First report of Fusarium wilt tropical race 4 in

Cavendish bananas caused by *Fusarium odoratissimum* in Colombia. *Plant Disease* 104(3), 994.

Getha, K., Vikineswary, S., Wong, W. H., Seki, T., Ward, A. and Goodfellow, M. 2005. Evaluation of *Streptomyces* sp. strain g10 for suppression of Fusarium wilt and rhizosphere colonization in pot-grown banana plantlets. *Journal of Industrial Microbiology and Biotechnology* 32(1), 24–32.

Guo, Z., Fan, H., Bai, T., Yang, P. and Zeng, L. 2015. Isolation and identification of banana vascular wilt in Yunnan and research on pathogencity of *Fusarium oxysporum* f. sp. *cubense* race 4. In: Proceedings of the Chinese Plant Protection Society Academic Annual Meeting, Changchun (in Chinese).

Hennessy, C., Walduck, G., Daly, A. and Padovan, A. 2005. Weed hosts of *Fusarium oxysporum* f. sp. *cubense* tropical race 4 in northern Australia. *Australasian Plant Pathology* 34(1), 115–7.

Hernández, Y., Magdama, F., Mostert, D., Bothma, S., Paredes Salgado, E. M., Terán, D., González, E., Angulo, R., Angel, L., Rodríguez, Y., Ortega, R., Viljoen, A. and Marys, E. E. 2023. First report of Fusarium wilt of Cavendish bananas caused by *Fusarium oxysporum* f. sp. *cubense* tropical race 4 in Venezuela. *Plant Disease* 107(10).

Heslop-Harrison, J. S. 2011. Genomics, banana breeding and superdomestication. *Proceedings of the V International Symposium on Banana: ISHS-ProMusa Symposium on Global Perspectives on Asian Challenges* 897, 55–62.

Heslop-Harrison, J. S. and Schwarzacher, T. 2007. Domestication, genomics and the future for banana. *Annals of Botany* 100(5), 1073–84.

Huang, S. P., Guo, T. X., Mo, J. Y., Li, Q. L., Tang, L. H., Chen, J., Wu, Y. D. and Wei, Q. 2016. Identification and distribution of *Fusarium oxysporum* f. sp. *cubense* physiological race in Guangxi. *Journal of Southern Agriculture* 47, 1326–31.

Huang, Y. H., Wang, R. C., Li, C. H., Zuo, C. W., Wei, Y. R., Zhang, L. and Yi, G. J. 2012. Control of Fusarium wilt in banana with Chinese leek. *European Journal of Plant Pathology* 134(1), 87–95.

Hung, T. N., Hung, N. Q., Mostert, D., Viljoen, A., Chao, C. P. and Molina, A. B. 2017. First report of Fusarium wilt on Cavendish bananas, caused by *Fusarium oxysporum* f. sp. *cubense* tropical race 4 (VCG 01213/16), in Vietnam. *Plant Disease* 102(2), 448.

Hwang, S. C. 1990. Somaclonal resistance in Cavendish banana to Fusarium wilt. In: Ploetz, R. C.. (Ed). *Fusarium Wilt of Banana*. 121–5.

Hwang, S. C. and Ko, W. H. 2004. Cavendish banana cultivars resistant to Fusarium wilt acquired through somaclonal variation in Taiwan. *Plant Disease* 88(6), 580–8.

Jain, S. and Maluszynski, M. 2004. Induced mutations and biotechnology in improving crops. *In Vitro Application in Crop Improvement*, 1st Edition, CRC Press, Florida, USA. 170-202.

Jie, L., Zifeng, W., Lixiang, C., Hongming, T., Patrik, I., Zide, J. and Shining, Z. 2009. Artificial inoculation of banana tissue culture plantlets with indigenous endophytes originally derived from native banana plants. *Biological Control* 51(3), 427–34.

Katan, T. and Di Primo, P. 1999. Current status of vegetative compatibility groups in *Fusarium oxysporum*: Supplement. *Phytoparasitica* 27, 273–7.

Kema, G. H. and Drenth, A. 2018. *Achieving Sustainable Cultivation of Bananas. Volume 1: Cultivation Techniques*. Burleigh Dodds Science Publishing Limited, 357.

Kema, G. H. and Drenth, A. 2020. *Achieving Sustainable Cultivation of Bananas. Volume 2: Germplasm and Genetic Improvement*. Burleigh Dodds Science Publishing, UK.

Kema, G. H., Drenth, A., Dita, M., Jansen, K., Vellema, S. and Stoorvogel, J. J. 2021. Fusarium wilt of banana, a recurring threat to global banana production. *Front. Plant Sci.* 11, 628888.

Khayat, E. 2020. Targeted improvement of Cavendish clones. In: *Achieving Sustainable Cultivation of Bananas. Volume 2: Germplasm and Genetic Improvement*. Burleigh Dodds Science Publishing, UK. 307–22.

Khayat, E., Duvdevani, A., Shchukin, A. and Lahav, E. 1998. Banana improvement program at Rahan meristem. *Proceedings of the II International Symposium on Banana: I International Symposium on Banana in the Subtropics* 490, 71–8.

Khayat, E. and Ortiz, R. 2011. Genetics of important. In: *Banana Breeding: Progress and Challenges*. CRC Press, Boca Raton, FL, 71–8.

Kidane, E. G. and Laing, M. D. 2010. Integrated control of fusarium wilt of banana (*Musa* spp.). *Proceedings of the International Conference on Banana and Plantain in Africa: Harnessing International Partnerships to Increase Research Impact* 879, 315–21.

Kloepper, J. W., Ryu, C. M. and Zhang, S. 2004. Induced systemic resistance and promotion of plant growth by *Bacillus* spp. *Phytopathology* 94(11), 1259–66.

Koenig, R. L., Ploetz, R. C. and Kistler, H. C. 1997. *Fusarium oxysporum* f. sp. *cubense* consists of a small number of divergent and globally distributed clonal lineages. *Phytopathology* 87(9), 915–23.

Lakshmanan, P., Selvaraj, P. and Mohan, S. 1987. Efficacy of different methods for the control of Panama disease. *Tropical Pest Management* 33(4), 373–4.

Leslie, J. F., Summerell, B. A. and Bullock, S. 2006. *The Fusarium Laboratory Manual*. Wiley Online Library.

Li, L.-F. and Ge, X.-J. 2017. Origin and domestication of cultivated banana. *Ecological Genetics and Genomics* 2, 1–2.

Lin, Y.-H., Su, C.-C., Chao, C.-P., Chen, C., Chang, C., Huang, J. and Chang, P. L. 2013. A molecular diagnosis method using real-time PCR for quantification and detection of *Fusarium oxysporum* f. sp. *cubense* race 4. *European Journal of Plant Pathology* 135(2), 395–405.

Ma, S.-S. and Shii, C.-T. 1972. *In vitro* formation of adventitious buds in banana shoot apex following decapitation. *Journal of the Chinese Society for Horticultural Science* 18, 135–42.

Martínez De La Parte, E., Meijer, H. J. G., Salacinas, M., Zuñiga-Burgos, V. A. and Kema, G. H. J. 2024. Endophytic colonization of non-host plants by the banana Fusarium wilt pathogen *Fusarium odoratissimum* contributes to inoculum persistence and modulates pathogenic fitness in banana. *Plant Pathology*.

Martinez De La Parte, E., Perez Vicente, L., Garcia-Bastidas, F., Bermúdez-Caraballoso, I., Schnabel, S., Meijer, H. J. G. and Kema, G. H. J. 2023. The vulnerability of Cuban banana production to Fusarium wilt caused by tropical race 4. *Phytopathology* doi:10.1094/PHYTO-04-23-0127-R.

Maryani, N. 2018. A complex relationship: banana & Fusarium wilt in Indonesia. PhD thesis. Wageningen University.

Maryani, N., Lombard, L., Poerba, Y. S., Subandiyah, S., Crous, P. W. and Kema, G. H. J. 2018. Phylogeny and genetic diversity of the banana Fusarium wilt pathogen *Fusarium oxysporum* f. sp. *cubense* in the Indonesian centre of origin. *Studies in Mycology* 92, 155–94.

McTaggart, A. R., James, T. Y., Shivas, R. G., Drenth, A., Wingfield, B. D., Summerell, B. A. and Duong, T. A. 2021. Population genomics reveals historical and ongoing recombination in the *Fusarium oxysporum* species complex. *Studies in Mycology* 99, 100132.

Meldrum, R. A., Daly, A. M., Tran-Nguyen, L. T. T. and Aitken, E. A. B. 2013. The effect of surface sterilants on spore germination of *Fusarium oxysporum* f. sp. *cubense* tropical race 4. *Crop Protection* 54, 194–8.

Menendez, T. and Shepherd, K. 1975. Breeding new bananas. *World Crops* May–June, 104–12.

Messiha, N. A. S., Van Diepeningen, A. D., Wenneker, M., van Beuningen, A. R., Janse, J. D., Coenen, T. G. C., Termorshuizen, A. J., van Bruggen, A. H. C. and Blok, W. J. 2007. Biological soil disinfestation (BSD), a new control method for potato brown rot, caused by *Ralstonia solanacearum* race 3 biovar 2. *European Journal of Plant Pathology* 117(4), 403–15.

Mohandas, S., Manamohan, M., Rawal, R. D., Chakraborty, S., Sreekantappa, H., Manjula, R. and Lakshmikantha, H. C. 2004. Interaction of *Fusarium oxysporum* f. sp. *cubense* with *Pseudomonas fluorescens* precolonized to banana roots. *World Journal of Microbiology and Biotechnology* 20(6), 651–5.

Molina, A., Fabregar, E., Ramillete, E., Sinohin, V. and Viljoen, A. 2011. Field resistance of selected banana cultivars against tropical race 4 of *Fusarium oxysporum* f. sp *cubense* in the Philippines. In: Proceedings of the Phytopathology IPPC Joint Meeting. Hawaii.

Molina, A., Fabregar, E., Sinohin, V. G., Herradura, L., Fourie, G. and Viljoen, A. 2008. Confirmation of tropical race 4 of *Fusarium oxysporum* f. sp. *cubense*, infecting Cavendish bananas in the Philippines. In: Proceedings of the Abstract of Presentation to the 2008 Centennial Meeting of the American Phytopathological Society.

Molina, A., Williams, R., Hermanto, C., Suwanda, M., Komolong, B. and Kokoa, P. 2010. Mitigating the threat of banana Fusarium wilt: understanding the agroecological distribution of pathogenic forms and developing disease management strategies. ACIAR, Australia, GPO Box 1571, 2010–12.

Molina, A., Xu, L., Roa, V., Bergh, I. V. D. and Borromero, K. 2005. Advancing banana and plantain R & D in Asia and the Pacific. Volume 13: Proceedings of the 3rd BAPNET Steering Committee Meeting, Guangzhou, China, 23–26 November 2004.

Moore, N., Pegg, K., Allen, R. and Irwin, J. 1993. Vegetative compatibility and distribution of *Fusarium oxysporum* f. sp. *cubense* in Australia. *Australian Journal of Experimental Agriculture* 33(6), 797–802.

Nakasato Tagami, G., van Westerhoven, A. C., Seidl, M., Van der Sluis, J., Cozzarelli, M., Balarezo, D., Pflücker, R., Marquez Rosillo, C., Clercx, L. and Kema, G. H. J. 2024. Aerial mapping of the Peruvian Chira valley banana production system to monitor the expansion of Fusarium wilt caused by tropical race 4. *PhytoFrontiers*. doi: 10.1094/PHYTOFR-07-23-0097-R.

Nel, B., Steinberg, C., Labuschagne, N. and Viljoen, A. 2007. Evaluation of fungicides and sterilants for potential application in the management of Fusarium wilt of banana. *Crop Protection* 26(4), 697–705.

Noordam, A. 1987. The yearround culture of tomatoes on rockwool in the Netherlands. *Proceedings of the Symposium on Horticultural Substrates and Their Analysis* 221, 357–62.

Novak, F., Afza, R., Duren, M. V. and Omar, M. 1990. Mutation induction by gamma irradiation of *in vitro* cultured shoot-tips of banana and plantain (*Musa* cvs). *Tropical Agriculture* 67, 21-8.

Novak, F., Afza, R., Morpurgo, R., Van Duren, M., Sacchi, M. and Khatri, A. 1995. Improvement of *Musa* through biotechnology and mutation breeding. *In Vitro Mutation Breeding of Bananas and Plantains* 57, 143–158.

O'Donnell, K., Kistler, H. C., Cigelnik, E. and Ploetz, R. C. 1998. Multiple evolutionary origins of the fungus causing Panama disease of banana: concordant evidence from nuclear and mitochondrial gene genealogies. *Proceedings of the National Academy of Sciences of the United States of America* 95(5), 2044–9.

O'Donnell, K., Ward, T. J., Geiser, D. M., Kistler, H. C. and Aoki, T. 2004. Genealogical concordance between the mating type locus and seven other nuclear genes supports formal recognition of nine phylogenetically distinct species within the *Fusarium graminearum* clade. *Fungal Genetics and Biology* 41(6), 600–23.

Ong, K. 1996. Fusarium wilt of banana in a Cavendish banana in a commercial farm in Malaysia. In: Frison, E. A., Horry, J. P. and De Waele, D. (Eds) *New Frontiers in Resistance Breeding for Nematode, Fusarium and Sigatoka (1995, Kuala 70 Lumpur, MY). Proceedings*. INIBAP, Montpellier.

Ordóñez, N., Salacinas, M., Mendes, O., Seidl, M. F., Meijer, H. J. G., Schoen, C. D. and Kema, G. H. J. 2019. A loop-mediated isothermal amplification (LAMP) assay based on unique markers derived from genotyping by sequencing data for rapid in planta diagnosis of Panama disease caused by tropical race 4 in banana. *Plant Pathology* 68(9), 1682–93.

Ordóñez, N., Seidl, M. F., Waalwijk, C., Drenth, A., Kilian, A., Thomma, B. P., Ploetz, R. C. and Kema, G. H. 2015. Worse comes to worst: bananas and Panama disease–when plant and pathogen clones meet. *PLoS Pathogens* 11. doi:10.1371/journal.ppat.1005197.

Ortiz, R., Ferris, R. S. B. and Vuylsteke, D. R. 1995. Banana and plantain breeding. In: Gowen, S. (Ed.) *Bananas and Plantains*. Springer Netherlands, Dordrecht, 110–46.

Pattison, A. B., Wright, C. L., Kukulies, T. L. and Molina, A. B. 2014. Ground cover management alters development of Fusarium wilt symptoms in Ducasse bananas. *Australasian Plant Pathology* 43(4), 465–76.

Pegg, K. G., Coates, L. M., O'Neill, W. T. and Turner, D. W. 2019. The epidemiology of fusarium wilt of banana. *Frontiers in Plant Science* 10, 1395.

Perrier, X., De Langhe, E., Donohue, M., Lentfer, C., Vrydaghs, L., Bakry, F., Carreel, F., Hippolyte, I., Horry, J. P., Jenny, C., Lebot, V., Risterucci, A. M., Tomekpe, K., Doutrelepont, H., Ball, T., Manwaring, J., de Maret, P. and Denham, T. 2011. Multidisciplinary perspectives on banana (*Musa* spp.) domestication. *Proceedings of the National Academy of Sciences of the United States of America* 108(28), 11311–8.

Persley, G. and De Langhe, E. 1987. Banana and plantain breeding strategies. In: Proceedings of an International Workshop Held at Cairns.State Mutual Book & Periodical Service, Ltd, Australia.

Pillay, M. and Tenkouano, A. 2011. *Banana Breeding: Progress and Challenges*. CRC Press, USA.

Ploetz, R. C. 1990. Variability in *Fusarium oxysporum* f. sp. *cubense*. *Canadian Journal of Botany* 68(6), 1357–63.

Ploetz, R. C. 1994. Panama disease: return of the first banana menace. *International Journal of Pest Management* 40(4), 326–36.

Ploetz, R. C. 2005a. Panama disease: an old nemesis rears its ugly head (Part 1: The beginnings of the banana export trades). *APSNet Feature Story*, August.

Ploetz, R. C. 2005b. Panama disease: an old nemesis rears its ugly head Part 2. The Cavendish era and beyond. *APSnet Feature*, October.

Ploetz, R. C. 2015a. Fusarium wilt of banana. *Phytopathology* 105(12), 1512–21.

Ploetz, R. C. 2015b. Management of Fusarium wilt of banana: a review with special reference to tropical race 4. *Crop Protection* 73, 7–15.

Puhalla, J. E. 1985. Classification of strains of *Fusarium oxysporum* on the basis of vegetative compatibility. *Canadian Journal of Botany* 63(2), 179–83.

Qi, P. 2001. Status report of banana Fusarium wilt disease in China. In: Molina, A. B., Nik Masdek, N. H. and Liew, K. W. (Eds) *Banana Fusarium Wilt Management: Toward Sustainable Cultivation*. INIBAP, Los Baños, 119–20.

Qi, Y. X., Zhang, X., Pu, J. J., Xie, Y. X., Zhang, H. Q. and Huang, S. L. 2008. Race 4 identification of *Fusarium oxysporum* f. sp. *cubense* from Cavendish cultivars in Hainan Province, China. *Australasian Plant Disease Notes* 3(1), 46–7.

Reyes-Herrera, P. H., Torres-Bedoya, E., Lopez-Alvarez, D., Burbano-David, D., Carmona, S. L., Bebber, D. P., Studholme, D. J., Betancourt, M. and Soto-Suarez, M. 2023. Genome sequence data reveal at least two distinct incursions of the tropical race 4 variant of Fusarium wilt into South America. *Phytopathology®* 113, 90–7.

Roux, N. 2004. Mutation induction in Musa - review. In: Jain, S. M. and Swennen, R. *Banana Improvement: Cellular, Molecular Biology, and Induced Mutations*. Sci Pub, Inc., Enfield 21–9.

Roux, N., Baurens, F.-C., Doležel, J., Hřibová, E., Heslop-Harrison, P., Town, C., Sasaki, T., Matsumoto, T., Aert, R., Remy, S., Souza, M. and Lagoda, P. 2008. Genomics of banana and plantain (*Musa* spp.), major staple crops in the tropics. In: Jain, S. M. and Swennen, R. *Genomics of Tropical Crop Plants*. Springer, 83–111.

Roux, N., Toloza, A., Dolezel, J. and Panis, B. 2004. Usefulness of embryogenic cell suspension cultures for the induction and selection of mutants in *Musa* spp. *Banana Improvement: Cellular, Molecular Biology and Induced Mutations*. Science Publishers Inc., 33–43.

Salacinas, M. 2019. Spot on: managing Panama disease of banana in the Philippines. PhD thesis. Wageningen University.

Salacinas, M., Meijer, H. J. G., Mamora, S. H., Corcolon, B., Mirzadi Gohari, A., Ghimere, B. and Kema, G. H. J. 2022. Efficacy of disinfectants against the causal agent of Panama disease in Cavendish bananas. *Plant Disease* 106(3), 966–74.

Saravanan, T., Muthusamy, M. and Marimuthu, T. 2003. Development of integrated approach to manage the Fusarial wilt of banana. *Crop Protection* 22(9), 1117–23.

Sardos, J., Breton, C., Perrier, X., Van den Houwe, I., Carpentier, S., Paofa, J., Rouard, M. and Roux, N. 2022. Hybridization, missing wild ancestors and the domestication of cultivated diploid bananas. *Frontiers in Plant Science* 13, 3668.

Schwab, J. D. 2020. Controlled and protected environment production of blueberries in the Midwest United States. MSc thesis. Kansas State University.

Shepherd, K. 1994. History and methods of banana breeding. *Infomusa* 3, 10–1.

Silva, S. O., Souza Junior, M. T., Alves, É. J., Silveira, J. R. S. and Lima, M. B. 2001. Banana breeding program at Embrapa. *Crop Breeding and Applied Biotechnology* 1(4), 399–436.

Smith, A., Ganesalingam, A., Lisle, C., Kadkol, G., Hobson, K. and Cullis, B. 2021. Use of contemporary groups in the construction of multi-environment trial datasets for selection in plant breeding programs. *Frontiers in Plant Science* 11, 623586.

Stover, R. 1962a. Fusarial wilt (Panama disease) of bananas and other *Musa* species. Fusarial wilt (Panama Disease) of Bananas and Other Musa species. Phytopathological paper No 4. Commonwealth Mycological Institute.

Stover, R. 1962b. Studies on Fusarium wilt of bananas: VIII. Differentiation of clones by cultural interaction and volatile substances. *Canadian Journal of Botany* 40(11), 1467–71.

Stover, R. and Buddenhagen, I. 1986. Banana breeding: polyploidy, disease resistance and productivity. *Fruits* 41, 175–91.

Su, H., Chuang, T. and Kong, W. 1977. Physiological race of Fusarial wilt fungus attacking Cavendish banana of Taiwan. *Taiwan Banana Research Institute, Special Publication* 2, 22.

Su, H., Hwang, S. and Ko, W. 1986. Fusarial wilt of Cavendish bananas in Taiwan. *Plant Disease* 70(9), 814–8.

Tao, C., Wang, Z., Liu, S., Lv, N., Deng, X., Xiong, W., Shen, Z., Zhang, N., Geisen, S., Li, R., Shen, Q. and Kowalchuk, G. A. 2023. Additive fungal interactions drive biocontrol of Fusarium wilt disease. *New Phytologist* 238(3), 1198–214.

Thangavelu, R., Palaniswami, A. and Velazhahan, R. 2004. Mass production of *Trichoderma harzianum* for managing fusarium wilt of banana. *Agriculture, Ecosystems and Environment* 103(1), 259–63.

Tripathi, J. N., Ntui, V. O., Ron, M., Muiruri, S. K., Britt, A. and Tripathi, L. 2019. CRISPR/Cas9 editing of endogenous banana streak virus in the B genome of *Musa* spp. overcomes a major challenge in banana breeding. *Communications Biology* 2, 1–11.

Van Den Berg, N., Berger, D. K., Hein, I., Birch, P. R., Wingfield, M. J. and Viljoen, A. 2007. Tolerance in banana to Fusarium wilt is associated with early up-regulation of cell wall-strengthening genes in the roots. *Molecular Plant Pathology* 8(3), 333–41.

van Westerhoven, A. C., Meijer, H. J. G., Seidl, M. F. and Kema, G. H. J. 2022. Uncontained spread of Fusarium wilt of banana threatens African food security. *PLoS Pathogens* 18(9), e1010769.

van Westerhoven, A. C., Meijer, H. J. G., Houdijk, J., Luntadila Matabuana, E., Martínez de la Parte, E., Seidl, M. F. and Kema, G. H. J. 2023. Dissemination of Fusarium wilt of banana in Mozambique caused by *Fusarium odoratissimum* Tropical Race 4. *Plant Dis.* doi: 10.1094/PDIS-07-22-1576-SC.

van Westerhoven, A. C., Aguilera-Galvez, C., Nakasato-Tagami, G., Shi-Kunne, X., Martinez de la Parte, E., Chavarro-Carrero, E., Meijer, H. J. G., Feurtey, A., Maryani, N., Ordóñez, N., Schneiders, H., Nijbroek, K., Wittenberg, A. H. J., Hofstede, R., García-Bastidas, F. A., Sørensen, A., Swennen, R., Drenth, A., Stükenbrock, E. H., Kema, G. H. J. and Seidl, M. F. 2024. Segmental Duplications Drive the Evolution of Accessory Regions in a Major Crop Pathogen. *New Phytologist* xx: xxx-xxx, in press.

Volkaert, H. 2011. Molecular analysis reveals multiple domestications of edible bananas. *Proceedings of the V International Symposium on Banana: ISHS-ProMusa Symposium on Global Perspectives on Asian Challenges* 897, 143–52.

Vuylsteke, D. 2000. Breeding bananas and plantains: from intractability to feasibility. *Proceedings of the I International Symposium on Banana: I International Conference on Banana and Plantain for Africa* 540, 1996, 149–56.

Welleman, J. C. C. and Verwer, F. L. 1983. Successful growing on Grodan rockwool in the Netherlands. *Proceedings of the International Symposium on Substrates in Horticulture Other than Soils In Situ* 150, 583–8.

Wu, K., Chen, W., Yang, S., Wen, Y., Zheng, Y., Anjago, W. M., Yun, Y. and Wang, Z. 2019. Isolation and identification of *Fusarium oxysporum* f. sp. *cubense* in Fujian Province, China. *Journal of Integrative Agriculture* 18(8), 1905–13.

Yadav, K., Damodaran, T., Dutt, K., Singh, A., Muthukumar, M., Rajan, S., Gopal, R. and Sharma, P. C. 2021. Effective biocontrol of banana Fusarium wilt tropical race 4 by a *Bacillus* rhizobacteria strain with antagonistic secondary metabolites. *Rhizosphere* 18, 100341.

Yuan, J., Raza, W., Shen, Q. and Huang, Q. 2012. Antifungal activity of *Bacillus amyloliquefaciens* NJN-6 volatile compounds against *Fusarium oxysporum* f. sp. *cubense*. *Applied and Environmental Microbiology* 78(16), 5942–4.

Zhang, H., Mallik, A. and Zeng, R. S. 2013. Control of Panama disease of banana by rotating and intercropping with Chinese chive (*Allium tuberosum* Rottler): role of plant volatiles. *Journal of Chemical Ecology* 39(2), 243–52.

Zheng, S. J., García-Bastidas, F. A., Li, X., Zeng, L., Bai, T., Xu, S., Yin, K., Li, H., Fu, G., Yu, Y., Yang, L., Nguyen, H. C., Douangboupha, B., Khaing, A. A., Drenth, A., Seidl, M. F., Meijer, H. J. G. and Kema, G. H. J. 2018. New geographical insights of the latest expansion of *Fusarium oxysporum* f. sp. *cubense* tropical race 4 into the Greater Mekong Subregion. *Frontiers in Plant Science* 9, 457.

Chapter 6

Towards sustainable management of Fusarium wilt of banana

André Drenth, The University of Queensland, Australia; and Gert H. J. Kema, Wageningen University, The Netherlands

1 Introduction

The rapid growth of the banana industry in Latin America in the latter part of the nineteenth century and early part of the twentieth century was made possible through improvements in logistics such as the building of railways in Central America, rapid sea transport to North America on specialized boats called reefers and an expanding railway and commercial distribution network throughout the USA. This system allowed for large amounts of bananas to rapidly and effectively be shipped from plantations in the tropics to consumers in the temperate zones, first in North America and then to Europe. Early attempts to ship banana varieties from the tropics to consumers in the temperate zones had been disappointing due to a lack of temperature control and uneven and early ripening of the fruit of some banana varieties. Once these issues were brought under control, the banana industry build an effective supply chain based on a single banana variety, 'Gros Michel', with great success due to its high productivity in plantation settings, large fruit, thick skin, extended green life, slow ripening, excellent transportability and good taste. These favourable

http://dx.doi.org/10.19103/AS.2022.0108.06

characteristics fuelled the development of a rapidly expanding monoculture of Gros Michel bananas grown in large-scale plantations near transport infrastructure in Latin America. Although ideal from a logistics and business perspective, enabling a constant supply and a consistent consumer experience, from a plant pathology perspective, it was a very high-risk strategy. The lack of genetic diversity in a crop grown in monoculture in large plantations over large areas is a prime example of how to create a high level of genetic vulnerability (National Research Council, 1972). The fact that genetic vulnerability is a real and major risk to food production is demonstrated by Fusarium wilt of banana (FWB), previously by the race 1 driven epidemic in Gros Michel in Latin America and currently by the dissemination of tropical race 4 (TR4).

2 The Fusarium race 1 epidemic in Gros Michel

FWB was initially caused by the so-called race 1, which originated in Southeast Asia (Ploetz and Pegg, 1997), and it has been hypothesized that causal agent(s) moved with asymptomatic bananas of the variety Silk from Asia to Jamaica sometime in the nineteenth century (Stover, 1962). From there, it slowly spread with contaminated planting material in the region, and the first major problems were noted in the Gros Michel plantations on the Atlantic coast of Panama and Costa Rica in the 1890s (Stover, 1962; Marquardt, 2001; McKenny, 1910). From this bridgehead, race 1 spread quickly to other banana-producing countries such as Suriname (1906), Trinidad and Tobago (1907), Cuba (1908), Guatemala (1910), Honduras (1916), Nicaragua (1919), Colombia (1929) and Venezuela (1930) in Central and South America (Ploetz, 2019) and started to have a devastating effect on the productivity of the large monoculture plantations of Gros Michel (Drenth and Kema, 2021; Marquardt, 2001; Ploetz, 2019; Soluri, 2005; Stover, 1962).

The massive destruction of the Gros Michel plantations led to various studies concerning the cause of the disease (Ashby, 1913), and a link was made to an earlier report by Bancroft (Bancroft, 1876), who first reported the disease from Australia and identified it as a Fusarium disease (Pegg et al., 1996). The epidemic caused havoc in plantations with the highly susceptible Gros Michel, and no effective control methods were identified despite large-scale efforts, such as flooding large plantations, to kill the fungus. Although this looked promising at first, the replanted fields very quickly succumbed to FWB again after replanting. Since no effective control methods were found, the banana companies at the time, and United Fruit Company in particular, saw one solution in moving to new areas, clearing the land and establishing new plantations. However, due to the classical problem of the lack of consultation of plant pathology expertise involved in critical decision making, they typically brought the disease with them in the planting material, equipment or soil and had to abandon these new plantations after a few years when they also collapsed due to FWB. This was a costly exercise that left a trail of economic, environmental and social destruction, and it has been estimated that over 40 000 ha of banana plantations had been abandoned by 1958 (Wardlaw, 1961). This shifting of cultivation and the need for clean land gave rise to corruption and political interference by the banana companies which turned the young democratic countries of Central America into what is often called 'banana republics' (Koeppel, 2008; Marquardt, 2001). It has been estimated that the FWB epidemic damage was over US$2.7 billion expressed in today's value (Ploetz, 2005), and this brought the Gros Michel-based banana industry to its knees in the late 1950s.

3 Infection and development of disease symptoms

Bananas are affected by a range of different plant diseases, of which FWB, also known as Panama disease, is one of the most important. Infection by the causal *Fusarium* spp. takes place in the secondary and tertiary roots (Wardlaw, 1961) which produce exudates that stimulate the germination of chlamydospores present in the soil (Ploetz and Pegg, 1999). The germ tubes of the spores penetrate the epidermal cells near the root tips and grow intercellularly along the elongation zone through the root cortex after which they enter the xylem vascular elements (Warman and Aitken, 2018). Once in the xylem, the fungus forms mycelium and microconidia to further colonize the xylem towards the rhizome. In resistant bananas, colonization is halted in the roots through the activation of a resistance response in the plant, preventing further ingress of the rhizome (Machardy and Beckman, 1981).

Once the fungus manages to move from the root into the rhizome, it can rapidly colonize the pseudostem through the formation of microconidia which are able to move quickly within the xylem vessels (Pegg et al., 2019; Trujillo, 1963; Machardy and Beckman, 1981). The upward movement of the fungus in the pseudostem is relatively quick, but the movement of microconidia upwards is restricted by perforation plates in the xylem. Recent research suggests that in addition to mycelial growth in the xylem, microconidia germinate, and their germ tubes facilitate passing through the perforation plates enabling recurrent conidia production for further upward movement (Warman and Aitken, 2018).

The pathogen acts more like a hemibiotroph during the early stages of infection, while it colonizes the vascular system in the pseudostem without causing external symptoms (Dita et al., 2018; Ploetz, 2019). However, at some stage, the plant starts to react to the invasion and tries to halt the fungus through the formation of gels, which enable the formation of phenolic compounds containing tyloses in an effort to contain the pathogen (Pegg et al., 2019). This process leads to the obstruction of the vascular system, which reduces the transport of water, resulting in the expression of external symptoms of water stress. Since bananas have a vascular capacity twice or thrice their need for growth, water stress symptoms only occur after significant progress of the disease and, hence, substantial reduction of water transport through the xylem vessels in the pseudostem (Beckman, 1990).

External symptoms of FWB include the formation of faint yellow streaks at the base of the petiole of the oldest leaf (Pegg et al., 2019) and chlorosis at the margins of the older leaves (Fig. 1a). These leaves eventually collapse at the junction of the petiole and the pseudostem, leaving a skirt of dead leaves draped around the plant (Fig. 1b). The pseudostem is also often split at the base (Fig. 1c). One of the most characteristic symptoms is the internal discolouration of the vascular system (Fig. 1d and e). Yellowish-to-brown and maroon-to-black lines may also be visible in the corm when cut (Fig. 1f). Discoloured vascular

Figure 1 Symptoms of Fusarium wilt in banana. (a) Yellowing and chlorosis at the margins of the oldest leaves; (b) leaves collapsed at the junction of the petiole and the pseudostem leaving a skirt of dead leaves draped around the plant; (c) splitting of the pseudostem at the base; (d) internal discolourations of the vascular system in the pseudostem; (e) close-up of vascular discolouration in the pseudostem; and (f) yellowish-to-brown and maroon-to-black lines in the corm when cut.

strands may also be visible in the rachis, and as a result of these combined symptoms affected plants rarely produce marketable bunches.

When the disease symptoms are fully expressed, and the banana plant is under significant water stress, the fungus also affects the parenchyma and the cortex and invades these tissues which are subsequently severely weakened. Here, large numbers of conidia and chlamydospores are produced and released into the soil when the plant collapses (Pegg et al., 2019). During the advanced stage of the disease, hyphae and sporodochia have also been observed protruding from stomata in the leaf sheaths, potentially providing an additional source of air and/or splash-borne inoculum (Warman and Aitken, 2018).

4 Efforts to manage Fusarium wilt of banana

4.1 Cultural control

The development of FWB in the field is to a large degree determined by the resistance level of the banana variety, the Fusarium race, and the environmental conditions prevailing in the plantation such as drought, flooding, heat stress and soil health, including microbial diversity and availability of nutrients as well as soil structure. Although the growth and proliferation of the pathogen occurs more rapidly under optimal conditions, the expression of FWB symptoms is accelerated under drought stress (Pegg et al., 2019). Drainage is important as wet soils and high rainfall have a negative effect on the production of bananas due to low oxygen levels in the soil, resulting in the death of root tips (Turner et al., 2007) and thus making them more prone to infection (Aguilar et al., 2000).

Optimal nutrition and appropriate fertilizer use promote plant health and development, which is important to maintain productivity. The use of nitrogen fertilizers and the pH of the soil have notable effects on the development of FWB (Simmonds, 1966). It has been observed for a long time that disease problems are higher in sandy soils with low pH. Disease-suppressive soils in Latin America tended to have a neutral or slightly higher than neutral pH (Stover, 1962; Segura et al., 2022). Hence, adjusting the pH to close to 7 was often recommended. Different soil types have been reported to have an impact on the development of disease in Central America. These soils were labelled at the time as 'short life' and 'long life' soils, and although there is a difference in the development of disease, it is not sufficient to control FWB (Stover, 1962). The mechanism behind what are now called suppressive soils has been a point of debate for a long time, and it is hypothesized that several abiotic soil factors affect saprophytic soil microflora which play a role in the suppression of soil-borne diseases (Cook and Baker, 1983; Mendes et al., 2011; Segura-Mena et al., 2021; Teixeira et al., 2021; Taniguchi et al., 2023). Although the management of nutrients in a plantation has an influence on the rate of development of disease, it is clear that changes in the pH and soil nutrients are not able to control FWB (Pegg et al., 2019).

4.2 Fallowing and crop rotation

Fallowing, either bare fallow or flood fallow, was trialled extensively in Latin America to revive FWB-affected plantations. Where possible in tropical lowlands, flooding reduced the pathogen population significantly but at the same time also reduced the populations of most microorganisms in the soil, creating a biological vacuum which was often quickly recolonized by Fusarium after the flood fallow period. In all cases, the pathogen population re-merged quickly after replanting with susceptible Gros Michel plants (Stover, 1962). It is unclear whether these infections were the result of spores which had survived in the soil or whether the fields were replanting with asymptomatic but infected plant materials. At that point in time, tissue culture of banana plants was not available, and hence only suckers from other plantations were used. Later research in Taiwan showed that 30–40% of asymptomatic suckers taken from an infected field harbour the fungus (Su et al., 1986). The flooding of thousands of hectares of infested land had large ecological and environmental impacts but did not result in effective long-term control (Stover, 1962).

4.3 Crop rotation and intercropping

Crop rotation is often used as an effective tool to reduce the impact of soil-borne diseases. However, in the case of FWB, the intercrop must be chosen wisely as it has been shown that rotation with, for example, sugarcane did not reduce disease incidence, while paddy rice reduced disease incidence significantly (Su et al., 1986). The use of groundcovers such as pinto peanut has been shown to have a small effect on disease reduction in the field (Pattison et al., 2014). From a productivity point of view, crop rotation and intercropping improve soil health and structure while reducing the population numbers of slow-multiplying pests, such as nematodes. Intercropping and vegetating the inter-rows also significantly reduce runoff and soil erosion, yielding significant environmental and sustainability benefits (Pattison et al., 2014). However, these factors are not sufficient to reduce the long-term impact of aggressive strains of the polycyclic causal agents of FWB in perennial banana plantations with very limited options for crop rotation (Ploetz, 2015).

4.4 Chemical control

Several soil fumigants have been trialled in the past with limited success, including methyl bromide against the so-called subtropical race 4, which is another Fusarium lineage, in South Africa (Herbert and Marx, 1990). Although some success has been achieved with annual crops, no effective control has been achieved in perennial crops such as bananas. Some results have been documented from India with the application of the highly systemic fungicide carbendazim (Lakshmanan et al., 1987). However, this product has now been banned in many countries around the world due to exposure risk.

Despite much research and screening of large numbers of products, it has recently been concluded that Fusarium wilts are in general not successfully controlled through the application of fungicides (Pegg et al., 2019). Fungicides may, however, play a role in eradication programs where symptomatic plants are injected in conjunction with a herbicide such as glyphosate, which is used to kill the plant and prevent any regrowth from the corm. However, it has been observed recently that the use of herbicides gives rise to senescence of the plant tissue which stimulates rapid colonization and the production of chlamydospores in the senescing plant tissues, thereby potentially increasing the inoculum load (Warman and Aitken, 2018; Pegg et al., 2019). Further research is needed to look at effective ways to kill banana plants while killing the fungus to prevent the accumulation of inoculum.

4.5 Biological control

Despite significant efforts over a long period of time and many claims of efficacy of biological control agents in laboratory or glasshouse environments, none have achieved effective control of FWD under field conditions (Bubici et al., 2019). There are many knowledge gaps in the development and application of biological control agents, and although some may reduce the incidence of disease to a certain degree, none have shown to be able to reduce the disease incidence to economic insignificant levels. Biological control is sometimes seen as a component of an integrated management approach, which comprises numerous different practices which collectively result in a significant reduction of the inoculum load in the soil and the incidence of disease to such a degree that a banana variety susceptible to FWD can be grown in a plantation setting (Dita et al., 2018). However, data from multiyear field trials to support such claims are not available (Bubici et al., 2019). A common problem is that there are unrealistic expectations concerning the efficacy of biological control agents for a polycyclic disease in a perennial crop like bananas. These production systems leave very little room for less-than-perfect efficacy, as even a low incidence of disease can quickly grow into an unmanageable problem. Although the efficacy of biological control agents and soil amendments is often backed up by experimental data, this is commonly obtained using potted banana plants over a short period of time, such that the data have little, if any, relationship to the efficacy one would experience in the field for this polycyclic disease in a perennial crop with a lifespan of 5 years or more (Ploetz, 2015).

4.6 Exclusion

Due to the difficulty in managing FWB using chemical, cultural or biological control, there are only a few alternatives left. Experience over the last century has shown that there are presently only two effective ways to control the disease. The first is exclusion of the pathogen and the second is large-scale deployment of effective resistance.

Preventing the introduction of the pathogen relies on a range of different activities and strategies to be able to enjoy producing bananas in the absence of the pathogen. The benefits of not having a disease, called exclusion benefits, have been calculated for a number of banana diseases in Australia, including FWB caused by TR4 (Cook et al., 2015). The value of the concept of exclusion benefits is to highlight the importance of the obvious fact that a disease is easiest to manage if you do not have it. It reflects the economic benefits of the absence of disease and gives insight into the financial benefits of quarantine and exclusion methods. However, the age-old problem with exclusion approaches has always been a poor understanding of risk in many industries and deploying scarce resources to prevent problems not yet experienced by industry or consumers. The authors of this chapter have observed that every plant disease starts somewhere small and at that point is ignored as it is of no significance. A little more attention is paid when the disease starts causing problems at a regional level somewhere in the world, at which stage it is often classed as an emerging disease. At this stage, investment – if any – in the disease problem is still very low as it is considered not a major problem. Once the disease starts spreading to multiple locations, a bit more attention is paid to the disease in the form of meetings and raising awareness, but resources to study the mode of spread and epidemiology of the disease are hard to come by. The next stage is an invasion of the pathogen in a country with a major production base at which point more attention is paid and some resources are allocated to research focussed on finding ways to 'manage' the disease, while little to no attention is paid to the epidemiology and eradication of the pathogen. Only when a disease starts spreading in a major production area and has some serious impact on production, industries push the panic button and resources start to flow to solve the problem, at which point it is too late and the growers have to deal with the consequences of a pathogen like FWB. This scenario has played like a broken record again and again in almost all banana-growing countries with regard to almost all banana diseases with disastrous consequences with regard to FWB caused by race 1 in the past and by TR4 at present (Drenth and Kema, 2021).

Exclusion can be implemented at different levels such as continent, country, regions, individual plantations and blocks within a plantation. The best approach is to have exclusion strategies in place at all levels, but this is often difficult. Often the exclusion is active one level below from where FWB is.

The spread of FWB in Taiwan has been studied in great detail (Su et al., 1986) and used as a recent case study (Pegg et al., 2019). Since FWB is a polycyclic disease with many generations during the production cycle, the progress of its epidemic spread follows a sigmoid disease progress curve. Considering this, it is very important to keep the amount of available inoculum as low as possible to keep the development of the disease in the lag phase.

Keeping the inoculum as low as possible and preventing further spread of the disease can be managed in a number of different ways. The recent experience after the invasion of TR4 into North Queensland in Australia has been to destroy the initial plantation and take out of production the whole plantation at this initial stage of invasion. The next phase involved taking out of production whole blocks when an infection was found on a second plantation. The reason for removing whole blocks, and not only a small section, or circle, around symptomatic plants is that the absence of symptoms in a block does not mean the absence of disease. The incubation period of the pathogen in bananas is quite long, especially at low levels of inoculum. While in the past, neighbouring plants in a circle may have been removed within a certain distance of the infected plants, this approach does not make a lot of sense in modern plantations with mechanized traffic in the inter-rows with tractors and other machinery which will effectively increase the risk of moving the pathogen along the rows and not between rows. The exclusion approach described earlier has slowed the spread of TR4 in Northern Australia significantly.

The destruction of infected plants to prevent the further spread of the pathogen has not been investigated in great detail and in some cases has given rise to the spread of the pathogen. Early attempts involved burying infected pseudostems (Su et al., 1986), which turned out to be very labour intensive and ineffective. In some cases, growers in Taiwan disposed of affected pseudostems by throwing them in irrigation ditches and thereby exacerbating the problems. In other countries, methods used for control of the bacterial wilt disease Moko, such as rice hull burning, were recommended (Molina et al., 2010). These recommended methods were untested and, due to insufficient heat penetration at depth and the handling and movement of plant material, people and soil, have been shown to be ineffective (Ploetz, 2015; Ploetz, 2019; Salacinas, 2019).

It is interesting to note that after more than a century of FWB invasions in banana plantations involving three different races, no highly effective destruction methods have been developed and tested which could be used to contain the disease. In the past, benzimidazole fungicides, including benomyl and carbendazim, which are highly systemic, have been shown to have an effect in the field (Lakshmanan et al., 1987) but are banned in most countries around the world for health and safety reasons. Although several fungicides are quite effective at inhibiting the growth of *Fusarium in vitro*, such as prochloraz and propiconazole, they have shown to be ineffective under field conditions, most likely due to their lower level of systemic ability within the banana plant.

A local exclusion strategy is currently being implemented in North Queensland, Australia. In 2015, TR4 was detected on one farm on which the plants were eradicated and the whole farm was taken out of production to reduce the risk of further spread. Following this, a schedule of inspections on nearby or connected farms was implemented. This system relies on the early detection of infected plants through regular surveys and testing of plants showing even the slightest wilting symptoms. If testing yields a positive result, the plant and a certain number of surroundings plants or in some cases whole fields are injected with a herbicide to prevent further production of inoculum and release into the soil. The positive plant and its immediate neighbours are carefully cut up above soil level and placed in strong plastic bags to prevent inoculum release in the environment. The plants are not dug up to prevent disturbing the soil and erosion during heavy rain, but the rhizome is gouged out and urea is placed in the cavity to promote decomposition. Urea is then applied in a high dose to the affected area prior to covering it with a strong plastic sheet. Urea is added in a high dose to the soil (1 kg/m^2) and to the bagged plant material as urea produces volatile ammonia after hydrolysis by urease, an enzyme which is present in many soil-borne organisms, root tips of bananas and Fusarium itself (Sequeira, 1963). In addition to the short-term production of ammonia, the accumulation of nitrite has also been shown to have an antifungal effect (Löffler et al., 1986). This method has been used to slow the spread of TR4 in North Queensland with some success as after 6 years the disease has only spread a small distance to some neighbouring plantations (Pegg et al., 2019).

5 Why is Fusarium wilt of bananas so difficult to control?

The epidemic of Fusarium race 1 in Gros Michel and the TR4 epidemic in Cavendish have made it clear that effective control of FWB in susceptible cultivars in a large-scale monoculture plantation environment is not achievable using biological, chemical or cultural measures (Pegg et al., 2019). In some other crops, Fusarium wilt can be managed, which raises the question of what the biological as well as anthropogenic reasons are behind this apparent inability to manage FWB in bananas. In this regard, we like to make the following statements:

1 The ability of a very small number of spores to start an epidemic in a perennial banana crop;
2 The presence of a polycyclic disease in a perennial cropping system requires a very high level of efficacy of any control measure to be effective over the life of a banana plantation;

3. The inability to eradicate Fusarium from the soil once an infestation has taken place;
4. A very long period of time an infested field remains contaminated. It has been reported that even after 40 years the soil is still infectious due to the ability of the causal fungus to survive in the soil short term as conidia and long term as chlamydospores in dead plant tissues or as an endophyte in a wide range of grasses and weeds which then act as a reservoir of inoculum over many years (Pegg et al., 2019; Ploetz, 2015);
5. The relatively long incubation time which enables the fungus to multiply in the plant without any external symptoms, thereby escaping early detection;
6. The ability of the fungus to be present in asymptomatic suckers which may be used as planting material;
7. The inability of intercropping to significantly reduce the disease incidence;
8. The ineffectiveness of soil fumigants to reduce the inoculum load in the soil. The fungus is present in the xylem of the banana plant and chlamydospores are produced in large numbers high in the pseudostem. These spores are encapsulated in the decaying plant tissue, which often renders soil fumigation ineffective. In addition, the spores can survive at depth in the field and start a new epidemic which means that options such as soil fumigation, solarization, anaerobic soil disinfestation, chemical fumigation and biofumigation have typically not been effective;
9. Underestimation of the impact of this soil-borne disease. The often small and local initial infection is not dealt with effectively as it is seen as a manageable problem by some in the industry until it keeps growing and destroys the industry in whole regions or countries. Clearly, the lessons from the impact of the race 1 epidemic have not been heeded nor enacted upon, such that we are repeating history;
10. The lack of effective eradication strategies once an invasion has taken place;
11. The lack of genetic diversity and absence of resistance to race 1 and TR4 in large-scale monoculture plantations of first Gros Michel and then Cavendish, respectively. Combined with an industry supply chain focused on a single banana variety this leads to an extremely high level of genetic vulnerability to existing and emerging pathogens, spreading around the world at an ever-increasing pace; and
12. The underinvestment in research in pathology, genetics and breeding at a global level has given rise to last-minute disaster management after

the invasion of a new pathogen has taken place. Research with a much longer delivery horizon is required to stay ahead of the pathogens.

6 The solution that created a future problem

The race 1 epidemic in Latin America gave rise to research into managing FWB with the aim to continue to grow a susceptible variety such as Gros Michel. As remarked by the United Fruit scientist John Johnston in 1923 'after 20 years of study and treatment, using all the customary methods used in the practice of disease control, have failed to yield a solution to the problem of Panama disease' (Soluri, 2005). The lack of effective control options led to the ongoing spread of the pathogen and the banana producers continually moving to new land, resulting in great economic, environmental and social costs. Based on the results from soil-based experiments on previously abandoned plantations, the United Fruit scientist Dr Vining Dunlap started a series of experiments using flood fallowing in 1939. After he claimed some initial success with this method, extensive flood-fallowing operations commenced in Honduras and elsewhere on thousands of hectares (Soluri, 2005).

Since managing FWB in plantations had failed, the only option at hand was shifting the cultivation of Gros Michel to new un-infested land. This strategy bought some time but ultimately failed due to the inability to secure clean land, the continuous recurrence of FWB in these new plantations and the increasing costs of infrastructure. A resistant replacement was needed, and the banana industry was looking for a banana variety that is resistant to FWB, has a high yield, has a long green life of the fruit enabling long-distance transport and is acceptable to the consumer. The first modest trials with the variety 'Cavendish' took place by the Standard Fruit company in 1943, and although there was excellent resistance to FWB, the bunches did not ripen naturally and required temperature control and the use of ethylene gas to ripen properly. It took until 1956 before Standard Fruit made the decision to convert all of its production to Cavendish. Initial market acceptance was poor, prices were low compared to Gros Michel and the thin-skinned Cavendish bananas suffered badly from bruising during transport. Although already used since the 1930s, boxes were introduced, and the fruit was cut, de-handed, washed and cooled in the plantation in packing sheds prior to shipping to overseas markets where self-service supermarkets preferred this new boxed product. The United Fruit company followed Standard Fruit's move to Cavendish 3–4 years later due to the increased costs of finding new clean land to grow Gros Michel (Soluri, 2005).

Although this switch from Gros Michel to Cavendish stopped the FWB problem in the short term, the intrinsic vulnerability of growing a single clone in monoculture on a global scale was not addressed and no diversification took place. This led one of the leading banana researcher at the time, Dr Harry

Stover, to state that the banana industries were 'extremely vulnerable to a new disease, especially a tropical race of Fusarium wilt that could devastate the basis of the industry – the Cavendish varieties' (Stover, 1986).

7 The rise of Fusarium wilt in bananas caused by tropical race 4

It is of interest to note that since Stover's statement the vulnerability of Cavendish plantations is under the spotlight again. Before this statement was made, plantations of Cavendish in Taiwan were already succumbing at an alarming rate to FWB caused by a new strain of the fungus (Su et al., 1977; Su et al., 1986) which soon started spreading in Southeast Asia and was devastating newly established Cavendish plantations in Indonesia and Malaysia and given the designation Tropical Race 4 (TR4) (Pegg et al., 1993). Soon after, this new race was found in China (1996), Northern Australia (1997), Philippines (2005) before it was found further afield in the Middle East (Lebanon, 2012), Africa (Mozambique, 2013) and Latin America (Colombia, 2019; Peru, 2021; Venezuela, 2023) (Drenth and Kema, 2021; Viljoen et al., 2020; van Westerhoven et al., 2022). Unfortunately, this time no clear replacement variety is waiting in the wings. At present, investments are being made to revive breeding programs and into novel techniques to engineer disease resistance which are undergoing field testing and are awaiting consumer acceptance (Dale et al., 2021).

8 The strong arm of resistance

The resistance of Cavendish against race 1 is durable. It has endured and has never been broken through mutations in race 1 strains, despite the huge selection pressure. Hence, resistance against FWB is extremely stable and durable. It is not the versatility of Fusarium which is causing the current problems, but the emergence and spread of a new strain that may have been existing for quite some time in the centre of origin (Ordonez et al., 2015; Maryani et al., 2019). In effect, it is now known that the strain affecting Cavendish may be considered a new species, *Fusarium odoratissimum* (Maryani et al., 2019). In the centre of origin, new pathogenic strains are expected to evolve and be able to successfully colonize the seeded wild bananas growing throughout the jungles of Southeast Asia. It has been hypothesized that this arms race with the host maintains sexual recombination in fungi under the red queen hypothesis originally stated in 1973 (Van Valen, 1973). Although it was long believed that FWB-causing Fusarium fungi are reproducing strictly asexual, recent research using multiple genome comparisons has shown that mating types are present and that sexual reproduction occurs within many phylogroups of Fusarium (McTaggart et al., 2021). Conclusions concerning the mode of reproduction are often based on fungi collected during disease outbreaks which significantly skew the data in favour of asexual reproduction while ignoring the process of sexual reproduction which most likely created these clones in the first place (Drenth et al., 2019).

Breeding of bananas started a century ago in 1922 in Trinidad and was aimed at creating a Gros Michel resistant to FWB (Wardlaw, 1961). At present there are 12 banana breeding programs around the world (Smith and Pillay, 2021), and different breeding objectives, strategies and approaches are being used from conventional breeding strategies involving evolutionary approaches (Amorin et al., 2021), reconstructive approaches (Bakry et al., 2021) as well as unconventional approaches involving somaclonal selection (Khayat, 2021), private and public genetic modification efforts (Dale et al., 2021).

Somaclonal variation has also been extensively exploited in bananas and has given rise to shorter-stature plants, improved fruit quality and higher productivity. The continuous selection of somaclones of Cavendish in many subsequent rounds has also given rise to clones with reduced susceptibility to TR4. The Taiwan Banana Research Institute has had some success with this approach, and several Giant Cavendish Tissue Culture Variants (GCTCV) clones have been produced over the years, such as GCTCV 218 (Formosana) and GCTCV 219 (Hwang and Ko, 2004). However, these plants are often taller and have a longer crop cycle and some have poor bunch characteristics. Although they are less susceptible to TR4, they still succumb to the disease and may in effect make it worse long term by multiplying the amount of available inoculum in the soil. In the belief that they have solved the problem, growers stop active removal of infected plants and continue business as before, which worsens the situation. As such, these somaclonal variants may at best produce a short-term solution to the problem. The other point to make here is that banana growers have to be careful that no planting material is taken for propagation from these as the fungus may still invade this material and thus may lead to the further spread of FWB and in effect make the problem bigger instead of smaller.

The race 1 epidemic in Latin America stimulated foundational breeding programs (Smith and Pillay, 2021) (Martínez de la Parte, 2023, Chapter 4, this volume). First, breeding started in Trinidad at the Imperial College of Tropical Agriculture where wild bananas were crossed with Gros Michel (Wardlaw, 1961). The United Fruit Company started its own breeding program in Honduras in 1959, and after a significant amount of pre-breeding among several diploid bananas several varieties were bred. However, acceptance of varieties from these programs has been minimal. Other breeding programs started after the introduction of black leaf streak disease in Africa.

The combination of a tropical crop cultivated by subsistence farmers and by large companies for export, the sterility of the existing varieties and the standardization of the supply chain since the 1960s focussed on Cavendish has led to stagnation on the variety improvement front and a lack of research funds for genetics and pre-breeding. Products from breeding programs make up a very small percentage of global production, and both Gros Michel and Cavendish did not arise from breeding programs but are very old landraces, identified at least 200 years ago, but potentially a lot longer, by Asian farmers in tropical rainforests. If breeding is to be successful, novel genetic approaches and wild banana germplasm need to be explored and investigated in much greater detail.

9 Conclusion

The expansion of large plantations to produce bananas for export based on a single variety, first Gros Michel, followed by Cavendish over the twentieth century, combined with increases in travel and trade and movement of plant material has given rise to the global spread of pathogens such as Yellow Sigatoka and Black leaf streak disease and FWB caused by race 1 and more recently by TR4 (Drenth and Kema, 2021). Due to the uniformity of the banana crop in most parts of the world and the lack of diversification, an extreme level of genetic vulnerability has been created which now risks derailing the global banana industry.

Over the last century, several major banana diseases, either originating from the centre of origin or new encounter pathogens, such as Moko in Latin America, have invaded many banana-growing regions around the world and have added significant costs, in the form of labour and agrochemical inputs, to the production of bananas (Soluri, 2005). However, unlike some of the other pathogens such as bacterial wilts (Blomme et al., 2017) and foliar diseases such as Black leaf streak (Guzman et al., 2019), there are no effective short- or long-term solutions to dealing with FWB in perennial banana plantations. Hence, the only effective options are exclusion and/or resistance. Exclusion has been effective for a significant amount of time to contain race 1 and TR4 to areas close to its centre of origin, but the overall lack of concerted action prior to the arrival of the disease has resulted in the spread of TR4, first within Asia, then to Australia, the Middle East, Africa and recently to Latin America. Now that the FWB TR4 pathogen has established itself in all banana-growing continents, it will continue to spread locally following the bridgehead effect (Drenth and Kema, 2021; Lombaert et al., 2010).

The impending spread of TR4 must be slowed down as at present no effective resistance exists in market-acceptable banana varieties. The lack of well-characterized germplasm and research in genetics and pre-breeding by industries and large producing and consuming countries has become very evident. The conservatism across the industry and its auxiliary stakeholders is remarkable, considering the impact of past plant disease epidemics in bananas (Drenth and Kema, 2021) and the global importance of the crop. Compared to other crops of similar economic significance, banana truly can be considered as an orphan crop when it comes to the genetic research effort.

An overview of genetic improvement in bananas has recently been published (Smith and Pillay, 2021). In addition to genetic improvement through classical plant breeding methods (Amorin et al., 2021) and reconstructive hybridization approaches (Bakry et al., 2021), other approaches are currently being explored such as mutagenesis and selection for resistant somaclonal variants (Khayat, 2021), as well as identifying and inserting resistance genes into Cavendish using a genetic modification approach through insertion of a resistance gene (Dale et al., 2021) or gene editing (Wang et al., 2020). All these public and private approaches are warranted to overcome the current TR4 pandemic.

10 Where to look for further information

- A good introduction and background to bananas and their production and fusarium wilt in bananas can be found at the Promusa website: https://www.promusa.org/Fusarium+oxysporum+f.+sp.+cubense.
- A series of research articles on the spread of fusarium wilt, epidemiology, detection and control options has recently been published as an e-book: Kema, G. H. J., Drenth, A., Dita, M., Jansen, K., Vellema, S. R. and Stoorvogel, J. eds. (2021). *Fusarium wilt of Banana, a Recurring Threat to Global Banana Production*. Lausanne: Frontiers Media SA. doi: 10.3389/978-2-88966-484-9. https://www.frontiersin.org/research-topics/7653/fusarium-wilt-of-banana-a-recurring-threat-to-global-banana-production#articles.
- Wageningen University hosts a website which provides insights and links to much of the current Fusarium research in bananas: https://fusariumwilt.org/index.php/en/about-fusarium-wilt/.

11 References

Aguilar, E. A., Turner, D. W. and Sivasithamparam, K. (2000). *Fusarium oxysporum* f. sp. *cubense* inoculation and hypoxia alter peroxidase and phenylalanine ammonia lyase activities in nodal roots of banana cultivars (Musa sp.) differing in their susceptibility to Fusarium wilt. *Australian Journal of Botany* 48(5): 589–596.

Amorin, E. P., Amorin, V. B. O., Silva, Md. S., Haddad, F., Ferreira, C. F., dos Santos Serejo, J. A. and Brazil, E. (2021). Developing hybrid banana varieties with improved properties. In: Kema, G. H. J. and Drenth, A. (Eds.), *Achieving Sustainable Cultivation of Bananas (Vol. 2) - Germplasm and Genetic Improvement.* Cambridge: Burleigh Dodds Science Publishing, pp. 323–337.

Ashby, S. F. (1913). Banana diseases in Jamaica. *Bulletin of the Department of Agriculture Jamaica* 2: 95.

Bakry, F., Horry, J. P. and Jenny, C. (2021). Making banana breeding more effective. In: Kema, G. H. J. and Drenth, A. (Eds.), *Achieving Sustainable Cultivation of Bananas (Vol. 2) - Germplasm and Genetic Improvement*. Cambridge: Burleigh Dodds Science Publishing, pp. 217–256.

Bancroft, J. (1876). *Votes and Proceedings 1877. Report of the Board Appointed to Enquire into the Cause of Disease Affecting Livestock and Plants*. Queensland, pp. 1011–1038.

Beckman, C. H. (1990). Host responses to the pathogen. In: Ploetz, R. C. (Ed.), *Fusarium Wilt of Banana*. St Paul, MN: APS Press, pp. 93–105.

Blomme, G., Dita, M., Jacobsen, K. S., Pérez Vicente, L., Molina, A., Ocimati, W., Poussier, S. and Prior, P. (2017). Bacterial diseases of bananas and enset: Current state of knowledge and integrated approaches toward sustainable management. *Frontiers in Plant Science* 8(1290): 1290.

Bubici, G., Kaushal, M., Prigigallo, M. I., Gómez-Lama Cabanás, C. and Mercado-Blanco, J. (2019). Biological control agents against fusarium wilt of banana. *Frontiers in Microbiology* 10(616): 616.

Cook, D. C., Taylor, A. S., Meldrum, R. A. and Drenth, A. (2015). Potential economic impact of Panama disease (tropical race 4) on the Australian banana industry. *Journal of Plant Diseases and Protection* 122(5–6): 229–237.

Cook, R. and Baker, R. (1983). The nature and paradise of biological control of plant pathogens. *The American Phytopathology Society* 539.

Dale, J., Tushemereirwe, W. and Harding, R. (2021). Genetic modification of bananas: The long road to farmers' fields. In: *Achieving Sustainable Cultivation of Bananas (Vol. 2): Germplasm and Genetic Improvement*. Cambridge: Burleigh Dodds Science Publishing, pp. 339–366.

Dita, M., Barquero, M., Heck, D., Mizubuti, E. S. G. and Staver, C. P. (2018). Fusarium wilt of banana: Current knowledge on epidemiology and research needs toward sustainable disease management. *Frontiers in Plant Science* 9(1468): 1468.

Drenth, A. and Kema, G. (2021). The vulnerability of bananas to globally emerging disease threats. *Phytopathology* 111(12): 2146–2161.

Drenth, A., McTaggart, A. R. and Wingfield, B. D. (2019). Fungal clones win the battle, but recombination wins the war. *IMA Fungus* 10(1): 18.

Guzman, M., Perez-Vicente, L., Carlier, J., Abadie, C., De Lapayere de Bellaire, L., Carreel, F., Marin, D. H., Romero, R. A., Gauhl, F., Pasberg-Gauhl. and Jones, D. R. (2019). Black leaf streak. In: *Handbook of Diseases of Banana, Abacá and Enset D. Jones*. Wallingford, Oxfordshire: CAB International, pp. 41–115.

Herbert, J. and Marx, D. (1990). Short-term control of Panama disease of bananas in South Africa. *Phytophylactica* 22(3): 339–341.

Hwang, S. C. and Ko, W. H. (2004). Cavendish banana cultivars resistant to fusarium wilt acquired through somaclonal variation in Taiwan. *Plant Disease* 88(6): 580–588.

Khayat, E. (2021). Targeted improvement of Cavendish clones. In: Kema, G. H. J. and Drenth, A. (Eds.), *Achieving Sustainable Cultivation of Bananas (Vol. 2) – Germplasm and Genetic Improvement*. Cambridge: Burleigh Dodds Science Publishing, pp. 307–321.

Koeppel, D. (2008). *Banana: The Fate of the Fruit That Changed the World*. New York: Hudson Street Press.

Lakshmanan, P., Selvaraj, P. and Mohan, S. (1987). Efficacy of different methods for the control of Panama disease. *Tropical Pest Management* 33(4): 373–374.

Löffler, H. J. M., Cohen, E. B., Oolbekkink, G. T. and Schippers, B. (1986). Nitrite as a factor in the decline of *Fusarium oxysporum* F. sp. dianthi in soil supplemented with urea or ammonium chloride. *Netherlands Journal of Plant Pathology* 92(4): 153-162.

Lombaert, E., Guillemaud, T., Cornuet, J. M., Malausa, T., Facon, B. and Estoup, A. (2010). Bridgehead effect in the worldwide invasion of the biocontrol harlequin ladybird. *PLoS ONE* 5(3): e9743.

Machardy, W. E. and Beckman, C. E. (1981). Vascular wilt Fusaria: infection and pathogenesis. In: Nelson, P. E., Tousson, T. A. and Cook, R. J. (Eds.), *Fusarium: Diseases, Biology and Taxonomy* London: Pennsylvania State Univ. Press. University Park. pp. 457.

Marquardt, S. (2001). "Green havoc": Panama disease, environmental change, and labor process in the Central American banana industry. *American Historical Review* 106(1): 49-80.

Maryani, N., Lombard, L., Poerba, Y. S., Subandiyah, S., Crous, P. W. and Kema, G. H. J. (2019). Phylogeny and genetic diversity of the banana Fusarium wilt pathogen *Fusarium oxysporum* f. sp. *cubense* in the Indonesian centre of origin. *Studies in Mycology* 92: 155-194.

McKenny, R. (1910). The Central American banana blight. *Science* 31(13): 750.

McTaggart, A. R., James, T. Y., Shivas, R. G., Drenth, A., Wingfield, B. D., Summerell, B. A. and Duong, T. A. (2021). Population genomics reveals historical and ongoing recombination in the *Fusarium oxysporum* species complex. *Studies in Mycology* 99: 100132.

Mendes, R., Kruijt, M., de Bruijn, I., Dekkers, E., van der Voort, M., Schneider, J. H., Piceno, Y. M., DeSantis, T. Z., Andersen, G. L., Bakker, P. A. and Raaijmakers, J. M. (2011). Deciphering the rhizosphere microbiome for disease-suppressive bacteria. *Science* 332(6033): 1097-1100.

Molina, A. Williams, R.C., Hermanto, C., Suwanda M., Komolong, B., and Kokoa, P. (2010). *Mitigating the Threat of Banana Fusarium Wilt: Understanding the Agroecological Distribution of Pathogenic Forms and Developing Disease Management Strategies*. Australian Centre for International Agricultural Research. Final report HORT/2005/136. pp 76.

National Research Council (1972). *Genetic Vulnerability of Major Crops*. Washington: National Academy of Sciences.

Ordonez, N., Seidl, M. F., Waalwijk, C., Drenth, A., Kilian, A., Thomma, B. P., Ploetz, R. C. and Kema, G. H. (2015). Worse comes to worst: Bananas and panama disease-when plant and pathogen clones meet. *PLoS Pathogens* 11(11): e1005197.

Pattison, A. B., Wright, C. L., Kukulies, T. L. and Molina, A. B. (2014). Ground cover management alters development of Fusarium wilt symptoms in Ducasse bananas. *Australasian Plant Pathology* 43(4): 465-476.

Pegg, K. G., Coates, L. M., O'Neill, W. T. and Turner, D. W. (2019). The epidemiology of fusarium wilt of banana. *Frontiers in Plant Science* 10(1395): 1395.

Pegg, K. G., Moore, N. Y. and Bentley, S. (1996). Fusarium wilt of banana in Australia: A review. *Australian Journal of Agricultural Research* 47(5): 637-650.

Pegg, K. G., Moore, N. Y. and Sorensen, S. (1993). Fusarium wilt in the Asian pacific region. In: Valmayor, R. V., Hwang, S. C., Ploetz, R. C., Lee, S. W. and Roa N. V. (Eds.), *Proceedings of the International Symposium on Recent Developments in Banana Cultivation Technology*. Laguna, Philippines, pp. 225-269.

Ploetz, R. and Pegg, K. (1997). Fusarium wilt of banana and Wallace's line: Was the disease originally restricted to his Indo-Malayan region? *Australasian Plant Pathology* 26(4): 239-249.

Ploetz, R. C. (2005). Panama disease, an old enemy rears its ugly head: Parts 1 &2. *Plant Health Progress* 7. DOI: 10.1094/PHP-2006-0308-01-RV.

Ploetz, R. C. (2015). Management of Fusarium wilt of banana: A review with special reference to tropical race 4. *Crop Protection* 73: 7–15.

Ploetz, R. C. (2019). Fusarium wilt. In: *Handbook of Diseases of Banana, Abacá and Enset D. Jones*. Wallingford, Oxfordshire: CAB International, pp. 207–228.

Ploetz, R. C. and Pegg, K. G. (1999). Fusarium wilt. In: Jones, D. R. (Ed.), *Diseases of Banana, Abaca and Enset*. Wallingford: CAB International Publishing, pp. 143–159.

Salacinas, M. (2019). *Spot On: Managing Panama Disease of Banana in the Philippines.*: PhD thesis. Wageningen University.

Segura, R. A., Stoorvogel, J. J. and Sandoval, J. A. (2022). The effect of soil properties on the relation between soil management and Fusarium wilt expression in Gros Michel bananas. *Plant and Soil* 471(1-2): 89–100.

Segura-Mena, R. A., Stoorvogel, J. J., García-Bastidas, F., Salacinas-Niez, M., Kema, G. H. J. and Sandoval, J. A. (2021). Evaluating the potential of soil management to reduce the effect of Fusarium oxysporum f. sp. cubense in banana (Musa AAA). *European Journal of Plant Pathology* 160(2): 441–455.

Sequeira, L. (1963). Effect of urea application on survival of Fusarium oxysporum f. cubence in soil. *Phytopathology* 53: 332–336.

Simmonds, N. W. (1966). *Bananas*. London: Longmans.

Smith, M. and Pillay, M. (2021). An overview of genetic improvement in bananas over the last century. In: Kema, G. H. J. and Drenth, A. (Eds.), *Achieving Sustainable Cultivation of Bananas (Vol. 2) - Germplasm and Genetic Improvement.* Cambridge: Burleigh Dodds Science Publishing, pp. 1–28.

Soluri, J. (2005). *Banana Cultures: Agriculture, Consumption, and Environmental Change in Honduras and the United States*. Austin: University of Texas Press.

Stover, R. H. (1962). *Fusarial Wilt (Panama Disease) of Bananas and Other Musa Species*. Kew, UK: Commonwealth Mycological Institute.

Stover, R. H. (1986). Disease management strategies and the survival of the banana industry. *Annual Review of Phytopathology* 24(1): 83-91.

Su, H. J., Chuang, T. Y. and Kong, W. S. (1977). Physiological race of fusarial wilt fungus attacking Cavendish banana of Taiwan. *The Taiwan Banana Research Institute* 2: 1–21.

Su, H. J., Hwang, S. C. and Ko, W. H. (1986). Fusarial wilt of Cavendish bananas in Taiwan. *Plant Disease* 70(9): 814–818.

Taniguchi, C. A. K., Artur, A. G., de Fátima Bruce da Silva, C., Queiroz, H. M., dos Santos, J. M. R. and de Freitas Andrade, H. A. (2023). Soil abiotic attributes related to fusarium wilt of bananas. *European Journal of Plant Pathology* 165(2): 349-361.

Teixeira, L., Heck, D., Nomura, E., Vieira, H. and Dita, M. (2021). Soil attributes, plant nutrition, and Fusarium wilt of banana in São Paulo, Brazil. *Tropical Plant Pathology* 46(4): 443–454.

Trujillo, E. E. (1963). Pathological-anatomical studies of Gros Michel bananas affected by Fusarium wilt. *Phyopathology* 53: 162–166.

Turner, D. W., Fortescue, J. A. and Thomas, D. S. (2007). Environmental physiology of the bananas (Musa spp.). *Brazilian Journal of Plant Physiology* 19(4): 463–484.

Van Valen, L. (1973). A new evolutionary law. *Evolutionary Theory* 1: 1–30.

van Westerhoven, A. C., Meijer, H. J. G., Seidl, M. F. and Kema, G. H. J. (2022). Uncontained spread of Fusarium wilt of banana threatens African food security. *PLoS Pathogens* 18(9): e1010769.

Viljoen, A., Ma, L.-J. and Molina, A. B. (2020). Chapter 8: Fusarium wilt (panama disease) and monoculture in banana production: Resurgence of a century-old disease. In: Ristaino, J. B. and Records, A. (Eds.), *Emerging Plant Diseases and Global Food Security*. St Paul, Minnesota, The American Phytopathological Society, pp. 159–184.

Wang, X., Yu, R. and Li, J. (2020). Using genetic engineering techniques to develop banana cultivars with fusarium wilt resistance and ideal plant architecture. *Frontiers in Plant Science* 11: 617528.

Wardlaw, C. W. (1961). *Banana Diseases – Including Plantains and Abaca* (Vol. 648). London: Longmans, Green and Co. Ltd.

Warman, N. M. and Aitken, E. A. B. (2018). The movement of *Fusarium oxysporum* f.sp. *cubense* (sub-tropical Race 4) in susceptible cultivars of banana. *Frontiers in Plant Science* 9(1748): 1748.

Chapter 7

Minor fungal diseases of banana

Luis Ernesto Pocasangre, Director of Research – EARTH University, Costa Rica; and Alfonso Martinuz, Especialista en Agricultura de la FAO para Nicaragua, Nicaragua

1 Introduction

Bananas are affected by both foliar and soilborne pathogens, which significantly reduce productivity and thereby the income of banana growers. The most important foliar disease is black Sigatoka, caused by *Mycosphaerella fijiensis* (now *Pseudocercospora fijiensis*), which needs at least one fungicide spray per week (up to 50 sprays per year), representing an average cost varying from US$1700 to 2400/ha/year (Pocasangre et al., 2015). The most important soilborne disease is Panama disease, caused by *Fusarium oxysporum* f.sp. *cubense*, consisting of Race 1, which affects banana cultivars such as Gros Michel, Lady Finger, and many of the Pome subgroup varieties, and Tropical Race 4, which affects all Cavendish cultivars and causes enormous economic loss in banana-producing countries worldwide.

In addition to the major diseases, there is a range of minor fungal and bacterial diseases attack bananas, affecting the quantity, quality, and market acceptance of the fruit in both export and local markets. There is a limited amount of information on these minor diseases, and little research effort has

http://dx.doi.org/10.19103/AS.2022.0108.07

been directed toward managing them. Some control of minor fungal diseases is achieved as a side effect of frequent fungicide applications to manage black Sigatoka. This chapter focuses on describing the following minor diseases of bananas:

- Cordana (*Cordana musae*);
- Deightoniella leaf spot (*Deightoniella torulosa*);
- Speckle (*Mycosphaerella musae*);
- Anthracnose (*Colletotrichum musae*);
- Cigar-end rot (*Verticillium theobromae*);
- Banana blast disease (*Pyricularia angulata*);
- Freckle disease (*Guignardia musae*); and
- Black cross leaf spot (*Phyllachora musicola*).

None of these diseases is usually considered a serious problem in the banana industry. However, in particular climatic conditions and with poor management of plantations, these minor diseases can cause significant economic losses. In addition, the withdrawal of several fungicides commonly used in the banana industry is likely to see some of these minor pathogens becoming a more significant problem in the future.

2 Cordana leaf spot

Zimmerman (1902) first reported Cordana leaf spot in bananas collected from Java and Indonesia. This study identified the causal agent as *Colletotrichum musae*, a fungus now more commonly known as *Cordana musae,* while a more recent name change has been proposed as *Neocordana musae* (Hernandez-Restrepo et al., 2015). While *Cordana musae* has a global distribution, another closely related species causing Cordana leaf spot, *Cordana johnstonii*, has a distribution limited to Southeast Asia.

Cordana leaf spot is usually a minor disease, but, under certain circumstances, it can cause defoliation in plantations (Stover, 1972). Cordana affects several banana cultivars. It has the ability to affect Cavendish cultivars, but the regular fungicide applications that control black Sigatoka in plantations growing bananas for export also control Cordana leaf spots. However, it is one of the most destructive fungal diseases affecting commercially grown Horn plantain and, in particular, false Horn plantain where the disease can destroy whole leaves. Cordana is a serious problem for smallholders growing plantains and other cooking bananas belonging to the Bluggoe type (Musa ABB) for local consumption. There are no studies evaluating the damage and economic impact of the disease.

2.1 Symptoms

Classical lesions caused by Cordana leaf spot cover the wide foliar lamina and are large, oval shaped, pale brown, and necrotic in the center, surrounded by a yellow halo (Jones, 2000). Initial symptoms resemble a big spot like a diamond with concentric necrotic rings (Fig. 1a). It is very common for lesions to coalesce to form a large area covering a large leaf area of the plant (Fig. 1b).

2.2 Host susceptibility

Cordana leaf spot particularly affects the plantain subgroup (Genome group AAB) including Horn, false Horn, and French plantain. The disease has also been reported in a large group of banana varieties, including Bluggoe (ABB), Pelipita types (ABB), Apple Banana (AAB), Mysore (AAB), Gros Michel (AAA), Cavendish cultivars (AAA), and diploids like Sucrier (AA). In Central America, Cordana leaf spot also significantly affects Gros Michel cultivars grown in highland agroforestry systems, involving intercropping with coffee, which provides an ideal environment for the development of disease.

2.3 Control

In commercial banana plantations, as well as in agroforestry and organic production systems, there is no specific control for Cordana. However, in export banana plantations, the weekly fungicide application for controlling black Sigatoka also controls Cordana. In the case of the cultivation of plantain, there are normally no applications of fungicide, and control is based on pruning affected leaves (deleafing) or cutting out the affected area. In organic banana production systems intercropped with coffee and/or cocoa, the application of Bordeaux mixture to control coffee rust also helps control Cordana leaf spot.

Figure 1 Cordana leaf spot caused by *Cordana musae*. (a) Big diamond spot on plantain (Musa AAB); (b) Cordana affecting the whole leaf of the plant on Bluggoe (Musa ABB).

3 Deightoniella leaf and fruit spot

Banana fruit speckle and black tip are caused by *D. torulosa*. The same fungus also causes leaf spot in old plantations where leaves are not removed. The disease is considered a relatively minor problem, particularly in Cavendish plantations that export bananas. *D. torulosa* is also known as a saprophytic colonizer of dead Musa leaves and flowers (Jones et al., 2000).

3.1 Symptoms

Lesions in the form of small black necrotic spots with a diameter of 1-2 mm first occur on the leaves, and these may grow into larger (25 mm) oval-shaped lesions surrounded by a black border (Fig. 2a). The lesions sometimes coalesce near the edge of the leaves, forming a curtain of necrotic tissue (Fig. 2b). Tissue-cultured Cavendish plants in greenhouse conditions are very susceptible to infection, resulting in leaf blight (Fig. 2c) (Leiva et al., 2013). Deightoniella leaf spot may be confused with leaf spot caused by other pathogens. Its symptoms can be distinguished from Cordana leaf spot by darker necrotic tissue in the center of the lesion, while morphological characteristics differentiate *D. torulosa* from pathogenic fungi. In addition to the lesions on the leaves, a speckle-like spotting of the leaf petiole may also be expressed.

Figure 2 Symptoms caused by *D. torulosa* at different stages. (a) and (b) leaf blight; (c) plantlets blight; (d) fruit black tip; (e) black spot on finger fruit (Photos Dr L. Perez Vicente).

In addition to leaf symptoms, disease symptoms are expressed on the fruit as pin spotting in the field (Fig. 2e). The fungus commonly occurs on dead flower parts and under wet conditions, and in case the fruit tip is damaged by sunburn or injury, it can infect the tip of the banana and cause black tip (Fig. 2d).

3.2 Host susceptibility

Deightoniella leaf spot has been reported mainly in cultivars belonging to the Cavendish subgroup such as Grande Naine, Williams, and Valery grown in large plantations. The disease has also been reported in Enset in parts of Ethiopia (Quimio and Tessera, 1996).

3.3 Control

There are no specific chemical control measures for Deightoniella leaf spot. However, there are several cultural practices for managing the disease aimed at reducing the inoculum load, such as cutting infected leaves and removing dead leaves when the disease is very severe. Another way to manage the disease is by reducing plant density to improve aeration and reducing relative humidity.

4 Speckle (*Mycosphaerella musae*)

Leaf speckle (common speckle, fruit speckle, tropical speckle) is caused by *Mycosphaerella musae*. Several other fungi have also been reported to cause various forms of speckle known under different names in different parts of the world. Speckle has been observed worldwide (Stover, 1969) and is common in Africa, Central America, and some parts of Oceania. It has also been recorded in Australia, The Cook Islands, the Federated States of Micronesia, The Marshall Islands, Palau, and The Solomon Islands. In smallholder settings, speckle is generally considered a disease of minor importance with little impact on yield. However, *Mycosphaerella musae* can be serious in commercial plantations if the disease is not controlled.

4.1 Symptoms

Initial symptoms of the disease include small water-soaked patches, which over time yield yellow to light brown irregular blotches on the lower surface of the leaf, darken to dark purple to black, and become visible on both sides of the leaf (Jones, 2000). Blotches coalesce to produce large bleached necrotic areas. The disease develops only on leaves over 5 weeks before flowering and does not appear above the fifth or sixth youngest leaf before flowering. When infection is severe, there is a loss of leaves, resulting in small bunches that are poorly filled and uneven ripening. Spread occurs via airborne ascospores produced in small black fruiting bodies (perithecia) on old leaves. After landing on a leaf, the spores germinate and the fungus grows on the surface of leaves for 5–6 weeks before infection. Infection typically occurs mainly on the shaded underside of the leaf where humidity is highest. The fungus can also infect the fruit, affecting the quality and market acceptance of the fruit (Fig. 3).

4.2 Host susceptibility

All the cultivars belonging to the Cavendish subgroup are susceptible to speckle, but field observations indicate that the cultivar Grande Naine is more susceptible than Valery. Speckle attacks several banana genotypes and has been reported to cause disease in Lady Finger (AAB), Ducasse (ABB), and Bluggoe types (ABB). The disease has also been reported in wild-type bananas (Pond, 1960). Disease development is faster in older and injured leaves.

4.3 Control

The most important control measure is to reduce the inoculum of the fungus by removing and destroying heavily speckled leaves from plants as well as destroying old leaves that are likely to contain spores of *M. musae*. It is important to remove shade and weeds around plants to keep humidity as low as possible. It is also important to desucker plants according to established good agricultural practices for bananas; this reduces humidity around the main plant as well as produces well-formed bunches. Another important practice is to reduce the density of plants again to improve airflow and reduce shading and humidity around each plant. Although there is no specific chemical control for speckle, fungicides used to control black Sigatoka will usually reduce speckle caused by *Mycosphaerella musae*.

Figure 3 Damage of speckle on Grande Naine. (a) Damage on the bunch; (b) damage on the hand.

5 Anthracnose

Anthracnose is caused by the fungus *Colletotrichum musae*. The disease is most prevalent in the postharvest and transport stages of banana production and seldom affects fruit on the plantation (Muirhead and Jones, 2000).

5.1 Symptoms

Typical lesions caused by Anthracnose are brown and black spots on the peel of banana fruit, especially when ripening. The spots increase in size and often coalesce to form an extensive sunken area on the peel of the finger (Fig. 4a). The sunken spots can become covered with a mass of pink to rusty-red spores. Infection can also take place at the cut end of individual fruit during packing, in which case the rot advances down the fruit stalk into the flesh of the fruit. When green fruit is infected, black scars form into a diamond shape. The pulp of the fruit remains unaffected. Anthracnose is also associated with crown rot of banana (Fig. 4b) but can also be caused by species of *Fusarium*.

5.2 Host susceptibility

Most commercial cultivars are affected by *Colletotrichum musae*. This includes all the Cavendish cultivars, Grande Naine, Williams, Valery, as well as baby bananas (Pisang mas (AA)). The disease is most frequently reported during the postharvest and transportation stages. Anthracnose is less frequent in plantain (Musa AAB) as well as in Prata cultivars (Musa AAB).

Figure 4 Damage caused by *Colletotrichum musae* in Grande Naine (Musa AAA). (a) Damage on finger; (b) crown rot in ripening banana.

5.3 Control

The most important control measure for Anthracnose is the protection of the bunch and avoiding physical damage during harvest, processing, and transporting of fruit. This is because the fungus cannot enter into green fruit directly but need tissue damage, wounds, or scars on the fruit to gain entry, infect, and colonize the ripening fruit. Another efficient measure to reduce the impact of the disease is monitoring the quality of packed fruit and rejecting hands and fingers with physical lesions which pose a risk to other fruits as they may start to sporulate during transport and ripening. Because the supply chain usually has the appropriate infrastructure, technology, and quality control systems, Anthracnose can be managed effectively and typically is not a major problem affecting export bananas. However, in local markets, where the storage, transport, and management of fruit are not given adequate attention, losses from Anthracnose can be very high.

6 Cigar-end rot

Cigar-end rot is a disease affecting banana fruit. There are two pathogens associated with the disease: *Musillium theobroma* (formerly *Verticillium theobromae*) and *Trachysphaera fructigena* (Zare et al., 2007; CABI, 2022). Cigar-end rot caused by *Musillium theobroma* has a pantropical distribution while cigar-end rot caused by *T. fructigena* is only found in West and Central Africa and Madagascar.

6.1 Symptoms

The black necrosis caused by the disease starts at the end of the immature fruit surrounding the dead floral parts and then extends back along the fruit for several centimeters. Necrosis can spread with the affected tissue and becomes more or less rounded in contrast to the angular shape of the immature fruit. The fruit pulp succumbs to dry rot, and the end of the fruit becomes dry and is frequently covered by ash gray fungal spores resembling a cigar tip, explaining the name of the disease (Fig. 5a). When the fruit is infected late in its development, only the tip is affected (Fig. 5a), but when the fruit is infected soon after the emergence of the bunch, a much larger part of the finger may be affected and typically more fingers are affected increasing the severity of the disease and affecting whole hands (Fig. 5b).

6.2 Host susceptibility

Cigar-end rot affects the fruit of most commercial cultivars of the subgroup Cavendish (Grande Naine, Williams, and Valery). Cigar end is also a serious

Figure 5 Cigar end on Grande Naine (Musa AAA). (a) Initial damage in tip fingers; (b) advanced damage on the whole finger and hand.

problem in Gros Michel (Musa AAA) and is frequently observed in false Horn plantain (Musa AAB) in Central and South America. The incidence of the disease is correlated with wet weather.

6.3 Control

It is essential to move and destroy (e.g. by burying) infected fruit or other plant material to prevent spores from spreading. Removal of dead flowers early may also reduce the inoculum load remove trash, which can accumulate when bunches are bagged. Fungicides to control black Sigatoka are also effective against this disease. To reduce infection, postharvest facilities need to be kept clean and regularly sanitized, and affected fingers should not be placed in delatexing tanks but removed prior to entering the postharvest facilities.

7 Banana blast disease

Banana blast disease was originally reported as being caused by the fungus *Pyricularia grisea*, the rarely seen sexual stage of *Magnaporthe grisea* explaining the common name of the disease. However, recent studies suggest *Pyricularia angulate* as the cause (Male et al., 2011; Ganesan et al., 2017). Banana blast is very common in nurseries of tissue-cultured plants (Ganesan et al., 2017). However, there is little information on and reports of *Pyricularia* affecting commercial banana plantations, and losses from *Pyricularia* in commercial plantations do not appear to be common. A disease of the fruit called pitting disease is caused by the same fungus and thus can be considered part of the same disease. Pitting used to be a major problem in expert plantations in Central America and Southeast Asia. Infection of the fruit causes blemishes reducing the marketability of the fruit while the ability of the disease to increase in severity during transport may cause further losses.

7.1 Symptoms

Typical symptoms of blast banana disease are minute oval-shaped brownish lesions with or without a yellow halo on the leaf. The spots may coalesce to affect the whole leaf. The fungus can also infect the pseudostem of younger plants, perhaps its most frequent symptom (Fig. 6). *Pyricularia* affects fruits near maturity, producing blemishes, which may increase in size and number when they are transported to the market. Symptoms appear initially as reddish spots on the peel that later become circular, necrotic, sunken spots, surrounded by a reddish halo, and, outside this, a greasy green halo.

7.2 Host susceptibility

Most commonly grown varieties including Cavendish (AAA) cultivars, Gros Michel (AAA), Plantains (AAB), and Silk (AAB) are affected by the disease.

Blast disease is very common in Enset cultivars in highland conditions in Ethiopia.

Figure 6 Damage of *Pyricularia* on the pseudostem on Grande Naine (Musa AAA).

7.3 Control

There are no specific control measures for managing *Pyricularia* in commercial banana plantations. If plants are affected, the best practice is pruning diseased leaves to reduce the inoculum load. If, on rare occasions, the fungus attacks the bunch, the fruit must be removed.

8 Freckle disease

Freckle on banana is caused by three different species of *Phyllosticta* namely *Phyllosticta musarum*, *Phyllosticta maculate*, and *Phyllosticta cavendishii* (Wong et al., 2012).

Freckle disease is common in Southeast Asia, and there are unconfirmed reports from Latin America and Africa. In the Philippines, it was the second most important disease after black leaf streak in commercial banana plantations of cultivars of the Cavendish subgroup (Corcolon and Raymundo, 2008) (Tsai and Liu, 1993). The disease is slow to spread within plantations and has a limited area of distribution (Ploetz et al., 2015). The main effect of the disease is that it blemishes the fruit, which reduces its marketability and may restrict market access.

8.1 Symptoms

Two types of leaf spotting occur, mainly on the upper surfaces of older leaves. The first consists of very small, dark-brown to black spots, less than 1 mm diameter; these give the leaf a sooty appearance and rough texture to the touch. The spots may cluster in lines and appear as streaks caused by drops of water containing spores running diagonally across the leaf, or from the midrib to the leaf edge, often along veins.

The second type of spotting consists of larger, individual, dark-brown to black spots, up to 4 mm in diameter. These spots may have gray or fawn centers and may aggregate to form large, rough, blackened areas or streaks. Severely affected leaves wither and die prematurely. Spots may also form on petioles, midribs, and transition leaves.

Fruit is also affected by freckle. As early as 2–4 weeks after bunch emergence, a few widely scattered spots may be seen or, in some cases, dense aggregates appear in the form of streaks or circles. The severity of the disease increases as the fruit matures. Individual spots first appear as minute reddish-brown flecks surrounded by a halo of dark-green, water-soaked tissue. In addition, large areas of the fruit surface may become black due to dense aggregations of spots. Although detracting from the appearance of the fruit, eating qualities are not affected (Drenth and Wong, Chapter 3).

8.2 Control

Depending on the species of *Phyllosticta* present, utilizing resistance traits in banana plants is an option. In areas where *Phyllosticta maculata* and *Phyllosticta musarum* are not pathogenic on cultivars of the AAA genomic group, planting of non-susceptible cultivars including Cavendish may be considered. Areas currently free of *Phyllosticta cavendishii* such as Africa and Latin America should consider strong biosecurity measures. Once the disease is present, cultural control methods such as removal of infected leaves and fruit and trimming of leaves before senescence may all help to reduce inoculum in the field. Early bagging and securing the bag tightly may prevent spores to enter via the peduncle. Frequent application of fungicides to manage black leaf streak inadvertently also controls freckle present on the leaves.

9 Black cross

Black cross leaf spot is caused by the ascomycete fungus *Phyllachora musicola*. It has been reported on cooking bananas in several countries in Oceania such as Samoa, Australia, New Guinea, and Tonga (Booth and Shaw, 1961). The disease is more serious in shady conditions and can cover the whole leaf (Gerlach, 1988). Cavendish varieties are resistant to this disease and as such black cross is not a problem in commercial plantations growing Cavendish cultivars.

9.1 Symptoms

The symptoms of black cross are expressed as a black four-pointed star at the underside of the leaf (Fig. 7) where the long arms are parallel to the veins on the leaf and the vertical arms are at right angles to the vein (Fig. 7). Crosses on the leaves can occur in large numbers (Fig. 8) and spores are formed at the underside of the leaves.

9.2 Host susceptibility

Black cross can attack most types of bananas including diploids and triploids. However, cultivars containing the B genome are in general more susceptible, such as cooking bananas like Bluggoe type (Musa ABB) as well as Lady Finger (Musa AAB). It is less pathogenic in the AAA genomic group, and cultivars belonging to the Cavendish subgroup have a good level of field resistance to this disease.

9.3 Control

In commercial plantations belonging to the Cavendish subgroup, no control is needed due to resistance to the disease. There is little information concerning

Figure 7 Symptoms of black cross on cooking banana in PNG (Photo credit Andre Drenth).

Figure 8 Leaf of cooking banana affected by black cross (Photo credit Andre Drenth).

the management of this disease in other varieties, but sanitation in the form of removing affected leaves will reduce the level of inoculum as the black crosses contain small fruiting bodies called perithecia which produce ascospores on the leaf surface. These spores are easily moved by wind and rain to new leaves to cause disease.

10 Conclusion

There are many plant diseases in the tropics, and Wellman (1968) listed a total of 180 plant diseases of bananas of which 9 can be considered major diseases (Drenth and Kema, 2021). It has to be noted that almost all diseases start as minor diseases somewhere in the world. Local and intercontinental spread then follows; its impact is amplified by the cultivation of a single banana variety, Cavendish, in large-scale monoculture settings with known susceptibility to eight of the major banana diseases (Drenth and Kema, 2021). Many minor diseases still have a regional distribution such that biosecurity measures are required to keep it that way. Other minor diseases are controlled in large-scale Cavendish plantations due to the frequent application of fungicides to control black Sigatoka and/or cultural methods such as removal of older infected leaves which controls several minor leaf diseases as well. However, if these controls are stopped such as during long periods of wet weather or other environmental or external factors, some of these minor diseases have the ability to have a significant impact on the volume of marketable fruit that can be produced. Currently, the banana industry is losing several molecules widely used for controlling major diseases, which also have a control of minor diseases, so it is expected that in the future several minor diseases will become a serious problem in specific places. Another factor that must affect the impact of minor diseases in the banana industry is climate change. For instance, Cordana is becoming an important disease in the organic banana production system in the dried region, where black Sigatoka is not causing significant damage but Cordana is taking advantage of the weather conditions to increase the damage. Finally, the most important factor that can increase the impact of minor diseases is the production system itself, because if we continue with the huge monoculture in combination with standard conventional practices of plantation management, plus the effect of climate changes as a result, it is expected that minor diseases can become a serious problem. For instance, Moko (*Ralstonia solanacearum*) is becoming a predominant disease because of the spread of the bacteria by contaminated water; more frequent hurricanes and tropical rains cause significant flooding in banana plantations, allowing movement of contaminated water , either causing infection in new plantations or increasing the problem in already-infected farms. In addition, there is little research to understand the biology and epidemiology of these minor diseases; as a result, there are no specific measures to control them. Finally, it is important to stress that plant breeding programs in bananas are focused on addressing major diseases like black Sigatoka, *Fusarium*, nematodes banana weevils, and others, but when new hybrids or resistant cultivar are deployed in the field, the farmers find that the new genetic material are susceptible to minor diseases. For instance, FHIA hybrids that were very resistant to black Sigatoka were very susceptible to Cordana in certain regions. Currently, there are several new strategies to breed bananas and it is expected that minor diseases can affect the new genetic planting material.

11 Where to look for further information

There are several sources of literature where information related to pest and diseases on banana can be found: www.ars.usda.gov; www.augura.com.co /biblioteca digital; www.CABI digital library; www.catie. ac.cr; corbana.co. cr; www. cirad.fr; www. earth.ac.cr; www.embrapa.br; www.epagri.sc.gov. br; www. fao.org; www.fhia.org.hn:www.iica.int; www.iniap.gob.ec; www.idiap.gob .pa; www. iniavenezuela.negocio .site; www. inta.gob.cr; www.musali.org: www .promusa.org.

12 References

Booth, C. and Shaw, D. E. (1961). Black cross-caused by new species of *Phyllochora* on bananas. *Papu New Guinea Agricultural Journal* 13: 157–159.

Buckel, I., Andernach, L., Schueffler, A., Piepinbring, M., Opatz, T. and Thines, E. (2017). Phytotoxic dioxolanones are potential virulence factors in the infection process of *Guignardia bidwellii*. *Scientific Reports* 7(8926): 8926. doi: 10: 10.1036/s4 1598-017-09157-6.

CABI (2022). *'Trachysphaera fructigena (Cigar End Rot of Banana)', CABI Compendium*. CABI Publishing International. doi: 10.1079/cabicompendium.54408.

Corcolon, B. M. and Raymundo, A. (2008). Estimating yield losses in banana due to freckle disease caused by *Phyllosticta musarum* (Cke.) van der Aa. *Philippine Journal of Crop Science* 33(2): 75–85.

Drenth, A. and Kema, G. (2021). The vulnerability of bananas to globally emerging disease threats. *Phytopathology* 111(12): 2146–2161.

Ganesan, S., Sing, H. S., Petikam, S. and Biswai, D. (2017). Pathological status of Pyricularia angulate causing blast and pitting disease of banana in Eastern India. *Plant Pathology Journal* 33(1): 9–20.

Gerlach, W. W. F. (1988). *Plant Diseases of Western Samoa. Samoan Germana Crop Protection Project. Western Samoa*, p. 215.

Jones, D. R. (2000). Black cross leaf spot in other leaf spots. In: *Diseases of Banana, Abaca and Enset*. Edited by D. R. Jones. CABI Publishing, pp. 97–99.

Jones, D. R., Lomerio, E. Q., Tesera, M. and Quimio, A. J. (2000). Black cross leaf spot in other leaf spots. In: *Diseases of Banana, Abaca and Enset*. Edited by D. R. Jones. CABI Publishing, pp. 102–103.

Jones, D. R. and Stover, R. H. (2000). Cigar end rot. In: *Diseases of Banana, Abaca and Enset*. Edited by D. R. Jones. CABI Publishing, pp. 175–177.

Leiva, M., Alvarado Capo, Y., Suarez, M. and Cruz, M. (2013). *Deightoniella torulosa* (Syd.) M.B. Ellis, causing leaf spot in acclimatized Grande naine (Musa, AAA) plants. Berkis R and Carrera K. *Revista de Proteccion Vegetal* 28: 145–148.

Male, M. F., Tan, Y. P., Vawdrey, L. L. and Shivas, R. G. (2011). Elucidation of the taxonomy and pathological status of Pyricularia associated with banana blast in Australia. *Australasian Plant Disease Notes* 6(1): 22–25. doi: 10.1007/s13314-011-0008-8.

Meredith, D. S. (1968). Freckle disease on bananas in Hawaii caused by *Phyllostictica musarum*. *Annals of Applied Biology* 62: 328–340.

Muirhead, I. F. and Jones, D. R. (2000). Anthracnose. In: *Diseases of Banana, Abaca and Enset*. Edited by D. R. Jones. CABI Publishing, pp. 199–203.

Ploetz, R. A., Kema, G. and Jun Ma, L. (2015). Impact of diseases in export and smallholder banana production. *Annual Review of Phytopathology* 53: 269–288.

Pocasangre, L. E., Vicente, P. and Ferris, L. H. (2015). *Banana Plant Disease Management in Organic Agriculture. Chapter Book. Plant Disease Management in Organic Agriculture*. St. Paul, MN: American Phytopathological Society (Australian Psychological Society), pp. 351–365.

Pond, W. (1960). Three leaf speckle diseases of the banana in Queensland. *Queensland Journal of Agricultural Science* 17: 271–309.

Quimio, A. J. and Tessera, M. (1996). *Diseases of Enset in Proceeding of International Workshop on Enset*. Ethiopia: Institute of Agricultural Research, pp. 188–203.

Stover, R. H. (1969). *Mychosphaerella* species associated with banana leaf spots. *Tropical Agricultural (Trinidad)* 46: 325–332.

Stover, R. H. (1994). *Mychosphaerella musae* and Cercospora non virulentum from Sigatoka leaf spots are identical. *Fruits* 49: 187–190.

Stover, R. H. (1972). *Diseases of Banana and Abaca*. Longman Publishers, Harlow, UK.

Tsai, Y. P. and Liu, S. H. (1993). Freckle disease of banana in Taiwan. In: *Proceeding of INIBAP-Basnet*, Los Baños, Philippines, pp. 298–307.

Wellman, F. L. (1968). More diseases on crops in the tropics than in the temperate zone. *Ceiba* 14: 17–28.

Wong, M. H., Crous, P. W., Henderson, J., Groenenwald, J. Z. and Drenth, A. (2012). *Phyllosticta* species associated with freckle disease of banana. *Fungal Diversity* 56(1): 173–187.

Wong, M.-H., Henderson, J. and Drenth, A. (2013). Identification and differentiation of *Phyllosticta* species causing freckle disease of banana using high resolution melting (HRM) analysis. *Plant Pathology* 62(6): 1285–1293.

Zimmerman, A. (1902). Uber einige tropischer Kulturpfalnzen beobactete Pilze. *Zentralblat fuer Bacteriologie ParasitenkundeInfektionskrankheiten und Hygiene* 8: 219 (abstract).

Part 2

Bacterial and phytoplasma diseases

Chapter 8

Moko bacterial wilt of banana

Mauricio Guzmán, Cropland Biosciences, Costa Rica; Marylin Sánchez, Mycelium Crop Science, Costa Rica; Ricardo Villalta, Independent Consultant, Costa Rica; Leonardo Pérez, Independent Consultant, Costa Rica; and Claudiana Carr, National Banana Corporation (CORBANA) and Cropland Biosciences, Costa Rica

1 Introduction

Moko caused by various strains of *Ralstonia solanacearum* is an important bacterial wilt disease of bananas and plantains affecting the global banana industry (Sequeira, 1998; Eden-Green, 2019). The highest impact of Moko has been in Central and South America and the Caribbean. Outside the Western Hemisphere, it has only been found in the Philippines and Malaysia (Thwaites et al., 1999; Robinson and Galan Sauco, 2010; Blomme et al., 2017; Eden-Green, 2019) (Table 1). The first known reference to the disease was in 1840 by Schomburgk on his travels to British Guiana (Martyn, 1934; Buddenhagen, 1961), now the Co-operative Republic of Guyana. In 1911, Rorer published the first detailed study of bacterial wilt of bananas in Trinidad. He mentioned that an epidemic of the disease in 1890 almost eliminated the plantain cultivar Moko (*Musa* ABB, subgroup Bluggoe), used as shade for young cacao trees as well as a source of income before cacao is ready to harvest (Rorer, 1911), giving rise to the popular name given to the disease (Stover, 1972; Frossard, 1987).

The plantain cultivar Moko is known in Central America as Guineo Cuadrado (González, 1987) and in South America and the Caribbean by several names

http://dx.doi.org/10.19103/AS.2022.0108.08

Table 1 Countries where Moko bacterial wilt has been reported and year of record

Region/country	Year of detection	References
Latin America and the Caribbean		
Trinidad and Tobago	1895	Rorer (1911)
Guyana	1840	Stover (1972); Phelps (1987)
Colombia	1954	Buddenhagen y Elsasser (1962); Lozano et al. (1969)
Peru	1954	Quinón and Aragaki (1963); French y Sequeira (1970)
Brazil	1954	Quinón and Aragaki (1963); Stover (1972)
Hawaii	1954	Quinón and Aragaki (1963); Stover (1972)
Nicaragua	1960	Stover (1972); Lehmann-Danziger (1987)
Grenada		Stover (1972); Hunt (1987)
Mexico	1960	Stover (1972); Phelps (1987)
Belize		Stover (1972); Black and Delbeke (1991)
Suriname		Stover (1972); Power (1976)
Guatemala		Stover (1972); Phelps (1987)
El Salvador		Stover (1972); Phelps (1987)
Costa Rica	1960	Stover (1972); Phelps (1987)
Panama	1960	Stover (1972); Phelps (1987)
Venezuela	?	Stover (1972); Phelps (1987)
Honduras	1961	Buddenhagen y Elsasser (1962); Stover (1972)
Jamaica	2004	EPPO (2004)
St Vincent	2007	EPPO (2007)
Ecuador	2013	Delgado et al. (2014)
Asia		
Philippines	1978	Rillo (1979)
Malaysia	2011	Zulperi and Sijam (2014)

such as Chato, Cachaco, Topocho, Poteau and Cacambou (Haddad and Borges, 1973; Frossard, 1987). It is grown in the region both for human consumption and as animal feed. However, due to its high susceptibility to bacterial wilt, plantations have been decimated, and some countries have banned planting the plantain cultivar Moko in banana-producing areas. Significant efforts have been made to replace it with other varieties more resistant to bacterial wilt such as Pelipita (*Musa* ABB). However, smallholder farmers still prefer the Moko cultivar and continue to plant it, creating a source of inoculum with the ability to spread to commercial banana and plantain plantations.

Moko disease goes under different local names and is known as 'Hereque' in Venezuela (Buddenhagen, 1961; Vivas et al., 2009), 'Maduraviche' in Colombia (Martínez and García, 2003) and 'Bugtok' in the Philippines when affecting fruits of 'Saba' (ABB) and 'Cardaba' (Molina, 1999; Molina, 2006; Blomme et al.,

2017; Eden-Green, 2019). Moko is endemic to Central and South America, and it is a highly infectious disease caused by various strains of the bacterium *R. solanacearum*. Almost all commercial varieties of bananas and plantains are susceptible to Moko. The pathogen is easily spread by natural means (e.g. by insects) and by humans through cultural practices used in commercial plantations. The incidence of Moko has increased in the last 15 years, giving rise to significant losses and increases in the cost to control this disease.

Once the disease enters a banana plantation, control becomes very difficult requiring labour-intensive and expensive management due to the ease with which the bacteria can spread, e.g. via insect transmission and cultural practices. Large corporations with extensive plantations have been able to successfully control the Moko disease by implementing very strict sanitary and biosecurity measures which come at a considerable cost (Sequeira, 1998; Eden-Green, 2019). However, small producers lack the knowledge and resources to implement the same level of control.

2 Disease symptoms

Moko can affect any of the commonly grown bananas and plantain cultivars with similar symptoms in all varieties. Some symptoms can be confused with other biotic and abiotic disorders. However, several are very specific to Moko, and a person with experience and training can make an accurate field diagnosis. Plants are affected at all stages of growth, but symptoms are most characteristic of young, rapidly growing plants (Buddenhagen, 1961; Stover, 1972). Characteristic symptoms were first described by Rorer (1911) and subsequently by other authors (Ashby, 1928; Wardlaw, 1935; Sequeira, 1958; Buddenhagen, 1961; Stover, 1972; González, 1987; Thwaites et al., 1999; Blomme et al., 2017 and Eden-Green, 2019).

2.1 External symptoms

2.1.1 Leaves

In young, fast-growing plants, the first, second or third leaf turns pale green, yellow-green or yellow and collapses near the junction of the petiole and lamina or towards the middle (Stover, 1972; González, 1987). This is one of the most characteristic symptoms of the disease (Fig. 1). Subsequently, the lower leaves also wilt and collapse. Collapse may occur while the leaves are still green or are only slightly off-colour (Buddenhagen, 1961). Next, the edge of the leaf becomes necrotic, followed by a yellow-orange halo that progresses through the entire leaf blade, causing rapid death of the leaf until the whole plant wilts in a few weeks (Fig. 1). The cigar leaf always appears withered or 'dormant' (its development delayed or stopped). The cigar leaf becomes opaque and

Figure 1 Typical external symptoms of Moko bacterial wilt in banana (a) cv. 'Valery' (*Musa* AAA, Cavendish subgroup) and plantain (b) cv. 'Dwarf Curraré' (*Musa* AAB, Plantain subgroup) in Costa Rica. Note the yellowing and collapses of young leaves near the junction of the petiole and lamina, which is one of the most characteristic symptoms of the disease. Photos: M. Guzmán.

sometimes necrotic at the base (Fig. 2d), especially in young suckers of 3 or 4 months of age (González, 1987). In plants which already have produced a bunch and are not producing any more leaves, the wilting of the existing leaves is hard to spot; sometimes only a slight yellowing in the youngest leaves is observed (González, 1987).

2.1.2 Suckers

Unlike Fusarium wilt, suckers are rapidly infected. Younger suckers blacken or necrotize and may warp or twist, showing stunted growth (Fig. 2a). The disease is characteristically seen in regrowth suckers 2–4 weeks after pruning (Buddenhagen, 1961; Stover, 1972). Suckers are black in colour and stunted and develop only one or a few leaves before wilting completely (Fig. 2b and c). Larger suckers express similar symptoms in their leaves as fast-growing plants (Fig. 2c and d). Affected leaves tend to stop growing, deform and widen at the base (Stover, 1972; González, 1987; Eden-Green, 2019). When bacterial infection takes place through the upper part of the plant, the external symptoms do not manifest as quickly on suckers which may remain asymptomatic for some time and symptoms only are expressed as the disease progresses in the plant over time. It is important, therefore, not to use suckers as propagating materials from fields affected by Moko as even asymptomatic plants may be infected.

Figure 2 Symptoms of Moko bacterial wilt in banana suckers cv. 'Grande Naine' (*Musa* AAA, Cavendish subgroup). Note the blackening, stunt growth and deformation of a young sucker still without true leaves (a), necrosis of young leaves (b), yellowing and collapse of young leaves (c) and cigar leaf appears withered and necrotic at the base (d) in young suckers. Photos: M. Guzmán and R. Villalta.

2.1.3 Fruit

The expression of symptoms of Moko in fruit is distinctive. In young bunches (1-2 weeks), development is arrested and young fingers and flowers turn black and rot (Fig. 3a). The fruit bud is small and may wither and/or become completely necrotic (Buddenhagen, 1961; Stover, 1972; González, 1987). In more developed bunches (7 weeks or more), some fingers ripen prematurely and the peel cracks and splits longitudinally, exposing the pulp (Stover, 1972; González, 1987; Eden-Green, 2019). An apparently healthy plant may show some yellow, split-cracked (Fig. 3b-d) and sometimes blackened fingers in bunches.

2.2 Internal plant symptoms

Infection by *R. solanacearum* is systemic. Once the bacteria enter from any part of the plant, they multiply and move through the vascular tissues, invading

Figure 3 External symptoms of Moko bacterial wilt in banana fruits cv. 'Grande Naine' (*Musa* AAA, Cavendish subgroup). Young bunch with black and rot flowers and fingers (a) and yellow, split-cracked fingers in bunches with advanced development (b–d). Photos: M. Guzmán and R. Villalta.

from the roots to the leaves and fruits and causing internal lesions in all organs. As a result, some vascular strands in roots, rhizome, pseudostem sheaths, leaf petioles, midrib, bunch stalk, finger peduncles and fruit peel turn light to dark brown (Figs. 4 and 5a-d). Extensive vascular discolouration is observed in the rhizome, especially in the stele adjacent to the cortex, and extending to the lateral buds (Figs. 4a and 6b). In the pseudostem, discoloured strands in plants without fruit are concentrated immediately outside the centre, becoming less frequent peripherally and centrally. The most extensive discolouration occurs in the sheaths of wilting leaves first (Buddenhagen, 1961; González, 1987; Thwaites et al., 1999). The same symptoms occur in plants carrying fruit, where vascular discolouration is more evident in the true stem and neighbouring sheaths (Figs. 4b and 6c). Only in advanced stages does vascular discolouration progress to the outer sheaths (González, 1987). This expression of symptoms contrasts that of Fusarium wilt, where discoloured vascular strands are concentrated at the

Figure 4 Internal symptom of vascular discolouration in different organs of the banana plant cv. 'Grande Naine' (*Musa* AAA, Cavendish subgroup): rhizome (a), pseudostem (b), leaf petiole (c and d) and fruit stalk (e and f). Photos: M. Guzmán.

periphery of the pseudostem (Buddenhagen, 1961; Stover, 1972; Eden-Green, 2019).

2.3 Internal fruit symptoms

In infected fruits, the pulp looks discoloured with brown rot that turns dry grey (Buddenhagen, 1961; Stover, 1972). This symptom is known as 'mummified fingers' (Figs. 5d–f and 6d). In some cases, the infection begins in the upper part of the plant via inoculation of the male of female floral parts. When this is

Figure 5 Internal symptoms of Moko bacterial wilt in banana fruits cv. 'Grande Naine' (*Musa* AAA, Cavendish subgroup). Vascular discolouration in the crown, peduncles and peel of fruit (a, b and c), dry rot of pulp and peel splitting (d, left healthy finger and right infected finger) and pulp dry rot (e and f). Photos: M. Sánchez and M. Guzmán.

the case, cutting the rhizome or lower part of the pseudostem may not reveal vascular discolouration, indicating that bacteria have not yet colonized these tissues (Stover, 1972; González, 1987).

Because internal symptoms of Moko, mainly in the rhizome and pseudostem, are somewhat similar in plants affected by other diseases, such as Fusarium wilt, vascular discolouration in these organs is not considered a characteristic symptom of the disease. Care must be taken before making a diagnosis based on this type of symptom. All varieties of bananas and plantains susceptible to Moko show the same external and internal symptoms (Fig. 6). In identifying the disease quickly and accurately, the most characteristic external symptoms are the following:

Figure 6 External and internal symptoms of Moko bacterial wilt in plantain cv. 'Dwarf Curraré' (*Musa* AAB, Plantain subgroup) in Costa Rica. Wilting (a), vascular discolouration in the rhizome (b), pseudostem and true stem (c) and fruits with pulp dry rot (d). Photos: R. Villalta, A. Madrigal and M. Guzmán.

1 The first, second or third leaf turns yellow-green and collapses at the junction of the petiole and lamina;
2 Split and premature ripening of some fingers with cracking of the peel and internal dry rot of fruit pulp; and
3 Blackened or wilting of young suckers.

3 Pathogen characteristics and diagnosis

The taxonomy and nomenclature of bacterial wilt pathogens has changed frequently. Frequent name changes occurred as methodology evolved from phenotypic, biochemical and molecular studies to genomics and functional genomics. The bacterial wilt pathogen causing Moko has been reclassified multiple times before it was transferred to the current genus, *Ralstonia* (Paudel et al., 2020). The genus *Ralstonia* belongs to the Phylum Proteobacteria,

Class β-Proteobaceria, Order Burkholderiales and Family Burkholderiaceae, which includes seven recognized species: *R. detusculanense*, *R. insidiosa*, *R. mannitolilytica*, *R. pickettii*, *R. pseudosolanacearum*, *R. syzygii* and *R. solanacearum* (García et al., 2019).

First named *Bacillus solanacearum* by Smith (1896), *R. solanacearum* was then successively known as *Bacterium solanacearum*, *Phytomonas solanacearum*, *Xanthomonas solanacearum*, *Pseudomonas solanacearum* (Paudel et al., 2020) and more recently, by application of DNA-based methods, *Burkholderia solanacearum* (Yabuuchi et al., 1992) and finally *R. solanacearum* (Yabuuchi et al., 1995; Blomme et al., 2017; Paudel et al., 2020). Plants affected by *R. solanacearum* include, but are not limited to, staple crops (potato), fruit crops (banana and tomato), oilseed crops (sunflower and groundnut), spice crops, fodder, flowers, forest trees (ironwood and eucalyptus), weeds and many ornamentals (Champoiseau et al., 2009; Mansfield et al., 2012; Paudel et al., 2020).

- *R. solanacearum* is a Gram-negative, rod-shaped, strictly aerobic bacterium that is 0.5–0.7 × 1.5–2.0 μm in size and a member of the nonfluorescent pseudomonad rRNA homology group II (Champoiseau et al., 2009; García et al., 2019). *R. solanacearum* is listed as the second most destructive plant pathogenic bacterium because of its wide-ranging economic impact (Mansfield et al., 2012).

3.1 Races and biovars

- *R. solanacearum* is considered a 'species complex' (RSSC) due to significant variation within the group (Fegan and Prior, 2005; Hong et al., 2012). The RSSC consists of many strains with different levels of pathogenicity towards different hosts. Some strains have a very wide host range, e.g. those from solanaceous plants occurring in tropical lowlands, while others have a restricted range.

Originally Buddenhagen et al. (1962) and Buddenhagen and Kelman (1964) distinguished three races of *R. solanacearum*:

- Race 1 affecting tobacco, tomato, many solanaceous and other weeds and certain diploids of bananas;
- Race 2 causing bacterial wilt of triploids of bananas and Heliconia; and
- Race 3 affecting potatoes and tomatoes but not highly virulent on other solanaceous crops.

Since 1964, two additional races have been proposed:

- Race 4 causing disease in ginger in the Philippines (Kam and Quimio, 1977); and

- Race 5 affecting mulberry (*Morus* spp.) in China (He et al., 1983).

Based on carbohydrate utilization, five biovars were created, based on the ability to utilize and/or oxidize several hexose alcohols and disaccharides (Hayward, 1964; He et al., 1983; Buddenhagen, 1986; Hayward, 1991; Champoiseau et al., 2009). According to this traditional classification, only Race 2, biovar 1 strains can cause Moko disease of bananas, though they are also pathogenic on solanaceous hosts such as tomatoes and potatoes (Buddenhagen and Kelman, 1964; Thwaites et al., 1999; Blomme et al., 2017; Eden-Green, 2019).

3.2 Moko strains

Several strains of Moko, also called pathotypes, have been described from different ecological niches and can be distinguished based on colony characteristics on a tetrazolium chloride (TZC) medium (Kelman, 1954), where the bacterium induces a purple-coloured formazan dye by the reduction of tetrazolium salts. Stover (1972), based on the works of Buddenhagen, describes the distinguishing features of four major pathotypes of *R. solanacearum* Race 2 designated as D, B, SFE and H.

- D (distortion): colonies large, irregular, fluidal, lace-like slime, often white or slight formazan (French and Sequeira, 1970; Stover, 1972; French, 1986; Thwaites et al., 1999). The D strain originated from wild Heliconia present in the Golfito area of Costa Rica (Buddenhagen, 1960; Sequeira and Averre, 1961; Stover, 1972), with low virulence on bananas and expressing only mild symptoms (Buddenhagen, 1961; Stover, 1972). According to Stover (1972), there is no wilting, plants remain stunted and new leaves are narrow and twisted and later blacken. Internal discolouration is the same as in wilt syndrome. Distortion is most pronounced on young plants.
- B (banana): probably by mutation from strain D, highly virulent on banana, cannot be distinguished from pathotype D in culture. No or slight oozing from male flower buds (Buddenhagen, 1960; Stover, 1972; Thwaites et al., 1999). Persisting in soils and spread mainly through tools and cultural practices (Eden-Green, 2019).
- SFR (small, fluidal, round): slight formazan, probably from Heliconia or strain B, highly virulent on banana and Bluggoe, oozes easily from flower bract scars, can invade the plant through flower scars and is well adapted to insect dispersal. The SFR strain appeared in a localized area in the Ulúa Valley in Honduras in the 1960s (Buddenhagen and Elsasser, 1962; Buddenhagen and Kelman, 1964; Stover, 1972), apparently originated in planting materials introduced from Venezuela (Sequeira, 1998). Now widely distributed in South and Central America, the Southern Caribbean

and Mexico (Stover, 1972; Phelps, 1987; Albuquerque et al., 2014; Eden-Green, 2019).

- H (Heliconia): probably by mutation from strain B in Costa Rica (Stover, 1972), looks like B in culture, oozing from male flower buds, transmitted by insects, first detected in Costa Rica, epidemic on Bluggoe but less virulent than SFR and non-virulent on banana (Stover, 1972; Frossard, 1987; Thurston, 1984).

Since 1970, other insect-transmitted strains have been described causing epidemics: SFR-C in Colombia and strain A (Amazon) in Peru and Brazil spreading in 'Chato' (ABB, Bluggoe subgroup) along the rivers in the Amazon basin (French and Sequeira, 1970; Frossard, 1987; Eden-Green, 2019). The R strain (red colony form) was described in Costa Rica as mildly pathogenic to heliconias and with very limited pathogenicity towards bananas (Sequeira and Averre, 1961; French, 1986).

In Honduras, Wood (1984) divided the population of the pathogen into three types: B, SFR and an afluidal variant looking like an avirulent type. SFR types are present only on Wakenaam island in Guyana, while B or D types were observed on the mainland (McDonald, 1985). In Surinam, only D strains are prevalent (Gangadin, 1984), and in Grenada, only SFR types were isolated (Cronshaw and Edmunds, 1980). In Trinidad, SFR seems to have replaced B strains (Frossard, 1987). The causal agent of Bugtok in the Philippines cannot be distinguished from B strains in culture (Frossard, 1987; Eden-Green, 1994; Thwaites et al., 1999; Eden-Green, 2019).

3.3 Phylotype-sequevar classification

The race–biovar classification system has been valuable and practical but proved unsatisfactory over time. There are no reliable laboratory tests to define the 'race' of an isolate because host ranges of strains are broad and often overlap and, with few exceptions, the biovars do not correspond to phylogenetically coherent groups (Champoiseau et al., 2009). Because of this, the race–biovar system has now, in most cases, been replaced by phylotype-sequevar hierarchical classification schemes that are widely accepted by the scientific community as best reflecting the genetic diversity of the species complex (Hong et al., 2012; Albuquerque et al., 2014; Blomme et al., 2017).

This phylogenetically based classification scheme was developed based on DNA sequence analysis (Fegan and Prior, 2005; Prior and Fegan, 2005), dividing the species complex into four phylotypes that broadly reflect the strains' ancestral relationships and geographical origins (Champoiseau et al., 2009):

- Phylotype I originated in Asia;
- Phylotype II originated in the Americas;
- Phylotype III originated in Africa; and
- Phylotype IV originated in Indonesia.

Phylotypes were subdivided into sequevars based on the sequence of the endoglucanasegene (Fegan and Prior, 2005; Prior and Fegan, 2005; Fegan and Prior, 2006). More recently, Safni et al. (2014), using a polyphasic taxonomic approach, has proposed dividing the *R. solanacearum* species complex into three species:

- *R. solanacearum* corresponding to phylotype II strains (including Moko strains);
- *R. pseudosolanacearum* corresponding to phylotypes I and III; and
- *R. syzygii* corresponding to phylotype IV.

Using proteomic and genomic approaches, Prior et al. (2016) have supported the proposed division of the *R. solanacearum* species complex by Safni et al. (2014) into these three species.

Strains of the Moko bacterial wilt from banana and *Heliconia* spp. are clustered into seven subgroups within Phylotype II sequevars IIA-6, IIA-24, IIA-41, IIA-53, IIB-3, IIB-4 and IIB-25 (Albuquerque et al., 2014; Blomme et al., 2017; Eden-Green, 2019). Subgroup IIB-3 includes the pathogen causing Bugtok disease in the Philippines, with a very low level of genetic diversity (Raymundo, 1998; Raymundo et al., 2005), supporting the theory that a single genotype of the banana pathogen was introduced from Central America (Eden-Green, 2019). According to Blomme et al. (2017), the polyphyletic nature of the *R. solanacearum* (phylotype II) strains causing Moko disease remains to be elucidated by an in-depth comparison of genomes.

3.4 Presence, isolation and culture of the pathogen

Because *R. solanacearum* is found in high cell densities in infected tissues, a common sign of the pathogen is a milky white bacterial ooze that accumulates on the surface of freshly cut stems, rhizomes or tubers (Denny, 2006; Champoiseau et al., 2009; García et al., 2019). In bananas and plantains, this ooze can be opaque dirty-white, grey or brown and may exude from the cut surfaces of the rhizome, pseudostem and the fruit stalk of diseased plants (Fig. 7a-c) (Ashby, 1929; Buddenhagen, 1961; Stover, 1972). Additionally, bacterial streaming can be observed within 15 minutes when cut sections from the rhizome, pseudostem or fruit stalk of infected plants are placed in a clear vial containing water to induce the secretion of a viscous, white-creamy slime distinctive of *R. solanacearum* (Fig. 7d). The streaming test can be a valuable,

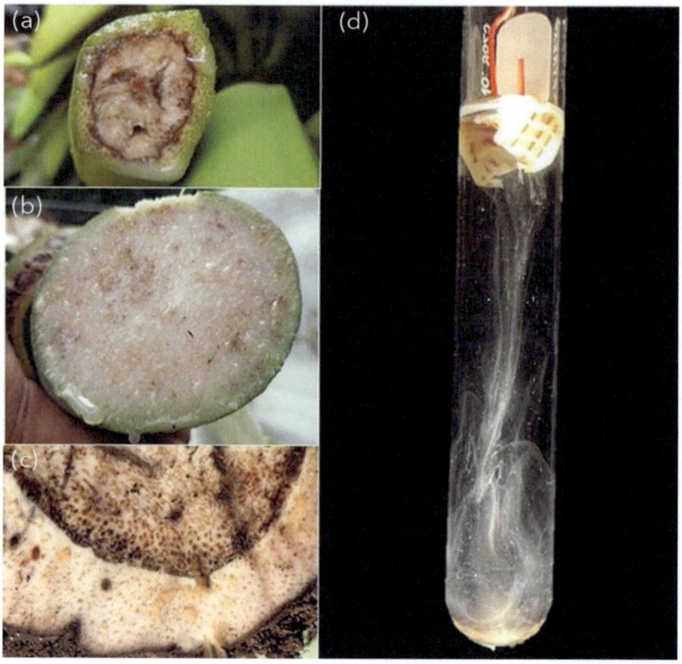

Figure 7 Bacterial ooze of *Ralstonia solanacearum* Race 2 from cuts in a finger, fruit stalk and rhizome (a, b, and c) and bacterial streaming emerging (d) from a small section of the pseudostem of a diseased plant of banana cv. 'Grande Naine' (*Musa* AAA, Cavendish subgroup). Photos: C. Carr and A. Nieuwenhuyse.

standard diagnostic tool for the quick detection of Moko bacterial wilt in bananas, and in cases when the bacterial cell density is low and the ooze does not form spontaneously on cut surfaces, the 'streaming' test can be a useful diagnostic tool, especially in the field (Allen et al., 2001; Denny, 2006; Momol and Champoiseau, 2009; García et al., 2019). Ooze and bacterial streaming are never observed in cases of Fusarium wilt (Buddenhagen, 1961; Stover, 1972).

- *R. solanacearum* can be isolated from water, soil or symptomatic plant tissues. Details of the recommended procedure, culture media and its composition can be found in Denny (2006) and García et al. (2019). Isolating the bacterium from infected plant tissues can provide a pure culture because the bacterial ooze produced by *R. solanacearum* is usually composed of pure bacterial cells (Denny, 2006). The bacterium causing Moko disease can be isolated from small pieces of tissue showing vascular discolouration excised from the rhizome, pseudostem, bunch stalk or fruit. The tissue must be surface sterilized and homogenized in sterile distilled water or 0.05 M potassium phosphate buffer using a sterile mortar (García et al., 2019). The pathogen can also be isolated from bacterial streaming in sterile distilled water. In both cases, the suspension obtained is streaked on standard semi-selective media such as casamino acid-peptone-glucose (CPG, Kelman, 1954), triphenyl tetrazolium chloride (TTC or TZC; Kelman, 1954) and modified Sequeira medium South Africa (SMSA, Elphinstone et al., 1996). SMSA medium is recommended when isolating the pathogen from asymptomatic host tissue, water or soil samples that tend to have low populations of *R. solanacearum* and higher levels of saprophytic bacteria and other contaminants (Denny, 2006; García et al., 2019). From these initial isolation plates, pure cultures of *R. solanacearum* can be generated using the standard dilution streaking technique and selecting a single colony (Fig. 8) (García et al., 2019).

For most strains, the optimal growth temperature is between 28°C and 32°C (Hayward, 1991; Schaad et al., 2001). On solid agar media, individual bacterial colonies are usually visible after 36–48 h of growth at 28°C (Champoiseau et al., 2009; García et al., 2019).

3.5 Pathogen identification

Trained people can diagnose Moko disease in the field based on visual symptoms. Culture-based methods, serological assays, biochemical tests and nucleic acid-based methods complement field diagnoses. Traditional and rapid methods used for initial screening under field conditions include the stem-streaming test and commercially available *R. solanacearum*-specific immunoassay strips (Denny, 2006; Kinyua et al., 2014; García et al., 2019). The culture-based method of streaking the bacterium on the semi-selective media TZC or SMSA is commonly employed for initial laboratory screening (Denny, 2006; Champoiseau, 2009; García et al., 2019) and allows differentiation between virulent and avirulent strains. In both media, virulent strains generate white colonies with pink centres, fluidal and irregularly shaped (Fig. 8), whereas colonies of avirulent strains are dark red, uniformly round, smaller and butyrous (buttery) or dry (Champoiseau et al., 2009; García et al., 2019).

Traditional and rapid diagnostic methods are generally reliable and easy to perform and are excellent decision-support tools for the management of the disease under field conditions. However, they have limitations when more precise and in-depth pathogen identification is required, e.g. for quarantine purposes and when conducting population and genetic diversity studies. For these purposes, highly sensitive and specific nucleic acid-based methods have been developed such as conventional polymerase chain reaction (PCR), quantitative PCR, microarray and loop-mediated isothermal amplification (Albuquerque et al., 2014; Cellier et al., 2017; Tran et al., 2016). García et al. (2019) have summarized the most common and widely used nucleic acid-based methods

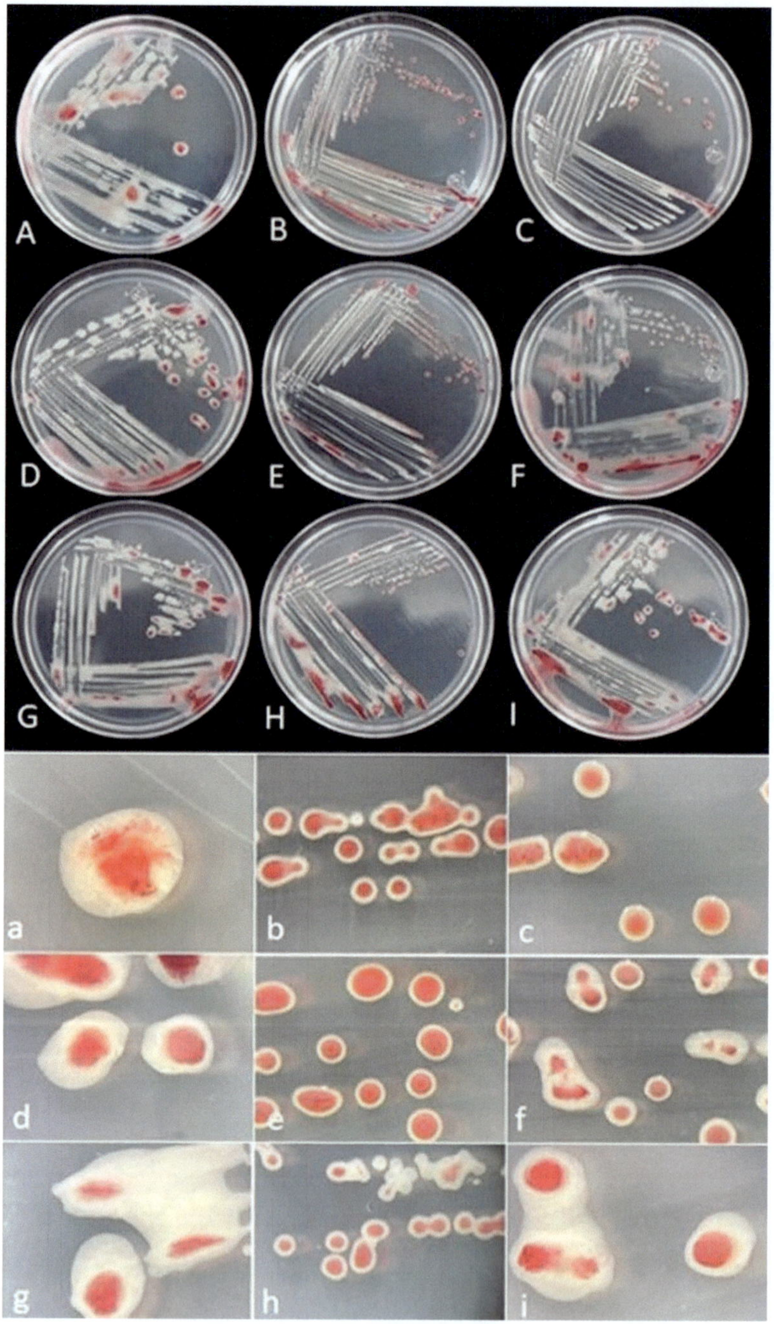

Figure 8 Cultures and colonies of virulent *R. solanacearum* strains on SMSA agar medium isolated from tomato (A–a), potato (B–b), Heliconia (C, D–c, d) and bananas (E–I and e–i). Photos: M. Sánchez and M. Guzmán.

currently available. The protocols for these methods must be standardized in each laboratory. The general protocol used in Costa Rica for PCR analysis of samples from plants suspected of being affected by Moko is as follows:

- Extraction of genetic material: the protocol described by Brandfass and Karlovsky (2008), with some modifications, is used for the extraction of DNA from plant tissue (rhizome, pseudostem or fruits) or cultures of the bacterium in semi-selective media.
- Detection of *R. solanacearum* from the extracted DNA: the presence of *R. solanacearum* is determined using a PCR with specific primers (Ito et al., 1997) 759/760 (759: 5′ GTC GCC GTC AAC TCA CTT TCC 3′; 760: 5′ GTC GCC GTC AGC AAT GCG GAA TCG 3′).
- Characterization of *R. solanacearum* strains: when amplified with 759/760 primers, the strains are further characterized through a multiplex PCR that includes specific primers to determine the phylotype (Nmult) and sequevar (Mus) designed by Fegan and Prior (2006). DNA amplification for sequevar 4 is characterized with two additional PCRs. The first with the BIOS1 primers (Pardo et al., 2019) and the second with duplex PCR primers 9 and 5 (Cellier et al., 2015) make it possible to accurately identify the strains affecting bananas.

Molecular techniques have become a valuable tool for diagnosing and studying pathogen populations; however, Koch's postulates are fundamental to confirming the pathogenicity of the isolated strains. Use of these postulates is especially important if strains are obtained from alternative hosts, water, soil or any other source other than diseased banana plants.

4 Disease cycle and epidemiology

4.1 Host range

R. solanacearum is of particular interest to the scientific community because of its global distribution, broad host range (over 250 plant hosts in 54 different plant families), virulence under diverse environmental conditions and its remarkable ability to retain pathogenicity genes in the absence of a host plant (Hayward, 1991; Paudel et al., 2020). Though the persistence of the bacterium in weed hosts has been questioned (Buddenhagen, 1986), there is sufficient evidence that infection of weeds plays a role in the Moko bacterial wilt disease cycle (Eden-Green, 2019). Table 2 shows a list of alternative hosts of *R. solanacearum* Race 2, the causal agent of Moko bacterial wilt disease in bananas and plantains, arranged by botanical family.

R. solanacearum has been isolated from wilted weeds in Costa Rica, but only the isolates from *Heliconia latispatha* and *H. caribaea* were pathogenic to

Table 2 List of alternative hosts of *Ralstonia solanacearum* Race 2, the causal agent of Moko bacterial wilt disease in bananas and plantains, by botanical family

Family	Host scientific name	References
Acanthaceae	*Blechum pyramidatum*	Prieto et al. (2012)
Apocynaceae	*Asclepias curassavica* L.	Belalcázar et al. (1968); Stover (1972); Granada (1996); Kelman (1953)
Araceae	*Colocasia esculenta*	Prieto et al. (2012)
	Xanthosoma roseum Schott.	Stover (1972)
	Ageratum conyzoides L.	Belalcázar et al. (1968); Granada (1996); Kelman (1953); Obregón et al. (2008)
	Ambrosia artemisiifolia L.	Belalcázar et al. (1968); Granada (1996); Kelman (1953)
	Bidens pilosa	Obregón et al. (2008)
	Bidens pilosa L.	Belalcázar et al. (1968); Granada (1996); Kelman (1953)
	Chaptalia nutans	Obregón et al. (2008)
	Emilia sonchifolia (L.) D.C.	Belalcázar et al. (1968); Granada (1996); Kelman (1953); Obregón et al. (2008)
	Galinsoga ciliata	Obregón et al. (2008)
	Galinsoga parviflora Cav.	Belalcázar et al. (1968); Granada (1996); Kelman (1953)
	Seneciodes cinerea	Obregón et al. (2008)
Balsaminaceae	*Impatiens balsamina* L.	Belalcázar et al. (1968); Granada (1996); Kelman (1953)
Boraginaceae	*Heliotropium* sp.	Obregón et al. (2008)
Brasicaceae	*Brassica campestris*	Belalcázar et al. (1968)
Campanulaceae	*Lobelia xalapensis*	Prieto et al. (2012)
Cannaceae	*Canna glauca* L.	Belalcázar et al. (1968); Granada (1996); Kelman (1953)
Commelinaceae	*Commelina diffusa* Burm. F.	Belalcázar et al. (1968); Granada (1996); Kelman (1953); Prieto et al. (2012)
	Commelina virginica	Obregón et al. (2008)
	Tripogandra cumanensis	Obregón et al. (2008)
	Tripogandra glandulosa	Obregón et al. (2008)
Compositae	*Bidens cynapiifolia* H.B.K.	Belalcázar et al. (1968); Granada (1996); Kelman (1953)
	Spilanthes acmella L.	Belalcázar et al. (1968); Granada (1996); Kelman (1953)
	Synedrella nodiflora (L.) *Gaertn*	Belalcázar et al. (1968); Granada (1996); Kelman (1953); Obregón et al. (2008)
Convolvulaceae	*Ipomoea trífida*	Obregón et al. (2008)
Cucurbitaceae	*Cucurbita maxima*	Prieto et al. (2012)

(Continued)

Table 2 *(Continued)*

Family	Host scientific name	References
Cyperaceae	*Cyperus* sp.	Obregón et al. (2008)
Euphorbiaceae	*Euphorbia heterophylla*	Obregón et al. (2008)
	Phyllanthus corcovadensis	Obregón et al. (2008)
	Ricinus communis	Stover (1972)
	Croton hirtus (L.) Herit.	Belalcázar et al. (1968); Granada (1996); Kelman (1953)
	Euphorbia graminea	Prieto et al. (2012)
	Euphorbia hirta L.	Belalcázar et al. (1968); Granada (1996); Kelman (1953); Obregón et al. (2008)
Fabaceae	*Desmodium* sp.	Obregón et al. (2008)
	Gliricidia sepium	Prieto et al. (2012)
Heliconiaceae	*Heliconia acuminata*	Sequeira y Averre (1961)
	Heliconia caribaea	Buddenhagen (1960)
	Heliconia latispatha	Buddenhagen (1960); Sequeira y Averre (1961)
	Heliconia imbricata	Sequeira y Averre (1961)
	Heliconia psittacorum	Prieto et al. (2012)
Labiataceae	*Salvia privodes*	Obregón et al. (2008)
Lamiaceae	*Salvia aff. lasiocephala*	Prieto et al. (2012)
	Stachys lamioides	Prieto et al. (2012)
Lythraceae	*Cuphea micrantha*	Prieto et al. (2012)
Malvaceae	*Sida* sp.	Obregón et al. (2008)
Musaceae	*Musa* sp.	Prieto et al. (2012)
Myrtaceae	*Psidium guajava*	Prieto *et al.* (2012)
Oxalidaceae	*Oxalis latifolia*	Prieto et al. (2012)
Phyllanthaceae	*Phyllanthus corcovadensis*	Belalcázar et al. (1968); Granada (1996); Kelman (1953); Prieto et al. (2012)
	Phyllanthus niruri L.	Belalcázar et al. (1968); Granada (1996); Kelman (1953)
Piperaceae	*Peperomia pellucida*	Obregón et al. (2008)
	Piper auritum H.B.K.	Stover (1972)
	Piper peltatum	Stover (1972)
	Piper sp.	Obregón et al. (2008); Prieto et al. (2012)
Poaceae	*Eleusine indica*	Prieto et al. (2012)
Polygalaceae	*Polygala paniculata* L.	Belalcázar et al. (1968); Granada (1996); Kelman (1953)
Polygonaceae	*Rumex crispus* L.	Belalcázar et al. (1968); Granada (1996); Kelman (1953)
Portulacaceae	*Portulaca oleracea*	Obregón et al. (2008); Prieto et al. (2012)

(Continued)

Table 2 (*Continued*)

Family	Host scientific name	References
Scrophulariaceae	*Scoparia dulcis* L.	Belalcázar et al. (1968); Granada (1996); Kelman (1953)
Solanaceae	*Browallia americana* L.	Belalcázar et al. (1968); Granada (1996); Kelman (1953)
	Capsicum sp.	Prieto et al. (2012)
	Datura stramonium	Belalcázar et al. (1968)
	Physalis angulata	Belalcázar et al. (1968); Granada (1996); Kelman (1953)
	Solanum caripense	Belalcázar et al. (1968)
	Solanum hirtum	Stover (1972)
	Solanum lycopersicum	Prieto et al. (2012)
	Solanum nigrum	Belalcázar et al. (1968); Belalcázar et al. (1968); Granada (1996); Kelman (1953); Stover (1972); Obregón et al. (2008); Prieto et al. (2012)
	Solanum torvum	Obregón et al. (2008)
	Solanum tuberosum	Prieto et al. (2012)
	Solanum umbellatum Mill.	Stover (1972)
	Solanum verbascifolium	Stover (1972)
Urticaceae	*Cecropia peltata*	Stover (1972)
	Pilea hyalina L.	Belalcázar et al. (1968); Granada (1996); Kelman (1953)
	Pilea sp.	Obregón et al. (2008)
	Plenax hirtus	Obregón et al. (2008)
Verbenaceae	*Lantana camara* L.	Belalcázar et al. (1968); Granada (1996); Kelman (1953)
	Lantana trifolia L.	Belalcázar et al. (1968); Granada (1996); Kelman (1953)
	Verbena litoralis H.B.K.	Belalcázar et al. (1968); Granada (1996); Kelman (1953)
Vitaceae	*Cissus sicyoides*	Obregón et al. (2008); Prieto et al. (2012)

bananas (Buddenhagen, 1960). The banana strain affecting young plants was found to be widespread in *Heliconia* spp. (*H. acuminata*, *H. imbricata* and *H. latispatha*) growing in virgin woodlands areas in Golfito on the Pacific coast of Costa Rica (Sequeira and Averre, 1961). Until early 1972, heliconias in the Caribbean zone of Costa Rica and other Central American areas were not infected by indigenous strains of *R. solanacearum*. However, once the banana strain is introduced into areas where heliconias are present, this genus can serve as a host (Stover, 1972; González, 1987) and heliconia plants can show symptoms of wilting and internal vascular discolouration (Fig. 9). According to González (1987), until 1972, the banana zone of Costa Rica was the only region of the Caribbean where Moko had not been detected. Until then, Almirante in Panama and Sarapiquí in Costa Rica had been the closest areas where Moko was found. In 1972, the disease appeared in several banana farms of the Caribbean coast of Costa Rica and, by the end of 1977, a large part of the country was affected by the disease.

In Colombia, Belalcázar et al. (1968) found 12 species of weeds that could carry the banana strain without showing external symptoms and four species expressing symptoms when inoculated: *Brassica campestris*, *Datura stramonium*, *Solanum caripense* and *Solanum nigrum*. Berg (1971) observed at least 39 known natural hosts of *R. solanacearum* in or near banana plantations

Figure 9 Symptoms of bacterial wilt (*Ralstonia solanacearum* Race 2) in plantations of heliconias in Costa Rica: *H. caribaea* (a and c), *H. psittacorum* (b) and growing of a wild heliconia (*H. latispatha*) in a banana plantation of Costa Rica (d), which can serve as an alternative host of the pathogen. Photos: M. Guzmán.

in Central America; however, not all these strains were pathogenic to bananas. Isolations from 45 weed species growing in banana plantations in Honduras showed that at least 10 were able to act as hosts of the SFR strain (Berg, 1971).

Belalcázar et al. (2004) conducted an extensive review of potential hosts of *R. solanacearum* Race 1, 2, or 3, and they reported 234 species distributed in 42 families and 114 genera. Of these, 26 weeds were identified as Race 2 hosts in subtropical climates where coffee and plantain are co-cultivated. A weed survey in banana and plantain plantations in Colombia found 27 weeds as pathogen hosts, of which eight of them were reported as a new host of *R. solanacearum* Race 2: *Chaptalia nutans*, *Seneciodes cinerea*, *Tripogandra glandulosa*, *Plenax hirtus*, *Peperomia pellucida*, *Tripogandra cumanensis*, *Desmodium* sp. and *Cissus sicyoides*. The last species (*Cissus sicyoides*) belongs to the botanical family Vitaceae, not previously reported as a host for *R. solanacearum* Race 2 (Obregón et al., 2008).

Prieto et al. (2012) have conducted surveys in areas where bananas, plantains, heliconias, bell pepper, potato and tomato are grown in Colombia to identify *R. solanacearum* hosts in weeds and crops. Special attention was given to banana and plantain plantations with a previous history of Moko disease. Plants were sampled within or around Moko foci, including symptomless plants. Sixty bacterial strains showing *R. solanacearum* characteristic colonies in SMSA medium were selected for further pathogenicity tests in seedlings of plantain cv. Dominico-Hartón (AAB, plantain subgroup). Twenty-six isolates induced Moko disease symptoms during the 60-day evaluation period. This study added 12 new hosts of *R. solanacearum* pathogenic to bananas, including nine weeds, and three crop plants: *Colocasia esculenta* L., *Cucurbita maxima* D. and *Psidium guajava* L.

There is now a long list of weeds and crops that have been reported as hosts of *R. solanacearum* Race 2 (Table 2), the causal agent of Moko, indicating that the bacteria can survive on a wide range of cultivated and wild hosts. However, early studies in Costa Rica have demonstrated that only isolates obtained from banana or *Heliconia* spp. were pathogenic to banana (Sequeira and Averre, 1961; Buddenhagen, 1960; Buddenhagen, 1961; Buddenhagen and Kelman, 1964; Eden-Green, 2019). It has been suggested that Moko disease probably originated from indigenous Heliconia plants in the jungles of Central America and possibly independently in northern South America (Sequeira and Averre, 1961; Buddenhagen and Kelman, 1964).

4.2 Infection, transmission and spread

Banana plant residues, contaminated soils and water and infected plants (including fruits, leaves and planting materials) are all potential sources of inoculum of the pathogen (Stover, 1972; González, 1987; Eden-Green, 2019).

Natural infections in bananas occur whenever the bacterium enter the roots directly or through wounds made by soil-inhabiting insects or nematodes (Kelman, 1953) or when it is carried into the vascular system of the inflorescence by insects or is introduced through wounds by means of insects or tools such as machetes (Buddenhagen and Elsasser, 1962; Thurston, 1984; Frossard, 1987).

Transmission through wounds created by agricultural tools used in banana plantations is common. Sequeira (1958) provided the first proof of wound transmission of the pathogen in 1958, half a century after the description of the disease. He showed that the bacterium was easily transmitted by pruning tools, particularly during de-suckering and de-leafing operations. Moko disease bacteria are readily transmitted from plant to plant on machetes used for pruning (Stover, 1972). It was later shown that wounds made with tools used in cultural practices are the main entry point for the bacteria in commercial banana plantations (Thurston, 1984; Eden-Green, 2019). Figure 10 shows the Moko bacterial wilt disease cycle and major pathways of transmission of *R. solanacearum* Race 2 in bananas and plantains.

Buddenhagen and Elsasser (1962) described the SFR strain on 'Bluggoe' in Honduras and demonstrated that insects were an important means of

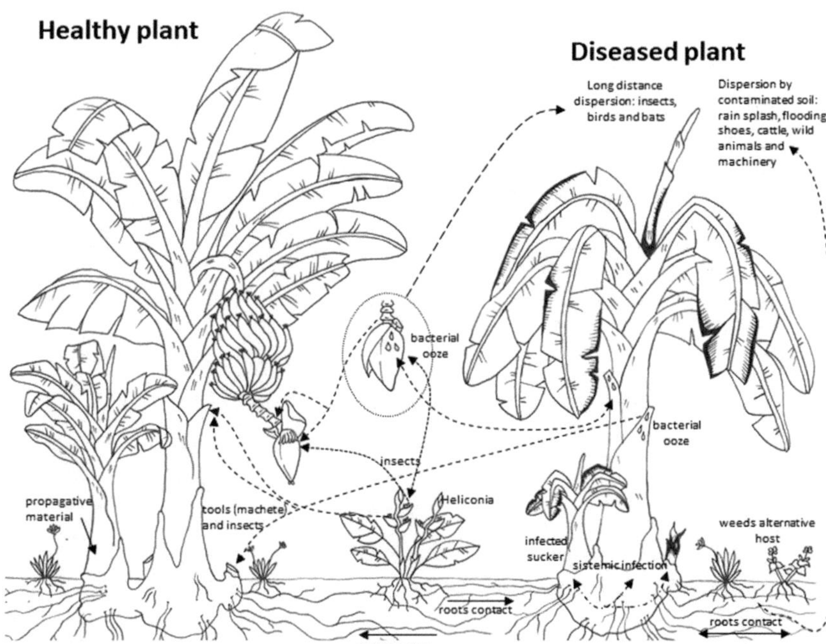

Figure 10 Moko bacterial wilt cycle and major pathways of transmission of *Ralstonia solanacearum* Race 2 in bananas and plantains. Drawing by Valentina Guzmán-Gómez.

bacterial spread. According to the authors, 50 species of insects were collected from inflorescences in one location, with up to 100 bees and wasp per hour frequenting a single male 'Bluggoe' inflorescence. About 5% of the bees (*Trigona corvina*) and wasps carried the bacteria, as determined by direct isolation. Moreover, large numbers of flying insects, mainly native bees (*Trigona* spp.), wasps (*Polybia* spp.), fruit flies (*Drosophila* spp.) and many other genera of flies, visit banana flowers and the wet surface of the peduncle where they suck the bacterial exudate from the male bract scars (Buddenhagen and Kelman, 1964; Stover, 1972; Thurston, 1984). In addition to insects, other animals such as birds, bats and rodents can disperse the bacteria when they feed on early-ripening fruits in the bunch (Fig. 11).

The source of bacteria for transmission by insects is the male part of the inflorescence. Bacterial ooze from cushions on diseased peduncles and from bases of bracts and bract scars of 'Bluggoe' plantain forms as milky white droplets for an approximately 10-day period, starting 15 days after

Figure 11 Native black bees (Trigona spp.), one of the main vectors of SFR strain of Moko, visiting banana flowers (a and b). Evidence of animal feeding in Cavendish (Musa AAA) banana fruits with symptoms of Moko (c, birds and d, bats and rodents) that can contribute to the dissemination of the pathogen. Photos: M. Guzmán.

infection (Buddenhagen and Elsasser, 1962). The B strain can also be insect transmitted to fruits through inflorescence, but this is much less common than the SFR strain because the B strain rarely oozes from infected flower buds (Stover, 1972). Insects can also transmit the pathogen from diseased plants through open wounds following sanitation and cultural practices such as leaf pruning (de-leafing), pruning of suckers and harvest of fruiting plants (Stover, 1972; Thurston, 1984; Eden-Green, 2019). According to Stover (1972), insect transmission of both B and SFR strains frequently occurs through freshly cut sucker surfaces, following pruning in commercial banana plantations.

Insect transmission to floral parts and inflorescence infections are rare on commercial banana plantations because regular removal of the male bell and bagging of bunches prevents bacterial infection through this route (Stover, 1972; González, 1987; Eden-Green, 2019). However, as Eden-Green (2019) indicated, these cultural practices are not implemented in smallholder plantings of 'Bluggoe', resulting in difficult-to-control epidemics of insect-transmitted Moko bacterial wilt in those plantings and beyond. The same author suggests that the rapidly spreading insect-transmitted epidemics probably account for the historic, devastating losses in 'Bluggoe' from Moko bacterial wilt in Guyana and Trinidad described by Rorer (1911). This may also explain the rapid spread of Moko in Honduras and within the Magdalena Valley in Colombia, where the disease spread over 90 km within 2 years, and the sporadic outbreaks in previously unaffected dessert banana plantations described by Buddenhagen and Elsasser (1962). Similar insect-borne epidemics of Moko bacterial wilt were reported in Peru (French and Sequeira, 1970) and Nicaragua (Lehmann-Danziger, 1987). Insect transmission most probably accounts for the extensive spread of Bugtok throughout central and southern Philippines since the first reports in 1965 (Eden-Green, 2019).

Root-to-root transmission is also possible as infected plants exude bacteria into the soil (Kelman and Sequeira, 1965; Frossard, 1987). Kelman and Sequeira (1965) concluded that bacteria may enter or egress at points of emergence of secondary roots. The egress from the roots of infected plants of large numbers of bacteria into the soil may play an essential role in spreading bacterial wilt to adjacent healthy plants. This is much more common with the B strain, which can survive in the soil for more extended periods (12–18 months), than the SFR strain, which rarely stays in the soil for more than 6 months (Stover, 1972). In infested soils, *R. solanacearum* may enter the roots through wounds made by soil-inhabiting insects and nematodes (Kelman, 1953). Sequeira (1958) suggested that *R. solanacearum* can enter banana roots directly, but this infection process is not common, suggesting wounds may be necessary. Nevertheless, positive infection by roots without mediating mechanical damage under experimental and field conditions has been observed (Sequeira, 1958; Buddenhagen and Elsasser, 1962; Kelman and Sequeira, 1965).

Buddenhagen and Kelman (1964) noted a correlation between the high incidence of Moko in bananas and infestation with nematodes, but this interaction still needs to be confirmed (Thurston, 1984). According to Eden-Green (2019), the main routes of survival and spread of the predominantly mechanically transmitted B strains of the Moko bacterial wilt pathogen are through roots following survival and persistence in the soil in infected banana residues, and possibly in association with weed hosts. Cultivated and wild alternative hosts are also a potential source of inoculum for the pathogen, as demonstrated by several researchers (Sequeira and Averre, 1961; Belalcázar et al., 1968; Berg, 1971; Belalcázar et al., 2004; Obregón et al., 2008; Prieto et al., 2012).

An efficient means of dissemination is via people who can carry infected materials such as suckers, rhizomes and sheaths or infested soil directly on shoes or indirectly via tractor wheels and cattle hooves (Frossard, 1987). The introduction of the pathogen can also occur via contaminated flood or irrigation water (Eden-Green, 2019). In Costa Rica, it has been observed that flooded areas are more prone to Moko infection because rivers can carry soil and plant material contaminated with the bacterium.

- *R. solanacearum* can survive in soils without a banana crop for some time, but, in the absence of amenable host plants, their survival in the soil is relatively poor (Sequeira, 1962; Granada and Sequeira, 1983; Eden-Green, 2019). High soil humidity favours the disease because it increases the survival and dissemination of the pathogen and the amount of infection (Thurston, 1984). Wardlaw (1972) reported experimental evidence that the bacterium survived less than 6 months in soil maintained at or below 65% water-holding capacity. Fallowing experiments achieved successful eradication after 12–18 months. Replanting trials under plantation conditions showed that the B strains survived fallow periods of less than 18 months (Sequeira, 1962), reduced to 9 months with dry season tillage (Sequeira, 1958), or 12 months in well-drained soils (Lozano et al., 1969). The B strain can survive longer in the soil than SFR strains (Thwaites et al., 1999). For this reason, fallow periods of 6 months and 12 months have been recommended for SFR and B strain infections, respectively (Sequeira, 1962; Stover, 1972).

Young plants develop symptoms earlier than more mature plants. In mature plants, the symptoms may appear from 6 weeks to 3 or more months after inoculation (Stover, 1972). In plants whose suckers were pruned with contaminated machetes or other tools used to remove unwanted suckers, symptoms appear 60–90 days later. This long incubation period complicates detection and control. Low temperatures also slow or reduce the development of the disease (Thurston, 1984).

5 Occurrence and impact

Moko bacterial wilt is present in many parts of the world. Reliable reports of the disease come from:

- North America: Mexico;
- Central America: Guatemala, Belize, Honduras, El Salvador, Nicaragua, Costa Rica and Panama;
- South America: Colombia, Venezuela, Guyana, Suriname, Brazil (Amazon basin and Sergipe), Peru and Ecuador; and
- the Caribbean: Trinidad, Grenada, Jamaica and St Vincent.

In Asia, Moko only occurs in commercial plantations of Cavendish in the Davao region of Mindanao in the Philippines and in peninsular Malaysia where the pathogen was also confirmed recently (Zulperi and Sijam, 2014; Zulperi et al., 2016). It is believed that the pathogen was introduced to the Philippines on propagating material of 'Valery' (*Musa* AAA Cavendish subgroup) imported from Honduras in 1968 (Rillo, 1979; Buddenhagen, 1994; Blomme et al., 2017; Eden-Green, 2019). Table 1 shows the countries where Moko disease has been reported, the year of detection and the supporting reference.

Moko bacterial wilt has caused significant losses to small banana and plantain producers in Latin America and the Caribbean (Robinson and Galan Sauco, 2010; González, 1987). Damage by the disease at the smallholder level is due to the loss of plants and area for planting, forcing a change in the crop to be cultivated.

In larger Central American export plantations, losses from Moko are relatively low at present. In Central America, the United Brands Co. (1979) reported losses of less than 1% of the plants in 12 150 ha of production. In Costa Rica, on a 350-ha farm of Cavendish banana with good management of the disease, the percentage of plants eliminated due to Moko cases ranged between 0.02% and 0.36% in a period between 2011 and 2021 (Table 3).

However, this low level of losses is only achieved at a high cost due to the strict prevention measures that must be implemented to keep the disease under control (Robinson and Galan Sauco, 2010). Measures to control Moko are expensive and hard to implement for small-scale producers of banana and plantain, such as those cultivating 'Bluggoe' (Robinson and Galan Sauco, 2010; Eden-Green, 2019). Stover and Simmonds (1987) indicated that the cost of preventive measures is much greater than the loss of plants. The most expensive task is the need for periodic inspection of each plant to ensure early

Table 3 Evolution of the number of cases of Moko bacterial wilt in a commercial Cavendish banana plantation of 350 ha in Costa Rica with good disease management practices and its impact in terms of plants eliminated and area affected by the eradication of the cases

Year	Number of Moko cases	Eradicated area (m^2)	Eradicated area (ha)	Total number of plants eradicated[a]	Total plants eradicated (%)
2011	12	771	0.08	130	0.02
2012	98	4816	0.48	809	0.14
2013	100	4806	0.48	807	0.14
2014	62	2944	0.29	495	0.08
2015	75	3600	0.36	605	0.10
2016	129	6402	0.64	1076	0.18
2017	104	4992	0.50	839	0.14
2018	93	4464	0.45	750	0.13
2019	119	5712	0.57	960	0.16
2020	260	12 480	1.25	2097	0.36
2021	90	4320	0.43	726	0.12
Total	**1142**	**55 307**	**5.53**	**10 233**	**1.57**

[a]Include diseased plants and the plants of the buffer zone.

detection of new outbreaks. They estimated that the cost of managing the disease amounted to US$400 000 per year (US$1 044 000 today) (US$32/ha/year; US$84 today), with plant losses increasing by 5% or more per year in the absence of preventive measures.

According to Stover and Simmonds (1987), the incidence of Moko in Central America was less than 10 cases in 500 ha, although it was two or three times higher in Honduras. In Costa Rica in 1998, the cost of eradicating a case of Moko of approximately 48 m^2 was estimated at US$234 (US$611 today), and the affected area had to remain in quarantine for 18 months with strict biosecurity measures; as each new case would extend the quarantine period (MAG, 1998). For 2021 in commercial farms affected by the disease, this cost is estimated between US$250–296 per case, including materials, disinfectants and labour (C. Vargas, Pers. Comm.).

The threat of Moko disease to export banana plantations and plantain for local consumption has prompted some countries to legislate to control the disease. Bunches affected by Moko can ripen prematurely in transport-producing ethylene affecting more fruits in the consignment. Since 1994, control of Moko in Costa Rica has been regulated by state law and made mandatory for all banana producers (Executive Decree No., 23810-MAG). The same decree prohibits the planting of highly susceptible cultivar Guineo Cuadrado (*Musa* ABB, subgroup Bluggoe) in any banana zone of the country as well as in places where cases of diseased plants are detected.

6 Management of the disease

Quarantine should be the first line of defence to avoid introducing Moko into a new country, region or farm. Where the pathogen is present, control strategies focused on early detection, isolation, eradication of sources of infection and the adoption of integrated management practices in an effort to limit the spread of the pathogen and rehabilitate previously affected areas (González, 1987; Eden-Green, 2019).

6.1 Exclusion

This concept should be understood as avoiding the disease entering an area free of it, including by dissemination from plant to plant, farm to farm or from one place to another (González, 1987). To achieve this, the movement of planting material like rhizomes and suckers from an infected zone to a Moko-free location must be avoided.

Within an infected site, movement of vegetative material for planting should also be avoided. Materials from disease-free areas or tissue culture plants must be used for replanting. Disinfecting footwear utilizing footbaths for personnel working in affected plantations and entering Moko-free locations is required. Footbaths should be constructed of plastic or concrete with a sponge impregnated with effective disinfectants such as quaternary ammonia, iodophors or glutaraldehyde. The tires of vehicles that travel through contaminated areas and those entering farms must also be cleaned and disinfected.

All sharp and pointed tools used in agricultural operations in affected plantations must be rigorously disinfected after use for each plant throughout the farm. Containers with disinfectants must be of adequate size for each tool so that the disinfectant covers their entire surface. Containers must be refilled with the disinfectant every day and cleaned to remove any tissue or organic matter that accumulates inside.

6.2 Identification and tracking of infected plants in the field

Early detection and diagnosis are essential to identify early outbreaks of disease. Well-trained personnel are needed to identify symptoms (known as Moko inspectors). Surveys to identify infected plants should preferably be done in weekly or biweekly cycles, from plant to plant, observing every part of the plant including suckers that may show any symptoms of the disease. Inspection (preferably biweekly) is particularly important before de-suckering to avoid possible dissemination by this means as well as after de-suckering and at harvest when fresh cuts in plants provide a source of inoculum and a court for infection.

6.3 Eradication

As a general recommendation, any suspected case of Moko recognized by a Moko inspector must be eradicated as soon as possible once confirmed. Eradication includes the infected plant and adjacent, apparently healthy plants to create a disease-free zone (buffer zone) with the objective of eliminating the source of infection promptly. Eradication methods may vary between countries and companies, although the minimum procedures to be implemented may be regulated by law, as is the case of Colombia and Costa Rica. Different herbicides, such as glyphosate, are applied by injection to kill the diseased plants. In addition, the soil needs to be disinfected with fumigants for rapid rehabilitation of contaminated areas and to reduce the risk of dispersal of the bacteria.

In 1976, methyl bromide was successfully used to disinfect soil and eradicate the pathogen from tissues of diseased plants. At that time, methyl bromide was used extensively in Central and South America through to the end of 2010 to eradicate Moko from Cavendish export plantations, mainly the most persistent pathotype B in soil. In 2015, methyl bromide was withdrawn in developing countries (in accordance with the Montreal protocol), partly on safety grounds and because it causes damage to the ozone layer. Dazomet and metam sodium were used as alternatives with good efficacy and no known damaging impact on the ozone layer (Cronshaw, 1998; Castañeda et al., 2002a; Castañeda et al., 2002b; Martínez and Guzmán, 2011). It is essential that approved application protocols are used with each soil fumigant to obtain the desired eradication effect.

According to González (1987), the first step to follow in the eradication of Moko cases in commercial Cavendish farms is to apply an insecticide, preferably with a motor pump, around the infected plant, in a radius of 15–20 m. To prevent insects from carrying inoculum from the infected plant, the insecticide should be sprayed from the outside in, using an insecticide approved for use on the crop. Vegetative material and cut plant material within the eradication area should also be sprayed with insecticide as many insects are attracted to exudates emanating from tissue and could spread the pathogen.

A radius of 5 m is marked around the diseased plant, which includes the area to be eradicated. In the case of isolated SFR strains, the radius of the eradication area could be reduced to 2.0 m (González, 1987). There is evidence in Colombia that *R. solanacearum* can be found in the soil up to a distance of 5 m from the diseased plant and up to 50 cm deep, depending on farm management practices (Obregón et al., 2011).

All plants within the demarcated area should be eliminated, starting with apparently uninfected ones and leaving the diseased plant to last. All the plants are cut into small pieces, including the rhizomes and roots, using thoroughly disinfected machetes and shovels for each plant. The remains of the diseased plant (whole plant) are placed in hermetically sealed plastic bags and treated with dazomet or metam sodium. The soil at the site of each rhizome, within a radius of 1.5–2 m, is removed to a depth of approximately 30 cm and treated with a soil fumigant such as dazomet (Basamid® 97MG, 120–150 g/ m^2) or metam sodium (44–60 g a.i./m^2) (Guzmán et al., 2019). Subsequently, the entire eradicated area is then covered with a 0.15-mm thick plastic tarp (preferably black), sealing the edges with soil to prevent the escape of the fumigant (Fig. 12). All tools and footwear of staff must be continually disinfected during the eradication process, especially every time the area is left. The tarp is removed 2–8 weeks later, respectively, depending on whether the SFR or B strain is present. The area can then be replanted with planting materials from disease-free areas.

In cases where eradication is only carried out by injecting herbicides, without fumigation of the soil, the eradication area may vary, and a fallow period should be added. Recommendations for the size of the eradication area and length of fallowing of buffer zones depend on the strains: 6 months and 10 m diameter around a single infected plant with the SFR strain and 12 months and 20 m diameter around a plant infected with the B strain (Sequeira, 1962; Stover, 1972). However, in Grenada, where the SFR strain is prevalent, a 6-month fallow period proved too short (Frossard, 1987).

In Colombia, the officially approved procedure includes an eradication area with a radius of 5 m, taking the diseased plant as a reference. All banana plants within the eradication area are injected with 20% glyphosate solution. Adult plants require a minimum dose of 50 mL distributed over five injection sites (10 mL per injection site) for optimal elimination. In addition, all suckers that are part of the production unit must be injected; in this case, the dose can vary between 10 mL and 30 mL, depending on their height. The injected plants must always remain standing and never cut into pieces. Instead, the bunches are cut from plants and stacked together with the debris remaining on the ground, applied with lime and then covered with a plastic tent. The eradicated area and its surroundings are maintained free of weeds for 6 months, and any regrowing plant is re-injected with glyphosate. However, Obregón et al. (2011) found that *R. solanacearum* can survive in banana soil and plant tissue for up to 6.5 months even after eradicating the plants with glyphosate.

Sequeira (1958) found that replanting of areas abandoned at Coto in Costa Rica due to severe *R. solanacearum* strain B infestation needed a fallow period of 2 or more years. More recent experiences in tropical lowlands in Costa Rica and other countries of Central America corroborate these observations. Chemical fumigation of infested soils is necessary for replanting Moko-affected areas where there is no extensive fallowing period.

Figure 12 Main steps in the process for eradication of a case of Moko in a plantain plantation cv. 'Curraré Enano' (*Musa* AAB, plantain subgroup): delimitation of the area to be eradicated with a perimeter ditch, footbath and container for boots and tools disinfecting (a), removal and cutting of the plants into small pieces with a machete and shovel (b), preparation of the metam sodium soil disinfectant solution (c), application of the soil disinfectant solution with a watering can and placement of the pieces of diseased plant in plastic bags (d), covering the affected with a plastic tarp and sealing the edges with soil (e) and final appearance of an eradicated case (f).

6.4 Cultural practices and integrated management

Once Moko establishes itself in a banana plantation, cultural practices are the main form of dissemination of the bacteria. Wardlaw (1972) estimated that almost 97% of the spread of strain B might be due solely to cultural practices.

Routine disinfection of cutting and digging tools is of particular importance in preventing Moko (Buddenhagen and Sequeira, 1958; Sequeira, 1958; Stover, 1972). Plantation workers often carry two of the same tool: one is left in a disinfectant solution while the other is being used. In infected areas, sterilization of tools is necessary between each plant. The addition of a marker dye (such as crystal violet) to the solution helps to monitor disinfection. Workers' footwear must also be disinfected when leaving contaminated areas. The use of well-constructed and designed footbaths is required. Tools and footwear can be disinfected with the same disinfectants. However, it is recommended to use higher doses for disinfecting footwear and tools for digging due to the presence of soil that reduces the effectiveness of the disinfectant.

A 10% formaldehyde solution (formalin) was used as an effective disinfectant until the 1990s but discontinued due to the risk to workers' health. Álvarez et al. (2015) recommended sodium hypochlorite as an alternative to formalin but is also toxic and corrosive and rapidly inactivates in the presence of soil and organic matter. Currently, iodophors, quaternary ammonium and glutaraldehyde are used as disinfectants for preventing Moko. They are effective at relatively low doses (1000–2000 ppm), making them safer for workers to handle. They are also inexpensive and widely available under different brand names. More recently, with the adoption of biosecurity measures for the prevention of Fusarium wilt tropical Race 4 in banana and plantain farms, doses of disinfectants have been increased to ensure the elimination of chlamydospores of Fusarium wilt. Table 4 indicates the doses of disinfectants recommended in Costa Rica for the prevention of Moko and Fusarium wilt (Guzmán et al., 2018; Pérez-Sánchez et al., 2019).

In Moko-infected areas, bunches should be bagged as soon as they emerge (premature bagging) to avoid transmission of the bacterium by insects. In addition, the plastic cover must be impregnated with an insecticide such as chlorpyrifos or bifenthrin to help repel insects attracted to the flowers. The plastic should be sufficiently transparent to allow observation of the bunch to facilitate timely detection of symptoms in fruits (necrotic or early ripening fingers) as this could indicate the presence of the disease. Cultural practices

Table 4 Disinfectant and doses recommended in Costa Rica for tools, vehicles and machinery in the prevention of Moko and Fusarium wilt tropical race 4 (Guzmán et al., 2018)

	Dose (ppm)	
Disinfectant	Tools	Footbaths, vehicles and machinery
Glutaraldehyde	1000	3000
Lodophors	2000	4000
Quaternary ammonium	2000	4000

such as deflowering in the field should be stopped in Moko-infected areas, and the removal of the male bud must be done manually.

6.5 Disease resistance

Commercial banana, plantain and cooking banana varieties are susceptible to B and SFR strains of Moko, with few exceptions. It is important to make a distinction between field resistance and intrinsic resistance to Moko (Frossard, 1987). Varieties that have persistent bracts and flowers provide fewer points for insects to acquire bacteria and are more difficult to infect by insects that transmit the disease (especially the SFR strain) than those varieties with dehiscent bracts and flowers, and these traits are recognized as having a high preventive value against Moko (Lozano et al., 1969). Some varieties less susceptible to transmission by insects could be vulnerable when the bacterium is injected into the plant (Stover, 1972) or inoculated into the soil, inducing damage to roots or transmitted by tools (Valencia-Valencia et al., 2014). This reflects the importance of considering the route of infection and the method of inoculation when evaluating resistance to Moko disease in different genetic materials.

The cultivar Pelipita (ABB) escapes infection because its male bracts and flowers are persistent. Since the early 1970s it has been promoted in Central and South America as a substitute for 'Bluggoe' plantain (ABB), which has been devastated by the SFR strain (Stover and Richardson, 1968; Stover, 1972; Eden-Green, 2019). 'Pelipita' also showed a high degree of resistance to three out of five strains of *R. solanacearum* from Colombia when roots were damaged with a knife and the bacteria inoculated into the soil (Valencia-Valencia et al., 2014). In contrast, Pelipita in Brazil was reported susceptible to Moko when the pathogen (10^8 CFU/mL) was injected into the pseudostem or the plant immersed in a suspension of the bacteria and roots were damaged with a knife (Mello Veras et al., 2002).

Horn plantain (AAB) appears to have some degree of resistance since the male axis on this variety degenerates early (Buddenhagen and Elsasser, 1962; Stover, 1972). However, Horn plantain becomes diseased following root infection or transmission by pruning tools (Buddenhagen and Elsasser, 1962). Other plantains such as 'False Horn' (AAB) and 'Dominco Harton' (AAB) cultivars are highly susceptible to root infection or transmission via infected tools. The 'Dwarf Cavendish' types that tend to retain male flowers and bracts are less prone to insect transmission (Eden-Green, 2019). The cultivar Yamgambi km5 (AAA, Ibota subgroup) was highly resistant to five strains of *R. solacearum* from Colombia, while the plantain hybrid FHIA 21 (AAAB) was found to be susceptible (Valencia-Valencia et al., 2014). However, Álvarez et al. (2015) reported a low incidence of Moko in FHIA 21 in plantations established in fields contaminated with *R. solanacearum* Race 2. These apparently contradictory results were later clarified by Pardo et al. (2019) who demonstrated that root wounds caused by a knife, before the inoculation of the bacteria into the soil, induced infection by the pathogen since the incidence of the disease was zero in plants whose roots were not damaged before inoculation.

In spite of the seriousness of Moko worldwide, genetic improvement programs for banana and plantain have no current projects nor are they selecting for resistance to bacterial wilt.

7 Conclusion and future trends

Moko disease causes significant losses in the banana- and plantain-producing regions of Latin America, the Caribbean and the Philippines. Quarantine should be the first line of defence for the exclusion of Moko disease from countries, regions or farms still free of the disease. There are still no effective curative treatments for affected plants nor resistance in commercially acceptable varieties, such that eradication remains the main control measure once the disease is detected. Management strategies should focus on early detection of diseased plants, isolation and eradication of sources of infection, including alternative hosts, and strict biosecurity measures to prevent the spread of the pathogen. Each new case detected on the farm must be quickly eradicated and the soil disinfected for prompt rehabilitation of planting areas.

The rapid increase in Moko cases observed in Central and South America over the past 15 years needs further study. More investment is required in research on the epidemiology, ecology, genetic and pathogenic variability of the pathogen as well as environmental conditions conducive to disease and the effect of climate change. There is a need to optimize strategies to prevent disease epidemics on farms as well as develop more effective procedures for eradication of Moko, ideally with less environmental impact and risk to workers' health.

Little effort has been placed on the genetic improvement of bananas and plantains towards resistance to Moko, and no significant progress in this direction is expected in the near future. Controlling Moko disease is labour intensive, expensive and difficult, especially when the insect-transmitted strain is present. The further spread of Moko to countries in the Caribbean, Asia or Africa, where bananas and plantains are a very important staple food, could have very serious consequences. Governments must be aware of the threat posed by this disease to sustainable banana and plantain production. More efforts are needed in biosecurity, in phytosanitary surveillance and in the training of growers, technicians and agronomists in the recognition of symptoms and effective management practices for the disease. The biosecurity measures adopted for the exclusion and management of Fusarium wilt tropical Race 4 may also contribute substantially to the prevention of Moko.

8 Where to look for further information

- PNUMA. Eliminación del bromuro de metilo en países en vías de desarrollo. Una historia de éxito. https://wedocs.unep.org/bitstream/handle/20.500.11822/28414/7674MethylBr_SP.pdf?sequence=3&isAllowed=y.
- Decreto N° 29.357/MAG - Estado de emergencia fitosanitaria por Moko y Sigatoka Negra en plantaciones bananeras. https://faolex.fao.org/docs/pdf/cos9565.pdf.
- https://www.promusa.org/.
- https://musanet.org/.

9 References

Albuquerque, G. M. R., Santos, L. A., Felix, K. C. S., Rollemberg, C. L., Silva, A. M. F., Souza, E. B., Cellier, G., Prior, P. and Mariano, R. L. 2014. Moko disease-causing strains of Ralstonia solanacearum from Brazil extend known diversity in paraphyletic phylotype II. *Phytopathology* 104(11): 1175–1182. doi: 10.1094/PHYTO-12-13-0334-R.

Allen, C., Kelman, A. and French, E. R. 2001. Brown rot. In: Loria, R., Franc, G. D. and Weingartner, D. P. (Eds.), *Compendium of Potato Diseases*. St. Paul, MN: American Phytopathological Society, pp. 11–13.

Álvarez, E., Pantoja, A., Gañán, L. and Ceballos, G. 2015. *Current Status of Moko Disease and Black Sigatoka in Latin America and the Caribbean, and Options for Managing Them*. CIAT Publication No. 404. Centro Internacional de Agricultura Tropical, Cali, Colombia; Food and Agriculture Organization of the United Nations, Rome, p. 40.

Argel, R. L. E., Bornacelly, H. H. and Hurtado, L. R. 2020. *Metodología para el Diagnóstico y la Erradicación del Moko (Ralstonia solanacearum) en el Cultivo de Banano. Centro de Investigaciones del Banano (CENIBANANO), Asociación Bananera de Colombia (AUGURA), Medellín, Colombia. 50 p.*

Ashby, S. F. 1928. The bacterial wilt of bananas and plantains. *The Journal of the Board of Agriculture of British Guiana* 1: 217–220.

Beaver, R. G. 1982. Use of picloram for eradication of banana diseased with Bunchy Top. *Plant Disease* 66(1): 906–907.

Belalcázar, S., Rosales, F. and Pocasangre, L. 2004. El "Moko" del plátano y banano y el rol de las plantas hospederas en su epidemiología. In: *ACORBAT 2004. Memorias XVI Reunión Internacional ACORBAT, 26 de Septiembre al 1 de Octubre de 2004*. Oaxaca, México, pp. 16–35.

Belalcázar, S., Uribe, G. and Thurston, H. 1968. Reconocimiento de hospedantes de Pseudomonas solanacearum E. F. Sm., en Colombia. *Fitotecnia Latinoamericana* 5: 89–99.

Belalcázar, C. S., Uribe, M. G. and Thurston, H. D. 1983. Reconocimiento de hospedantes a Pseudomonas solanacearum. *Colombia. Rev. Inst. Colombiano Agropec* 3: 37–46.

Berg, L. A. 1971. Weed hosts of the SFR strain of Pseudomonas solanacearum, causal organism of bacterial wilt of bananas. *Phytopathology* 61(10): 1314–1315.

Black, R. and Delbeke, A. 1991. Moko disease (Pseudomonas solanacearum) of Musa in Belize. *Tropical Science* 31: 347–353.

Blomme, G., Dita, M., Jacobsen, K. S., Pérez-Vicente, L., Molina, A., Ocimati, W., Poussier, S. and Prior, P. 2017. Bacterial diseases of bananas and Enset: Current state of knowledge and integrated approaches toward sustainable management. *Frontiers in Plant Science* 8: 1290. doi: 10.3389/fpls.2017.01290.

Brandfass, C. and Karlovsky, P. 2008. Upscaled CTAB-based DNA extraction and real-time PCR assays for Fusarium culmorum and F. graminearum DNA in plant material with reduced sampling error. *International Journal of Molecular Sciences* 9(11): 2306–2321. doi: 10.3390/ijms9112306.

Buddenhagen, I. W. 1960. Strains of Pseudomonas solanacearum, in indigenous hosts in banana plantations of Costa Rica, and their relationship to bacterial wilt of bananas. *Phytopathology* 50: 660–664.

Buddenhagen, I. W. 1961. Bacterial wilt of bananas: History and known distribution. *Tropical Agriculture (Trinidad)* 38: 107–121.

Buddenhagen, I. W. 1986. Bacterial wilt revisited. In: Persley, G. J. (Ed.), *Bacterial Wilt Disease in Asia and the South Pacific: Proceedings of the an International Workshop Held at PCARRD*, Los Baños, Philippines, 8–10 October 1985. ACIAR Proceedings No. 13. Canberra, Australia: Australian Centre for International Agricultural Research, pp. 126–143.

Buddenhagen, I. W. 1994. Banana diseases caused by bacteria. In: Ploetz, R. C., Zentmyer, G. A., Nishijima, W. T., Rohrbach, K. G. and Ohr, H. D. (Eds.), *Compendium of Tropical Fruit Diseases*. St Paul, MN: APS Press, pp. 15–17.

Buddenhagen, I. and Kelman, A. 1964. Biological and physiological aspects of bacterial wilt caused by Pseudomonas solanacearum. *Annual Review of Phytopathology* 2(1): 203–230.

Buddenhagen, I. W. and Elsasser, T. A. 1962. An insect-spread bacterial wilt epiphytotic of Bluggoe banana. *Nature* 194(4824): 164–165.

Buddenhagen, I. W. and Sequeira, L. 1958. Disinfectants and tool disinfection for prevention of spread of bacterial wilt of bananas. *Plant Disease Reporter* 42: 1399–1404.

Buddenhagen, I., Sequeira, L. and Kelman, A. 1962. Designation of races in Pseudomonas solanacearum (Abstr.). *Phytopathology* 52: 726.

Castañeda, D. A., Morales, J. G., Elejalde, G. J. and Mejia, G. A. 2002a. Alternatives to the use of methyl bromide in banana production in Colombia. In: Batchelor, T. and Bolivar, J. (Eds.), *Proceedings of the International Conference on Alternatives to Methyl Bromide - The Remaining Challenges, Seville, Spain*. Brussels: European Commission.

Castañeda, D. A., Sánchez, G. J., Elejalde-Gaviria, G. J., Patiño, L. F., Morales, J. G. and Mejia-Meza, G. A. 2002b. Alternatives to the use of methyl bromide in banana production in Colombia. In: *ACORBAT 2002. Memorias XV Reunión Internacional ACORBAT, 27 de Octubre al 2 de Noviembre de 2002*. Colombia: Cartagena de Indias, pp. 207–211.

Cellier, G., Moreau, A., Chabirand, A., Hostachy, B., Ailloud, F. and Prior, P. 2015. A duplex PCR assay for the detection of Ralstonia solanacearum phylotype II strains in Musa spp. *PLoS ONE* 10(3): e0122182. doi: 10.1371/journal.pone.0122182.

Champoiseau, P. G. 2009. Protocols for Ralstonia solanacearum culture preservation. Available at: https://plantpath.ifas.ufl.edu/rsol/CulturePreservation.html.

Champoiseau, P. G., Jones, J. B. and Allen, C. 2009. *Ralstonia solanacearum* Race 3 biovar 2 causes tropical losses and temperate anxieties. *Plant Health Progress*. doi: 10.1094/PHP-2009-0313-01-RV.

Cronshaw, D. K. 1998. Basamid granular for the control OK Moko disease. In: *ACORBAT 1998. Memorias XII Reunión Internacional ACORBAT, 27 de Octubre al 2 de Noviembre de 1996*. San José: confederación unitaria de trabajadores, pp. 312–315.

Cronshaw, D. K. and Edmunds, J. E. 1980. Note on the identification and distribution of moko disease in Grenada. *Tropical Agrículture (Trinidad)* 57: 171–172.

Delgado, R., Morillo, E., Buitrón, J., Bustamante, A. and Sotomayor, I. 2014. First report of Moko disease caused by Ralstonia solanacearum race 2 in plantain (Musa AAB) in Ecuador. *New Disease Reports* 30(1): 23. doi: 10.5197/j.2044-0588.2014.030.023.

Denny, T. P. 2006. Plant pathogenic Ralstonia species. In: Gnanamanickam, S. S. (Ed.), *Plant-Associated Bacteria*. Dordrecht, The Netherlands: Springer, pp. 573–644.

Eden-Green, S. J. 1994. Diversity of Pseudomonas solanacearum and related bacteria in South East Asia: New directions for Moko disease. In: Hayward, A. C. and Hartman, G. L. (Eds.), *Bacterial Wilt: The Disease and Its Causative Agent, Pseudomonas solanacearum*. Wallingford, UK: CAB International, pp. 25–34.

Eden-Green, S. J. 2019. Moko bacterial wilt and Bugtok. In: Jones, D. R. (Ed.), *Diseases of Banana, Abacá and Enset*. Wallington, UK: CABI Publishing, pp. 314–323.

Elphinstone, J. G., Hennessy, J., Wilson, J. K. and Stead, D. E. 1996. Sensitivity of different methods for the detection of *Pseudomonas solonacearum* in potato tuber extracts. *EPPO/OPEPP Bulletin* 23: 663–678.

EPPO. 2004. First report of Moko disease of banana in Jamaica. Available at: https://gd.eppo.int/reporting/ article-1555.

EPPO. 2007. New data on quarantine pests and pests of the EPPO. Available at: https://gd.eppo.int/reporting/article-1233. Alert Publications, List.

Fahy, P. C. and Persley, G. J. 1983. *Plant Bacteria/Diseases*. Australia: Academic Press.

Fegan, M. and Prior, P. 2005. How complex is the "*Ralstonia solanacearum* species complex"? In: Allen, C., Prior, P. and Hayward, A. C. (Eds.), *Bacterial Wilt Disease and the Ralstonia solanacearum Species Complex*. St. Paul, MN: American Phytopathological Society Press, pp. 449–461.

Fegan, M. and Prior, P. 2006. Diverse members of the *Ralstonia solanacearum* species complex cause bacterial wilts of banana. *Australian Plant Pathology* 35: 93–101. doi: 10.1186/s12864-016-2413-z.

French, E. R. 1986. Interaction between strains of *Pseudomonas solanacearum*, its hosts and the environment. In: Persley, G. J. (Ed.), *Bacterial Wilt Diseases in Asia and the South Pacific: Proceedings of the an International Workshop Held at PCARRD*, Los Baños, Philippines, 8–10 October 1985. ACIAR Proceedings No. 13. Canberra, Australia: Australian Centre for International Agricultural Research, pp. 99–104.

French, E. R. and Sequeira, L. 1970. Strains of *Pseudomonas solanacearum* from Central and South America: A comparative study. *Phytopathology* 60(3): 506–512.

Frossard, P. 1987. Moko disease threatens the West Indies. In: Walmsley, D. (Ed.), *Seminar Proceedings on Improving Citrus and Banana Production in the Caribbean Through Phytosanitation*, St. Lucia, W.I., 2-5 December 1986. CTA, Center for Ageing Research and Development in Ireland, Wageningen: CTA 2 December 1986. Seminar on

Improving Citrus and Banana Production in the Caribbean through Phytosanitation, Castries, Sainte-Lucie, pp. 115–120.

Fucikovsky, L. Z. and Santos, M. O. 1993. Advance of bacterial wilt in bananas in Mexico. In: Hartman, G. L. and Hayward, A. C. (Eds.), *Bacterial Wilt: Proceedings of the an International Conference Held at Kaohsiung*, Taiwan, 28–31 October 1992. ACIAR Proceedings No. 45. Canberra, Australia: Australian Centre for International Agricultural Research, pp. 341–342.

Gangadin, S. 1984. Race, biotype and strain characterisation of the Moko disease pathogen in Suriname. *De Surinaamse Landbouw* 32: 75–82.

García, R. O., Kerns, J. P. and Thiessen, L. 2019. Ralstonia solanacearum species complex: A quick diagnostic guide. *Plant Health Progress* 20(1): 7–13.

González, M. 1987. *Enfermedades del Cultivo del Banano. Universidad de Costa Rica, Facultad de Agronomía*. Escuela de Fitotecnia, Programa de Comunicación Agrícola. Asociación Bananera Nacional, p. 102.

Granada, G. 1996. Hospederos de Pseudomonas solanacearum Raza 2 en condiciones de la zona platanera del departamento del Quindío. In: Colombiana, C. (Ed.), *de Investigación Agropecuaria, Armenia*, Colombia, p. 96.

Granada, G. A. and Sequeira, L. 1983. Survival of Pseudomonas solanacearum in soil, rhizosphere and plant roots. *Canadian Journal of Microbiology* 29(4): 433–440.

Guzmán, M., Pérez, L., Villalta, R., Carr, C. and Sandoval, J. 2018. Fusarium wilt tropical race 4 (Foc TR4) a serious threat to the production of bananas and plantains: Recommendations to prevent the entry of the pathogen into banana plantations. *Corbana* 44: 1–8.

Guzmán, M., Sánchez, M. and Villalta, R. 2019. Moko o marchitamiento bacteriano del banano y el plátano. Hoja Divulgativa N° 19. Eje de Fitoprotección, Dirección de Investigaciones, CORBANA 2p.

Haddad, G. O. and Borges, F. O. 1973. *Los bananos en Venezuela. Estudio y descripción de clones de plátano y cambur.* Dirección de Investigaciones, Ministerio de Agricultura y Cría, p. 103.

Hayward, A. C. 1964. Characteristics of Pseudomonas solanacearum. *Journal of Applied Bacteriology* 27(2): 265–277. doi: 10.1111/j.1365-2672.1964.tb04912.x.

Hayward, A. C. 1991. Biology and epidemiology of bacterial wilt caused by Pseudomonas solanacearum. *Annual Review of Phytopathology* 29: 65–87.

He, L. Y., Sequeira, L. and Kelman, A. 1983. Characteristics of strains of Pseudomonas solanacearum from China. *Plant Disease* 67(12): 1357–1361.

Hong, J. C., Norman, D. J., Reed, D. L., Momol, M. T. and Jones, J. B. 2012. Diversity among Ralstonia solanacearum strains isolated from the southeastern United States. *Phytopathology* 102(10): 924–936.

Hunt, P. 1987. Current strategies for Moko control in Grenada: Technical and logistical constraints. In: *Seminar Proceedings. Improving Citrus and Banana Production in the Caribbean Through Phytosanitation, St Lucia, W.I., 2-5 September 1986*. Wageningen: CTA/Center for Ageing Research and Development in Ireland, pp. 121–129.

Ito, S., Ushijima,Y., Fujii, T., Tanaka, S., Kameyalwaki, M., Yoshiwara, S. and Kishi, F. 1998. Detection of viable cells of Ralstonia solanacearum in soil using a semi-selective medium and a PCR technique. *Journal of Phytopathology* 146(8–9): 379–384.

Kam, E. Y. and Quimio, A. J. 1977. Characteristics of Pseudomonas solanacearum strains that infect ginger (Zingiber officinale) in the Philippines. *Philippines Phytopathology* 13: 5–6.

Kelman, A. 1953. The bacterial wilt caused by Pseudomonas solanacearum. *North Carolina Agricultural Experiment Station Technical Bulletin* 99: 1–194.

Kelman, A. 1954. The relationship of pathogenicity of Pseudomonas solanacearum to colony appearance in a tetrazolium medium. *Phytopathology* 44: 693–695.

Kelman, A. and Sequeira, L. 1965. Root-to-root spread of Pseudomonas solanacearum. *Phytopathology* 55: 304–309.

Kinyua, Z. M., Miller, S. A., Ashlina, C. and Subedi, N. 2014. Bacterial wilt disease Ralstonia solanacearum: Standard operating procedure for use in diagnostic laboratories. Available at: https://ipmil.cired.vt.edu/wp-content/uploads/2014/06/SOP-Ralstonia-solanacerum-EastAfricaFinal-Apr2014-2.pdf.

Lehmann-Danziger, H. 1987. The distribution of Moko disease in Central and South America and its control on plantains and bananas. In: *Seminar Proceedings. Improving Citrus and Banana Production in the Caribbean Through Phytosanitation. St Lucia, W.I., 2–5 September 1986.* Wageningen, The Netherlands: CTA/Center for Ageing Research and Development in Ireland, pp. 130–152, 153–155.

Lozano, J. C. and Sequeira, L. 1970. Differentiation of races of Pseudomonas solanacearum by a leaf infiltration technique. *Phytopathology* 60(5): 833–838.

Lozano, J. C., Thurston, H. D. and Galvez, G. E. 1969. Control del 'Moko' del platano y banano causado por la bacteria Pseudomonas solanacearum. *Agricultura Tropicale* 25: 315–324.

Manicka, G., N. S., Lokeswari, T. S. and Nandini, K. R. 1979. Bacteria! Wilt of banana in Southern India. *Plant Diseases Reporter* 63: 525–528.

Mansfield, J., Genin, S., Magori, S., Citovsky, V., Sriariyanum, M., Ronald, P., Dow, M., Verdier, V., Beer, S. V., Machado, M. A., Toth, I., Salmond, G. and Foster, G. D. 2012. Top 10 plant pathogenic bacteria in molecular plant pathology. *Molecular Plant Pathology* 13(6): 614–629. doi: 10.1111/j.1364-3703.2012.00804.x.

Martínez, G. A. and García, R. F. 2003. Manejo de la enfermedad del Moko o Ereke en el cultivo del plátano para la Orinoquía colombiana. Boletín Divulgativo No. 9. Corporación Colombiana de Investigación Agropecuaria (Corpoica). *Meta, Colombia*. Villavicencio: Japan International Cooperation Agency (JICA), p. 20.

Martinez, I. and Guzmán, M. 2011. *Moko o marchitamiento bacteriano del banano y plátano (Ralstonia solanacearum Yabuuchi et al.). Proyecto demostrativo con implementación de BPA en el cultivo del banano. Plegable divulgativo N° 1.* Costa Rica: Corporación Bananera Nacional Dirección de Investigaciones, Sección de Fitopatología.

Martyn, E. B. 1934. A note on plantain and banana diseases in British Guiana with special references to wilt. *The Journal of the Board of Agriculture of British Guiana* 5: 120–183.

McDonald, F. D. 1985. Identification and distribution of strains of Pseudomonas solanacearum causing Moko disease in Guyana. *Tropical Agrículture (Trinidad)* 62: 113–114.

Mello Veras, S. de, Gasparotto, L., Boher, B., Gato, A. and M. G. 2002. Availação da bananeira cultivar Pelipita á Ralstonia solanacearum raça 2. Coletânea dos trabalhos da CDSV/AM. *Ministério da Agricultura, Pecuária e Abastecimento, Delegacia Federal de Agricultura no Amazonas, Manaus, Brazil*, pp. 94–96.

Molina, A. B. 1999. Fruit rot diseases of cooking banana in Southeast Asia. *Infomusa* 8: 29–30.

Molina, A. B. 2006. Managing bacterial wilt/fruit rot disease of banana in Southeast Asia. In: Karamura, E., Osiru, M., Blomme, G., Lusty, C. and Picq, C. (Eds.), *Proceedings of the Banana Xanthomonas Wilt Regional Preparedness and Strategy Development Workshop, Held in Kampala, Uganda, 14–18 February 2005*. INIBAP Regional Office for Asia and the Pacific, pp. 26–31.

Momol, P. G. C. and Champoiseau, T. M. 2009. Bacterial wilt of tomato. Available at: https://plantpath.ifas.ufl.edu/rsol/RalstoniaPublications_PDF/BacterialWiltTomato_Revised_PDF.pdf.

Obregón, M., Rodríguez, P., Morales, J. and Salazar, M. 2008. Hospedantes de Ralstonia solanacearum en plantaciones de banano y plátano en Colombia. *Revista Facultad Nacional de Agronomía Medellín* 2: 4518–4526.

Obregón, M., Rodríguez, P. A. and Salazar, M. 2011. Supervivencia de Ralstonia solanacearum en suelo y tejido de plantas de banano en Urabá, Colombia. *Fitosanidad* 15(2): 91–97.

Pardo, J. M., López-Álarez, D., Ceballos, G., Álvarez, E. and Cuellar, W. J. 2019. Detection of Ralstonia solanacearum phylltype II, race 2 causing Moko disease and validation of genetic resistance observed in the hybrid plantain FHIA-21. *Tropical Plant Pathology* 44(4): 371–379. doi: 10.1007/s40858-019-00282-3.

Paudel, S., Dobhal, S., Alvarez, A. M. and Arif, M. 2020. Taxonomy and phylogenetic research on Ralstonia solanacearum species complex: A complex pathogen with extraordinary economic consequences. *Pathogens* 9(11): 886. doi: 10.3390/pathogens9110886.

Pérez-Sánchez, L., Sánchez, M., Carr, C., Villalta, M. and Guzmán, M. 2019. *Medidas de Bioseguridad para la Prevención del Marchitamiento por Fusarium Raza 4 Tropical (Foc R4T) en Fincas Bananeras. Boletín N° 6, Dirección de Investigaciones, CORBANA. 39 p.*

Phelps, R. H. 1987. The status of Moko disease in the Caribbean. In: *Seminar Proceedings. Improving Citrus and Banana Production in the Caribbean Through Phytosanitation, St Lucia*, W.I., 2–5 September 1986. Wageningen, The Netherlands: CTA/Center for Ageing Research and Development in Ireland, pp. 100–107.

Power, R. H. 1976. Moko: A new bacterial disease on banana and plantain in Suriname. *De Surinaamse Landbouw* 24: 85–92.

Prieto, J., Morales, J. and Salazar, M. 2012. Identification of new hosts for Ralstonia solanacearum (Smith) race 2 from Colombia. *Protección Veg* 3: 151–161.

Prior, P., Ailloud, F., Dalsing, B. L., Remenant, B., Sánchez, B. and Allen, C. 2016. Genomic and proteomic evidence supporting the division of the plant pathogen Ralstonia solanacearum into three species. *BMC Genomics* 17: 90. doi: 10.1186/s12864-016-2413-z.

Prior, P. and Fegan, M. 2005. Recent developments in the phylogeny and classification of Ralstonia solanacearum. *Acta Horticulturae* 695: 127–136.

Quinón, V. and Aragaki, M. 1963. Bacterial wilt of bird-of-Paradise caused by Pseudomonas solanacearum. *Phytopathology* 53: 1115–1116.

Raymundo, A. K. 1998. Analysis of genetic variation of a population of banana infecting strains of Ralstonia solanacearum. In: Prior, P., Allen, C. and Elphinstone, J. (Eds.), *Bacterial Wilt Disease: Molecular and Ecological Aspects*. Heidelberg: Springer, pp. 56–60.

Raymundo, A. K., Orlina, M. E., Lavina, W. A. and Opina, N. L. 2005. *Comparative Genome Plasticity of Tomato and Banana Strains of Ralstonia solanacearum in the Philippines*. Allen Press.

Rillo, A. R. 1979a. Bacterial wilt of banana in the Philippines. *FAO Plant Protection Bulletin* 27: 105–108.

Rillo, A. R. 1979b. Le fletrissement bacterian du banan ier aux Philippines. *Phyto Food and Agriculture Organization* 27 105–108.

Robinson, J. R. and Galan Sauco, V. 2010. *Bananas and Plantains* (2nd edn.). Wallingford, Oxfordshire: CABI Publishing International.

Rorer, J. B. 1911. A bacterial disease of bananas and plantains. *Phytopathology* 1: 45–49.

Safni, I., Cleenwerck, I., De Vos, P., Fegan, M., Sly, L. and Kappler, U. 2014. Polyphasic taxonomic revision of the Ralstonia solanacearum species complex: Proposal to emend the descriptions of Ralstonia solanacearum and Ralstonia syzygii and reclassify current R. syzygii strains as Ralstonia syzygii subsp. syzygii subsp. nov., R. solanacearum phylotype IV strains as Ralstonia syzygii subsp. indonesiensis subsp. nov., banana blood disease bacterium strains as Ralstonia syzygii subsp. celebesensis subsp. nov. and R. solanacearum phylotype I and III strains as Ralstonia pseudosolanacearum sp. nov. *International Journal of Systematic and Evolutionary Microbiology* 64(9): 3087–3103. doi: 10.1099/ijs.0.066712-0.

San Juan, M. 1979. Endemic diseases of bananas in the Philippines. In: *Proceedings of Symposium of Philippine Phytopathology*. University of the Philippines at Los Baños, Laguna, Philippines, December 14–15, 1977, pp. 1917–1977.

Schaad, N. W., Jones, J. B. and Chun, W. 2001. *Laboratory Guide for the Identification of Plant Pathogenic Bacteria* (3rd edn.). St. Paul, MN: American Phytopathological Society.

Sequeira, L. 1962. Control of bacterial wilt of bananas by crop rotation and fallowing. *Tropical Agriculture (Trinidad)* 39: 211–217.

Sequeira, L. 1998. Bacterial wilt: The missing element in international banana improvement programs. In: Prior, P., Allen, C. and Elphinstone, J. (Eds.), *Bacterial Wilt Disease: Molecular and Ecological Aspects*. Heidelberg: Springer, pp. 6–14.

Sequeira, L. and Averre, C. W. 1961. Distribution and Pathogenicity of Strains of Pseudomonas solanacearum from Virgins Soils in Costa Rica. *Plant Disease Report* 45: 435–440.

Sequeira, L. C. 1958. Bacterial wilt of bananas: Dissemination of the pathogen and control of the disease. *Phytopathology* 48: 64–69.

Smith, E. F. 1896. *A Bacterial Disease of the Tomato, Eggplant, and Irish Potato (Bacillus solanacearum n. sp.)*. Washington: United States Department of Agriculture, Division of Vegetable Physiology and Pathology. Bulletin No. 12, pp. 1–28.

Stover, R. H. 1972. *Banana, Plantain and Abaca Diseases*. Kew, UK: Commonwealth Mycological Institute, p. 316.

Stover, R. H. and Richardson, D. L. 1968. Pelipita, an ABB Bluggoe-type plantain resistant to bacterial and fusarial wilts. *Plant Disease Reporter* 52: 901–903.

Stover, R. H. and Simmonds, N. W. 1987. *Bananas* (3rd edn.). Harlow, UK: Longmans, p. 468.

Thurston, H. D. 1984. Tropical plant diseases. *American Phytopathology Società St. Paul, USA*, p. 208.

Thwaites, R., Eden-Green, S. J. and Black, R. 1999. Diseases caused by bacteria. In: Jones, D. R. (Ed.), *Diseases of Banana, Abacá and Enset*. Wallington, UK: CABI Publishing, pp. 213–240.

Tran, T. M., Jacobs, J. M., Huerta, A., Milling, A., Weibel, J. and Allen, C. 2016. Sensitive, secure detection of race 3 biovar 2 and native U.S. strains of Ralstonia solanacearum. *Plant Disease* 100(3): 630–639.

Valencia-Valencia, L., Álvarez-Cabrera, E. and Castaño-Zapata, J. 2014. Resistencia de 34 Genotipos de Plátano (Musa AAB) y Banano (Musa AAA) a Cinco Cepas de Ralstonia solanacearum Raza 2 (Smith). *Revista Agronomía* 22(2): 21–34.

Vivas, V., Urdaneta, I., Rangel, S. and Hernández, J. 2009. Caracterización e incidencia de Ralstonia solanacearum Smith en plantas de Musa AAB en el Sector "El Roble", sur del Lago de Maracaibo, Venezuela. *Revista UDO Agrícola* 9(2): 383–392.

Wardlaw, C. W. 1935. *Diseases of the Banana and of the Manila Hemp Plant*. London: Macmillan, p. 615.

Wardlaw, C. W. 1972. *Banana Diseases Including Plantains and Abaca* (2nd edn.). London: Longmans, p. 648.

Wood, A. C. 1984. Moko disease: Atypical symptoms induced by afluidal variants of Pseudomonas solanacearum in banana plants. *Phytopathology* 74(8): 972–976.

Yabuuchi, E., Kosako, Y., Hano, I., Hotta, H. and Nishiuchi, Y. 1995. Transfer of two Burkholderia and an Alcaligenes species to Ralstonia gen. nov.: Proposal of Ralstonia pickettii (Ralston, Palleroni, and Douderoff 1973) comb. nov., Ralstonia solanacearum (Smith, 1896) comb. nov. and Ralstonia eutropha (Davis 1969) comb nov. *Microbiology and Immunology* 39(11): 897–904. doi: 10.1111/j.1348-0421.1995.tb03275.x.

Yabuuchi, E., Kosako, Y., Oyaizu, H., Yano, I., Hotta, H., Hashimoto, Y., Ezaki, T. and Arakawa, M. 1992. Proposal of Burkholderia gen. nov. and transfer of seven species of the genus Pseudomonas homology group II to the new genus, with the type species Burkholderia cepacia (Palleroni and Holmes 1981) comb. nov. *Microbiology and Immunology* 36(12): 1251–1275. doi: 10.1111/j.1348-0421.1992.tb02129.x.

Zulperi, D. and Sijam, K. 2014. First report of Ralstonia solanacearum Race 2 biovar 1 causing Moko disease of banana in Malaysia. *Plant Disease* 98(2): 275.

Zulperi, D., Sijam, K., Ahmad, Z. A. M., Awang, Y., Ismail, S. I., Asib, N. and Hata, E. M. 2016. Genetic diversity of Ralstonia solanacearum phylotype II sequevar 4 strains associated with Moko disease of banana (Musa spp.) in Peninsular Malaysia. *European Journal of Plant Pathology* 144(2): 257–270.

Chapter 9

Banana Blood disease

Jane Ray, Northern Territory Government, Australia; Siti Subandiyah, Universitas Gadjah Mada, Indonesia; and André Drenth, The University of Queensland, Australia

1 Introduction

Blood disease is a bacterial wilt of banana caused by *Ralstonia syzygii* subsp. *celebesensis* (Safni et al., 2014). It causes distinctive symptoms characterised by reddish-brown vascular staining, wilting and senescing of leaves, and pulp rot of green fruit, making them inedible (Tjahjono and Eden-Green, 1988). The name 'Blood disease' comes from the discoloured sap that oozes from fresh cuts on diseased pseudostems and fruit. Blood disease was first reported just over 100 years ago from the Selayar islands and southern Sulawesi and until about 30 years ago remained confined to that area (Rijks, 1916; Gäumann, 1921; Ray et al., 2021). Since the late 1980s, the disease has spread and is now found on several of the larger islands of the Indonesian archipelago and was recently also identified from peninsular Malaysia, causing significant economic losses over a large and increasing geographic area (Teng et al., 2016; Safni et al., 2018; Ray et al., 2021).

The banana variety Kepok (syn: Nipah) is popular and grown widely in Indonesia and Malaysia by smallholder farmers, providing a source of income and food (Buddenhagen, 2009; Teng et al., 2016). The Kepok (ABB) variety is consumed raw as fruit or cooked and is highly susceptible to Blood disease

http://dx.doi.org/10.19103/AS.2022.0108.09

affecting livelihoods in the region (Robinson, 1996; Ray et al., 2021). Experiments also demonstrated that Cavendish, the most commonly grown and traded banana variety globally, is susceptible, and at this stage, no resistance has been identified as yet within the genus *Musa* (Ray et al., 2022b). The pathogen invades the banana plant in a systemic fashion, and suckers originating from a contaminated mat may become infected and develop symptoms (Ray et al., 2022a; Gäumann, 1921; Hadiwiyono, 2011). Recently, Ray (2022a, 2022b) experimentally determined that Blood disease spreads from diseased to healthy plants predominantly through mechanical transmission by insects, animals, and tools within and between banana fields. Long distance spread was demonstrated to be able to occur through infection of asymptomatic suckers used for planting material (Ray et al., 2022a) as hypothesised previously (Gäumann, 1921; Eden-Green, 1994b).

There are no known effective chemical or biological control methods. Once a banana plant is infected, there is no cure, and the plant will eventually die. The only available disease management is through cultural control methods, such as the use of clean planting material, avoiding infection, and removing sources of infection, which highlights the importance of area-wide disease management and efforts to prevent local and long-distance spread of Blood disease. The rapid spread in the recent past is of concern and has given rise to a major emerging threat to banana production in Southeast Asia and beyond (Buddenhagen, 2009; Ray et al., 2021; Ray et al., 2022a; Hermanto and Emilda, 2013).

2 Disease symptoms

Infection of banana plants can occur in different ways which may produce differences in the sequence of expression of symptoms. When infection occurs through the male bell, the first signs of disease are the appearance of small droplets of ooze on the male bell which is followed by retention and desiccation of the bracts on the male bell (Fig. 1a and b) (Ray et al., 2021; Ray et al., 2022a). During periods of high moisture availability, the male bell may ooze bacteria approximately 2 weeks after infection for several days (Ray et al., 2022a; Hadiwiyono, 2011). The pathogen progresses internally from the bell via the vascular tissue of the lower rachis, staining it internally reddish brown while moving towards the fruit. From that point onwards, discolouration of the fruit pulp and internal vascular staining of the rachis, peduncle, and true stem progress simultaneously (Fig. 2a, b and c) (Ray et al., 2021). Foliar symptoms, expressed as leaf wilt and necrosis, occur after the pathogen has reached the corm and colonised the leaf sheaths that make up the pseudostem. As the disease progresses, the bell and lower rachis become necrotic (Fig. 1c) (Ray et al., 2021).

Figure 1 Symptom progression of Blood disease following infection through the male bell. (a) Droplets of bacteria oozing from a Cavendish male bell. (b) Cavendish male bell with desiccated and retained bracts. (c) Kepok banana inflorescence with advanced symptoms showing a necrotic male bell and rachis with green fruit and rotten pulp.

During the early stages of infection, the fruit looks externally unaffected, the skin remains green while the vascular bundles within the pedicel and fruit skin become stained, and the fruit pulp discolours and rots (Fig. 2a). Diseased fruit is inedible, and the texture and colour of the pulp varies (Fig. 3). Pulp texture can be dry, wet, or slimy and can be solid, or hollow forming a cavity or any combination of these (Tjahjono and Eden-Green, 1988). The colour of the pulp can be cream, yellow, light brown, reddish, dark brown, or a combination of these colours, and the pulp sometimes has a dark halo just under the fruit skin (Fig. 3). For most varieties, all fruits in the bunch express symptoms simultaneously (Eden-Green, 1994b).

Once the pathogen reaches the corm, the vascular staining progresses through the leaf sheaths, usually colonising those toward the true stem first (Fig. 2c) (Gäumann, 1922a; Stover and Espinoza, 1992; Ray et al., 2021). Vascular staining can be seen in the petiole (Fig. 2d). Close inspection reveals that some of the more recently colonised xylem vessels of the leaf sheaths express vascular discolouration of a lighter colour and intermittent vascular discolouration (Ray et al., 2021). Infected banana leaves express wilting, sometimes showing chlorosis prior to collapsing and becoming necrotic, dry, and brown (Fig. 4) (Ray et al., 2021). The leaves hang down against the stem leaving the bunch exposed (Fig. 4a) (Gäumann, 1922a). In the later stages of the disease, the fruit can appear to ripen unevenly (Fig. 4a), and in some cases, the fruit skin may split (Tjahjono and Eden-Green, 1988; Eden-Green and

Figure 2 Vascular staining caused by *Ralstonia syzygii* subsp. *celebesensis* in Kepok banana plants. (a) Staining in the rachis, fruit skin, and pedicle. (b) peduncle. (c) staining of the pseudostem concentrated toward the true stem. (d) Vascular staining extends into the petiole.

Sastraatmadja, 1990). The whole plant eventually collapses (Ray et al., 2021; Ray et al., 2022a). When the disease progresses to the corm, it may result in the subsequent infection and expression of symptoms in the suckers (Ray et al., 2022a).

Infection can also occur through the roots and by tool transmission during de-suckering or de-leafing (Ray et al., 2022a; Baharuddin, 1994). In such cases, the disease causes the same symptoms as for infection through the male bell,

Figure 3 Banana fruits infected with *Ralstonia syzygii* subsp. *celebesensis* appear normal on the outside but the fruit pulp is rotten and inedible. The pulp rot varies in colour and texture. (a, b) Symptoms in Kepok banana fruit. (c, d) Symptoms in Cavendish banana fruit.

except that the banana leaves generally express earlier and stronger chlorosis and later become black and water soaked or brown and dry (Fig. 4b) (Ray et al., 2022a). Very young shoots may turn black and rot. Young plants, when infected, usually wilt and collapse before shooting, although infection can occur at any growth stage (Ray et al., 2022a). Suckers attached to an infected mother plant are subject to a variable latency period and may become infected at any growth stage, and in some instances, healthy bunches are produced (Ray et al., 2022a; Eden-Green, 1994b).

Blood disease and *Fusarium* wilt caused by *Fusarium oxysporum* f.sp. *cubense* (Foc) can be confused as they express similar symptoms in field-grown banana plants. Both diseases cause the leaves to yellow and wilt and the vascular strands in the pseudostem to stain. The difference is that *R. syzygii* subsp. *celebesensis* causes rot of the fruit pulp and vascular staining concentrated in the true stem and in the leaf sheaths toward the centre of the pseudostem (Ray et al., 2021). In contrast, Foc causes vascular staining in the leaf sheaths toward the outside of the pseudostem and does not cause the fruit pulp to discolour and rot (Ploetz et al., 2003).

Figure 4 Banana plants showing symptoms of Blood disease. (a) Leaves wilt, become necrotic and hang down against the pseudostem leaving the bunch exposed; the bell is necrotic and the bunch shows uneven fruit ripening. (b) The arrow denotes a diseased necrotic mother plant, all shoots attached to the corm show leaf wilt, chlorosis, scorch, and necrosis of the leaves.

3 Pathogen characteristics and diagnosis

There are three major bacterial wilt diseases of banana expressing a range of similar symptoms. Two of the diseases are members of the *Ralstonia solanacearum* species complex (*R. solanacearum sensu lato*). Moko (Bugtok) caused by *R. solanacearum sensu stricto* and Blood disease caused by *R. syzygii* subsp. *celebesensis* (Ray et al., 2021; Safni et al., 2014). The third disease is banana *Xanthomonas* wilt caused by *Xanthomonas vasicola* pv. *musacearum* (Blomme, 2023). Distinguishing between the diseases caused by *Xanthomas* and *Ralstonia* spp. is based on the different growth morphology of these bacteria on Petri plates after isolation. Distinguishing between Moko and Blood disease caused by *Ralstonia* spp. is more complex and benefits from understanding the current taxonomy.

3.1 Taxonomy and characterisation of Ralstonia solanacearum sensu lato

Blood disease and Moko's causal organisms reside with the diverse taxon *R. solanacearum sensu lato*, classified within the family Burkholderiaceae, order

Burkholderiales, class Betaproteobacteria of the phylum Proteobacteria (Yabuuchi et al., 1995). Taxonomists and plant pathologists have historically struggled to create a functional classification system for the diverse taxon *R. solanacearum sensu lato* responsible for wilt diseases in over 200 different plant species in more than 50 different families (Kelman, 1953). Over time, this group has been classified in various ways using several techniques adding to the confusion (Table 1).

Ralstonia solanacearum sensu lato was initially classified into five races based on host ranges, then into five biovars based on carbohydrate utilisation (Buddenhagen et al., 1962; He, 1983; Hayward, 1964; Cook et al., 1989). The taxon was subsequently separated into multi-locus genotypes (MLG) and two divisions based on restriction fragment length polymorphism (RFLP) groupings (Table 1) (Cook et al., 1989). Several studies based on sequencing of gene regions supported the divisions and separated the strains into four broad genetic groups called phylotypes (Table 1) (Taghavi et al., 1996; Fegan and Prior, 2005; Remenant et al., 2010). These phylotypes were further divided into sequevars or sequence variants that are groups of isolates clustering based on partial endoglucanase gene sequences (Fegan and Prior, 2005). This sequevar system can discriminate between some distinct bacterial diseases within *R. solanacearum sensu lato*, but not those within phylotype IV, including Blood disease (Fegan and Prior, 2005).

A more recent taxonomic revision of *R. solanacearum sensu lato* divided the taxon into three species corresponding to the phylotypes: *R. solanacearum* (phylotype II), *R. pseudosolanacearum* (phylotypes I and III), and *R. syzygii* (phylotype IV) (Table 1) (Safni et al., 2014). Using this approach *Ralstonia syzygii* was further separated into three subspecies namely, *R. syzygii* subsp. *celebesensis* (Blood disease of banana), *R. syzygii* subsp. *Syzygii* (Sumatra disease of clove), and *R. syzygii* subsp. *indonesiensis* (bacterial wilt of Solanaceae and clove) (Safni et al., 2014). This taxonomic revision is currently widely accepted by the scientific community (Prior et al., 2016).

Ralstonia solanacearum sensu lato has a bipartite genomic structure consisting of two large replicons, a ~3.6-Mb chromosome and a ~1.5–2.1-Mb megaplasmid (Remenant et al., 2010; Salanoubat et al., 2002). The chromosome length is similar for all members of *R. solanacearum sensu lato* (Remenant et al., 2011). Whereas the size of the megaplasmid for *R. syzygii* subsp. *celebesensis* at ~1.6 Mb and *R. syzygii* subsp. *syzygii* at ~1.7 Mb are significantly smaller than all other strains within *R. solanacearum sensu lato* at ~2.1Mb (Remenant et al., 2011).

Strains of *Ralstonia solanacearum sensu lato* also exhibit distinctive growth properties when grown on casamino acid peptone glucose medium amended with 2,3,5-triphenyltetrazolium chloride (TZC) in Petri plates, facilitating selection of the target organism (Kelman, 1954). In particular, *R. syzygii* subsp.

Table 1 Historic classification and current taxonomy and nomenclature of the *Ralstonia solanacearum* species complex, including host range and geographic origin.

Category	Data	***Ralstonia solanacearum* species complex (*R. solanacearum sensu lato*)**						Reference
Species	Molecular, DNA-DNA hybridiz. & *MALDI-TOF-MS	*R. pseudosolanacearum*		*R. solanacearum*	*Ralstonia syzygii*			(Safni et al., 2014)
Subspecies					*R. syzygii* subsp. *indonesiensis*	*R. syzygii* subsp. *celebesensis*	*R. syzygii* subsp *syzygii*	(Safni et al., 2014)
Phylotype	partial ITS, *hrpB*, & *egl* gene seq.	Phylotype I	Phylotype III	Phylotype II	Phylotype IV	Phylotype IV	Phylotype IV	(Fegan and Prior, 2005; Prior and Fegan, 2005)
Sequevar	*egl* gene seq.	12, 13, 14, 15, 16, 17, 18, 34, 44, 48	19, 20, 21, 22, 23	1, 2, 3, 4, 5, 6, 7,28, 36, 39, 24, 35, 54, 55, 56, 57, 41, 50	8, 11, 10	10	9	(Fegan and Prior, 2005; Santiago et al., 2017; Xu et al., 2009; Prior and Fegan, 2005)
Multi-locus genotypes (MLG)	**RFLP groups	8, 9, 10, 11, 12, 13, 14, 15, 16, 17, 18, 19, 20, 21, 22, 23	?	1, 2, 3, 4, 5, 6, 7, 24, 25, 26, 27, 28, 29, 30, 31, 32, 33, 34	?	?	?	(Cook et al., 1989; Prior and Fegan, 2005)
Division	RFLP & 16S rRNA gene seq.	Division 1	?	Division 2a	Division 2b	Division 2b	Division 2b	(Taghavi et al., 1996; Cook et al., 1989)
Biovar	Carbohydrate utilisation	3, 4, 5	N2, 1, 2T	N2, 1, 2, 2T	2, 1, 2T	N/A	N/A	(Fegan and Prior, 2005; Fegan and Prior, 2006; Hayward, 1964; Cook et al., 1989)
Biotype	Carbohydrate utilisation	8	9,10,11	1,2,3,4,5,6,7	11			(Hayward, 1964; Fegan and Prior, 2005)
Race	Host range	1, 4, 5	?	1, 2, 3	?	N/A	N/A	(Buddenhagen et al., 1962; He, 1983)
Host range	Host	Wide range, Solanaceae & other (mulberry, ginger)	Wide range, Solanaceae & other	Banana, potato, Solanaceae	Solanaceae & clove	Banana	Clove	(Buddenhagen et al., 1962; He, 1983; Safni et al., 2014)
Common name		Bacterial wilt	Bacterial wilt	Moko, Bacterial wilt	Bacterial wilt	Blood disease	Sumatra disease	
Probable origin	Sampling location	Asia	Africa	America	Indonesia, Australia, Japan	Indonesia	Indonesia	(Prior and Fegan, 2005)

Exception: ARCH0732, Biovar 2, Division 1, Ph IV, Seq 11 (Taghavi et al., 1996; Prior, 2005).

[a] Matrix-assisted laser desorption/ionisation-time of flight mass spectrometry (MALDI-TOF-MS).

[b] Restriction fragment length polymorphism (RFLP).

celebesensis produces distinctive non-fluidal red colonies with a white margin after 4–5 days (Fig. 5a) (Ray et al., 2021). In contrast, *R. syzygii* subsp. *syzygii* does not grow under these conditions, and virulent isolates of *R. syzygii* subsp. *indonesiensis* and most other *Ralstonia* strains causing bacterial wilt grow more rapidly, producing fluidal colonies after 2–3 days (Kelman, 1954; Roberts et al., 1990).

Ralstonia syzygii subsp. *celebesensis* can be differentiated from strains of *R. solanacearum sensu stricto* that cause various wilt diseases, including Moko disease of banana through biochemical tests and host range. Bacterial strains causing Moko disease cause wilt on tomato, eggplant, and capsicum plants whilst *R. syzygii* subsp. *celebesensis* is not pathogenic on these hosts (Baharuddin, 1994; French and Sequeira, 1970). Isolates of *R. syzygii* subsp. *celebesensis* are different to *R. solanacearum* in that they are unable to reduce nitrate or oxidase glucose, sucrose, mannose, or ribose (Eden-Green, 1994a). Additionally, *R. syzygii* subsp. *celebesensis* utilised only 51% of the carbon sources compared to *R. solanacearum,* which was able to utilise 97% of those tested (Eden-Green, 1994a).

3.2 Sample selection, isolation and storage

Sample selection, collection, and transport are vital elements for obtaining a bacterial isolate from symptomatic banana material to definitively identify *R. syzygii* subsp. *celebesensis* using molecular diagnostics. The selection of banana plant materials in the field is important as the causal bacteria is slow growing. Isolations from vascular staining within: (1) fruit skin or pedicel are preferred, followed by (2) petiole, then (3) the outermost symptomatic layers of the pseudostem (Ray et al., 2021). Sample collection should consist of several intact fruits, sections of the petiole, or pseudostem tissue, all with visible vascular staining. Place the samples into a plastic bag with a dry paper towel. Protect samples from extreme heat and sunlight but keep them at room temperature during transportation and until the bacterial isolation is complete (Ray et al., 2021). Most *Ralstonia* spp. are tropical pathogens theorised to be cold sensitive and may die when refrigerated.

A detailed description of the isolation of *R. syzygii* subsp. *celebesensis* from plant tissues has been published (Ray et al., 2021; Rincón-Flórez et al., 2022) and is briefly outlined here. For isolation, select tissue with vascular staining, preferably from the fruit's skin, the pedicle, or the petiole. If only pseudostem tissue is available, then isolate from tissues showing recent vascular staining that is lighter in colour or intermittent as isolations are difficult or impossible from older infections or rotten tissues (Ray et al., 2021). Place a small amount of chopped vascular stained tissue in sterile water and incubate for 10 min.

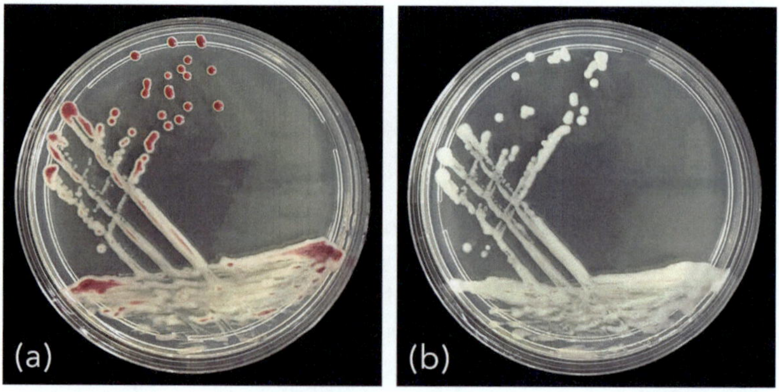

Figure 5 Culture morphology of *Ralstonia syzygii* subsp. *celebesensis* after 5 days at 28°C. (a) Growth on casamino acid peptone glucose medium amended with tetrazolium chloride (½ TZC) in Petri plate. (b) Growth on casamino acid peptone glucose medium (CPG) in Petri plate.

Streak a small amount of the liquid onto 1/2 TZC + C media (10 g L^{-1} peptone, 1 g L^{-1} casein hydrolysate, 5 g L^{-1} D-glucose, 16 g L^{-1} Bacto agar, 25 mg L^{-1} TTC, 100 mg L^{-1} C) in Petri plates (Kelman, 1954; Ray et al., 2021). After 3-5 days, select colonies with characteristic morphology to obtain pure cultures for identification or storage (Fig. 5a) (Ray et al., 2021). Pure cultures can be grown on casamino acid peptone glucose medium (CPG; 10 g L^{-1} peptone, 1 g L^{-1} casein hydrolysate, 5 g L^{-1} D-glucose, 16 g L^{-1} Bacto agar) for biochemical testing, DNA extraction, and storage (Fig. 5b) (Ray et al., 2021; Rincón-Flórez et al., 2022). Fresh cultures can be stored in sterile water at room temperature in the dark for approximately 1–3 years or long term on cryobeads at $^{-}80$°C (Ray et al., 2021; Schaad et al., 2001).

3.3 *Molecular assays to detect* **Ralstonia syzygii** *subsp.* **celebesensis**

Two specific molecular diagnostic assays based on standard PCR and analysis through gel electrophoresis are reported to detect *R. syzygii* subsp. *celebesensis*. An assay using the 121 primer set amplifies a 317-bp product from the megaplasmid (intergenic region), and the BDB2400 primer set amplifies a 131-bp product from the chromosome (Das, 2004; Tan, 2003; Kubota et al., 2011). These primer sets were initially validated on less than 23 isolates of *R. syzygii* subsp. *celebesensis* originating from a limited geographic area (Das, 2004; Kubota et al., 2011). A recent rigorous validation of these assays involving specificity, sensitivity, and robustness using 339 samples, including 174 isolates of *R. syzygii* subsp. *celebesensis* from diverse geographic origins demonstrated that the assay based on the BDB2400 primer set lacked specificity as it failed to detect 38 *R. syzygii* subsp. *celebesensis* isolates originating from the Sulawesi region of Indonesia. It was concluded that the 121 assay is specific, sensitive, reliable, and fit for purpose (Rincón-Flórez et al., 2022). In addition to this PCR-based assay, they also developed a real-time assay to detect *R. syzygii subsp. celebesensis* using a DUF4043 primer set and probe which amplifies a 57-bp product from the chromosome (Rincón-Flórez et al., 2022). This assay has also been rigorously validated (Rincón-Flórez et al., 2022).

For a definitive identification such as that required for a countries 'first report', two different reliable molecular assays that target different parts of the genome are recommended. This dual assay approach provides greater confidence in the identification, overcoming risks of false negative or false positive results associated with a genetic mutation within the population affecting a primer target site. Consequently, a definitive identification can be obtained for *R. syzygii subsp. celebesensis* using the standard PCR-based 121 assay targeting the megaplasmid combined with the real-time DUF4043 assay targeting the chromosome, and by ensuring that all amplifications include both positive and negative controls (Rincón-Flórez et al., 2022).

4 Disease cycle and epidemiology

Short distance or plant-to-plant dispersal of Blood disease is associated with mechanical transmission from diseased to healthy plants through the movement of the bacterium by insects, animals, and tools or via contaminated soil or water (Fig. 6) (Ray et al., 2022b; Ray et al., 2022a; Buddenhagen, 2009; Stover and Espinoza, 1992; Gäumann, 1922a). Long-distance dispersal is predominantly through the movement of suckers derived from diseased parent plants (Fig. 6) and also may occur through the movement of diseased fruit (Gäumann, 1922a; Stover and Espinoza, 1992; Tjahjono and Eden-Green, 1988; Ray et al., 2022a).

4.1 Inoculum sources

R. syzygii subsp. *celebesensis* is highly infective, and a low inoculation concentration of 10^2 CFU.mL^{-1} is adequate to initiate disease (Ray et al., 2022b; Ray et al., 2022a). Approximately, 2 weeks after inoculation of the male bell and during periods of high moisture availability, the bract scars and flower cushions become infectious (Ray et al., 2022a; Hadiwiyono, 2011). These plant parts ooze infectious sap and remain infectious for an undetermined period but likely several days (Ray et al., 2022a; Hadiwiyono, 2011). Infections become systemic, and symptomatic plant parts of the fruit, pseudostem, bunch peduncle, and rachis when cut yield infectious sap capable of initiating disease (Hadiwiyono,

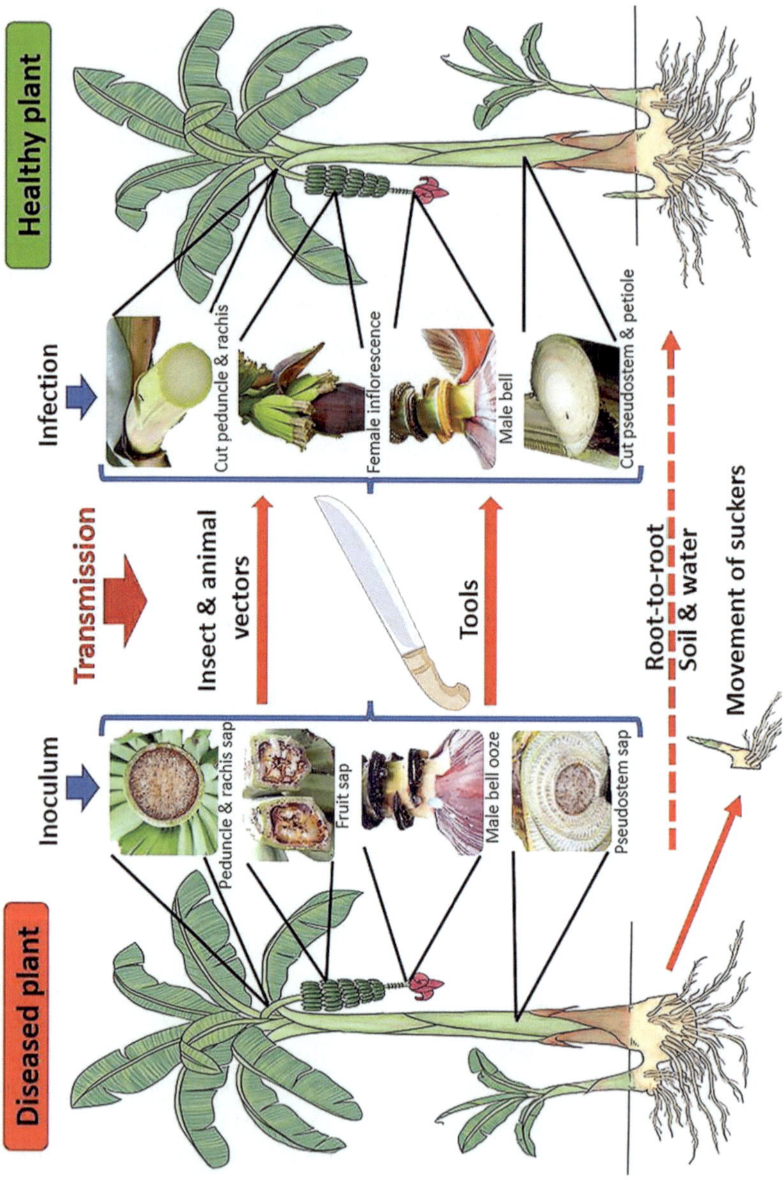

Figure 6 Disease cycle of *Ralstonia syzygii* subsp. *celebesensis*, the causal agent of Blood disease of banana. Banana plants drawn by Nur Izza F. Himmah.

2011; Ray et al., 2022a). The role and importance of below-ground plant parts such as the roots as sources of inoculum remains unclear (Ray et al., 2022a).

4.2 Transmission

The male and female parts of the banana inflorescence, various fresh-cut plant parts, and the roots when wounded act as infection courts (Ray et al., 2022a; Ray et al., 2022b; Baharuddin, 1994). Infection occurs through pruning and harvesting wounds, including cuts to petioles, pseudostems, bunch peduncles, and the corm (Ray et al., 2022a). It has recently been experimentally demonstrated that infection occurs through open xylem vessels occurring at the female flowers, the male bell bract scar, the male bell flower cushion, the rachis when denavelled, and de-flowered fruit (Ray et al., 2022b).

Transmission of *R. syzygii* subsp. *celebesensis* occurs from the movement of sap and ooze from any infectious plant part on the bodies of insects, birds and bats, or through tools to any susceptible plant part (Ray et al., 2022a; Ray et al., 2022b). *Ralstonia syzygii* subsp. *celebesensis* is highly tool transmissible, and plant xylem exposed through pruning or harvesting activities by growers or contract harvesters, reveal inoculum sources, and infection courts (Ray et al., 2022a). Transmission of *R. syzygii* subsp. *celebesensis* has been shown to occur from diseased to healthy banana roots but does not appear to occur readily (Ray et al., 2022a). Transmission may also occur through water and soil contaminated with *R. syzygii* subsp. *celebesensis*, although preliminary research suggests that the role of water and soil is minor in transmission, which may be due to the lack of flagella in this bacterium in contrast to Ralstonia solanacearum sensu stricto causing Moko {Ray, 2022 #442; Eden-Green and Sastraatmadja, 1990).

4.3 Survival or overwintering

Ralstonia syzygii subsp. celebesensis survives in living host (*Musa* spp.) tissue. The bacteria's survival in other potential hosts, their roots or the rhizosphere is unknown, and no natural infection has been reported beyond *Musa* spp. Symptoms of wilt developed following inoculation of *R. syzygii subsp. celebesensis* in various potted plants, including *Heliconia* spp. *Canna* sp., *Strelitzia* sp., and some weeds (Baharuddin, 1994). This is different compared to the role alternative hosts play in the survival of other *Ralstonia* spp. including Moko disease (Granada and Sequeira, 1983). Gäuman (1923) reported that *Ralstonia syzygii* subsp. *celebesensis* survives in soil for 6–12 months. No subsequent research on soil survivability, soil transmission, infectivity, or amelioration treatments exist to support or expand upon these findings.

4.4 Epidemiology

There are epidemic and endemic phases of banana Blood disease. The epidemic phase is characterised by invasion into an area previously free of the disease (Buddenhagen, 2009). In this scenario, many susceptible banana plants occur in close proximity resulting in the build-up of very high inoculum levels, giving rise to high levels of disease incidence in a wide range of banana varieties (Gäumann, 1923; Buddenhagen, 2009). Over time, some bananas die out as suckers become infected, and non-productive plants are removed to make way for other crops, resulting in reduced host density. This lower host density results in a lesser inoculum load, causing reduced disease incidence characteristic of the endemic phase (Buddenhagen, 2009; Ray et al., 2021).

4.5 Host range

Ralstonia syzygii subsp. *celebesensis* infects many *Musa* spp., both cultivated and wild, while some variation in host susceptibility has been reported (Tjahjono and Eden-Green, 1988; Gäumann, 1921). Gäumann (1921) tested over 100 varieties of *Musa* spp. and found them all susceptible. Various non-*Musa* spp. hosts grown in pots developed symptoms of wilt following inoculation of their wounded roots with a solution of *R. syzygii* subsp. *celebesensis* (Baharuddin, 1994). Out of the 20 different plant species inoculated, symptoms were recorded in *Heliconia revolta*, *Heliconia collinsiena*, *Canna indica*, *Strelitzia reginae*, *Solanum nigrum*, *Datura stramonium*, and *Asclepias currassiva* (Baharuddin, 1994). Even though *Solanum lycopersicum* and *S. melogena* are susceptible to the causal agent of Moko *R. solanacearum sensu stricto*, they are not susceptible to infection by *R. syzygii* subsp. *celebesensis* (Baharuddin, 1994; Eden-Green, 1994a). There are no reports of *R. syzygii* subsp. *celebesensis* having been isolated from naturally infected non-*Musa* host plant species.

5 Occurrence and impact

Blood disease was first reported on a small island named Kayaudi, part of the Selayar island group south of Sulawesi in Indonesia in 1905 (Rijks, 1916). The disease destroyed commercial plantations of Kepok (Bluggoe subgroup) bananas on the island, and the loss of productivity caused the plantations to be abandoned (Rijks, 1916). In response to this, the Dutch authorities sent Gäumann to investigate the distribution, biology, and control of Blood disease (Gäumann, 1921; Gäumann, 1922b; Gäumann, 1923). His investigations revealed that Blood disease originated from the forests of mainland southern Sulawesi where wild banana varieties are common (Gäumann, 1921; Gäumann,

1922b; Gäumann, 1923). The epidemic observed in the Selayar islands was hypothesised to have occurred through the introduction of the disease from southern Sulawesi (Gäumann, 1923).

Gäumann determined that Blood disease was limited to south Sulawesi (Gäumann, 1922a). After Gäumann's investigations, the Dutch authorities enacted quarantine preventing the movement of plant materials from the region. As a result, the disease was successfully contained in this region for over 60 years, until it was detected in West Java in 1987 (Tjahjono and Eden-Green, 1988). From West Java, the disease spread rapidly over the next three decades, and is now recorded from 25 of Indonesia's 34 provinces, and has recently spread to peninsular Malaysia (Ray et al., 2021; Teng et al., 2016).

The economic impact of Blood disease occurring in Indonesia and Malaysia is difficult to quantify and disease control measures are minimal or non-existent. Although, significant crop losses have been reported in both countries where the highly susceptible cooking banana variety Kepok (Syn:Nipah) is popular (Ray et al., 2021; Teng et al., 2016; Rijks, 1916; Supriadi, 2005). Indonesia is a large banana producing country with an annual production of around 7MT (FAOSTAT, 2018). Here, Blood disease is one of the most economically significant plant diseases, and losses due to the epidemics that swept through Sumbawa and Lombok in 1999 and Sumba in 2012 approached 100% (Supriadi, 2005; Ray et al., 2021).

6 Management of Blood disease

Resistance to *R. syzygii* subsp. *celebesensis* is unknown in the *Musa* genus. Gäumann tested over 100 banana varieties in field trials near Majene and reported some variation in susceptibility, although all varieties tested eventually succumbed to the disease (Gäumann, 1921). Recently, Ray (2021) confirmed Blood disease from 18 different banana varieties during field surveys. Varieties with retained bracts or that lack a male bell were thought to be resistant (Buddenhagen, 2009) but they are more likely to 'avoid' infection due to the inaccessibility of this infection court to insects, birds, and bats that can inadvertently transfer bacteria (Buddenhagen, 2009; Ray et al., 2022b). It is unknown if banana varieties with partial resistance to *R. syzygii* subsp. *celebesensis* exist: infection not progressing from a parent plant to the sucker via the corm such as has been reported in Bugtok disease of cooking banana caused by a strain of *Ralstonia solanacearum* (Soguilon et al., 1995). Should such partial resistance occur within *Musa* for *R. syzygii* subsp. *celebesensis* it may enable additional control options such as just removal of the diseased stem from the corm rather than the whole corm and all attached shoots.

Control of Blood disease currently relies on cultural methods, as there are no effective chemical or biological controls (Hermanto and Emilda, 2013; Buddenhagen, 2009). Managing local dispersal of Blood disease through cultural control relies on reducing the inoculum load and preventing infection. Prevention of long-distance disease spread relies on restricting the movement of banana planting material and fruits from diseased areas, which can be achieved through quarantine restrictions (Eden-Green and Sastraatmadja, 1990; Ray et al., 2021). In areas free of the disease, banana plantings should use tissue culture or certified disease-free bits and suckers obtained from disease-free areas.

In areas where Blood disease occurs, management options include:

- sanitation of tools used for bunch harvesting, pruning, and de-suckering,
- denavelling using a forked stick soon after the last hand is formed,
- minimisation of leaf pruning, de-suckering and de-flowering,
- bagging of bunches with sealed cloth bags at bunch emergence,
- use of clean planting materials, such as those obtained from tissue culture,
- if tissue culture is not feasible, source disease-free suckers and bits,
- early detection of diseased banana plants followed by prompt destruction of the whole mat,
- destruction of banana mat *in situ* using a systemic herbicide, and
- delay replanting at the location following the destruction or death of diseased banana plants.

These disease control strategies are appropriate for the individual farm unit. Although, area-wide management would lead to better overall disease suppression and may eventually lead to local disease area freedom. Development, validation, and extension of effective disease management options for Blood disease are urgently required in areas where the disease occurs to prevent further economic losses and geographic spread.

7 Conclusion and future trends

Recent studies have provided a greater understanding of the distribution, biology, diagnostics, and epidemiology of *R. syzygii* subsp. *celebesensis* (Ray et al., 2022a; Ray et al., 2022b; Ray et al., 2021; Rincón-Flórez et al., 2022). These studies also highlight areas in need of further research such as developing and testing disease management strategies on a large scale, evaluating the bacteria's transmission and survivability in soil and water, and determine if any resistance within the genus *Musa* exists through large-scale screening of edible banana varieties as well as wild germplasm. It also remains undetermined if strains with different biological properties exist in parts of Indonesia not yet investigated.

Further research on disease control would benefit smallholder and commercial growers in Southeast Asia and beyond. Areas of further study include: (1) methods to destroy banana plants *in situ* using herbicides and delivery options to kill the larger Kepok banana plants, (2) effective tool sanitation techniques, (3) assessment of comparative efficacy of insect-proof cloth bags, denavelling timing, and techniques, and (4) to develop soil amelioration techniques to reduce fallow time between banana plantings to prevent replant disease transmission.

After development and testing of disease mitigation and control methodologies in areas where Blood disease exists extension and dissemination of this information is urgently needed. Additionally, areas still free of the disease would benefit from restrictions on the movement of planting material from locations with the disease. Action is urgently required to prevent the further geographic expansion of this destructive disease in Southeast Asia and beyond.

8 Where to look for further information

A general introduction and background to banana diseases can be found in the book 'Handbook of diseases of banana, abaca and enset' (Jones, 2019). For more detailed information on the bacterial diseases that affect bananas see the review article 'Bacterial diseases of bananas and enset: current state of knowledge and integrated approaches toward sustainable management' (Blomme et al., 2017). For a current single source of detailed information on Blood disease see the PhD thesis 'Biology and epidemiology of banana Blood disease' (Ray, 2022).

9 References

Baharuddin, B. (1994). *Pathological, biochemical and serological characterization of the Blood disease bacterium affecting banana and plantain (Musa spp.) in Indonesia*. PhD thesis, University of Göttingen, Cuvillier Verlag, Germany.

Blomme, G., Dita, M., Jacobsen, K. S., Pérez Vicente, L., Molina, A., Ocimati, W., Poussier, S. and Prior, P. (2017). Bacterial diseases of bananas and enset: Current state of knowledge and integrated approaches toward sustainable management. *Frontiers in Plant Science* 8(1290): 1–25.

Blomme, G. (2023). Banana Xanthomonas Wilt. In: Drenth, A. and Kema, G. H. J. (Eds.), *Achieving sustainable cultivation of bananas: Volume 3. Diseases and pests*. Cambridge: Burleigh Dodds Science Publishing.

Buddenhagen, I. W. (2009). Blood bacterial wilt of banana: History, field biology and solution. *Acta Horticulturae* 828: 57–68.

Buddenhagen, I. W., Sequeira, L. and Kelman, A. (1962). Designations of races in *Pseudomonas solanacearum*. *Phytopathology* 52: 726.

Cook, D., Barlow, E. and Sequeira, L. (1989). Genetic diversity of *Pseudomonas solanacearum*: detection of restriction fragment length polymorphisms with DNA probes that specify virulence and the hypersensitive response. *Molecular Plant-Microbe Interactions* 1: 113–121.

Das, S. H. J. (2004). *Molecular diagnostics of the Blood disease bacterium*. Honours thesis, The University of Queensland Brisbane, Australia.

Eden-Green, S. (1994a). Diversity of *Pseudomonas solanacearum* and related bacteria in south east Asia: new directions for Moko disease. In: Hayward, A. C. and G. L. H. (Eds.), *Bacterial wilt: the disease and its causative agent, Pseudomonas solanacearum*. Wallingford: CAB International, pp. 25-34.

Eden-Green, S. (1994b). *Musa disease fact sheet No. 3. Banana Blood disease*. Montpellier: INIBAP Parc Scientifique Agropolis.

Eden-Green, S. and Sastraatmadja, H. (1990). Blood disease in Indonesia. *FAO Plant Protection Bulletin* 38(38): 49-50.

FAOSTAT. (2018). *FAOSTAT Database*. Rome: Food and Agriculture Organization of the United Nations.

Fegan, M. and Prior, P. (2005). How complex is the *Ralstonia solanacearum* species complex. In: Allen, C., Prior, P. and Hayward, A. C. (Eds.), *Bacterial Wilt Disease and the Ralstonia solanacearum species Complex*. St. Paul, MN: APS Press, pp. 449-461.

Fegan, M. and Prior, P. (2006). Diverse members of the *Ralstonia solanacearum* species complex cause bacterial wilts of banana, *Australasian Plant Pathology*. 35: 93-101.

French, E. R. and Sequeira, L. (1970). Strains of *Pseudomonas solanacearum* from Central and South America: A comparative study. *Phytopathology* 60(3): 506-512.

Gäumann, E. A. (1921). Onderzoekingen over de bloedziekte der bananen op Celebes I. *Mededeelingen van het Instituut voor Plantenziekten* 50: 1-47.

Gäumann, E. A. (1922a). Blood disease banana. *The Review of Applied Mycology* 1: 225-227.

Gäumann, E. A. (1922b). Over ziekten van de pisangs. *Teysmania* 33: 1-14.

Gäumann, E. A. (1923). Onderzoekingen over de bloedziekte der bananen op Celebes II. *Mededeelingen van het Instituut voor Plantenziekten* 59: 1-45.

Granada, G. A. and Sequeira, L. (1983). Survival of *Pseudomonas solanacearum* in soil, rhizosphere, and plant roots. *Canadian Journal of Microbiology* 29(4): 433-440.

Hadiwiyono, H. (2011). Blood bacterial wilt disease of banana: the distribution of pathogen in infected plant, symptoms, and potentiality of diseased tissues as source of infective inoculums. *Nusantara Bioscience* 3(3): 112-117.

Hayward, A. C. (1964). Characteristics of *Pseudomonas solanacearum*. *Journal of Applied Bacteriology* 27(2): 265-277.

He, L. Y. (1983). Characteristics of strains of *Pseudomonas solanacearum* from China. *Plant Disease* 67(12): 1357-1361.

Hermanto, C. and Emilda, D (2013). Bunch management of banana to control Blood disease. *Australasian Plant Pathology* 42(6): 653-658.

Jones, D. R. (2019). *Handbook of diseases of banana, abaca and enset*. CABI, Wallingford, UK.

Kelman, A. (1953). The bacterial wilt caused by *Pseudomonas solanacearum*. *Technical Bulletin of North Carolina Agricultural Experiment Station* 99: 1-194.

Kelman, A. (1954). The relationship of pathogenicity of *Pseudomonas solanacearum* to colony appearance in a tetrazolium medium. *Phytopathology* 44(12): 693-695.

Kubota, R., Schell, M. A., Peckham, G. D., Rue, J., Alvarez, A. M., Allen, C. and Jenkins, D. M. (2011). In silico genomic subtraction guides development of highly accurate, DNA-based diagnostics for *Ralstonia solanacearum* race 3 biovar 2 and Blood disease bacterium. *Journal of General Plant Pathology* 77(3): 182-193.

Ploetz, R. C., Thomas, J. E. and Slabaugh, W. R. (2003). Diseases of banana and plantain. In: Ploetz, R. C. (Ed.), *Diseases of tropical fruit crops*. Cambridge: CABI Publishing, pp. 73–134.

Prior, P., Ailloud, F., Dalsing, B. L., Remenant, B., Sanchez, B. and Allen, C. (2016). Genomic and proteomic evidence supporting the division of the plant pathogen *Ralstonia solanacearum* into three species. *BMC Genomics* 17(90): 90.

Ray, J. D. (2022). Biology and epidemiology of banana Blood disease. PhD thesis. The University of Queensland, Queensland, Australia.

Ray, J. D., Subandiyah, S., Prakoso, A. B., Rincón-Flórez, V. A., Carvalhais, L. C. and Drenth, A. (2022a). Transmission of Blood disease in banana. *Plant Disease* 106(8): 2155-2164.https://doi.org/10.1094/PDIS-10-21-2373-RE.

Ray, J. D., Subandiyah, S., Rincon-Florez, V. A., Prakoso, A. B., Carvalhais, L. C. and Drenth, A. (2022b). Susceptibility of the banana inflorescence to Blood disease. *Phytopathology* 112(4): 803–810.

Ray, J. D., Subandiyah, S., Rincon-Florez, V. A., Prakoso, A. B., Mudita, I. W., Carvalhais, L. C., Markus, J. E. R., O'Dwyer, C. A. and Drenth, A. (2021). Geographic expansion of banana Blood disease in Southeast Asia. *Plant Disease* 105(10): 2792–2800.

Remenant, B., Coupat-Goutaland, B., Guidot, A., Cellier, G., Wicker, E., Allen, C., Fegan, M., Pruvost, O., Elbaz, M., Calteau, A., Salvignol, G., Mornico, D., Mangenot, S., Barbe, V., Médigue, C. and Prior, P. (2010). Genomes of three tomato pathogens within the *Ralstonia solanacearum* species complex reveal significant evolutionary divergence. *BMC Genomics* 11(1): 379.

Remenant, B., de Cambiaire, J. C., Cellier, G., Jacobs, J. M., Mangenot, S., Barbe, V., Lajus, A., Vallenet, D., Medigue, C., Fegan, M., Allen, C. and Prior, P. (2011). *Ralstonia syzygii*, the Blood disease bacterium and some Asian *R. solanacearum* strains form a single genomic species despite divergent lifestyles. *PLoS ONE* 6(9): e24356.

Rijks, A. B. (1916). Rapport over een onderzoek naar de pisangsterfte op de saleiereilanden. *Mededeelingen van het Laboratorium voor Plantenziekten* 21: 1–16.

Rincón-Flórez, V. A., Ray, J. D., Carvalhais, L. C., O'Dwyer, C. A., Subandiyah, S., Zulperi, D. and Drenth, A. (2022). Diagnostics of banana Blood disease. *Plant Disease* 106(3): 947–959.

Roberts, S. J., Eden-Green, S. J., Jones, P. and Ambler, D. J. (1990). *Pseudomonas syzygii*, sp. nov., the cause of Sumatra disease of cloves. *Systematic and Applied Microbiology* 13(1): 34–43.

Robinson, J. C. (1996). *Banana and plantains*. UK: CAB International Wallingford.

Safni, I., Cleenwerck, I., De Vos, P., et al. (2014). Polyphasic taxonomic revision of the *Ralstonia solanacearum* species complex: proposal to emend the descriptions of *Ralstonia solanacearum* and *Ralstonia syzygii* and reclassify current *R. syzygii* strains as *Ralstonia syzygii* subsp. *syzygii* subsp. nov., *R. solanacearum* phylotype IV strains as *Ralstonia syzygii* subsp. *indonesiensis* subsp. nov., banana Blood disease bacterium strains as *Ralstonia syzygii* subsp. *celebesensis* subsp. nov. and *R. solanacearum* phylotype I and III strains as *Ralstonia pseudosolanacearum* sp. nov. *International Journal of Systematic and Evolutionary Microbiology* 64(9): 3087–3103.

Safni, I., Subandiyah, S. and Fegan, M. (2018). Ecology, epidemiology and disease management of *Ralstonia syzygii* in Indonesia. *Frontiers in Microbiology* 9: 419.

Santiago, T., Lopes, C., Caetano-Anollés, G. and Mizubuti, E. (2017). Phylotype and sequevar variability of *Ralstonia solanacearum* in Brazil, an ancient centre of diversity of the pathogen. *Plant Pathol* 66: 383–392.

Salanoubat, M., Genin, S., Artiguenave, F., Gouzy, J., Mangenot, S., Arlat, M., Billault, A., Brottier, P., Camus, J. C., Cattolico, L., Chandler, M., Choisne, N., Claudel-Renard, C., Cunnac, S., Demange, N., Gaspin, C., Lavie, M., Moisan, A., Robert, C., Saurin, W., Schiex, T., Siguier, P., Thébault, P., Whalen, M., Wincker, P., Levy, M., Weissenbach, J. and Boucher, C. A. (2002). Genome sequence of the plant pathogen *Ralstonia solanacearum*. *Nature* 415(6871): 497–502.

Schaad, N. W., Jones, J. B. and Chun, W. (2001). *Laboratory Guide for the Identification of Plant Pathogenic Bacteria*. American Phytopathological Society (APS Press), St Paul, USA.

Soguilon, C., Magnaye, L. and Natural, M. (1995). Bugtok disease of banana. *Musa Disease Fact Sheet No. 6*. Montpellier: International Network for the Improvment of Banana and Plantain, Parc Scientifique Agropolis.

Stover, R. H. and Espinoza, A. (1992). Blood disease of bananas in Sulawesi. *Fruits* 47(5): 611–613.

Supriadi. (2005). Present status of Blood disease in Indonesia. In: Allen, C., Prior, P. and Hayward, A. C. (Eds.), *Bacterial Wilt disease and the Ralstonia solanacearum species complex*. St. Paul, MN: APS Press, pp. 395–404.

Taghavi, M., Hayward, C., Sly, L. I. and Fegan, M. (1996). Analysis of the phylogenetic relationships of strains of *Burkholderia solanacearum*, *Pseudomonas syzygii*, and the Blood disease bacterium of banana based on 16S rRNA gene sequences. *International Journal of Systematic and Evolutionary Microbiology* 46(1): 10–15.

Tan, P. (2003). *The Blood disease bacterium: exploiting genetic diversity for the development of a molecular diagnostic test*. Honours thesis, University of Queensland, Brisbane, Australia.

Teng, S. K., Aziz, N. A. A., Mustafa, M., Laboh, R., Ismair, I. S., Sulaiman, S. R., Azizan, A. A. and Devi, S. (2016). The occurrence of Blood disease of banana in Selangor, Malaysia. *International Journal of Agriculture and Biology* 18: 92–97.

Tjahjono, B. and Eden-Green, S. (1988). Blood disease of banana in Indonesia. *5th Int. Congress of Plant Pathology.(Abst. of paper)*. Kyoto, Japan.

Xu, J., Pan, Z., Prior, P., Xu, J., Xu, J., Zhang, Z., Zhang, H., Zhang, L., He, L. and Feng, J. (2009). Genetic diversity of *Ralstonia solanacearum* strains from China. *European Journal of Plant Pathology* 125: 641–653.

Yabuuchi, E., Kosako, Y., Yano, I., Hotta, H. and Nishiuchi, Y. (1995). Transfer of two *Burkholderia* and an *Alcaligenes* species to *Ralstonia* gen. nov. Proposal of *Ralstonia pickettii* (Ralston, Palleroni and Doudoroff 1973) comb. nov., *Ralstonia solanacearum* (Smith 1896) comb. nov. and *Ralstonia eutropha* (Davis 1969) comb. nov. *Microbiology and Immunology* 39(11): 897-904.

Chapter 10

Xanthomonas wilt of banana

Guy Blomme, The Alliance of Bioversity International and The International Center for Tropical Agriculture (CIAT), Ethiopia; Elizabeth Kearsley, BlueGreen Labs, Belgium; and Walter Ocimati, The Alliance of Bioversity International and The International Center for Tropical Agriculture (CIAT), Uganda

1 Introduction

Over the past two decades, Xanthomonas wilt of banana (BXW) has impacted banana growing regions across East and Central Africa. BXW affects mostly small-holder farmers, who represent ~90% of the banana production systems, and severely impacts their livelihood and food security (Tripathi et al., 2009; Ploetz et al., 2015; Blomme et al., 2017a, 2020a). Economic losses further resonate along the banana value chain, resulting in substantial losses at national levels (Di Cori et al., 2018). Additionally, the impact of BXW extends far beyond food security and economic losses. Both the presence of the disease and the need for controlling measures have affected various ecosystem services, including regulatory services such as soil erosion control, water quality and pollination enhancement in fields and landscapes, and supporting and cultural services (Ocimati et al., 2018a). Reports of farmers moving away from banana production as their main production system has also resulted in extensive land-use change (Ocimati et al., 2020).

BXW was first recorded on enset in Ethiopia (Castellani, 1939; Yirgou and Bradbury, 1968) and was later found to have been transmitted to bananas (Yirgou and Bradbury, 1974). In 2001 BXW was recorded outside of Ethiopia in Uganda and the Eastern Democratic Republic of Congo (Tushemereirwe et al., 2003; Ndungo et al., 2004), and it has since spread to the rest of the African

http://dx.doi.org/10.19103/AS.2022.0108.10

Great Lakes countries comprising Tanzania, Rwanda (Reeder et al., 2007), Burundi (Carter et al., 2010), and Kenya (Carter et al., 2010). Since these first reports, BXW has spread to various regions within these countries and while the introduction of disease management practices has shown to reduce infection levels (Kubiriba et al., 2012; Nkuba et al., 2015), eradication has rarely been achieved, often leading to resurgence events (Blomme et al., 2014; Shimwela et al., 2017; Nakato et al., 2018a; Ocimati et al., 2019). The combination of ease of spread of the pathogen, the susceptibility of all edible banana cultivars grown in East and Central Africa and various human factors have contributed to the dissemination of this pathogen (Blomme et al., 2017a; Ocimati, 2019; Ocimati et al., 2019).

Long-term management practices, taking into account the feasibility for farmers to uphold them, combined with extensive support from other stakeholders are the main way forward to reduce infection levels and keep them down. Effective management also requires a shift from reactive to proactive interventions, moving toward a focus on preventive practices to reduce the risk of both new infections and re-infections. In recent years, research efforts have managed to implement this shift, focusing on improving effective and practical control measures (e.g. introducing the single diseased stem removal (SDSR) control package, Blomme et al., 2017b), improving the ease of use of existing preventive measures (e.g. feasibility of tool sterilization; Ocimati et al., 2021) and moving toward a long-term solution in the form of resistant cultivars (Tripathi et al., 2021) and the accessibility to early warning systems (Ocimati et al., 2019; Selvaraj et al., 2019, 2020). Focus has furthermore shifted toward addressing issues that have emerged regarding the implementation of BXW management practices, ranging from practical to socioeconomic and political issues (McCampbell et al., 2018), and presenting suitable solutions for the small-holder farmer. Here we present the state of knowledge on BXW epidemiology, available and upcoming technologies for BXW management and the implementation thereof.

2 Xanthomonas wilt of banana epidemiology

2.1 Causal agent

Xanthomonas wilt is caused by a rod-shaped, aerobic, Gram-negative bacterium with a single polar flagellum (Bradbury, 1986) in the Gamma subclass of Proteobacteria (Smith et al., 2008). As the causal agent of BXW of enset and banana, it was initially described in Ethiopia as *Xanthomonas musacearum* (Yirgou and Bradbury, 1968, 1974) and shortly after classified as *X. campestris* pv. *musacearum* (Xcm) (Young et al., 1978). Further characterization based on genetic and genomic evidence, however, suggests the pathogen is of the species *X. vasicola* (Aritua et al., 2007; Studholme et al., 2010; Wasukira et al.,

2012, 2014). The reclassification as a new pathovar *X. vasicola* pv. *musacearum* (Xvm) was proposed (Aritua et al., 2008) and accepted (Studholme et al., 2020).

Xvm is genetically a highly monomorphic pathogen with only very limited diversity found among isolates across East and Central Africa (Wasukira et al., 2012; Nakato et al., 2019b, 2021). A recent study by Nakato et al. (2021) revealed six closely related clades of Xvm in Ethiopia, four more than previously reported (Wasukira et al., 2012). This suggests that a higher genetic diversity of Xvm could be present in Ethiopia from which a restricted set of genotypes escaped its ancestral home and caused BXW outbreaks across the East African Great Lakes region (Nakato et al., 2021). No variable characteristics or differences in virulence factors have so far been reported among different sub-lineages or isolates of Xvm (Wasukira et al., 2017).

2.2 Infection

Xvm is a vascular pathogen (Yirgou and Bradbury, 1968, 1974) that enters a banana or enset host plant through wounds exposing internal tissue. After inoculation, the bacteria spread systemically within the plant (Ssekiwoko et al., 2010; Ocimati et al., 2013a). Monitoring the systemicity of Xvm through the banana plant, Karamura et al. (2016) observed that on average the inoculum takes 35–42 days to reach the corm, with symptom expression already occurring after 2–3 weeks. An average incubation period of 3 weeks is also reported by Blomme et al. (2017a), although it varies with cultivar, plant growth stage, mode of disease transmission and environmental conditions.

Xvm is, however, not fully systemic within a banana mat. Healthy-looking plants, able to produce edible bunches and healthy-looking lateral shoots, develop in mats that contain an infected plant or plants (Tushemereirwe et al., 2003; Ssekiwoko et al., 2010; Ocimati et al., 2013a, 2015). While the mechanism behind this incomplete systemicity is not fully understood, it appears that Xvm does not efficiently spread once in the central cylinder and cortex of the corm (Ssekiwoko et al., 2006a; Ocimati et al., 2013a,b) or further into corm roots (Ocimati et al., 2013b). Nevertheless, a random spread pattern from infected mother plants to lateral suckers has been observed, and symptom expression on lateral shoots can be observed from 81 days after initial inoculation (Ocimati et al., 2013a; Karamura et al., 2016). Hence, the risk of lateral infection thereby does exist.

Latent infections of Xvm have frequently been observed, with infections able to persist in parts of a mat for over 2 years without causing visible symptoms (Ocimati et al., 2013a, 2015). These further complicate the matter and present issues for management practices (Ocimati et al., 2013a,b, 2015).

2.3 Spread

Insect vectors form the primary mode of natural transmission of BXW (Tinzaara et al., 2006; Rutikanga et al., 2015; Shimwela et al., 2017). By moving between inflorescences of infected and healthy plants, insects often come into contact with the abscission sites of male flowers and bracts where xylem vessels are unobstructed, thereby providing a direct pathway for the Xvm bacteria to the vascular system in the floral stalk/peduncle of the plant (Ocimati et al., 2013a; Nakato et al., 2014). Insect-mediated transmission through female inflorescences only occurs rarely as these scars are shown to dehydrate more rapidly (Nakato et al., 2018a) and the low inoculum levels that are introduced generally do not induce symptoms (Nakato et al., 2014).

The ecology and behavioral habits of insect populations directly impact the prevalence and distance of this transmission pathway. For example, fruit flies will only travel the distance of a few neighboring plants during their lifespan, whereas wasps and bees will travel further, increasing the range of transmission (Smith et al., 2008; Rutikanga et al., 2015). In addition, farms at higher altitudes and those experiencing lower temperatures will be less affected by insect-mediated transmission simply due to lower insect activity (Mwangi et al., 2006; Tripathi et al., 2009; Rutikanga et al., 2015). Natural transmission can further occur through nectar-collecting and fruit-eating birds and bats (Buregyeya et al., 2008, 2014; Blomme et al., 2014), although this is generally less common. Soil-related dispersion of Xvm is of limited importance due to short survival periods of Xvm in soil in the absence of the host plant (Mwebaze et al., 2006; Welde-Michael et al., 2008).

Imprudent practices by people also contribute to disease spread. Firstly, the use of contaminated farming tools plays a critical role in Xvm transmission (Yirgou and Bradbury, 1974; Ploetz et al., 2015). Xvm bacteria have been shown to survive for up to 6 days and 20 days on non-stainless steel and stainless steel knives, respectively (Buregyeya et al., 2008). Hence, consecutive management of host plants using contaminated tools provides for optimal transmission within a farm. Contaminated tools also serve as a transmission pathway between farms as tools are often shared within communities, or even across large distances through hired labor (Nakato et al., 2013a; Blomme et al., 2020b). Secondly, the use of infected, often-asymptomatic planting material is another important mode of transmission (Karamura et al., 2008). Most farmers (~90%) source new planting material in the form of suckers from their own or neighboring farms with an associated risk of transferring (latently) infected plants (Tinzaara et al., 2013). Xvm may also spread through the trade of banana bunches and leaves used to wrap these bunches (Nakato et al., 2013a). Transmission through trade occurs when knives used to peel fruits or cut leaves by the recipient are later used for farm management practices in healthy banana fields or when traders indiscriminately use their cutting tools to harvest bunches across farms without sterilization between plants or farms.

Finally, other local activity on farms, such as roaming animals, have also been reported to contribute to disease transmission (Tinzaara et al., 2013; Blomme et al., 2014).

A combination of the various modes of spread has contributed to the overall dissemination across East and Central Africa following patterns referred to as the 'invasive bridgehead effect' (Lombaert et al., 2010), reported for various banana diseases (Drenth and Kema, 2021). This effect describes the patterns of long-distance spread followed by secondary spread from newly established sites. The transmission of BXW to new regions over long distances results from movement of infected planting materials, trade of bunches, contaminated tools used by traders, possibly through bats and birds (with a large flying radius), followed by local spread via flying insects, farmer-owned garden tools and browsing animals (Blomme et al., 2020b).

2.4 Symptoms

BXW symptom expression can occur as early as 2–4 weeks after infection (Tripathi et al., 2009; Karamura et al., 2016) but can be prolonged up to 16 months depending on conditions (Ocimati et al., 2013a). Symptoms characteristically include leaf yellowing and wilting caused by blocking of vascular bundles in the pseudostem (Fig. 1a), dry rot of the male inflorescence (Fig. 1b), premature and uneven ripening of bunches (Fig. 1b) and internal browning of the fruit pulp (Fig. 1c) (Tripathi et al., 2009; Ocimati et al., 2013a; Nakato et al., 2014). Additional stem discoloration and bacterial ooze in cut pseudostems (Fig. 1d) or leaf petioles are also common (Tinzaara et al., 2006; Blomme et al., 2017a).

2.5 Host range, tolerance and susceptibility

Banana, plantain (*Musa* spp.) and enset (*Ensete ventricosum*) are the only known natural hosts of Xvm, although the level of tolerance and susceptibility has been observed to vary among genotypes (Ssekiwoko et al., 2006b; Kebede and Gemmeda, 2017; Nakato et al., 2019a). All edible bananas grown in affected regions in Africa (e.g. Cavendish (AAA), East African Highland cooking/beer bananas (AAA), 'Kayinja' and 'Bluggoe' (ABB), 'Sukari Ndizi' (AAB), AAA dessert types and AAB plantain types) are susceptible to the pathogen, while resistance has been identified in the wild diploid banana cultivar *Musa balbisiana* (Ssekiwoko et al., 2006b, 2015; Tripathi et al., 2019).

The mechanisms of resistance of *M. balbisiana* have been of substantial interest for researchers (Ssekiwoko et al., 2015; Tripathi et al., 2019). Most

Figure 1 Symptoms of BXW on banana hosts: (a) yellowing and wilting of leaves; (b) wilting of male bud bracts and premature ripening of the fruit; (c) fruit pulp discolouration and (d) yellow ooze on cut pseudostem. Source: Photos by Guy Blomme.

recently, through transcriptome-wide differential gene expression between *M. balbisiana* and the highly susceptible 'Pisang Awak', Tripathi et al. (2019) observed an upregulation of genes associated with the response to biotic stress and the activation of both PAMP-triggered defense and R protein-mediated defense in *M. balbisiana*. In susceptible banana cultivars, Xvm disables the plants' detection system and downstream defense reactions (Ssekiwoko et al., 2015). Nakato et al. (2019a) nevertheless have identified a range in cell-mediated tolerance among the various susceptible genotypes. By inoculating young plants of 72 different *Musa* accessions, they showed that genotypes belonging to the *M. acuminata* subsp. *zebrina* and Yalim belonging to the subspecies *banksii* are less susceptible, showing reduced symptom expression and severity.

Banana cultivars also show variation in their susceptibility to initial infection due to differences in inflorescence morphology (Changa et al., 2019). Within certain genotypes, the vulnerable infection sites for insect-mediated transmission are not formed or exposed as they lack the male inflorescence, have persistent male flowers and bracts and or form dry wounds after abscission of bracts or flowers (Karamura et al., 2008; Tripathi and Tripathi, 2009; Changa et al., 2019). It has been shown that ABB cultivars (e.g. 'Pisang

Awak' or 'Bluggoe') are particularly prone to insect-vectored transmission (Nkuba et al., 2015), while several of the cultivars within the East African Highland banana (EAHB) group are less susceptible to inflorescence infection, due to the presence of persistent male bracts and flowers (Tripathi et al., 2009; Karamura et al., 2010). Persistent male bracts and flowers of 'Dwarf Cavendish' can be seen in Fig. 2.

Xvm has also been found to infect alternative hosts including the perennials *Canna indica* and sugarcane, and annual crops such as maize and sorghum (Ssekiwoko et al., 2006b; Aritua et al., 2008; Karamura et al., 2015; Chala et al., 2016; Ocimati et al., 2018b). These species could play roles in the short-term survival and possible perpetuation of Xvm within landscapes.

Figure 2 Banana cultivar differences in inflorescence morphology: (a) a Dwarf Cavendish (AAA) plant with persistent male bracts and flowers, able to escape insect-vectored Xvm transmission; (b) a 'Pisang awak' (ABB) plant with non-persistent male bracts and flowers, creating insect-vectored entry pathways for Xvm into the vascular system of the plant's peduncle or flower stalk and (c) insect vectors (bees) on open wounds on the inflorescence after drop off of male flowers. Source: Photos by Guy Blomme.

3 Xanthomonas wilt management: prevention and control

3.1 Disease detection and diagnosis

Early detection and accurate diagnosis of BXW is a critical first step in disease management, allowing for a targeted selection of control measures and the rapid implementation thereof. The ability to take fast action at the first signs of new infections eases the application of control measures, achieves fast disease subsidence rate and reduces the risk of further disease transmission. Optimal selection of disease control measures does depend on reliable and specific diagnosis of the causal pathogen as symptoms observed in the field can be confused with other biotic or abiotic stresses. Generally, the initial detection of BXW is based on visual assessment of symptoms. While often it stops there, multiple diagnostic tools are available to verify the on-farm assessment. Microbiological, serological and molecular diagnostic methods can be used, with varying degrees of accuracy, accessibility, ease of use and cost. These techniques remain limited in both their temporal and spatial extent, with detection and diagnosis based on single point measurements at farm or community levels, with limited early warning potential. New developments in large-scale disease detection through integrated remote sensing and machine learning approaches (Selvaraj et al., 2020) could bridge this gap and prove complementary to the established detection/diagnostic methods.

3.1.1 Visual detection

Visual assessment of symptoms is the most practical means available to farmers in East and Central Africa for early BXW detection. Effective disease detection is based on monitoring at a plant-by-plant basis at regular intervals (Blomme et al., 2017a). Weekly or biweekly field inspections are generally recommended based on the average time of symptom expression, which can vary depending on local circumstances (e.g. cultivar, farming system, altitude) (Ssekiwoko et al., 2006a; Welde-Michael et al., 2008; Ocimati et al., 2013a,b; Blomme et al., 2017a). Accurate BXW diagnosis through visual inspection should be based on the presence of yellow ooze on cut pseudostems and fruit pulp discoloration. Leaf yellowing and drying symptoms can be confused with Fusarium wilt symptoms (Tushemereirwe et al., 2003; Blomme et al., 2017a). As BXW is currently only present in Africa, any confusion with Moko/Bugtok disease or Blood disease that are only present in Latin America and/or Asia is not relevant at present (Blomme et al., 2017a).

3.1.2 Microbiological diagnosis

Xvm can be isolated on several culture media, including on yeast extract peptone glucose agar (YPGA; Mwangi et al., 2007), cellobiose cephalexin agar (Tripathi et al., 2007) and yeast extract tryptone sucrose agar (Tripathi et al., 2007). These media can be complemented with 5-fluorouracil, cephalexin and cycloheximide to produce semi-selectivity (Tripathi et al., 2007). Culturing Xvm colonies for identification does take several days (Mwebaze et al., 2006; Mwangi et al., 2007), making this method very time consuming and less convenient for large sample batches (Karamura et al., 2016). Microbiological diagnosis is also limited to research facilities.

3.1.3 Serological diagnosis

Xvm can be identified with an enzyme-linked immunosorbent assay (ELISA) using specific antibodies (Nakato et al., 2013b). Specifically, polyclonal antibody detection through a direct antigen coating (DAC)-ELISA-based method offers a rapid and reliable result (Nakato et al., 2013b). Rapid, on-site detection of Xvm can be achieved using a polyclonal antibody-based lateral flow device (Karamura et al., 2016), which can be used on-site to detect latent infections or new Xvm infections before symptoms are presented (Karamura et al., 2016). However, the use of these approaches has to date been limited to research due to the associated high costs.

3.1.4 Molecular diagnosis

Polymerase chain reaction (PCR)-based diagnosis is the most reliable method available, able to distinguish Xvm from closely related strains not pathogenic to banana and enset. Pathovar-specific primers are used to amplify Xvm strains (Adriko et al., 2012), with ongoing development of markers with increased specificity further advancing this diagnostic tool (Nakato et al., 2018b). This approach requires costly equipment and a high expertise, and is as such limited to certified research laboratory facilities.

3.1.5 Remote sensing and machine learning-based approaches

Disease detection using digital imaging combined with machine learning classification approaches provide a great tool for disease surveillance and early warning systems. The robust identification of multiple banana diseases, including BXW, has been achieved through deep learning classification of both real-time field images (Selvaraj et al., 2019), drone and high-resolution satellite images (Selvaraj et al., 2020). Currently, the accuracy of these classifications range around 70–99% (Selvaraj et al., 2019, 2020). Integrated multi-level sensing systems (satellites, drones and ground-based digital imagery) would be necessary for monitoring banana crop health at different spatial scales. This approach also requires costly equipment and facilities, making it cost prohibitive at present.

At field level, a smartphone app, called Tumaini, able to detect symptoms of five banana diseases, including BXW, and one pest is currently available on the Google Play Store. This mobile phone app helps farmers to accurately identify BXW disease and provides disease control solutions. The processing and mapping of positive app observations on the PestDisPlace platform is being put in place, and once operational, this platform will serve as an early warning system and will be accessible to app users. Upon detection and confirmation of the disease, or eventually when presence of the disease in neighboring villages or landscapes is indicated on the PestDisPlace platform maps, farmers could then also reach out to research and extension agencies for additional support.

3.2 Control measures

3.2.1 Cultural control packages, complete diseased mat uprooting and single diseased stem removal

The cultural practices of complete diseased mat uprooting (CDMU) and, more recently, SDSR form the bedrock of BXW management across East and Central Africa (Blomme et al., 2017a,b). Both practices center around the removal of Xvm inoculum from the infected field and are combined with preventive measures (discussed further on) to reduce the risk of disease transmission, namely timely removal of the male inflorescence buds using a forked stick, disinfection of tools and the use of clean planting material (Karamura et al., 2008; Tripathi et al., 2009; Blomme et al., 2014, 2019; Ocimati et al., 2021).

CDMU is the originally recommended control package in which, as the name states, the entire infected mat is uprooted and destroyed (Tushemereirwe et al., 2003). While a very effective method at reducing or even eliminating disease inoculum, low adoption levels have been reported due to its high labor demand and associated high labor cost and time consumption (Mwangi et al., 2007; Blomme et al., 2014; Blomme et al., 2021). Both the removal and the destruction of the infected plant debris through burning or burying, as initially postulated, are very physically demanding. Composting of the plant debris without burying is a less labor-intensive option (Mwebaze et al., 2006; Kubiriba et al., 2014). However, uprooted mats should be kept as intact as possible before they are disposed of at a compost heap, as each additional cut in a pseudostem will create additional ooze that could potentially lead to infections within the field, if e.g. picked up by insects or browsing animals. The need to replace infected mats with new clean planting material adds another hurdle, being both difficult to access and expensive. Blomme et al. (2021) identified the acquisition of clean planting material as the largest cost associated with the CDMU package for small-holder farmers in Rwanda, representing 78% of the total cost. Production losses resulting from this practice are also high, since healthy bunch-producing plants within an infected mat also need to be removed (Blomme et al., 2021). Moreover, it can take a long time for the field to be restored to pre-infection production levels, up to 20 months at the mid- to high-altitude sites in East and Central Africa (Blomme et al., 2014), resulting in a low return on investment, potentially too low for small-holder farmers.

The updated control package based on SDSR is presented as a suitable alternative for CDMU in most circumstances (Fig. 3; Blomme et al., 2017b). The success of SDSR can be explained by the incomplete systemic spread of Xvm bacteria, by which not all the physically interconnected lateral shoots within an infected mat are necessarily colonized (Ssekiwoko et al., 2006a; Ocimati et al., 2013a, 2015; Blomme et al., 2017b). Visibly diseased plants within a mat are cut off at soil level, while all symptomless suckers are allowed to remain. This method is far less labor intensive and time consuming than CDMU (Blomme et al., 2021), while it remains highly effective in disease reduction and elimination (Kubiriba et al., 2012; Blomme et al., 2017b). Yield losses are also minimized as healthy plants remain available for production of bunches, and the cost associated with new planting material is avoided (Blomme et al., 2021). Blomme et al. (2021) showed that SDSR could be effectively applied for a fraction of the time and cost of CDMU, reportedly 88% less time and a 96% lower cost, making it the more attractive control package for most farmers.

Incomplete removal of the inoculum and 'invisible' latent infections do pose a risk, although the remaining low levels of Xvm inoculum in a mat

Figure 3 Application of the SDSR control package: (a) single diseased stem removal; (b) early male bud removal using a forked stick and (c) sterilizing a machete using soap and water. Source: Photos by Guy Blomme and Walter Ocimati.

only occasionally lead to symptomatic plants (Ocimati et al., 2013a, 2015). Effective application of the SDSR control package requires regular attention from the farmer through close monitoring of the field and continuous removal of new symptomatic plants (Blomme et al., 2017b; Ntamwira et al., 2019). It is recommended to apply SDSR at regular intervals during the first phase of this management practice. Blomme et al. (2019) suggest at least weekly monitoring during the first 3–4 months of application. The frequency of field monitoring can be reduced when most of the Xvm inoculum seems to be removed and potential new symptoms only present sporadically (Blomme et al., 2019). Long incubation periods of BXW up to 24 months (Ocimati et al., 2013a, 2015) and the occurrence of small outbreaks at irregular intervals over a period of many months (Blomme et al., 2021) indicate that long-term monitoring is essential. Using the SDSR control package, field disease incidence has been shown to reduce drastically over time from up to 80% incidence to below 2% within 3 months, and below 1% within 5–10 months (Blomme et al., 2017b). Starting from lower initial disease incidences of around 3–9%, Blomme et al. (2021) reported even faster declines in disease incidence resulting in incidences of <1% after one month and <0.5% within 3 months. Ocimati et al. (2015) does note that if new infections can be prevented, the level of Xvm inoculum in the affected mats will continue to decline with subsequent generations and the productivity in the fields can be fully restored.

Collective action of both the CDMU and SDSR control package, i.e. all affected farmers in a community incorporate the control measures, drastically improves the effectiveness of the measures (Blomme et al., 2019). Isolated application allows for re-infections to occur from neighboring farms (Rutikanga et al., 2016; Blomme et al., 2019). When applying SDSR, the prolonged presence of the inoculum during the early stages of control also increases the possibility of disease transmission (Ocimati et al., 2015), although a high frequency of application of SDSR during this initial control phase could mitigate this risk (Blomme et al., 2019).

Selection of either CDMU or SDSR often depends on the farming system and incidence of disease. CDMU is generally recommended for larger, well-managed and market-oriented banana production systems that are able to invest in targeted disease eradication (Blomme et al., 2014, 2017a,b). SDSR is more appropriate and recommended for less intensively managed banana cropping systems, often small-holder subsidence farmers, that benefit more from this cost-effective treatment (Blomme et al., 2014, 2017b). CDMU could also prove cost-effective for small-holder farming when the BXW infection is still new with only few plants affected, providing a window for full eradication (Blomme et al., 2014, 2017a,b). Complementary application of CDMU and SDSR is also advisable for small scale-setting, with SDSR used as a monitoring approach after CDMU application (Blomme et al., 2017b).

3.2.2 Variations on basic control packages

3.2.2.1 Uprooting entire diseased fields and fallowing

Roguing of all banana mats in highly infected fields followed by fallowing or crop rotation cycles using non-host plants has been recommended (Turyagyenda et al., 2008). While the CDMU control package does not specifically define an interval between the uprooting of diseased mats and the planting of new and healthy suckers, this approach advises a banana-free fallowing period of 6–12 months to ensure no Xvm inoculum remains in the soil (Turyagyenda et al., 2008; Rutikanga et al., 2016). Several pitfalls are however associated with this recommendation. Firstly, although an Xvm-inoculum free field can be achieved after the fallow period, re-infestations can easily occur from neighboring infected fields. Only through collective action and participation of the entire community will the risk of re-infection be substantially reduced. Secondly, smallholder farmers are generally not keen on imposing a fallow period due to low land availability and the loss of income. While crop rotation with non-host crops could alleviate this loss (Mwangi et al., 2006), the high cash value obtained from the banana crop is not easily replaced by another crop and thus impacts the livelihood of the farmer. Thirdly, complete elimination of Xvm inoculum is not ensured if the field is not closely monitored. For example, remnant corm pieces resulting from incomplete uprooting could lead to new shoot emergence, or the presence of a dominantly occurring perennial weed *Canna indica* (Rutikanga et al., 2016), found to act as a host, could render the fallow period unsuccessful.

3.2.2.2 Chemical destruction of diseased mats

Uprooting of entire disease mats (i.e. CDMU control) is very labor intensive and time consuming. Destruction of the mat via injection of herbicides (e.g. 2,4-D and glyphosate) substantially eases this task of killing and subsequently removing the infected mat (Blomme et al., 2008) and is equally effective as manual roguing (Blomme et al., 2008; Kubiriba et al., 2014). Although the cost of labor can be substantially reduced, the perceived additional cost of herbicides and applicators and the potentially limited availability of these chemicals impede adoption. The prolonged presence of the injected diseased mat in the field also obstructs farmers to move forward with planting new disease-free crops, holding up the availability of productive land as well as negatively affecting browsing cattle and small ruminants (Blomme et al., 2008, 2014).

3.2.3 Biological control

Biological control using antagonistic bacteria has been shown to be an efficient practice in the control of several *Xanthomonas* species (Marin et al., 2019).

Biocontrol of Xvm, however, remains in an exploratory phase and no in situ application has been reported. Experimental studies have shown suppression of Xvm after inoculating plants with antagonistic isolates from the rhizosphere and phyllosphere of the enset hosts (Abayneh, 2010) and from banana tissue (Mahuku, 2016), but none of these have been tested for their effect on BXW in the field.

3.3 Preventive measures

3.3.1 De-budding of the male inflorescence

Removal of the male bud (i.e. de-budding) just after the formation of the last hand of the bunch is a highly effective preventive technique for all banana bacterial wilts (Blomme et al., 2017a). This technique prevents the formation of the primary infection site for insect vectors (male inflorescence part/male bract and flower wounds) and thus prevents insect vector transmission of BXW. The use of a wooden forked stick for twisting the male peduncle is highly recommended (Fig. 3b; Blomme et al., 2005) to avoid cross-contamination that would occur when e.g. using metal cutting tools (e.g. machetes).

3.3.2 Tool sterilization

Regular sterilization of tools during field management is necessary to prevent disease transmission. Effective sterilization can be achieved using household bleach (sodium hypochlorite, NaOCl, 3.5%), by heating tools in fire up to a minute or until too hot to touch, by inserting tools into boiling water for at least 1 min and by cleaning with household soaps and detergents (Ocimati et al., 2021). Other methods, including insertion of the tool into hot ash, washing with cold water or the use of botanicals for their antibiotic properties, have proven less effective, only reducing Xvm populations and are thus not recommended (Ocimati et al., 2021). Nevertheless, not all sterilization options are practical. Farmers have reported that repetitive excessive heating of metal tool blades ultimately damages the tools, while household bleach can be too expensive for regular use (Ochola et al., 2014; Shimwela et al., 2016; Blomme et al., 2019; Ocimati et al., 2021). Ocimati et al. (2021) recommend sterilization through washing with soap/detergents (Fig. 3c) and tool insertion into boiling water for at least 1 min as more practical and affordable options.

3.3.3 Clean planting material

The use of clean planting material when setting up new fields or when replacing uprooted infected plants is critical to prevent bringing in Xvm inoculum and (re-)infecting a field (Sivirihauma et al., 2017; Jacobsen et al., 2019). Clean planting material in the form of tissue cultured plantlets, macro-propagated plantlets or suckers from clean mother gardens is recommended. However, planting material certification programs are basically non-existent for bananas in East and Central Africa (Jacobsen et al., 2019), and no regulated or practical framework is available for small-holder farmers (McCampbell et al., 2018). Most farmers have to rely on suckers sourced from their own field or neighboring farms (Tinzaara et al., 2013) with an associated risk of transmitting pests and disease (Tinzaara et al., 2013).

Careful screening and selection of symptomless suckers, even from infected fields (although preferably from disease-free mats), could help overcome the burden of lack of certified planting material with a minimized risk, resulting in productive crop establishment (Sivirihauma et al., 2017). Similarly, as SDSR, this practice is based on the incomplete systemic movement of Xvm within a mat (Ocimati et al., 2013a, 2015) by which healthy suckers can be present in an infected mat. The most concerning issue with this practice is the potential presence of latent infections in the transplanted suckers, requiring close monitoring during subsequent plant growth and applying SDSR in case of diseased shoots/plants (Sivirihauma et al., 2017). Several experimental studies have shown the efficacy of this practice as evidenced by low incidence levels of BXW disease, even with material sourced from highly infected fields (Ocimati et al., 2015; Sivirihauma et al., 2017).

The use of symptomless suckers from BXW-affected fields should be avoided when moving at larger spatial scales. Since a disease-free status of these suckers cannot be ensured, the transmission to new regions where BXW is not present or prevalent brings with it a high risk of BXW disease introduction. The selection of clean planting material from recognized disease-free sources should be prioritized (Jacobsen et al., 2019).

3.3.4 Amendments within the small-holder banana farming system

The structure of the small-holder banana farming system can affect the ease by which BXW spreads and small adaptations can be made by the farmer to reduce this risk. Firstly, the selection of banana cultivars that are less susceptible to insect-mediated Xvm transmission can substantially reduce the risk of infection (Blomme et al., 2017a; Changa et al., 2019, see Section 2.5). Farmers could, for example, opt for East African Highland bananas (AAA) and plantains (AAB) with persistent or semi-persistent neuter flowers and floral buds as potential alternatives for the more insect vector transmission susceptible ABB and AAB dessert cultivars (Tripathi et al., 2009; Karamura et al., 2010). The farmers' needs with regard to crop production will ultimately affect this decision. However, even incorporating a higher diversity of cultivars, increasing the presence of escape cultivars (i.e. not susceptible to insect vector transmission), can reduce the risk of establishment and spread of BXW (Ocimati et al., 2018b; Uwamahoro et al., 2019; Blomme et al., 2020b). Similarly, increasing crop diversity through intercropping of bananas with non-host crops is another practice that may reduce transmission (Vidal et al., 2017). This increased diversity can namely affect the microclimate, diversify the insect community or act as a physical barrier for spread. Increased management associated with intercropping should, however, be done with care as increased field visits and the use of garden tools also increase disease transmission risk (Ocimati et al., 2019; Blomme et al., 2017b; Uwamahoro et al., 2019). Accordingly, free roaming animals should be avoided in fields with possible BXW infections and kept away from infected plant debris left from BXW roguing.

3.3.5 Development of resistant cultivars

The development of bacterial wilt-resistant banana cultivars has received significant funding and may play a key role in BXW preventive management (Tripathi et al., 2020). The development of resistant cultivars through conventional breeding is limited by the lack of natural resistance found across the edible *Musa* germplasms (Blomme et al., 2014; Nakato et al., 2019a). Resistance to Xanthomonas wilt has only been identified in *M. balbisiana* (Ssekiwoko et al., 2006b; Kebede and Gemmeda, 2017; Nakato et al., 2019a), a wild relative of banana. The mechanisms of resistance toward Xvm and its genetics are not well understood and introgression of other unwanted traits of *M. balbisiana*, mainly with regard to fruit quality, reduce its appeal as a useful source for breeding. The tolerance recently observed in two subspecies of *M. acuminata* (Nakato et al., 2019a) may provide new options for conventional breeding approaches. Further issues of long generation times, various levels of ploidy, sterility of most cultivars and limited genetic variability complicate breeding of resistant banana cultivars (Tripathi et al., 2004; Smith and Pillay, 2020).

Alternatively, genetic engineering for resistance has come up with promising results. The integration of transgenes encoding for plant ferredoxin-like protein (pflp) and hypersensitive response assisting protein (hrap) isolated from sweet pepper (*C. annuum*), both defense genes intensifying a hypersensitive response (Lin et al., 1997; Chen et al., 2000), has led to the successful establishment of resistance to Xvm in banana cultivars (Tripathi et al., 2010; Namukwaya et al., 2012). Similarly, transgenetic expression of the rice pattern recognition receptor, Xa21 gene, selected for its broad-spectrum resistance against the rice pathogen *Xanthomonas oryzae* pv. *oryzae* (Xoo) (Wang et al., 1996), has resulted in enhanced resistance to Xvm (Tripathi et al., 2014). Laboratory, screenhouse and initial field testing has shown a high level of resistance to BXW without any known adverse effects on plant physiology or performance (Tripathi et al., 2010, 2014; Namukwaya et al., 2012).

Recent advances in CRISPR/Cas9-based genome editing, which provides the capability of creating precise alterations in the plant genome, may accelerate the development of resistant cultivars (Tripathi et al., 2020). Recently, robust CRISPR/Cas9-based genome editing of banana has been established (Naim et al., 2018; Kaur et al., 2018; Ntui et al., 2020; Tripathi et al., 2020), and the first advances toward targeted genome manipulations for BXW resistance have been made (Tripathi et al., 2021). Tripathi et al. (2021) targeted the downy mildew resistance 6 (DMR6) gene, which is a suppressor of plant immunity upregulated during pathogen-infection (Low et al., 2020). Enhanced BXW resistance was achieved in MusaDMR6 mutants of banana without any known adverse effects on plant growth.

The rolling out of resistant cultivars from genetic engineering or genome editing should not be expected in the short term, as research is still ongoing concerning the long-term efficacy of the produced clones, and comprehensive risk assessments and safety evaluations are needed (e.g. Jin et al., 2017). Moreover, most countries across East and Central Africa do not have an established regulatory framework in place with regard to genetically modified crops, which further delays implementation.

3.3.6 Risk assessments, epidemiological modeling and early warning systems

Access to accurate and timely assessments of the risk of new BXW infections and transmission is critical for risk mitigation through the timely implementation of preventive strategies and early disease control. Epidemiological models are critical tools for risk assessment, providing predictions based on either an empirical or mechanistic understanding of disease transmission combined with current data on disease incidences. Current BXW modeling efforts vary in complexity and predictive power focusing on various levels of detail on modes of spread, environmental and climatological factors, variations of farming systems and management options (Kweyunga and Tumwiine, 2017; Nakakawa et al., 2017; Shimwela et al., 2017; Kweyunga et al., 2018; Xing et al., 2020). Visual representations of modeling results through mapping of at-risk regions form highly informative and easily interpretable communication tools for various stakeholders (Shimwela et al., 2017; Ocimati et al., 2019). Finally, these models can be used for more than risk assessment alone and aid in the identification of optimal control frameworks and cost-effective management practices for various regions (Nakakawa et al., 2017).

Timely risk assessments and the usefulness of predictive modeling, however, highly depend on the accuracy and timely availability of data on

BXW incidences across East and Central Africa. Data collection is generally done through surveillance by trained staff across specific regions of interest, but detailed long-term monitoring across large spatial scales is difficult to achieve. Recent advances combining digital imaging with disease diagnostics through machine learning may improve the accessibility of up-to-date disease incidence data (Selvaraj et al., 2019, 2020; Adewopo et al., 2021; see Section 3.1). The recent development and implementation of BXW-focused mobile phone applications, 'Tumaini' and 'ICT4BXW', form a first step toward closely involving farmers with data collection efforts, while at the same time providing access to real-time information and early warning systems (Selvaraj et al., 2019, 2020; Adewopo et al., 2021).

4 Implementation of Xanthomonas wilt management

Various iterations of the BXW management options have been disseminated to farmers by research and development agencies with variable degrees of success achieved across the impacted East and Central Africa regions (Blomme et al., 2014, 2017a,b; Tinzaara et al., 2019). Substantial successes have been obtained through full application of the main cultural control packages CDMU and SDSR (i.e. complemented with the preventive measures of male bud removal and the use of clean garden tools and clean planting material), resulting in effective disease containment, reduced disease incidence and severity, and the recovery of field production levels (Kubiriba et al., 2014; Nakakawa et al., 2017; Kikulwe et al., 2019). Incomplete application of these packages, however, often through omission of preventive measures (Blomme et al., 2017a), has rendered the management of BXW difficult and ineffective in some regions (Jogo et al., 2013). The adoption of all components of the recommended control packages by local farmers is influenced by the know-how of the farmer, practical ease of use of the techniques, availability of resources (time, labor, money), the perceived effectiveness of the separate practices, the needs of the farm, the intensity of dissemination efforts and the level of trust farmers have in the extension services (Jogo et al., 2013; McCampbell et al., 2018; Blomme et al., 2019; Ntamwira et al., 2019).

The coverage, content and reach of extension services can make or break BXW management (Ocimati et al., 2021), including the sensitization and training of farmers, the introduction of bylaws and task forces surveying areas and enforcing rigorous action when needed (McCampbell et al., 2018). Weak extension systems with a low knowledge base give rise to low adoption levels by farmers and incomplete or incorrect implementation of the measures at the farm level. A better understanding of the struggles the farmers face and the factors influencing farmers' adoption decisions could guide the improvement of extension efforts (Jogo et al., 2013; McCampbell et al., 2018). Accordingly, the distribution of funds to various stakeholders in terms of research versus extension services and aid for farmers should be better aligned to address these issues (Di Cori et al., 2018).

Extension services tailored to regional and national needs have proven to be important through the Roots, Tubers and Bananas scaling project (Rietveld, 2021) in which specific scaling strategies of SDSR were developed for Burundi, Uganda and eastern Democratic Republic (DR) of Congo. In Uganda, multiple governmental agencies and programs dealing with the control of BXW have been in place for over a decade, and the scaling of innovations such as SDSR could follow existing strategies. In Burundi, reaching and direct training of substantial percentages of households, presenting them with cost-effective and feasible management practices to control BXW (i.e. SDSR), proved sufficient to motivate farmers. In eastern DR Congo, providing direct training in remote areas proved difficult and additional extension efforts via radio campaigns were needed. Presenting local leaders with positive results of SDSR also prompted the implementation of local bylaws at several sites, mandating the application of SDSR. Additionally, testing new engagement strategies in eastern DR Congo showed that the employment of organized cutting units to inspect, train farmers and jointly cut diseased plants within the communities, rather than relying on farmer volition, proved more effective to reach community-wide BXW management and could be incorporated in national extension strategies. Identifying innovation bottlenecks according to cultural settings is also critical to improve implementation of BXW management.

4.1 Implementation issues for small-holder farmers

4.1.1 Practical constraints

Practically, the implementation of BXW management is not always straightforward for small-holder farmers due to the limited availability of resources (time, labor, money) and difficulties combining disease control practices with general farm activities. Throughout Section 3, multiple practical constraints for farmers have already emerged, including incomplete or inaccessible tool sterilization options, limited availability of clean planting material and the high labor demand, time constraints and cost of full control packages. Furthermore, the time required to achieve a return on investment from BXW management (i.e. a disease-free, productive crop) can be too long for resource-constrained farmers, to sustain their livelihood in the short term.

Farmers have been known to take shortcuts with BXW management. For example, by saving time through dropping some of the control practices (e.g. de-budding) that form part of the control package, which frees up time to perform other more lucrative activities with a faster return on investment. The time farmers spend on applying control practices decreases over time once the visible infection levels in the field have reduced.

Both the time-consuming activity of long-term monitoring of their fields and the perceived reduction in need to apply BXW management measures after initial incidence numbers have dropped lead to farmers abandoning the application of effective management practices (McCampbell et al., 2018). In larger households with more labor potential available, more components of the control packages are generally adopted (Jogo et al., 2013).

Timely application of the recommended control packages is also not always easy to achieve. For example, the removal of the male bud is very time-specific and is often performed too late for effective prevention of contact with insects (Muhangi et al., 2006). Management of BXW can also interfere with the timing of other farm practices. In farms where the banana plants are intercropped with annual crops, regular cutting of diseased plants (SDSR) during cropping seasons can damage these intercrops (Blomme et al., 2017b; Iradukunda et al., 2019). Household needs will generally determine which practices are prioritized.

4.1.2 Access to information and training

Having reliable access to information and training is an important factor for the small-holder farmer. While several forms of extension services such as awareness campaigns, farmers' field schools and community actions have proven to be effective (Kubiriba et al., 2012; Nkuba et al., 2015; Tinzaara et al., 2016), several recent case studies revealed that reliable and continued access to knowledge extension services remains too low, leaving farmers with knowledge gaps (Nakato et al., 2016; McCampbell et al., 2018; Blomme et al., 2020b). This issue is even exacerbated for women. While women are often more involved in the day-to-day management of banana plantations (Kikulwe et al., 2018; Kawarazuka et al., 2020), sensitization and training programs have mostly been attended by male farmers as head of their household (Jogo et al., 2013; Iradukunda et al., 2019).

Training and awareness campaigns are also important to rectify often-wrongful perceptions of BXW measures by farmers, which can drive the decision-making process for adoption (Jogo et al., 2013; Blomme et al., 2019). For example, the regular re-appearance of diseased plants when using the SDSR control package can be perceived as a failure of the applied measures and can be very discouraging for farmers (Blomme et al., 2019). Farmers can also perceive a delay in implementation of measures as beneficial with the goal of safeguarding the short-term production of other crops, especially when disease incidence is low (Blomme et al., 2017b). Providing detailed information about the most cost-effective combinations of management, the benefits of early application of control methods and expected timeframes for return on investment is important to clarify this misconception (Nakakawa et al., 2017; Gotor et al., 2020).

4.1.3 Difficulties realizing collective action

Mobilizing entire communities and realizing collective action for the management of BXW has been done with varying levels of success (McCampbell et al., 2018; Ocimati et al., 2020). Specifically, in small-holder farming communities, area-wide management is more difficult to achieve due to the varying levels of importance of bananas for farmers' livelihood, more limited market-oriented production and the often weak extension support (Rutikanga et al., 2016; Blomme et al., 2019).

4.2 Integrated approaches toward sustainable management

Integrated participatory approaches are often recognized as the best way forward to improve extension support for local small-holder farmers. Effective extension support has been achieved through various approaches that depend on the presence of external stakeholders, including community sensitization for action, participatory development communication and farmer field schools (Kubiriba et al., 2012; Blomme et al., 2014; Tinzaara et al., 2016). However, continued regular presence of external support is not sustainable, generally related to funding constraints, and often pulled back when infection levels in a region decline (Tinzaara et al., 2019), often followed by a resurgence of the disease. Participatory approaches still rely on support from external stakeholders, but add a two-way stream of information by which farmers can communicate issues and contribute in the development of modified techniques targeted toward the specific needs of the community (Jogo et al., 2013). Partially transferring responsibility of communication to the community increases the level of participation and accountability within the community and can enhance wider adoption of BXW management.

Several iterations of participatory approaches have been developed. For example, Nakato et al. (2016) assessed the Community Knowledge Worker (CKW) initiative to improve the flow of information and connections among stakeholders. Local leaders were hereby trained as CKWs and acted as a communication node between local farmers and external experts. Using mobile phone applications, they could provide farmers with direct access to agricultural information through an agriculture call center and simultaneously collect data and surveys on local BXW infestations and management issues for researchers. Another approach by Tinzaara et al. (2019), namely the learning and experimentation approaches for farmers, centered around the two-way sharing of innovations between facilitators and farmers and the systematic experimentation and validation of technologies. More basic approaches such as exchange visits can also be very effective, opening up communication channels between farmers and communities and enhancing the adoption of improved practices (Kikulwe et al., 2019).

Enhancing participatory approaches through citizen science with information, communication and technology (ICT)-based tools has also recently been investigated. Through the development of mobile phone applications (e.g. Tumaini app, ICT4BXW; Selvaraj et al., 2019, 2020; Adewopo et al., 2021), farmers are able to digitally access current information while at the same time contributing through data collection. Real-time information from farmers on BXW infection levels can substantiate risk assessments of disease transmission and contribute to early warning systems. While showing great potential, digital extension tools are far from being widely implemented, with small-holder farmers still having limited access to smartphones and poor network coverage often occurring in many areas (Adewopo et al., 2021).

Finally, calls are being made to enhance the scope of extension services and integrate broader schemes of issues that farmers face. This could take various forms, including integrating marketing and strengthening agricultural trading to improve livelihoods (Tinzaara et al., 2019), integrating the issue of BXW into overarching food security projects (Mbure et al., 2018) or combining disease management needs with other environmental issues (Ocimati et al., 2018a; Blomme et al., 2020b). Establishing such integrated approaches will require substantial engagement of a wide spectrum of stakeholders, combined with elaborate coordination and collaboration.

5 Conclusion and future trends in research

Research efforts are currently heavily focused on improving the implementation of control measures, evaluating various modes of extension efforts and the development of improved preventive options. The successful development and deployment of resistant cultivars will be a game-changer for preventive BXW management and continued research on breeding, and genetic engineering for resistance (including cis-gene approaches and CRISPR) is expected in the coming years. Further advancements in communications to strengthen participatory extension services are also expected. While dependent on the availability of technology in often remote places, smartphone applications are expected to directly improve a two-way stream of information between farmers and other stakeholders. Applying citizen science approaches for data collection on disease presence through smartphone applications is expected to improve disease surveillance across regions. Combining this ground-truth data with airborne and satellite imaging and machine learning diagnostic analysis are expected to strengthen early warning systems.

6 Where to look for further information

The following articles provide a good overview:

- Blomme, G., Jacobsen, K., Ocimati, W. and Beed, F. D. (2014). Fine-tuning banana Xanthomonas wilt control options over the past decade in East and Central Africa, Eur. J. Plant Pathol., 139, 265–281.
- Blomme, G., Dita, M., Jacobsen, K. S., Vicente, L. P., Molina, A., Ocimati, W., Poussier, S. and Prior, P. (2017). Bacterial diseases of bananas and enset: current state of knowledge and integrated approaches toward sustainable management, Front. Plant Sci., 8, 1290.
- Blomme, G. and Ocimati, W. (2018). Xanthomonas bacterial wilt, in D. R. Jones (ed.), Chapter 5: Diseases Caused by Bacteria and Phytoplasmas, Handbook of Diseases of Banana, Abacá and Enset, pp. 296–313. Wallingford: CAB International. ISBN: 9781780647197.
- McCampbell, M., Schut, M., Van den Bergh, I., van Schagen, B., Vanlauwe, B., Blomme, G., Gaidashova, S., Njukwe, E. and Leeuwis, C. (2018). Xanthomonas Wilt of Banana (BXW) in Central Africa: opportunities, challenges, and pathways for citizen science and ICT-based control and prevention strategies, NJAS - Wagening. J. Life Sci., 86–87, 89–100.
- Tinzaara, W., Karamura, E., Blomme, G., Jogo, W., Ocimati, W. and Kubiriba, J. (2013). Communication approaches for sustainable management of banana Xanthomonas wilt in East and Central Africa, in G. Blomme et al. (eds), Banana systems in the humid highlands of Sub-Saharan Africa: enhancing resilience and productivity, pp. 224–234. CAB International. ISBN: 9781780642314.

Key research in BXW is performed by the following organizations:

- The Alliance of Bioversity and CIAT;
- The International Institute of Tropical Agriculture;
- National Agricultural Research Organisation, Uganda;
- Institut National des Etudes et Recherches Agronomique, DR Congo;
- Rwanda Agriculture Board, Rwanda; and
- Institut des Sciences Agronomiques du Burundi, Burundi.

7 References

Abayneh, T. B. (2010). Evaluation of biological control agents against bacterial wilt pathogen (*Xanthomonas campestris* pv. *musacearum*) of Ensete (*Ensete ventricosum*), Master thesis, School of Graduate Studies of Addis Ababa University, 66 p.

Adewopo, J. B., McCampbell, M., Mwizerwa, C. and Schut, M. (2021). A reality check for digital agricultural extension tool development and use, *Rural* 21(01/21), 23–25.

Adriko, J., Aritua, V., Mortensen, C. N., Tushemereirwe, W. K., Kubiriba, J. and Lund, O. S. (2012). Multiplex PCR for specific and robust detection of *Xanthomonas campestris* pv. *musacearum* in pure culture and infected plant material: specific Xcm diagnosis, *Plant Pathol.* 61(3), 489–497.

Aritua, V., Nanyonjo, A., Kumakech, F. and Tushemereirwe, W. (2007). Rep-PCR reveals a high genetic homogeneity among Ugandan isolates of *Xanthomonas campestris* pv. *Musacearum*, *Afr. J. Biotech.* 6, 179–183.

Aritua, V., Parkinson, N., Thwaites, R., Heeney, J. V., Jones, D. R., Tushemereirwe, W., Crozier, J., Reeder, R., Stead, D. E. and Smith, J. (2008). Characterization of the Xanthomonas sp. causing wilt of enset and banana and its proposed reclassification as a strain of X. Vasicola, *Plant Pathol.* 57, 070918211612008-070918211612???

Arthur, W., Geoffrey, T. and Jerome, K. (2017). PCR validation of predicted virulence factors in a collection of *Xanthomonas campestris* (pv. *musacearum* and *vascolurum*) strains, *Afr. J. Biotechnol.* 16(12), 558–572.

Blomme, G., Dita, M., Jacobsen, K. S., Pérez Vicente, L., Molina, A., Ocimati, W., Poussier, S. and Prior, P. (2017a). Bacterial diseases of bananas and enset: current state of knowledge and integrated approaches Toward sustainable management, *Front. Plant Sci.* 8, 1290.

Blomme, G., Ocimati, W., Sivirihauma, C., Vutseme, L., Mariamu, B., Kamira, M., van Schagen, B., Ekboir, J. and Ntamwira, J. (2017b). A control package revolving around the removal of single diseased banana stems is effective for the restoration of Xanthomonas wilt infected fields, *Eur. J. Plant Pathol.* 149(2), 385–400.

Blomme, G., Dusingizimana, P., Ntamwira, J., Kearsley, E., Gaidashova, S., Rietveld, A., Van Schagen, B. and Ocimati, W. (2021). Comparing effectiveness, cost- and time-efficiency of control options for Xanthomonas wilt of banana under Rwandan agro-ecological conditions, *Eur. J. Plant Pathol.* 160(2), 487–501.

Blomme, G., Jacobsen, K., Ocimati, W., Beed, F., Ntamwira, J., Sivirihauma, C., Ssekiwoko, F., Nakato, V., Kubiriba, J., Tripathi, L., Tinzaara, W., Mbolela, F., Lutete, L. and Karamura, E. (2014). Fine-tuning banana Xanthomonas wilt control options over the past decade in East and Central Africa, *Eur. J. Plant Pathol.* 139(2), 271–287.

Blomme, G., Mukasa, H., Ssekiwoko, F. and Eden-Green, S. (2005). On-farm assessment of banana bacterial wilt control options, *Afr. Crop Sci. Conf. Proc.* 7, 317–320.

Blomme, G., Ocimati, W., Felde, A. Z., Amwonya, D. and Kantungeko, D. (2020a). A literature review on yield gaps of various root, tuber and banana crops as a background for assessing banana yield reductions due to pests and diseases at a field site in Western Burundi, *Afr. J. Agric. Res.* 16, 1169–1183.

Blomme, G., Ocimati, W., Amato, S., Felde, A., Kamira, M., Bumba, M., Bahati, L., Amini, D. and Ntamwira, J. (2020b). Banana pest risk assessment along banana trade axes running from low to high altitude sites, in the Eastern DR Congo and in Burundi, *Afr. J. Agric. Res.* 16(9), 1253–1269.

Blomme, G., Ocimati, W., Sivirihauma, C., Lusenge, V., Bumba, M. and Ntamwira, J. (2019). Controlling Xanthomonas wilt of banana: influence of collective application, frequency of application, and social factors on the effectiveness of the Single Diseased Stem Removal technique in eastern Democratic Republic of Congo, *Crop Prot.* 118, 79–88.

Blomme, G., Turyagyenda, L. F., Mukasa, H. and Eden-Green, S. (2008). The effectiveness of different herbicides in the destruction of banana Xanthomonas wilt infected plants. Special Issue, *Afr. Crop Sci. J.* 16(1), 103–110.

Bradbury, J. F. (1986). *Guide to Plant Pathogenic Bacteria*, pp. 198–260, CAB International Mycological Institute, Kew.

Buregyeya, H. G., Tusiime, J., Kubiriba, J. and Tushmereirwe, W. K. (2008). Evaluation of distant transmission of banana bacterial wilt in Uganda, Conference on Banana and Plantain in Africa: Harnessing International Partnerships to Increase Research Impact, Mombasa, October 5–9.

Buregyeya, H., Kubiriba, J., Tusiime, G., Kityo, R., Ssekiwoko, F. and Tushemerierwe, W. K. (2014). Role of birds & bats in Long Distance transmission of banana bacterial wilt in Uganda, *Int. J. Agric. Innov. Res.* 2, 2319–1473.

Carter, B. A., Reeder, R., Mgenzi, S. R., Kinyua, Z. M., Mbaka, J. N., Doyle, K., Nakato, V., Mwangi, M., Beed, F., Aritua, V., Lewis Ivey, M. L., Miller, S. A. and Smith, J. J. (2010). Identification of *Xanthomonas vasicola* (formerly *X. campestris* pv. *musacearum*), causative organism of banana Xanthomonas wilt, in Tanzania, Kenya and Burundi, *Plant Pathol.* 59(2), 403.

Castellani, E. (1939). 'Su un marciume dell' Ensete, *L'Agric. Coloniale Firenze* 33, 297–300.

Chala, A., Kebede, T. and Blomme, G. (2016). Natural occurrence and pathogenicity of Xanthomonas bacteria on selected plants, *Afr. J. Biotechnol.*, 15(39), 2146–2155.

Changa, C. M., Tushemereirwe, W., Ssekiwoko, F., Kubiriba, J., Mudonyi, B., Vanhaelewyn, L. and Tusiime, G. (2019). Banana field resistance to insect-vector transmission of bacterial wilt caused by *Xanthomonas campestris* pv. *musacearum*, *Afr. J. Agric. Res.* 14(1), 24–33.

Chen, C. H., Lin, H. J., Ger, M. J., Chow, D. and Feng, T. Y. (2000). The cloning and characterization of a hypersensitive response assisting protein that may be associated with the harpin-mediated hypersensitive response, *Plant Mol. Biol.* 43(4), 429–438.

Di Cori, V., Kikulwe, E., Kozicka, M. and Gotor, E. (2018). *Understanding the Economic Impact of BXW and Its Management Practices in East and Central Africa*, p. 16, Bioversity International, Rome, Italy, ISBN: 978-92-9255-116-2.

Drenth, A., and Kema, G. (2021). The vulnerability of bananas to globally emerging disease threats. *Phytopathology* 111(12):2146–2161. doi: 10.1094/PHYTO-07-20-0311-RVW. Epub 2021 Dec 7. PMID: 34231377.

Gotor, E., Di Cori, V., Pagnani, T., Kikulwe, E., Kozicka, M. and Caracciolo, F. (2020). Public and private investments for banana Xanthomonas Wilt control in Uganda: the economic feasibility for smallholder farmers, *Afr. J. Sci. Technol. Innov. Dev.* 14(1), 135–146, doi: 10.1080/20421338.2020.1816616.

Iradukunda, F., Bullock, R., Rietveld, A. and van Schagen, B. (2019). Understanding gender roles and practices in the household and on the farm: implications for banana disease management innovation processes in Burundi, *Outlook. Agr.* 48(1), 37–47.

Jacobsen, K., Omondi, B. A., Almekinders, C., Alvarez, E., Blomme, G., Dita, M., Iskra-Caruana, M.-L., Ocimati, W., Tinzaara, W., Kumar, P. L. and Staver, C. (2019). Seed degeneration of banana planting materials: strategies for improved farmer access to healthy seed, *Plant Pathol.* 68(2), 207–228.

Jin, Y., Goodman, R. E., Tetteh, A. O., Lu, M. and Tripathi, L. (2017). Bioinformatics analysis to assess potential risks of allergenicity and toxicity of HRAP and PFLP proteins in genetically modified bananas resistant to Xanthomonas wilt disease, *Food Chem. Toxicol.* 109(1), 81–89.

Jogo, W., Karamura, E., Tinzaara, W., Kubiriba, J. and Rietveld, A. (2013). Determinants of farm-level adoption of cultural practices for banana Xanthomonas Wilt control in Uganda, *J. Agric. Sci.* 5(7), 70–81.

Karamura, E. B., Turyagyenda, F. L., Tinzaara, W., Blomme, G., Ssekiwoko, F., Eden-Green, S., Molina, A. and Markham, R. (2008). *Xanthomanas Wilt of Banana in East and Central Africa*. Diagnostic and Management Guide, pp. 60, Bioversity International, Rome, Italy.

Karamura, E., Kayobyo, G., Tushmereirwe, W., Benin, S., Blomme, G., Eden Green, S. and Markham, R. (2010). Assessing the impacts of banana bacterial wilt disease on banana (Musa spp.) productivity and livelihoods of Ugandan farm households, *Acta Hortic.* 879(879), 749–755.

Karamura, G., Smith, J., Studholme, D., Kubiriba, J. and Karamura, E. (2015). Comparative pathogenicity studies of the *Xanthomonas vasicola* species on maize, sugarcane and banana, *Afr. J. Plant Sci.* 9(9), 385–400.

Karamura, G., Ochola, D., Smith, J., Kubiriba, J. and Mwaka, G. (2016). 'Evaluation of a lateral flow device for in-field detection of Banana Xanthomonas Wilt and its application in tracking the systemicity of *Xanthomonas campestris* pv. *musacearum*', *Afr. J. Agric. Res.*, 11(25), 2250–2256.

Kaur, N., Alok, A., Shivani, , Kaur, N., Pandey, P., Awasthi, P. and Tiwari, S. (2018). CRISPR/Cas9- mediated efficient editing in phytoene desaturase (PDS) demonstrates precise manipulation in banana cv. Rasthali genome, *Funct. Integr. Genomics* 18(1), 89–99.

Kawarazuka, N., Damtew, E., Mayanja, S., Okonya, J. S., Rietveld, A., Slavchevska, V. and Teeken, B. (2020). A gender perspective on pest and disease management from the cases of roots, tubers and bananas in Asia and Sub-Saharan Africa, *Front. Agron.* 2, 7.

Kebede, T. and Gemmeda, L. (2017). 'Evaluation of banana genotype resistant to Xanthomonas wilts disease (*Xanthomonas campestris* pv. *musacearum*) in south east of Ethiopia', *Afr. J. Microbiol. Res.*, 11(15), 603-612.

Kikulwe, E. M., Okurut, S., Ajambo, S., Gotor, E., Ssali, R. T., Kubiriba, J. and Eldad, K. (2018). 'Does gender matter in effective management of plant disease epidemics? Insights from a survey among rural banana farming households in Uganda', *J. Dev. Agric. Econ.*, 10(3), 87–98.

Kikulwe, E. M., Kyanjo, J. L., Kato, E., Ssali, R. T., Erima, R., Mpiira, S., Ocimati, W., Tinzaara, W., Kubiriba, J., Gotor, E., Stoian, D. and Karamura, E. (2019). Management of banana Xanthomonas Wilt: evidence from impact of adoption of cultural control practices in Uganda, *Sustainability* 11(9), 2610.

Kubiriba, J., Bagamba, F., Rockfeller, E. and Tushemereirwe, W. K. (2012). 'The changing spread dynamics of banana Xanthomonas wilt (BXW) in Uganda, *Uganda J, Agric. Sci.* 13, 53–60.

Kubiriba, J., Muthomi, J., Ndungo, V., Johnson, K., Rockefeller, E., Ivan, R., Wilberforce, T. and Fina, O. (2014). Strategies for rehabilitation of banana fields infested with *Xanthomonas campestris* pv. *musacrearum*, *J. Crop. Prot.* 3, 21–29.

Kweyunga, E. H. and Tumwiine, J. (2017). 'A mathematical model for the vector transmission and control of Banana Xanthomonas Wilt', *J. Math. Res.*, 9(4), 101–113.

Kweyunga, E. H., Tumwiine, J. and Karamura, E. (2018). Modeling the dynamics of Banana Xanthomonas Wilt transmission incorporating infectious force in both asymptomatic and symptomatic stages, *J. Adv. Math. Comput. Sci* 29(3), 1–17.

Lin, H.-J., Cheng, H.-Y., Chen, C.-H., Huang, H.-C. and Feng, T.-Y. (1997). Plant amphipathic proteins delay the hypersensitive response caused by harpin (Pss) and Pseudomonas syringae pv. Syringae, *Physiol. Mol. Plant Pathol.* 51(6), 367–376.

Lombaert, E., Guillemaud, T., Cornuet, J. M., Malausa, T., Facon, B. and Estoup, A. (2010). Bridgehead effect in the worldwide invasion of the biocontrol Harlequin Ladybird, *PLoS ONE* 5(3), e9743.

Low, Y. C., Lawton, M. A. and Di, R. (2020). Validation of barley 2OGO gene as a functional orthologue of Arabidopsis DMR6 gene in Fusarium head blight susceptibility, *Sci. Rep-UK* 10, 9935.

Mahuku, G. (2016). Endophytic bacteria associated with banana and their potential for controlling banana Xanthomonas wilt, Master thesis, Makerere University, Kampala, Uganda.

Marin, V. R., Ferrarezi, J. H., Vieira, G. and Sass, D. C. (2019). Recent advances in the biocontrol of *Xanthomonas* spp., *World J. Microbiol. Biotechnol.* 35(5), 72.

Mbure, G., Nzawele, D. B. and Byamungu, J. (2018). Strengthening local systems to control Xanthomonas wilt in eastern democratic Republic of Congo, *Acta Hortic.* 1196(1196), 237–246.

McCampbell, M., Schut, M., Van den Bergh, I., van Schagen, B., Vanlauwe, B., Blomme, G., Gaidashova, S., Njukwe, E. and Leeuwis, C. (2018). Xanthomonas Wilt of Banana (BXW) in Central Africa: opportunities, challenges, and pathways for citizen science and ICT-based control and prevention strategies, *NJAS: Wageningen Journal of Life Sciences* 86–87(1), 89–100.

Muhangi, J., Nankinga, C., Tushemereirwe, W. K., Rutherford, M., Ragama, P., Nowakunda, K. and Abeyasekera, S. (2006). Impact of awareness campaigns for banana bacterial wilt control in Uganda, *Afr. Crop Sci. J.* 14, 175–183.

Mwangi, M., Tinzaara, W., Vigheri, N., Flora, N., Philip, R. and Ranajit, B. (2006). Comparative study of Banana Xanthomonas wilt spread in mid and high altitudes of the Great Lakes Region of Africa, *Tropentag*, Conference on International Agricultural Research for Development, University of Bonn, p. 5.

Mwangi, M., Mwebaze, M., Bandyopadhyay, R., Aritua, V., Eden-Green, S., Tushemereirwe, W. and Smith, J. (2007). Development of a semi-selective medium for isolating *Xanthomonas campestris* pv. *musacearum* from insect vectors, infected plant material and soil, *Plant Pathol.* 56(3), 383–390.

Mwebaze, J., Tusiime, G., Teshemereirwe, W. and Kubiriba, J. (2006). The survival of *Xanthomonas campestris pv. musacearum* in soil and plant debris, *Afr. Crop Sci. J.* 14, 121–127.

Naim, F., Dugdale, B., Kleidon, J., Brinin, A., Shand, K., Waterhouse, P. and Dale, J. (2018). Gene editing the phytoene desaturase alleles of Cavendish banana using CRISPR/Cas9, *Transgenic Res.* 27(5), 451–460.

Nakakawa, J., Mugisha, J. Y. T., Shaw, M. W., Tinzaara, W. and Karamura, E. (2017). Banana *Xanthomonas* Wilt infection: the role of debudding and roguing as control options within a mixed cultivar plantation, *Int. J. Math. Math. Sci.* 2017, 1–13.

Nakato, G. V., Beed, F. D., Ramathani, I. and Opio, F. (2013a). Risk of banana Xanthomonas wilt spread through trade, *J. Crop Prot.* 2, 151–161.

Nakato, G. V., Akinbade, S. A., Kumar, L., Bandyopadhyay, R. and Beed, F. (2013b). Development of ELISA for the detection of *Xanthomonas campestris* pv. *musacearum*, the causal agent of BXW: banana Xanthomonas wilt. In: *Banana Systems in the Humid Highlands of Sub-Saharan Africa: Enhancing Resilience and Productivity,*

G. Blomme, P. van Asten and B. H. Vanlauwe (Eds), CABI Publishing, Wallingford, Oxfordshire, pp. 93–101.

Nakato, G. V., Ocimati, W., Blomme, G., Fiaboe, K. K. M. and Beed, F. (2014). Comparative importance of infection routes for banana Xanthomonas wilt and implications on disease epidemiology and management, *Can. J. Plant Pathol.* 36(4), 418–427.

Nakato, G. V., Beed, F., Bouwmeester, H., Ramathani, I., Mpiira, S., Kubiriba, J. and Nanavati, S. (2016). Building agricultural networks of farmers and scientists via mobile phones: case study of banana disease surveillance in Uganda, *Can. J. Plant Pathol.* 38(3), 307–316.

Nakato, V., Mahuku, G. and Coutinho, T. (2018a). *Xanthomonas campestris* pv. *musacearum*: a major constraint to banana, plantain and enset production in central and east Africa over the past decade, *Mol. Plant Pathol.* 19(3), 525–536.

Nakato, G. V., Wicker, E., Coutinho, T. A., Mahuku, G. and Studholme, D. J. (2018b). A highly specific tool for identification of *Xanthomonas vasicola* pv. *musacearum* based on five Xvm-specific coding sequences, *Heliyon* 4(12), e01080.

Nakato, G. V., Christelová, P., Were, E., Nyine, M., Coutinho, T. A., Doležel, J., Uwimana, B., Swennen, R. and Mahuku, G. (2019a). Sources of resistance in *Musa* to *Xanthomonas campestris* pv. *musacearum*, the causal agent of banana Xanthomonas wilt, *Plant Pathol.* 68(1), 49–59.

Nakato, G. V., Fuentes Rojas, J. L., Verniere, C., Blondin, L., Coutinho, T., Mahuku, G. and Wicker, E. (2019b). A new multi locus Variable Number of Tandem Repeat Analysis Scheme for epidemiological surveillance of *Xanthomonas vasicola* pv. *musacearum*, the plant pathogen causing bacterial wilt on banana and enset, *PLoS ONE* 14(4), e0215090.

Nakato, G. V., Studholme, D. J., Blomme, G., Grant, M., Coutinho, T. A., Were, E. M., Wicker, E. and Mahuku, G. (2021). SNP-based genotyping and whole genome sequencing reveal previously unknown genetic diversity in *Xanthomonas vasicola* pv. *musacearum*, causal agent of banana Xanthomonas wilt, in its presumed Ethiopian origin, *Plant Pathol.* 70(3), 534–543.

Namukwaya, B., Tripathi, L., Tripathi, J. N., Arinaitwe, G., Mukasa, S. B. and Tushemereirwe, W. K. (2012). Transgenic banana expressing Pflp gene confers enhanced resistance to Xanthomonas wilt disease, *Transgenic Res.* 21(4), 855–865.

Ndungo, V., Bakelana, K., Eden-Green, S. and Blomme, G. (2004). An outbreak of banana Xanthomonas wilt (*Xanthomonas campestris* pv. *musacearum*) in the Democratic Republic of Congo, *Infomusa* 13, 43–44.

Nkuba, J., Tinzaara, W., Night, G., Niko, N., Jogo, W., Ndyetabula, I., Mukandala, L., Ndayihazamaso, P., Niyongere, C., Gaidashova, S., Rwomushana, I., Opio, F. and Karamura, E. (2015). 'Adverse impact of banana Xanthomonas wilt on farmers' livelihoods in Eastern and Central Africa', *Afr. J. Plant Sci.*, 9(7), 279–286.

Ntamwira, J., Blomme, G., Bahati, L. and Ocimati, W. (2019). Effect of timing of diseased plant cutting, altitude and banana cultivar on efficacy of singly removing Xanthomonas wilt infected banana plants, *Eur. J. Plant Pathol.* 154(2), 477–489.

Ntui, V. O., Tripathi, J. N. and Tripathi, L. (2020). Robust CRISPR/Cas9 mediated genome editing tool for banana and plantain (*Musa* spp.), *Curr. Plant Biol.* 21, 100128.

Ochola, D., Jogo, W., Odongo, M., Tinzaara, W., Onyango, M. and Karamura, E. B. (2014). 'Household dynamics influencing effective eradication of Xanthomonas wilt in smallholder banana systems in Ugunja division – Kenya', *Afr. J. Agric. Res.*, 9(26), 2031–2040.

Ocimati, W., Ssekiwoko, F., Karamura, E., Tinzaara, W., Eden-Green, S. and Blomme, G. (2013a). Systemicity of *Xanthomonas campestris* pv. *musacearum* and time to disease expression after inflorescence infection in East African highland and Pisang Awak bananas in Uganda, *Plant Pathol.* 62(4), 777–785.

Ocimati, W., Ssekiwoko, F., Karamura, E. B., Tinzaara, W. and Blomme, G. (2013b). Does *Xanthomonas campestris* pv. *musacearum* colonize banana cord root tissue? In: *Acta Hortic.* International Symposium on Banana: ISHS-ProMusa Symposium on Bananas and Plantains: Towards Sustainable Global Production 986(986), 103–109.

Ocimati, W., Nakato, G. V., Fiaboe, K. M., Beed, F. and Blomme, G. (2015). Incomplete systemic movement of *Xanthomonas campestris* pv. *musacearum* and the occurrence of latent infections in Xanthomonas wilt-infected banana mats, *Plant Pathol.* 64(1), 81–90.

Ocimati, W., Groot, J. C. J., Tittonell, P., Taulya, G. and Blomme, G. (2018a). Effects of Xanthomonas wilt and other banana diseases on ecosystem services in banana-based agroecosystems, *Acta Hortic.* 1196(1196), 19–32.

Ocimati, W., Were, E., Groot, J. C. J., Tittonell, P., Nakato, G. V. and Blomme, G. (2018b). Risks posed by intercrops and weeds as alternative hosts to *Xanthomonas campestris* pv. *musacearum* in banana fields, *Front. Plant Sci.* 9, 1471.

Ocimati, W. (2019). Sustainability of banana-based agroecosystems affected by Xanthomonas wilt disease of banana. PhD dissertation, Wageningen University, p.234.

Ocimati, W., Bouwmeester, H., Groot, J. C. J., Tittonell, P., Brown, D. and Blomme, G. (2019). The risk posed by Xanthomonas wilt disease of banana: mapping of disease hotspots, fronts and vulnerable landscapes, *PLoS ONE* 14(4), e0213691.

Ocimati, W., Groot, J. J. C., Tittonell, P., Taulya, G., Ntamwira, J., Amato, S. and Blomme, G. (2020). Xanthomonas wilt of banana drives changes in land-use and ecosystem services across infected landscapes, *Sustainability* 12(8), 3178.

Ocimati, W., Tazuba, A. F. and Blomme, G. (2021). Farmer friendly options for sterilizing farm tools for the control of Xanthomonas Wilt disease of banana, *Front. Agron.* 3, 655824.

Ploetz, R. C., Kema, G. H. J. and Ma, L. J. (2015). Impact of diseases on export and smallholder production of banana, *Annu. Rev. Phytopathol.* 53, 269–288.

Reeder, R. H., Muhinyuza, J. B., Opolot, O., Aritua, V., Crozier, J. and Smith, J. (2007). Presence of banana bacterial wilt (*Xanthomonas campestris* pv. *musacearum*) in Rwanda, *Plant Pathol.* 56(6), 1038–1038.

Rietveld, A. M. (2021). Broadening the scaling of banana Xanthomonas wilt (BXW) management in East and Central Africa. *End-of-Project Report, CGIAR Research Program on Roots, Tubers and Bananas*, 12 p.

Rutikanga, A., Night, G., Tusiime, G., Ocimati, W. and Blomme, G. (2015). Spatial and temporal distribution of insect vectors of *Xanthomonas campestris* pv. *musacearum* and their activity across banana cultivars grown in Rwanda. Proceedings of the 7th Congress on Plant Protection. Plant Protection Society of Serbia, IOBC-EPRS, IOBC-WPRS, Belgrade, pp. 139–153.

Rutikanga, A., Sivirihauma, C., Ocimati, W., Night, G., Murekezi, C., Ndungo, V., Mugiraneza, T., Rurangwa, E. and Blomme, G. (2016). Breaking the cycle of *Xanthomonas campestris* pv. *musacearum* in infected fields through the cultivation of annual crops and disease control in adjacent fields, *J. Phytopathol.* 164(9), 659–670.

Selvaraj, M. G., Vergara, A., Ruiz, H., Safari, N., Elayabalan, S., Ocimati, W. and Blomme, G. (2019). AI-powered banana diseases and pest detection, *Plant Methods* 15(1), 92.

Selvaraj, M. G., Vergara, A., Montenegro, F., Alonso Ruiz, H., Safari, N., Raymaekers, D., Ocimati, W., Ntamwira, J., Tits, L., Omondi, A. B. and Blomme, G. (2020). Detection of banana plants and their major diseases through aerial images and machine learning methods: a case study in DR Congo and Republic of Benin, *ISPRS J. Photogramm.* 169, 110–124.

Shimwela, M. M., Ploetz, R. C., Beed, F. D., Jones, J. B., Blackburn, J. K., Mkulila, S. I. and van Bruggen, A. H. C. (2016). Banana Xanthomonas wilt continues to spread in Tanzania despite an intensive symptomatic plant removal campaign: an impending socio-economic and ecological disaster, *Food Secur.* 8(5), 939–951.

Shimwela, M. M., Blackburn, J. K., Jones, J. B., Nkuba, J., Narouei-Khandan, H. A., Ploetz, R. C., Beed, F. and van Bruggen, A. H. C. (2017). Local and regional spread of banana Xanthomonas wilt (BXW) in space and time in Kagera, Tanzania, *Plant Pathol.* 66(6), 1003–1014.

Sivirihauma, C., Ocimati, W., Vutseme, L., Ntamwira, J., Bahati, L. Blomme, G. (2017). 'Symptomless banana suckers sourced from Xanthomonas wilt infected fields are a viable alternative for seed within infected banana-based landscapes lacking access to clean planting materials', *Afr. J. Agric. Res.*, 12(31), 2490–2498.

Smith, J. J., Jones, D. R., Karamura, E., Blomme, G. and Turyagyenda, F. L. (2008). *An Analysis of the Risk from Xanthomonas campestris pv. Musacearum to Banana Cultivation in Eastern, Central and Southern Africa*, Bioversity International, Montpellier, France.

Smith, M. and Pillay, M. (2020). An overview of genetic improvement in bananas over the last century. In: *Achieving Sustainable Cultivation of Bananas Volume 2: Germplasm and Genetic Improvement*, Kema, G. H. J. and Drenth, A. (Eds), Burleigh Dodds Science Publishing Limited, Cambridge, UK.

Ssekiwoko, F., Turyagyenda, L. F., Mukasa, H., Eden-Green, S. and Blomme, G. (2006a). Systemicity of *Xanthomonas campestris* pv. *musacearum* in flower-infected banana plants, XVII ACORBAT International meeting; Banana: A sustainable business, October 15–20, 2006, Joinville – Santa Catarina – Brazil. pp. 789–793.

Ssekiwoko, F., Tushemereirwe, W., Batte, M., Ragama, P. and Kumakech, A. (2006b). Reaction of banana germplasm to inoculation with *Xanthomonas campestris pv. musacearum*, *Afr. Crop Sci. J.* 14(2), 151–155.

Ssekiwoko, F., Turyagyenda, L. F., Mukasa, H., Eden-Green, S. and Blomme, G. (2010). Spread of *Xanthomonas campestris* pv. *musacearum* in banana (*Musa* spp.) plants following infection of the male inflorescence, *Acta Hortic.* 879(879), 349–356.

Ssekiwoko, F., Kiggundu, A., Tushemereirwe, W. K., Karamura, E. and Kunert, K. (2015). *Xanthomonas vasicola* pv. *musacearum* down-regulates selected defense genes during its interaction with both resistant and susceptible banana, *Physiol. Mol. Plant Pathol.* 90, 21–26.

Studholme, D. J., Kemen, E., MacLean, D., Schornack, S., Aritua, V., Thwaites, R., Grant, M., Smith, J. and Jones, J. D. (2010). Genome-wide sequencing data reveals virulence factors implicated in banana Xanthomonas wilt, *FEMS Microbiol. Lett.* 310(2), 182–192.

Studholme, D. J., Wicker, E., Abrare, S. M., Aspin, A., Bogdanove, A., Broders, K., Dubrow, Z., Grant, M., Jones, J. B., Karamura, G., Lang, J., Leach, J., Mahuku, G., Nakato, G. V., Coutinho, T., Smith, J. and Bull, C. T. (2020). Transfer of *Xanthomonas campestris* pv. *arecae* and *X. campestris* pv. *musacearum* to *X. vasicola* (Vauterin) as *X. vasicola* pv.

arecae comb. nov. and *X. vasicola* pv. *musacearum* comb. nov. and Description of *X. vasicola* pv. *vasculorum* pv. Nov, *Phytopathology®* 110(6), 1153–1160.

Tinzaara, W., Gold, C. S., Ssekiwoko, F., Bandyopadhyay, R., Abera, A. and Eden-Green, S. J. (2006). Role of insects in the transmission of banana bacterial wilt, *Afr. Crop Sci. J.* 14(2), 105–110.

Tinzaara, W., Karamura, E. B., Blomme, G., Jogo, W., Ocimati, W., Rietveld, A., Kubiriba, J. and Opio, F. (2013). Why sustainable management of Xanthomonas Wilt of Banana in East and Central Africa has been elusive, *Acta Hortic.* 986(986), 157–164.

Tinzaara, W., Karamura, E. B., Kubiriba, J., Ochola, D., Ocimati, W., Blomme, G. and Ssekiwoko, F. (2016). The banana Xanthomonas wilt epidemic in east and central Africa: current research and development efforts, *Acta Hortic.* 1114(1114), 267–274.

Tinzaara, W., Ssekiwoko, F., Kikulwe, E. M. and Karamura, E. (2019). 'Effectiveness of learning and experimentation approaches for farmers as a community based strategy for banana Xanthomonas wilt management', *J. Agric. Ext. Rural Dev.*, 11(7), 128–138.

Tripathi, L., Tripathi, J. N. and Tushmereirwe, W. K. (2004). Strategies for resistance to bacterial wilt disease of bananas through genetic engineering, *Afr. J. Biotechnol.* 3, 688–692.

Tripathi, L., Tripathi, J. N., Tushemereirwe, W. K. and Bandyopadhyay, R. (2007). Development of a semi-selective medium for isolation of *Xanthomonas campestris* pv. *musacearum* from banana plants, *Eur. J. Plant Pathol.* 117(2), 177–186.

Tripathi, L., Mwangi, M., Abele, S., Aritua, V., Tushemereirwe, W. K. and Bandyopadhyay, R. (2009). Xanthomonas wilt: a threat to banana production in east and central Africa, *Plant Dis.* 93(5), 440–451.

Tripathi, L. and Tripathi, J. (2009). Relative susceptibility of banana cultivars to *Xanthomonas campestris* pv. *musacearum*, *Afr. J. Biotechnol.* 8, 5343–5350.

Tripathi, L., Mwaka, H., Tripathi, J. N. and Tushemereirwe, W. K. (2010). Expression of sweet pepper Hrap gene in banana enhances resistance to *Xanthomonas campestris* pv. *musacearum*, *Mol. Plant Pathol.* 11(6), 721–731.

Tripathi, J. N., Lorenzen, J., Bahar, O., Ronald, P. and Tripathi, L. (2014). Transgenic expression of the rice Xa21 pattern-recognition receptor in banana (*Musa* sp.) confers resistance to *Xanthomonas campestris* pv. *musacearum*, *Plant Biotechnol. J.* 12(6), 663–673.

Tripathi, L., Tripathi, J. N., Shah, T., Muiruri, K. S. and Katari, M. (2019). Molecular basis of disease resistance in banana progenitor *Musa balbisiana* against *Xanthomonas campestris* pv. *musacearum*, *Sci. Rep.* 9(1), 7007.

Tripathi, L., Ntui, V. O. and Tripathi, J. N. (2020). CRISPR/Cas9-based genome editing of banana for disease resistance, *Curr. Opin. Plant Biol.* 56, 118–126.

Tripathi, J. N., Ntui, V. O., Shah, T. and Tripathi, L. (2021). CRISPR/Cas9-mediated editing of *DMR6* orthologue in banana (*Musa* spp.) confers enhanced resistance to bacterial disease, *Plant Biotechnol. J.* 19(7), 1291–1293, doi: 10.1111/pbi.13614.

Turyagyenda, L. F., Blomme, G., Ssekiwoko, F., Karamura, E., Mpiira, S. and Eden-Green, S. (2008). Rehabilitation of banana farms destroyed by Xanthomonas wilt in Uganda, *J. Appl. Bio.Sci.* 8, 230–235.

Tushemereirwe, W., Kangire, A., Smith, J., Ssekiwoko, F., Nakyanzi, M., Kataama, D., Musiitwa, C. and Karyaija, R. (2003). An outbreak of bacterial wilt on banana in Uganda, *Infomusa* 12, 6–8.

Uwamahoro, F., Berlin, A., Bylund, H., Bucagu, C. and Yuen, J. (2019). Management strategies for banana Xanthomonas wilt in Rwanda include mixing indigenous and improved cultivars, *Agron. Sustain. Dev.* 39(2), 22.

Vidal, T., Boixel, A., Durand, B., de Vallavieille-Pope, C., Huber, L. and Saint-Jean, S. (2017). Reduction of fungal disease spread in cultivar mixtures: impact of canopy architecture on rain-splash dispersal and on crop microclimate, *Agric. Forest Meteorol.* 246, 154-161.

Wang, G. L., Song, W. Y., Ruan, D. L., Sideris, S. and Ronald, P. C. (1996). The cloned gene Xa21 confers resistance to multiple *Xanthomonas oryzae* pv. *oryzae* isolates in transgenic plants, *Mol. Plant Microbe Interact.* 9(9), 850–855.

Wasukira, A., Tayebwa, J., Thwaites, R., Paszkiewicz, K., Aritua, V., Kubiriba, J., Smith, J., Grant, M. and Studholme, D. J. (2012). Genome-wide sequencing reveals two major sub-lineages in the genetically monomorphic pathogen *Xanthomonas campestris* pathovar *musacearum*, *Genes* 3(3), 361–377.

Wasukira, A., Coulter, M., Al-Sowayeh, N., Thwaites, R., Paszkiewicz, K., Kubiriba, J., Smith, J., Grant, M. and Studholme, D. J. (2014). Genome sequencing of *Xanthomonas vasicola* pathovar *vasculorum* reveals variation in plasmids and genes encoding lipopolysaccharide synthesis, Type-IV pilus and Type-III secretion effectors, *Pathogens* 3(1), 211–237.

Wasukira, A., Tusiime, G. and Kubiriba, J. (2017). 'PCR validation of predicted virulence factors in a collection of *Xanthomonas campestris* (pv. *musacearum* and *vascolurum*) strains', *Afr. J. Biotechnol.*, 16(12), 558–572.

Welde-Michael, G., Bobosha, K., Addis, T., Blomme, G., Mekonnen, S. and Mengesha, T. (2008). Mechanical transmission and survival of bacterial wilt on enset, *Afr. Crop Sci. J.* 16(1), 97–102.

Xing, Y., Hernandez Nopsa, J. F., Andersen, K. F., Andrade-Piedra, J. L., Beed, F. D., Blomme, G., Carvajal-Yepes, M., Coyne, D. L., Cuellar, W. J., Forbes, G. A., Kreuze, J. F., Kroschel, J., Kumar, P. L., Legg, J. P., Parker, M., Schulte-Geldermann, E., Sharma, K. and Garrett, K. A. (2020). Global cropland connectivity: a risk factor for invasion and saturation by emerging pathogens and pests, *Bioscience* 70(9), 744–758.

Yirgou, D. and Bradbury, J. F. (1968). Bacterial wilt of enset (*Enset ventricosum*) incited by *Xanthomonas musacearum* sp. n., *Phytopathol.* 58, 111–112.

Yirgou, D. and Bradbury, J. F. (1974). A note on wilt of banana caused by the enset wilt organism *Xanthomonas musacearum*, *East Afr. Agric. For. J.* 40(1), 111–114.

Young, J. M., Dye, D. W., Bradbury, J. F., Panagopoulos, C. G. and Robbs, C. F. (1978). A proposed nomenclature and classification for plant pathogenic bacteria, *N. Z. J. Agric. Res.* 21(1), 153–177.

Chapter 11

Bacterial soft rot of banana

Luis Pérez-Vicente, Institute of Plant Health Research (INISAV), Cuba; and Armando García-Suárez, Central Plant Quarantine Laboratory (LCCV) of Central Unit of Laboratories (ULCSA), Cuba

1 Introduction

Bananas, plantains, abaca and enset are affected by several different pathogenic bacteria and phytoplasmas, which are efficiently transmitted through banana propagation materials, irrigation water, tools during cultural practices, insects, animals and persons moving inside plantations (Blomme et al., 2017). They can have a significant impact on banana and plantain production due to their effect on plant development and the cost of management. Although often minor diseases, bacterial soft rots can erupt under wet, anoxic conditions and cause significant losses.

Bacterial pseudostem and rhizome wet rots of bananas and plantains were previously described as head rot but are now known to be different diseases caused by related but distinct pathogens. Rhizome soft rot and head rot are caused by *Pectobacterium carotovorum* subsp. *carotovorum* (Jones, 2018; Hauben et al., 1998 [synonym *Erwinia carotovora*]), while pseudostem wet rot and rhizome rot are caused by *Dickeya paradisiaca* (Dickey and Victoria, 1980; Samson et al., 2004 [synonym *Erwinia chrysanthemi* pv. *chrysanthemi* (Burkholder et al., 1953) Dye 1969*; E. chrysanthemi* pv. *paradisiaca* (Victoria & Barros, 1969) Dickey & Victoria 1980]). Both bacteria, and particularly *D. paradisiaca*, are frequently present in plants infected by *Fusarium oxysporum*

http://dx.doi.org/10.19103/AS.2022.0108.11

f. sp. *cubense*, the causal agent of Fusarium wilt of banana. *Dickeya zeae* and *Dickeya dadantii* have recently been reported as causing soft rot diseases in bananas in China (Zhang et al., 2014; Liu et al., 2016).

2 Disease symptoms

Pectobacterium carotovorum subsp. *carotovorum* and *D. paradisiaca* belong to the group of bacteria known as soft rotting 'Erwinia'. They cause maceration of affected tissues through exuding enzymes that degrade cell walls (Toht et al., 2003) and lead to characteristic putrefaction. Symptoms include plants with poor growth and development, wilting, wet and dry pseudostem and rhizome rots, plant tipping and stench (Stover, 1972; Wardlaw, 1972; Rivera, 1978; Dickey and Victoria, 1980; Jones, 2018; Blomme et al., 2017; Pérez-Vicente et al., 2020). Both diseases have been confused due to the relative similarity of symptoms and the concurrence of the causal agents in symptomatic plants. However, there is sufficient symptomatic, etiological and epidemiologic evidence to consider them as similar but different diseases caused by related but different pathogens (Rivera, 1978; Dickey and Victoria, 1980; Rivera and Ezavin, 1989; Elphinstone, 2008; Li et al., 2011).

2.1 *Head rot or rhizome soft rot by* Pectobacterium carotovorum *subsp.* carotovorum

Head rot is commonly found in newly planted (3–5 months old) fields in wet soils with poor drainage and high subterranean water tables, high pH, during hot humid weather and prolonged rain periods favoring the facultatively anaerobic nature of the pathogens involved (Stover, 1972; Jones, 2018).

The disease is more intense in young suckers and ratoons, which produce an intense unpleasant odor. Affected plants show poor emergence of shoots, retarded growth, dwarfing, marginal necrosis or scorching of the lower leaves, yellowing and wilting and sometimes the young newly emerging leaves fail to open and appear brown and show rotting, leading to arresting plant growth (Stover, 1972; Rivera, 1978; Nagaraj et al., 2012; Rhagavi et al., 2019; Pérez-Vicente et al., 2020). After rainy periods in the lower lying and wet parts of plantations, affected plants develop marginal necrosis of leaves and intense leaf yellowing and wilting. Leaves finally die and hang around the pseudostem (Fig. 1). Inside, the pseudostem tissues are decomposed and become translucid. The rhizome shows pockets of wet and soft tissues of cream to dark brown color surrounded by a darker border and produces a fetid odor. This wet pocket can cover parts or all of the rhizome surface. Mature plants affected by the disease seldom exhibit outward symptoms but are easily blown or pushed over. When the plant falls over due to the splitting of the rhizome, fractions of

Figure 1 Head rot or rhizome soft rot by *Pectobacterium carotovorum* subsp. *carotovorum*: (a) and (b) First external symptom of wilting of leaves in the cultivar Grand Nain (AAA) in a wet soil; (c) and (d) wilted plants of cultivar Burro CEMSA (Bluggoe, ABB) grown in a low-lying wet soil; (e) advanced internal wet rot of pseudostem; (f) and (g) rhizome soft of cultivars Grand Nain (subgroup Cavendish, AAA) and Burro CEMSA (subgroup Bluggoe, ABB), respectively. Source: Photos by L. Pérez-Vicente.

infected tissue remain adhered to the pseudostem and the rest of the rhizome remains anchored by the roots. This phenomenon is called tip-overs or snap-offs in reference to the broken corm (Stover, 1972). The fall of plants affected by soft rot pathogens can be differentiated from other causes because of the complete necrosed rhizome and roots which are exposed when the plant falls. The bacteria can be transmitted to the daughter suckers of the clump.

2.2 Pseudostem wet rot and rhizome rot by Dickeya paradisiaca

Two disease syndromes are caused by *D. paradisiaca*: (i) one related to a soft rot of the pseudostem progressing from the rhizome upward when infected plant material is planted or from petioles of pruned leaves progressing downward, indicating transmission by machete and other tools. (ii) The other is related to a peripherical externally visible necrosis around the union of the rhizome with the pseudostem which internally exhibits necrosis with a cavern expressing a spongy aspect (Rivera and Ezavin, 1989; Blomme et al., 2017), associated with injuries during the cultural practice of pruning of buds and suckers.

Plants developing from an infected rhizome or sucker show slow growth, chlorotic and flaccid leaves and a sheath rot that disseminates upward from the base of the plant to the rest of pseudostem (Stover, 1959, 1972; Llanos

Figure 2 Cavendish affected plants by *D. paradisiaca* in a hot wet soil. Note poor developing, leaves yellowing and death of the plants. Source: Photo by L. Pérez-Vicente.

et al., 1967; Rivera and Ezavin, 1989). These plants develop young chlorotic leaves with necrotic margins and small buds. Some plants eventually collapse and die (Fig. 2). Symptoms can also appear as translucid spots on leaf sheaths which progress from petioles of pruned leaves downward, indicating a pathogen transmitted by machete and/or other cutting tools. Rotting advances downward and toward the inside of the leaf sheaths in different parts of the pseudostem or in the base of the leaves, weakening the pseudostem and giving rise to collapse when the plant bends over due to the bunch weight or wind action (Fig. 3) (Rivera and Ezavin, 1989; Blomme et al., 2017). A fetid amber fluid emerges from affected tissues when pressure is applied. Frequently, the young affected plants do not flower (Rivera, 1978). Additional symptoms are cream to dark brown patches and collar rot of the rhizome at the soil level. These patches develop into cavities that resemble damage caused by the banana black weevil (*Cosmopolites sordidus* Gemar; Martínez et al., 1989) (see Chapter 18). Splitting of the pseudostem at the soil level is a common symptom of advanced stages of infection in the cultivars Robusta and Grand Nain. In such cases, plants are prone to topple over when pushed, by wind action or the bunch weight.

The bacterial pathogen can grow in tissue culture plantlets generated from infected donor plants. The bacteria grow well in the tissue culture multiplication media, causing concern in the tissue culture banana laboratories in Cuba for a period (Hernández-Pérez et al., 1994). In the nurseries during the hardening of plants, the incidence can reach up to 5% (Thomas et al., 2011; 2015), whereas in India it was considered a major limiting factor in this phase of the micropropagation process (Thangavelu, 2009).

Figure 3 Pseudostem wet rot and rhizome necrosis by *D. paradisiaca*: (a) Burro CEMSA (Bluggoe subgroup, ABB), heavily affected field; (b) Burro CEMSA-affected plant bend over at the pseudostem lesion; (c) external symptoms of pseudostem wet rot; (d) sucker affected with necrotic lesions; (e) external watery rot lesion; (f) internal progress of infection; (g) internal pseudostem rot symptoms in a longitudinal and transversal section view, respectively; (h) rhizome necrosis in plantains (AAB). Source: Photos by L. Pérez-Vicente.

3 Pathogen characteristics and diagnosis

Pectobacterium carotovorum subsp. *carotovorum* and *Dickeya* belong to Gamma Proteobacteria and the family Pectobacteriaceae (Toht et al., 2003). Both produce pectolytic enzymes that can be readily shown by a potato tuber inoculation test or crystal violet pectate (CVP) culture media (Hauben et al., 1998).

On the basis of 16S rDNA sequences, soft rotting members of the genus *Erwinia* were reclassified as *P. carotovorum* subsp. *atrosepticum*, *P. carotovorum* subsp. *carotovorum* and *P. chrysanthemi* (Hauben et al., 1998). Later, *P. chrysanthemi* was transferred to *Dickeya* (Samson et al., 2004). These bacteria are opportunistic invaders of wounds and are geographically widely distributed.

Soft rot bacteria can be diagnosed using microbiological, biochemical and serological methods such as indirect immunofluorescence and enzyme-linked immunosorbent assay (ELISA), as well as different variants of polymerase chain reaction (PCR)-based assays (Pérombelon and Van der Wolf, 2002; Potrykus et al., 2014; NPC Seed Certification Sub-Committee, 2017). These PCR assays have been developed to differentiate *P. carotovorum* subsp. *carotovorum* and *P. carotovorum* subsp. *atrosepticum*, from *Dickeya* spp. in potato seed production and can reliably differentiate the different species of soft rot pathogens in *Musa*. Diagnosis based on PCR is more specific, fast and sensitive than microbiological methods.

P. carotovorum subsp. *carotovorum* and *D. paradisiaca* identification methods include morphological, physiological, biochemical and pathological features. Several methods including indirect immunofluorescence, ELISA, metabolic enzyme profiling and different PCR-based methods are used as they achieve higher levels of accuracy and sensitivity than traditional methods.

3.1 *Diagnosis of* Pectobacterium carotovorum *subsp.* carotovorum

Pectobacterium carotovorum subsp. *carotovorum* is a Gram-negative, straight rod, 0.5–1.0 × 1.0–3.0 µm, motile by peritrichous flagella, facultatively anaerobic, oxidase-negative, catalase-positive, ferments glucose and reduces nitrate (Brenner et al. 1994; Bradbury, 1986). The main characteristic that distinguishes the soft rotting *Pectobacterium* and *Dickeya* subsp. from other species of *Erwinia* is their ability to produce large quantities of pectolytic enzymes that enable them to macerate parenchymatous tissue of a wide range of host plant species.

Pectobacterium carotovorum subsp. *carotovorum* can be isolated using the semi-selective diagnostic medium CVP (Hyman et al., 2001) and incubated at differential temperatures to identify the different soft rotting strains (Pérombelon & Hyman, 1986). On CVP, only soft rotting *Pectobacterium* can form characteristic cavities after 48 h, whereas *P. carotovorum* subsp. *atrosepticum* (which causes black leg disease in potato) generally forms cavities at 27 °C only, *P. carotovora* subsp. *carotovora* and *Dickeya chrysanthemi*, also do so up to temperatures of 33.5 °C and 37 °C, respectively (Pérombelon & Hyman, 1986).

Serological (IFI and ELISA tests) and different types of PCR diagnostic methods with a higher level of sensitivity and specificity are used to diagnose soft rot bacteria. PCR variants with specific primers have been described (Kang et al., 2003; Potrykus et al., 2014; Dadaşoğlu and Kotan, 2017). A selective nested PCR protocol using universal rice primers which facilitates a reported specific detection of 2–4 CFU mL^{-1} of *P. carotovorum* subsp. *carotovorum* strains from host tissues has been developed (Kang et al., 2003).

3.2 Dickeya paradisiaca

The causal bacterium of pseudostem wet rot and rhizome rot was previously named *E. chrysanthemi* pv. *chrysanthemi* (Burkholder et al., 1953; Dye 1969), *E. carotovora* subsp. *paradisiaca* (Victoria and Barros, 1969), *Erwinia paradisiaca* (Fernández-Borrero and López-Ouque, 1970), *E. chrysanthemi* (Fernández and López, 1970; Rivera, 1978), *E. chrysanthemi* pv. *paradisiaca* Dickey and Victoria (Dickey and Victoria, 1980) and then transferred to the current name *D. paradisiaca* (Samson et al., 2004).

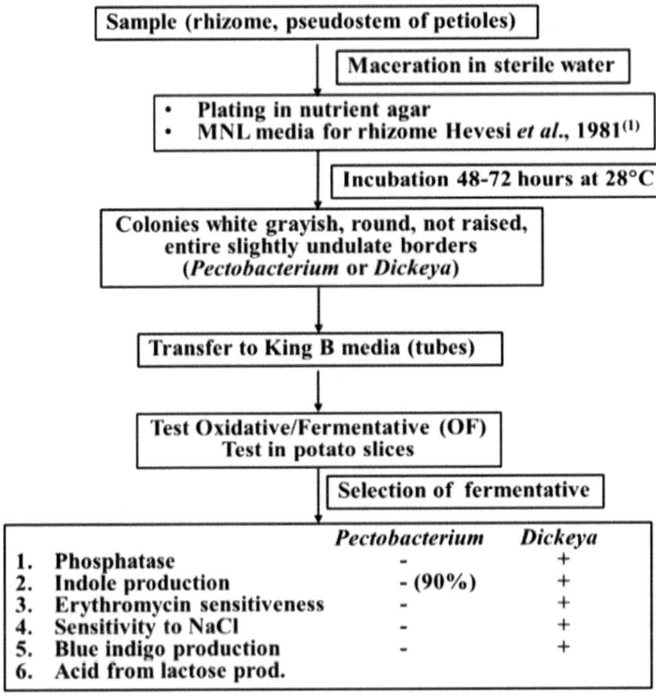

Figure 4 Isolation overview for microbiological identification of *P. carotovorum* and *D. paradisiaca* from samples of banana tissues and soil. MNL media is a selective media with bacto-peptone and beef extract 4 g each; yeast extract 1 g; dextrose and NaCl 2.5 g each; agar 20 g; pH 7.2; water to 1L. After sterilization is added: (1) 1.5 mL of a crystal violet (4% solution in sterile water) and (2) 10 mL of sodium novobiocin solution (at 10 g mL^{-1} of sterile water) by 100 mL of warm sterile media (Hevesi et al., 1981). *D. paradisiaca* colonies in MNL media, have a defined mauve rose color in the center with blueish, irregular, diffuse borders and decolorate media that surround the colony. For the rest of test, consult Lelliot and Stead (1987) and Schaad et al. (2001).

Dickeya paradisiaca can be isolated from leaves, rhizomes and roots. Plant tissue samples are washed in sterile water and macerated in a sterile recipient (Fig. 4).

In a study with more than 200 *Dickeya* ssp. isolates from different hosts, Elphinstone (2008) developed CVPM media (modification of the CVP media) that allows selectively isolation of all *Dickeya* spp. from different hosts.

Dickeya paradisiaca is a Gram-negative aerobic rod with peritrichous flagella. Some principal phenotypic characteristics of strains as described by Dickey and Victoria (1980) and EPPO (1990) appear in Table 1. The main useful characteristics to differentiate *D. paradisiaca* from *D. zeae* and *D. dadantii* reported causing soft rot diseases in banana in China appears in Table 2 (Zhang et al., 2014; Liu et al., 2016).

Table 1 Some principal phenotypic characteristics of *D. paradisiaca* as described by Dickey and Victoria (1980) and EPPO (1990)

Culture media	Description
Nutrient agar (after 48 h)	White to slightly gray, with irregular borders, fine granulated growth and after 4 days exhibit a defined, raised center. Cells from a 24-h-old nutrient agar culture had average dimensions of 0.6 × 1.7 m. The cells were motile and peritrichous; four to six flagella commonly discernible. The colonies on nutrient agar after 24 h at 27°C are convex, slightly to moderately irregular and undulate, pale cream-colored, butyrous and averaged 1.5–2.5 mm in diameter
In YDC (yeast extract-dextrose-calcium carbonate)	Can produce a non-diffusible blue pigment. Colonies are convex or somewhat umbonate, irregular, undulate, light tan, butyrous and 2.5–3.5 mm in diameter
Kelman's media	They appear olive green with irregular margins with characteristic tiny black inclusions visible with a dissecting microscope
PDA	Young colonies are either circular, convex, smooth and entire or sculptured with irregular margins, depending on the moisture content of the growth medium. After 4–5 days, both colony types resemble a fried egg, with a pinkish, round, raised center and lobed periphery, which later becomes feathery or almost coralloid (EPPO, 1990)
Positive	Negative
Facultative anaerobic	Cytochrome oxidase
Pectate degradation	Casein hydrolysis
Potato soft rot	Gelatine liquefaction
Gas from glucose	Phenylalanine deamination
Catalase	Gluconate oxidation
Phosphatase	Blue pigment on YDC agar
8-Galactosidase	Arginine decarboxylase
Indole	Growth on 5% NaCl
Nitrate reduction	Deoxyribonuclease
Growth at 36°C and 39°C	Urease
Susceptibility to:	Lecithinase
Penicillin G (2U)	
Erythromycin (15 g)	

To differentiate phenotypes of *D. chrysanthemi* from *D. paradisiaca*, one may consult the assays used by the NPC Seed Certification Sub-Committee (2017; Table 3).

Different PCR-based methods have been developed over time and are currently routinely used for disease diagnosis and pathogen monitoring. *D. paradisiaca* from banana and plantain can be detected from tissues, soil and water with a PCR assay using primers LF (5′-TTCGTCTAGAGGCCCAGGAC-3′)

Table 2 Main differential biochemical characters among *Dickeya* spp. associated with banana soft rots

Biochemical character	*D. dadantii* & *D. zeae*	*D. paradisiaca*
Arginine dihydrolase	+	+
Lactose	+	(−)
Inulin	−	−
D-arabinose	+	+
Mannitol	+	−
Myo-Inositol	+	−
Casein	+	−

Legend from Stoica C. Regnum Prokaryotae. Genus Dickeya. Available online at https://www.tgw1916.net (accessed January 2022):
+: positive 90–100% of strains.
−: negative 90–100% of strains.
(+): positive 75–89% of strains.
(−): negative 75–89% of strains.
Source: Zhang et al. (2014) and Liu et al. (2016).

Table 3 Essays to differentiate phenotypes of *D. chrysanthemi* from *D. paradisiaca*

Species name	*D. chrysanthemi*	*D. paradisiaca*	*D. solani*	*D. zeae*
Reference	Samson et al. (2005)	Samson et al. (2005)	van der Wolf et al. (2014)	Samson et al. (2005)
Growth at 39°C	+	+	Weak	+
D-Arabinose	−	+	+	+
Melibiose	+	Variable	+	+
Rafinose	+	Variable	+	+
Manitol	+	−	+	+
β-Gentiobiose	−	+	Weak	−

Source: NPC Seed Certification Sub-Committee (2017).

and LR (5′-TCAGCTTGTTCCGGATTGTT-3′) designed from the 16S–23S rDNA coding spacer regions. The amplification products were a unique DNA fragment of approximately 171 bp from all banana *Dickeya* sp. (*P. chrysanthemi*) isolates tested (Li et al., 2011).

4 Disease cycle and epidemiology

Pectobacterium carotovorum subsp. *carotovorum* and *D. paradisiaca* are vascular wilt pathogens, colonizing the xylem and becoming systemic within plants where they degrade succulent fleshy plant organs such as roots, rhizomes, stem cuttings and thick leaves. These two bacterial pathogens can remain latent in plants and can thus be spread in cuttings used as planting materials and planting operations including peeling and heat-treating rhizomes

Figure 5 Scheme of the cycle of infection of *P. carotovorum* subsp. *carotovorum* and *D. paradisiaca* in banana and plantain.

for nematode control (Stover, 1959). Both bacteria are also transmitted on tools used for cropping activities (EPPO, 2003). The soft rot bacteria can survive as endophytes, epiphytes and saprophytes in plants, on plant surfaces and in the soil and groundwater (Toht et al., 2003). The bacterium can also be transmitted through infected tissue culture plantlets. An overview of infection pathways for both diseases is shown in Fig. 5.

There is an association between soil wetness and disease incidence. In irrigated areas of Honduras, nearly twice as many head rot cases were recorded in a region with high rainfall as compared to a region with low rainfall (Wardlaw, 1972). Wet and cool soils with high water tables favor outbreaks of the disease. It is common to find infected plants in the lowest sections of plantations and after heavy rain periods. In Cuba, the disease has been frequently observed in Bluggoe (ABB) and Pisang Awak (ABB) cultivars in wet and low-lying sections of fields. However, there is no evidence of differences in susceptibility among cultivars under favorable soil and weather conditions for disease development.

Dickeya paradisiaca has a low saprophytic ability in soil (Rivera et al., 1983). Under fallow in the absence of its host plant, it survives between 30 and 40 days at soil depths between 0 and 30 cm. From this relatively low ability to survive as a saprophyte in the soil, it can be concluded that the principal manner of dissemination is infected planting material and tools used for cropping practices.

5 Occurrence and impact

Often considered minor diseases of banana and plantain, both soft rot diseases are currently widely distributed in banana-producing countries. Their economic impact is related to the cultural practices, environmental conditions and soil characteristics in plantations. Reports of outbreaks and the impact of individual bacterial soft rot diseases have not always been accurate, due to the similarity of disease symptoms and the concurrence of both pathogens in some cases.

5.1 *Rhizome soft rot or head rot by P. carotovorum subsp. carotovorum*

The disease was first described in detail in Honduras in 1948, as the cause of poor establishment of newly planted rhizomes, particularly in high rainfall areas and in locations in Central America submitted to flood fallowing in an effort to reduce Fusarium wilt incidence (Stover, 1959, 1972; Wardlaw, 1972). The disease was serious in flood-fallowed Gros Michel plantations 6 months after planting with common incidences of 15 diseased plants/ha and at times reaching 50% in localized areas. In Cavendish varieties the incidence of disease was lower, but up to 30% of newly planted corms have been reported to succumb to the disease in wet, cool weather. Losses of mature plants are greatest during the first l–2 years after planting and the disease incidence usually declines with age. However, under conditions suitable for disease, high levels of incidence could be found in localized areas in plantations more than 5 years old (Stover, 1972).

In Cuba, there are early reports of concurrent rhizome soft rot symptoms in banana plants affected by Fusarium wilt (Johnston, 1915, 1918). In Jamaica, Shillingford (1974) reported severely diseased pseudostems of Cavendish plants affected by bacterial soft rot of rhizomes which broke at the soil level prior to the fruit reaching maturity.

5.2 *Pseudostem wet rot and rhizome necrosis by* Dickeya paradisiaca

Pseudostem wet rot and rhizome necrosis by *D. paradisiaca* was first described in the Cauca Valley in Colombia, where it caused severe losses in almost 2000 ha of plantains (Llanos, 1967; Fernández y López, 1970). The pathogen can cause 80–90% production losses within 5 years after initial infection in a plantation (Dickey and Victoria, 1980). Pseudostem wet rot incidences of 3% and 6% were found during a disease survey in the Chinchiná and Palestina municipalities of Colombia, respectively, due to the lack of use of healthy planting material and lack of tool disinfection measures (Alarcón-Restrepo and Castaño-Zapata, 2006).

Pseudostem wet rot and rhizome necrosis have caused significant losses in Central America and the Caribbean in Cavendish bananas, plantains and cooking bananas (Llanos, 1967; Ordosgoitti et al., 1974; Rivera, 1978; Rivera et al., 1980; Rivera and Ezavin, 1989). *D. paradisiaca* has been highly prevalent in Cuba in the past due to transmission through infected planting material and contaminated tools used for pruning of senescent and Sigatoka leaf spot infected leaves and for pruning of suckers (Rivera, 1978; Pérez-Vicente, 2003).

In Cuba during the 1970s, *D. paradisiaca* caused major losses in Cavendish (AAA) and plantain (AAB) plantations with incidences of between 50% and 75% (Rivera, 1978). During the 1980s, there were reported incidences of *D. paradisiaca* of 95% and 89% in Cavendish (AAA) and plantains (AAB) plantations, respectively, in the Jovellanos municipality of Matanzas province (Martínez et al., 1989). In the 1980s, the implementation in large banana and plantain enterprises of an integrated management program based on the use of certified healthy planting material from nurseries reduced disease incidence to negligible levels. Currently, the disease is widely distributed in Cuba among smallholders growing bananas (AA, AAA, ABB), plantains, (AAB), cooking bananas (ABB), landrace cultivars as well as FHIA hybrids (AAAB, AABB). The severity depends on the origin of the planting material and the cultivation practices in the field.

Pseudostem and rhizome rot by *D. paradisiaca* is also considered an important disease in nurseries and banana plantations in India (Chattopadhyay and Mukherjee; 1986; Sigh, 1990; Thangavelu, 2009; Patel et al., 2010; Thammaiah et al., 2010; Nagaraj et al., 2012; Arun et al., 2012). The pathogen causes significant disease problems in the cultivars Grand Nain, Rhasthali (Silk, AAB) and Nendran (subgroup Plantain AAB; Nagaraj et al., 2012; Rhagavi et al., 2019). There are reports of incidence as high as 30–35% in Cavendish plantations with a high level of toppling of plants due to pseudostem and rhizome rots in Bangalore and Kolar districts of Karnataka states in India (Nagaraj et al., 2012). The level of incidence of rhizome rot became alarmingly high in India and as a result was included as a priority trait in breeding for banana (Sathiamoorthy, 1994).

The disease is also present in Israel (Volcani and Zutra, 1967) and in Papua New Guinea (Tomlinson et al., 1987). During 2009 and 2010 in the Guangdong province, China, severe outbreaks of soft rot caused by *Dickeya* sp. affecting 82% of the fields and 40–50% of plants of an ABB banana dessert cultivar were reported (Li et al., 2011).

6 Management of the disease/pest

Short-distance dissemination is linked to transmission from diseased to healthy plants by contaminated tools during the cultural practices of de-leafing and the rhizome injuries caused during pruning of suckers when no disinfection is applied between plants. Insects belonging to the curculionid *Metamasius hemipterus* and the banana weevil borer, *C. sordidus*, have also been associated with the presence and spread of the disease in the field (Rivera,

1978; Belalcázar et al., 1991). Long-distance dissemination is linked to the movement of infected planting material and the use of tools contaminated in infected fields. Contamination of tissue culture plantlets has been a cause of concern in tissue culture laboratories when meristems are obtained from infected plants in the field (Jones, 2018; Hernández-Pérez et al., 1994; Blomme et al., 2017).

Disease management of bacterial soft rots is based on a range of exclusion, sanitation and disinfection measures and cultural practices.

1 Exclusion: avoid using planting material from infected areas. It is advisable to use healthy planting material from certified nurseries and tissue culture facilities. Management of bacterial soft rot diseases in Cavendish bananas and plantains in many countries is based on ELISA indexing of meristem donors' plants for tissue culture multiplication which has shown to be highly effective (Pérez-Vicente, 2003). In addition, meristems used for tissue culture should be disinfected with 3% sodium hypochlorite prior to implanting in the media. No antibiotics should be used in tissue culture media as they may mask the presence of these bacteria which grow well on tissue culture media. It has been shown that the systematic elimination of contaminated cultures in tissue culture can eliminate bacteria after the fifth or sixth multiplication step (Hernández et al., 1994).
2 Avoiding establishing new fields during periods of the highest rainfall in soils that do not have sufficient drainage.
3 Providing balanced nutrition and proper irrigation.
4 Improving soil drainage and destruction of plant residues in affected fields.
5 Application of copper fungicide to the rhizome and buds at planting.
6 Disinfection of tools and equipment after each plant with 3% sodium hypochlorite or 2500 g mL^{-1} benzalkonium chloride (quaternary ammonia) (Pérez-Vicente and Martínez-de La Parte, 2015).
7 Removal of rotted plants and disinfection of the soil underneath.
8 Pseudostem sheaths should not be cut, and elimination of old and *Pseudocercospora* spp.-affected leaves must be cut, leaving a 10 cm long portion of the leaf petioles (Fernández and López, 1970). In some places, smallholders spray pseudostems with copper fungicides 0.5% Cu^{2+} during the rainy season.

Resistance is often a major cornerstone of disease management. In an inoculation study, 25 cultivars and *Musa* species belonging to genome groups AA, BB, AAA, AAB and ABB were shown to be susceptible to disease after inoculation (Padrón, 1983). It has to be noted that the incidence of disease based on the natural infection of disease in the field is often erratic as it is affected by multiple factors such as the transmission, source and health of the planting material, cropping and sanitation practices on the farm, and the cropping history of the fields among others.

The bacteria grow well in most tissue culture media and can be present in the tissue culture plantlets. Selecting healthy plant meristem donors, indexed by ELISA or PCR and systematic elimination protocols in tissue cultures facilities can remove the pathogens from the production system after the sixth multiplication step (Hernández et al., 1994). The general use of healthy tissue culture plantlets in previously affected production fields combined with the systematic disinfection of tools in most cases results in negligible levels of disease in plantations.

7 Future trends and conclusion

Phylogenetic studies carried out with *Pectobacterium* and *Dickeya* spp. have allowed the development of more specific and sensitive diagnostic methods for bacterial soft rot pathogens. It is to be expected that these tools will also enable more detailed epidemiological and disease control studies. Many of the published data on the epidemiology and etiology of these two pathogens are derived from research and experiences in vegetables, potatoes and ornamentals. Plant pathology research on soft rot bacteria in bananas and plantains has received relatively limited attention and funding because of their limited importance in large-scale production systems that routinely implement best management practices. Hence, soft rot diseases are more prevalent among smallholders that do not implement exclusion and tool disinfection practices and may have limited access to clean planting material.

There is a lack of insight into the presence of resistance in *Musa* germplasm, and existing information is basically related to general observations on cases presented under different production conditions. More detailed studies on the reaction of *Musa* germplasm to *D. paradisiaca* are required to be able to identify and deploy effective resistance.

8 Where to look for further information

- Pérombelon, M. C. M. and Kelman, A. (1983). Ecology of soft rot erwinias. *Annual Review of Phytopathology* 18: 361–387.
- Schaad, N. W., Jones, J. B. and Chun, W. (2001). *Laboratory Guide for Identification of Plant Pathogenic Bacteria*. Third Edition. For the Bacteriology Committee of the American Phytopathological Society. APS Press, p. 381.

9 References

Alarcón-Restrepo, J. J. and Castaño-Zapata, J. (2006). Reconocimiento fitosanitario de las principales enfermedades del plátano Dominico Hartón, (*Musa* A B Simmonds). *Agronomy* 14(1): 65–79.

Arun, W. A., Bohra, P., Umesha, K., Chandrashekar, S. C., Sathyanarayana, B. N. and Sreeramu, B. S. (2012). Successful rescue and field establishment of native banana varieties severely affected by rhizome rot. *Journal of Agriculture and Rural Development in the Tropics and Subtropics* 113(2): 147–154.

Belalcázar, S., Merchán, V. M., Mayorga, M., Londoño, M. G., Pulido, J. I. and García, F. (1991). *El Cultivo del Plátano en el Trópico*. CIID, ICA, INIBAP y Comité Departamental de Cafeteros del Quindío.

Blomme, G., Dita, M., Jacobsen, K. S., Pérez-Vicente, L., Molina, A., Ocimati, W., Poussier, S. and Prior, P. (2017). Bacterial diseases of bananas and enset: Current state of knowledge and integrated approaches toward sustainable management. *Frontiers in Plant Science* 8: 1290. DOI: 10.3389/fpls.2017.01290.

Bradbury, J. F. (1986). *Guide to Plant Pathogenic Bacteria*. CAB International Mycological, Institute, Farham House, Slough, p. 332.

Brenner, D. J., Krieg, N. R. and Staley, J. T. (1994). Bergey's manual of systematic bacteriology (Vol. 2). In *The Proteobacteria, Part B. Gammaproteobacteria* (2nd edn.). Springer.

Chattopadhyay, P. K. and Mukherjee, N. (1986). A pseudostem rot of banana due to Erwinia chrysanthemi pv. *paradisiaca*. *Current Science* 55: 789–790.

Dadaşoğlu, F. and Kotan, R. (2017). Identification and characterization of *Pectobacterium carotovorum*. *The Journal of Animal and Plant Sciences* 27(2): 647–654.

Dickey, R. S. and Victoria, J. I. (1980). Taxonomy and emended description of strains of *Erwinia* isolated from *Musa paradisiaca* Linnaeus. *International Journal of Systematic Bacteriology* 30(1): 129–134.

Elphinstone, J. (2008). *Revised Investigation of Erwinia chrysanthemi (Dickeya dianthicola and Other Dickeya spp.) Able to Infect Potatoes*. Project Report, Potato Council. Agriculture & Horticulture Development Board (AHDB).

EPPO. (1990). *Data Sheets on Quarantine Pests*. Erwinia chrysanthemi, p. 5.

EPPO. (2003). Diagnostic protocols for regulated pests. *Ralstonia solanacearum* PM 7 21(1). *OEPP/EPPO Bulletin* 34: 173 –178.

Fernández, B. D. and López, D. G. (1970). Pudrición acuosa del pseudotallo del plátano (*Musa paradisiaca*) causada por *Erwinia chysanthemi* n. sp. *Cenicafe* 21: 1–44.

Fernández-Borrero, O. and López-Ouque, S. (1970). Pudrición acuosa del pseudotallo del plátano *Musa paradisiaca* causada por *Erwinia paradisiaca*. n. sp. *Cenicafe* 21: 3–44.

Hauben, L., Moore, E. R., Vauterin, L., Steenackers, M., Mergaert, J., Verdonck, L. and Swings, J. (1998). Phylogenetic position of phytopathogens within the Enterobacteriaceae. *Systematic and Applied Microbiology* 21(3): 384–397.

Hernández-Pérez, R., Herrera-Isla, L., Pichardo-Moya, T. and Alvarado-Capó, Y. (1994). Diagnóstico de *Erwinia chrysanthemi* Burk *et al.*, en el proceso de micropropagación del plátano (*Musa* spp.). *Centro Agrícola* 21(I): 62–67.

Hevesi, M., Rivera, N. and Perez, L. (1981). Medio selectivo para aislar y conservar a *Erwinia chrysanthemi*. *Ciencia y Técnica en la Agricultura Ser Prot Plant* 4: 61–70.

Hyman, L. J., Sullivan, L., Toth, I. K. and Pérombelon, M. C. M. (2001). Modified crystal violet pectate medium (CVP) based on a new polypectate source (Slendid) for the detection and isolation of soft rot erwinias. *Potato Research* 44(3): 265–270.

Johnston, J. R. (1915). La enfermedad del plátano en Cuba. Estación Experimental Agronómica. Cuba. *Circular* 47: 1–3.

Johnston, J. R. (1918). El plátano y sus enfermedades. Algunos inconvenientes de cultivo en Cuba, *Musa paradisiaca* y *Musa sapientum*. *Revista Agricultura, Comercio y Trabajo* 1: 419–421.

Jones, D. R. (2018). Rhizome and pseudostem bacterial rots. In: Jones, D. R. (Ed.), *Handbook of Diseases of Banana, Abaca and Enset*. CAB International, Wallingford, UK, pp. 329–338.

Kang, H. W., Kwon, S. W. and Go, S. J. (2003). PCR-based specific and sensitive detection of *Pectobacterium carotovorum* ssp. *carotovorum* by primers generated from a URP-PCR fingerprinting-derived polymorphic band. *Plant Pathology* 52: 127–133.

Kelman, A. (1983). Citation classic. *Current Contents* 9: 20.

Li, P., Lin, B., Shen, H. and Pu, X. (2011). Species-specific detection of *Dickeya* sp. (*Pectobacterium chrysanthemi*) in infected banana tissues, soil and water. *African Journal of Biotechnology* 10(74): 16774–16780.

Liu, Q., Xiao, W., Wu, Z., Li, S., Yuan, Y. and Li, H. (2016). Identification of *Dickeya dadantii* as a causal agent of banana bacterial sheath rot in China. *Journal of Plant Pathology* 98(3): 503–510.

Llanos, C. (1967). Una nueva enfermedad del plátano en el Valle del Cauca: la bacteriosis. *Journal of Tropical Agriculture* 23: 806–812.

Martínez, N., Rivera, N., Cabrera, R. y Waves, W. (1989). Detección y Comportamiento de la Enfermedad Necrosis del Cormo del Plátano en la Empresa V. I. en Matanzas, L. Ciencia y Técnica en la Agricultura, *Serie Protección de Plantas* 12(1): 111–120.

Nagaraj, M. S., Umashankar, N., Palanna, K. B. and Khan, A. N. A. (2012). Etiology and management of tip-over disease of banana by using biological agents. *International Journal of Advanced Biological Research* 2(3): 483–486.

Ordosgoitti, F., Santos, R. and Hadded, C. (1974). La pudrición acuosa del pseudotallo del plátano y su presencia en tres regiones plataneras de Venezuela. *Agron Trop* 24: 247–258.

Padrón, J. (1983). Varietal response of banana to wet rot of plantain (*Erwinia chrysanthemi* Burk et al.). *Cultivos Tropicales* 5(1): 75–82.

Patel, P. R. and Shukla, A. (2010). Management of rhizome rot in banana. In: Mustaffa, M. M., Shiva, K. N., Padmanabhan, B. and Mayil Vagnan, M. (Eds.), *Abstracts, Global Conference on Banana, Tiruchirapalli, India*. Association for Improvement in Production and Utilisation of Banana (AIPUB) and the National Research Centre for Banana (NRCB).

Pérez-Vicente, L. (2003). Manejo integrado de plagas y enfermedades en bananos y plátanos en Cuba. In: Rivas, G. and Rosales, F. (Eds.), *Actas del Taller "Manejo Convencional y Alternativo de la Sigatoka Negra, Nematodos y Otras Plagas Asociadas al Cultivo de Musáceas", Celebrado en Guayaquil, Ecuador* (Vols. 11–13). de agosto, 37–54.

Pérez-Vicente, L., Fernández-González, E. and Javer-Higgison, E. (2020). Chapter 35: Banana and plantain diseases of Cuba and Caribbean basin: Impact, epidemiology and management advances. In: Uma, S., Vaganan, M. M. and Agrawal, A. (Eds.), *Bananas and Plantains: Leading-Edge Research and Development (Vol. 1):*

Diversity, Improvement and Protection. ICAR-National Research Centre for Banana, Tiruchirappalli, India, 529–562.

Pérez-Vicente, L. and Martínez-de La Parte, E. (2015). Efecto de un desinfectante de superficies a base de amonios cuaternarios sobre *Fusarium oxysporum* f. sp. cubense, *Dickeya paradisiaca* y *Colletotrichum musae*. Resúmenes III Congreso Latinoamericano de Bananos y plátanos 17–22 agosto, 2015 Turupá, Santa Catarina Brasil.

Pérombelon, M. C. M. and Hyman, L. J. (1986). A rapid method for identifying and quantifying soft rot erwinias directly from plant material based on their temperature tolerance and sensitivity to erythromycin. *Journal of Applied Bacteriology* 60(1): 61–66.

Pérombelon, M. C. M. and Van der Wolf, J. M. (2002). Methods for the detection and quantification of *Erwinia carotovora* subsp. *atroseptica* (*Pectobacterium carotovorum* subsp. *atrosepticum*) on potatoes: A laboratory manual. Scottish Crop Research Institute Occasional Publication No. 10 (rev. version), p. 82.

Potrykus, M., Sledz, W., Golanowska, M., Slawiak, M., Binek, A., Motyka, A., Zoledowska, S., Czajkowski, R. and Lojkowska, E. (2014). Simultaneous detection of major blackleg and soft rot bacterial pathogens in potato by multiplex polymerase chain reaction. *Annals of Applied Biology* 165(3): 474–487.

Rhagavi, G., Muthamilan, M., Nakkeeran, S., Kumaravadivel, N., Sivakumar, U. and Suganthi, A. (2019). Molecular detection of the causative agent of soft rot of (*Pectobacterium carotovorum* subsp. *carotovorum*) in banana (*Musa* sp.). *International Journal of Current Microbiology and Applied Sciences* 8(11): 1854–1868.

Rivera, N. (1978). Estudio comparativo de dos nuevas enfermedades bacterianas en áreas plataneras de Cuba. *Agrotecnia de Cuba* 10: 35–44.

Rivera, N., Albornoz, A. and Gil, C. (1983). Estudio de la sobrevivencia de *Erwinia chrysanthemi* pv. *paradisiaca* en suelo. *Ciencia Técn. Agric. Ser. Prot. Plantas* 6(2): 71–82.

Rivera, N. and Ezavin, M. (1989). Necrosis del cormo del plátano causada *por Erwinia chrysanthemi*. *Ciencia Técnica Agric. Ser. Prot. Plantas* 18: 59–69.

Rivera, N., García, A. and Gil, C. (1980). Caracterización patotípica de aislamientos de *Erwinia chrysanthemi* procedentes de plátano y maíz. *Ciencia Técn. Agric. Ser. Prot. Plantas* 3: 47–60.

Samson, R., Legendre, J. B., Christen, R., Fischer-Le Saux, M., Achouak, W. and Gardan, L. (2004). Transfer of *Pectobacterium chrysanthemi* (Burkholder *et al.* 1953) Brenner *et al.* 1973 and *Brenneria paradisiaca* to the genus *Dickeya* gen. nov. as *Dickeya chrysanthemi* comb. nov and *Dickeya paradisiaca* comb. nov and delineation of four novel species, *Dickeya dadantii* sp nov., *Dickeya dianthicola* sp nov., *Dickeya dieffenbachiae* sp nov and *Dickeya zeae* sp nov. *International Journal of Systematic and Evolutionary Microbiology* 55: 1415–1427.

Sathiamoorthy, S. (1994). *Musa* improvement in India. In: Jones, D. R. (Ed.), *The Improvement and Testing of Musa: A Global Partnership*. INIBAP, France, 188–200.

Seed, N. P. C. and Certification Sub-Committee (2017). *Protocol for Detection of Dickeya and Pectobacterium in Potato Tubers, Stems, or Irrigation Water*, p. 16.

Shillingford, C. A. (1974). Bacterial rhizome rot in Jamaica. *Plant Disease Reporter* 58: 214–218.

Stover, R. H. (1959). Bacterial and rhizome rot of bananas. *Phytopathology* 49: 290–292.

Stover, R. H. (1972). *Banana, Plantain and Abaca Diseases*. Clay Mathematics Institute, Kew, UK.

Thammaiah, N., Kanamadi, V. C., Shirol, A. M., Kulkarni, M. S. and Swamy, G. S. K. (2010). Management of Tip over Disease of Banana cv. Naine, G. Association for Improvement in Production and Utilisation of Banana (AIPUB) and the National Research Centre for Banana (NRCB). In: Mustaffa, M. M., Shiva, K. N., Padmanabhan, B. and Mayil Vagnan, M. (Eds.), *Abstracts, Global Conference on Banana*, Tiruchirapalli, India, p. 106.

Thangavelu, R. (2009). Management of fungal and bacterial diseases in micropropagated plants. In: *Proceedings of the 2nd National Conference on Banana, Jaigon, Maharastra, India*. Association for the Improvement of Production of Banana (AIPUB) and National Research Centre for Banana (NPCB), Thiruchirapalli, pp. 101–107.

Thomas, J. E. (2015). *MusaNet. Technical Guidelines for the Safe Movement of Musa Germplasm* (3rd edn.). Bioversity International, Rome, p. 60.

Thomas, P., Goplakrishnan, C. and Krishnareddy, M. (2011). Soft rot inciting *Pectobacterium carotovorum*. *Plant Cell, Tissue and Organ Culture* 105(3): 423–429.

Toht, I. K., Bell, K. S., Holeva, M. C. and Birch, P. R. J. (2003). Soft rot Erwiniae: From genes to genomes. *Molecular Plant Pathology* 4(1): 17–30.

Tomlinson, D. L., King, G. A. and Ovia, A. (1987). Bacterial corm and rhizome rot of banana in Papua New Guinea caused by *Erwinia chrysanthemi*. *Tropical Pest Management* 33(3): 196–199.

Van der Wolf, J. M., Nijhuis, E. H., Kowalewska, M. J., Saddler, G. S., Parkinson, N., Elphinstone, J. G., Pritchard, L., Toth, I. K., Lojkowska, E., Potrykus, M., Waleron, M., de Vos, P., Cleenwerck, I., Pirhonen, M., Garlant, L., Helias, V., Pothier, J. F., Pfluger, V., Duffy, B., Tsror, L. and Manulis, S. (2014). *Dickeya solani* sp nov., a pectinolytic plant-pathogenic bacterium isolated from potato (*Solanum tuberosum*). *International Journal of Systematic and Evolutionary Microbiology* 64(3): 768–774.

Victoria, J. L. and Barros, O. (1969). Etiología de una nueva enfermedad bacterial del plátano *Musa paradisiaca* L. En: *Colombia. Revista Inst. Colomb. Agropecu* (Vol. 4). ICA, 173–190.

Volcani, Z. and Zutra, D. (1967). Bacterial soft rot in Israel. *Reviews in Applied Mycology* 46: 288.

Wardlaw, C. W. (1972). London, UK. Longman. 878 pp. *Banana Diseases: Including Plantain and Abaca*.

Zhang, J., Shen, H., Pu, X., Lin, B. and Hu, J. (2014). Identification of Dickeya zeae as a causal agent of bacterial soft rot in banana in China. *Plant Disease* 98(4): 436–442.

Chapter 12

Phytoplasma diseases of banana plants

Lilia C. Carvalhais[1], The University of Queensland, Australia; and Richard Davis[2], Australian Department of Agriculture, Fisheries and Forestry, Australia

1 Introduction

Formerly known as mycoplasma-like organisms, phytoplasmas were first described in plants in 1967 through electron microscopy (Doi et al., 1967). They are bacteria that lack cell walls and hence are pleiomorphic. Their cell size ranges from 200 to 800 nm and they inhabit the plant phloem and tissues of phloem-feeding insects, which act as vectors (Namba, 2019; Weintraub and Beanland, 2006; Doi et al., 1967). Hundreds of plant species have been reported to be infected by phytoplasmas, including wild, ornamental and economically relevant crop species (Maejima et al., 2014). Phytoplasma diseases of banana were unknown until two reports were published 10 years apart, one in 1999 and the other in 2009. Both documented plant symptoms including stunting, and the associated '*Candidatus* Phytoplasma' species was identified as '*Candidatus* Phytoplasma asteris', from the aster yellows group (16SrI). The first record was from Maoming, Guangdong Province in China, where phytoplasmas were detected through nested PCR and restriction fragment length polymorphism (RFLP) analysis in phloem tissue from plants

[1] Centre for Horticultural Science, Queensland Alliance for Agriculture and Food Innovation, The University of Queensland, Australia.
[2] Science and Surveillance Group, Biosecurity Plant Division, Department of Agriculture, Fisheries and Forestry, Australia.

http://dx.doi.org/10.19103/AS.2022.0108.12

that also expressed banana bunchy top disease symptoms (Li et al., 1999). The second report was from Uttar Pradesh Aligarh district in northern India, where phytoplasmas were detected in leaf tissues of stunted banana plants using a PCR test that targeted a region of the 16S rRNA coding gene followed by DNA sequence analysis (Singh et al., 2009). These reports did not specify the banana cultivar or genomic group affected. Later, '*Ca.* P. asteris' was associated with symptoms of banana elephantiasis disease in Suriname and Colombia in South America (Aliaga et al., 2018). This disease had previously been reported in Panama, Dominican Republic, Honduras and Costa Rica, but a phytoplasmal aetiology was not identified (Stover, 1972). Banana elephantiasis has been described as a rare disorder, expressing symptoms such as an enhanced growth at the base of the pseudostem, within the connecting tissues between the pseudostem and rhizome, which causes ruptures of the outer sheaths (Fig. 1a). The exposed tissue around the expanded base of the pseudostem acquires a reddish yellow to dark brown colour due to oxidation. Typically, the rhizome surfaces out of the soil and resembles the structure of an onion bulb sometimes displaying a conical shape (Fig. 1b). It is not unusual that the pseudostem breaks and collapses at its narrowest section, due to rotting of the pseudostem tissue at the base where it connects to the corm. In this case, a pineapple-like stump is usually left (Fig. 1c). Reduced formation of suckers, displaying necrotic leaf tips, and smaller fruit size have been observed (Aliaga et al., 2018; Arango-Bernal and Fernadez-Borrero, 1984). Losses in production of 9–72% in commercial AAA and AAB genome group banana crops have been documented in a few regions in Colombia between 2016 and 2017 (Aliaga et al., 2018).

A phytoplasma-associated disease significantly affecting banana plants emerged in Papua New Guinea (PNG). It is a wilt disease that has a devastating impact on cooking banana plants (ABB genomic group) (Davis et al., 2012). The disease was first noticed in the Bogia district of the Madang Province and has spread within PNG from there and recently was also identified in the adjacent country, Solomon Islands (Davis et al., 2015). Losses have been extensive, reaching thousands of plant deaths in areas such as the Markham Valley in Morobe Province (Birte Komolong, PNG National Agricultural Research Institute, personal communication). Symptoms on banana are associated with a group of closely related and novel phytoplasmas, tentatively named as banana wilt associated phytoplasma (BWAP) (Davis et al., 2012). In the northwest of PNG, a phytoplasma suspected to be native to the island of New Guinea associated with a serious lethal wilt disease of both coconut palms (*Cocos nucifera*) and banana plants has recently been classified as the new species '*Ca.* P. noviguineense' (Miyazaki et al., 2018). The disease in coconut is named Bogia coconut syndrome. Elsewhere in PNG, further work has pointed to an identical disease of cooking banana plants which is associated with phytoplasmas

Figure 1 (a) Banana variety Pisang Awak (genomic group ABB) displaying symptoms of elephantiasis disease associated with '*Candidatus* Phytoplasma asteris'. The bases of the pseudostem have enhanced growth and the outer sheaths are ruptured. (b) Banana variety Dominico-Harton (genomic group AAB, plantain subgroup) with rhizome surfacing showing a conical shape. (c) Banana variety Gros Michel (genomic group AAA) showing pineapple-like stump left when pseudostem broke and collapsed in its narrowest section. Photo credit: Miguel Dita, Alliance Biodiversity International-International Center for Tropical Agriculture.

closely related to, but not part of the taxon '*Ca.* P. noviguineense' (Jones et al., 2021; Davis et al., 2015).

There is also a report of an aster yellows group (16SrI-B) phytoplasma in Indonesia, associated with a very low incidence of symptoms that are very similar to those of BWAP in the banana varieties Manggala (ABB) and Nangka king (AAA-AAB) in Tasikmalaya and Banjar, Java, Indonesia. The symptoms of this phytoplasma disease include leaf yellowing, underdeveloped fruit, discontinuous discolouration along the vascular tissue with black and wet necrotic pockets (Sibarani et al., 2019).

As the banana wilt diseases in PNG and Solomon Islands are the most significant in terms of yield losses for smallholders producing bananas and have been subject to most research, only these diseases are considered further in this chapter.

2 Disease symptoms

Cooking banana varieties affected by phytoplasmas in PNG and the Solomon Islands show obvious yellowing of leaves and/or leaf death (Fig. 2a, b, d, e). Fruit bunches, if they do occur, miss fingers in many hands while the fingers which are present are poorly filled (Fig. 2c) and few suckers are produced.

Figure 2 (a) Banana plants expressing yellowing symptoms on leaves and leaf death; (b), (d), (e). Close-up of banana leaves showing yellowing, with a typical orange tinge; (c) Banana bunch from a symptomatic plant with reduced number of hands and fingers, which are distorted, small and poorly filled. Photo credit: André Drenth, University of Queensland.

Figure 3 (a) Discontinuous streaking on vascular tissue along pseudostem. (b) Transversal section of the pseudostem showing vascular discolouration. (c) Close-up of wet necrotic pockets on vascular tissue. (d) Vascular discolouration along leaf midrib. Photo credit: André Drenth, University of Queensland.

No signs of discolouration are observed in the interior of the fingers. Internal pseudostem symptoms are highly distinctive and quite different from those of other fungal and bacterial wilt diseases of banana. Phytoplasma infection is characterised by a discontinuous vascular streaking in the pseudostem, appearing as small (one to several cm only) regions of bright yellow to dark brown vascular tissues with indistinct margins, usually wet, forming necrotic pockets (Fig. 3a–c). Indistinct vascular streaking can also extend into the midrib of older leaves at advanced stages of the disease (Fig. 3d). Necrosis of parts of the male inflorescence is also a symptom unique to this disease, with stunted male flowers showing a black colour, as well as the anthers showing discolouration (Fig. 4a–d).

Figure 4 (a) Longitudinal section of a banana inflorescence showing discoloured necrotic immature male flowers. (b) Banana inflorescence with some stunted necrotic male flowers exposed after bracts have been removed. (c) Banana inflorescence of two different varieties, one that was infected by phytoplasmas (left) and a healthy plant from another variety (right). (d) Discoloured necrotic anthers were exposed after petals were removed. Photo credit: André Drenth, University of Queensland.

3 Pathogen characteristics and taxonomic placement

Multiplication of phytoplasma cells only occurs in the plant phloem and in the haemolymph of insects. Within plants, phytoplasmas typically attach to the membrane of the sieve elements and colonise different organs in an uneven fashion, usually reaching higher densities in sink tissues (Christensen et al., 2004; Huang et al., 2020). In banana, necrosis of the phloem, yellowing and wilting of the leaves are the most common symptoms associated with phytoplasmas.

Phytoplasmas cannot be cultured in standard laboratory media and only multiply in living host plant tissue which has hampered research progress and, hence, our understanding of these plant pathogens. Microscopy and serology-based detection methods have been developed in the past but more recently PCR-based assays are most commonly used for diagnostics purposes (Marcone and Ragozzino, 1996; Musetti et al., 2002; Hodgetts et al., 2014). In the early 1990s, analysis of the 16S rRNA gene sequence revealed that phytoplasmas form a monophyletic group within the class Mollicutes (Namba et al., 1993). Phytoplasma genomes are considerably smaller than those of other bacteria, varying in size from 0.474 kb to 1116 kb (Kumari et al., 2019). The first classification system of phytoplasmas was based on RFLP analyses of a 1.25 kb amplicon of the 16S rRNA gene digested with 17 different restriction enzymes (Lee et al., 1998). From these analyses, similarity coefficients were calculated and used to identify distinct groups and subgroups (Zhao and Davis, 2016). This system is still being used in addition to the near-full-length 16S rRNA gene sequence analysis mainly through a virtual RFLP analysis software known as *i*PhyClassifier (Zhao et al., 2012). Although the validity of the 16Sr groups defined by RFLP has been confirmed as being congruent with clades obtained by phylogenetic analysis of the full 16S rRNA gene (Wei et al., 2007; Weintraub and Jones, 2009), the information content will never be as informative as the DNA sequence it is derived from.

The novel taxon '*Ca.* P. noviguineense' (Miyazaki et al., 2018) was established using rRNA gene sequences from infected coconut palms, betelnut palms and banana plants from the Madang Province where Bogia coconut syndrome occurs, in addition to using one sequence from a banana plant in the Western Province where Bogia coconut syndrome had not been reported previously (Miyazaki et al., 2018). When discovered, the phytoplasmas from northwest PNG identified as '*Ca.* P. noviguineense' and BWAPs from eastern PNG and Solomon Islands did not belong to any currently recognised 16S rRNA group (Davis et al., 2015). It is important to note, however, that the closest matches (with similarity coefficients below the group/subgroup cut-off point) belonged to phytoplasmas associated with devastating lethal yellowing diseases of coconut palms elsewhere in the world in groups 16SrIV and 16SrXXII. More recently, '*Ca.* P. noviguineense' has been assigned to the new group, 16SrXXXVIII (Wei and Zhao 2022). Also of importance is that a comparison of 16S rRNA gene sequences available on GenBank indicates that the BWAPs from eastern PNG and Solomon Islands are sufficiently different to fall outside '*Ca.* P. noviguineense' and possibly represent another novel '*Candidatus* Phytoplasma' taxon. The16S rRNA gene sequences of BWAPs from the Solomon Islands archipelago (Genbank accession numbers KP642660-3) are 96.5–96.6% similar to the sequence of the '*Ca.* P. noviguineense' reference strain (LC228755). The cut-off point for inclusion within a '*Ca.* Phytoplasma' sp. was 97.5% similarity, according to prevailing phytoplasma nomenclature rules at the time (IRPCM, 2004), and this threshold has now been raised to 98.65% similarity (Bertaccini et al., 2022).

4 Diagnosis

Reliable detection and identification of phytoplasmas are crucial for early containment and management of phytoplasma-associated diseases. Nested PCR and real-time PCR are the most used detection methods as they allow a high level of sensitivity required for detection of phytoplasmas occurring in low titre (Hodgetts et al., 2007; Nejat and Vadamalai, 2013). Both methods often target conserved sequences of the 16S, 23S rRNA genes or their intergenic spacer regions (ITS) (Smart et al., 1996; Hodgetts et al., 2009).

Other diagnostic assays targeting certain 16Sr groups or specific phytoplasma groups (e.g. apple proliferation group) have been developed, as well as fast in-field methods, such as the loop-mediated isothermal amplification (LAMP) (Bekele et al., 2011b; Hodgetts et al., 2011; Lu et al., 2016).

Although it can be advantageous to use LAMP and obtain fast diagnosis in the field, careful attention needs to be paid to sample preparation and the use of controls which makes confirmation of results by an alternative method in the laboratory necessary. DNA sequences other than ribosomal genes have also been targeted for detection of phytoplasmas, including genes that encode protein translocase subunits (*secY*, *secA*), the elongation factor Tu gene (*tuf*), genes encoding ribosomal proteins (*rplV/rpl22* and *rpsC/rps3*), and a gene encoding the 60 kDa chaperonin (*cpn60*) (Dumonceaux et al., 2014; Jovic et al., 2021; Makarova et al., 2012; Martini et al., 2007; Bekele et al., 2011a). These methods were mostly developed for use on infected plant material, but some have also been applied to insect tissue (Dumonceaux et al., 2014; Martini et al., 2007). A comprehensive comparison between them has not been carried out hitherto. Assays based on genes encoding ribosomal subunits and their spacer regions are frequently used as universal methods given how ubiquitous and conserved these genes are. Often these non-rRNA gene-based methods are used either to acquire a finer differentiation between phytoplasma strains, for the diagnosis of specific phytoplasma groups or as alternatives for confirmation of the diagnosis (Martini et al., 2019; Bekele et al., 2011a).

5 Disease cycle and epidemiology

Phytoplasmas are inoculated straight into phloem vessels of plants by polyphagous insects from the order Hemiptera, such as leafhoppers, planthoppers and psyllids (Weintraub and Beanland, 2006). The direct delivery of phytoplasmas into the vascular system also allows evasion of plant immunity responses (Huang et al., 2020). Phytoplasmas often trigger changes in plant development via degradation of plant transcription factors resulting in abnormalities such as phyllody which seem to enhance plant colonisation by insect vectors as more feeding sites are available (Toruno et al., 2016; Sugio et al., 2011).

A general disease cycle of phytoplasmas on banana is given in Fig. 5. The vector, depicted here as a leafhopper, feeds on the phloem sap of a banana plant infected with a phytoplasma in a process called acquisition feeding (Fig. 5b). A latency period then follows, during which phytoplasma cells multiply within the vector and move from the gut to the hemolymph and finally to the salivary glands (Fig. 5c). Subsequently, when the infected insect feeds from the phloem sap of a healthy plant, phytoplasma cells are transferred to its phloem in a process known as inoculation feeding (Fig. 5d). Phytoplasmas then multiply in phloem vessels and when a certain titer is reached symptoms start to be expressed in the plant (Fig. 5e). Whilst sucker production is reduced in infected plants, most of those that do grow appear symptomless both externally and internally. However, both symptomatic and asymptomatic

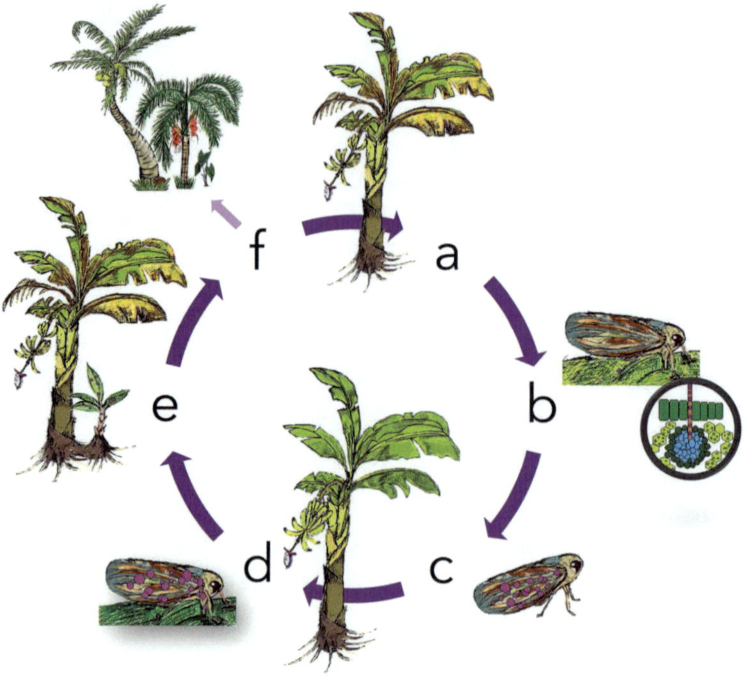

Figure 5 General disease cycle of phytoplasmas on banana. (a) Banana plant showing yellowing symptoms. (b) Acquisition feeding: leafhopper sucking phloem sap from phytoplasma-infected banana plant. (c) Latency period, when phytoplasmas multiply inside leafhopper, moving from the gut to the hemolymph and finally to the salivary glands. (d) Inoculation feeding, when phytoplasma-infected leafhopper delivers phytoplasmas cells present in the saliva to the phloem vessels. (e) Phytoplasma multiplication and symptom expression in the infected banana plant, with asymptomatic banana sucker harbouring low numbers of phytoplasma cells. (f) From left to right: coconut palm, another principal host, and two possible alternative hosts: betelnut palm and taro.

suckers arising from infected mother plants are known to test positive for phytoplasma (Davis et al., 2012). Coconut and other palms such as betelnut (*Areca catechu*) are also main hosts of phytoplasmas pathogenic to banana (Fig. 5f), while taro (*Colocasia esculenta*) was deemed a key alternative host for '*Ca.* P. noviguineense' based on nested PCR and sequence analysis (Gurr, 2020). It is unclear what the native host of this phytoplasma is but in the Madang Province there is a record of '*Ca.* P. noviguineense' detected in *Hydriastele costata*, a native palm that is commonly used in house construction (P. Kokoa, PNG Chief Plant Protection Officer, personal communication).

Whilst nothing is so far known about vectors of the phytoplasmas affecting banana in eastern PNG and the Solomon Islands, significant advances have been made in northwest PNG where Bogia coconut syndrome occurs. Six insect taxa belonging to four families, Derbidae, Lophopidae, Flatidae and

Ricaniidae, were identified as putative vectors in localised 'hot spots' where both coconut palms and banana plants were diseased (Lu et al., 2016). This work included experiments with an artificial feeding medium to verify the ability of phytoplasma transmission by living insects (Gurr, 2020). The planthoppers *Zophiuma pupillata* and *Lophops saccharicida* were identified as potential vector species of importance. Subsequently, field-collected individuals of these two vector species were capable of transmitting phytoplasma to banana plants in caged transmission trials (Gurr, 2020).

Banana phytoplasmas have spread within PNG and to the Solomon Islands most likely through movement of banana planting material (suckers) by people, which is a common practice in Melanesian countries and elsewhere. Betelnuts and taro corms are commodities that also regularly travel long distances in PNG and Solomon Islands, representing a biosecurity risk to unaffected regions. Tissue culture banana plantlets would also pose a risk of long-distance transmission, if infected. Indexing for phytoplasmas in bananas used for planting material should significantly decrease the likelihood of new incursions or the spread of phytoplasma-associated diseases.

6 Occurrence and impact

A lethal wilt of cooking banana plants was first noticed in PNG in the early 1990s by growers in the Bogia region of Madang Province in the northwest of the country (Lastus Kuniata, Ramu Agri Industries, personal communication). By the mid-2000s, identical symptoms were observed at the eastern end of the country on Buka Island in what is now known as the Autonomous Region of Bougainville (P. Kokoa, PNG Chief Plant Protection Officer, personal communication). Later studies verified a widespread but patchy incidence of the BWAP phytoplasma across PNG and into the Solomon Islands archipelago (Davis et al., 2012; Davis et al., 2015). Simultaneously in northwest PNG in the early 1990s, a new lethal wilt disease of coconut palms that became known as Bogia coconut syndrome appeared and the growers themselves were the first to suspect a connection between the coconut and banana diseases (P. Kokoa, PNG Chief Plant Protection Officer, personal communication).

The current geographic distribution of phytoplasmas causing wilt disease of cooking banana plants in PNG and Solomon Islands is provided in Fig. 6. Important biological differences may reflect the taxonomic distinction between the BWAPs from the Solomon Islands to the east of PNG and those that fall within '*Ca.* P. noviguineense' in the west (Davis et al., 2015; Jones et al., 2021). For example, it is noteworthy that no phytoplasma-like palm disease has been found so far in these locations in the east (Fig. 6).

The areas that have been affected the most and sustaining ongoing losses in banana production due to a disease associated with phytoplasmas

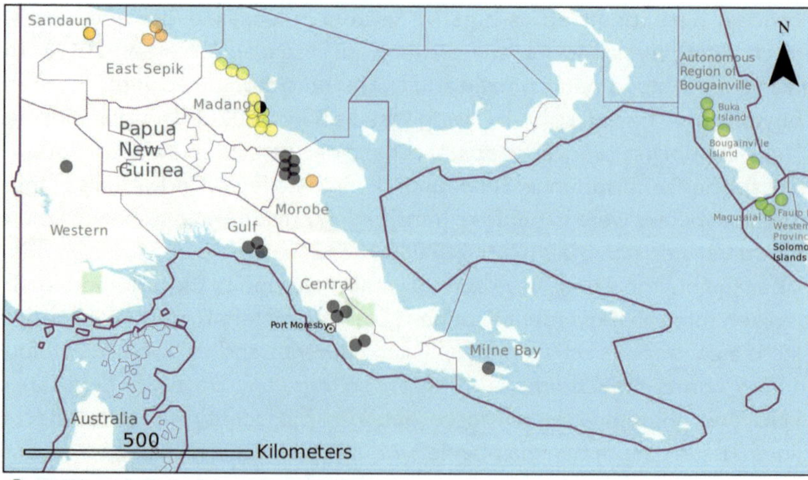

- BWAP falls outside '*Ca. P. noviguineense*'; asymptomatic coconuts
- BWAP that falls within '*Ca. P. noviguineense*'; symptomatic coconuts
- Debepare phytoplasma (falls within *Ca. P. noviguineense*') and other *Ca. P. noviguineense* occur together; symptomatic coconuts
- Debepare phytoplasma (falls within *Ca. P. noviguineense*'); asymptomatic coconuts
- Unknown BWAP; asymptomatic coconuts

Figure 6 Map of Papua New Guinea and Solomon Islands indicating approximate locations of recorded occurrences of phytoplasmas. The Debepare phytoplasma was first found in banana in 2010 in the village of Debepare, Western Province, associated with BWAP symptoms (and no visible symptoms in coconut palms). This isolate was later shown to be part of the novel taxon '*Ca.* P. noviguineense' described by Miyazaki et al. (2018). The locations of the three records of Debepare phytoplasmas around Kerema in the Gulf Province are not completely accurate (precise coordinates unavailable). At the time of the design of this map, no records of Bogia Coconut Syndrome were known from anywhere outside of Madang Province. Map data © OpenStreetMap contributors, Microsoft, Esri Community Maps contributors, Map layer by Esri. Map credit: Jed Calvert, University of Queensland.

are the Madang and Morobe Provinces in PNG. In this region banana growers are subsistence smallholders growing multiple crops, which makes losses difficult to quantify. Given that banana is a staple crop in PNG and second top commodity, with 1.3 million tonnes produced in 2021, significant losses in banana production in any region in PNG have a major impact on food and income security for the local population (FAO, 2023).

7 Management of the disease

7.1 Cultivar susceptibility and available host resistance

In PNG and the Solomon Islands, only the widely grown cooking bananas (ABB genomic group) have been severely affected by the wilt disease associated with phytoplasmas. The majority of these varieties fall in the Kalapua subgroup in PNG (Davis et al., 2012), but the ABB dessert banana known as Yawa is also known to be significantly affected in the Markham Valley and elsewhere (National Agricultural Research Institute, PNG, unpublished data). No commercial-scale monoculture plantations of Cavendish varieties exist in this region but records of detections in the Madang Province of the BWAP phytoplasma in symptomless banana plants of the Cavendish subgroup (AAA genomic group) growing adjacent to dying cooking banana plants exist (P. Kokoa, PNG Chief Plant Protection Officer, personal communication). Caged plant studies also confirmed infection of externally symptomless Cavendish subgroup banana plants by '*Ca*. P. noviguineense', but in these experiments no attempt was made to determine if there were any internal symptoms (Gurr, 2020).

7.2 Cultural control

There are no proven curative options to control phytoplasma-associated diseases of field crops, which leaves growers with preventative measures and destruction of infected plants. The best approach to prevent phytoplasma-associated diseases from spreading is the use of pathogen-free planting material and monitoring and early detection followed up by rogueing of affected plants. Early containment and follow-up eradication strategies are reliant on early identification of disease symptoms. Having diagnostic tests available for the detection of phytoplasmas early in the asymptomatic phase is important for the identification of infected plants and their subsequent removal to limit further spread.

An integrated disease management strategy is needed for growers to combat phytoplasma wilt diseases of banana plants in PNG and the Solomon Islands. This would involve biosecurity measures to protect yet unaffected regions, such as what is currently being implemented by authorities in the Solomon Islands, where the outbreak is new and confined to islands adjacent to PNG (F. Tsatsia, Director, Biosecurity Solomon Islands, personal communication). In regions already affected, a strict rogueing policy is needed and such phytosanitary measures require knowledge of disease symptoms in bananas and potential alternative hosts. This could be combined with strategies to control the insect vectors, based on information on their biology and ecology, which is currently lacking. There is also a need to gain more knowledge on the rate of development of systemic infection and on the distribution of phytoplasmas in infected banana plants. This is mostly to inform disease testing protocols for imported germplasm with respect to timing of testing of new plantlets grown from tissue culture. The availability of clean planting material following rigorous phytoplasma indexing assessments is crucial to prevent new infections.

8 Conclusion and future trends

Frequent movement of people in the modern world has significantly increased the spread of plant diseases, with climate change and continuing expansion of world commerce expected to aggravate this problem even more (Ristaino et al., 2021). Therefore, research focused on the prevention of emerging diseases and mitigation practices to control new outbreaks is paramount to global food security.

Wilt disease associated with phytoplasmas on banana in PNG and Solomon Islands is an emerging disease problem and its impact in areas where it occurs has been devastating. Plant diseases caused by phytoplasmas are especially impactful in vegetatively propagated crops with a perennial lifestyle, such as banana. This means that smallholder growers may be reluctant to rogue diseased plants bearing fruit which hampers efforts to reduce the disease in an area. To avoid further spread and continuous yield losses, it is crucial to raise awareness on the importance of rogueing and the use of clean planting material among growers in the areas affected by this disease. Clean planting material can only be reliably obtained from plants known to be disease-free, which can be challenging if effective phytoplasma indexing methods are not available and accessible. Basic research on the biology of banana phytoplasmas is crucial to understanding the patterns of disease transmission and emergence of new pathogen variants. Comprehensive and long-term disease surveillance also needs to be scaled to have reliable insights into the geographic distribution of the pathogen and enable effective monitoring systems. Emerging diseases are often not attracting a lot of research and only local investments are available. This allows them to turn into severe outbreaks that are more challenging and substantially more expensive to control given the insufficient availability of epidemiological data (Ristaino et al., 2021; Drenth and Kema Chapter 6 of this book). By then investments in research typically improve but the required mitigation efforts are much greater, such as in the case of the majority of major banana diseases, including black leaf streak, yellow Sigatoka, Moko, bunchy top, Fusarium wilt race 1 and Fusarium wilt tropical race 4 (Drenth and Kema, 2021). Ongoing funding and consolidated international collaboration aiming at shared research and access to data are fundamental to reduce the impact of plant disease threats.

9 Where to look for further information

Further general information on phytoplasmas in banana and other plants can be found in the following literature sources:

- Jones, D. (2018). Diseases caused by bacteria and phytoplasmas. In: Jones, D. (ed). *Handbook of Diseases of Banana, Abaca and Enset*. Boston, MA: CABI, pp. 342–346

- Olivier, C. Y., Dumonceaux, T. J. and Pérez-López, E. (2020). *Sustainable Management of Phytoplasma Diseases in Crops Grown in the Tropical Belt: Biology and Detection*. Switzerland: Springer.
- Rao, G. P., Bertaccini, A., Fiore, N. and Liefting, L. W. (2018). *Phytoplasmas: Plant Pathogenic Bacteria-I: Characterisation and Epidemiology of Phytoplasma-Associated Diseases*. Singapore: Springer.

10 References

Aliaga, F., Hopp, E., Álvarez, E. and Lopez-Lavalle, L. A. B. (2018). First report of a '*Candidatus* Phytoplasma asteris' isolate associated with banana elephantiasis disease in Colombia. *New Disease Reports* 37(1): 12.

Arango-Bernal, L. G. and Fernandez-Borrero, O. (1984). *Elefantiasis o Embalconamiento del Plátano y del Banano*. Report. Colombia: Corporación Colombiana de Investigación Agropecuaria.

Bekele, B., Abeysinghe, S., Hoat, T. X., Hodgetts, J. and Dickinson, M. (2011a). Development of specific secA-based diagnostics for the 16SrXI and 16SrXIV phytoplasmas of the Gramineae. *Bulletin of Insectology* 64: S15–S16.

Bekele, B., Hodgetts, J., Tomlinson, J., Boonham, N., Nikolić, P., Swarbrick, P. and Dickinson, M. (2011b). Use of a real-time LAMP isothermal assay for detecting 16SrII and XII phytoplasmas in fruit and weeds of the Ethiopian Rift Valley. *Plant Pathology* 60(2): 345–355.

Bertaccini, A., Arocha-Rosete, Y., Contaldo, N., Duduk, B., Fiore, N., Montano, H. G., Kube, M., Kuo, C. H., Martini, M., Oshima, K., Quaglino, F., Schneider, B., Wei, W. and Zamorano, A. (2022). Revision of the 'Candidatus Phytoplasma' species description guidelines. *International Journal of Systematic and Evolutionary Microbiology* 72(4): 005353.

Christensen, N. M., Nicolaisen, M., Hansen, M. and Schulz, A. (2004). Distribution of phytoplasmas in infected plants as revealed by real-time PCR and bioimaging. *Molecular Plant-Microbe Interactions* 17(11): 1175–1184.

Davis, R. I., Henderson, J., Jones, L. M., McTaggart, A. R., O'Dwyer, C., Tsatsia, F., Fanai, C. and Rossel, J. B. (2015). First record of a wilt disease of banana plants associated with phytoplasmas in Solomon Islands. *Australasian Plant Disease Notes* 10(1): 14.

Davis, R. I., Kokoa, P., Jones, L. M., Mackie, J., Constable, F. E., Rodoni, B. C., Gunua, T. G. and Rossel, J. B. (2012). A new wilt disease of banana plants associated with phytoplasmas in Papua New Guinea (PNG). *Australasian Plant Disease Notes* 7(1): 91–97.

Doi, Y., Teranaka, M., Yora, K. and Asuyama, H. (1967). Mycoplasma-or PLT group-like microorganisms found in the phloem elements of plants infected with mulberry dwarf, potato witches' broom, aster yellows, or paulownia witches' broom. *Japanese Journal of Phytopathology* 33(4): 259–266.

Drenth, A. and Kema, G. (2021). The vulnerability of bananas to globally emerging disease threats. *Phytopathology* 111(12): 2146–2161.

Dumonceaux, T. J., Green, M., Hammond, C., Perez, E. and Olivier, C. (2014). Molecular diagnostic tools for detection and differentiation of phytoplasmas based on chaperonin-60 reveal differences in host plant infection patterns. *PLoS ONE* 9(12): e116039.

FAO (2023). FAOSTAT. *IGO* Extracted from: https://www.fao.org/faostat/en/#rankings/commodities_by_country.

Gurr, G. (2020). Bogia coconut syndrome and related Phytoplasma syndromes in Papua New Guinea: Developing biological knowledge and a risk management strategy. Report for Australian Centre for International Agricultural Research. Report no HORT/2012/087, 10 September. Bruce (ACT): Australia.

Hodgetts, J., Ball, T., Boonham, N., Mumford, R. and Dickinson, M. (2007). Use of terminal restriction fragment length polymorphism (T-RFLP) for identification of phytoplasmas in plants. *Plant Pathology* 56(3): 357–365.

Hodgetts, J., Boonham, N., Mumford, R. and Dickinson, M. (2009). Panel of 23S rRNA gene-based real-time PCR assays for improved universal and group-specific detection of phytoplasmas. *Applied and Environmental Microbiology* 75(9): 2945–2950.

Hodgetts, J., Johnson, G., Perkins, K., Ostoja-Starzewska, S., Boonham, N., Mumford, R. and Dickinson, M. (2014). The development of monoclonal antibodies to the secA Protein of Cape St. Paul wilt disease phytoplasma and their evaluation as a diagnostic tool. *Molecular Biotechnology* 56(9): 803–813.

Hodgetts, J., Tomlinson, J., Boonham, N., Gonzalez-Martin, I., Nikolić, P., Swarbrick, P., Yankey, E. N. and Dickinson, M. (2011). Development of rapid in-field loop-mediated isothermal amplification (LAMP) assays for phytoplasmas. *Bulletin of Insectology* 64: S41–S42.

Huang, W. J., Reyes-Caldas, P., Mann, M., Seifbarghi, S., Kahn, A., Almeida, R. P. P., Béven, L., Heck, M., Hogenhout, S. A. and Coaker, G. (2020). Bacterial vector-borne plant diseases: Unanswered questions and future directions. *Molecular Plant* 13(10): 1379–1393.

IRPCM. (2004). '*Candidatus* Phytoplasma', a taxon for the wall-less, non-helical prokaryotes that colonise plant phloem and insects. *International Journal of Systematic and Evolutionary Microbiology* 54(4): 1243–1255.

Jones, L. M., Pease, B., Perkins, S. L., Constable, F. E., Kinoti, W. M., Warmington, D., Allgood, B., Powell, S., Taylor, P., Pearce, C. and Davis, R. I. (2021). '*Candidatus* Phytoplasma dypsidis', a novel taxon associated with a lethal wilt disease of palms in Australia. *International Journal of Systematic and Evolutionary Microbiology* 71(5): 004818.

Jovic, J., Marinkovic, S., Jakovljevic, M., Krstić, O., Cvrković, T., Mitrović, M. and Toševski, I. (2021). Symptomatology, (co)occurrence and differential diagnostic PCR identification of '*Ca.* Phytoplasma solani' and '*Ca.* Phytoplasma convolvuli' in field bindweed. *Pathogens* 10(2): 160.

Kumari, S., Nagendran, K., Rai, A. B., Singh, B., Rao, G. P. and Bertaccini, A. (2019). Global status of phytoplasma diseases in vegetable crops. *Frontiers in Microbiology* 10: 1349.

Lee, I.-M., Gundersen-Rindal, D. E., Davis, R. E. and Bartoszyk, I. M. (1998). Revised classification scheme of phytoplasmas based an RFLP analyses of 16S rRNA and ribosomal protein gene sequences. *International Journal of Systematic Bacteriology* 48(4): 1153–1169.

Li, H., Qiu, B., Li, L. and Shi, C. (1999). Amplification of phytoplasma 16S rDNA from banana bunchy top disease, RFLP and sequence analysis of 16S rDNA fragment. *Wei Sheng Wu Xue Bao* 39(4): 315–320.

Lu, H. Y., Wilson, B. A. L., Ash, G. J., Woruba, S. B., Fletcher, M. J., You, M., Yang, G. and Gurr, G. M. (2016). Determining putative vectors of the Bogia Coconut Syndrome Phytoplasma using loop-mediated isothermal amplification of single-insect feeding media. *Scientific Reports* 6: 35801.

Maejima, K., Oshima, K. and Namba, S. (2014). Exploring the phytoplasmas, plant pathogenic bacteria. *Journal of General Plant Pathology* 80(3): 210–221.

Makarova, O., Contaldo, N., Paltrinieri, S., Kawube, G., Bertaccini, A. and Nicolaisen, M. (2012). DNA barcoding for identification of '*Candidatus* Phytoplasmas' using a fragment of the elongation factor Tu gene. *PLoS ONE* 7(12): e52092.

Marcone, C. and Ragozzino, A. (1996). Comparative ultrastructural studies on genetically different phytoplasmas using scanning electron microscopy. *Petria* 6(2): 125–136.

Martini, M., Lee, I. M., Bottner, K. D., Zhao, Y., Botti, S., Bertaccini, A., Harrison, N. A., Carraro, L., Marcone, C., Khan, A. J. and Osler, R. (2007). Ribosomal protein gene-based phylogeny for finer differentiation and of phytoplasmas. *International Journal of Systematic and Evolutionary Microbiology* 57(9): 2037–2051.

Martini, M., Quaglino, F. and Bertaccini, A. (2019). Multilocus genetic characterization of phytoplasmas. In: Assunta, B., Oshima, K., Kube, M. and Rao, G. P. (Eds.), *Phytoplasmas: Plant Pathogenic Bacteria-III*. Singapore: Springer, pp. 161–200.

Miyazaki, A., Shigaki, T., Koinuma, H., Iwabuchi, N., Rauka, G. B., Kembu, A., Saul, J., Watanabe, K., Nijo, T., Maejima, K., Yamaji, Y. and Namba, S. (2018). '*Candidatus* Phytoplasma noviguineense', a novel taxon associated with Bogia coconut syndrome and banana wilt disease on the island of New Guinea. *International Journal of Systematic and Evolutionary Microbiology* 68(1): 170–175.

Musetti, R., Loi, N., Carraro, L. and Ermacora, P. (2002). Application of immunoelectron microscopy techniques in the diagnosis of phytoplasma diseases. *Microscopy Research and Technique* 56(6): 462–464.

Namba, S. (2019). Molecular and biological properties of phytoplasmas. *Proceedings of the Japan Academy. Series B, Physical and Biological Sciences* 95(7): 401–418.

Namba, S., Oyaizu, H., Kato, S., Iwanami, S. and Tsuchizaki, T. (1993). Phylogenetic diversity of phytopathogenic Mycoplasmalike organisms. *International Journal of Systematic Bacteriology* 43(3): 461–467.

Nejat, N. and Vadamalai, G. (2013). Diagnostic techniques for detection of Phytoplasma diseases: Past and present. *Journal of Plant Diseases and Protection* 120(1): 16–25.

Ristaino, J. B., Anderson, P. K., Bebber, D. P., Brauman, K. A., Cunniffe, N. J., Fedoroff, N. V., Finegold, C., Garrett, K. A., Gilligan, C. A., Jones, C. M., Martin, M. D., MacDonald, G. K., Neenan, P., Records, A., Schmale, D. G., Tateosian, L. and Wei, Q. (2021). The persistent threat of emerging plant disease pandemics to global food security. *Proceedings of the National Academy of Sciences of the United States of America* 118(23): e2022239118.

Sibarani, S. M. A., Joko, T. and Subandiyah, S. (2019). Detection and identification of banana-associated phytoplasma using nested-PCR method. *Jurnal Perlindungan Tanaman Indonesia* 23(1): 148–155.

Singh, B. R., Aminuddin, A., Al-Khedhairy, A. A., Al-Qurainy, F. and Musarrat, J. (2009). Molecular diagnostics and phylogenetic analysis of '*Candidatus* Phytoplasma asteris' (16SrI-Aster yellow group) infecting banana (Musa spp.). *African Journal of Biotechnology* 8(21): 5819–5824.

Smart, C. D., Schneider, B., Blomquist, C. L., Guerra, L. J., Harrison, N. A., Ahrens, U., Lorenz, K. H., Seemüller, E. and Kirkpatrick, B. C. (1996). Phytoplasma-specific PCR primers based on sequences of the 16S-23S rRNA spacer region. *Applied and Environmental Microbiology* 62(8): 2988–2993.

Stover, R. H. (1972). Non-infectious diseases and abnormalities of uncertain cause. In: Stover, R. H. (Ed.), *Banana, Plantain and Abaca Diseases.* Bucks, England: Commonwealth Agricultural Bureaux, pp. 243–244.

Sugio, A., Kingdom, H. N., MacLean, A. M., Grieve, V. M. and Hogenhout, S. A. (2011). Phytoplasma protein effector SAP11 enhances insect vector reproduction by manipulating plant development and defense hormone biosynthesis. *Proceedings of the National Academy of Sciences of the United States of America* 108(48): E1254–E1263.

Toruno, T. Y., Stergiopoulos, I. and Coaker, G. (2016). Plant-pathogen effectors: Cellular probes interfering with plant defenses in spatial and temporal manners. *Annual Review of Phytopathology* 54: 419–441.

Wei, W., Davis, R. E., Lee, I. M. and Zhao, Y. (2007). Computer-simulated RFLP analysis of 16S rRNA genes: Identification of ten new Phytoplasma groups. *International Journal of Systematic and Evolutionary Microbiology* 57(8): 1855–1867.

Wei, W. and Zhao, Y. (2022). Phytoplasma taxonomy: nomenclature, classification, and identification. *Biology* 11: 1119.

Weintraub, P. G. and Beanland, L. (2006). Insect vectors of phytoplasmas. *Annual Review of Entomology* 51: 91–111.

Weintraub, P. G. and Jones, P. (2009). *Phytoplasmas: Genomes, Plant Hosts and Vectors*. Wallingford, UK: CABI Publishing.

Zhao, Y. and Davis, R. E. (2016). Criteria for Phytoplasma 16Sr group/subgroup delineation and the need of a platform for proper registration of new groups and subgroups. *International Journal of Systematic and Evolutionary Microbiology* 66(5): 2121–2123.

Zhao, Y., Lee, I., Wei, W., Shao, J., Suo, X. and Davis, R. E. (2012). The iPhyClassifier II: An update to the online tool for Phytoplasma identification and classification. *Phytopathology* 102(7): 143–143.

Part 3

Viral diseases

Chapter 13

Banana bunchy top virus

John E. Thomas, The University of Queensland, Australia

1 Introduction

Bananas (*Musa* sp.) are one of the most important food crops worldwide and are the most valuable fruit product in international trade, alone comprising more than 14% of total fruit value and worth US$14.45 billion in 2017 (ITC, 2017). Despite this, only 15% of world production is exported, the remainder consumed locally, often as a staple food.

The crop is subject to severe losses due to several serious plant diseases (Ploetz et al., 2015; Drenth and Kema, 2021). Among these, banana bunchy top disease (BBTD), caused by banana bunchy top virus (BBTV), is the most important of the virus diseases, with infected plants rarely producing a bunch (Kumar et al., 2015). Over the last century, BBTD has continually expanded its geographic range causing significant losses. The impact of the disease is exacerbated through the lack of effective resistance, ready transmission through

http://dx.doi.org/10.19103/AS.2022.0108.13

infected planting material, and the high impact of the disease on Cavendish cultivars, which currently dominate the world export trade. Cavendish cultivars are estimated to constitute 46% of total world banana production (Vézina, 2017).

Although BBTD was first reported in 1889 (Darnell-Smith, 1924) and was one of the very early plant diseases to be ascribed a viral cause (Magee, 1927), identification of the virus itself had to wait another 60 years (Wu and Su, 1990c). Since then, rapid advances have been made in understanding the molecular features of the virus (Burns et al., 1995; Karan et al., 1997; Stainton et al., 2012; Stainton et al., 2015) and its interaction with its plant host (Amin et al., 2011; Niu et al., 2009) and vector (Watanabe and Bressan, 2013). However, despite these advances, the disease continues to spread to new countries (Thomas, 2019), and current control still relies on quarantine, clean planting material, and inspection and eradication, much as it did a century ago (Magee, 1927).

This chapter gives an overview of the historical aspects of the disease and the early studies and outlines the symptoms of the disease and biological characteristics of the pathogen related to host range, transmission, and spread. This is followed by an overview of control strategies which involve the detection of the pathogen, its epidemiology and resistance in the host, and how this can be used in management strategies.

2 Historical aspects and early studies

BBTD was first recorded in 1889, from Fiji, but it was likely to have been present there for at least 10 years prior. Over the next 5 years, by 1895, plantation production was slashed by over 80% (Darnell-Smith, 1924; Magee, 1953) primarily due to BBTD. The disease appeared not to be endemic, but its origin was a mystery, due to the paucity of surveys and disease records around the world at that time. Other early records include Egypt (1901), Australia, and Sri Lanka (1913), and a similar disease in abaca (*Musa textilis*) in the Philippines (1910) (Magee, 1953). The disease had then not been reported from the Americas and most of Africa, supported by some surveys, nor from southeast Asia or China, but there with little if any documentary evidence (Magee, 1953).

The cause of BBTD was initially elusive, with fungi (Knowles and Jepson, 1912, cited in Magee, 1927), bacteria (Darnell-Smith, 1924), nematodes (Nowell, 1925 and Fahmy, 1924, cited in Magee, 1927), soil, and nutritional factors (Darnell-Smith, 1924; Darnell-Smith and Tryon, 1923) all considered as possible causal agents. Eventually, through a thorough and elegant study, Magee (1927) showed that BBTD was an aphid-transmitted virus disease, and based on knowledge of the modes of transmission, he devised an effective control program that was implemented in Australia and led to the restoration of the Australian industry which had collapsed in affected areas after the introduction of BBTD in 1913 (Magee, 1936).

3 Symptoms

The symptoms of BBTD are distinctive and generally manifest as successively smaller, erect, and rather brittle leaves, with chlorotic, upturned margins, giving the infected plant a bunched or rosetted appearance (Fig. 1a and b). Symptoms first appear on the second, or later, leaf produced after aphid inoculation and do not develop on leaves formed prior to the first symptomatic leaf. In many

Figure 1 Symptoms caused by BBTV in banana (a) Infected suckers showing marginal leaf chlorosis, stunting, and rosetting in a Cavendish cultivar (AAA); (b) cv. Ladyfinger plants (AAB) showing marginal chlorosis, a bunched leaf aspect, and leaf upcurling; (c) dark green streaks and flecks on a leaf petiole (Cavendish banana); (d) dark green dots and dashes on the minor leaf veins, forming 'hooks' as they enter the midrib (Cavendish banana).

cultivars, including those of the Cavendish subgroup (AAA), the first symptoms are evident on close inspection as dark green streaks on the leaf petiole (Fig. 1c) and dark green streaks or dots on the minor veins, often on the basal area of the leaf lamina, the streaks forming 'hooks' as they enter the pale midrib (Fig. 1d). These leaf symptoms are best seen from the underside of the leaf in transmitted light. Plants infected late in the cropping cycle may produce a small bunch with distorted fruit (Fig. 2a and b), or if a very late infection, just dark green streaks on the tips of the bracts (Fig. 2c). However, most infected plants will not produce a bunch in the current or subsequent years and will become moribund in advanced stages of the disease (Magee, 1927). Plants produced from infected planting material display symptoms in all leaves, are very stunted, and do not produce fruit (Magee, 1927).

Strains of BBTV exist which cause differing degrees of symptom expression. Mild strains, producing only very limited leaf streaking, and symptomless strains have both been reported from Taiwan.

Figure 2 (a, b) Small, malformed bunches and (c) dark green streaks formed at the tip of a flower bract, formed on Cavendish cultivar plants infected late in the cropping cycle.

Although the symptoms caused by BBTV are very characteristic, they are mimicked by the related abaca bunchy top virus (ABTV; Sharman et al., 2008), which has been recorded from the Albay province in the Philippines and Sarawak (Malaysia). ABTV and BBTV can each infect both banana and abaca (*M. textilis*). Although few specific surveys have been done, the distribution of ABTV appears to be limited. Mixed infections of the two viruses can occur (Galvez et al., 2020) and recombination between ABTV and BBTV has been identified (Stainton et al., 2012).

4 Properties of banana bunchy top virus

BBTD is caused by members of the species BBTV (genus *Babuvirus,* family Nanoviridae), which have a circular single-stranded multicomponent DNA genome. The six genome components are each packaged in a separate 17-19 nm diameter isometric virus particle (Fig. 3), equal to the smallest virus particles from any organism. The genome components are each 1.0-1.1 kb in length, with the same basic structure, and with the exception of DNA-R, an open reading frame (ORF) encoding a single protein. DNA-R codes for the master replication initiation protein (Rep), which initiates replication of all six components, but also contains a shorter, internal ORF of unknown function. The other DNA components and the proteins they encode are DNA-C (Clink), DNA-M (movement protein), DNA-N (nuclear shuttle protein), DNA-S (capsid protein), and DNA-U3 (unknown function and potentially expressed in some BBTV isolates only) (Thomas, 2021). Replication is via a rolling circle mechanism,

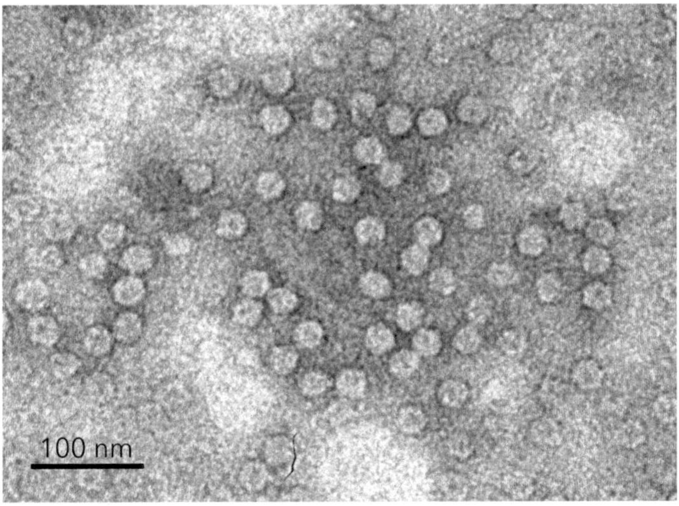

Figure 3 Purified virions of BBTV negatively contrasted with ammonium molybdate. Bar represents 100 nm (courtesy K. Crew and M. Vance).

and all components share two conserved common regions (major common region and stem-loop common region). The latter contains an origin of replication and a nonanucleotide sequence (TATTATT/AC) which is conserved among nanovirids and geminivirids (Bell et al., 2002).

In addition to these genomic components, some BBTV isolates contain additional 1.0-1.1 kb self-replicating components called alphasatellites. These alphasatellites are dependent on BBTV for their movement within the plant, encapsidation, and transmission, but their function, if any, is not known and they appear not to affect symptom expression (Bell et al., 2002; Briddon et al., 2018). They are phylogenetically distinct and are more closely related to the alphasatellites of members of the genus *Nanovirus* than to the DNA-R molecules of BBTV (Gronenborn et al., 2018). Further details of the biochemical and molecular properties of BBTV can be found in Thomas (2019).

Two broad groups of BBTV isolates have been recognized, based on the phylogenetic analysis of the nucleotide sequences of all six genome components (Stainton et al., 2012; Karan et al., 1994). These groups have been designated the 'South Pacific' and 'Asian' groups (Karan et al., 1994) or more recently the 'Pacific-Indian Oceans' (IPO) and the 'South-East Asian' (SEA) groups, respectively (Kumar et al., 2015). Despite these sequence differences, and the preponderance of alphasatellites in the Asian/SEA group, no obvious biological differences are apparent.

Suppression of RNA silencing in BBTV infections has been reported. Niu et al. (2009) concluded that both the movement protein and the coat protein are silencing suppressers (the movement protein being the stronger), acting at different positions in the silencing pathway. However, Amin et al. (2011) concluded that the movement protein and the cell cycle link protein are suppressors of RNA silencing. Both groups have identified the movement protein as a major pathogenicity determinant and suppressor of RNA silencing, though differing activities have been identified for the other BBTV proteins. Although the experimental systems used by both groups were very similar, they used BBTV isolates from different groups (SEA – Niu et al., 2009; IPO – Amin et al., 2011). There is evidence that the two groups have followed separate evolutionary paths (Hu et al., 2007), and it is possible that they have also evolved somewhat different strategies for RNA silencing (Amin et al., 2011).

5 Transmission

BBTV is transmitted through infected vegetative planting material, including tissue culture plantlets, and by an aphid vector, but not by mechanical transmission or through the soil (Magee, 1927; Drew et al., 1989).

In 1925, the banana aphid, *Pentalonia nigronervosa* (Fig. 4), was demonstrated to be the vector of BBTV (Magee, 1927). This aphid species was first described on banana by Coquerel (1859). A similar aphid was independently described from *Caladium* and named *P. caladii* (van der Goot, 1917). Subsequently, these species were considered synonymous (Hardy, 1931) or forms of the same species (*f. typica* and *f. cladii*; Eastop, 1966). More recently, it has been demonstrated, using morphological, biological, and molecular methods, that *P. nigronervosa* and *P. caladii* should indeed be regarded as distinct species (Bhadra and Agarwala, 2012; Bhadra and Agarwala, 2010; Foottit et al., 2010). The host ranges of the two species overlap, though the former has a host preference for *Musa* spp. and the latter for other species in the order Zingiberales and the family Aracaeae. Foottit (2010) reported *P. nigronervosa* from *Musa* and *Heliconia* and *P. caladii* from *Caladium, Colocasia, Cucurma, Cyrtosperma, Alpinia, Hedychium, Musa,* and *Zingiber*. The feeding and breeding preferences of the two aphid species among banana cultivars and other hosts are also variable (Rahma et al., 2021; Suparman et al., 2017; Ngatat et al., 2017; Suparman and Setyawati, 2011; Bhadra and Agarwala, 2010).

Although both aphid species have been shown to be vectors of BBTV (Greenwell and Bressan, 2011), most research of BBTV vector transmission has

Figure 4 A heavy infestation of banana aphids (*Pentalonia nigronervosa*) on a young banana leaf. Inset: a single banana aphid.

not distinguished between these two species. In almost all cases, though, the aphids have been taken from banana, where the vast majority of aphids are *P. nigronervosa*. Early research (Magee, 1940) showed that the transmission of BBTV was of the circulative, non-propagative mode, with extended acquisition and inoculation times required, a latent period, and retention of infectivity by aphids for nearly two weeks, including continuing infectivity after moults. This and subsequent works have demonstrated the following transmission parameters (Magee, 1940; Hu et al., 1996; Anhalt and Almeida, 2008; Jebakumar et al., 2018): minimum acquisition access period 45 min–17 h; minimum inoculation access period 15 min–1.5 h; retention of infectivity after removal from virus source 13–20 days. A minimum latent period of at least 'a few hours to 2 days' was recognized by Magee (1940) and reported as a minimum of 20 h by Anhalt and Almeida (2008). All life stages of the aphid are capable of transmission and reported efficiencies for single aphids are in the range 10–67% (Hu et al., 1996; Magee, 1927; Thiribhuvanamala et al., 2001; Wu and Su, 1990a). The virus neither replicates in the vector (Hafner et al., 1995; Watanabe and Bressan, 2013) nor is transmitted to the parthenogenetic offspring (Magee, 1940; Watanabe and Bressan, 2013). The growth rate of *P. nigronervosa* was highest at 25°C (Robson et al., 2007a), and similarly, virus transmission rates were higher at 25-30°C, than at 20°C, and decreased or ceased at 10–16°C (Anhalt and Almeida, 2008; Magee, 1940; Wu and Su, 1990a).

A recent report suggests that *Hysteroneura setariae* (rusty plum aphid), usually found on Poaceae, may also transmit BBTV (Sila et al., 2019). The aphids *Myzus persicae*, *Aphis gossypii*, *Rhopalosiphum padi*, and the whitefly *Bemisia tabaci* have been shown not to transmit BBTV (Yasmin et al., 2001).

BBTV is also readily transmitted via vegetative propagation; either using conventional bits and suckers and all meristematic eyes or suckers in an infected mat will ultimately become infected (Magee, 1927). It is also transmitted through tissue culture (Ramos and Zamora, 1990; Wu and Su, 1991; Drew et al., 1989), though transmission rates of the virus to plantlets are not always 100% (Thomas et al., 1995; Drew et al., 1989).

6 Host range and cultivar susceptibility

The progenitors of edible diploid and triploid bananas are the wild, seeded diploid species in the Eumusa section, *M. acuminata* ssp. alone (contributing the A genome) or in hybrid combination with *M. balbisiana* (contributing the B genome), with an occasional small contribution from *M. schizocarpa* (S genome). Species in the Australimusa section (T genome) also play a minor role (Carreel, 1994; D'Hont et al., 2000).

Most, if not all, edible banana cultivars are susceptible to BBTV, though differing in the degrees of ease of infection and intensity of symptoms (Ngatat

Table 1 The known host range of banana bunchy top virus

Host Order and Family	Host Species	Reference
Zingiberales		
Musaceae	*Musa acuminata* ssp. *banksii*	(Magee, 1927)
	M. acuminata ssp. *halabanensis*	(Rahayuniati et al., 2021a)
	M. acuminata ssp. *longipetiolata*	(Rahayuniati et al., 2021a)
	M. acuminata ssp. *malaccensis*	(Rahayuniati et al., 2021a)
	M. acuminata ssp. *siamea*	J. Thomas, M. Vance, K. Crew, unpublished
	M. acuminata ssp. *sumatrana*[a]	(Rahayuniati et al., 2021a)
	M. acuminata ssp. *burmannica*	(Thomas, 2019)
	M. acuminata ssp. *zebrina*	(Thomas, 2019)
	M. balbisiana	(Espino et al., 1993)
	M. beccarii	(Sachter-Smith, 2015)
	M. coccinea	(Thomas, 2019)
	M. itinerans var. itinerans	(Sachter-Smith, 2015)
	M. jackeyi	(Thomas, 2019)
	M. lateritia	(Sachter-Smith, 2015)
	M. lutea	(Sachter-Smith, 2015)
	M. ornate	(Thomas, 2019)
	M. textilis	(Magee, 1927)
	M. thomsonii	(Sachter-Smith, 2015)
	M. troglodytarum	(Dubousquet et al., 2018)
	M. velutina[a]	(Thomas and Dietzgen, 1991)
	Ensete superbum	(Selvarajan and Balasubramanian, 2013)
	E. ventricosum	(Magee, 1927)
Zingiberaceae	*Alpinia galangal*[a]	(Rahayuniati et al., 2021b)
	A. purpurata[a]	(Dubousquet et al., 2018)
	A. zerumbet	(Pinili et al., 2013)
	Cucurma longa[a]	(Rahayuniati et al., 2021b)
	Hedychium coronarium	(Su et al., 1993)
	Kaempferia galanga	(Rahayuniati et al., 2021b)
	Zingiber officinale	(Rahayuniati et al., 2021b)
Cannaceae	*Canna indica*	(Su et al., 1993; Pinili et al., 2013)
Heliconiaceae	*Heliconia aurantiaca*	(Hamim et al., 2017)
Alismetales		
Araceae	*Colocasia esculenta*	(Pinili et al., 2013)

[a] Natural infection has been recorded.

et al., 2017; Espino et al., 1993; Mware, 2016; Niyongere et al., 2011; Sachter-Smith, 2015). Cultivars in the Cavendish subgroup (AAA) are among the most susceptible and severely affected.

All wild and ornamental *Musa* species tested are also susceptible (Table 1). The related genus *Ensete* is also susceptible: *E. superbum* (Selvarajan and Balasubramanian, 2013), *E. ventricosum* (Magee, 1927). The record of BBTV from *M. acuminata* ssp. *sumatrana* appears to be the first report of BBTV from a wild banana species in its natural habitat (Rahayuniati et al., 2021a).

BBTV was initially thought to be confined to hosts in the genera *Musa* and *Ensete* (Dale, 1987; Magee, 1927), but recent research has provided unequivocal evidence that alternative hosts outside the Musaceae do exist. Infection by aphid inoculation and confirmation of infection by molecular assays or back inoculation to banana has been demonstrated for the hosts noted in Table 1. *A. purpurata* (Dubousquet et al., 2018) and *A. galangal* and *Cucurma longa* (Rahayuniati et al., 2021b) were recorded as natural field infections. Conflicting results in past and current research where these hosts were not confirmed (Rahayuniati et al., 2021b; Hu et al., 1996; Magee, 1927; Geering and Thomas, 1997) may be due to the use of virus isolates from different subgroups or differing feeding preferences and transmission characteristics of the two aphid vector species, *P. nigronervosa* and *P. caladii*. The need to distinguish between the two vector species has only recently become apparent (Foottit et al., 2010) and consequently the precise vector species used had not been defined in many of the earlier experiments.

The significance of these alternative hosts of BBTV and *P. nigronervosa*, and the role of *P. caladii* in the transmission of BBTV in the natural ecosystem, is not known and is an important area that requires investigation.

7 Origin of the pathosystem, distribution and international spread

7.1 Possible origin of banana bunchy top disease

Although BBTD was first recognized in Fiji, and was soon afterward found in Egypt, Sri Lanka, and Australia, it is apparent that the disease did not originate in these countries, but rather was imported in infected planting material (Magee, 1953).

Several lines of evidence point to an origin for BBTV in south-east Asia. The major vector of BBTV, *P. nigronervosa* is thought to have originated in south-east Asia. Although all *Musa* species and cultivars appear to be susceptible to BBTV, *M. balbisiana* and cultivars containing the B genome appear in general to be more field tolerant and less easily infected with the virus (Thomas, 2019) as would be expected for a species that has evolved in close association with a pathogen. The center of origin for *M. balbisiana* is thought to be the more northerly areas of south-east Asia from eastern India, through Bangladesh, the northern parts of Myanmar, Thailand, Laos and Vietnam, and southern China (Perrier et al., 2011). The maximum sequence diversity for BBTV is found in Vietnam and southern China (Rao et al., 2017; Islam et al., 2010; Bell et al., 2002; Hu et al., 2007; Stainton et al., 2015). Although the IPO group of BBTV isolates has not been recorded from south-east Asia, it has been shown through recombination and reassortment analysis that they must have co-existed at some point (Stainton et al., 2012). Also, mild or symptomless isolates, lacking DNA-N, and an example of reassortment between SEA and IPO groups of BBTV have both been found in Taiwan (Hu et al., 2007). Alphasatellites are more frequently found associated with Asian isolates of BBTV, especially from Vietnam and Taiwan (Bell et al., 2002; Horser et al., 2001), and it has been noted that the nucleotide sequences of their major common regions (CR-Ms) are phylogenetically closer to DNA-R of the SEA rather than the IPO group, implying that the Asian isolates may be older (Hu et al., 2007). Finally, the other two known members of the genus *Babuvirus*, ABTV (Sharman et al., 2008) and cardamom bushy dwarf virus (CBDV; Mandal et al., 2013), have also been described from this region and share the banana aphid vector.

7.2 Distribution and international spread of banana bunchy top virus

BBTV now has a wide and expanding geographic distribution, including countries in Asia, sub-Saharan Africa, and Oceania (Thomas, 2019), but at this stage is absent from the Americas. It has most recently been detected for the first time in Tanzania (Mpoki et al., 2022) and Uganda (Ocimati et al., 2021). Detailed molecular analysis suggests that the current distribution of BBTV has resulted from a limited number of introductions of infected planting material, or infective aphids, mostly from the diversity hotspots in India and south-east Asia (Stainton et al., 2015). The current global distribution could be accounted for by as few as 14 movements between eight pairs of location within the last 1000 years, but primarily during the past 300 years. South-east Asia was the most probable virus origin, but the Indian sub-continent was the 'most connected location'. Phylogenetic clustering of isolates suggested limited introductions to individual countries or regions (Stainton et al., 2015; Kumar et al., 2011).

8 Detection

There are a variety of diagnostic assays available for BBTV, many commonly used and based on conventional strategies, and some more novel in their approach and yet to be widely applied.

Polyclonal and monoclonal antibodies have been employed in ELISA to detect BBTV since the 1990s, and provide robust, reliable assays, especially in situations where limited laboratory facilities are available. These assays have sufficient sensitivity to detect BBTV in field plants, including from pre-symptomatic tissues, tissue culture plantlets, and individual infective aphids (Wu and Su, 1990b, K. Crew, M. Vance, N. Tran, J. Thomas, unpublished; Dietzgen and Thomas, 1991; Thomas and Dietzgen, 1991; Geering and Thomas, 1996). The polyclonal antibodies and some individual monoclonal antibodies cross-react with the other viruses from the genus *Babuvirus*, ABTV (Sharman et al., 2008), and CBDV (Mandal, 2010), but not with viruses in the related genus *Nanovirus* (Katul et al., 1993). Polyclonal antiserum against expressed coat protein of BBTV has been prepared and used in a gold-nano particle-based dot immunobinding assay to detect the virus in field samples (Majumder et al., 2013). Single-chain variable fragment (scFv) monoclonal antibodies have been prepared against expressed coat protein of BBTV, and in plate-trapped antigen ELISA, and were able to detect the virus from field-infected banana samples (Shilpa, 2013). These scFv antibodies were also used in dot immunobinding assays on polyvinylidene Fluoride (PVDF) membranes to detect purified BBTV coat protein (Namratha et al., 2018). Another antibody-based detection assay developed for BBTV is impedimetric detection using CdSe quantum dots (Majumder et al., 2020).

A variety of nucleic acid-based assays have been developed for BBTV. PCR (Xie and Hu, 1995; Vikram et al., 2020; Mansoor et al., 2005) and real-time PCR (Chen and Hu, 2013; Watanabe et al., 2013) were shown to have significantly greater sensitivity than ELISA. Extracts from banana can have inhibitory effects in PCR, but these can be avoided using immunocapture PCR (Sharman et al., 2000) or simplified extraction procedures (Thompson and Dietzgen, 1995; Selvarajan et al., 2015). Lateral flow devices have been developed to detect PCR products, with a reported sensitivity ten times greater than detection by gel (Wei et al., 2014). Other nucleic acid-based detection assays for BBTV detection include loop-mediated isothermal amplification (LAMP) (Peng et al., 2012; Zhang et al., 2018) and recombinase polymerase amplification (Kapoor et al., 2017).

9 Epidemiology and modeling

Natural spread of BBTV at a local level is predominantly via its main aphid vector, *P. nigronervosa*. This aphid has a worldwide distribution and occurs wherever bananas are grown. It prefers a relatively moderate optimum temperature of 25°C for its lifecycle (Robson et al., 2007a) and is generally found in protected parts of the banana plant, including the base or throat of the pseudostem, under leaf sheaths, and under favorable conditions on the young leaves of emerging suckers (Robson et al., 2006; Johnson, 1963; Kolkaida, 1954). Aphid numbers are adversely affected by high rainfall but prosper during the warmer months, especially so in sub-tropical areas and at higher altitudes (Niyongere et al., 2013; Magee, 1927). Aphid numbers are promoted in poorly managed plantings where de-leafing and de-suckering are neglected, leaving increased protected habitats (Niyongere et al., 2013).

The rate of spread of BBTD is faster in the warmer seasons (Varghese et al., 2020; Allen, 1978a), partly due to more favorable conditions for the development of the aphid vector populations and their efficiency of virus transmission, and partly due to the increased growth rate of the banana plants. Symptoms only appear on the second, or later, leaf produced after infection (Magee, 1927; Niyongere et al., 2013; Allen, 1978a), and Magee (1927) was only able to transmit BBTV from symptomatic leaves. In sub-tropical Australia, plants can produce two new leaves in 19 days in summer but can take 125 days in winter (Allen, 1978a).

Initial epidemiological studies in the sub-tropical growing region of eastern Australia assumed the presence of a single susceptible host (*Musa* spp.) and a single aphid vector species (*P. nigronervosa*) (Magee, 1927; Allen, 1978a,b; Allen, 1987). Detailed modeling studies on individual disease outbreaks in northern NSW highlighted the relatively short distances attributed to aphid vector transmission. The average distance of spread from the predicted source of infection was 15.5 m, with about two thirds of new infections occurring within 20 m and 99% within 86 m (Allen, 1987, 1978b). The chance of infection in a new plantation within the first year was 88% if established alongside a BBTV-infected planting, but only 27% if separated by 50–1000m and less than 5% if more than 1000m apart (Allen and Barnier, 1977). Later studies in the Philippines (Opina and Milloren, 1996; Smith et al., 1998) and Burundi (Niyongere et al., 2013) have also supported this predominantly restricted spread by aphid vectors. Allen (1987) considered the incubation period of the disease (infection to first symptoms) to be the time taken to produce two new leaves. He estimated the latent period of the disease (incubation period of the disease plus the time taken for the next infection to occur from this source plant) to be the average the time taken for 3.7 leaves to be produced since infection. This is the average time, derived from observations of actual epidemics; though theoretically, the minimum latent period could be less than this and be affected by the size of aphid vector populations. This period includes the time taken for symptoms to develop in the infected plant, plus the time taken for the aphid to acquire an infective dose of the virus, then move to another plant. This period is seasonally dependent and averaged 57.8 days in sub-tropical Australia. A stochastic, spatio-temporal computer simulation was developed based on these studies and used to evaluate control measures (Allen, 1987). The disease control program is now based on an inspection interval of four weeks in BBTV-infected plantings, during which time no more than four new leaves should have been produced, and the risk of re-transmission from infected plants minimized.

More recently, further models have been developed, using recent inspection and eradication data from NSW and Queensland in Australia (Thomas, 2018: ; C. A. Gilligan, pers. comm.; Varghese et al., 2020). Researchers from the University of Cambridge developed a stochastic, spatially explicit, epidemiological compartmental model to describe the spread of BBTV at within-farm level and a mechanistic, spatio-temporal framework for modeling individual-to-individual transmission (contact model) at the landscape scale. Inference on the model parameters was performed using Bayesian computation methods involving Markov chain Monte Carlo (MCMC) coupled with data augmentation techniques. Input data included the GPS coordinates of each infected plant and its time of infection. This study showed that the current control program was based on sound epidemiological principles, and that relaxation of any of the components (inspection frequency or efficiency, eradication efficiency) would result in loss of control and eventual rapid spread of the disease. Varghese et al. (2020) also developed a stochastic network-based model using Bayesian computation methods with MCMC, based on sub-sections of a plantation rather than individual plant data. This model indicated seasonality in all parameters and could predict specific high-risk areas of the plantation warranting more frequently and focused inspections.

The epidemiological significance of the recent recognition of at least one more vector species (*P. caladii*) and additional alternative hosts (Section 5) is uncertain, and most of these studies have been done under artificial, experimental conditions. However, the widespread, natural occurrence of BBTV in the alternative host *Alpinia purpurata* has been reported (Wong et al., 2019).

10 Resistance

As noted in Section 6, banana cultivars and wild germplasm are known to differ widely in their reaction to BBTV, both in degree of symptom expression and ease of infection by aphid inoculation. For example, the cultivar Gros Michel (AAA) exhibits less severe symptoms and lower levels of field infection than many other cultivars (Niyongere et al., 2011; Ngatat et al., 2017; Magee, 1948: A. D. W. Geering and J. E. Thomas, unpublished). A similar situation has been reported for *M. acuminata* ssp. *malaccensis* (AA; Sachter-Smith, 2015). It has frequently been observed that cultivars which show less severe symptoms or lower levels of field infection often contain a B genome component (Ngatat et al., 2017, 2022; Niyongere et al., 2011; Espino et al., 1993), though this situation is neither consistent nor exclusive. Definitive evidence of immunity to BBTV has been elusive. However, recent studies in the Philippines have shown that many accessions of the wild seeded banana *M. balbisiana* were not able to be infected after repeated aphid inoculation under controlled conditions in the screenhouse followed by field exposure to high intensity inoculum pressure for up to four years (Dela Cueva et al., 2023), suggesting immunity to infection derived from the B genome. Similarly, the difficulty in infecting cv. Gros Michel (AAA) and *M. acuminata* ssp. *malaccensis* suggests that resistance mechanisms may be involved in the A genome as well as the B genome.

There are many reports of 'resistance' in the literature but a caveat here is that these experiments have sometimes been based on field observations under natural infection and/or were based on visual observations for symptoms only (Mshani et al., 2008; Espino et al., 1993). The genetic basis for these differences in susceptibility is unknown. The virus titre in source plants can influence the efficiency of infection of plants by aphids (Jebakumar et al., 2018), and some cultivars reported as resistant to BBTV have been shown to be susceptible when increased numbers of aphids were used for inoculation (Thomas, 2019). Interestingly, some banana cultivars, including Veimama (AAA, Cavendish subgroup) have been observed, very infrequently, to recover from typical symptoms and to become symptomless (Magee, 1948; K. Crew and J. Thomas, unpublished).

An alternative to natural sources of resistance in the genus *Musa* has been sought through a number of transgenic approaches.

For example, the expression of defective Rep protein (encoded by DNA-R) including mutant, antisense, partial Rep genes and inverted repeat of partial Rep gene, and theorized to out-compete native virus Rep, have shown promise in glasshouse studies (Borth et al., 2011) but appear not to have achieved field application.

In glasshouse trials in India, the RNAi strategy using intron hairpin RNA transcripts based on BBTV Rep coding sequence was used with banana cv. Rasthali (AAA) (Shekhawat et al., 2012) and cv. Virupakshi (AAB) (Elayabalan et al., 2013). Transformed plants were reported to be resistant after aphid inoculation with BBTV. However, there have been no reports of field trials with these transformants. A similar strategy was used by Mware (2016) in Australia with Cavendish (AAA) with DNA-R or DNA-N, but without success. However, the same strategy using DNA-M (movement protein) RNA transcripts was successful, with 85% of inoculated plants symptomless and with no detectable virus after 10 months. Although successful in the glasshouse, field trials with these lines in Malawi were less successful, experiencing field infection rates of over 40% (Kato, 2019). A novel strategy based on the INPACT system (Dugdale et al., 2014) has been developed for BBTV (Kato, 2019). This system involves transformation of the plant with a cassette containing a split 'suicide gene' (barnase, a cytotoxic RNase from *Bacillus amyloliquefaciens*) with the 3′ and 5′ segments inverted. These are flanked by introns, followed by the BBTV origin of replication. The system is activated upon infection by BBTV, whereupon the Rep nicks the cassette at the origins of replication, allowing release and circularization of the suicide gene. With processing of the transcript, the introns and origins of replication are removed and the now active barnase gene translated. The presence of the barnase results in virus-activated cell death, thereby halting virus replication. Alone, this strategy was ineffective for the control of BBTV in glasshouse studies, but when combined as a 'dual cassette' with a hairpin construct targeting DNA-M, 7/9 transformed banana lines displayed no symptoms when challenged with BBTV. When indexed by PCR, three of these lines produced strong amplicons, but amplicons for the remaining four were only faintly visible (Kato, 2019). These constructs are yet to be field tested.

11 Management strategies

The use of insecticides to manage BBTD through control of the aphid vector has met with limited success. *P. nigronervosa* is often present in inaccessible parts of the plant, such as between the leaf sheaths on the pseudostem and on the base of the pseudostem up to several centimeters below the soil surface (Magee, 1927). This makes the use of contact treatments problematic. Foliar applications of the insecticide imidacloprid did not provide fully systemic coverage, leaving newly emerged leaves unprotected (Robson et al., 2007b). Likewise, pseudostem injections of imidacloprid were not fully effective at killing aphids on the youngest expanded leaf (K. Crew and J. Thomas, unpublished).

Biopriming and biohardening of plants prior to BBTV infection with formulations of endophytic (*Bacillus subtilis, B. pumilus*) and rhizospheric (*Pseudomonas fluorescens*) bacteria has shown some promise for BBTD control in India (Jebakumar and Selvarajan, 2018; Kavino et al., 2008, 2007; Harish et al., 2008, 2009). These studies used cvs Grand Nain (AAA), Robusta (AAA), or Virupakshi (AAB) and demonstrated lower rates of infection and delayed symptom expression in glasshouse and field studies with application of the formulations prior to virus inoculation. However, despite the apparent benefit of this approach, widespread adoption by farmers does not appear to have occurred.

In the Philippines, gamma irradiation has been attempted on tissue culture plantlets banana cv Lakatan (AAA, syn. 'Pisang Berangan') (Damasco et al., 2006) and abaca (Dizon et al., 2012) with the aim of developing BBTV resistance. Additionally, somaclonal variants, which often occur in tissue culture, were evaluated for BBTV resistance and one putatively resistant line was selected for further evaluation (Damasco et al., 2008). However, there have been no further reports of this work to enable their potential to be evaluated.

As noted earlier, immunity to BBTV in edible bananas is not known, and differences among cultivars in tolerance or resistance to infection can be difficult to exploit. Potentially useful cultivars might not be adaptable to local environmental conditions or meet consumer acceptance. Thus, control has traditionally been based on cultural practices using local cultivars.

Seminal work on BBTD was conducted by Magee in the early 1920s (Magee, 1927), and these studies included a strategy for rehabilitation of the devastated Australian banana industry through cultural control practices. These strategies included two major principles:

- Exclusion of the disease from unaffected and lightly affected areas; and
- Eradication of infected plants from both lightly and heavily-affected areas.

Control measures were enforced under State legislation in the affected States of New South Wales and Queensland, and included:

- Registration of all commercial banana plantings;
- Establishment of quarantine zones;
- Regular inspections of all banana plantings, commercial and non-commercial, by trained inspectors;
- Immediate destruction of infected plants;
- Restrictions on, and permits for, supply, movement, and planting of suckers;
- Use of disease-free planting material; and
- Extension and education program to promote grower awareness.

These measures enabled the rehabilitation of the Australian industry and with some adjustments still form the basis of disease control today. In the 1930s, overproduction and complacency allowed a resurgence of BBTD, but this was brought under control again by refocusing on the program (Magee, 1936). Total eradication has only been achieved in a couple of districts, but to this day BBTD remains at a low, manageable level across the industry. Across the NSW sub-tropical industry, the average incidence from 2010 to 2013 was ca 0.07% (Pekin, 2015). The Australian industry has now almost completely relocated from the affected sub-tropical areas to tropical north Queensland, but through the ongoing implementation of these control measures BBTD has remained confined to the subtropics providing growers in the tropics with substantial benefit from absence of the disease (Cook et al., 2012).

Feral bananas and plants growing in smallholdings and backyards can be significant sources of infection. The work of Smith et al. (1998) in Mindanao, the Philippines, showed that despite regular intensive rogueing in commercial plantations, clear gradients of infection were present originating from nearby affected smallholdings and providing an ongoing source of reinfection.

Despite the success in Australia, coordinated control programs across regions or countries have generally been difficult to instigate and maintain, especially without enforced legislative support. However, a simplified strategy was tested in several African countries (Benin, Burundi, Malawi), based on long-term consistent rogueing and the use of low-risk replanting material (Omondi et al., 2020). In Benin and Burundi, trials started in regions with relatively high levels of infection. Over a two-year period, BBTD incidence was rapidly reduced from an initial monthly detection rate of 20–50% in the managed fields to ongoing rates of 5–10%. In a high BBTD incidence area in Malawi, banana-free buffer zones of at least 100m surrounded fields where rogueing was implemented and healthy planting material was used. Here, monthly disease incidence was kept below 5% over a three-year period. In contrast, control plots planted adjacent to BBTD-infected bananas and not subject to control showed a cumulative disease rate of 90% after three years. For replanting, these experiments used symptomless planting material sourced from within the fields. With the relatively low levels of field infection, the risk of inadvertent selection of latent, symptomlessly infected material was reduced. Thus, with a concerted effort a sustainable level of banana production was maintained in the face of significant levels of BBTD in the landscape.

Critical in these programs of disease control is the continued education through extension programs of farmers and agricultural specialists in BBTD awareness (Omondi et al., 2020; Abiola et al., 2020). Such programs are only successful when there is a high level of buy-in, compliance, and technical support, but many factors can work against success. Ongoing commitment and co-operation across whole communities can often be difficult to achieve, sometimes resulting in partial implementation of the control strategy with compromised effectiveness. By-passing the clean planting material step by using untested material, for reasons of expediency or profit, can also occur. Nevertheless, disease mitigation and recovery of production in the face of BBTD epidemics is possible, across a range of resource inputs, as long as control strategies are well-considered and there is ongoing commitment.

12 Future trends in research

The outstanding need in addressing the impact of BBTD is to secure durable resistance in banana cultivars that meet the requirements of both large-scale commercial producers and smallholders. Promising advances have been made in the areas of both the search for natural resistance and in the application of RNAi and other transgenic approaches. There is currently no information reported on the genetic basis of observed natural resistance, neither the number of independent sources nor the number or characteristics of the allele(s) involved. Despite these promising observations on natural resistance, there have been limited collecting missions in the center of origin of *Musa* in south-east Asia, with the aim of screening wild germplasm. These efforts need to continue, especially as this germplasm is under threat due to disease pressure, the clearing of natural habitats for development, and the effect of climate change on ecological niches. Although promising results have been obtained for RNAi-based resistance in glasshouse studies (Mware, 2016), these results have yet to be convincingly transferred to the field (Kato, 2019). The strategy to combine RNAi-based resistance with INPACT technology warrants further investigation under field conditions. It is surprising though that the level of resistance demonstrated for INPACT + RNAi lines with DNA-M hairpin constructs (Kato, 2019) was less than that observed for RNAi lines with DNA-M hairpin constructs alone (Mware, 2016). Research across the broad area of resistance is the subject of ongoing research, and needs continued focused support.

The epidemiology of BBTV appears to be more complicated than originally assumed. However, it remains to be seen how important non-*Musa* hosts and aphid vectors other than *P. nigronervosa* are in natural ecosystems, as opposed to glasshouse studies. Field studies in this area are urgently needed. This could be the key to failure to be able to eradicate BBTV from production areas, despite reduction of disease incidence to very low levels.

13 Where to look for further information

13.1 Further reading

- Alliance Bioversity CIAT website; https://www.promusa.org/Bunchy+top.
- Alliance for Banana Bunchy Top Disease Control in Africa; https://www.bbtvalliance.org/.
- Thomas, J. E., Gronenborn, B., Harding, R. M., Mandal, B., Grigoras, I., Randles, J. W., Sano, Y., Timchenko, T., Vetten, H. J., Yeh, H.-H., Ziebell, H. and ICTV Report Consortium. (2021). ICTV virus taxonomy profile: Nanoviridae. *Journal of General Virology* 102:1544.
- Thomas, J. E. (2015). *MusaNet Technical Guidelines for the Safe Movement of Musa Germplasm*. 3rd edn. Bioversity International, Rome, Italy.

14 References

Abiola, A., Zandjanakou-Tachin, M., Aoudji, K. N. A., Avocevou-Ayisso, C. and Kumar, P. L. (2020). Adoption of roguing to contain banana bunchy top disease in South-East Bénin: Role of farmers' knowledge and perception. *International Journal of Fruit Science* 20(4): 720–736.

Allen, R. N. (1978a). Epidemiological factors influencing the success of roguing for the control of bunchy top disease of bananas in New South Wales. *Australian Journal of Agricultural Research* 29(3): 535–544.

Allen, R. N. (1978b). Spread of bunchy top disease in established banana plantations. *Australian Journal of Agricultural Research* 29(6): 1223–1233.

Allen, R. N. (1987). Further studies on epidemiological factors influencing control of banana bunchy top disease, and evaluation of control measures by computer simulation. *Australian Journal of Agricultural Research* 38(2): 373–382.

Allen, R. N. and Barnier, N. C. (1977). *The Spread of Bunchy Top Disease Between Banana Plantations in the Tweed River District During 1975–76*, 27–28.

Amin, I., Ilyas, M., Qazi, J., Bashir, R., Yadav, J. S., Mansoor, S., Fauquet, C. M. and Briddon, R. W. (2011). Identification of a major pathogenicity determinant and suppressors of RNA silencing encoded by a South Pacific Isolate of Banana bunchy top virus originating from Pakistan. *Virus Genes* 42(2): 272–281.

Anhalt, M. D. and Almeida, R. P. (2008) Effect of temperature, vector life stage, and plant access period on transmission of banana bunchy top virus to banana. *Phytopathology* 98(6): 743–748.

Bell, K. E., Dale, J. L., Ha, C. V., Vu, M. T. and Revill, P. A. (2002). Characterisation of Rep-encoding components associated with banana bunchy top Nanovirus in Vietnam. *Archives of Virology* 147(4): 695–707.

Bhadra, P. and Agarwala, B. K. (2010). A comparison of fitness characters of two host plant-based congeneric species of the banana aphid, *Pentalonia nigronervosa* and *P. caladii*. *Journal of Insect Science* 10: 140.

Bhadra, P. and Agarwala, B. K. (2012). On the morphological and genotypic variations of two congeneric species of banana Aphid *Pentalonia* (Homoptera: Aphididae) from India. *Advances in Life Sciences* 2(3): 75–81.

Borth, W., Perez, E., Cheah, K., Chen, Y., Xie, W. S., Gaskill, D., Khalil, S., Sether, D., Melzer, M., Wang, M., Manshardt, R., Gonsalves, D. and Hu, J. S. (2011). Transgenic banana plants resistant to *Banana bunchy top virus* infection. *Acta Horticulturae* (897): 449–457.

Briddon, R. W., Martin, D. P., Roumagnac, P., Navas-Castillo, J., Fiallo-Olivé, E., Moriones, E., Lett, J. M., Zerbini, F. M. and Varsani, A. (2018). Alphasatellitidae: A new family with two subfamilies for the classification of geminivirus- and Nanovirus-associated alphasatellites. *Archives of Virology* 163(9): 2587–2600.

Burns, T. M., Harding, R. M. and Dale, J. L. (1995). The genome organization of banana bunchy top virus: Analysis of six ssDNA components. *Journal of General Virology* 76(6): 1471–1482.

Carreel, F. (1994). Etude de la diversite genetique des bananiers (genre Musa) a l'aide des marqueurs RFLP. *National Agronomique Paris-Grigon*.

Chen, Y. and Hu, X. P. (2013). High-throughput detection of banana bunchy top virus in banana plants and aphids using real-time TaqMan (R) PCR. *Journal of Virological Methods* 193(1): 177–183.

Cook, D. C., Liu, S., Edwards, J., Villalta, O. N., Aurambout, J. P., Kriticos, D. J., Drenth, A. and De Barro, P. J. (2012). Predicting the benefits of banana bunchy top virus exclusion from commercial plantations in Australia. *PLoS ONE* 7(8): e42391.

Coquerel, C. (1859). Note sur quelques insectes de Madagascar et de Bourbon. *Annales de la Société Entomologique de France (3ième séries)* 7: 239–260.

Dale, J. L. (1987). Banana bunchy top: An economically important tropical plant virus disease. *Advances in Virus Research* 33: 301–325.

Damasco, O. P., Dizon, T. O., Tayobong, R. R. P., Dela Cueva, F. S., Sison, M. L., Cruz, D. and Mendoza, E. M. (2008). Banana bunchy top virus resistant somaclonal variant of banana (Musa sp.) cv. lakatan. *Philippine Journal of Crop Science* (Supplement 1): 90–91.

Damasco, O. P., Estrella, J. B., Caymo, L. S., Dizon, T. O., Rabara, R. C., Cruz Jr, F. S. and Mendoza, E. M. T. (2006). Banana bunchy top virus (BBTV) resistance in cultivar "Lakatan" developed via gamma irradiation of shoot tips. *Phillipine Journal of Crop Science* 31(3): 21–34.

Darnell-Smith, G. P. (1924). "Bunchy top" disease in banana. *Queensland Agricultural Journal* 21: 169–179.

Darnell-Smith, G. P. and Tryon, H. (1923). Banana bunchy top disease. *Queensland Agricultural Journal* 19: 32–33.

Dela Cueva, F. M., Perez, N. A. M., Benjamin, A. F. A., Yanos, L. A., Gueco, L. S. and Thomas, J. (2023). Resistance of Musa balbisiana accessions of the Philippines to banana bunchy top virus. *Plant Disease*.

D'Hont, A., Paget-Goy, A., Escoute, J. and Carreel, F. (2000). The interspecific genome structure of cultivated banana, *Musa* spp. revealed by genomic DNA in situ hybridization. *Theoretical and Applied Genetics* 100(2): 177–183.

Dietzgen, R. G. and Thomas, J. E. (1991). Properties of virus-like particles associated with banana bunchy top disease in Hawaii, Indonesia and Tonga. *Australasian Plant Pathology* 20(4): 161–165.

Dizon, T. O., Damasco, O. P., Lobina, I. T., Pinili, M. S., Lalusin, A. G. and Natsuaki, K. T. (2012). Induction of putative resistant lines of abaca (Musa textilis nee) to banana bunchy top virus and banana bract mosaic virus through in vitro mutagenesis. *Journal of ISSAAS* 18(1): 87–99.

Drenth, A. and Kema, G. (2021). The vulnerability of bananas to globally emerging disease threats. *Phytopathology* 111(12): 2146–2161.

Drew, R. A., Moisander, J. A. and Smith, M. K. (1989). The transmission of banana bunchy-top virus in micropropagated bananas. *Plant Cell Tissue and Organ Culture* 16(3): 187–193.

Dubousquet, V., Massé, D. and Crew, K. (2018). Banana bunchy top virus in non-banana hosts in French Polynesia. In: *Banana Scientific Symposium*. Queensland Department of Agriculture and Fisheries, Cairns, Queensland.

Dugdale, B., Mortimer, C. L., Kato, M., James, T. A., Harding, R. M. and Dale, J. L. (2014). Design and construction of an in-plant activation cassette for transgene expression and recombinant protein production in plants. *Nature Protocols* 9(5): 1010–1027.

Eastop, V. (1966). A taxonomic study of Australian Aphidoidea (Homoptera). *Australian Journal of Zoology* 14(3): 399–592.

Elayabalan, S., Kalaiponmani, K., Subramaniam, S., Selvarajan, R., Panchanathan, R., Muthuvelayoutham, R., Kumar, K. K. and Balasubramanian, P. (2013). Development of Agrobacterium-mediated transformation of highly valued hill banana cultivar Virupakshi (AAB) for resistance to BBTV disease. *World Journal of Microbiology and Biotechnology* 29(4): 589–596.

Espino, R. C., Magnaye, L. V., Johns, A. P. and Juanillo, C. (1993). Evaluation of Philippine banana cultivars for resistance to bunchy-top and fusarium wilt. In: Valmayor, R. V., Hwang, S. C. and Ploetz, R. (Eds.), *Proceedings of the International Symposium on Recent Developments in Banana Cultivation Technology*, Chiuju, Pingtung, Taiwan, 14–18 December 1992. Los Banos, Laguna, Philippines: INIBAP/ASPNET, pp. 89–102.

Fahmy, T. (1924). Plant diseases of Egypt. Minerals and agriculture in Egypt: *Bulletin* 30: 1–10.

Foottit, R. G., Maw, H. E. L., Pike, K. S. and Miller, R. H. (2010). The identity of *Pentalonia nigronervosa* Coquerel and *P. caladii* van der Goot (Hemiptera: Aphididae) based on molecular and morphometric analysis. *Zootaxa* 2358(1): 25–38.

Galvez, L. C., Barbosa, C. F. C., Koh, R. B. L. and Aquino, V. M. (2020). Loop-mediated isothermal amplification (LAMP) assays for the detection of abaca bunchy top virus and banana bunchy top virus in abaca. *Crop Protection* 131: 105101.

Geering, A. D. W. and Thomas, J. E. (1996). A comparison of four serological tests for the detection of banana bunchy top virus in banana. *Australian Journal of Agricultural Research* 47(3): 403–412.

Geering, A. D. W. and Thomas, J. E. (1997). Search for alternative hosts of banana bunchy top virus in Australia. *Australasian Plant Pathology* 26(4): 250–254.

Greenwell, A. M. and Bressan, A. (2011). Transmission of *banana bunchy top virus* (BBTV) by multiple clones of *Pentalonia nigronervosa* Coquerel and *Pentalonia caladii* van der Goot In: *Entomological Society of America Annual Meeting*. Entomological Society of America, Reno, Nevada, USA, p. D0079.

Gronenborn, B., Randles, J. W., Knierim, D., Barrière, Q., Vetten, H. J., Warthmann, N., Cornu, D., Sileye, T., Winter, S. and Timchenko, T. (2018). Analysis of DNAs associated with coconut foliar decay disease implicates a unique single-stranded DNA virus representing a new taxon. *Scientific Reports* 8(1): 5698.

Hafner, G. J., Harding, R. M. and Dale, J. L. (1995). Movement and transmission of banana bunchy top virus DNA component one in bananas. *Journal of General Virology* 76(9): 2279–2285.

Hamim, I., Green, J. C., Borth, W. B., Melzer, M. J., Wang, Y. N. and Hu, J. S. (2017). First report of banana bunchy top virus in Heliconia spp. on Hawaii. *Plant Disease* 101(12): 2153.

Hardy, G. H. (1931). Aphididae in Australia. *The Proceedings of the Royal Society of Queensland* 43: 31–36.

Harish, S., Kavino, M., Kumar, N., Balasubramanian, P. and Samiyappan, R. (2009). Induction of defense-related proteins by mixtures of plant growth promoting endophytic bacteria against Banana bunchy top virus. *Biological Control* 51(1): 16–25.

Harish, S., Kavino, M., Kumar, N., Saravanakumar, D., Soorianathasundaram, K. and Samiyappan, R. (2008). Biohardening with plant growth promoting rhizosphere and endophytic bacteria induces systemic resistance against Banana bunchy top virus. *Applied Soil Ecology* 39(2): 187–200.

Horser, C. L., Karan, M., Harding, R. M. and Dale, J. L. (2001). Additional Rep-encoding DNAs associated with banana bunchy top virus. *Archives of Virology* 146(1): 71–86.

Hu, J. M., Fu, H. C., Lin, C. H., Su, H. J. and Yeh, H. H. (2007). Reassortment and concerted evolution in banana bunchy top virus genomes. *Journal of Virology* 81(4): 1746–1761.

Hu, J. S., Wang, M., Sether, D., Xie, W. and Leonhardt, K. W. (1996). Use of polymerase chain reaction (PCR) to study transmission of banana bunchy top virus by the banana aphid (*Pentalonia nigronervosa*). *Annals of Applied Biology* 128(1): 55–64.

Islam, M. N., Naqvi, A. R., Jan, A. T. and Haq, Q. M. R. (2010). Genetic diversity and possible evidence of recombination among *banana bunchy top virus* (BBTV) isolates. *International Research Journal of Microbiology* 1(1): 1–12.

ITC. (2017). *International Trade Centre News*, 12 June 2018. https://www.intracen.org/layouts/2coltemplate.aspx?pageid=47244640256&id=47244673949.

Jebakumar, R. M., Balasubramanian, V. and Selvarajan, R. (2018). Virus titre determines the efficiency of Pentalonia nigronervosa (Aphididae: Hemiptera) to transmit banana bunchy top virus. *Virusdisease* 29(4): 499–505.

Jebakumar, R. M. and Selvarajan, R. (2018). Biopriming of micropropagated banana plants at pre- or post-BBTV inoculation stage with rhizosphere and endophytic bacteria determines their ability to induce systemic resistance against BBTV in cultivar Grand Naine. *Biocontrol Science and Technology* 28(11): 1074–1090.

Johnson, J. (1963). Biology of the bunchy top aphid Pentalonia nigronervosa Coq. *Agricultural Research in Kerala* 2: 45–51.

Kapoor, R., Srivastava, N., Kumar, S., Saritha, R. K., Sharma, S. K., Jain, R. K. and Baranwal, V. K. (2017). Development of a recombinase polymerase amplification assay for the diagnosis of banana bunchy top virus in different banana cultivars. *Archives of Virology* 162(9): 2791–2796.

Karan, M., Harding, R. M. and Dale, J. L. (1994). Evidence for two groups of banana bunchy top virus isolates. *Journal of General Virology* 75(12): 3541–3546.

Karan, M., Harding, R. M. and Dale, J. L. (1997). Association of banana bunchy top virus DNA components 2 to 6 with banana bunchy top disease. *Molecular Plant Pathology*: 1–16. http://www.bspp.org.uk/mppol/1997/0624karan.

Kato, M. (2019). *Molecular Strategies for Resistance to Circular Single-Stranded DNA Viruses*. Brisbane, QLD: Queensland University of Technology.

Katul, U., Vetten, H. J., Maiss, E., Makkouk, K. M., Lesemann, D. and Casper, R. (1993). Characterisation and serology of virus-like particles associated with faba bean necrotic yellows. *Annals of Applied Biology* 123(3): 629–647.

Kavino, M., Harish, S., Kumar, N., Saravanakumar, D., Damodaran, T. and Samiyappan, R. (2007). Potential implications of biopriming in banana (*Musa* spp.) plantlets against banana bunchy top virus (BBTV). *Journal of Plant Interactions* 2(3): 149–158.

Kavino, M., Harish, S., Kumar, N., Saravanakumar, D. and Samiyappan, R. (2008). Induction of systemic resistance in banana (*Musa* spp.) against *banana bunchy top virus* (BBTV) by combining chitin with root-colonizing *Pseudomonas fluorescens* strain CHA0. *European Journal of Plant Pathology* 120(4): 353–362.

Kolkaida, A. (1954). A study of the banana aphid, Pentalonia nigronervosa Coq. (Hemiptera Homoptera: Aphididae). *Bulletin Societé Fouad 1st., Entomologie. S, A.A* 38: 231–250.

Kumar, P. L., Hanna, R., Alabi, O. J., Soko, M. M., Oben, T. T., Vangu, G. H. and Naidu, R. A. (2011). Banana bunchy top virus in sub-Saharan Africa: Investigations on virus distribution and diversity. *Virus Research* 159(2): 171–182.

Kumar, P. L., Selvarajan, R., Iskra-Caruana, M.-L., Chabannes, M. and Hanna, R. (2015). Biology, etiology, and control of virus diseases of banana and plantain. In: Gad, L. and Nikolaos, I. K. (Eds.), *Advances in Virus Research*. Academic Press, pp. 229–269.

Magee, C. J. (1936). Bunchy top of bananas - Rehabilitation of the banana industry of N.S.W. *The Journal of the Australian Institute of Agricultural Science* 2: 13–16.

Magee, C. J. (1953). Some aspects of the bunchy top disease of banana and other *Musa* spp. *Journal and Proceedings of the Royal Society of New South Wales* 87: 3–18.

Magee, C. J. P. (1927). Investigation on the bunchy top disease of the banana, Bulletin No. 30, p. 86.

Magee, C. J. P. (1940). Transmission studies on the banana bunchy-top virus. *The Journal of the Australian Institute of Agricultural Science* 6: 109–110.

Magee, C. J. P. (1948). Transmission of banana bunchy top to banana varieties. *The Journal of the Australian Institute of Agricultural Science* 14: 18–24.

Majumder, S., Bhattacharya, B., Singh, P. K. and Johari, S. (2013). Detection of banana bunchy top virus using impedance spectroscopy. *Sensor Letters* 11(11): 2055–2059.

Majumder, S., Bhattacharya, B., Singh, P. K., Johari, S., Singh, B. and Rahman, R. (2020). Impedimetric detection of Banana bunchy top virus using CdSe quantum dots for signal amplification. *SN Applied Sciences* 2(4): 7.

Mandal, B. (2010). Advances in small isometric multicomponent ssDNA viruses infecting plants. *Indian Journal of Virology* 21(1): 18–30.

Mandal, B., Shilpi, S., Barman, A. R., Mandal, S. and Varma, A. (2013). Nine novel DNA components associated with the foorkey disease of large cardamom: Evidence of a distinct Babuvirus species in *Nanoviridae*. *Virus Research* 178(2): 297–305.

Mansoor, S., Qazi, J., Amin, I., Khatri, A., Khan, I. A., Raza, S., Zafar, Y. and Briddon, R. W. (2005). A PCR-based method, with internal control, for the detection of banana bunchy top virus in banana. *Molecular Biotechnology* 30(2): 167–170.

Mshani, V. E., Soko, M. M., Banda, D. L. N., Mwenebanda, B. M. L., Msosa, T. K. and Mwamlima, L. H. (2008). *Evaluation of Banana and Plantain Hybrid Cultivars in Banana Bunchy Top Affected Areas*.

Mware, B. O. (2016). *Development of Banana Bunchy Top Virus Resistance in Bananas: Rnai Approach*. Brisbane, QLD: Queensland University of Technology.

Namratha, N., Narayan, B., Chougal, A. and Chidanand, A. R. (2018). Development of Rapidot immunoassays for the detection of banana bunchy top virus. *Journal of Agricultural Science and Food Research* 9(4): 1000245.

Ngatat, S., Hanna, R., Kumar, P. L., Gray, S. M., Cilia, M., Ghogomu, R. T. and Fontem, D. A. (2017). Relative susceptibility of Musa genotypes to banana bunchy top disease in Cameroon and implication for disease management. *Crop Protection* 101: 116–122.

Ngatat, S., Hanna, R., Lienou, J., Ghogomu, R. T., Nguidang, S. P. K., Enoh, A. C., Ndemba, B., Korie, S., Fotso Kuate, A., Nanga Nanga, S., Fiaboe, K. K. M. and Kumar, P. L. (2022). Musa germplasm A and B genomic composition differentially affects their susceptibility to banana bunchy top virus and its aphid vector, *Pentalonia nigronervosa*. *Plants* 11(9): 1206.

Niu, S., Wang, B., Guo, X., Yu, J., Wang, X., Xu, K., Zhai, Y., Wang, J. and Liu, Z. (2009). Identification of two RNA silencing suppressors from banana bunchy top virus. *Archives of Virology* 154(11): 1775–1783.

Niyongere, C., Ateka, E., Losenge, T., Blomme, G. and Lepoint, P. (2011). Screening Musa genotypes for banana bunchy top disease resistance in Burundi. *Acta Horticulturae* 897: 439–447.

Niyongere, C., Losenge, T., Ateka, E. M., Ntukamazina, N., Ndayiragije, P., Simbare, A., Cimpaye, P., Nintije, P., Lepoint, P. and Blomme, G. (2013). Understanding banana bunchy top disease epidemiology in Burundi for an enhanced and integrated management approach. *Plant Pathology* 62(3): 562–570.

Ocimati, W., Tazuba, A. F., Tushemereirwe, W. K., Tugume, J., Omondi, B. A., Acema, D., Were, E., Onyilo, F., Ssekamate, A. M., Namanya, P., Kubiriba, J., Erima, R., Okurut, A. W., Kutunga, D. and Blomme, G. (2021). First report of banana bunchy top disease caused by Banana bunchy top virus in Uganda. *New Disease Reports* 44(2): e12052.

Omondi, B. A., Soko, M. M., Nduwimana, I., Delano, R. T., Niyongere, C., Simbare, A., Kachigamba, D. and Staver, C. (2020). The effectiveness of consistent roguing in managing banana bunchy top disease in smallholder production in Africa. *Plant Pathology* 69: 1754.

Opina, O. S. and Milloren, H. J. L. (1996). Dynamics of banana bunchy top within commercial populations of banana (*Musa* sp.) cv Lakatan. *Philippine Phytopathology* 32: 75–82.

Pekin, J. (2015). Final report banana bunchy top virus - Phase 2.

Peng, J., Zhang, J., Xia, Z., Li, Y., Huang, J. and Fan, Z. (2012). Rapid and sensitive detection of Banana bunchy top virus by loop-mediated isothermal amplification. *Journal of Virological Methods* 185(2): 254–258.

Perrier, X., De Langhe, E., Donohue, M., Lentfer, C., Vrydaghs, L., Bakry, F., Carreel, F., Hippolyte, I., Horry, J. P., Jenny, C., Lebot, V., Risterucci, A. M., Tomekpe, K., Doutrelepont, H., Ball, T., Manwaring, J., de Maret, P. and Denham, T. (2011). Multidisciplinary perspectives on banana (*Musa* spp.) domestication. *Proceedings of the National Academy of Sciences of the United States of America* 108(28): 11311–11318.

Pinili, M. S., Nagashima, I., Dizon, T. O. and Natsuaki, K. T. (2013). Cross-transmission and new alternate hosts of *Banana bunchy top virus*. *Tropical Agriculture and Development* 57(1): 1–7.

Ploetz, R. C., Kema, G. H. J. and Ma, L. J. (2015). Impact of diseases on export and smallholder production of banana. *Annual Review of Phytopathology* 53: 269–288.

Rahayuniati, R. F., Hartono, S., Somowiyarjo, S., Subandiyah, S. and Thomas, J. E. (2021a). Characterization of banana bunchy top virus on Sumatra (Indonesia) wild banana. *Biodiversitas Journal of Biological Diversity* 22(3): 1243–1249.

Rahayuniati, R. F., Subandiyah, S., Hartono, S., Somowiyarjo, S., Kurniawan, R. E. K., Prakoso, A. B., Crew, K., Vance, M. E., Ray, J. D. and Thomas, J. E. (2021b). Recent distribution and diversity analysis on banana bunchy top virus of banana and alternative host in Indonesia. *Tropical Plant Pathology* 46(5): 506–517.

Rahma, S., Maryana, N. and Hidayat, P. (2021). Host preference of *Pentalonia nigronervosa* Coquerel and *P. caladii* van der Goot (Hemiptera: Aphididae) on various host plants. In: *International E-Conference on Sustainable Agriculture and Farming System*. IOP Publishing.

Ramos, C. S. and Zamora, A. B. (1990). Elimination of banana bunchy top infection from banana (*Musa* sp) cv Lakatan by heat pretreatment and meristem culture. *Philippine Journal of Crop Science* 15: 119–123.

Rao, X. Q., Wu, Z. L., Zhou, L., Sun, J. and Li, H. P. (2017). Genetic diversity of banana bunchy top virus isolates from China. *Acta Virologica* 61(2): 217–222.

Robson, J. D., Wright, M. G. and Almeida, R. P. P. (2006). Within-plant distribution and binomial sampling of Pentalonia nigronervosa (Hemiptera: Aphididae) on banana. *Journal of Economic Entomology* 99(6): 2185–2190.

Robson, J. D., Wright, M. G. and Almeida, R. P. P. (2007a). Biology of Pentalonia nigronervosa (Hemiptera, Aphididae) on banana using different rearing methods. *Environmental Entomology* 36(1): 46–52.

Robson, J. D., Wright, M. G. and Almeida, R. P. P. (2007b). Effect of Imidacloprid foliar treatment and banana leaf age on Pentalonia nigronervosa (Hemiptera, Aphididae) survival. *New Zealand Journal of Crop and Horticultural Science* 35(4): 415–422.

Sachter-Smith, G. L. (2015). *In Search of Naturally Occurring Host Resistance to Banana Bunchy Top Virus*. Manoa, HI: Master of Science, University of Hawai'i at Mānoa.

Selvarajan, R. and Balasubramanian, V. (2013). Natural occurrence of Banana bunchy top virus in Ensete superbum in India. *Indian Journal of Virology* 24(1): 97–98.

Selvarajan, R., Balasubramanian, V. and Sasireka, T. (2015). A simple, rapid and solvent free nucleic acid extraction protocol for detection of banana bunchy top virus by polymerase chain reaction and loop-mediated isothermal amplification. *European Journal of Plant Pathology* 142(2): 389–396.

Sharman, M., Thomas, J. E. and Dietzgen, R. G. (2000). Development of a multiplex immunocapture PCR with colourimetric detection for viruses of banana. *Journal of Virological Methods* 89(1–2): 75–88.

Sharman, M., Thomas, J. E., Skabo, S. and Holton, T. A. (2008). Abacá bunchy top virus, a new member of the genus Babuvirus (family Nanoviridae). *Archives of Virology* 153(1): 135–147.

Shekhawat, U. K. S., Ganapathi, T. R. and Hadapad, A. B. (2012). Transgenic banana plants expressing small interfering RNAs targeted against viral replication initiation gene display high-level resistance to banana bunchy top virus infection. *Journal of General Virology* 93(8): 1804–1813.

Shilpa, B. (2013). *Development of Single Chain Fragment Variable Monoclonal Antibody against Banana Bunchy Top Virus Coat Protein*. Dharwad: University of Agricultural Sciences.

Shimwela, M. M., Mahuku, G., Mbanzibwa, D. R., Mkamilo, G., Mark, D., Mosha, H. I., Pallangyyo, B., Fihavango, M., Oresanya, A., Ogunsanya, P. and Kumar, P. L. (2022). First report of banana bunchy top virus in banana and plantain (*Musa* spp.) in Tanzania. *Plant Disease* 106(4): 1312.

Sila, S., Abadi, A. L. and Mudjiono, G. (2019). Transmission of banana bunchy top virus (BBTV) by Hysteroneura setariae (Thomas) aphids from Paspalum conjugatum Bergius weeds. *Bio-Science Research* 16(1): 36–44.

Smith, M. C., Holt, J., Kenyon, L. and Foot, C. (1998). Quantitative epidemiology of banana bunchy top virus disease and its control. *Plant Pathology* 47(2): 177–187.

Stainton, D., Kraberger, S., Walters, M., Wiltshire, E. J., Rosario, K., Halafihi, M., Lolohea, S., Katoa, I., Faitua, T. H., Aholelei, W., Taufa, L., Thomas, J. E., Collings, D. A., Martin, D. P. and Varsani, A. (2012). Evidence of inter-component recombination, intra-component recombination and reassortment in banana bunchy top virus. *Journal of General Virology* 93(5): 1103–1119.

Stainton, D., Martin, D. P., Muhire, B. M., Lolohea, S., Halafihi, M., Lepoint, P., Blomme, G., Crew, K. S., Sharman, M., Kraberger, S., Dayaram, A., Walters, M., Collings, D. A., Mabvakure, B., Lemey, P., Harkins, G. W., Thomas, J. E. and Varsani, A. (2015). The global distribution of Banana bunchy top virus reveals little evidence for frequent recent, human-mediated long distance dispersal events. *Virus Evolution* 1(1): vev009.

Su, H.-J., Wu, R.-Y. and Tsao, L.-Y. (1993). Ecology of banana bunchy-top virus disease. In: Valmayor, R. V., Hwang, S. C. and Ploetz, R. (Eds.), *Proceedings of the International Symposium on Recent Developments in Banana Cultivation Technology*. Los Banos, Philippines: International Network for the Improvement of Banana and Plantain/ ASPNET, pp. 308–312.

Suparman, S., Gunawan, B., Pujiastuti, Y., Arsi, A. and Cameron, R. R. (2017). Alternative hosts of banana Aphid Pentalonia nigronervosa Coq. (Hemiptera: Aphididae), the vector transmitting banana bunchy top virus. *Journal of Advanced Agricultural Technologies* 4(4): 354–359.

Suparman, N. and Setyawati, N. (2011). Preferensi dan Kecocokan Inang *Pentalonia nigronervosa* Coquerel (Hemiptera: Aphididae) terhadap Berbagai Varietas Pisang (Preference and suitability of *Pentalonia nigronervosa* Coquerel (Hemiptera: Aphididae) to various banana varieties.). *Indonesian Journal of Entomology* 8(2): 73–84.

Thiribhuvanamala, G., Sabitha, D., Ganapathy, T. and Manickam, K. (2001). Transmission studies of banana bunchy top virus. *Madras Agricultural Journal* 88(10/12): 738–739.

Thomas, J. (2019). Banana bunchy top. In: Jones, D. (Ed.), *Handbook of Diseases of Banana, Abacá and Enset*. Wallingford, Oxfordshire, UK: CAB International, pp. 362–376.

Thomas, J. E. (2018). Final report: Banana bunchy top virus control data.

Thomas, J. E. (2021). Banana bunchy top virus (Nanoviridae). In: Bamford, D. H. and Zuckerman, M. (Eds.), *Encyclopedia of Virology* (4th edn.). Oxford: Academic Press, pp. 169–175.

Thomas, J. E. and Dietzgen, R. G. (1991). Purification, characterization and serological detection of virus-like particles associated with banana bunchy top disease in Australia. *Journal of General Virology* 72(2): 217–224.

Thomas, J. E., Smith, M. K., Kessling, A. F. and Hamill, S. (1995). Inconsistent transmission of banana bunchy top virus in micropropagated bananas and its implication for germplasm screening. *Australian Journal of Agricultural Research* 46(3): 663–671.

Thompson, D. and Dietzgen, R. G. (1995). Detection of DNA and RNA plant viruses by PCR and RT-PCR using a rapid virus release protocol without tissue homogenisation. *Journal of Virological Methods* 54(2–3): 85–95.

van der Goot, P. (1917). Zur Kenntnis der Blattläuse Javas. *Contributions à la Faune des Indes Néderlandaises* 1: 1–301.

Varghese, A., Drovandi, C., Mira, A. and Mengersen, K. (2020). Estimating a novel stochastic model for within-field disease dynamics of banana bunchy top virus via approximate Bayesian computation. *PLoS Computational Biology* 16(5): e1007878.

Vézina, A. (2017). *The Hidden Side of Banana Diversity*.

Vikram, S., Ravishankar, C. and Afaque, Q. (2020). Sensitive and closed tube plant DNA virus detection via PCR. *Research Journal of Biotechnology* 15(6): 111–116.

Watanabe, S. and Bressan, A. (2013). Tropism, compartmentalization and retention of banana bunchy top virus (*Nanoviridae*) in the aphid vector *Pentalonia nigronervosa*. *Journal of General Virology* 94(1): 209–219.

Watanabe, S., Greenwell, A. M. and Bressan, A. (2013). Localization, concentration, and transmission efficiency of banana bunchy top virus in four asexual lineages of Pentalonia aphids. *Viruses* 5(2): 758–776.

Wei, J. T., Liu, H. X., Liu, F., Zhu, M., Zhou, X. and Xing, D. (2014). Miniaturized paper-based gene sensor for rapid and sensitive identification of contagious plant virus. *ACS Applied Materials and Interfaces* 6(24): 22577–22584.

Wong, M., Dubousquet, V. and Vairaa, M. (2019). Field infection of *Alpinia purpurata* with banana bunchy top virus in French Polynesia. In: *9th Interntional Geminivirus Symposium and 7th. ssDNA Comparative Virolog Workshop*. University of California, Davis, CA.

Wu, R.-Y. and Su, H.-J. (1990a). Transmission of banana bunchy top virus by aphids to banana plantlets from tissue culture. *Botanical Bulletin of Academia Sinica* 31: 7–10.

Wu, R. Y. and Su, H. J. (1990b). Production of monoclonal antibodies against banana bunchy top virus and their use in enzyme-linked immunosorbent assay. *Journal of Phytopathology* 128(3): 203–208.

Wu, R. Y. and Su, H. J. (1990c). Purification and characterization of banana bunchy top virus. *Journal of Phytopathology* 128(2): 153–160.

Wu, R. Y. and Su, H. J. (1991). Regeneration of healthy banana plantlets from banana bunchy top virus-infected tissues cultured at high temperature. *Plant Pathology* 40(1): 4–7.

Xie, W. S. and Hu, J. S. (1995). Molecular cloning, sequence analysis, and detection of banana bunchy top virus in Hawaii. *Phytopathology* 85(3): 339–347.

Yasmin, T., Khalid, S. K., Soomro, M. H. S., Malik, S. A., Shah, H. and Ahmad, I. (2001). Specificity of host-pathogen interaction of banana bunchy top disease. *Journal of Biological Sciences* 1(4): 212–213.

Zhang, J., Borth, W., Lin, B., Melzer, M., Shen, H., Pu, X., Sun, D., Nelson, S. and Hu, J. (2018). Multiplex detection of three banana viruses by reverse transcription loop-mediated isothermal amplification (RT-LAMP). *Tropical Plant Pathology* 43(6): 543–551.

Chapter 14

Banana streak disease

Andrew D. W. Geering, Queensland Alliance for Agriculture and Food Innovation, The University of Queensland, Australia

1 Introduction

Banana streak is the most widely distributed viral disease of bananas and plantains in the world. This disease was first identified in the Niéky lowlands of Côte d'Ivoire in 1958, where concern over sudden outbreaks of the disease led to the very first research program to investigate symptomatology, yield impacts and methods of eradication (Lassoudiere, 1974; Lassoudiere, 1979). Yot-Dauthy and Bové (1966) recognized that the disease was different from typical mosaic disease caused by cucumber mosaic (CMV) as the pathogenic agent could not be mechanically transmitted to tobacco plants, and therefore they introduced a new name for the disease, 'la mosaïque en tirets', which translates to 'streak mosaic'. Ironically, the disease had been illustrated in an earlier textbook on banana pathology by Wardlaw (1961) but incorrectly attributed to infection by CMV.

Benham Lockhart was the first to provide definitive evidence of a viral aetiology for banana streak disease (Lockhart, 1986) and to later propose that the virus belonged to a new group of reverse-transcribing viruses in the family *Caulimoviridae* (Lockhart, 1990; Medberry et al., 1990), later named 'badnavirus' (bacilliform DNA virus). He, in collaboration with Neil Olszewski, Roger Hull and members of their research teams, was also the first to make the paradigm-shifting discovery that banana streak virus (BSV) DNA was integrated into the nuclear genome of *Musa*, and this was the likely origin of infections in new

http://dx.doi.org/10.19103/AS.2022.0108.14

banana and plantain hybrids produced by the international banana breeding programs (La Fleur et al., 1996; Ndowora et al., 1999; Harper et al., 1999). Other notable milestones in the early history of research were sequencing of the banana streak OL virus (BSOLV) genome (Harper and Hull, 1998) and the revelation that banana streak disease is caused by a cryptic species complex of different badnaviruses (Geering et al., 2000).

The various species of BSV have the potential to reduce both the yield and quality of fruit from infected banana plants, but the magnitude of this impact is very much dependent on the combination of virus and host genotypes and prevailing environmental conditions (Dahal et al., 2000). In many cases, the impacts of the BSVs on the production of bananas are only mild. However, the problem of activation of integrated BSV DNA has caused great disruptions to the international banana breeding programs and to nurseries and germplasm collections utilising tissue culture for propagation.

2 Disease symptoms

The most characteristic symptoms of banana streak disease are narrow, chlorotic, interveinal flecks, streaks or spindle-shaped lesions on the leaf lamina, which turn necrotic as the leaf ages (Fig. 1). These streaks are often confined to segments of the leaf, with the remainder of the leaf appearing asymptomatic. Sometimes, the chlorosis is more extensive, forming blotches or broad stripes. The streaking is often sporadic in appearance, with some layers of leaves being entirely asymptomatic. The environmental factors affecting symptom expression are poorly defined, with contrasting reports of symptom expression being favoured by both cooler (Dahal et al., 1998; Thottappilly et al., 1998) and hotter (Daniells et al., 2001; Lockhart, 1986) temperatures. There is a strong effect of plant phenology on symptom expression, with the incidence of symptomatic leaves steadily increasing as the plant matures, reaching a maximum at the time of bunch initiation (Daniells et al., 2001). Recently emerged leaves with chlorotic streaking have the highest virus titres (Daniells et al., 2001), but as the leaf ages and the streaks become strongly chlorotic or necrotic, the virus titre declines, probably as a consequence of cell apoptosis (Ngo et al., 2020).

Many other types of symptoms are occasionally observed. The leaf sheaths may split at the base and detach from the pseudostem (Fig. 2), and the internal parts of the pseudostem and the cigar leaf can become necrotic. The peel of the fruit may be thinner and more difficult to separate from the pulp. The phyllotaxy of the plant may be altered, with the leaves emerging along a single vertical plane, much like the appearance of a Traveller's Palm (*Ravenala madagascariensis*). The crown of the plant may become choked due to the shortening of the internodes in the true stem and the bunch may erupt part way up the pseudostem.

Figure 1 Variation in leaf streak symptoms caused by banana streak virus. (a) *Musa* AAA group 'Williams' infected by banana streak CA virus; (b) a *Musa* AAB group IRFA hybrid infected by banana streak IM virus; and (c) *Musa* AAB group 'Mysore' infected by banana streak MY virus. Photographs: A. D. W. Geering.

Figure 2 Splitting of the leaf sheaths and black streaking on the pseudostem of a banana streak virus-infected banana plant. Photograph: M. Chabannes.

3 Pathogen characteristics and diagnosis

Banana streak disease is caused by a complex of nine officially recognized and one tentative viral species in the genus *Badnavirus* (Table 1), all of which have bacilliform-shaped virions, 30 × 130–150 nm in size (Teycheney et al., 2020). There is often not a practical need to distinguish to species level, in which case the common name 'banana streak virus' or the acronym 'BSV' is used. These viruses have circular, non-covalently closed, double-stranded (ds) DNA genomes of 7.2–7.6 kbp (Table 1). All protein-coding capacity is contained within one DNA strand, designated the positive strand, and there are three open reading frames (ORFs). The function of the ORF1 gene product is unknown; ORF2 encodes a multifunctional virion-associated protein; ORF3 encodes a large polyprotein, which is cleaved into movement, capsid, aspartic protease and replicase [reverse transcriptase (RT) with RNase H1 (RH1) activity] proteins through the action of the aspartic protease (Geering, 2021). During replication, the virus genome alternates between dsDNA and single-stranded RNA through cycles of transcription and reverse transcription (Geering, 2021). Reverse transcription and virion assembly is thought to occur within electron-dense inclusion bodies, otherwise called viroplasms, which are located in the cytoplasm of parenchymatous cells in the leaf and are 1.2–1.5 μm in diameter (Ngo et al., 2020).

Table 1 The badnaviruses associated with banana streak disease

Virus	Acronym	Derivation of two letter code[a]	NCBI Reference Sequence[b]	Genome size (bp)[b]	Integrated/A or B genome
Banana streak OL virus	BSOLV	OL = Obino l'Ewai	NC_003381	7389	Yes/B
Banana streak GF virus	BSGFV	GF = Goldfinger	NC_007002	7263	Yes/B
Banana streak IM virus	BSIMV	IM = Imové	NC_015507	7769	Yes/B
Banana streak MY virus	BSMYV	MY = Mysore	NC_006955	7650	Yes/B
Banana streak VN virus	BSVNV	VN = Viet Nam	NC_007003	7801	No
Banana streak UA virus	BSUAV	UA = Uganda A	NC_015502	7519	No
Banana streak UI virus	BSUIV	UI = Uganda I	NC_015503	7458	No
Banana streak UL virus	BSULV	UL = Uganda L	NC_015504	7401	No
Banana streak UM virus	BSUMV	UM = Uganda M	NC_015505	7532	No
Banana streak CA virus	BSCAV	CA = Cavendish	NC_015506	7408	No

[a]Imové is French creole for 'it's bad', reflecting the impact of the virus on the CIRAD banana breeding program in Guadeloupe. The other codes refer to the cultivar or country in which the virus was first found.
[b]Data for exemplar isolates of the species.

Unlike retroviruses, badnaviruses do not actively integrate into the genome of the host plant as part of their replication cycle, but nevertheless there are integrated forms of badnaviral DNA in banana and many other plant genomes (Diop et al., 2018). The mechanism of integration is still a matter of conjecture but may have been associated with non-homologous end joining repair of dsDNA breaks in the chromosomes, a process that often leads to the capture of extraneous DNA as a filler material (Gorbunova and Levy, 1997). To be inherited from one generation to the next, the viral DNA must first have integrated into the chromosomes of germline cells. The imprecise nature of non-homologous end joining has meant that many of the badnaviral DNA fragments that have been captured in the plant genome are fragmented and often rearranged when compared to their ancestral viral genome. Estimates of the age of the integration events range from hundreds of thousands to millions of years ago, well before the banana was domesticated (Gayral et al., 2010; Chabannes et al., 2021). Following integration, the selection pressures on the viral DNA change and mutations occur in places that render the sequence replication-defective (e.g. disrupted ORFs). The mutations tend to be random in nature, so that if there are multiple copies of integrated viral DNA within the plant, it is possible to estimate the ancestral viral genome sequence by creating a consensus of the integrated sequences (Geering et al., 2014).

The various BSV species are indistinguishable based on symptomatology as expressed in banana plants, vector specificity and virion morphology and can only reliably be identified using molecular diagnostic assays. The main criterion used to demarcate species in the genus *Badnavirus* is less than 80% nucleotide sequence identity in the RT-RH1 coding region of ORF3 (Teycheney et al., 2020). To assist with classification, an RT-RH1 sequence alignment including all officially recognized species can be downloaded from the International Committee on Taxonomy of Viruses Resources page for the *Caulimoviridae* (Teycheney et al., 2020).

Immunocapture polymerase chain reaction (IC-PCR) is a popular diagnostic method for the BSVs (Harper et al., 1999a; Le Provost et al., 2006; Geering et al., 2000; Thomas, 2015), but usage of this method is constrained by a worldwide shortage of antibodies. In the IC-PCR method, the PCR tubes are coated with BSV-specific antibodies, which capture and immobilize the virions on the side of the tube; the remaining cell contents are then washed away. The DNA in the bound virions is then released during the first heat denaturation step of the PCR. The primary reason for using this variation of the traditional PCR method is to enrich encapsidated viral DNA and at the same time eliminate viral DNA that is integrated into the nuclear genome of the plant, otherwise called endogenous viral DNA. However, there is non-specific adsorption of plant DNA to the walls of the tube, and precautions need to be taken to avoid this from happening. The magnitude of non-specific adsorption of plant DNA is both time and temperature dependent and limiting the immunocapture incubation step to 3 h at room temperature minimizes this problem (Le Provost et al., 2006). After the immunocapture step, a DNase I treatment can be done to eliminate all unencapsidated DNA, while the virion-associated DNA is protected by the capsid shell and remains intact (Thomas, 2015). Finally, non-specific adsorption of plant DNA to the PCR tube can be monitored by running a PCR assay for a highly repetitive DNA element in the banana genome either as a duplex or as a parallel assay with the virus-specific assay (Le Provost et al., 2006).

There are few sources and a limited range of BSV antibodies for use in IC-PCR. This reagent shortage is partly due to the technical difficulties associated with purifying BSV for antiserum production but also because many of the BSVs occur in tropical, developing countries, where the equipment needed to do the virus purifications is not widely available. These problems can be circumvented by using chemically synthesized peptides that mimic the major linear epitopes in the N-terminal intrinsically disordered domain of the BSV capsid protein as immunogens (Vo et al., 2016). The antibodies that are generated using these peptides are suitable for use in IC-PCR but not enzyme-linked immunosorbent assay (ELISA).

Degenerate PCR primers have been designed for the universal detection of all badnavirus species, including the BSVs. The two most widely used pairs of PCR primers for this purpose are BADNA1A/4 (Geering et al., 2005a; Harper et al., 2002) and BadnaFP/RP (Yang et al., 2003), and both target the RT region of ORF3. The two reverse primers, BADNA4 and BadnaRP, are essentially the same, whereas the forward primers, BADNA1 and BadnaFP, are distinct but partially overlapping. The sequence of the BadnaFP primer published by Yang et al. (2003) has an error, which has been corrected in later publications (James et al., 2011a). Many new badnavirus species have recently been described and claims of universality of the PCR assays are no longer valid, although these assays are still suitable for all currently recognized BSV species.

Conventional PCR using a total plant DNA extract as template can be done for the BSV species that are not integrated in the *Musa* genome, such as banana streak CA virus (BSCAV), banana streak VN virus (BSVNV), banana streak UI virus (BSUIV), banana streak UL virus (BSULV), and banana streak UM virus (BSUMV) (Table 1). Furthermore, banana streak GF virus (BSGFV), banana streak IM virus (BSIMV), banana streak MY virus (BSMYV) and BSOLV are only integrated into the B genome of *Musa*, and therefore if testing is done on bananas with an AA or AAA genotype such as 'Cavendish' bananas, then again, conventional PCR is a suitable method of diagnosis. Care must be taken with the design of the PCR primers to ensure species-specificity, as there is a background of unrelated endogenous badnaviral elements in the A genome, which could cause non-specific amplification products if the primers correspond to conserved sequences (Geering et al., 2005a).

Attempts have been made to develop quantitative PCR (qPCR) methods that discriminate circular, episomal forms of the BSV genome from corresponding endogenous viral elements that are present in the B genome of the banana (Delanoy et al., 2003; Ricciuti et al., 2021). In the case of the BSOLV qPCR assay developed by Delanoy et al. (2003), the discriminatory power of the assay is based on structural aberrations in the endogenous viral DNA compared to the episomal viral DNA, with a single amplicon of 1.3 kb expected for episomal BSOLV DNA, compared to three amplicons of c. 2.7, 7.7

and 15-16 kb for endogenous BSOLV DNA (Ricciuti et al., 2021). An alternative detection strategy was investigated by Ricciuti et al. (2021), who hypothesized that the concentration of endogenous viral DNA in a total DNA extract from healthy plants of a nominated hybrid cultivar would be constant, and any statistically significant deviation above this baseline concentration would indicate infection. However, against expectations, the baseline concentration of viral DNA was shown to be very variable between replicate healthy plants, ruling out this method of qPCR for use in diagnosis. The explanation given for this baseline variability was the generation of extrachromosomal copies of viral DNA by the transcription of the integrated DNA, followed by occasional reverse transcription.

After PCR, the next most popular diagnostic method for the BSVs is rolling circle amplification (RCA), utilizing phi29 DNA polymerase. In the first iteration of this method (James et al., 2011a), a commercial RCA kit (TempliPhi™) was used, but the random hexamer mixture had to be supplemented with degenerate badnavirus primers, including the aforementioned BadnaFP/RP primers, to improve the sensitivity of the assay to a point where the BSVs could be reliably detected. The two major disadvantages of this original RCA method were the cost of the commercial kit, and the lack of target specificity as non-specific amplification products were generated from other circular DNA molecules within the plant such as mitochondrial DNA. Further improvements to the RCA method have been made, which have increased the sensitivity and specificity of the assay, as well as decreased costs, by allowing assay reaction mixtures to be made from individual constituents rather than relying upon a preprepared commercial kit (Sukal et al., 2019). It is important to emphasize that purified virus cannot be used as a template for RCA, as the discontinuities in each strand of the encapsidated viral DNA cause premature termination of DNA polymerization. For this reason, an immunocapture enrichment step cannot be combined with RCA. The template for RCA is the covalently closed, minichromosomal viral DNA, which is present within the plant cell nucleus and is represented within total plant DNA extracts.

A less commonly used method for BSV diagnosis is ELISA, but the sensitivity of detection using this method is generally less than PCR or RCA. As with IC-PCR, the worldwide shortage of BSV antibodies prevents more widespread adoption of ELISA for diagnosis. Commercial ELISA kits for BSMYV and BSOLV are currently available from the Leibniz Institute DSMZ-German Collection of Microorganisms and Cell Cultures and are effective for the detection of these species (Ngo et al., 2020), although the antibodies have limited serological cross-reactivity with other BSV species.

4 Disease cycle and epidemiology

Banana streak disease has an unusual epidemiology as new infections can arise following the activation of viral sequences that are integrated into the nuclear genome of the banana plant. This phenomenon was discovered by Ndowora et al. (1999) and Harper et al. (1999b), who sought to explain why the AAB plantain cultivar 'Obino L'Ewai', as well as tetraploid hybrid progeny plants produced when this cultivar was used as a female parent in breeding experiments, became infected with BSOLV after initiation into tissue culture. A locus in the nuclear genome of 'Obino L'Ewai' was characterized, which contains the complete genome sequence of BSOLV, but the sequence is interrupted by a large, scrambled region containing fragments of non-contiguous and inverted viral sequences. Direct sequence duplications were also observed at the termini of the locus as well as at either end of the scrambled region. It was proposed that through two homologous recombination events, the scrambled region could be excised and the ends of the locus joined to create a circular, transcriptionally active form of the viral genome. Subsequent investigations linked this locus to the B genome of banana (Geering et al., 2001).

Independent confirmation that integrated BSOLV DNA in 'Obino L'Ewai' was responsible for new infections was provided by Chabannes et al. (2013), who discovered just a single genetic locus containing integrated BSOLV DNA in the diploid genome of *M. balbisiana* 'Pisang Klutung Wulung' (PKW). This locus is heterozygotic, with one allele (eBSOLV-1) identical in sequence and structure to that found in 'Obino L'Ewai', while the second (eBSOLV-2) has a deletion of 309 bp at the junction of ORF3 and the intergenic region, which would render the allele replication-defective. When PKW was crossed with a tetraploid *Musa acuminata* male parent, infection only developed in the proportion of the progeny population inheriting the eBSOLV-1 allele.

The phenomenon of activation of integrated viral sequences to cause new infections has now been extended to two more badnavirus species, banana streak GF virus (BSGFV) and banana streak IM virus (BSIMV), both of which are associated with the B genome of *Musa*. Like BSOLV, there is only a single, heterozygotic locus containing BSGFV DNA in the PKW genome, and only one allele (eBSGFV-7) at this locus is replication-competent, while the other (eBSGFV-9) is replication-defective because of the accumulation of non-synonymous and null mutations (Gayral et al., 2008; Chabannes et al., 2013). In contrast, there are two structurally identical alleles of BSIMV DNA at a single locus in PKW, and while these two alleles show very minor differences in sequence, both appear capable of producing infectious viral sequences (Chabannes et al., 2013). It is now possible to genotype a particular banana or plantain cultivar to determine if it is carrying a replication-competent allele of endogenous BSOLV, BSGFV or BSIMV, and therefore prone to infection from the activation of these sequences (Gayral et al., 2010; Chabannes et al., 2013; Gayral et al., 2008; Umber et al., 2016).

BSMYV DNA is also integrated into the B genome of *Musa* (Geering et al., 2005b), but there is no evidence for the activation of this DNA to cause infection arising from the international *Musa* breeding programs. However, BSMYV infections are very rarely observed in 'Lady Finger' (AAB) and 'Goldfinger' (AAAB) plants in subtropical Australia, and the most plausible explanation for these infections is the activation of cognate endogenous viral sequences (ADW Geering and JE Thomas, unpublished results).

Tissue culture propagation is a well-documented trigger for the activation of the eBSOLV-1 and eBSGV-7 alleles to cause new infections (Dallot et al., 2001; Côte et al., 2010; Umber et al., 2022). However, eBSIMV is not triggered in the same way, with no infection observed after 10 subculture cycles of cell culture of four hybrid cultivars (AAB and ABB) and two *M. balbisiana* (BB) accessions that harbour this DNA (Umber et al., 2022). Other abiotic stresses likely cause the activation of the eBSVs (Umber et al., 2022), but the nature of these stresses is poorly defined. The presence of a replication-competent eBSV allele does not necessarily condemn the plant to becoming infected with the virus. For example, two AAB group plantain cultivars, 'Kelong Mekintu' and 'Black Penkelon', do not become infected with BSGFV when propagated by tissue culture, even though they both contain the eBSGFV-7 allele (Côte et al., 2010; Umber et al., 2022). The rates of activation of eBSOLV-1 by tissue culture vary according to the background genetic composition of the plant, with some hybrid cultivars being more prone to infection than others, and infection in diploid *M. balbisiana* is entirely repressed (Umber et al., 2022). Finally, activation of the eBSV alleles reaches a maximum rate between the fifth and tenth subculture steps of cell culture and sharply declines thereafter (Umber et al., 2022).

A model has been proposed by Staginnus and Richert-Pöggeler (2006) to explain how the expression of the eBSVs is regulated by the plant. A facet of this model is that the eBSV sequences are transcribed at low levels in some cells, and the RNA transcripts act as precursors for the formation of small RNA molecules, which in turn guide transcriptional and post-transcriptional gene silencing of the virus and the corresponding eBSV loci (Duroy et al., 2021). Both interspecific hybridization and tissue culture are known to drastically change DNA methylation patterns, which could release the transcriptional silencing on the virus and provide a cellular environment that is more favourable for viral replication (Azizi et al., 2020; Greaves et al., 2015). Moreover, tissue culture propagation promotes DNA recombination, thereby facilitating the reconstitution and release of a functional viral genome from the plant genome. A final component in the model proposed by Staginnus and Richert-Pöggeler (2006) is that the BSVs encode a silencing suppressor protein, which prevents the reinstatement of the epigenetic controls on the virus by the plant, but currently there is no empirical evidence to support this hypothesis.

It must be emphasized that the integration of badnaviral DNA is not unique to *M. balbisiana* but also occurs in *M. acuminata*, *M. basjoo*, *M. itinerans*, *M. laterita*, *M. ornata* and *M. schizocarpa* (Chabannes et al., 2021; Diop et al., 2018; Geering et al., 2005a; D'Hont et al., 2012). However, the integrated badnaviral sequences in *Musa* species other than *M. balbisiana* are thought to be replication-defective because the sequences are incomplete or excessively rearranged and mutated to allow reconstitution of an autonomously replicating viral genome (D'Hont et al., 2012; Chabannes et al., 2021). There are also many BSV species that are not integrated in any *Musa* genome, and these viruses likely originate from alternative host plant species. For example, the main infection reservoir for BSCAV is thought to be sugarcane (Muller et al., 2011; Braithwaite et al., 2004), a plant that often is grown in mixed farming systems with banana.

Once a plant becomes infected with BSV, the disease cycle is like most other badnavirus species (Fig. 3). It is expected that all BSVs are transmitted in a semi-persistent manner by mealybugs (Pseudococcidae), although empirical evidence of transmission is lacking for many combinations of virus and mealybug species. Known vectors include the citrus mealybug *Planococcus citri*, which transmits BSOLV, BSGFV and BSMYV (Geering et al., 2005b; Meyer et al., 2008); the vine mealybug *P. ficus*, which transmits BSOLV and BSGFV

Figure 3 Banana streak disease cycle, illustrating different virus transmission pathways: mealybug-vectored transmission from sugarcane to banana or between bananas and activation of endogenous banana streak viruses in the B genome of banana. Illustration by L.-S. Pretorius.

(Meyer et al., 2008); the pineapple mealybug *Dysmicoccus brevipes*, which transmits BSOLV (Kubiriba et al., 2001; Meyer et al., 2008); and the grey sugarcane mealybug *Saccharicoccus sacchari*, which transmits one or more species of BSV, the identities of which were not precisely determined (Kubiriba et al., 2001; Lockhart and Olszewski, 1993). Badnaviruses in general have weak vector specificity, with at least 14 different mealybug species capable of transmitting the cacao swollen shoot group of viruses (Muller, 2021). First instar nymphs (crawlers) seeking feeding spots are considered the dispersive stage of the mealybug vector and are most likely responsible for dissemination in the field of the BSVs. The first instar nymphs can be carried to the surrounding plants by ants or by wind but rarely from actively walking the entire distance across leaf bridges (Grasswitz and James, 2008; Barrass et al., 1994; Daane et al., 2012). Ants have a symbiotic relationship with mealybugs and help to defend them against predators and parasitoids, as well as preventing disease by cleaning up their excrement, called 'honey dew', which is sweet and sticky (Mgocheki and Addison, 2010).

The rate of spread of the BSVs by mealybugs is thought to be extremely slow and when epidemics have occurred, particularly in *Musa* AAA genotypes, this is thought to be a consequence of multiplication of the virus by the division of infected plants through propagation practices by humans. In a field trial run by Daniells et al. (2001), equal numbers of healthy and infected plants were established, but no natural spread of the virus was observed over a 24-month period.

5 Occurrence and impact

Collectively, the various species of BSV are the most widely geographically distributed of all the viruses that infect banana. The BSVs that are integrated in the B genome of banana, such as BSOLV, BSMYV, BSGFV and BSIMV, could reasonably be expected to occur in every country where hybrid bananas or plantains (*Musa* AB, AAB and ABB group cultivars) are grown in significant quantities. The remaining species of BSV that are not integrated into the banana genome have more restricted geographic ranges. BSVNV has only been recorded from Vietnam, China, and Cuba (Javer-Higginson et al., 2014; Lheureux et al., 2007; Zhuang et al., 2011). BSUAV, BSUIV, BSULV and BSUMV are seemingly confined to East Africa (James et al., 2011b; Harper et al., 2005; Karanja et al., 2008; Chabannes et al., 2021). BSCAV is only recorded from Australia (Geering et al., 2000), China, Guadeloupe, Kenya and Mauritius (Jaufeerally-Fakim et al., 2006). However, the main field host of BSCAV is thought to be sugarcane, and as this plant species is an asymptomatic host and vegetatively propagated, the distribution of this virus is probably much greater than currently recognized due to a lack of sampling effort in sugarcane.

The economic impacts of the various species of BSV are poorly described, and data have mainly been collected in an opportunistic fashion. The only purposefully designed field trial to measure the effects of infection on plant growth and yield was that of Daniells et al. (2001), using the host-pathogen combination of BSCAV and *Musa* AAA group Cavendish subgroup cv. 'Williams'. In the plant crop, the only yield parameters that were affected by infection were the time periods between planting and bunch emergence and harvest, which were delayed by 26 days and 18 days, respectively. In the first ratoon crop, the infected plants were slightly stunted and produced 7% lighter bunches and 5% shorter fingers and the cropping cycle was again extended, leading to an overall reduction in yield per annum of 11%. It was concluded that the plants in the ratoon crop were less able to tolerate infection by BSCAV than those of the plant crop because of greater plant competition and generally less favourable conditions for plant growth. A similar magnitude of yield impact by an identified species of BSV infecting a Cavendish-type banana ('Poyo') growing in Côte d'Ivoire has been reported, with the mean weight of bunches reduced by 8% (Lassoudiere, 1979). In common with the Australian study, the yield impacts only became obvious in the first ratoon crop.

Arguably, the greatest impact of the BSVs has been indirect through disruption to the international banana breeding programmes and the *Musa* germplasm collections. During the 1990s, BSV became alarmingly common in new, synthetic hybrids of *M. acuminata* and *M. balbisiana* produced by plant breeders, and the presence of the virus curtailed the commercialization of many promising new cultivars. A moratorium was placed on the distribution of BSV-infected plants held at the International *Musa* Germplasm Transit Centre in Belgium, and this effectively locked up about 300 of the nearly 1500 accessions held in the collection, with a disproportionate impact on the plantain cultivars (Thomas et al., 2015). This moratorium has recently been relaxed.

6 Management of the disease

Control of banana streak disease in dessert bananas with an AAA genotype is reasonably straightforward. The major point of intervention for disease control is in the nursery, and it is important that plants from which propagules are sourced are tested to demonstrate freedom from infection. Even though symptoms may be sporadic in appearance, infected plants are very rarely asymptomatic for the entirety of the cropping cycle (Daniells et al., 2001), and therefore it should be possible to establish a clean nursery block through close observation of symptoms alone, even if virus testing is not possible. The rate of virus spread by mealybugs is so slow that the treatment of the crop with insecticides is not advisable as a control option for the disease for either economic or environmental reasons. The waxy coating of the mealybug also repels contact insecticides. The virus can easily be eradicated from a crop by rogueing diseased plants, and it is important that all parts of the corm and associated suckers are either physically removed or killed by injecting the plant with a herbicide such as glyphosate. The virus does not persist in the soil for any length of time nor can the plant become infected by mechanical inoculation, and therefore it is possible to replant immediately.

Banana streak disease is more difficult to control when the source of infection lies within the genome of the plant. Only a narrow gene pool of *M. balbisiana* has been used for plant breeding, and new accessions of this species are being collected in the hope that some are free of replication-competent endogenous badnaviral sequences (Caruana et al., 2019). The 'Honduras' accession of *M. balbisiana* is devoid of endogenous BSIMV but does contain a mixture of replication-competent and replication-defective alleles of endogenous BSOLV and BSGFV. These alleles do segregate in seedling progeny arising from self-crosses of the parent plant, and success has been obtained in breeding an improved line of *M. balbisiana* that only contains replication-defective eBSVs (Umber et al., 2016). In tetraploid hybrid *Musa* genotypes, there is evidence for homeologous chromosome pairing between the A and B genomes during meiosis, a phenomenon that facilitates chromosome translocations through recombination and the occasional production of gametes in which the part of the B genome carrying the endogenous badnaviral sequences has been swapped with the paired part of the A genome, which is devoid of these sequences (Noumbissié et al., 2016). In addition to the methods mentioned earlier, CRISPR/Cas9 genome editing may be used to inactivate replication-competent endogenous badnaviral sequence in the *M. balbisiana* genome (Tripathi et al., 2019).

Conventional screening for resistance to the BSVs has rarely been done, nor considered necessary, as the probability of secondary infection in the field is so small. Nevertheless, there are indications that some *Musa* genotypes are more tolerant of BSV infection than others. For example, four wild *Musa* AA accessions and five *Musa* AA/AAA cultivars were challenged with BSCAV by James et al. (2021) under glasshouse conditions, and significant variability in disease severity was observed. Wild-type *M. acuminata* subsp. *zebrina* (AA) became infected but remained asymptomatic for the 5-month duration of the experiment, whereas *M. acuminata* ssp. *siamea* 'Khae Phrae' (AA) plants stopped growing at 10-weeks post-inoculation and developed cigar leaf necrosis, causing three of the five inoculated plants to die.

7 Conclusion and future trends

Badnaviruses are by far the largest and most economically important group of viruses in the family *Caulimoviridae*, but knowledge about the replication cycle of these viruses is lagging well behind that of the more heavily studied virus, cauliflower mosaic virus. Research on badnaviruses has been hampered by a lack of model host plant systems, for which sophisticated molecular toolboxes are available. By using infectious clones of BSV (James et al., 2021), it is now possible to do reverse genetics on the virus to determine the functions of individual genes such as ORF1 and to investigate in fine detail the interactions between the virus, host plant and vector. There is also the possibility to introduce the virus into more amenable experimental host plants such as rice by agroinoculation (Bouhida et al., 1993).

There are many knowledge gaps in the epidemiology of the BSVs. Conclusions about the rates of spread of the viruses in the field and yield penalties from infection are based upon a small number of studies, which understandably represent only a very narrow spectrum of the different permutations of environmental conditions and plant and virus genotypes. Decisions about quarantine and the risks associated with distributing BSV-infected germplasm have been made based upon a very low knowledge base.

Great advances have been made in understanding how new infections can develop from eBSVs in the banana genome, and methods have now been developed to counter this problem in the *Musa* breeding programs. Activation of these sequences appears to be an undesirable consequence of modern plant breeding practices and tissue culture propagation. If the only impact of the endogenous badnaviral sequences on the host plant was negative, it would seem likely that these sequences would have been eliminated over time through purifying selection. However, the opposite holds, with the endogenous badnaviral sequences persisting over thousands of plant generations, well before the plant was domesticated by humans and the genomes of cultivated banana varieties fixed because of plant sterility. Significant questions remain as to what benefits these endogenous badnaviral sequences might impart, especially the high copy number, replication-defective integrated badnaviral sequences in the A genome of *Musa*. A popular theory is that these sequences induce RNAi-based resistance against the cognate infectious form of the virus. While this hypothesis remains untested, it is unlikely to be their only function as the copy number of the sequences in the *Musa* genome is far greater than that required to induce efficient gene silencing. Furthermore, many of the endogenous badnaviral sequences appear to be remnants of ancestral viruses that have long gone extinct. There are many examples of endogenous retroviruses contributing to the evolution of animals by providing novel protein-coding genes or transcription regulatory elements (Chuong, 2018), and it is likely that the endogenous badnaviral sequences act in a similar manner.

Diagnosis of the BSVs no longer presents the challenge it once did, and in countries such as Australia, which has a well-organized banana industry dominated by Cavendish-type bananas, banana streak disease is rarely ever seen. Through a wider adoption of clean planting material schemes, banana streak disease will continue to decline in importance elsewhere in the world.

8 Where to look for further information

Additional information on these viruses can be found at:

- CABI Compendium, https://www.cabidigitallibrary.org/journal/cabicompendium.
- International Committee on Taxonomy of Viruses, https://ictv.global/.
- Thomas J. E. (2015). *Technical Guidelines for the Safe Movement of Musa Germplasm*. 3rd Edition. (Bioversity International: Rome), https://cgspace.cgiar.org/handle/10568/118459.

9 References

Azizi, P., Hanafi, M. M., Sahebi, M., Harikrishna, J. A., Taheri, S., Yassoralipour, A. and Nasehi, A. (2020). Epigenetic changes and their relationship to somaclonal variation: a need to monitor the micropropagation of plantation crops. *Functional Plant Biology* 47(6): 508–523.

Barrass, I. C., Jerie, P. and Ward, S. A. (1994). Aerial dispersal of first- and second-instar longtailed mealybug, *Pseudococcus longispinus* (Targioni Tozzetti) (Pseudococcidae: Hemiptera). *Australian Journal of Experimental Agriculture* 34(8): 1205–1208.

Bouhida, M., Lockhart, B. E. L. and Olszewski, N. E. (1993). An analysis of the complete sequence of a sugarcane bacilliform virus genome infectious to banana and rice. *Journal of General Virology* 74(1): 15–22.

Braithwaite, K. S., Geijskes, R. J. and Smith, G. R. (2004). A variable region of the sugarcane bacilliform virus (SCBV) genome can be used to generate promoters for transgene expression in sugarcane. *Plant Cell Reports* 23(5): 319–326.

Caruana, M.-L., Ly, N.-S., Haevermans, T., Jenny, C., Perrier, X., Sardos, J. and Chabannes, M. (2019). Screening of wild seedy *Musa balbisiana* biodiversity to fill up phylogenetic gaps and fuel future breeding programs. PE 1108. In: *Plant and Animal Genome Conference (PAG)*, San Diego, USA, pp. 311–311.

Chabannes, M., Baurens, F. C., Duroy, P. O., Bocs, S., Vernerey, M. S., Rodier-Goud, M., Barbe, V., Gayral, P. and Iskra-Caruana, M. L. (2013). Three infectious viral species lying in wait in the banana genome. *Journal of Virology* 87(15): 8624–8637.

Chabannes, M., Gabriel, M., Aksa, A., Galzi, S., Dufayard, J. F., Iskra-Caruana, M. L. and Muller, E. (2021). Badnaviruses and banana genomes: a long association sheds light on *Musa* phylogeny and origin. *Molecular Plant Pathology* 22(2): 216–230.

Chuong, E. B. (2018). The placenta goes viral: retroviruses control gene expression in pregnancy. *PLoS Biology* 16(10): e3000028.

Côte, F. X., Galzi, S., Folliot, M., Lamagnère, Y., Teycheney, P. Y. and Iskra-Caruana, M. L. (2010). Micropropagation by tissue culture triggers differential expression of infectious endogenous *banana streak virus* sequences (eBSV) present in the B genome of natural and synthetic interspecific banana plantains. *Molecular Plant Pathology* 11(1): 137–144.

D'Hont, A., Denoeud, F., Aury, J. M., Baurens, F. C., Carreel, F., Garsmeur, O., Noel, B., Bocs, S., Droc, G., Rouard, M., Da Silva, C., Jabbari, K., Cardi, C., Poulain, J., Souquet, M., Labadie, K., Jourda, C., Lengellé, J., Rodier-Goud, M., Alberti, A., Bernard, M., Correa, M., Ayyampalayam, S., Mckain, M. R., Leebens-Mack, J., Burgess, D., Freeling, M., Mbéguié-A-Mbéguié, D., Chabannes, M., Wicker, T., Panaud, O., Barbosa, J., Hribova, E., Heslop-Harrison, P., Habas, R., Rivallan, R., Francois, P., Poiron, C., Kilian, A., Burthia, D., Jenny, C., Bakry, F., Brown, S., Guignon, V., Kema, G., Dita, M., Waalwijk, C., Joseph, S., Dievart, A., Jaillon, O., Leclercq, J., Argout, X., Lyons, E., Almeida, A., Jeridi, M., Dolezel, J., Roux, N., Risterucci, A. M., Weissenbach, J., Ruiz, M., Glaszmann, J. C., Quétier, F., Yahiaoui, N. and Wincker, P. (2012). The banana (*Musa acuminata*) genome and the evolution of monocotyledonous plants. *Nature* 488: 213–217.

Daane, K. M., Almeida, R. P. P., Bell, V. A., Walker, J. T., Botton, M., Fallahzadeh, M., Mani, M., Miano, J. L., Sforza, R., Walton, V. M. and Zaviezo, T. (2012). Biology and management of mealybugs in vineyards. In: Bostanian, N. J., Vincent, C. and Isaacs, R. (Eds). *Arthropod Management in Vineyards: Pests, Approaches, and Future Directions*. Dordrecht: Springer Netherlands, pp. 271–307.

Dahal, G., Hughes, J. d'A., Thottappilly, G. and Lockhart, B. E. L. (1998). Effect of temperature on symptom expression and reliability of banana streak badnavirus detection in naturally infected plantain and banana (*Musa* spp.). *Plant Disease* 82(1): 16–21.

Dahal, G., Ortiz, R., Tenkouano, A., Hughes, J. d'A., Thottappilly, G., Vuylsteke, D. and Lockhart (2000). Relationship between natural occurrence of banana streak Badnavirus and symptom expression, relative concentration of viral antigen, and yield characteristics of some micropropagated *Musa* spp. *Plant Pathology* 49(1): 68–79.

Dallot, S., Acuña, P., Rivera, C., Ramírez, P., Côte, F., Lockhart, B. E. and Caruana, M. L. (2001). Evidence that the proliferation stage of micropropagation procedure is determinant in the expression of *Banana streak virus* integrated into the genome of the FHIA 21 hybrid (*Musa* AAAB). *Archives of Virology* 146(11): 2179–2190.

Daniells, J. W., Geering, A. D. W., Bryde, N. J. and Thomas, J. E. (2001). The effect of *Banana streak virus* on the growth and yield of dessert bananas in tropical Australia. *Annals of Applied Biology* 139(1): 51–60.

Delanoy, M., Salmon, M., Kummert, J., Frison, E. and Lepoivre, P. (2003). Development of real-time PCR for the rapid detection of episomal *Banana streak virus* (BSV). *Plant Disease* 87(1): 33–38.

Diop, S. I., Geering, A. D. W., Alfama-Depauw, F., Loaec, M., Teycheney, P. Y. and Maumus, F. (2018). Tracheophyte genomes keep track of the deep evolution of the *Caulimoviridae*. *Scientific Reports* 8(1): 572.

Duroy, P.-O., Seguin, J., Ravel, S., Rajeswaran, R., Laboureau, N., Salmon, F., Delos, J. M. E., Pooggin, M., Iskra Caruana, M. L. and Chabannes, M., (2021). Infective endogenous viral elements are tightly controlled by RNA silencing pathways in banana. In: *EpiCATCH COST Action Conference on "Epigenetic Mechanisms of Crop Adaptation to Climate Change"*, Florence, Italy, p.13.

Gayral, P., Blondin, L., Guidolin, O., Carreel, F., Hippolyte, I., Perrier, X. and Iskra-Caruana, M. L. (2010). Evolution of endogenous sequences of *Banana streak virus*: what can we learn from banana (*Musa* sp.) evolution? *Journal of Virology* 84(14): 7346–7359.

Gayral, P., Noa-Carrazana, J. C., Lescot, M., Lheureux, F., Lockhart, B. E., Matsumoto, T., Piffanelli, P. and Iskra-Caruana, M. L. (2008). A single *Banana streak virus* integration event in the banana genome as the origin of infectious endogenous pararetrovirus. *Journal of Virology* 82(13): 6697–6710.

Geering, A. D. W. (2021). Badnaviruses (*Caulimoviridae*). In: Bamford, D. H. and Zuckerman, M. (Eds), *Encyclopedla of Vlrology* (4th edn.), Amsterdam: Elsevier/Academic Press, Netherlands. pp.158–168.

Geering, A. D. W., Maumus, F., Copetti, D., Choisne, N., Zwickl, D. J., Zytnicki, M., McTaggart, A. R., Scalabrin, S., Vezzulli, S., Wing, R. A., Quesneville, H. and Teycheney, P. Y. (2014). Endogenous florendoviruses are major components of plant genomes and hallmarks of virus evolution. *Nature Communications* 5: 5269.

Geering, A. D. W., McMichael, L. A., Dietzgen, R. G. and Thomas, J. E. (2000). Genetic diversity among *Banana streak virus* isolates from Australia. *Phytopathology* 90(8): 921–927.

Geering, A. D. W., Olszewski, N. E., Dahal, G., Thomas, J. E. and Lockhart, B. E. (2001). Analysis of the distribution and structure of integrated *Banana streak virus* DNA in a range of *Musa* cultivars. *Molecular Plant Pathology* 2(4): 207–213.

Geering, A. D. W., Olszewski, N. E., Harper, G., Lockhart, B. E. L., Hull, R. and Thomas, J. E. (2005a). Banana contains a diverse array of endogenous badnaviruses. *Journal of General Virology* 86(2): 511–520.

Geering, A. D. W., Pooggin, M. M., Olszewski, N. E., Lockhart, B. E. and Thomas, J. E. (2005b). Characterisation of Banana streak Mysore virus and evidence that its DNA is integrated in the B genome of cultivated *Musa*. *Archives of Virology* 150(4): 787–796.

Gorbunova, V. and Levy, A. A. (1997). Non-homologous DNA end joining in plant cells is associated with deletions and filler DNA insertions. *Nucleic Acids Research* 25(22): 4650–4657.

Grasswitz, T. R. and James, D. G. (2008). Movement of grape mealybug, *Pseudococcus maritimus*, on and between host plants. *Entomologia Experimentalis et Applicata* 129(3): 268–275.

Greaves, I. K., Gonzalez-Bayon, R., Wang, L., Zhu, A., Liu, P. C., Groszmann, M., Peacock, W. J. and Dennis, E. S. (2015). Epigenetic changes in hybrids. *Plant Physiology* 168(4): 1197–1205.

Harper, G. and Hull, R. (1998). Cloning and sequence analysis of banana streak virus DNA. *Virus Genes* 17(3): 271–278.

Harper, G., Dahal, G., Thottappilly, G. and Hull, R. (1999a). Detection of episomal banana streak badnavirus by IC-PCR. *Journal of Virological Methods* 79(1): 1–8.

Harper, G., Hart, D., Moult, S. and Hull, R. (2002). Detection of *Banana streak virus* in field samples of bananas from Uganda. *Annals of Applied Biology* 141(3): 247–257.

Harper, G., Hart, D., Moult, S., Hull, R., Geering, A. and Thomas, J. (2005). The diversity of *Banana streak virus* isolates in Uganda. *Archives of Virology* 150(12): 2407–2420.

Harper, G., Osuji, J. O., Heslop-Harrison, J. S. and Hull, R. (1999). Integration of banana streak badnavirus into the *Musa* genome: molecular and cytogenetic evidence. *Virology* 255(2): 207–213.

James, A. P., Geijskes, R. J., Dale, J. L. and Harding, R. M. (2011a). Development of a novel rolling-circle amplification technique to detect *Banana streak virus* that also discriminates between integrated and episomal virus sequences. *Plant Disease* 95(1): 57–62.

James, A. P., Geijskes, R. J., Dale, J. L. and Harding, R. M. (2011b). Molecular characterisation of six badnavirus species associated with leaf streak disease of banana in East Africa. *Annals of Applied Biology* 158(3): 346–353.

James, A. P., Kidanemariam, D. B., Hamill, S. D., Dale, J. L. and Harding, R. M. (2021). Infectivity of an infectious clone of banana streak CA virus in A-genome bananas (*Musa acumlnata* ssp.). *Viruses* 13(6): 1071.

Javer-Higginson, E., Acina-Mambole, I., González, J. E., Font, C., González, G., Echemendía, A. L., Muller, E. and Teycheney, P.-Y. (2014). Occurrence, prevalence and molecular diversity of banana streak viruses in Cuba. *European Journal of Plant Pathology* 138: 157–166.

Jaufeerally-Fakim, Y., Khorugdharry, A. and Harper, G. (2006). Genetic variants of banana streak virus in Mauritius. *Virus Research* 115(1): 91–98.

Karanja, L., Wangai, A., Harper, G. and Pathak, R. S. (2008). Molecular identification of *Banana streak virus* isolates in Kenya. *Journal of Phytopathology* 156(11–12): 678–686.

Kubiriba, J., Legg, J. P., Tushemereirwe, W. and Adipala, E. (2001). Vector transmission of *Banana streak virus* in the screenhouse in Uganda. *Annals of Applied Biology* 139(1): 37–43.

La Fleur, D. A., Lockhart, B. and Olszewksi, N. E. (1996). Portions of the banana streak badnavirus genome are integrated in the genome of its host *Musa* sp. *Phytopathology* 86: S100.

Lassoudiere, A. (1974). La mosaïque dite "a tirets" du bananier 'Poyo' en Côte d'Ivoire. *Fruits* 29: 349–357.

Lassoudiere, A. (1979). Mise en évidence des répercussions économiques de la mosaïque en tirets du bananier en Côte d'Ivoire. Possibilités de lutte par éradication. *Fruits* 34(1): 3–34.

Le Provost, G., Iskra-Caruana, M. L., Acina, I. and Teycheney, P. Y. (2006). Improved detection of episomal *Banana streak vIruses* by multiplex immunocapture PCR. *Journal of Virological Methods* 137(1): 7–13.

Lheureux, F., Laboureau, N., Muller, E., Lockhart, B. E. L. and Iskra-Caruana, M.-L. (2007). Molecular characterization of banana streak acuminata Vietnam virus isolated from *Musa acuminata siamea* (banana cultivar). *Archives of Virology* 152: 1409–1416.

Lockhart, B. E. L. (1986). Purification and serology of a bacilliform virus associated with banana streak disease. *Phytopathology* 76(10): 995–999.

Lockhart, B. E. L. (1990). Evidence for a double-stranded circular DNA genome in a second group of plant viruses. *Phytopathology* 80(2): 127–131.

Lockhart, B. E. L. and Olszewski, N. E. (1993). Serological and genomic heterogeneity of banana streak Badnavirus: implications for virus detection in *Musa* germplasm. In: Ganry, J. (Ed.). *Breeding Banana and Plantain for Resistance to Diseases and Pests*. Montpellier, France: CIRAD, in collaboration with INIBAP (International Network for the Improvement of Banana and Plantain), pp.105–113.

Medberry, S. L., Lockhart, B. E. and Olszewski, N. E. (1990). Properties of Commelina yellow mottle virus's complete DNA sequence, genomic discontinuities and transcript suggest that it is a pararetrovirus. *Nucleic Acids Research* 18(18): 5505–5513.

Meyer, J. B., Kasdorf, G. G. F., Nel, L. H. and Pietersen, G. (2008). Transmission of activated-episomal *Banana streak OL (badna)virus* (BSOLV) to cv. Williams banana (*Musa* sp.) by three mealybug species. *Plant Disease* 92(8): 1158–1163.

Mgocheki, N. and Addison, P. (2010). Spatial distribution of ants (Hymenoptera: Formicidae), vine mealybugs and mealybug parasitoids in vineyards. *Journal of Applied Entomology* 134(4): 285–295.

Muller, E. (2021). Cacao swollen shoot virus (*Caulimoviridae*). In: Bamford, D. H. and Zuckerman, M. (Eds). *Encyclopedia of Virology* (4th edn.). Oxford: Academic Press, pp. 274–284.

Muller, E., Dupuy, V., Blondin, L., Bauffe, F., Daugrois, J. H., Nathalie, L. and Iskra-Caruana, M. L. (2011). High molecular variability of sugarcane bacilliform viruses in Guadeloupe implying the existence of at least three new species. *Virus Research* 160(1–2): 414–419.

Ndowora, T., Dahal, G., LaFleur, D., Harper, G., Hull, R., Olszewski, N. E. and Lockhart, B. (1999). Evidence that badnavirus infection in *Musa* can originate from integrated pararetroviral sequences. *Virology* 255(2): 214–220.

Ngo, T. H., Webb, R., Crew, K. S., Vance, M. E., Thomas, J. E. and Geering, A. D. W. (2020). Identification of putative viroplasms within banana cells infected by banana streak MY virus. *Journal of General Virology* 101(12): 1305–1312.

Noumbissié, G. B., Chabannes, M., Bakry, F., Ricci, S., Cardi, C., Njembele, J.-C., Yohoume, D., Tomekpe, K., Iskra-Caruana, M.-L., D'Hont, A. and Baurens, F.-C. (2016). Chromosome segregation in an allotetraploid banana hybrid (AAAB) suggests a translocation between the A and B genomes and results in eBSV-free offsprings. *Molecular Breeding* 36(4): 38.

Ricciuti, E., Laboureau, N., Noumbissié, G., Chabannes, M., Sukhikh, N., Pooggin, M. M. and Iskra-Caruana, M. L. (2021). Extrachromosomal viral DNA produced by transcriptionally active endogenous viral elements in non-infected banana hybrids impedes quantitative PCR diagnostics of banana streak virus infections in banana hybrids. *Journal of General Virology* 102(11): 1–14.

Staginnus, C. and Richert-Pöggeler, K. R. (2006). Endogenous pararetroviruses: two-faced travelers in the plant genome. *Trends in Plant Science* 11(10): 485–491.

Sukal, A. C., Kidanemariam, D. B., Dale, J. L., Harding, R. M. and James, A. P. (2019). Assessment and optimization of rolling circle amplification protocols for the detection and characterization of badnaviruses. *Virology* 529: 73–80.

Teycheney, P. Y., Geering, A. D. W., Dasgupta, I., Hull, R., Kreuze, J. F., Lockhart, B., Muller, E., Olszewski, N., Pappu, H., Pooggin, M. M., Richert-Pöggeler, K. R., Schoelz, J. E., Seal, S., Stavolone, L., Umber, M. and ICTV Report Consortium (2020). ICTV virus taxonomy profile: *Caulimoviridae*. *Journal of General Virology* 101(10): 1025–1026.

Thomas, J. E. (2015). *MusaNet Technical Guidelines for the Safe Movement of Musa Germplasm* (3rd edn.). Rome: Bioversity International.

Thomas, J., Iskra Caruana, M.-L., Kumar, P. L., Roux, N., Chabannes, M., Massart, S., Mathieu, Y., Chase, R. and Van Den Houwe, I. (2015). Position paper on a strategy to distribute banana (*Musa*) germplasm with endogenous banana streak virus genomes. Available at: https://agritrop.cirad.fr/579485/1/Position%20paper%20on%20BSV%20distribution%20strategies_versionWEB.pdf (accessed 21 March 2023).

Thottappilly, G., Dahal, G. and Lockhart, B. E. L. (1998). Studies on a Nigerian isolate of banana streak Badnavirus: I. Purification and enzyme linked immunosorbent assay. *Annals of Applied Biology* 132(2): 253–261.

Tripathi, J. N., Ntui, V. O., Ron, M., Muiruri, S. K., Britt, A. and Tripathi, L. (2019). CRISPR/Cas9 editing of endogenous banana streak virus in the B genome of *Musa* spp. overcomes a major challenge in banana breeding. *Communications Series Biology* 2(1): 46.

Umber, M., Pichaut, J.-P., Farinas, B., Laboureau, N., Janzac, B., Plaisir-Pineau, K., Pressat, G., Baurens, F., Chabannes, M., Duroy, P., Guiougou, C., Delos, J., Jenny, C., Iskra-Caruana, M., Salmon, F. and Teycheney, P. (2016). Marker-assisted breeding of *Musa balbisiana* genitors devoid of infectious endogenous banana streak virus sequences. *Molecular Breeding* 36(6): 74.

Umber, M., Pressat, G., Fort, G., Plaisir Pineau, K., Guiougiou, C., Lambert, F., Farinas, B., Pichaut, J. P., Janzac, B., Delos, J. M., Salmon, F., Dubois, C. and Teycheney, P. Y. (2022). Risk assessment of infectious endogenous banana streak viruses in Guadeloupe. *Frontiers in Plant Science* 13: 951285.

Vo, J. N., Campbell, P. R., Mahfuz, N. N., Ramli, R., Pagendam, D., Barnard, R. and Geering, A. D. W. (2016). Characterization of the banana streak virus capsid protein and mapping of the immunodominant continuous B-cell epitopes to the surface-exposed N terminus. *Journal of General Virology* 97(12): 3446–3457.

Wardlaw, C. W. (1961). *Banana Diseases*. Edinburgh: Longmans, Green and Co.

Yang, I. C., Hafner, G. J., Revill, P. A., Dale, J. L. and Harding, R. M. (2003). Sequence diversity of South Pacific isolates of Taro bacilliform virus and the development of a PCR-based diagnostic test. *Archives of Virology* 148(10): 1957–1968.

Yot-Dauthy, D. and Bové, J.-M. (1966). Mosaïque du bananier. Identification et purification de diverses souches du virus. *Fruits* 21: 449–466.

Zhuang, J., Wang, J.-h., Zhang, X. and Liu, Z.-x. (2011). Molecular characterization of *Banana streak virus* isolate from *Musa Acuminata* in China. *Virologica Sinica* 26: 393–402.

Chapter 15

Other viral pathogens of banana

Andrew D. W. Geering, Queensland Alliance for Agriculture and Food Innovation, The University of Queensland, Australia

1 Introduction

Banana (*Musa* spp.) and abacá (*Musa textilis*) are susceptible to infection by a broad range of viruses, which have received little attention compared to banana bunchy top virus and the banana streak complex of viruses, because they are not widely distributed [e.g. banana bract mosaic virus (BBrMV)], because they do not cause appreciable yield declines [e.g. banana mild mosaic virus (BanMMV)] or because they are only sporadically a problem [e.g. cucumber mosaic virus (CMV)]. These viruses are disseminated through vegetative propagative practices and may reach a very high level of incidence, if not uniformly infect some cultivars. In this chapter, the characteristics of these minor viral pathogens of banana, plantain and abacá are discussed.

2 Cucumber mosaic virus

The first description of infectious chlorosis or mosaic disease of banana is credited to Charles Magee, who reported the disease in banana

http://dx.doi.org/10.19103/AS.2022.0108.15

plants growing in northern New South Wales, Australia, in May 1929 and demonstrated the infectious nature of the disease by transmitting it to healthy banana plants using the banana aphid *Pentalonia nigronervosa* (Magee, 1930). In further studies, Magee successfully transmitted the suspected virus from banana to cucurbit and other indicator hosts such as tobacco and tomato using several different aphid species. Based on symptomatology, he correctly concluded that the disease was caused by CMV, which at the time was known as cucumis virus 1 (Magee, 1940). CMV has now been reported as a pathogen of both banana and abacá (Furuya et al., 2006) from many countries around the world.

2.1 Impact and host range of cucumber mosaic virus

On a global scale, CMV is only a minor pathogen of banana and is generally sporadic in appearance, but occasionally severe epidemics and serious yield losses are reported. As with all plant viruses, CMV was named after the first host plant species in which it was recorded and the name belies its importance as a pathogen of many other plant species on all arable continents, including vegetables (e.g. tomato, capsicum, cucurbits and lettuce), pulse legumes (e.g. lupin, lentil and chickpea) and a wide variety of ornamental plant species and weeds (CABI, 2022). With 1071 host plant species in 521 genera from 100 families, CMV has one of the widest host ranges of any plant virus (Yoon et al., 2019). Infection in banana is often a spillover of virus epidemics that occur in more susceptible plant species that grow under or near the bananas.

2.2 Disease symptoms of cucumber mosaic virus

The symptoms of CMV infection in banana vary from a mild leaf mottle to plant mortality (Fig. 1). At the severe end of the disease spectrum, sheath leaf rot and internal necrosis of the pseudostem are observed. Plant stunting, deformed fruit and leaves and varying degrees of leaf chlorosis, mosaic patterns and necrotic streaking may also be observed (Chou et al., 2009). The severity of disease symptoms is influenced by both the virus and plant genotypes as well as by the prevailing environmental conditions. On the Deccan plateau of southern India, the heart rot syndrome in 'Basrai' bananas (*Musa* AAA group, Cavendish subgroup) only occurs during the cooler winter growth season, whereas during the warmer summer growth season, milder mosaic symptoms predominate (Mali and Rajegore, 1980). There is also circumstantial evidence that CMV synergistically interacts with BanMMV to cause leaf necrosis, compared to when the plant is infected with CMV by itself, when milder mosaic symptoms are observed (Caruana and Galzi, 1998).

Figure 1 Various *Musa* AAA group cultivars infected with cucumber mosaic virus (photographs: A. D. W. Geering (a) and Scot C. Nelson (b and c).

2.3 Taxonomy of cucumber mosaic virus, strain variation and satellite RNAs

CMV is classified in the genus *Cucumovirus*, which is one of six genera in the family *Bromoviridae* (Bujarski et al., 2019). It has isometric virions with a diameter of 29 nm and a genome of 8623 nt that is divided between three linear, positive-sense, single-stranded RNA segments labelled RNA 1, 2 and 3 in order of decreasing size (NCBI RefSeq: NC_002034, NC_002035 and NC_001440, respectively). RNA 1 is monocistronic and codes for a replication-associated protein, whereas the remaining two RNA segments are bicistronic, with the RNA-dependent RNA polymerase (RdRp) and an RNA-silencing suppressor protein coded by RNA 2 (protein 2b) and the movement protein (MP) and coat protein (CP) coded by RNA 3. Bicistronic RNAs pose a problem for the plant's protein translational machinery, as the scanning ribosome will detach from the RNA once the stop codon of the first open reading frame (ORF) is reached and will not reinitiate translation at the start codon of the second ORF. To overcome this problem, CMV produces two sub-genomic RNAs, labelled RNA 4a and 4b, each of which contains the downstream ORF of RNA 2 and 3, respectively.

Apart from the core genomic RNA components, extra-viral RNA components known as satellite RNAs are sometimes observed, which bear no sequence homology to CMV but are dependent on this virus for replication and packaging into virions. These satellite RNAs can alter symptom severity but in an unpredictable manner. Sometimes, the presence of a satellite RNA makes symptoms worse (Takanami, 1981; Xu and Roossinck, 2000), but at other times, the symptoms may be ameliorated (Mossop and Francki, 1979). Satellite RNAs have been observed associated with at least two CMV isolates from banana, but in these instances, the satellite RNA appeared either to mitigate disease symptoms (Gafny et al., 1996) or to not have any impact on the disease (Chou et al., 2009).

CMV is a very variable virus and is divided into two subgroups, I and II, based on phylogenetic relationships using the CP gene and differential reactions with selected monoclonal antibodies (Hsu et al., 2000). Subgroup I isolates of the virus are further divided into IA and IB, and IB isolates are hypothesised to have evolved in eastern Asia (Hirsch and Moury, 2021). Virtually all isolates of CMV that have been recovered from banana belong to subgroup I (Kebede and Majumder, 2021; Fidan and Koç, 2019; Singh et al., 1995; Khan et al., 2011; Vishnoi et al., 2013; Chou et al., 2009; Kebede and Majumder, 2020; Ayo-John et al., 2008). This preponderance of subgroup I isolates in banana does not necessarily indicate greater susceptibility of the plant to this virus subgroup. About 80% of all CMV isolates discovered in the world belong to subgroup I (Hirsch and Moury, 2021). Furthermore, subgroup II isolates are thought to have lower temperature optima for replication, hence a greater prevalence in more temperate regions of the world (Hirsch and Moury, 2021).

2.4 Diagnosis of cucumber mosaic virus

Diagnosis of CMV infections is quite straightforward. Commercial enzyme-linked immuosorbent assay (ELISA) kits for this virus are available from a range of companies, and these kits provide sensitive and specific virus detection and are easy to use. Furthermore, some companies offer ELISA kits that distinguish the two subgroups of viruses. For even greater convenience, lateral flow strips (immunostrips) can also be purchased, providing a 5-min diagnosis. One important consideration when doing any type of serological assay for CMV is to use a citrate buffer-based tissue extraction solution, as when using more conventional phosphate buffers, the virus appears to precipitate in the presence of the plant extract, with a subsequent loss in sensitivity of the assay (Francki and Hatta, 1980). A suitable tissue extraction solution for CMV contains 0.05 M trisodium citrate, pH 8.0, 0.5 mM ethylenediaminetetraacetic acid, 2% polyvinylpyrrolidone, 0.05% Tween-20 and 0.5% monothioglycerol (Sharman et al., 2000b).

Numerous other options exist to diagnose CMV infections, including reverse transcription polymerase chain reaction (RT-PCR) (Hu et al., 1995; Singh et al., 1995), recombinase polymerase amplification (Srivastava et al., 2019), Luminex xTAG technology (Bald-Blume et al., 2017) and RT-loop-mediated isothermal amplification (Zhang et al., 2018; Hasiów-Jaroszewska et al., 2018). A multiplex immunocapture PCR (M-IC-PCR) for the simultaneous detection of CMV, banana bunchy top virus and BBrMV in crude sap extracts has also been developed (Sharman et al., 2000b). Under a transmission electron microscope, the isometric virions of CMV are quite distinctive as they often have a characteristic 'doughnut-like' appearance due to the penetration of the negative stain into the centre of the virion (Hatta and Francki, 1979; Vishnoi et al., 2013). CMV is mechanically transmissible and can be inoculated onto biological indicator plants such as *Nicotiana tabacum*, *N. glutinosa*, *Chenopodium amaranticolor*, *Chenopodium quinoa* and *Vigna unguiculata* (Chou et al., 2009). Other banana-infecting viruses that produce superficially similar disease symptoms will not infect these plant species.

2.5 Epidemiology and control of cucumber mosaic virus

CMV is transmitted in a non-persistent manner by at least 80 different aphid species (CABI, 2022). Virus acquisition by the aphid can occur in as little as 5 s, and transmission may be equally brief. However, the aphid only remains viruliferous for a matter of hours and the virus is lost when the aphid moults or repeatedly probes healthy plant tissue. Migratory winged aphids (alatae) are poor at distinguishing a suitable host plant through visual or olfactory cues alone and undertake exploratory probes before choosing to settle, moving short distances if a suitable host plant is not immediately found (Döring, 2014). It is during this host-finding phase that aphids transmit viruses such as CMV, and aphid species that are incapable of colonising the plant species in question may still be important as virus vectors (Berlandier et al., 1997). These characteristics of non-persistent transmission mean that the use of insecticides to control the vectors is ineffective, as the insecticides do not kill the aphids fast enough to prevent acquisition and transmission of the virus.

Alternative host plant species play an important role in the epidemiology of CMV in banana. In Morocco, where most bananas are produced under cover in plastic greenhouses (Galán Saúco, 2020), epidemics of CMV are exacerbated by intercropping with muskmelon *Cucumis melo* (Bouhida and Lockhart, 1990). In banana gardens in southern India, CMV can be found in weedy *Celosia* and *Datura* spp. (Mali and Rajegore, 1980). It therefore follows that one of the main pillars of disease control is the eradication of virus reservoirs in or near the banana plantations. As with all other viral pathogens of banana, it is also very important to begin the crop with clean planting materials and to rogue infected plants as they occur, ensuring that no suckers remain.

CMV is an exceptionally well-studied plant virus, with comprehensive information available for many aspects of its biology, but there are still knowledge gaps relating to the epidemiology of this virus in banana. The major alternative hosts of the virus will vary between regions and need to be identified to develop region-specific disease control programs. The question of whether the spread of CMV in banana is monocyclic or polycyclic (i.e. secondary spread occurs between the banana plants through the action of aphids) has not been determined. Yield impacts from infection have not been quantified nor the effects of infection on the quality of the fruit. There are well-studied examples of synergism between CMV and co-infecting viruses in other crop species (Murphy and Bowen, 2006; Wege and Siegmund, 2007; Wang et al., 2002), and this phenomenon warrants further investigation in banana.

3 Banana bract mosaic virus

Banana bract mosaic disease (syn. Kokkan disease in India) of banana was first reported in India in 1966 and in the Philippines in 1979 (Magnaye and Espino, 1990; Samraj et al., 1966). Preliminary evidence for a viral aetiology for this disease was provided by Magnaye and Espino (1990), who successfully transmitted the infective agent using aphids, and then by Espino et al. (1990), who observed flexuous and filamentous virions in enriched plant extracts under a transmission electron microscope. BBrMV was tentatively assigned to the genus *Potyvirus* by Bateson and Dale (1995) using a universal potyvirus RT-PCR assay followed by amplicon sequencing and a reference genome sequence of a Philippines banana virus isolate was generated by Ha et al. (2008). The presence of BBrMV in southern India was confirmed by Rodoni et al. (1997) and Thomas et al. (1997).

BBrMV likely originated in India, as the Southeast Asian and eastern Pacific virus isolates form a subclade that is nested within a larger clade of Indian virus isolates (Rodoni et al., 1997; Siljo et al., 2012; Balasubramanian and Selvarajan, 2014; Selvarajan and Balasubramanian, 2023; Zhang et al., 2016). The geographic distribution of BBrMV now extends to islands in the Pacific Ocean and South America, and this trans-oceanic spread of the virus would have occurred as a result of human-mediated transport of plants or plant propagules of banana or ornamental host plant species.

3.1 Impact and host range of banana bract mosaic virus

BBrMV is found in India, Sri Lanka, the Philippines, Thailand, Viet Nam, Western Samoa, the USA (Hawaii) and Ecuador (Rodoni et al., 1999; Quito-Avila et al., 2013; Wang et al., 2010; Thomas et al., 1997). While BBrMV is common in India and the Philippines, records elsewhere are limited to single disease reports, suggesting that the virus is still restricted in distribution in these countries. BBrMV is ubiquitous in the AAB cooking banana cultivars 'Saba' and 'Cardaba' throughout the Philippines and also infects 'Cavendish' plantations to varying extents in Mindanao (Magnaye and Espino, 1990). BBrMV is an important pathogen of abacá (*M. textilis*) in south and south-eastern Luzon and in the Eastern Visayas, where it rivals the abacá strain of sugarcane mosaic virus (SCMV) in prevalence and is often confused with this virus (Cruz et al., 2016; Sharman et al., 2000a).

BBrMV infection reduces the yield of bananas by 20–50%, depending on the cultivar in question and environmental conditions such as soil fertility (Selvarajan and Jeyabaskaran, 2006; Thomas et al., 1999a; Gasper, 2001). Infection with this virus has a small impact on the number of hands per bunch and the number of fingers per hand, but a very large impact on the length, girth and weight of the fingers (Selvarajan and Jeyabaskaran, 2006; Gasper, 2001). In a small percentage of severely diseased bananas, the cigar leaf becomes necrotic, the meristem dies and a bunch is not thrown. Economic losses associated with bract mosaic disease in 'Nendran' plantains in India are estimated at US$48 million per annum (Selvarajan and Balasubramanian, 2023).

In addition to abacá, the alternative hosts of BBrMV are ornamental bananas (*Musa ornata* and hybrids) (Madhavan et al., 2022), cardamom (*Elettaria cardamomum*) (Siljo et al., 2012) and purple ginger (*Alpinia purpurata*) (Wang et al., 2010; Zhang et al., 2016). The possible roles of these alternative host plants in viral epidemics in banana are discussed later.

3.2 Taxonomy of banana bract mosaic virus

BBrMV is classified in the genus *Potyvirus*, which is one of 12 genera in the family *Potyviridae* (Inoue-Nagata et al., 2022). It has flexuous and filamentous virions that are approximately 725 nm long and a positive sense, single-stranded RNA genome of 9711 nt (NCBI RefSeq: NC_009745). The genome of BBrMV is typical of a potyvirus, with a 9378 nt-long ORF, which encodes a polyprotein that is cleaved into ten mature proteins at semi-conserved sites by three self-encoded proteases (Inoue-Nagata et al., 2022). The BBrMV capsid has only a single capsomer, the CP, although three or more major protein bands are observed when purified virions are run on a sodium dodecyl sulphate-polyacrylamide gel, possibly resulting from protein degradation during purification (Thomas et al., 1997). A second small ORF called *pipo* is located in a +2 translational reading frame to the polyprotein ORF, and through a transcriptional slippage mechanism, RNA transcripts are generated that encode a P3N–PIPO fusion protein (Olspert et al., 2015; Chung et al., 2008).

3.3 Disease symptoms of banana bract mosaic virus

In the initial stages of the disease in banana, continuous greenish or brownish spindle-shaped streaks or long, irregularly scattered, continuous and discontinuous stripes appear along the leaf petiole (Magnaye and Espino, 1990). When the bunch emerges, similar discolorations are observed along the peduncle and on the bracts of the inflorescence (Fig. 2a). Leaf mosaic symptoms are often absent but when present appear as distinct discontinuous streaks or spindle-shaped lesions running parallel to the primary veins (Fig. 2b-c). If the outer leaf sheaths of the pseudostem are peeled away, pinkish or reddish-brown spindle-shaped streaks to stripe discolorations of varying lengths along the lower leaf sheaths are distinctive (Fig. 2d). These streaks can turn necrotic in some cultivars such as 'Nendran' (*Musa* AAB group). The bunches that are produced by BBrMV-infected plants are undersized and the fingers are shorter and thinner, abnormally curved and brittle and can exhibit mosaic patterns (Gasper, 2001). Leaf phyllotaxy may be disrupted, with the leaves arranged along a single vertical plane (Traveller's Palm appearance). Symptom expression is temperature dependent, and during the hot, humid summer season in Kerala, India (March to May), the symptoms of infection in banana ameliorate or disappear altogether (Selvarajan and Balasubramanian, 2023).

Abacá plants infected by BBrMV also have mosaic patterns on the flower bracts and chlorotic streaking or spindle-shaped lesions on the leaves and

Figure 2 Various *Musa* AAB group cultivars infected with banana bract mosaic virus: (a), bract mosaic symptoms on inflorescence (photograph: J. E. Thomas); (b) and (c), foliar symptoms (photograph: A. D. W. Geering) and (d), red streaks and necrosis on pseudostem (photograph: R. Selvarajan).

petioles (Lalusin and Villavicencio, 2015). The leaves may be severely distorted (stringy in appearance), and veins originating from the midrib are raised in older leaves.

3.4 Diagnosis of banana bract mosaic virus

BBrMV is most likely to be confused with CMV and SCMV based on disease symptoms. Identification of the virus can be quickly narrowed down to the genus level by the examination of a sap extract under a transmission electron microscope, as potyvirus virions are distinctive in size and shape and occur at a reasonably high titre. For a definitive diagnosis, a serological or molecular diagnostic assay is required. Highly specific monoclonal and polyclonal antibodies to BBrMV have been prepared using either CP expressed in *Escherichia coli* or virions purified from plants to immunise laboratory animals (Thomas et al., 1997; Rodoni et al., 1999; Selvarajan et al., 2020b). These antibodies have been successfully deployed in a variety of serological formats such as ELISA, Western blot, IC-PCR and lateral flow strip. A double-antibody sandwich ELISA reagent set is commercially available from Agdia Inc. (Elkhart, Indiana).

Rapid and sensitive species-specific RT-PCR and RT-LAMP assays have been developed for BBrMV (Sankaralingam et al., 2006; Koh et al., 2020; Wang et al., 2010; Madhu Kovileri et al., 2023; Siljo et al., 2014) and are a convenient and simple alternative to the aforementioned serological detection assays. Potyvirus universal RT-PCR assays (Pappu et al., 1993; Langeveld et al., 1991) have also been applied to the diagnosis of BBrMV (Bateson and Dale, 1995; Thomas et al., 1997; Balasubramanian and Selvarajan, 2012), but the amplicon must be sequenced to allow the identification of the virus to species level. Finally, BBrMV is one of three target viruses in the M-IC-PC developed by Sharman et al. (2000b).

3.5 Epidemiology and control of banana bract mosaic virus

As with all potyviruses, BBrMV is transmitted in a non-persistent manner by aphids and the vector specificity is weak, meaning a broad range of aphid species are potentially capable of transmitting the virus. Documented vectors include *Aphis gossypii*, *P. nigronervosa* and *Rhopalosiphum maidis* (Magnaye and Espino, 1990; Muñez, 1992; Madhavan et al., 2022). Attempts to mechanically transmit the virus between bananas or from banana to other plant species have been unsuccessful. BBrMV is seed transmitted at a high rate in fertile, diploid bananas, but this mode of transmission is mainly a concern for the *Musa* breeding programs and germplasm conservation programs, as commercial cultivars are sterile and vegetatively propagated (Selvarajan et al., 2020a).

There are many questions surrounding the epidemiology of BBrMV, such as the rate and timing of secondary spread of the virus by aphids in the field and the role of alternative host plant species as sources of viral inoculum. BBrMV is a pathogen of cardamom (*Elettaria cardamomum*) in southern India and is most prevalent in crops where either banana is grown adjacent to or immediately before a cardamom crop (Siljo et al., 2012). Close genetic relationships between some banana and cardamom virus isolates have been observed, supporting the hypothesis of exchange of the virus between these two crops. In Hawaii, BBrMV has only ever been detected in *Alpinia purpurata*, and there are no records of it spreading into bananas in this region (Wang et al., 2010; Zhang et al., 2016). However, the Hawaiian virus isolates are most closely related to banana and abacá virus isolates from Southeast Asia, suggesting cross-species transmission at one time in the past (Zhang et al., 2016).

As with all viral pathogens of banana, the cornerstone of a control program for BBrMV is the use of planting material that is certified to be free of the virus through rigorous testing of mother plants. Infected plants in a crop should be rogued, ensuring that the whole plant is either killed or removed to prevent suckers sprouting from the corm. There are no known sources of immunity to BBrMV, although some banana cultivars potentially have lower susceptibility or greater tolerance to infection (Estelitta et al., 1996). Much more systematic experimentation needs to be done before recommendations can be made as to possible disease resistant cultivars. As a rule, insecticides are ineffective at controlling non-persistently transmitted aphid-borne viruses, and their usage is counterproductive because of off-target effects on beneficial insects such as parisitoid wasps and pollinators. Applications of higher rates of fertiliser in the plant crop help to mitigate the impacts of the disease in the plant crop, but no benefits are received in later ratoon crops (Selvarajan et al., 2009) and the overuse of fertilisers should be discouraged for economic and environmental reasons.

4 Sugarcane mosaic virus

SCMV is one of the most economically important plant viruses in the world, affecting a range of monocotyledonous crops, including sugarcane, sorghum, maize and various turf and pasture grass species. This virus is highly variable, with many distinct strains recognised, but only one, the abacá strain (SCMV-abacá), is known to infect *Musa* species, including *M. textilis* (abacá or Manilla hemp) and *Musa* AAA group cultivars.

4.1 Occurrence and impact of sugarcane mosaic virus

SCMV-abacá is restricted to the Philippines, where *M. textilis* is native, and where a mosaic-like disease of this plant species was first noticed in 1925 (Lalusin and Villavicencio, 2015). World War II hastened the spread of SCMV-abacá, as many abacá plantations were abandoned during this period, allowing the pathogen to spread unabated (Lalusin and Villavicencio, 2015). After the war, prices for abacá fibre soared, causing a frenzy of new plantings. However, due to inadequate farmer education and overall disorganisation within the industry, actions were not taken to ensure that the planting materials were free of virus and that existing severely diseased plantations were destroyed after they had reached the end of their productive life (Reinking, 1955).

SCMV-abacá is still regarded as a very serious constraint to abacá production in the Philippines, particularly in the provinces of Camarines Norte, Camarines Sur, Albay, Sorsogon and Catanduanes (Raymundo et al., 2002). An economic analysis of the impact of SCMV alone on the abacá industry has not been done, but about 5.2% of abacá plants in the Bicol region in the Philippines were affected by a viral disease of one type or another (either bunchy-top or mosaic) in 1991, causing an estimated fibre yield loss of more than 800 000 kg valued at about ₱18 million (Raymundo et al., 2002). Both the yield and tensile strength of abacá fibre is reduced as a result of the mosaic disease complex (dela Cruz and Raymundo, 2009).

4.2 Taxonomy of sugarcane mosaic virus

SCMV is a member of the genus *Potyvirus* and as such shares many characteristics with BBrMV such as virion morphology, genome organisation and replication strategy (see Section 3.2). The exemplar virus isolate for SCMV is from maize grown in China (NCBI Reference Sequence: NC_003398). The genome of an isolate of SCMV-abacá has not been sequenced in entirety, only a c. 2 kb segment from the 3′ end (GenBank: AY222743) (Gambley et al., 2004). The N-terminal region of the CP of SCMV-abacá has an atypical amino acid repeat motif, which is thought to determine host specificity (Desbiez et al., 2014), supporting the recognition of this virus as a distinct biological strain of SCMV (Gambley et al., 2004).

4.3 Disease symptoms of sugarcane mosaic virus

The symptoms of SCMV-abacá infection in abacá are chlorotic, interveinal dashes, stripes and spindle-shaped lesions on the leaves and chlorotic mottling on the leaf sheaths and pseudostem (Lalusin and Villavicencio, 2015). The chlorotic areas on the leaf develop rusty brown borders and extend from the midrib to the leaf margin. BBrMV and CMV also infect abacá in the Philippines and cause overlapping disease symptoms. It is difficult to interpret much of the scientific literature on abacá mosaic disease, as the diagnostic technologies either were not yet available or were not utilised to allow differentiation of these three viruses.

4.4 Diagnosis of sugarcane mosaic virus

SCMV-abacá can be separated from other viruses associated with abacá mosaic disease (CMV and BBrMV) using maize as a biological indicator plant (Eloja and Tinsley, 1963). However, care must be taken to choose one of the older maize cultivars, as many modern cultivars have been bred to have some form of resistance to SCMV.

Several companies sell high-quality ELISA reagents for SCMV, but virtually all are created using antibodies generated against sugarcane or maize strains of the virus, which are weakly serologically related to SCMV-abacá (Eloja and Tinsley, 1963; Gambley et al., 2004). Although the sensitivity of detection of SCMV-abacá would not be as great as with the homologous antigen, these ELISA reagents should still allow the detection of SCMV-abacá. The CP of SCMV-abacá has been expressed in *E. coli* and used to prepare antibodies against the virus (Gambley et al., 2004).

A multitude of RT-PCR assays have been developed for SCMV, but these would need to be validated against SCMV-abacá before they could be used, as this is a divergent strain of the virus. A triplex RT-PCR assay has been developed for the simultaneous detection of BBrMV, SCMV-abacá and a *Musa* gene as an assay quality control (Galvez et al., 2020).

4.5 Epidemiology and control of sugarcane mosaic virus

SCMV-abacá is transmitted in a non-persistent manner by aphids such as *R. maidis* and *Aphis gossypii* (Gambley et al., 2004). The various strains of SCMV occupy different ecological niches, so it is not possible to extrapolate host range data from typical sugarcane or maize strains to the abacá strain. The only known natural hosts of SCMV-abacá are *Musa* spp. (banana and abacá), although *Zea mays* is an experimental host (Eloja and Tinsley, 1963). During the post-World War II period, corn was often grown as a secondary crop to abacá to supplement farmers' income, and it is perhaps a telling observation that when corn was interplanted with abacá, the productive life of the plantation was only 4 to 18 months, compared to 2½ to 6 years when the corn was planted apart from the abacá (Reinking, 1955).

There are several recent reports of SCMV infecting *Canna* species and hybrids (Li et al., 2019; Tang et al., 2016; Betancourth et al., 2020), and experiments to determine whether these virus isolates could infect *Musa* spp. are warranted. Reciprocal aphid inoculations of SCMV-abacá onto *Canna indica* resulted in infection (Celino, 1940), although this result needs to be substantiated using modern diagnostic techniques.

Any disease management plan for SCMV-abacá must begin with the use of clean planting materials and the rogueing of infected plants. Disease management is hindered by neighbouring farmers abandoning plantations when the disease incidence becomes too high, creating large, external sources of inoculum that are not easily managed. The use of insecticides to control aphid vectors is ineffective and harmful to the environment (Celebrar et al., 1970). There are good prospects for breeding resistance to the various viruses that infect abacá, as this plant species is diploid and seeded and cross-compatible with several *Musa* species including *M. acuminata*, *M. balbisiana*, *M. lolodensis* and *M. borneensis*, although the F1 hybrid population is often partially or highly sterile (Lalusin and Villavicencio, 2015). Two hybrid accessions created by crossing *M. balbisiana* 'Pacol' with *M. textilis* 'Abuab' have strong resistance against abacá bunchy top disease and partial resistance against the mosaic diseases (Parac et al., 2019; Parac et al., 2021). As abacá is grown for fibre and not for food, developing transgenic resistance against the virus may be another option.

5 Banana mild mosaic virus

BanMMV is widely distributed around the world (De Clerck et al., 2017), a consequence of the fact that symptoms of infection are typically very mild or absent, and therefore the virus has escaped the notice of farmers and scientists alike. The discovery of BanMMV was serendipitous and occurred during routine screening of *Musa* germplasm accessions and field plants for unidentified viruses by electron microscopic examination of enriched sap extracts (viral minipreps) (Thomas et al., 1999b; De Clerck et al., 2017). This virus has received very little research attention, commensurate with its perceived lack of economic importance, and hence there is very little information known about this virus.

5.1 Occurrence and impact of banana mild mosaic virus

BanMMV most commonly occurs in AB/AAB genotype cooking bananas and plantains, rarely in East African highland and other minor AA/AAA genotype cultivars, and has never been recorded in Cavendish subgroup cultivars (Reichel et al., 2003; De Clerck et al., 2017). There have been no studies to measure the yield impact caused by BanMMV, but there are no overt effects of infection on bunch or fruit characteristics. Some banana cultivars, such as 'Ducasse' (AAB group, syn. 'Pisang Awak'), appear to be uniformly infected with the virus, but this has not detracted from utilization of these cultivars.

BanMMV often occurs in mixed infections with other viruses, mainly banana streak OL virus but sometimes also CMV (De Clerck et al., 2017; Reichel et al., 2003; Caruana and Galzi, 1998), and the issue of synergistic interactions between these viruses requires investigation. In Guadeloupe, unusually severe disease symptoms (necrosis) have been noticed in 'Cavendish' plants infected with both CMV and BBrMV compared to CMV alone (Caruana and Galzi, 1998).

5.2 Taxonomy of banana mild mosaic virus

BanMMV is classified within the genus *Banmivirus* in the family *Betaflexiviridae*. Virions are flexuous and filamentous, about 580 nm long and 14 nm wide (Thomas et al., 1999b), and contain a single-stranded genomic RNA of 7352 nt in the case of the exemplar isolate (NCBI Reference Sequence: NC_002729) of the species (Gambley and Thomas, 2001). The genome of BanMMV contains five ORFs: ORF 1 encodes the RNA-dependent RNA polymerase; ORFs 2-4 encode the triple gene block MPs and ORF 5 encodes the CP (Gambley and Thomas, 2001).

5.3 Disease symptoms of banana mild mosaic virus

Banana plants infected by BanMMV are often asymptomatic (Thomas et al., 1999b). Symptoms, when they occur, can be transitory in appearance and disappear as the plant matures. Fine chlorotic or silvery streaks are noticeable on the leaves of some cultivars infected by the virus, for example 'Daluyao' and 'Pisang Seribu', respectively, grading to a very mild mosaic in other cultivars such as 'Ducasse' (Fig. 3) (Thomas et al., 1999b).

Figure 3 *Musa* AAB group 'Ducasse' plant infected with banana bract mosaic virus.

5.4 Diagnosis of banana mild mosaic virus

Currently available diagnostic methods for BanMMV are RT-PCR or high-throughput sequencing (Hanafi et al., 2022; Thomas, 2015; Teycheney et al., 2007; Hanafi et al., 2020). Polyvalent antibody preparations have been prepared against BanMMV, which can be used in IC-RT-PCR (Teycheney et al., 2007), but these antibodies do not offer sufficient sensitivity or specificity of detection for deployment in other serological assay formats such as ELISA. One of the major challenges with primer design for RT-PCR is the very large sequence diversity between virus isolates. Some divergent isolates of BanMMV have only 75% nt identity with the reference genome sequence for the species across the entire virus genome (Hanafi et al., 2020). The recommended RT-PCR method for BanMMV is a form of 3′ RACE, whereby the reverse primer is an anchored oligo-dT primer (Poty-1 primer) that binds to the 3′ poly(A) tail of the genomic RNA, and the forward primer (BanMMV CP9 primer) is a degenerate primer incorporating inosine nucleosides, which binds to a conserved sequence motif within the CP gene (Hanafi et al., 2020).

5.5 Epidemiology and control of banana mild mosaic virus

No vectors of BanMMV have been identified, although there is circumstantial evidence of field spread as aggregated patterns of infection have been observed in the CIRAD *Musa ex situ* collection of live plants maintained in Guadeloupe (Teycheney et al., 2005a). The only known host plants of BanMMV are cultivated *Musa*, and infections are maintained by vegetative propagation practices.

BanMMV can be eliminated from *Musa* germplasm using a combination of heat therapy and meristem tip culture, although the treatment success rate is only 73% (Hanafi et al., 2022). There are concerted efforts to eliminate BanMMV from the ITC *Musa* germplasm collection where feasible. One of the problems with using tissue culture to eliminate BanMMV is that there is a concomitant risk of activating endogenous banana streak virus sequences, and therefore one problem of infection may only be replaced by a second.

6 Musa ornata-associated banmivirus

A second tentative banmivirus species called Musa ornata associated banmivirus (MOaBV) has been identified in the corms and leaves of *M. ornata* accession ITC0528 by high-throughput sequencing (Hanafi et al., 2022). At 7364 nt, the genome of this new virus is almost identical in length to that of BanMMV, and the arrangement of ORFs is likewise similar. Currently, the only way to distinguish the two viruses is by sequence analysis, as the genome sequences share 62.7% nt identity (Hanafi et al., 2022). Other than the aforementioned sequence data, nothing is known about the biology of MOaBV, but it is reasonable to assume that it would share many features with BanMMV.

7 Banana virus X

As its name suggests, banana virus X (BVX) is an enigmatic virus, only known from bananas growing on the island of Guadeloupe in the French West Indies (Teycheney et al., 2005b). BVX belongs to the family *Betaflexiviridae* but has not yet been assigned to a virus genus as only the 3′ half of its genome has been sequenced (GenBank: AY710267). The discovery of BVX was serendipitous during routine RT-PCR screening of the *ex situ Musa* germplasm collection of CIRAD for BanMMV (Teycheney et al., 2005a). BVX was identified in two distinct mother plants of one accession (*Musa* ABB group cv. 'Ney Mannan'), both of which were asymptomatically infected. A degenerate oligonucleotide RT-PCR assay has been developed, which enables simultaneous detection of BVX and BanMMV (Teycheney et al., 2007). Using this diagnostic assay, BVX has been detected in asymptomatic 'Cavendish' and 'Figue Pomme' plants from several farms around Guadeloupe, suggesting that the virus is widely distributed and maybe even indigenous to the island (Péréfarres et al., 2009). Again, all plants were asymptomatic, suggesting that BVX is non-pathogenic to banana.

8 Banana die-back virus

Isometric virions, 28–30 nm in diameter, have been associated with severely stunted banana cv. 'Valery' (*Musa* AAA group, Cavendish subgroup) plants in Ibadan, Nigeria (Hughes et al., 1998). A serological relationship was established between this virus, named banana die-back virus, and several nepoviruses including tobacco ringspot virus (Hughes et al., 1998). Moreover, this virus was mechanically transmissible to *V. unguiculata* TVu-76 and *Nicotiana occidentalis* but could not be back-transmitted to banana; hence proof of pathogenicity was not provided. No other information on this virus is available.

9 Conclusion and future trends

A recommendation that is common to all disease management plans for viral pathogens of banana is to use clean planting material and to rogue infected plants from the crop as they occur. However, this is not always an option as some banana cultivars such as 'Ducasse' are uniformly infected with BanMMV, but the question needs to be raised as to whether the virus is pathogenic and control measures are warranted.

High-throughput sequencing has accelerated the discovery of plant viruses, particularly those that latently infect the plant. It is almost certain that there are more undocumented viruses that infect banana, for example, uncharacterised closterovirid virions have been noticed in some *Musa* germplasm (De Clerck et al., 2017). Biological characterisation is not keeping pace with the rate of virus discovery, posing challenges to biosecurity authorities, as there is insufficient knowledge about the biology of the viruses to make informed decisions about the need to impose quarantine regulations. Counterintuitively, some plant viruses impart benefits to plants such as drought tolerance and are only pathogenic in some circumstances (González et al., 2021). Research is required to investigate the yield impact of several of the viruses mentioned in this chapter, but this has not been a priority for funding agencies as the viruses mainly occur in traditional landraces and not in the major commercial cultivars.

There are significant knowledge gaps concerning the epidemiology of all the plant viruses mentioned in this chapter, even for CMV as it pertains to banana. It is a concern that BBrMV is still expanding in geographic distribution, and in the case of Hawaii, its introduction appears to have been facilitated by the movement of alternative host plant species of the virus. The same risk applies to SCMV-abacá, which is likely to have a broader host range than just cultivated *Musa*.

10 Where to look for further information

Additional information on these viruses can be found at:

- CABI Compendium, https://www.cabidigitallibrary.org/journal/cabicompendium.
- International Committee on Taxonomy of Viruses, https://ictv.global/.
- Thomas, J. E. (2015). *Technical Guidelines for the Safe Movement of Musa Germplasm. 3rd Edition*. (Bioversity International: Rome), https://cgspace.cgiar.org/handle/10568/118459.

11 References

Ayo-John, E. I., d'Arros, H. J., Ekpo, E. J. A. and Shoyinka, S. A. (2008). A survey in Southern Nigeria reveals the presence of *Cucumber mosaic virus* subgroup I in *Musa* crops. *Fruits* 63(3): 135–143.

Balasubramanian, V. and Selvarajan, R. (2012). Complete genome sequence of a banana bract mosaic virus isolate infecting the French plantain cv. Nendran in India. *Archives of Virology* 157(2): 397–400.

Balasubramanian, V. and Selvarajan, R. (2014). Genetic diversity and recombination analysis in the coat protein gene of *Banana bract mosaic virus*. *Virus Genes* 48(3): 509–517.

Bald-Blume, N., Bergervoet, J. H. W. and Maiss, E. (2017). Development of a molecular assay for the detection of *Cucumber mosaic virus* and the discrimination of its subgroups I and II. *Journal of Virological Methods* 243: 35–43.

Bateson, M. F. and Dale, J. L. (1995). Banana bract mosaic virus: characterisation using Potyvirus specific degenerate PCR primers. *Archives of Virology* 140(3): 515–527.

Berlandier, F. A., Thackray, D. J., Jones, R. A. C., Latham, L. J. and Cartwright, L. (1997). Determining the relative roles of different aphid species as vectors of cucumber mosaic and bean yellow mosaic viruses in lupins. *Annals of Applied Biology* 131(2): 297–314.

Betancourth, C., Salomón, C., Moreno Henao, J. M., Montaño, S., Salazar, C., Uribe, P., Martínez, A., Muñoz, L. and Cuervo, M. (2020). First report of Sugarcane mosaic virus in achira (*Canna edulis* Ker.) in Nariño, Colombia. *Revista U.D.C.A. Actualidad and Divulgación Científica* 23(1): e1461.

Bouhida, M. and Lockhart, B. E. (1990). Increase in importance of cucumber mosaic virus infection in greenhouse-grown bananas in Morocco. *Phytopathology* 80(10): 981.

Bujarski, J., Gallitelli, D., García-Arenal, F., Pallás, V., Palukaitis, P., Reddy, M. K., Wang, A. and ICTV Report Consortium (2019). ICTV Virus Taxonomy Profile: *Bromoviridae*. *Journal of General Virology* 100(8): 1206–1207.

CABI (2022). *Cucumber Mosaic Virus (Cucumber Mosaic).* CABI Publishing International C.A.B. International.

Caruana, M. L. and Galzi, S. (1998). Identification of uncharacterised filamentous viral particles on banana plants. *Acta Horticulturae* (490): 323–336.

Celebrar, P. R., Gonzales, C. S., Eloja, A. L. and Gavarra, M. R. (1970). Further studies on three methods of weed control and insecticide spraying as control measures against the abaca mosaic disease. *Philippine Journal of Plant Industry* 35: 79–86.

Celino, M. S. (1940). Experimental transmission of the mosaic of Abaca, or Manila hemp plant (*Musa textilis* Née). *Philippine Agriculturist* 29(5): 379–403.

Chou, C.-N., Chen, C.-E., Wu, M.-L., Su, H. and Yeh, H. (2009). Biological and molecular characterization of Taiwanese isolates of cucumber mosaic virus associated with Banana mosaic disease. *Journal of Phytopathology* 157(2): 85–93.

Chung, B. Y. W., Miller, W. A., Atkins, J. F. and Firth, A. E. (2008). An overlapping essential gene in the *Potyviridae*. *Proceedings of the National Academy of Sciences* 105(15): 5897–5902.

Cruz, F. C. S., Belen, G. B. and Alviar, A. N. (2016). Serological and molecular detection of mixed bunchy top and mosaic virus infections in abaca (*Musa textilis* Nee). *Philippine Agricultural Scientist* 99(1): 88–98.

De Clerck, C., Crew, K., Van den Houwe, I., McMichael, L., Berhal, C., Lassois, L., Haissam Jijakli, M., Roux, N., Thomas, J. and Massart, S. (2017). Lessons learned from the virus indexing of *Musa* germplasm: insights from a multiyear collaboration. *Annals of Applied Biology* 171(1): 15–27.

dela Cruz, C. S. and Raymundo, A. D. (2009). Decline in fiber yield and tensile strength of abaca due to mosaic and cost-returns analysis under varying disease-influenced production scenarios. *Philippine Journal of Crop Science* 34: 75–87.

Desbiez, C., Chandeysson, C. and Lecoq, H. (2014). A short motif in the N-terminal part of the coat protein is a host-specific determinant of systemic infectivity for two potyviruses. *Molecular Plant Pathology* 15(2): 217–221.

Döring, T. F. (2014). How aphids find their host plants, and how they don't. *Annals of Applied Biology* 165(1): 3–26.

Eloja, A. L. and Tinsley, T. W. (1963). Abacá mosaic virus and its relationship to sugar-cane mosaic. *Annals of Applied Biology* 51(2): 253–258.

Espino, T. M., Exconde, S. B., Zipagan, F. B. and Espino, R. R. C. (1990). Banana bract mosaic, a new disease of banana II. Isolation and purification for monoclonal antibody production. *The Philippine Agriculturist* 73(1): 61–68.

Estelitta, S., Suma, A. and Kadhakrishnan, T. C. (1996). Field screening of banana germplasm against kokkan disease. *Journal of Tropical Agriculture* 34: 68–69.

Fidan, H. and Koç, G. (2019). Occurrence, ecology and phylogeny of banana streak badnavirus (BSV) and cucumber mosaic cucumovirus (CMV) in *Musa* sp. production areas of the Mediterranean coastline of Turkey. *Applied Ecology and Environmental Research* 17(3): 5935.

Francki, R. I. B. and Hatta, T. (1980). Cucumber mosaic virus: variation and problems of identification. *Acta Horticulturae* (110): 167–174.

Furuya, N., Dizon, T. O. and Natsuaki, K. T. (2006). Molecular characterization of *Banana bunchy top virus* and *Cucumber mosaic virus* from Abaca (*Musa textilis* Nee). *Journal of Agricultural Sciences Tokyo University of Agriculture* 51(2): 92–101.

Gafny, R., Wexler, A., Mawassi, M., Israeli, Y. and Bar-Joseph, M. (1996). Natural infection of banana by a satellite-containing strain of cucumber mosaic virus: nucleotide sequence of the coat protein gene and the satellite RNA. *Phytoparasitica* 24(1): 49–56.

Galán Saúco, V. (2020). Broad overview of the subtropical banana industry. *Acta Horticulturae* 1272: 1–12.

Galvez, L. C., Koh, R. B. L., Barbosa, C. F. C. and Aquino, V. M. (2020). Multiplex reverse transcription-polymerase chain reaction for simultaneous detection of banana bract mosaic virus (BBrMV) and sugarcane mosaic virus (SCMV) in abaca. *Canadian Journal of Plant Pathology* 42(4): 572–583.

Gambley, C. F. and Thomas, J. E. (2001). Molecular characterisation of Banana mild mosaic virus, a new filamentous virus in *Musa* spp. *Archives of Virology* 146(7): 1369–1379.

Gambley, C. F., Thomas, J. E., Magnaye, L. V. and Herradura, L. (2004). Abacá mosaic virus: a distinct strain of Sugarcane mosaic virus. *Australasian Plant Pathology* 33(4): 475–484.

Gasper, S. R. (2001). Effect of Kokkan disease caused by banana bract mosaic virus on the growth and yield of banana. M.Sc. Thesis, Kerala Agricultural University: Vellayani, India.

González, R., Butković, A., Escaray, F. J., Martínez-Latorre, J., Melero, Í., Pérez-Parets, E., Gómez-Cadenas, A., Carrasco, P. and Elena, S. F. (2021). Plant virus evolution under strong drought conditions results in a transition from parasitism to mutualism. *Proceedings of the National Academy of Sciences of the United States of America* 118(6): e2020990118.

Ha, C., Coombs, S., Revill, P. A., Harding, R. M., Vu, M. and Dale, J. L. (2008). Design and application of two novel degenerate primer pairs for the detection and complete genomic characterization of potyviruses. *Archives of Virology* 153(1): 25–36.

Hanafi, M., Rong, W., Tamisier, L., Berhal, C., Roux, N. and Massart, S. (2022). Detection of banana mild mosaic virus in *Musa* in vitro plants: high-throughput sequencing presents higher diagnostic sensitivity than (IC)-RT-PCR and identifies a new Betaflexiviridae species. *Plants* 11(2): 226.

Hanafi, M., Tahzima, R., Ben Kaab, S., Tamisier, L., Roux, N. and Massart, S. (2020) Identification of divergent isolates of banana mild mosaic virus and development of a new diagnostic primer to improve detection. *Pathogens* 9(12): 1045.

Hasiów-Jaroszewska, B., Kasprowicz, M. and Borodynko-Filas, N. (2018). Rapid detection of *Cucumber mosaic virus* isolates representing distinct phylogenetic subgroups by reverse transcription, loop-mediated isothermal amplification. *Journal of Plant Diseases and Protection* 125(2): 227–230.

Hatta, T. and Francki, R. I. B. (1979). Enzyme cytochemical method for identification of cucumber mosaic virus particles in infected cells. *Virology* 93(1): 265–268.

Hirsch, J. and Moury, B. (2021). Cucumber mosaic virus (*Bromoviridae*). In: Bamford, D. H. and Zuckerman, M. (Eds), *Encyclopedia of Virology* (4th edn.). Oxford: Academic Press, pp. 371–382.

Hsu, H. T., Barzuna, L., Hsu, Y. H., Bliss, W. and Perry, K. L. (2000). Identification and subgrouping of *Cucumber mosaic virus* with mouse monoclonal antibodies. *Phytopathology* 90(6): 615–620.

Hu, J. S., Li, H. P. and Barry, K. (1995). Comparison of dot blot, ELISA, and RT-PCR assays for detection of two cucumber mosaic virus isolates infecting banana in Hawaii. *Plant Disease* 79(9): 902–906.

Hughes, J. D., Speijer, P. R. and Olatunde, O. (1998). Banana die-back virus: a new virus infecting banana in Nigeria. *Plant Disease* 82(1): 129–129.

Inoue-Nagata, A. K., Jordan, R., Kreuze, J., Li, F., López-Moya, J. J., Mäkinen, K., Ohshima, K., Wylie, S. J. and ICTV Report Consortium (2022). ICTV Virus Taxonomy Profile: *Potyviridae*. *Journal of General Virology* 103(5): 1–2.

Kebede, Y. and Majumder, S. (2020). Molecular detection and first report of *Cucumber mosaic virus* infecting 'Cavendish' banana plants in Ethiopia. *Journal of Plant Diseases and Protection* 127(3): 417–420.

Kebede, Y. and Majumder, S. (2021). Molecular characterization of *Cucumber mosaic virus* sub group IB infecting Cavendish banana plants in Ethiopia. *Journal of Plant Protection Research* 61(4): 392–401.

Khan, S., Jan, A. T., Aquil, B. and Haq, Q. M. R. (2011). Coat protein gene based characterization of cucumber mosaic virus isolates infecting banana in India. *Journal of Phytology* 3: 94–101.

Koh, R. B. L., Barbosa, C. F. C., Aquino, V. M. and Galvez, L. C. (2020). Rapid, simple detection of banana bract mosaic virus in abaca using a one-step reverse transcription loop-mediated isothermal amplification (RT-LAMP) assay. *Journal of General Plant Pathology* 86(6): 433–441.

Lalusin, A. G. and Villavicencio, M. L. H. (2015). Abaca (*Musa textilis* Nee) breeding in the Philippines. In: Cruz, V. M. V. and Dierig, D. A. (Eds), *Industrial Crops: Breeding for BioEnergy and Bioproducts*. New York: Springer, pp. 265–289.

Langeveld, S. A., Dore, J. M., Memelink, J., Derks, A. F., van der Vlugt, C. I., Asjes, C. J. and Bol, J. F. (1991). Identification of potyviruses using the polymerase chain reaction with degenerate primers. *Journal of General Virology* 72(7): 1531–1541.

Li, Y., Xia, F., Wang, Y., Yan, C., Jia, A. and Zhang, Y. (2019). Characterization of a highly divergent *Sugarcane mosaic virus* from *Canna indica* L. by deep sequencing. *BMC Microbiology* 19(1): 260.

Madhavan, S., Balasubramanian, V., Ramajayam, D., Raju, D. V. S., Prasad, K. V. and Selvarajan, R. (2022). Occurrence of *Banana bract mosaic virus* on *Musa ornata* Roxb based hybrids in India. *VirusDisease* 33(4): 397–403.

Madhu Kovileri, M., Nair, S. and Loius, V. (2023). One-step reverse transcription-loop mediated isothermal amplification (RT-LAMP) for closed-tube colorimetric detection of banana bract mosaic virus in banana (*Musa* spp.). *3 Biotech* 13(5): 131.

Magee, C. J. (1930). A new virus disease of bananas. *Agricultural Gazette of New South Wales* 41(12): 929 p.

Magee, C. J. P. (1940). Transmission of infectious chlorosis or heart-rot of the Banana and its relationship to Cucumber mosaic. *Journal of the Australian Institute of Agricultural Science* 6(1): 44–47.

Magnaye, L. V. and Espino, R. R. C. (1990). Note: banana bract mosaic, a new disease of banana. I. Symptomatology. *The Philippine Agriculturist* 73: 55–59.

Mali, V. R. and Rajegore, S. B. (1980). A cucumber mosaic virus disease of banana in India. *Journal of Phytopathology* 98(2): 127–136.

Mossop, D. W. and Francki, R. I. B. (1979). Comparative studies on two satellite RNAs of cucumber mosaic virus. *Virology* 95(2): 395–404.

Muñez, A. R. (1992). Symptomatology, transmission, and purification of banana bract mosaic virus (BBMV) in 'Giant Cavendish' banana. M.Sc. Thesis, University of the Philippines Los Baños: Los Baños, Phillipines.

Murphy, J. F. and Bowen, K. L. (2006). Synergistic disease in pepper caused by the mixed infection of *Cucumber mosaic virus* and *Pepper mottle virus*. *Phytopathology* 96(3): 240–247.

Olspert, A., Chung, B. Y., Atkins, J. F., Carr, J. P. and Firth, A. E. (2015). Transcriptional slippage in the positive-sense RNA virus family *Potyviridae*. *EMBO Reports* 16(8): 995–1004.

Pappu, S. S., Brand, R., Pappu, H. R., Rybicki, E. P., Gough, K. H., Frenkel, M. J. and Niblett, C. L. (1993). A polymerase chain reaction method adapted for selective amplification and cloning of 3' sequences of potyviral genomes: application to dasheen mosaic virus. *Journal of Virological Methods* 41(1): 9–20.

Parac, E. P., Lalusin, A. G., Pangga, I. B. and Cruz, F. C. S. (2021). Characteristics of selected hybrids of abaca (*Musa textilis* Nee) with resistance to bunchy top. *Philippine Agricultural Scientist* 103: 1–12.

Parac, E. P., Lalusin, A. G. and Pangga, I. B. (2019). Response of selected hybrids of abaca (*Musa textilis* Nee) to bunchy top and mosaic diseases in the field. *Philippine Journal of Crop Science* 44: 109.

Péréfarres, F., Le Provost, G., Acina, I., Lockhart, B. E. L., Allah Dghim, A., Iskra-Caruana, M. L., Candresse, T. and Teycheney, P. Y. (2009). Detection, incidence and diversity of banana streak viruses, banana mild mosaic virus and banana virus X in Guadeloupe. *Acta Horticulturae* 828: 205–212.

Quito-Avila, D. F., Ibarra, M. A., Alvarez, R. A., Ratti, M. F., Espinoza, L., Cevallos-Cevallos, J. M. and Peralta, E. L. (2013). First report of *Banana bract mosaic virus* in 'Cavendish' banana in Ecuador. *Plant Disease* 97(7): 1003–1003.

Raymundo, A. D., Bajet, N. B., Sumalde, A., Cipriano, B. P., Borromeo, R., Garcia, B. S., Tapalla, P. and Fabellar, N. (2002). Mapping the spread of abaca bunchy-top and mosaic diseases in the Bicol and Eastern Visayas Regions. Philippines. *Philippine Agricultural Scientist* 84: 352–361.

Raymundo, A. D., Cipriano, B. P., Garcia, S. I. and Tapalla, P. B. (2002). Economic losses in abaca due to bunchy-top and mosaic virus diseases in the Bicol and Eastern Visayas region. *Journal of Tropical Plant Pathology* 38: 31–34.

Reichel, H., Martínez, A. K., Arroyave, J. A., Sedano, R., Morales, F. J., Duterme, O., Kummert, J. and Lepoivre, P. (2003). First report of *Banana mild mosaic virus* isolated from plantains (*Musa* AAB) in Colombia. *Plant Disease* 87(9): 1150.

Reinking, O. A. (1955). Review of Abaca Mosaic Control Program in the Philippines, 1950–1954. Agriculture Division, United States of America Operations Mission to the Philippines, Manila.

Rodoni, B. C., Ahlawat, Y. S., Varma, A., Dale, J. L. and Harding, R. M. (1997). Identification and characterization of banana bract mosaic virus in India. *Plant Disease* 81(6): 669–672.

Rodoni, B. C., Dale, J. L. and Harding, R. M. (1999). Characterization and expression of the coat protein-coding region of banana bract mosaic potyvirus, development of diagnostic assays and detection of the virus in banana plants from five countries in Southeast Asia. *Archives of Virology* 144(9): 1725–1737.

Samraj, J., Menon, M. R., Christudas, S. P. and Satyarajan, P. K. (1966). Kokkan, a new disease of banana (*Musa paradisiaca* Linn.). *Agricultural Research Journal of Kerala* 4: 116.

Sankaralingam, A., Baranwal, V. K., Ahlawat, Y. S., Devi, R. and Ramiah, M. (2006). RT-PCR detection and molecular characterization of *Banana bract mosaic virus* from the pseudostem and bract of banana. *Archives of Phytopathology and Plant Protection* 39(4): 273–281.

Selvarajan, R. and Balasubramanian, V. (2023). Chapter 24. Genetic analysis of banana bract mosaic virus and its management. In: Gaur, R. K., Patil, B. L. and Selvarajan, R. (Eds), *Plant RNA Viruses*. Academic Press, pp. 495–523.

Selvarajan, R., Balasubramanian, V., Jeyabaskaran, K. J., Pandey, S. D. and Mustaffa, M. M. (2009). Management of bract mosaic disease through higher doses of fertilizers in banana cv. Ney Poovan (AB). *Indian Journal of Horticulture* 66(3): 303–305.

Selvarajan, R., Balasubramanian, V., Priyanka, P., Jebakumar, R. M., Selvam, K. P. and Uma, S. (2020a). Evidence of seed transmission of *Banana bract mosaic virus* in *Musa* synthetic diploid H-201, a possible threat to banana breeding. *European Journal of Plant Pathology* 156(4): 1181–1187.

Selvarajan, R., Kanichelvam, P. S., Balasubramanian, V. and Sethurama Subramanian, S. (2020b). A rapid and sensitive lateral flow immunoassay (LFIA) test for the on-site detection of banana bract mosaic virus in banana plants. *Journal of Virological Methods* 284: 113929.

Selvarajan, R. and Jeyabaskaran, K. J. (2006). Effect of banana bract mosaic virus (BBrMV) on growth and yield of cultivar Nendran (Plantain, AAB). *Indian Phytopathology* 59(4): 496–500.

Sharman, M., Gambley, C. F., Oloteo, E. O., Abgona, R. V. J. and Thomas, J. E. (2000a). First record of natural infection of abaca (*Musa textilis*) with banana bract mosaic Potyvirus in the Philippines. *Australasian Plant Pathology* 29(1): 69.

Sharman, M., Thomas, J. E. and Dietzgen, R. G. (2000b). Development of a multiplex immunocapture PCR with colourimetric detection for viruses of banana. *Journal of Virological Methods* 89(1–2): 75–88.

Siljo, A., Bhat, A. I. and Biju, C. N. (2014). Detection of *Cardamom mosaic virus* and *Banana bract mosaic virus* in cardamom using SYBR Green based reverse transcription-quantitative PCR. *VirusDisease* 25(1): 137–141.

Siljo, A., Bhat, A. I., Biju, C. N. and Venugopal, M. N. (2012). Occurrence of *Banana bract mosaic virus* on cardamom. *Phytoparasitica* 40(1): 77–85.

Singh, Z., Jones, R. A. C. and Jones, M. G. K. (1995). Identification of cucumber mosaic virus subgroup 1 isolates from banana plants affected by infectious chlorosis disease using RT-PCR. *Plant Disease* 79(7): 713–716.

Srivastava, N., Kapoor, R., Kumar, R., Kumar, S., R K, S., Kumar, S. and Baranwal, V. K. (2019). Rapid diagnosis of *Cucumber mosaic virus* in banana plants using a fluorescence-based real-time isothermal reverse transcription-recombinase polymerase amplification assay. *Journal of Virological Methods* 270: 52–58.

Takanami, Y. (1981). A striking change in symptoms on cucumber mosaic virus-infected tobacco plants induced by a satellite RNA. *Virology* 109(1): 120–126.

Tang, W., Xu, X. H., Sun, H. W., Li, F., Gao, R., Yang, S. K., Lu, X. B. and Li, X. D. (2016). First report of *Sugarcane mosaic virus* Infecting *Canna* spp. in China. *Plant Disease* 100(12): 2541–2541.

Teycheney, P. Y., Laboureau, N., Iskra-Caruana, M. L. and Candresse, T. (2005a). High genetic variability and evidence for plant-to-plant transfer of *Banana mild mosaic virus*. *Journal of General Virology* 86(11): 3179–3187.

Teycheney, P. Y., Marais, A., Svanella-Dumas, L., Dulucq, M. J. and Candresse, T. (2005b). Molecular characterization of banana virus X (BVX), a novel member of the *Flexiviridae* family. *Archives of Virology* 150(9): 1715–1727.

Teycheney, P. Y., Acina, I., Lockhart, B. E. and Candresse, T. (2007). Detection of *Banana mild mosaic virus* and Banana virus X by polyvalent degenerate oligonucleotide RT-PCR (PDO-RT-PCR). *Journal of Virological Methods* 142(1–2): 41–49.

Thomas, J. E. (2015). *MusaNet Technical Guidelines for the Safe Movement of Musa Germplasm* (3rd edn.). Rome: Bioversity International.

Thomas, J. E., Geering, A. D. W., Gambley, C. F., Kessling, A. F. and White, M. (1997). Purification, properties, and diagnosis of banana bract mosaic potyvirus and its distinction from abaca mosaic potyvirus. *Phytopathology* 87(7): 698–705.

Thomas, J. E., Iskra-Caruana, M. L., Magnaye, L. V., Jones, D. R. (1999a). Bract mosaic. In: Jones, D. R. (Ed.). *Diseases of Banana, Abacá and Enset*. Wallingford: CABI Publishing, pp. 253–256.

Thomas, J. E., Lockhart, B. E. L. and Iskra-Caruana, M. L. (1999b). Banana mild mosaic. In: Jones, D. R. (Ed.), *Diseases of Banana, Abacá and Enset*. Wallingford: CABI Publishing, pp. 275–278.

Vishnoi, R., Kumar, S. and Raj, S. K. (2013). Molecular characterization of a *Cucumber mosaic virus* isolate associated with mosaic disease of banana in India. *Phytoparasitica* 41(5): 545–555.

Wang, I. C., Sether, D. M., Melzer, M. J., Borth, W. B. and Hu, J. S. (2010). First report of *Banana bract mosaic virus* in flowering ginger in Hawaii. *Plant Disease* 94(7): 921–921.

Wang, Y., Gaba, V., Yang, J., Palukaitis, P. and Gal-On, A. (2002). Characterization of synergy between *Cucumber mosaic virus* and potyviruses in cucurbit hosts. *Phytopathology* 92(1): 51–58.

Wege, C. and Siegmund, D. (2007). Synergism of a DNA and an RNA virus: enhanced tissue infiltration of the Begomovirus *Abutilon mosaic virus* (AbMV) mediated by *Cucumber mosaic virus* (CMV). *Virology* 357(1): 10–28.

Xu, P. and Roossinck, M. J. (2000). Cucumber mosaic virus D satellite RNA-induced programmed cell death in tomato. *The Plant Cell* 12(7): 1079–1092.

Yoon, J.-Y., Palukaitis, P. and Choi, S.-K. (2019). Chapter 1: Host range. In: Palukaitis, P. and García-Arenal, F. (Eds), *Cucumber Mosaic Virus*. St Paul: The American Phytopathological Society, pp.15–18.

Zhang, J., Borth, W., Lin, B., Melzer, M., Shen, H., Pu, X., Sun, D., Nelson, S. and Hu, J. (2018). Multiplex detection of three banana viruses by reverse transcription loop-mediated isothermal amplification (RT-LAMP). *Tropical Plant Pathology* 43(6): 543–551.

Zhang, J., Borth, W. B., Lin, B., Dey, K. K., Melzer, M. J., Shen, H., Pu, X., Sun, D. and Hu, J. S. (2016). Deep sequencing of banana bract mosaic virus from flowering ginger (*Alpinia purpurata*) and development of an immunocapture RT-LAMP detection assay. *Archives of Virology* 161(7): 1783–1795.

Part 4

Nematode pests

Chapter 16

Nematodes in banana

Mieke Daneel, Agricultural Research Council – Tropical and Subtropical Crops/North-West University, South Africa; and Akhona Mbatyoti, Agricultural Research Council – Tropical and Subtropical Crops, South Africa

1 Introduction

Banana and plantain (*Musa*) is the fourth most important global food crop and is produced in 130 tropical and subtropical countries (FAO, 2020). Banana production has increased in the last few decades from an estimated 69 million metric tonnes in 2000–2002 to 116 million metric tonnes in 2017–2019 with an approximate value of US$31 billion (FAO, 2020). However, just 15% of total global banana production is traded on the international market where it represents a considerable portion of export revenue for many Latin American and Caribbean countries. It is an essential source of income and employment for many households in these regions. Production for export accounts for about 22 million tonnes (FAO, 2021), while the remainder is consumed domestically, providing a substantial source of nutrition and food security for more than 400 million people, mostly in producing countries in Africa, Asia, and Latin America (FAO, 2020; Voora et al., 2020). Sub-Saharan African countries, which produce one-third of the world's bananas, depend on banana both as their staple food and as a source of income.

Various banana genotypes are available, but the main variety worldwide is Cavendish (AAA) as it is the only export variety and accounts for approximately 47% of global production (Kema et al., 2021). In sub-Saharan Africa, especially

http://dx.doi.org/10.19103/AS.2022.0108.16

East and West Africa, other banana varieties such as plantains and eastern highland cooking bananas (EAHB) are used more frequently and make up about 70% of the crop grown in the continent (Dale et al., 2017; Kikulwe and Asindu, 2020). Elsewhere in Asia, banana is not consumed as a staple food and a much more diverse range – from starchy to sweet banana material – is consumed (Dale et al., 2017).

Yields in different parts of the world differ greatly (Table 1). Reasons for these differences are the use of different banana types and varieties, which are divided between Cavendish type bananas and 'plantains and others' (FAOSTAT, 2022). Another important factor is the use of two different production systems – commercial versus smallholder farming systems. While in commercial systems, fertilizer, irrigation, and pesticides are recommended to be applied regularly, this is very different in smallholder farming systems. In the latter, yields are drastically lower (Table 1) due to no or very limited fertilizer applications, drought, and poor soil conditions as well as bacterial and fungal diseases. In both farming systems, nematodes also play a considerable role in reducing yields.

Table 1 Yield data (t/ha/year) on the average values for 2018–2020 for Cavendish and plantain and others per country (FAOSTAT, 2022)

Country	Banana	Plantain and others	Country	Banana	Plantain and others
Central and Latin America			**Australasia**		
Brazil	14.68	–	China	33.38	–
Costa Rica[a]	47.51	10.50	Taiwan	23.35	–
Dominican Republic[a]	42.06	21.11	India	35.30	–
Honduras[a]	51.64	12.16	Indonesia[a]	55.71	–
Colombia[a]	26.08	8.52	The Philippines	32.55	11.92
Ecuador[a]	37.89	5.88	Malaysia	14.41	–
Haiti	4.89	6.49	Australia[a]	29.55	–
Europe					
Spain[a]	44.22	–	Papua New Guinea	16.11	–
Africa					
Benin	4.87	–	Uganda	–	4.24
Cote d'Ivoire[a]	44.70	3.76	Kenya	22.50	12.06
Cameroon	13.03	14.22	Zambia	4.45	–
Ghana	11.28	11.15	Malawi	30.82	10.29
Nigeria	–	6.25	South Africa[a]	60.91	–

[a] Indicates mainly commercial farming systems especially for Cavendish bananas.

2 Plant-parasitic nematodes

Nematodes are major economic pests that need to be managed to ensure sustainable fruit production. It is estimated that parasitism by nematodes results in yield losses of 30–60% of bananas globally (Sikora et al., 2018). According to Tripathi et al. (2015), plant-parasitic nematodes cause losses of up to 70% in plantains and cooking bananas in Africa. In South Africa, damage by plant-parasitic nematodes was estimated to result in yield losses of 19% (Willers, 1998), while yield losses of 30–50% were projected in Costa Rica and Panama (Davide, 1995).

Banana roots are attacked simultaneously by several plant-parasitic nematode species, with the most important being *Radopholus similis*, *Pratylenchus goodeyi*, *Pratylenchus coffeae*, and *Helicotylenchus multicinctus* but also including sedentary endo-parasites such as *Meloidogyne* and *Rotylenchulus reniformis*. There are many other species associated with banana roots, but they are currently not considered serious pests for banana (Gowen et al., 2005; Daneel and De Waele, 2017).

2.1 Burrowing nematode

The most damaging and widespread banana nematode is *R. similis*, commonly known as burrowing nematode.

2.1.1 Life cycle and symptoms

Burrowing nematodes are a migratory endo-parasite that spend their entire life cycle of about 20–25 days inside the root and rhizome tissue of the banana plant (Sikora et al., 2018) (Fig. 1a). Females and all juvenile stages are infective and cause damage to the root tissue, while males when present do not cause damage. Eggs are deposited in the root tissue and the females lay one to six eggs per day.

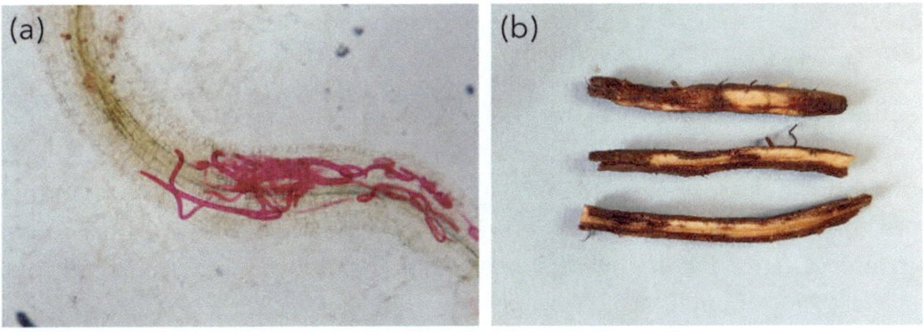

Figure 1 (a) *R. similis* coloured using the Byrd et al. (1983) method in the roots of bananas. (b) *R. similis* damage in the cortex of the roots.

Figure 2 Plants fallen over due to *R. similis* in combination with wind damage.

Figure 3 Pseudostem with typical burrowing nematode damage showing leaves not releasing from around the pseudostem.

Burrowing nematodes enter the soft roots at the root tip or in the region where hair roots are produced, rather than through hardened, old, or decaying roots. It normally takes less than 24 h to enter the roots. They do not normally leave the root tissue, except when moving out of decaying roots or in case of overcrowding. They then move to healthy roots and infect adjacent plant material (Haegeman et al., 2010).

As the nematodes move inside the root tissue, lesions enlarge and coalesce and symptoms in the roots are expressed as necrosis and reddish lesions in the internal tissue, including root cortex and, rarely, stele of the roots (Fig. 1b). Burrowing nematode damage in the roots restricts nutrient and water uptake, resulting in yellowing of the leaves, stunted plant growth and lack of vigour, reduction in the number and size of leaves, leaves not releasing from around the pseudostem, smaller bunches, longer time between ratoons, and toppling of the plants due to poor anchorage in the soil (Gowen et al., 2005; Quénéhervé, 2008) (Figs. 2 and 3). The presence of nematodes also results in a shorter plantation life (Coyne et al., 2022).

2.1.2 Distribution and spread

R. similis occurs throughout the banana-producing areas worldwide and prefers temperature ranges between 24°C and 32°C, with an optimum temperature for reproduction of 30°C and altitudes less than 1400 m above sea level (Kashaija et al., 1994; Sarah et al., 1996; Coyne and Kidane, 2018). *R. similis* is also commonly found in subtropical regions such as South Africa (Daneel and De Waele, 2017) and Australia (Pattison et al., 2002). Several biotypes of *R. similis* exist and their pathogenicity differs depending on their rate of reproduction (Tarté et al., 1981; Kaplan et al., 2000; Cobon and Pattison, 2003).

Although *R. similis* has many alternative hosts (Sarah et al., 1996), *Musa* and *Ensete* are two of the most important, which have helped the distribution of this nematode throughout the world with plant material, for example, suckers and rhizomes (O'Bannon, 1977). The use of tissue culture planting material is limiting its distribution and it is probably also the reason why *R. similis* is not present in areas such as Israel, the Canary Islands, the Cape Verde Islands, Cyprus, Crete, Mauritius, Taiwan, the highlands of Eastern Africa, most of South Africa, Mozambique, and parts of southern Mexico (Wehunt and Edwards, 1968; Stover, 1972; Gowen and Quénéhervé, 1990). Another reason for its limited spread within certain countries such as South Africa is the presence of strict quarantine regulations prohibiting the movement of sucker material between regions and even farms (Daneel and De Waele, 2017).

2.1.3 Control strategies

In tropical regions, *R. similis* can produce up to four generations per year, while in subtropical regions this is significantly less due to a colder winter period in which both plant and nematode are much less active. The potential for crop losses is therefore much higher in tropical regions, requiring more extensive control activities underpinned by a larger amount of research on this topic (Araya and Lakhi, 2004; Jesus et al., 2015; Salguero et al., 2016).

Burrowing nematodes are controlled using different strategies, depending on commercial or smallholder farming systems. In commercial systems, chemical control is the main control strategy, although biological control and cultural practices are also used. The number of applications of nematicides or biologicals depends on whether plantations are situated in tropical or subtropical regions, the amount of nematodes present, and the additional control strategies implemented in management strategies. For organic farmers, other options have to be followed, including the addition of organic material such as banana waste material, manures, and mulches (Ngosong et al., 2019). In smallholder farming systems, often no control is practised, although the use of paired or hot water-treated rhizomes is recommended (Coyne et al., 2010).

2.2 Root lesion nematodes

Another migratory endo-parasitic nematode infecting banana roots and rhizomes is *Pratylenchus* or root lesion nematode. Two species commonly found are *P. coffeae* and *P. goodeyi*.

2.2.1 Life cycle and symptoms

Pratylenchus are migratory endo-parasitic nematodes that feed on the cortex of banana, plantain, and *Ensete* roots, resulting in brown to purplish necrotic lesions in the cortical and epidermal tissue where they feed and multiply. All stages of these nematodes can invade the roots while eggs are laid in the root tissues or in the soil. After entering the roots, lesion nematodes migrate between and within the cells in the cortex, parallel to the stele. Cavities that coalesce are caused by feeding on neighbouring cells (Pinochet, 1978). The damage described by Pinochet (1978) is similar to *R. similis* damage described by Blake (1961, 1966), except for the absence of cell enlargement or increase in size of cell nucleus or nucleolus. The average life cycle from egg to egg is about 27 days at a temperature range of 25–30°C.

Symptoms caused by lesion nematode damage are due to reduced nutrient and water uptake resulting in chlorosis, stunting, lengthening of the vegetative cycle, reduction in leaf size and number, smaller bunches, and toppling. Above- and below-ground symptoms of lesion nematode damage are similar to damage caused by burrowing nematodes.

2.2.2 Distribution and spread

Pratylenchus, similarly to *R. similis*, is generally spread through the distribution of infected planting material (Gowen et al., 2005).

P. coffeae has a wide host range and distribution in most tropical and subtropical countries, but its impact is more localized than *R. similis* (Sikora et al., 2018). In some areas such as the Pacific Islands, it is the most important pest species of banana, while in other areas it is more specific to certain varieties. In Thailand, it affects Pisang Awak (ABB), while in Honduras it is the most important nematode species affecting Cavendish cultivars (AAA). In Ghana, losses of up to 60% in plantain production have been reported (Bridge et al., 1997; Burke et al., 2015). In Gabon, *P. coffeae* is the most dominant species (Loubana et al., 2007). In East Africa, the species is becoming a more important pest (Seshu-Reddy et al., 2007; Luambano et al., 2019), while in West Africa, *P. coffeae* has already replaced *R. similis* (Coyne, 2009). In South Africa, *P. coffeae* is endemic and found in most banana-producing areas in contrast to *R. similis*, which is not present in all areas (Daneel and De Waele, 2017).

P. goodeyi is mostly limited to the Afromontane highlands of Africa and is usually associated with higher altitudes and lower temperature ranges of 16–21°C (Kashaija et al., 1994; Pinochet et al., 1995; Price and Bridge, 1995). In Uganda and Kenya, the species is found only at altitudes of 1600 m above sea level or higher. In a study conducted in Cameroon, *P. goodeyi* was found in altitudes ranging from 800 m or more above sea level (Loubana et al., 2007). *P. goodeyi* is a major pest of highland bananas (EAHB) in East Africa and *Ensete* in Ethiopia (Kashaija et al., 1994; Bridge et al., 1995). The species is the main nematode pest in the Canary Islands (Guiran de and Vilardebo, 1962). *P. goodeyi* has been detected in association with banana in Egypt, Madeira, Crete, and Australia (Oteifa, 1962; Vovlas et al., 1994; Bridge et al., 1997). It is, however, worth noting that *P. goodeyi* distribution is shifting as this species has been detected more recently in hot tropical areas in Kenya (Seshu-Reddy et al., 2007) and Tanzania (Luambano et al., 2019).

2.2.3 Control strategies

Control strategies for lesion nematodes are similar to burrowing nematodes as it is also a migratory endo-parasite of banana that spends most of its life cycle inside the roots of the banana plant.

2.3 Spiral nematodes

Spiral nematodes or Hoplolaimidae is a large family consisting of many genera, among which *Helicotylenchus* is probably the most widely distributed. *Helicotylenchus* has many species but, in banana, *H. multicinctus* is the most damaging.

2.3.1 Life cycle and symptoms

Most spiral nematodes are ectoparasites, but *H. multicinctus* is considered a migratory endoparasite since the juveniles, adults, and eggs can be found in

Figure 4 Spiral nematode damage in roots, showing dark necrotic lesions in the outer cortex.

the cortex of the plant root (Strich-Harari and Zuckerman, 1964). They feed on the cytoplasm of cells close to the epidermis (Blake, 1966) and cause cellular damage that is seen as small and shallow necrotic lesions in the outer cortex of the roots (Quénéhervé and Cadet, 1985) (Fig. 4). Females lay eggs in roots that hatch within 5–6 days; they have three more moults into third, fourth juvenile, and then adult stage. The different life stages are very similar to the adult except for size. Their life cycle is approximately 32 days in optimal conditions (Southey, 1986).

Typical plant-parasitic nematode damage symptoms include stunted plants and reduction in plant and bunch size. Toppling can be seen in severe *H. multicinctus* damage, but this is rare since it is mostly found in combination with other plant-parasitic nematodes.

2.3.2 Spread and distribution

H. multicinctus is commonly found in banana-growing areas and is regarded as an important parasite of bananas, especially in subtropical regions where conditions are suboptimal for banana production (McSorley and Parrado, 1986; Quénéhervé, 2009). However, it is difficult to estimate crop losses caused by this species since *H. multicinctus* is very seldom found on its own. In Uganda, a study was conducted where *R. similis* and *H. multicinctus,* both as a single and mixed population, were evaluated for nematode damage and only *R. similis* had a significant correlation with root damage (Barekye et al., 2005). In South Africa, *H. multicinctus* and *Meloidogyne* are the most common nematodes found in banana; however, damage is limited (Daneel et al., 2015; Daneel and De Waele, 2017). Coyne et al. (2018) found that *H. multicinctus* is commonly associated with *Musa* and when in combination with *Meloidogyne*, it can cause severe damage to plantains, particularly in West and Central Africa.

2.3.3 Control strategies

Control strategies are not specifically targeted at spiral nematodes but rather for the nematode complex found in the plantations. Strategies include those described above.

2.4 Root-knot nematodes

Meloidogyne or root-knot nematodes are also commonly reported in association with banana and plantain; however, the genus is not considered an important pest of these crops. Root-knot nematodes are mostly found together with other plant-parasitic nematodes such as *R. similis* or *Pratylenchus* spp., which result in more severe damage symptoms than *Meloidogyne* can cause alone (De Waele and Davide, 1998).

2.4.1 Life cycle and symptoms

Root-knot nematode infestations result in gall formation on the roots, as can be seen in Fig. 5. The life cycle of *Meloidogyne* ranges from 4 to 6 weeks, depending on climatic conditions and species present (De Waele and Davide, 1998). It consists of an egg, four juvenile stages, of which the J2 stage is the only mobile stage, while the other juvenile stages stay sedentary and swell while developing into the adult female. Some juveniles can develop into vermiform males. The J2 moves towards the roots where it penetrates the root tip, or it moves close to other penetration sites or where minor wounds are present. It enters the endodermis and when entering the stele, it induces multinucleate giant cells on which the J2 feed. Surrounding this tissue, it induces the multiplication of cortical cells, resulting in gall formation (De Waele and Davide, 1998).

Above-ground symptoms are similar to other banana nematodes. However, Rabie (1991) reported that, in combination with stress factors such as drought and/or cold temperatures, root-knot nematodes were involved in inflicting

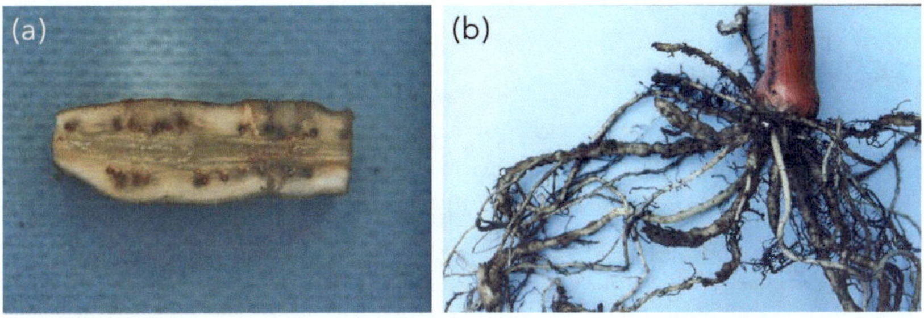

Figure 5 (a) Root-knot nematode damage on inside of roots. (b) Root-knot nematode damage on small tissue culture plants.

symptoms described as 'false panama disease'. The symptoms, like panama disease, included progressive dying back of older leaves starting at the tip. Further investigation showed that galls occurred in the primary and secondary roots and distortion and bifurcations were visible due to severe infection with rhizomes showing brown-purple to reddish-brown discolouration.

2.4.2 Spread and distribution

Although *Meloidogyne* is found worldwide, root-knot nematodes are only predominant in banana when the climate is too cold for *R. similis*, or burrowing nematodes have not been introduced or *P. goodeyi* is not present, which includes subtropical areas such as South Africa, Mediterranean areas such as Crete and Lebanon, tropical regions situated at higher altitudes such as the Andes in Colombia, and in lowlands in Africa where *R. similis* has not been introduced (e.g. Ghana, Tanzania). In Southeast Asia, root-knot nematodes are often the most common and abundant nematode species of many native diploid and triploid varieties grown and they are also predominant in Cavendish bananas in the Philippines (De Waele and Davide, 1998). In South Africa, root-knot nematodes are the most common plant-parasitic nematodes recorded to parasitize roots and rhizomes of mostly younger banana plants, although damage is limited to these younger plants (Daneel et al., 2015; Daneel and De Waele, 2017).

As for other banana nematode parasites, root-knot nematodes can be distributed through infected planting material. However, due to their extremely wide range of host plants, they are present in most agricultural areas and can infect bananas directly.

2.4.3 Control strategies

As for other species, root-knot nematode damage is not treated on its own, but the entire nematode complex is managed.

2.5 Species complexes

The two lesion nematodes do not commonly occur together, although there have been instances where both *Pratylenchus* species were found to have infected banana (Araya et al., 2002). *Pratylenchus* does occur in mixed populations with *Meloidogyne* spp., *H. multicinctus*, and *R. similis*, and the most dominant species differ per country as can be seen, for example, in the study by Loubana et al. (2007) where the species complex differed between Cote d'Ivoire, Gabon, and Cameroon.

Nematode damage in the roots restricts nutrient and water uptake, resulting in leaf yellowing, stunted plant growth and lack of vigour, reduction in the number and size of leaves, retarded growth, smaller bunches, longer time between ratoons, and toppling due to poor anchorage (Gowen et al., 2005; Quénéhervé, 2008).

3 Control of nematodes

To ensure sustainable banana and plantain production, plant-parasitic nematodes need to be managed in both commercial and smallholder farming systems. Besides chemical (Quénéhervé, 2009; Salguero et al., 2016; Coyne et al., 2018) and biological control (Shanti and Rajendran, 2006; Koffi et al., 2013; Olaniyi, 2014; El-Nagdi et al., 2015; Su et al., 2017), other alternative strategies to control or manage nematodes include the use of clean planting material, heat treatment of planting material, clean soil, cover crops, or alternate cropping, fallowing, flooding, resistant or tolerant plant material, transgenic resistance, and the addition of organic material (Quénéhervé, 2008; Coyne et al., 2010, 2018; Tabarant et al., 2011). These practices can be used on their own, but more often in combination to improve their effectiveness. More details on sustainable management strategies are presented elsewhere in this book.

3.1 Alternate cropping systems

Most smallholder farmers do not fallow or use cover cropping due to insufficient land and the potential loss due to lower or no production (Tripathi et al., 2015). Another challenge is that most nematode species associated with banana have a wide host range; therefore, a plant might be resistant to one species but not another. Examples of cover crops that have been successful in suppressing plant-parasitic nematodes include *Crotolaria juncea* and *Tagetes* spp. (Wang and Hooks, 2009). Furthermore, a wide range of plant species with resistance to *Meloidogyne* exist, but in banana production systems this might not be helpful, since *Meloidogyne* is only of limited importance in terms of causing nematode damage in banana. The literature suggests that long-term rotation with sugarcane reduced *R. similis* populations (Robinson and Galán Saúco, 2010), while other plants can also be planted between two consecutive banana crops such as Guinea grass + Siratro (Colbran, 1964), Pangola grass (Smith and Thames, 1969), Sudan grass (Milne and Keetch, 1976; Ternisien and Melin, 1989), horse bean (Zem and Lordello et al., 1983), and Jarra digit grass (Hall et al., 1993).

Sarah (1989) recommended an extended natural fallow followed by growing groundnut for a 3–4-month period. Alternatively, the weed, *Chromolaena odorata,* could be used as a cover crop to reduce *R. similis* populations from the soil before the replanting of dessert bananas (Sarah, 1989). However, in Cameroon, in a maize and groundnut alternating cropping system, heavy *R. similis* infestations were observed in the subsequent plantain crop (Price, 1994). Quénéhervé (2009) found that when alternating crops such as maize and okra, groundnut and soybean, and sweet potato and amaranth were planted, only the latter combination was able to suppress *R. similis* for almost 18 months. Conversely, significant yield differences were observed for all the alternate cropping systems compared to a permanent plantain crop.

There is currently no easy solution available to reduce the population densities of plant-parasitic nematodes in banana and plantain roots and rhizomes. Effective control of plant-parasitic nematodes also depends on many factors, including climate, nematode community complex, and *Musa* material, to name just a few.

3.2 Agro-ecosystems biodiversity

Changes in the distribution of nematode species may be used as an alternative approach to reduce the potential impact of the nematode complex in bananas. For example, *P. goodeyi* is less damaging to banana than *R. similis* (Talwana et al., 2003; Coyne et al., 2018), so by replacing *R. similis* with *P. goodeyi,* root damage might be reduced.

Similarly, *H. multicinctus* is less damaging than *R. similis* (Barekye et al., 2005; Coyne et al., 2013), while both species are found in the same niche (Tixier et al., 2006, 2008; Quénéhervé, 2008; Duyck et al., 2009). If the contribution of each species to this nematode complex can be changed, banana root damage might be reduced (Duyck et al., 2009, 2011). By using the cultivar FB920, Tixier et al. (2008) were able to show that *H. multicinctus* could outcompete *R. similis,* while the total nematode numbers of the two species combined were lower as well. This could result in lower damage and less need for pesticides.

Changing agro-ecosystems can increase biodiversity, but it is important to understand the interactions between the different nematode species in such niches as well as gaining knowledge about plant–nematode interactions. By modifying the abundance of host plants, the balance of nematode species can be changed (Duyck et al., 2009). Other factors that play a critical role in species composition and niche partitioning are type of soil, climatic factors, and cultivars (Duyck et al., 2009, 2011; Quénéhervé et al., 2011). Research points to information on using multiple banana varieties as another option to enhance biodiversity and reduce nematode damage (Quénéhervé et al., 2011), while Poeydebat et al. (2017) looked at plant richness and the effect on nematode damage. However, the number of nematodes is not always reduced, as was shown by Poeydebat et al. (2017), where higher plant biodiversity increased the number of plant-parasitic nematodes. Quénéhervé et al. (2011), on the other hand, found that by planting varietal mixtures of banana material, *R. similis* damage could be reduced because it was replaced by *H. multicinctus*. It is, however, suggested that re-creating favourable conditions for plant biodiversity will increase global sustainability and resilience of banana agro-ecosystems (Tixier et al., 2008).

3.3 Chemical control

In commercial banana systems, which are mostly, but not exclusively, used for exported bananas, chemical control has been used successfully for decades. However, older nematicides are class I chemicals, indicating severe toxicity to humans, other organisms, and the environment (Chitwood, 2003; Desaeger et al., 2020). Problems with groundwater contamination, male sterility, and other effects (Taylor, 2003) have been observed over many years and pressure from consumers eventually led to the withdrawal of these products (Desaeger et al., 2020). New nematicides have been developed with a lower environmental impact, but they are very costly and not feasible for use in smallholder farming systems. Although chemical control provides immediate results and is highly effective, it is only a temporary solution and follow-up applications are necessary in tropical regions. Therefore, chemical control products will at least partly need to be substituted by a range of cultural practices, including clean planting material (*in vitro* plants), rotation crops, biological control, fallow, and organic amendments. New application technologies (e.g. nanomaterials) are being investigated that might provide additional control but must comply with new regulatory requirements in terms of quality and safety of the international banana trade, while new pesticides with improved safety profiles will add to the spectrum of available products.

3.4 Resistance breeding

One of the main pillars of an IPM strategy is resistance breeding. Challenges faced by conventional banana breeding programmes, besides sterility and the lack of seed in the triploid plant material, include the fact that a wide variety of plant-parasitic nematode species attack an extensive variety of *Musa* hybrids planted worldwide, in addition to a combination of several pests and diseases attacking the banana plant concurrently. Transgenic crop improvement strategies may provide an alternative to classical breeding, as has been shown by Tripathi et al. (2015) in plantain field trials against *Helicotylenchus* and *R. similis* infections. While new research is attempting to incorporate banana genes (cisgenes) into Cavendish (Dale et al., 2017), the ultimate aim is to modify the Cavendish genome using gene editing to generate the necessary traits such as nematode resistance. Genetic modification, as indicated elsewhere in this book series (Dale et al., 2020), is probably the most cost-effective way to achieve effective disease resistance for smallholder farmers (Ahmad et al., 2012).

A big challenge in banana nematode control is the concurrent infection by several nematode species in the root system and as yet no hybrid has shown complete resistance to nematode complexes as required to provide good control (Pinochet, 1988; De Waele and Elsen, 2002; Lorenzen et al., 2010). Although resistance to *R. similis* has been identified (Quénéhervé et al., 2009), these materials are still moderate hosts to other species such as *Pratylenchus*, which is becoming a significant pest affecting the banana across sub-Saharan Africa (Coyne, 2009). It is also worth mentioning that species complexes differ considerably within and among locations, thus limiting resistance as a control strategy (Coyne et al., 2013).

Well-known sources of resistance against *R. similis* include 'Pisang Jari Buaya' (AA) (PYB) and 'Yangambi km5' (AAA) (Pinochet and Rowe, 1978; Wehunt et al., 1978; Sarah et al., 1992; Price, 1994), while 'PYB' has been introgressed to 'FHIA-01' (AAAB), which is also resistant to *R. similis* (Pinochet and Rowe, 1979; Viaene et al., 2003). 'Yangambi km5' is reported to be resistant to both *R. similis* and *P. goodeyi* (Fogain and Gowen, 1998). Other sources of resistance have been found in several genome groups in *M. acuminata* and *M. balbisiana*, in both wild and cultivated diploids (Moens et al., 2005; Dochez et al., 2006; Gaidashova et al., 2008; Quénéhervé et al., 2009). In Uganda, a conventionally bred hybrid, 'Kiwangaazi', was released, showing tolerance to *R. similis* (Nowakunda et al., 2015).

3.5 Genetic modification

An alternative way to control nematodes could be by use of genetically modified (GM) plant material, which could confer resistance to nematodes. However, broad-spectrum resistance is required to manage multiple nematode species that pose a threat to banana and plantain crops. Transgenic approaches have considerable potential for sustainable improvement of banana production (Dale et al., 2020). Across banana-producing areas, the banana plants used for production are sterile, with no or limited wild material being available, and material is propagated using tissue culture material or sucker material, reducing gene flow as an issue.

RNA interference (RNAi) based defences have been developed and have shown promise against *Meloidogyne* spp. (Huang et al., 2006; Yadav et al., 2006; Ibrahim et al., 2011; Papolu et al., 2013), *R. similis* (Haegeman et al., 2009; Li et al., 2015a, b), and *Pratylenchus* spp. (Joseph et al., 2012; Tan et al., 2013).

Tripathi et al. (2015) reported that both a peptide and a cystatin provided a high level of resistance in plantain to *R. similis* and *H. multicinctus* with noticeable effects over an extended time period as both nematode species were unable to maintain their population densities in the roots until the crop reached harvest. Nevertheless, the issue of safety is a constraint when promoting transgenic plant material and needs to be considered. Hu et al. (2020) suggested that possibilities such as 'gene-deletor' techniques are or might become available in due course.

While in transgenic GMO approaches, external genes are incorporated from outside the *Musa* genome, new research aims to incorporate banana genes (cisgenes) into Cavendish (Dale et al., 2017, 2021). These approaches have resulted in banana varieties with enhanced levels of pro-vitamin A and increased resistance to *Fusarium* wilt. Such approaches can also be used to modify the Cavendish genome to generate traits such as nematode resistance. Genetic modification may in the short-term be the only way to achieve nematode resistance to enable smallholder farmers to grow banana with no additional chemical inputs (Ahmad et al., 2012).

4 Future trends in research

With predicted temperature increases, it is suggested that climate change will have a significant impact on agriculture and food production (IPCC, 2007; Knueppel et al., 2009). There is great concern regarding how this will affect crop production, including pests and diseases. Coyne et al. (2018) state that 'given our limited knowledge in general on species distribution, temperature thresholds, and survival under different climates, it will be much guess work and many assumptions'.

Climate change has shown that it has the ability to alter the distribution of nematode species. For example, *R. similis* was previously not found in altitudes of 1400 m above sea level, but has recently been detected in higher altitudes in Uganda (Talwana et al., 2016; Erima et al., 2017).

A study in Martinique confirmed that temperature, rainfall, and soil type influenced the abundance of the six nematode species found in bananas, which further confirms that changing temperatures and variations in rainfall patterns as a result of climate change can have a considerable impact on nematode communities (Duyck et al., 2009). For example, the abundance of *H. multicinctus* decreased with increasing rainfall (Duyck et al., 2009). Similar behaviour in relation to *H. multicinctus* populations was also observed on plantains in the Caribbean (Hutton, 1978). By contrast, *P. coffeae* numbers increased with increased rainfall (Duyck et al., 2009). In Uganda, increased populations of *R. similis*, *H. multicinctus*, and *Meloidogyne* spp. were observed with increases in temperature (Erima et al., 2017). In Cameroon, *R. similis* and

Hoploilaimus sp. infestations in banana were more prevalent during low-rainfall and high-temperature periods (Neba et al., 2014).

Changes in nematode species distribution and impact due to climate change will have a significant effect on farmers producing banana and plantain and more specifically smallholder famers, which dominate in sub-Saharan Africa (Coyne et al., 2018). Although climate change is a challenge for nematode management, it can also be used to our advantage. For instance, if nematode complexes change due to higher or lower temperatures and some of these newly introduced species are less pathogenic or damaging to the banana root system, it might be favourable to embrace this change while cultivar selection could be adapted to the advantage of the less damaging species. As mentioned earlier, *P. goodeyi* is less damaging than *R. similis*, and is found in higher altitudes than *R. similis*. However, the distribution of both species has changed as studies conducted with *P. goodeyi* in Kenya and Tanzania (Seshu-Reddy et al., 2007; Luambano et al., 2019) and with *R. similis* in Kenya (Nyang'au et al., 2021) have shown. To ensure that damage does not accelerate, it would be beneficial to manipulate the rhizosphere so that *P. goodeyi* is allowed to dominate species in higher altitudes, for example, by using specific *Musa* material preferred by *P. goodeyi* (Price, 1994).

On the other hand, studies have shown that *P. coffeae* is found in most places in mixed populations with *R. similis*. Again, it might be beneficial to enhance the occurrence of *P. coffeae* and suppress that of *R. similis* by using other banana varieties. However, some studies have shown that *P. coffeae* is as damaging as *R. similis* or has simply replaced *R. similis* (Coyne, 2009). In such cases, other strategies will have to be followed.

5 Conclusion

It is important to realize that effective and sustainable techniques for reducing nematode problems must be developed in addition to the above-mentioned cultural practices. Methods include changing the nematode community structure into a less damaging nematode complex, which can be achieved by changing banana varieties or planting alternate crops. By altering the root rhizosphere, conditions may be favourable for the survival and reproduction of some nematode species at the cost of others. This manipulation may allow less damaging species to dominate, resulting in reduced nematode damage. Therefore, more research is necessary to allow realistic predictions in this regard and for a wide variety of climatic and soil conditions. However, convincing farmers to apply such an approach will be challenging because it represents a different way of thinking.

Increasing biodiversity can also potentially reduce nematode problems since higher biodiversity generally limits pest and disease infections (Quénéhervé et al., 2011; Poeydebat et al., 2017). It is important to recognize that choice of plant species in crop rotation and intercropping approaches is likely an important factor.

Climate change is resulting in shifting distribution patterns of the major nematode species infecting banana, as has been proven by several publications (Seshu-Reddy et al., 2007; Luambano et al., 2019; Coyne et al., 2018). Perhaps this disastrous reality can be used to our advantage while manipulating the nematode community into a less damaging one.

Future research should focus on genetic modification as a priority in genetic improvement of the banana. Including or manipulating certain selected traits in the banana and plantain genomes could reduce the need for additional inputs of harsh chemicals. Adopting the CRISPR technique, for example, might help protect these crops against devastating pests and diseases since researchers can use the gene-editing tool to boost the banana's defences while retaining the desired productivity and consumer traits.

Improving biodiversity to reduce pest and disease problems should also receive sufficient attention. Biodiversity can be enriched by planting a combination of cover crops, intercropping, mixed crops, and planting a variety of selections of *Musa*. This diversity above-ground will definitely influence biodiversity in the rhizosphere, which can be improved even more by the addition of mulches, manures, and organic amendments, of which an abundance of material is available that still needs to be investigated for its efficacy on nematode control. A good example is mulch of *Tithonia diversifolia*, which reduced nematode numbers and improved yield (Coyne et al., 2005).

Increased biodiversity could be complemented with the manipulation of damaging nematode communities by providing feeding material to promote the less damaging nematode species. Furthermore, tools available in other industries such as infrared photography can be incorporated into management programmes to predict, anticipate, and manage nematode problems, specifically in commercial plantations. Improved monitoring approaches are key to reducing pest and disease problems and may lead producers to change their attitude towards farming practices. It is envisaged that a range of commercial biological control agents and softer chemical control agents will also become available in the near future for inclusion in IPM programmes. With climate change putting more pressure on natural resources such as water and increasing temperatures, cultivars/hybrids that will be suitable for these conditions are urgently required.

6 Where to look for further information

- Daneel, M. S. and De Waele, D. (2017). Nematode pests of bananas. In: Fourie, H., Spaull, W. V., Jones, R., Daneel, M. S. and De Waele, D. (Eds.), *Nematology in South Africa: A View from the 21st Century*. Cham, Switzerland: Springer, pp. 359–372.

- Quénéhervé, P. (2009). Integrated management of banana nematodes. In: Ciancio, A. and Mukerji, K. G. (Eds.), *Integrated Management of Fruit Crops and Forest Nematodes*. Switzerland: Springer, pp. 3-61.
- Sikora, R. A., Coyne, D. and Quénéhervé, P. (2018). Nematode parasites of bananas and plantains. In: Sikora, R. A., Coyne, D., Hallmann, J. and Timper, P. (Eds.), *Plant-Parasitic Nematodes in Subtropical and Tropical Agriculture*. 3rd edition. Wallingford, UK: CAB International, pp. 617-657.

7 References

Ahmad, P., Ashraf, M., Younis, M., Hu, X., Kumar, A., Akram, N. A. and Al-Qurainy, F. 2012. Role of transgenic plants in agriculture and biopharming. *Biotechnology Advances* 30(3): 524–540.

Araya, M., De Waele, D. and Vargas, R. 2002. Occurrence and population densities of nematode parasites of banana (*Musa* AAA) roots in Costa Rica. *Nematropica* 32: 21–33.

Araya, M. and Lakhi, A. 2004. Response to consecutive nematicide applications using the same product in *Musa* AAA, cv. Grande Naine originated from *in vitro* propagative material and cultivated on a virgin soil. *Nematologia Brasileira* 28(1): 55–61.

Barekye, A., Kashajia, I. N., Tushemereirwe, W. K. and Adipala, A. 2005. Comparison of damage levels caused by *Radopholus similis* and *Helicotylenchus multicinctus* on bananas in Uganda. *Annals of Applied Biology* 137: 273–278.

Blake, C. D. 1961. Root rot of bananas caused by *Radopholus similis* (Cobb) and its control in New South Wales. *Nematologica* 6(4): 295–310.

Blake, C. D. 1966. The histological changes in banana roots caused by *Radopholus similis* and *Helicotylenchus multicinctus*. *Nematologica* 12(1): 129–137.

Bridge, J., Fogain, R. and Speijer, P. 1997. *The Root Lesion Nematodes of Bananas. Pratylenchus coffeae (Zimmermann, 1898) Filipjev & Schuurman Stekhoven, 1941, Pratylenchus goodeyi Sher & Allen, 1953. Musa Pest Fact Sheet No 2*. Montpellier, France: INIBAP.

Bridge, J., Pricey, N. S. and Kofi, P. 1995. Plant parasitic nematodes of plantain and other crops in Cameroon, West Africa. *Fundamental and Applied Nematology* 18(3): 251–260.

Burke, M., Scholl, E. H., Bird, D. McK., Schaff, J. E., Coleman, S., Crowell, R., Diener, S., Gordon, O., Graham, S., Wang, X., Windham, E., Wright, G. M. and Opperman, C. H. 2015. The plant parasite *Pratylenchus coffeae* carries a minimal nematode genome. *Nematology* 17: 1–17.

Byrd, D. W., Kirkpatrick, T. and Barker, K. R. 1983. An improved technique for clearing and staining plant tissues for detection of nematodes. *Journal of Nematology* 15: 42–43.

Chabrier, C. and Quénéhervé, P. 2003. Control of the burrowing nematode (*Radopholus similis* Cobb) on banana: Impact of the banana field destruction method on the efficiency of the following fallow. *Crop Protection* 22(1): 121–127.

Chitwood, D. J. 2003. Nematicides. In: Plimmer, J. R. (Ed.), *Encyclopedia of Agrochemicals* (Vol. 3). New York: John Wiley & Sons, pp. 1104–1115.

Cobon, J. A. and Pattison, T. 2003. Variability in host range and pathogenicity of Australian *Radopholus* spp. on *Musa* cultivars. *Australasia Nematology Newsletter* 14: 11–20.

Colbran, R. C. 1964. Cover crops for nematode control in old banana land. *Queensland Journal of Agricultural Science* 21: 233–236.

Coyne, D. 2009. Pre-empting plant-parasitic nematode losses on *Musa* spp. *Acta Horticulturae* 828: 227–236.

Coyne, D. and Kidane, S. 2018. Nematode pathogens. In: Jones, D. (Ed.), *Diseases of Banana, Abac´a and Enset* (2nd edn.). Wallingford, UK: CABI Publishing.

Coyne, D., Rotimi, O., Speijer, P., De Schutter, B., Dubois., T., Auwerken, A. T., and De Waele, D. 2005. Effects of nematode infection and mulching on plantain (*Musa* spp., AAB-group) yield of ratoon cycles and plantation longevity in southeastern Nigeria. *Nematology* 7: 531–541.

Coyne, D. L., Cortada, L., Dalzell, J. J., Claudius-Cole, A. O., Haukeland, S., Luambano, N. and Talwana, H. 2018. Plant-parasitic nematodes and food security in sub-Saharan Africa. *Annual Review of Phytopathology* 56: 381–403.

Coyne, D. L., Omowumi, A., Rotifa, I. and Afolami, S. O. 2013. Pathogenicity and damage potential of five plant-parasitic nematode species on plantain (*Musa* spp., AAB genome) cv. *Nematology* 15(5): 589–599.

Coyne, D. L., Wasukira, A., Dusabe, J., Rotifa, I. and Dubois, T. 2010. Boiling water treatment: A simple, rapid and effective technique for producing healthy banana and plantain (*Musa* spp.) planting material. *Crop Protection* 29(12): 1478–1482.

Dale, J., Paul, J.-Y., Dugdale, B. and Harding, R. 2017. Modifying bananas: From transgenics to organics? *Sustainability* 9(3): 333.

Dale, J., Tushemereirwe, W. and Harding, R. 2020. Genetic modification of bananas: The long road to farmers' fields. In: Kema, G. and Drenth, A. (Eds.), *Achieving Sustainable Cultivation of Bananas (Vol. 2): Germplasm and Genetic Improvement*. United Kingdom: Burleigh Dodds Science Publishing, pp. 339–366.

Daneel, M. S., De Jager, K., Van den Bergh, I., De Smet, M. and De Waele, D. 2015. Occurrence and pathogenicity of plant-parasitic nematodes on commonly grown banana cultivars in South Africa. *Nematropica* 45: 118–127.

Daneel, M. S. and De Waele, D. 2017. Nematode pests of bananas. In: Fourie, H., Spaull, W. V., Jones, R., Daneel, M. S. and De Waele, D. (Eds.), *Nematology in South Africa: A View From the 21st Century*. Cham, Switzerland: Springer, pp. 359–372.

Davide, R. G. 1995. Overview of nematodes as a limiting factor on *Musa* production. In: Frison, E. A., Horry, J. P. and de Waele, D. (Eds.), *New Frontiers in Resistance Breeding for Nematode, Fusarium and Sigatoka*. Montpellier, France: International Network for the Improvement of Banana and Plantain, pp. 27–31.

De Waele, D. and Davide, R. G. 1998. *The Root-Knot Nematodes of Banana Meloidogyne incognita (Kofoid & White, 1919) Chitwood, 1949 and Meloidogyne javanica (Treub, 1885) Chitwood, 1949 M. Pest Factsheet No 3.*

De Waele, D. and Elsen, A. 2002. Migratory endoparasities: *Pratylenchus* and *Radopholus* species. In: Starr, J. J., Cook, R. and Bridge, J. (Eds.), *Plant Resistance to Parasitic Nematodes*. Wallingford, UK: CABI Publishing, pp. 175–206.

Desaeger, J., Wram, C. and Zasada, I. 2020. New reduced-risk agricultural nematicides - Rationale and review. *Journal of Nematology* 52: e2020-91.

Dochez, C., Dusabe, J., Whyte, J., Tenkouano, A., Ortiz, R. and De Waele, D. 2006. New sources of resistance to *Radopholus similis* in *Musa* germplasm from Asia. *Australasian Plant Pathology* 35: 381–485.

Drenth, A., Kema, G. and Stergiopoulos, I. 2016. With the familiar Cavendish banana in danger, can science help it survive? *The Conversation*. Available at: http://

theconversation.com/with-the-familiar-cavendish-banana-in-danger-canscience-help-it-survive-64206.

Duyck, P.-F., Pavoine, S., Tixier, P., Chabrier, C. and Quénéhervé, P. 2009. Host range as an axis of niche partitioning in the plant-feeding nematode community of banana agroecosystems. *Soil Biology and Biochemistry* 41(6): 1139–1145.

Duyck, P.-F., Pavoine, S., Tixier, P., Vinatier, F., Loubana, P.-M., Chabrier, C. and Quénéhervé, P. 2011. Niche partitioning based on soil type and climate at the landscape scale in a community of plant-feeding nematodes. *Soil Biology and Biochemistry* 44: 49–55.

El-Nagdi, W. N. A., Abd El-Khair, H. and El-Ghonaimy, A. M. 2015. Field application of biological control on root-knot nematode and fusarium root rot fungus in banana Cv. Grand naine. Middle East. *Journal of Agricultural Research* 4: 545–555.

Erima, R., Kubiriba, J., Komutunga, E., Nowakunda, K., Namanya, P., Seruga, R., Nabulya, G., Ahumuza, E. and Tushemereirwe, W. K. 2017. Banana pests and diseases spread to higher altitudes due to increasing temperature over the last 20 years. *African Journal of Environmental Science and Technology* 11(12): 601–608.

FAO. 2020. *Bananas*. Available at: http://www.fao.org/economic/est/est-commodities/bananas/en/.

FAO. 2021. *Banana Market Review – Preliminary Results 2020*. Rome.

Fernandez, E., Mena, J., Gonzalez, J. and Marquez, M. E. 2005. Banana-root-system: Towards a better understanding for its productive management. *Proceedings of the an International Symposium Held in San José*, Costa Rica, 3–5 November 2003, 193–200.

Fogain, R. and Gowen, S. R. 1998. Yangambi km5 (Musa, A. A. A. Ibota Subgroup): A possible source of resistance to Radopholus similis and Pratylenchus goodeyi. *Fundamental and Applied Nematology* 21: 75–80.

Food and Agriculture Organization of the United Nation. 2022. *FAOSTAT*.

Gaidashova, S. V., Uwimpuhwe, B. and Karamura, E. B. 2008. Identification of banana varieties with resistance to nematodes in Rwanda. *African Crop Science Journal* 16(1): 27–33.

Gowen, S. and Quénéhervé, P. 1990. Nematode parasites of bananas, plantains and abaca. In: Luc, M., Sikora, R. A. and Bridge, J. (Eds.), *Plant Parasitic Nematodes in Subtropical and Tropical Agriculture*. Wallingford, UK: CAB International, pp. 431–460.

Gowen, S. R., Quénéhervé, P. and Fogain, R. 2005. Nematode parasites of bananas and plantains. In: Luc, M., Sikora, R. A. and Bridge, J. (Eds.), *Plant Parasitic Nematodes in Subtropical and Tropical Agriculture*. Wallingford, UK: CAB International, pp. 611–643.

Guiran de, G. and Vilardebo, A. 1962. Le bananier aux îles Canaries. IV. Les nématodes parasites du bananier. *Fruits* 17: 263–277.

Haegeman, A., Elsen, A., De Waele, D. and Gheysen, G. 2010. Emerging molecular knowledge on *Radopholus similis*, an important nematode pest of banana. *Molecular Plant Pathology* 11(3): 315–323.

Haegeman, A., Vanholme, B. and Gheysen, G. 2009. Characterization of a putative endoxylanase in the migratory plant-parasitic nematode *Radopholus similis*. *Molecular Plant Pathology* 10(3): 389–401.

Hall, T. J., Walduck, G. D. and Walker, R. W. 1993. Register of Australian herbage plant cultivars. *Australian Journal of Experimental Agricultural* 33: 674–676.

Hu, C., Yang, Q., Shao, X., Dong, T., Bi, F. C., Li, C., Deng, G., Li, Y., Yi, G. and Dou, T. 2020. The application of the 'Gene-deletor' technology in banana. *Plant Cell, Tissue and Organ Culture* 140: 105–114.

Huang, G. Z., Allen, R., Davis, E. L., Baum, T. J. and Hussey, R. S. 2006. Engineering broad root-knot resistance in transgenic plants by RNAi silencing of a conserved and essential root-knot nematode parasitism gene. *Proceedings of the National Academy of Sciences of the United States of America* 103(39): 14302–14306.

Hutton, D. G. 1978. Influence of rainfall on some plantain nematodes in Jamaica. *Nematropica* 8: 34–39.

Ibrahim, H. M. M., Alkharouf, N. W., Meyer, S. L. F., Aly, M. A. M., Gamal El-Din, Ael K., Hussein, E. H. A. and Matthews, B. F. 2011. Post-transcriptional gene silencing of root-knot nematode in transformed soybean roots. *Experimental Parasitology* 127(1): 90–99.

InAfrica 2016. Bananas from Africa. Available at: http://inafrica24.com/modernity/bananas-from-africa/.

IPCC. 2007. 2007: Synthesis report. Contribution of working groups I, II and III to the fourth assessment report of the Intergovernmental Panel on Climate Change Core Writing Team. In: Pachauri, R. K. and Reisinger, A. (Eds.), *Climate Change*. Geneva, Switzerland: IPCC, p. 104.

Jesus, F. N., Damasceno, J. C. A., Barbosa, D. H. S. G., Malheiro, R., Pereira, J. A. and Soare, A. C. F. 2015. Control of the banana burrowing nematode using sisal extract. *Agronomy for Sustainable Development* 35(2): 783–791.

Joseph, S., Gheysen, G. and Subramaniam, K. 2012. RNA interference in *Pratylenchus coffeae*: Knock down of Pc-pat-10 and Pc-unc-87 impedes migration. *Molecular and Biochemical Parasitology* 186(1): 51–59.

Kaplan, D. T., Thomas, W. K., Frisse, L. M., Sarah, J. L., Stanton, J. M., Speijer, P. R., Marin, D. H. and Opperman, C. H. 2000. Phylogenetic analysis of geographically diverse *Radopholus similis* via rDNA sequence reveals a monomorphic motif. *Journal of Nematology* 32(2): 134–142.

Kashaija, I. N., Speijer, P. R., Gold, C. S. and Gowen, S. R. 1994. Occurrence, distribution and abundance of plant parasitic nematodes of bananas in Uganda. *African Crop Science Journal* 2: 99–104.

Kema, G. H. J., Drenth, A., Dita, M., Jansen, K., Vellema, S. and Stoorvogel, J. J. 2021. Fusarium Wilt of Banana, a Recurring Threat to Global Banana Production. *Frontiers in Plant Science* 11 – 2020. https://doi.org/10.3389/fpls.2020.628888.

Kikulwe, E. M. and Asindu, M. 2020. A contingent valuation analysis for assessing the market for genetically modified planting materials among banana producing households in Uganda. *GM Crops and Food* 11(2): 113–124. DOI: 10.1080/21645698.2020.1720498.

Knueppel, D., Demment, M. and Kaisser, L. 2009. Validation of household food insecurity access scale in rural Tanzania. *Public Health Nutrition* 13: 360–367.

Koffi, M. C., Vos, C., Draye, X. and Declerck, S. 2013. Effects of *Rhizophagus irregularis* MUCL 41833 on the reproduction of *Radopholus similis* in banana plantlets grown under in vitro culture conditions. *Mycorrhiza* 23(4): 279–288.

Li, Y., Wang, K., Xie, H., Wang, D. W., Xu, C. L., Huang, X., Wu, W. J. and Li, D. L. 2015a. Cathepsin B cysteine proteinase is essential for the development and pathogenesis of the plant parasitic nematode *Radopholus similis*. *International Journal of Biological Sciences* 11(9): 1073–1087.

Li, Y., Wang, K., Xie, H., Wang, Y. T., Wang, D. W., Xu, C. L., Huang, X. and Wang, D. S. 2015b. A nematode calreticulin, Rs-CRT, is a key effector in reproduction and pathogenicity of *Radopholus similis*. *PLoS ONE* 10(6): e0129351.

Lorenzen, J., Tenkouano, A., Bandyopadhyay, R., Vroh, B., Coyne, D. and Tripathi, L. 2010. Overview of banana and plantain (*Musa* spp.) improvement in Africa: Past and future. *Acta Horticulturae* 879: 595–603.

Loubana, P. M., Sarah, J. L., Sakwe, N. P., Mavoungou, J. F., Gnonhouri, P., Boisseau, M. and Fourét, E. 2007. Study of genetic diversity of plant parasitic nematodes on bananas and plantains in Central and West Africa. *Proceedings of the African Crop Science Conference* (Vol. 8), pp. 783–786.

Luambano, N. D., Kashando, B. E., Masunga, M. M., Mwenisongole, A. E., Mziray, M. F., Mbaga, J. E., Polini, R. M. and Mgonja, D. M. 2019. Status of *Pratylenchus coffeae* in banana-growing areas of Tanzania. *Physiological and Molecular Plant Pathology* 105: 102–109.

McSorley, R. and Parrado, J. L. 1986. Nematological Review: *Helicotylenchus multicinctus* on bananas: An international problem. *Nematologica* 16: 73–91.

Milne, D. L. and Keetch, D. P. 1976. Some observations on the host plant relationships of *Radopholus similis* in natal. *Nematropica* 6: 13–17.

Moens, T., Araya, M., Swennen, R. and De Waele, D. 2005. Screening of *Musa* cultivars for resistance to *Helicotylenchus multicinctus*, *Meloidogyne incognita*, *Pratylenchus coffeae* and *Radopholus similis*. *Australasian Plant Pathology* 34(3): 299–309.

Neba, A. G., Ndonwo, A. S. and Mimba, A. 2014. Impacts of climate variability and change in banana yields in the CDC-Delmonte Banana Project, Tiko, South-West Region, Cameroon. *Journal of the Cameroon Academy of Science* 11(2&3): 131–142.

Ngosong, C., Justing, O. and Tening, A. S. 2019. Mulching: A sustainable option to improve soil health. In: Panpatte, D. G. and Jhala, Y. K. (Eds.), *Soil Fertility Management for Sustainable Development*, pp. 231–249.

Nowakunda, K., Barekye, A., Ssali, R. T., Namaganda, J., Tushemereirwe, W. K., Nabulya, G., Erima, R., Akankwasa, K., Hilman, E., Batte, M. and Karamura, D. 2015. 'Kiwangaazi' (syn 'KABANA 6H') Black Sigatoka Nematode and Banana Weevil Tolerant 'Matooke' Hybrid Banana Released in Uganda. *HortScience* 50(4): 621–623.

Nyang'au, D., Atandi, J., Cortada, L., Nchore, S., Maina Mwangi, M. and Coyne, D. 2021. Diversity of nematodes on banana (*Musa* spp.) in Kenya linked to altitude and with a focus on the pathogenicity of *Pratylenchus goodeyi*. *Nematology* 24(2): 137–147.

O'Bannon, J. H. 1977. Worldwide dissemination of *Radopholus similis* and its importance of crop production. *Journal of Nematology* 9(1): 16–25.

Olaniyi, M. O. 2014. Effects of mycorrhizal inoculant and organic mulches on nematode damage to cooking banana. *Journal of Biology, Agriculture and Healthcare* 4: 81–86.

Oteifa, B. A. 1962. Species of root-lesion nematodes commonly associated with economic crops in the Delta of the U. A. R. *Plant Disease Reporter* 46: 712–714.

Papolu, P. K., Gantasala, N. P., Kamaraju, D., Banakar, P., Sreevathsa, R. and Rao, U. 2013. Utility of host delivered RNAi of two FMRF amide like peptides, *flp*-14 and *flp*-18, for the management of root knot nematode, *Meloidogyne incognita*. *PLoS ONE* 8(11): e80603.

Pattison, A. B., Stanton, J. M., Cobon, J. A. and Doogan, V. J. 2002. Population dynamics and economic threshold of the nematodes *Radopholus similis* and *Pratylenchus goodeyi* on banana in Australia. *International Journal of Pest Management* 48(2): 107–111.

Pinochet, J. 1978. Histopathology of the root lesion nematode, *Pratylenchus coffeae* on plantains, Musa AAB. *Nematologica* 24(3): 337–340.

Pinochet, J. 1998. Comments on the difficulty in breeding bananas and plantains for resistance to nematodes. *Revue de Nematologie* 11: 3-5.

Pinochet, J., Fernandez, C. and Sarah, J. L. 1995. Influence of temperature on *in vitro* reproduction of *Pratylenchus coffeae, P. goodeyi* and *Radopholus similis*. *Fundamental and Applied Nematology* 18: 391-392.

Pinochet, J. and Rowe, P. R. 1978. Reaction of two banana cultivars to three different nematodes. *Plant Disease Reporter* 62: 727-729.

Pinochet, J. and Rowe, P. R. 1979. Progress in breeding for resistance to *Radopholus similis* on bananas. *Nematropica* 9: 76-77.

Poeydebat, C., Carval, D., Tixier, P. and Chabrier, C. 2017. Does plant richness alter multitrophic soil food web and promote plant-parasitic nematodes. *Applied Soil Ecology* 117-118: 137-146.

Price, N. S. 1994. Field trial evaluation of nematode susceptibility within *Musa*. *Fundamental and Applied Nematology* 17: 391-396.

Price, N. S. and Bridge, J. 1995. Pratylenchus goodeyi (Nematoda: Pratylenchidae): A plant-parasitic nematode of the montane highlands of Africa. *Journal of African Zoology* 109: 435-442.

Quénéhervé, P. 2008. Integrated management of banana nematodes. In: Ciancio, A. and Mukerji, K. G. (Eds.), *Integrated Management of Fruit Crops and Forest Nematodes*. Cham: Springer Science, pp. 3-61.

Quénéhervé, P. 2009. Integrated management of banana nematodes. In: Ciancio, A. and Mukerji, K. G. (Eds.), *Integrated Management of Fruit Crops and Forest Nematodes*. Switzerland: Springer, pp. 3-61.

Quénéhervé, P., Barrière, V., Salmon, F., Houdin, F., Achard, R., Gertrude, J.-C., Marie-Luce, S., Chabrier, C., Duyck, P. and Tixier, P. 2011. Effect of banana crop mixtures on the plant-feeding nematode community. *Applied Soil Ecology* 49: 40-45.

Quénéhervé, P. and Cadet, P. 1985. Etude de la dynamique de l'infestion en nématodes transmis par les rhizomes du bananier cv Poyo en Côte d'Ivoire. *Revue de Nématologie* 8: 257-263.

Quénéhervé, P., Salmon, F., Topart, P. and Horry, J. P. 2009. Nematode resistance in bananas: Screening results on some new *Mycosphaerella* resistant banana hybrids. *Euphytica* 165(1): 137-143.

Rabie, E. C. 1991. *Die invloed van Meloidogyne javanica en M. incognita op die voorkoms van Vals Panaman siekte by piesangs*. MSc Dissertation, University of Pretoria, Pretoria.

Robinson, J. C. and Galán Saúco, V. 2010. *Bananas and Plantains*. Wallingford, UK: CAB International.

Salguero, D., Rudon, G., Blanco, R., Moya, C., Ramclamy, W., Medina, L., Azofeifa, D. and Araya, M. 2016. Effect of different nematicide applications per year on banana (*Musa* AAA) root nematode control and crop yield. *Journal of Applied Biological Sciences* 101: 9598-9609.

Sarah, J. L., Blavignac, F., Sabatini, C. and Boisseau, M. 1992. Une methode de laboratoire pour le criblage varietal des bananiers vis-a-vis de la resistance aux nematodes. *Fruits* 47: 559-564.

Sarah, J. P. 1989. Nematological review: Banana nematodes and their control in Africa. *Nematropica* 19: 199-216.

Sarah, J. P., Pinochet, J. and Stanton, J. 1996. The burrowing nematode of bananas. In: *Radopholus similis Cobb, 1913. Musa Fact Sheet Nr 1*. Montpellier, France: INIBAP.

Seshu Reddy, K. V., Prasad, J. S., Speijer, P. R., Sikora, R. A. and Coyne, D. L. 2007. Distribution of plant-parasitic nematodes on *Musa* in Kenya. *Infomusa* 16: 18–23.

Shanti, A. and Rajendran, G. 2006. Biological control of lesion nematode in banana. *Nematologia Mediterranea* 34: 69–75.

Shelton, A. M., Sances, F. V., Hawley, J., Tang, J. D., Boune, M., Jungers, D., Collins, H. L. and Farias, J. 2000. Assessment of insecticide resistance after the outbreak of diamondback moth (Lepidoptera: Plutellidae) in California in 1997. *Journal of Economic Entomology* 93(3): 931–936.

Sikora, R. A., Coyne, D. and Quénéhervé, P. 2018. Nematode parasites of bananas and plantains. In: Sikora, R. A., Coyne, D., Hallmann, J. and Timper, P. (Eds.), *Plant-Parasitic Nematodes in Subtropical and Tropical Agriculture* (3rd edn.). Wallingford, UK: CAB International, pp. 617–657.

Smith, L. R. and Thames, W. H. 1969. Plantain nematodes and preliminary crop rotation studies for their control in the Dominican Republic. In: *Proceedings of the Symposium on Tropical Nematology*, Rio Piedras, Puerto Rico, pp. 84–89.

Southey, J. F. 1986. Principles of sampling for nematodes. In: Southey, J. F. (Ed.), *Laboratory Methods for Work With Plant and Soil Nematodes*. London: HMSO, pp. 1–4.

Stover, R. H. 1972. Banana, Plantain and Abaca Diseases. United Kingdom: Commonwealth Mycological Institute, p. 316.

Strich-Harari, D. and Zuckerman, B. M. 1963. The life stages of *Helicotylenchus multicinctus* (Cobb) in banana roots. *Nematologica* 9(3): 347–353.

Su, L., Shen, Z., Ruan, Y., Tao, C., Chao, Y., Li, R. and Shen, Q. 2017. Isolation of antagonistic endophytes from banana roots against *Meloidogyne javanica* and their effects on soil nematode community. *Frontiers in Microbiology* 8: 2070.

Tabarant, P., Villenave, C., Risede, J.-M., Roger-Estrade, J. and Dorel, M. 2011. Effects of organic amendments on plant-parasitic nematode populations, root damage, and banana plant growth. *Biology and Fertility of Soils* 47(3): 341–347.

Talwana, H., Sibanda, Z., Wanjohi, W., Kimenju, W., Luambano, N., Massawe, C., Manzanilla Lopez, R. H., Davies, K. G., Hunt, D. J., Sikora, R. A., Coyne, D. L., Gowen, S. R. and Kerry, B. R. 2016. Agricultural nematology in East and Southern Africa: Problems, management strategies and stakeholder linkages. *Pest Management Science* 72(2): 226–245.

Talwana, H., Speijer, P. R., Gold, C. S., Swennen, R. and De Waele, D. 2003. A comparison of the effects of the nematodes *Radopholus similis* and *Pratylenchus goodeyi* on growth, root health and yield of an East African highland cooking banana (*Musa* AAA-group). *International Journal of Pest Management* 49(3): 199–204.

Tan, J. A. C. H., Jones, M. G. K. and Fosu-Nyarko, J. 2013. Gene silencing in root lesion nematodes (*Pratylenchus* spp.) significantly reduces reproduction in a plant host. *Experimental Parasitology* 133(2): 166–178.

Tarté, R., Pinochet, J., Gabrielli, C. and Ventura, O. 1981. Differences in population increase, host preference and frequency of morphological variants among isolates of the banana race of *Radopholus similis*. *Nematropica* 11: 43–52.

Taylor, A. L. 2003. Nematocides and nematicides – A history. *Nematropica* 33: 225–232.

Ternisien, E. and Melin, P. 1989. Etude des rotations culturales en bananeraie. II. Impact des cultures de rotations sur la production bananière et l'état sanitaire du sol. *Fruits* 44: 445-454.

Tixier, P., Risède, J.-M., Dorel, M. and Malezieux, E. 2006. Modelling population dynamics of banana plant-parasitic nematodes: A contribution to the design of sustainable cropping systems. *Ecological Modelling* 198(3-4): 321-331.

Tixier, P., Salmon, F., Chabrier, C. and Quénéhervé, P. 2008. Modelling pest dynamics of new crop cultivars: The FB920 banana with the *Helicotylenchus multicinctus-Radopholus similis* nematode complex in Martinique. *Crop Protection* 27(11): 1427-1431.

Tripathi, L., Babirye, A., Roderick, H., Tripathi, J. N., Changa, C., Urwin, P. E., Tushemereirwe, W. K., Coyne, D. and Atkinson, H. J. 2015. Field resistance of transgenic plantain to nematodes has potential for future African food security. *Scientific Reports* 5: 8127.

Viaene, N., Duran, L. F., Rivera, J. M., Duenas, J., Rowe, P. and De Waele, D. 2003. Responses of banana and plantain cultivars, lines and hybrids to the burrowing nematode *Radopholus similis*. *Nematology* 5(1): 85-98.

Voora, V., Larrea, C. and Bermudez, S. 2020. *Global Market Report: Bananas Sustainable Commodities Marketplace*. Available at: https://www.iisd.org/system/files/publications/ssi-global-market-report-banana.pdf (Series 2019).

Vovlas, N., Avgelis, A., Goumas, D. and Frisullo, S. 1994. A survey of banana diseases in sucker propagated plantation in Crete. *Nematologia Mediterranea* 22: 101-107.

Walker, P. T., Hebblethwaite, M. J. and Bridge, J. 1984. Project for banana pest control and improvement in Tanzania. EEC report for the government of Tanzania. London: Produced by Tropical Development and Research Institute, p. 129.

Wang, K. H. and Hooks, C. R. 2009. Plant-parasitic nematodes and their associated natural enemies within banana (Musa spp.) plantings in Hawaii. *Nematropica* 39: 57-74.

Wehunt, E. J. and Edwards, D. I. 1968. Radopholus similis and other nematode species on bananas. In: Smart, G. C. and Perry, V. G. (Eds.), *Tropical Nematology*. Gainesville: University of Florida Press, pp. 1-19.

Wehunt, E. J., Hutchinson, D. J. and Edwards, D. I. 1978. Reaction of banana cultivars to the burrowing nematode (*Radopholus similis*). *Journal of Nematology* 10(4): 368-370.

Willers, P. 1998. Nematological research in subtropical industries. *Neltropica Bulletin* 299: 12-13.

Yadav, B. C., Veluthambi, K. and Subramaniam, K. 2006. Host-generated double stranded RNA induces RNAi in plant-parasitic nematodes and protects the host from infection. *Molecular and Biochemical Parasitology* 148(2): 219-222.

Zem, A. C. and Lordello, L. G. E. 1983. *Estudos sobre Hospedeiros de Radopholus similis e Helicotylenchus multicinctus. Reuniao Brasileira de Nematologia, Piracicaba (BR)* (Vol. 7). Brazil: Sociedade Brasileira de Nematologia Publicaçao, pp. 175-187.

Chapter 17

Towards sustainable management of nematodes in banana

Anthony B. Pattison and Jennifer A. Cobon, Department of Agriculture and Fisheries, Australia; Mario Araya-Vargas, AMVAC Chemical Corporation, Costa Rica; and Christian Chabrier, CIRAD, France

1 Introduction

The world production of bananas and plantain in 2019 was 116 MT from 5.1 Mha (http://www.fao.org/faostat/en/#data/QC), grown on all continents with tropical and subtropical environments. Bananas are cultivated in a range of production systems, which include intensive monoculture plantation systems receiving high inputs to produce high outputs of up to 65 t/ha/yr for international export markets. However, up to 85% of global production is for local markets and largely produced in subsistence smallholder systems, which are low yielding <10 t/ha/yr, but are essential for the livelihood and food security of smallholder farmers. Intensive banana plantation production systems are dominated by monocultures of 'Cavendish' cultivars, whereas many smaller subsistence style productions systems rely on a diverse range of

http://dx.doi.org/10.19103/AS.2022.0108.17
Published by Burleigh Dodds Science Publishing Limited, 2024.

Figure 1 Symptoms of plant-parasitic nematodes causing a direct damage to banana growth by (a) slowing vegetative growth, resulting in 'choking', (b) reduced bunch size, (c) necrotic root systems, and (d) toppling of plants (Photos A. Pattison).

banana and plantain cultivars to supply fruit for a multitude of uses, including fresh fruit, cooking and beverage manufacture.

Plant-parasitic nematodes present a major constraint to production of bananas and plantains. There are a wide range of endoparasitic and ectoparasitic nematodes that feed on the roots of banana, with 146 species belonging to 43 genera associated with bananas throughout the world (Gowen and Queneherve, 1990). The plant-parasitic nematodes in banana soils can have direct impacts on the banana plant, such as reduced plant growth, reduced yield, and toppling (Fig. 1a-d). Direct impacts of nematodes on bananas are mostly due to endoparasitic nematode, which enter the plant and feed inside the root tissue, such as *Radopholus similis*, *Pratylenchus goodeyi*, *P. coffeae*, *Helicotylenchus multicinctus* and *Meloidogyne* spp. (Fig. 2a-f). Plant-parasitic nematodes can also have indirect impacts, by slowing production cycles, reducing crop uniformity, increasing crop inputs, developing complexes with secondary fungi and making plants more susceptible to environmental stresses. The indirect impacts are often associated with ectoparasitic nematodes, which

Published by Burleigh Dodds Science Publishing Limited, 2024.

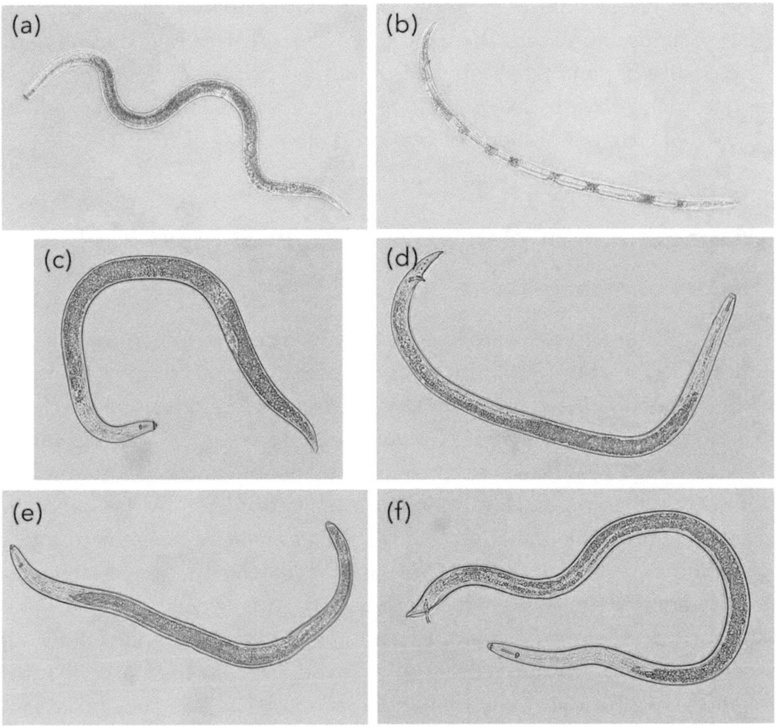

Figure 2 Principal plant-parasitic nematodes isolated from the roots of bananas (a) *Radopholus similis* female (b) and male, (c) *Pratylenchus goodeyi* female (d) and male, (e) *Helicotylenchus multicinctus* female and (f) male (Photos courtesy of Wayne O'Neill).

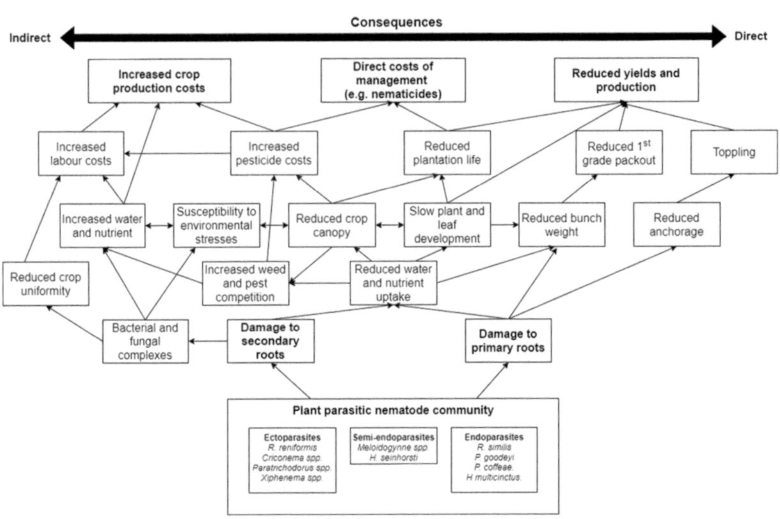

Figure 3 Conceptual model of the direct and indirect impacts of plant-parasitic nematodes on banana production.

feed on root tissue from outside the plant, subtly decreasing productivity as their abundance increases. The consequences of plant-parasitic nematodes feeding on the banana plant form a continuum of direct and indirect impacts, ranging from the obvious, like plant toppling to subtle decreases in productivity due to loss of crop uniformity and increased labour needs (Fig. 3).

2 Management of plant-parasitic nematodes

2.1 History of management

Plant-parasitic nematodes were known to parasitise the roots of bananas from as early as 1891 (Cobb, 1915). The extent of nematode infestations was found to be widely associated with most banana plantations around the world (Colbran and Saunders, 1961; Blake, 1961; Gowen and Queneherve, 1990; Gowen et al., 2005; Gowen, 1995). However, their impact on production was not widely recognised until commercial monoculture plantation systems for export were developed. In these plantations, plant-parasitic nematodes were recognised as yield limiting pests as early as the 1950s and research commenced to establish control systems (Quénéhérve, 1993). Reduced productivity due to plant-parasitic nematodes became more evident in South and Central America following the replacement of Gros Michel cultivars, with Cavendish cultivars that were resistant to Fusarium wilt Race 1. Early management efforts relied on applications of fumigant nematicides such as 1,2-dibromo-3-chloropropane (DBCP), being applied using hand-held injectors for soil treatment (Quénéhérve, 1993). However, it was also recognised that planting material was a major source of spread and transmission of plant-parasitic nematodes. The systems for treating planting material relied upon removing visibly infected material and subjecting planting material to a hot water treatment (Blake, 1961; Colbran, 1964b).

The introduction of non-fumigant organophosphate and carbamate nematicides to manage the abundance of plant-parasitic nematodes in commercial Cavendish plantations became widespread during the 1970s. These non-fumigant nematicides were cholinesterase inhibitors affecting the nematodes' nervous system, which either caused direct death or paralysis, preventing the nematodes from feeding, resulting in starvation (Bunt, 1987). Field experiments demonstrated yield increases in the order of 5–50% using nematicides in most commercial banana plantations in the majority of production regions around the world (Quénéhérve, 1993; Broadley, 1979b; Broadley, 1974; Araya, 1999; Chavez et al., 2020; Jaramillo et al., 2019; Salguero et al., 2016). Furthermore, treatment of propagation material started to involve paring of planting material, followed by immersion in a 250 ppm fenamiphos solution for 10 min to reduce *R. similis* (Broadley, 1979a). Continuous use of nematicides in the banana monoculture systems gave rise to enhanced microbial degradation of organophosphate and carbamate nematicides by the early 1990s (Pattison et al., 2000a; Racke and Coats, 1988; Moens et al., 2004; Anderson, 1989). In addition, by the 2000s nematicide contamination around banana plantations was being reported, raising questions concerning the ongoing application of nematicides in many parts of the world (Castillo et al., 2000; Charlier et al., 2009). Environmental and safety concerns led to the deregistration of some nematicides and restrictions in market acceptability of fruit destined for export, requiring the need for alternative and integrated nematode management systems (Stirling and Pattison, 2008; Quénéhérve, 1993). Since that time integrated nematode management systems have been continually being improved through new insights, development of novel products, and aided by banana growers' desire to improve soil health, coupled with pressure from retailers requiring certified reductions in the use of nematicides (such as Global GAP or Fair Trade).

2.2 Development of thresholds and testing services for management decision-making

Monitoring nematode damage and linking that to the abundance of particular nematode species is an important part of integrated nematode management programs (Stanton et al., 2001). For bananas, monitoring plant-parasitic nematodes can be done by extracting the nematodes from soils and roots or examining the damage on roots caused by the nematodes. Broadley (1979c) developed a disease index based on the severity of lesions in the root cortex and Stanton et al. (2001) found that this was a more precise method for assessing the nematode status in a plantation than nematode counts from root and soil samples. Assessment of root damage in bananas caused by nematodes was found to be rapid and easy to use. In a survey of banana grower practices in Australia in 2003, two-thirds of banana growers were aware of the root disease index method for monitoring nematode damage and nearly half were either using it regularly or having it done by a commercial monitoring service (Stirling and Pattison, 2008). For those countries focussed on production of bananas for export, quality assurance and management certification systems, such as Rainforest Alliance, Global G.A.P, Sedex etc., nematode counts are required to justify applications of nematicides for banana production.

3 Cultural controls

3.1 Quarantine and biosecurity

Not all plant-parasitic nematodes damaging bananas occur in all banana producing countries, regions or plantations. For this reason, exclusion through quarantine and biosecurity can be useful in preventing incursions of new and exotic plant-parasitic nematodes. Singh et al. (2013) listed eight regulated plant-parasitic nematodes for banana, including:

- Helicotylenchus microcephalus, H. multicinctus;
- Heterodera oryzicola;
- Hoplolaimus pararobustus;
- Pratylenchus coffeae, P. goodeyi;
- Radopholus similis; and
- Xiphinema ifacolum.

In addition, *Meloidogyne javanica, Scutellonema minutum,* and *Tylenchorhynchus nudus* were listed as being damaging to bananas but were not regarded as nematode pests requiring regulation (Singh et al., 2013).

Internal country regulations and on-farm biosecurity practices can also prevent introduction of plant-parasitic nematodes onto new banana plantations. For example, with the expansion of bananas to north Queensland from subtropical areas in the early 1990s considerable effort was put into preventing the spread of nematodes into new production areas. Initially, regulations regarding the movement and use of banana planting material (Plant Protection Act 1989, Queensland) were used to discourage growers distributing planting material from plantations infested with *R. similis*. This led to the establishment of dedicated planting material nurseries on farms to control introduction of nematodes into new production areas. Some owners

Figure 4 Signage advising farm is a nematode-free area and the need to wash vehicles before entering (Photo A. Pattison).

Published by Burleigh Dodds Science Publishing Limited, 2024.

of new plantations introduced their own quarantine procedures to prevent the introduction of *R. similis* on machinery and footwear (Fig. 4) (Stirling and Pattison, 2008). Educational programmes in addition to regulation played a major role in preventing spread through increased awareness of the potential for nematode carryover on planting material.

3.2 Clean planting material

Prior to the development of *in vitro* banana planting material through tissue culture, most banana planting material was sourced from fields that were being taken out of production because of low productivity. This is still commonly practised in many smallholder systems (Tenkouano et al., 2006). Such planting material was often infested with nematodes and therefore treated with either hot water or a nematicide before use. The commercial availability of tissue culture derived banana plants as a source of planting material reduced the need for banana

Figure 5 Large-scale production of (a) *in vitro* banana plantlets and (b) acclimatisation facility in Taiwan (Photo A. Pattison).

growers to use rhizome material sourced on-farm (Fig. 5a, b). The quality of tissue-cultured material is often monitored under industry managed quality assurance schemes, such as for example QBAN in Australia (Hamill et al., 2009). A survey conducted in north Queensland indicated that 87% of banana growers believed they used clean planting material to manage the spread of nematodes, but only 25% were using tissue culture derived plantlets (Stirling and Pattison, 2008).

3.3 Nematode persistence

Regular renewal of banana fields is not common in many banana exporting countries, with plantation fields in production for 30–40 years. In Queensland, Australia, most banana growers renew fields every 10 years, with a break of 6-24 months between successive banana crops. The fallow period traditionally has consisted of weeds and volunteer bananas, rather than a non-host crop to *R. similis* (Stirling and Pattison, 2008). *R. similis* can survive in decomposing root and corm material for up to 6 months (Blake, 1969). The survival of *R. similis* on volunteer bananas growing in the fallow period between cropping cycles is a major cause of carryover of nematodes to the next crop. Chemical destruction of banana plants by injecting glyphosate into the pseudostems of banana plants was six times more effective at reducing survival of *R. similis* compared with mechanical destruction (Chabrier and Queneherve, 2003). The cropping data showed a 29% increase in production following the second crop cycle in areas where bananas were eradicated using a glyphosate injection (Chabrier and Queneherve, 2003).

Soil water has a significant impact on *R. similis* persistence in banana fields. Water run-off is a major source of dispersal and reinfection in banana plantations with *R. similis* (Chabrier and Quénéhervé, 2008). Soil moisture conditions in the fallow period significantly affected the survival of *R. similis*, which are not well adapted to surviving in saturated soils (Chabrier et al., 2010b). It also turned out that males, although having a reduced capacity to feed on banana roots, survived better without a host than females (Chabrier et al., 2010a).

3.4 Crop rotation

Fallowing and rotation with non-host crops has been investigated in many countries for the reduction of plant-parasitic nematodes prior to planting bananas (Ternisien and Ganry, 1990). Rotation crops for bananas can be grown for additional income or used as a green manure fallow crop to supplement organic matter inputs and protect the soil. Common rotations for bananas used for additional income include papaya, sugarcane and pineapple. The inclusion of sugarcane in the banana production system requires careful selection of nematode resistant sugarcane cultivars to efficiently reduce *R. similis* populations. Pineapple, a non-host to *R. similis*, requires intensive weeding to be productive, but efficiently supresses the abundance of *R. similis* in soil, making it suitable for rotation with bananas (Chabrier and Queneherve, 2003; Zhong et al., 2016).

The adoption of non-host fallow crops as a nematode management practice by banana growers requires that they fit within the banana production system. Early screening found grasses that were resistant to *R. similis* like green panic (*Panicum maximum*) (Colbran, 1964a) and Jarra grass *(Digitaria milanjiana)* (Hall et al., 1993), but they were not adopted by banana growers due to the difficulty in establishment or weed potential in the subsequent banana crop. Further screening of the multiplication of *R. similis* on a wide range of crops demonstrated that several grasses, including, signal grass (*Urochloa decumbens*), Rhodes grass (*Chloris gayana*), Bahia grass (*Paspalum notatum*) and various brassicas were relatively poor hosts of the nematode and were useful rotation crops (Pattison et al., 2000b; Pattison, 2021) (Fig. 6a, b). This has led to recommendations of different fallow crops and cultivars for different environments and time periods (Pattison, 2021).

Figure 6 Rotation crops used to reduce plant-parasitic nematodes in bananas such as (a) Brassicas and (b) a sunn hemp and sorghum mix. (Photo A. Pattison).

Weed control is also an essential component of effective nematode management. Weeds mostly belonging to three families – the Euphorbiaceae, Poaceae, and Solanaceae – have been found to be significant reservoirs of plant-parasitic nematodes including *R. similis* in banana fields (Quénéhervé et al., 2006).

4 Biological control

Biological control of plant-parasitic nematodes in bananas can have a direct impact on the nematodes through reducing their populations, or an indirect effect by increasing the plant's ability to tolerate nematode infestation (Table 1). Direct effects can be either due to antagonism based on the synthesis of lytic enzymes, toxic insecticidal crystal proteins, volatile compounds or parasitism (Table 1).

Table 1 Examples of biological controls of plant-parasitic nematodes of bananas using specific organisms that have a direct or indirect impact on nematodes or practices that alter the general soil microbial community

		Specificity of biological methods	
		Specific organisms	General microbial community
Impact on plant-parasitic nematodes	Direct	• Purpureocillium (Paecilomyces) lilacinus (Rumbos et al., 2008; Mendoza et al., 2007; Khan et al., 2006; Paz Filho et al., 2019; Fernandez et al., 2005) • Pasteuria penetrans (Devrajan et al., 2003) • Pochonia chlamydosporia (Barbosa et al., 2019)	• Mulch (Wairegi and van Asten, 2010; Gaidashova et al., 2010) • Soil Organic Matter (Wang and Hooks, 2009) • Organic amendments (Tabarant et al., 2011b; Esnard et al., 1998; Pattison et al., 2011) • Plant extracts (Kosma et al., 2011; Jesus et al., 2015) • Plant diversity (Poeydebat et al., 2017)
	Indirect	• Bacillus spp. (Mendoza and Sikora, 2009) • Fusarium oxysporum (Mendoza and Sikora, 2009; Chaves et al., 2009; Paparu et al., 2009; Sikora et al., 2008) • Pseudomonas fluorescens (Kavino et al., 2010) • Glomus spp. (Seenivasan, 2017) • Trichoderma spp. (Sikora et al., 2008)	• Nutrients (Wairegi and van Asten, 2010) • Biostimulants (D'Addabbo et al., 2019) • Humic acid (Seenivasan and Senthilnathan, 2018) • pH (Almeida et al., 2018)

Indirect antagonism is expressed through competition for nutrients, inducing systemic resistance or the release of nematode modulating molecules, which is included among the indirect mechanisms (Table 1) (Gamalero and Glick, 2020).

4.1 Specific biocontrol organisms

The effectiveness of specific organisms at reducing the relative abundance of plant-parasitic nematode populations is typically density dependent, being more effective at high nematode density (Stirling, 2014). Three specific biocontrol organisms are widely reported for control of plant-parasitic nematodes in banana: *Purpureocillium lilacinus* (formerly *Paecilomyces lilacinus), Pasteuria penetrans,* and *Pochonia chlamydosporia* (Table 1). However, these organisms were most effective against root-knot nematodes species in banana, with limited efficacy on endoparasitic nematodes. Gamalero and Glick (2020) suggested that once nematodes are established within the root systems of plants, they are difficult to control. As most economically important plant-parasitic nematodes of bananas are endoparasitic, much of the research on specific biological control organisms has focused on plant endophytic fungi and bacteria. This has included *Fusarium oxysporum* isolates (Mendoza and Sikora, 2009; Chaves et al., 2009; Paparu et al., 2009; Sikora et al., 2008) (Table 1), which tend to increase plant growth in the presence of nematodes in controlled glasshouse experiments, as well as lead to a suppression of nematodes (Li et al., 2015). New studies on soil micro-organisms associated with nematode decline have revealed new microbial species with antagonistic potential against plant-parasitic nematodes (Topalović et al., 2020). For example, *Bacillus* spp. isolates, which produce nematicidal compounds, were found to induce a plant-mediated responses, leading to the suppression of root-knot nematodes (Topalović et al., 2020).

4.2 General biological control

General biological control is based on the provision of better soil conditions, retarding the build-up of plant-parasitic nematodes and increasing plant tolerance to damage (Table 1). This occurs due to non-specific microbial and plant interactions in the soil, which maintain a low abundance of plant-parasitic nematodes (Poeydebat et al., 2017). The effectiveness of general biological control is related to total soil microbial biomass and is enhanced by practices that conserve soil organic matter (Stirling, 2014). Organic matter promotes nematode antagonistic organisms, stimulating their competitiveness against plant-parasitic nematodes (Thoden et al., 2011) and strengthening their positive interactions with plants (Stirling, 2014). Pattison et al. (2011) found that the application of amendments high in C and N, such as banana residue, grass hay, mill mud (a by-product from sugarcane mills) and compost suppressed plant-parasitic nematodes through developing a more favourable environment for antagonistic organisms. Similarly, in Guadeloupe sugarcane bagasse, sugarcane sludge and plant residues decreased the abundances of plant-parasitic nematodes in bananas by up to 96% (Tabarant et al., 2011a; Tabarant et al., 2011b).

The cost of importation of organic matter required to sustain and maintain the general suppression of plant-parasitic nematodes in bananas can be prohibitive, if an available source is not readily available. Alternatively, organic matter can be obtained from residues of crops grown *in situ* (Stirling, 2014). Shade tolerant ground covers such as carpet grass (*Axonopus spp.*) and forage peanut (*Arachis pintoi*) can be grown around banana plants and in inter-rows and managed in a way such that they do not compete with the crop for water and nutrients. This practice was not detrimental to banana root growth but improved the physical and biological status of banana soil (Djigal et al., 2012). Intercropping with legumes, such as sunn hemp (*Crotalaria juncea*), was also found to reduce *R. similis* abundance and improve root growth of bananas relative to monoculture bananas with bare inter-rows (Chitamba et al., 2014). Intercropping with vegetable crops and inclusion of bananas in agroforestry production systems are viable options for managing plant-parasitic nematodes for smallholder growers, allowing them to include cropping diversity into their production systems (Blomme et al., 2020; Ocimati et al., 2021).

The crop residues that accumulate on the soil surface as a result of de-leafing, de-suckering and harvesting operations typically amount to 10-25 t dry matter/ha. In export banana plantations, this material was traditionally moved into the inter-row, leaving the soil bare around the plant to facilitate nematicide application. However, allowing the crop residues to accumulate around the base of the banana plant, increased soil carbon (C) levels, while reducing the abundance of plant-parasitic nematodes (Pattison et al., 2008). The biochemical characteristics, particularly the C:N ratio, of organic residues added to banana soils, strongly influenced the nematode community composition, and therefore the potential damage caused by plant-parasitic nematodes (Chauvin et al., 2015). In general, practices that tend to reduce microbial biomass, activity, and diversity, such as monocultures, low soil pH, high fertiliser, and agrochemical inputs, tend to result in greater abundance of plant-parasitic nematodes (Almeida et al., 2018; Pattison et al., 2008; Declerck et al., 1998; Wairegi et al., 2010; Silva et al., 2018).

5 Chemical control

As outlined in Section 2.1, chemical nematicides were often the principal management method for plant-parasitic nematodes in commercial and export banana plantations until the new millennia. Most of the products available were organophosphates or carbamates, which are being phased out due to their toxicity and non-compliance with new regulations. However, unlike other crop protectants there have not been any suitable replacement chemicals released to the banana industry due to the difficulties in finding highly selective products that target nematodes, higher environmental and safety standards, and the cost of registration (Desaeger et al., 2020). The cost of bringing a new chemical active ingredient to market is increasing every year and is now estimated to have surpassed US$250 -300 million. The majority of research done on new nematicides is focused on their risk to the environment and human health, including potential impacts on a wide range of non-target organisms (Desaeger et al., 2020). Despite these significant hurdles several new nematicides have become available to growers and several more are being developed by the agrochemical industry. The new products tend to be more selective, often only targeting nematodes, are less toxic and safer to apply (Desaeger et al., 2020). Ideally, a nematicide will only affect plant-parasitic nematodes, work consistently, will not leave residue in the soil or plants, is easy and safe to apply and is inexpensive (Desaeger et al., 2020). The new generation of nematicides have a much better profile in terms of operator safety and selectivity. An overview of the most significant new chemical nematicides to emerge in the last decade was provided by Desaeger et al. (2020) and with relevance to banana include the following:

- Tioxazafen (Nemastrike®, Monsanto/Bayer) is a systemic nematicide belonging to the chemical class oxadiazoles. Its mode of action may be linked to disruption of ribosomal activity of nematodes. However, recent safety concerns may limit its further use.
- Spirotetramat (Movento®, Monsanto/Bayer) is a spirocyclic tetramic acid derivative, which has unique translocation properties within the entire vascular system of the plant. To be active against nematodes, the compound needs to be hydrolyzed to the active spirotetramat-enol form.
- Fluensulfone (Nimitz®, Adama) is a member of the class of 1,3-thiazoles, a new group of chemical nematicides. The mode of action of fluensulfone is as a fatty acid beta oxidation inhibitor.
- Fluopyram (Verango®, Velum®, Indemnify®, Bayer CropScience) was initially registered as a fungicide. The nematicidal activity of fluopyram was not known or disclosed until later and was first registered as a nematicide in Honduras for banana under the name Verango® (Desaeger et al., 2020).
- Fluazaindolizine (Salibro®, DuPont/Corteva) was registered in 2021. Fluazaindolizine is a member of the class of carboxamides, which

specifically targets nematodes. Its mode of action is unknown; however, it is distinct from carbamates, organophosphates or any other known nematicide groups.

The development of new nematicides often begins with the bacterial-feeding nematode *Caenorhabditids elegans* as a model using high-throughput screening, which means that its suppression might not translate well to plant-parasitic nematodes. In product development, the root-knot nematode (*Meloidogyne* spp.) is often used as a model and efficacy may not necessarily translate to control of important endoparasitic nematodes in bananas.

Natural products accounted for almost 70% of all new active ingredient registrations, with biological nematicides becoming more important for the future (Desaeger et al., 2020). Some essential oils derived from plants show efficacy and may lead to the development of new nematicidal compounds (Fierascu et al., 2020). The mechanism responsible for the nematicidal action of essential oils is unclear, but maybe due to a permeability changes of nematode cell membranes or development of neurotoxins (Fierascu et al., 2020). Extracts from other plants like *Agave sisalana* and *Azadirachta indica* have also been shown to be effective at reducing the abundance of *R. similis* in the roots of banana plants and may offer alternatives to suppress nematodes (Jesus et al., 2015; Kosma et al., 2011).

Finding novel biochemical targets within the nematode can help to identify more selective nematicides and develop more options for growers to counteract resistance and enhanced degradation issues. Chen et al. (2020) identified 65 nematicidal compounds belonging to a diverse group of chemicals classified as amides, esters, ketones, thioethers, sulfones, hydrazones, tioxazafen analogues, and organometallic complexes. While many of the products may never be registered for the management of plant-parasitic nematodes of bananas, they highlight the search for replacements of the older organophosphates and carbamate nematicides. Furthermore, metabolomics and transcriptomics methods are powerful tools to help understand the toxicity and mode of action of new compounds on plant-parasitic nematodes (Chen et al., 2020).

6 Genetic resistance

Host resistance is an effective means of nematode control in many crops to limit plant-parasitic nematodes. However, in bananas and plantains, there has been little progress in breeding for resistance to nematodes because of sterility and the genetic complexity within the genus *Musa* (Seenivasana, 2018) and the low priority given to this trait in breeding programmes in the past due to the availability of chemical nematicides (Haegeman et al., 2010). Sources of resistance in bananas and plantains have been identified in *Musa* germplasm growing in the wild, landraces, local banana cultivars, and hybrids in breeding programmes (Viaene et al., 2003; Dochez et al., 2006; Quénéhervé et al., 2008). However, very few new cultivars with plant-parasitic nematode resistance from breeding programmes are grown in farmer fields today.

Along with the identification of resistant germplasm, an increased understanding is emerging about the chemical and genetic background of resistance. The identification of resistant and tolerant sources of *Musa* to *R. similis* has led to the identification of chemical markers in resistant cultivars (Seenivasana, 2018). Plant defence responses and secondary metabolites produced by bananas with the ability to suppress plant-parasitic nematodes have also been identified (Kazuki et al., 2019). Changes in the lignin and phenol profiles were identified in banana genotypes resistant to *R. similis* infection. Increased lignification was found to be a general defence response, whereas increased phenol concentration was found to be more specific to nematode infection (Dhakshinamoorthy et al., 2014; Hölscher et al., 2014; Wuyts et al., 2007). The identification of nematode transcriptomes may provide insights into the parasitism process and provide new avenues for more efficient control measures. Genes involved in the parasitism process, as well as genes essential for nematode survival, could provide useful information for inclusion in development and deployment of resistance (Jacob et al., 2008).

7 Integrated pest and disease management

Development of integrated plant-parasitic nematode management systems commenced with the combination of soil sanitation practices and clean planting material (Mateille et al., 1988; Quénéhérve, 1993). Early attempts used flood fallowing as a soil disinfestation method before replanting bananas, which reduced nematodes, but inconsistent results, lack of water, sloping fields and costs prevented large scale adoption (Mateille et al., 1988). More recent integrated nematode management systems include multiple practices based on; crop rotation, reduced tillage, rationalisation of fertiliser applications, manure incorporation, careful application of pesticides and crop water management. A combination of farm practices have a greater effect as part of a whole system to induce soil suppressiveness to reduce plant-parasitic nematode abundance (Silva et al., 2018) by changing soil microbial parameters linked to general suppression (Garbeva et al., 2004). The incorporation of such practices to manage plant-parasitic nematodes on bananas ultimately depends on consumer demands and resources available to banana growers.

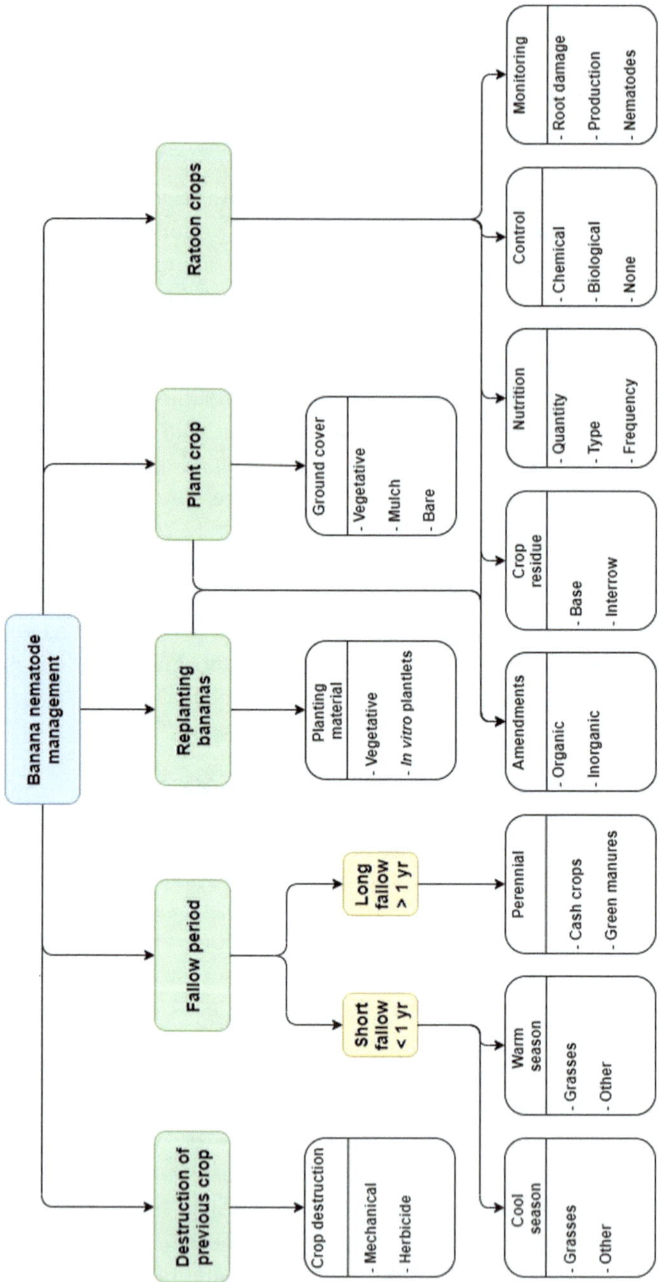

Figure 7 Management decisions made at different stages of the banana cropping systems that influence the abundance and impact of plant-parasitic nematodes.

8 Examples of integrated nematode control in different countries

8.1 Australia

The gradual implementation of an integrated nematode management programme has resulted in a steady decrease in the amount of nematicide used in the Australian banana industry (Stirling and Pattison, 2008). It is estimated that less than one-third of banana growers use nematicides, instead of relying on an integrated approach to manage plant-parasitic nematodes (Fig. 7). This begins with the thorough destruction of affected banana crops followed by a fallow period where non-host crops are grown. Before replanting bananas, soil improvements are made that include adjustment of drainage and soil pH. When replanting tissue culture plantlets are widely used, purchased through an accredited scheme to ensure quality. Once plants become established, ground cover vegetation is maintained, with crop residues being kept around the base of the banana plants. There has been a significant rationalisation of nitrogen fertiliser inputs in north Queensland's banana industry due to fertiliser run-off regulations to protect the Great Barrier Reef. This had the added benefit in aiding the development of nematode suppressive soils. Roots are regularly monitored for nematode damage with accredited private and government diagnostic services available to banana growers (Fig. 7).

8.2 Costa Rica

Costa Rica has a long history of commercial banana cultivation, taking place for more than 100 years, with most plantations growing bananas for more than 20 years. Costa Rica produces bananas mostly for the export trade to the USA and the European Union, with all farms requiring a certification system for production and environmental sustainability (Rainforest Alliance, Global G.A.P., SCS Global Services Standard, Sedex, ETI). Within these certification systems, the use of phytosanitary products for pest and disease control are highly regulated. Plant-parasitic nematodes became a constraint to banana production in Costa Rica following the replacement of Gros Michel (*Musa* AAA) with Cavendish cultivars (*Musa* AAA) (Araya and Vargas, 2018; Araya et al., 2002; Araya-Vargas et al., 1995). The damage caused by plant-parasitic nematodes led to a greater emphasis on improving root health to overcome yield reductions. Improving the drainage within plantations, through primary, secondary, and tertiary structures, removed root growth impediments caused by saturated soils. The improved root growth resulted in increased tolerance of banana plants to plant-parasitic nematodes.

The introduction of Integrated Pest Management (IPM) led to monitoring for plant-parasitic nematodes (Araya-Vargas and Cheves-Vargas, 1998; Araya-Vargas, 1998). All Costa Rican banana farms producing export bananas monitor nematodes, at least every four months. Any application of chemical nematicides needs to be based on data from a monitoring program, and only when the nematode density exceeds the economic threshold of 5000-8000 plant-parasitic nematodes per 100 g of root can a nematicide be applied (Araya-Vargas, 2004). New chemical groups, such as fluopyran and avermectins are now being made available to replace organophosphate and carbamate nematicides. In addition, banana grower's pay strict attention to providing optimal growing conditions through ongoing drainage system maintenance, application of organic amendments and lime to adjust soil pH. Further alternatives are being explored such as the use of biological control agents (Salas, 2007) and differing sources of organic matter (Vargas and Araya, 2009). However, initial results do not match the efficacy of chemical control, slowing their uptake.

It is now recommended that Costa Rican banana plantations renew 5–10% of their cultivated area each year, commencing 10 years after planting, in areas where production has declined. In areas designated for renewal crop destruction commences with glyphosate injection of the previous crop. Most banana growers fallow for 1–3 months, taking the opportunity to improve drainage, loosen subsoil, and adjust soil pH before replanting with tissue-cultured banana plantlets. Growers recommence nematode monitoring four months after planting following the same procedures as described for the production areas.

8.3 French West Indies

The development of a banana plantation renewal programme in the French West Indies has allowed a dramatic reduction on the impact of nematodes, with a 50% reduction in abundance of *R. similis* in plantations and 12-fold reduction in nematicide application, with no reductions in crop productivity. The injection of glyphosate into the pseudostem to destroy existing banana plants was key to commencing the renewal programme. Initially, a 'clean fallow' (regular herbicide applications) was implemented, but erosion concerns saw this replaced with rotation crops, mainly sugarcane or *Urochloa* spp. Schemes with specific criteria were established to encouraged grower adoption, involving subsidised cost of certified tissue culture plants and interception drains to prevent recontamination through run-off water. The implementation of this integrated cropping systems has meant that *R. similis* has become a minor constraint to banana production, resulting in a decline in nematicide use from 84.4 tonnes in 1997 to 6.6 tonnes in 2009 (Chabrier and Quénéhervé, 2010) and then less than 2.8 tonnes of active ingredient in 2015. Over the same period, average banana yields increased slightly from 30 tonnes/ha to 34 tonnes/ha.

8.4 Ivory Coast

Commercial banana plantations on the Ivory Coast resembled the banana production system of the French West Indies of the 1980s. One of the differences was that the social pressure to reduce pesticide use was much lower and there were no support schemes to encourage growers to move towards a more integrated cropping system. Hence, few plantations have adopted rotation crops in their banana cropping system. Nevertheless, banana plantations in production for more than 30 years, usually adopt a one-year fallow period every 6 to 15 years. Growers frequently plant *Tithonia diversifolia* as cover plant to reduce *R. similis* abundance (Gnonhouri et al., 2020). The majority of the plantations which have been renovated this way in the last 15 years have been re-planted with tissue culture plants, with special attention to avoid the reintroduction of nematode via infected plant material or via irrigation water. The result is that only a half of commercial plantations are treated with nematicide and receive 1–2 applications/year depending on nematode monitoring results.

8.5 Subtropical bananas

Banana production in subtropical regions is typically limited to steep sloping land or semi-arid desert regions to avoid frost and chilling. Many of the plant-parasitic nematodes found in banana production in tropical regions are of concern in subtropical regions, such as *R. similis*, *Meloidogyne* spp., and *Helicotylenchus multicinctus*. However, due to the environmental differences affecting nematode development and growth of bananas, plant-parasitic nematodes often regarded as secondary nematodes in tropical regions, can have a greater impact on production in subtropical regions. Additionally, the endoparasitic nematode *Pratylenchus goodeyi*, which has its distribution closely linked to altitude and temperature, is found in subtropical banana production areas in eastern Australia and the Canary Islands as well as the highlands of Africa.

In many subtropical banana-production areas, the lack of infrastructure and mechanisation means that regular plantation renewal is not commonly practised, and fields remain in continuous production for 20 years or more (Dita et al., 2020). Due to the reduced productivity of subtropical banana plantations, the application of nematicides is often uneconomic, and plants show a weaker response to their applications as compared to the tropics (Pattison et al., 2002). When banana fields are renewed, there is the opportunity to include nematode resistant rotation crops and replanting with tissue culture planting material. Shade tolerant ground covers or intercropping is also recommended to retain soil around banana plants, particularly on sloping land, with the selection of nematode resistant ground cover species seen as a priority. In semi-arid areas where crop density is increased to reduce wind damage and increase water efficiency, the inclusion of ground covers in plantations is problematic. Other agronomic practices aimed at improving the general soil and plant health of the banana plant, such as applications of organic amendments, retention of crop residue, weed, pest and disease control, and maintaining appropriate nutrient levels can all help to reduce the impacts of plant-parasitic nematodes in subtropical banana regions.

9 Conclusion and future trends

The management of plant-parasitic nematodes in banana systems needs to change in response to growing consumer demands for safer food and lessening impacts of agriculture on global ecosystems. There has been an increase in the understanding of microbial-plant-pathogen interactions, which may help the banana industry alleviate problems caused by plant-parasitic nematodes in plantations. Studies on micro-organisms, organic manures, and physicochemical soil aspects indicate suppressiveness can be achieved through improving soil management practices. The shortage of effective chemical nematicides and the demand for sustainable production brings new challenges in the management of plant-parasitic nematodes (Silva et al., 2018), particularly for commercial banana plantations. Banana production and management of plant-parasitic nematodes face additional challenges, such as climate change and changes in farming technology in efforts to better manage plant-parasitic nematodes.

9.1 Challenges

9.1.1 Climate change and impacts

Climate change represents a major challenge for banana producers globally, with declining yields expected in some areas by 2050 (Varma and Bebber, 2019). Changing climatic conditions are likely to impose additional stress on crops like bananas. Banana crops with a compromised root system, due to the feeding and damage caused by plant-parasitic nematodes, are more vulnerable to any impacts of climatic stress and may suffer a greater impact on productivity and viability of the plantation (Brentu et al., 2004).

9.1.2 Loss of chemicals

Loss of current nematicides registered for use in bananas is a likely scenario in many commercial banana producing countries. The availability and efficacy of new generation nematicides is still to be determined. Many of the new nematicide products are being developed for efficacy against root-knot nematodes in annual cropping systems. Further investigations are required to determine how the products fit into a perennial system where migratory endoparastic nematodes are the biggest problem. It is likely that new nematicides will be integrated with farm cultural practices and/or biological products and used as a 'corrective treatment' rather than the principal method of management.

9.3 Future technologies

9.3.1 Precision agriculture

Precision agriculture and decision support systems are being developed for export banana plantations with yield maps linked to other available information for the plantation (Stoorvogel et al., 2004; Leroux et al., 2018). However, the asynchronous nature of banana crops makes applications of precision agriculture difficult (Lamour et al., 2020). With the need to utilise farm inputs more effectively, precision agricultural principles are being developed, with correlations to soil properties (Freitas et al., 2016). Plant-parasitic nematodes have a two-way impact on applications of precision agriculture. Firstly, mapping can identify areas of high plant-parasitic nematodes abundance, which allows targeted nematode management interventions. Secondly, plant-parasitic nematodes represent a major source of variability in banana fields having direct and indirect impacts on plant production making yield predictions difficult (Fig. 3).

9.3.2 Specific biocontrols

There are many nematode biocontrol mechanisms, especially parasitism, induced systemic resistance, and volatile organic compounds using microbial consortia or specific bacterial or fungal isolates, for nematode management (Topalović et al., 2020). The specific organisms need to be chosen to take advantage of the increased knowledge in chemical communications between bananas, plant-parasitic nematodes, and other soil micro-organisms. A comprehensive understanding of the plant-beneficial microbial consortia and the mechanisms underlying disease suppression may help to develop synthetic microbial communities for biocontrol of plant-parasitic nematodes (Topalović et al., 2020). Genome sequencing of potential biocontrols may increase knowledge of enzymatic pathways to allow selection of organisms that have specific functions for biological control of plant-parasitic nematodes.

9.3.3 Plant microbiomes

There is a growing understanding that plants rely on their root microbiome when they are under attack by pathogens and parasites. Therefore, rhizosphere organisms can act as a first line of defence against nematode invasion, while the root endophytic organisms can act as a second line of defence against successfully invaded nematodes (Topalović et al., 2020). Understanding how farm management decisions can enhance the diversity and regulatory function of the banana microbiome is essential in developing plant-parasitic nematode suppressive banana production systems. The banana microbiome potentially plays a role in alleviation of abiotic stress to plants. Therefore, enhancing the microbiome around the time of an abiotic perturbation may enhance the microbial or plant induced protection, reducing the additional impact on plant stress caused by plant-parasitic nematodes.

9.3.4 Chemical signalling

There has been an increase in the knowledge of chemical signalling in the study of plant-pathogen-microbe interactions. Nematode's sense environmental signals using different sensory organs, such as amphids. These sensory organisms pick up chemical cues released from the roots which guide plant-parasitic nematodes to the host roots, (Topalović et al., 2020). It is possible to apply this information to develop interruption systems, reducing the likelihood of plant-parasitic nematodes becoming established with the root systems of bananas. The identification of effectors, pathogen-associated molecular patterns (PAMPs), and damage-associated molecular patterns (DAMPs) to trigger immune response in banana include the secretion of anti-nematode enzymes, the production of anti-nematode compounds, cell wall reinforcement, production of reactive oxygen species and nitric oxide, and hypersensitive response-mediated cell death (Kazuki et al., 2019). The induced plant responses to plant-parasitic nematodes in bananas represents novel methods for nematode management, with minimal environmental impact.

An anti-nematode phenolic compound, the phenylphenalenone anigorufone has been identified in resistant banana cultivars at *R. similis* infection sites (Dhakshinamoorthy et al., 2014; Hölscher et al., 2014). Anigorufone was reported to have high nematicidal activity because of the formation of large lipid–anigorufone complexes in the bodies of *R. similis*. Furthermore, the flavonoids that have anti-nematicidal activity mostly belong to the classes of flavonol kaempferol, which inhibited egg hatching of *R. similis* (Wuyts et al., 2006). Wuyts et al. (2006) also found that the flavaonols, kaempferol, quercetin, and myricetin were repellents and nemastatic to *M. incognita* juveniles.

Improvements in RNA interference technology allows the development of biopesticides which can prevent losses from plant-parasitic nematodes (Fletcher et al., 2020; Haegeman et al., 2010). Additionally, CRISPR technology has the potential to improve resistance and immune response of bananas to plant-parasitic nematodes through targeted manipulation of the plant's genome.

9.3.5 Nematode symbionts

Plant-parasitic nematodes are now known to have their own microbial communities attaching to the surface of nematodes or as endosymbionts, putatively aiding the parasitism of plants (Topalović et al., 2020; Brown, 2018). *R. similis* was found to harbour a Wolbachia-like bacterium. Wolbachia is a mutualist, as the nematodes cannot survive without this endosymbiont, which produce essential metabolites for the nematode (Haegeman et al., 2010). It has been suggested that horizontal gene transfer from nematode endosymbionts to the nematode host, assisted the nematodes to acquire plant cell wall degrading enzymes, giving them an evolutionary advantage (Haegeman et al., 2011; Brown, 2018).

9.4 Integrated technology

The maintenance of a functioning root system to sustain banana production is changing from highly chemical dependent practices to a more integrated and ecological systems approach. Plant-parasitic nematodes in bananas will remain a constraint to production but by integrating better farm management options their impact can be reduced. Increased understanding of the biochemical interactions of the plant-nematode-microbiome demonstrated that farm practices that conserved soil biodiversity improved suppression of plant-parasitic nematodes. The challenge in managing plant-parasitic nematodes is to integrate practices that allow production to continue in the face of greater consumer demands for reduced pesticides with improved environmental outcomes. This requires changing banana production system from high-input monocultures to diverse agroecological systems, which utilise beneficial soil microbial interactions, resulting in greater tolerance and resilience to plant-parasitic nematodes.

10 Where to look for further information

10.1 Websites

- Lambert, K. and Bekal, S. (2002). Introduction to plant-parasitic nematodes. *The Plant Health Instructor*. https://doi.org/10.1094/PHI-I-2002-1218-01 (apsnet.org).

10.2 General books

- Ciancio, A. and Mukerji, K. G. (2009). Integrated Management of Plant Pests and Diseases. *Integrated Management of Fruit Crops and Forest Nematodes*. Springer.

- Luc, M., Sikora, R. A. and Bridge, J. (2005). *Plant Parasitic Nematodes in Subtropical and Tropical Agriculture*. CABI Publishing, Wallingford, UK.
- Perry, R. N. and Moens, M. (2006). *Plant Nematology*. CABI, CAB International.
- Shurtleff, M. C. and Averre, C. W. (2000). *Diagnosing Plant Diseases Caused by Nematodes*. The American Phytopathological Society. APS PRESS, St. Paul, Minnesota.
- Sikora, R. A., Coyne, D., Hallmann, J. and Timper, P. (2018). *Plant Parasitic Nematodes in Subtropical and Tropical Agriculture*. 3rd Edition. CABI, CAB International.
- Stirling, G., Hayden, H., Pattison, T. and Stirling, M. (2016). *Soil Health, Soil Biology, Soilborne Diseases and Sustainable Agriculture*. CSIRO Publishing, pp 280.
- Stirling, G. R. (2014). *Biological Control of Plant-parasitic Nematodes: Soil Ecosystem Management in Sustainable Agriculture*. 2nd Edition. CABI.
- Waele, D. D. and Elsen, A. (2007). Challenges in tropical plant nematology. *Annual Review of Phytopathology* 45(1): 457–485. doi:10.1146/annurev.phyto.45.062806.094438.

10.3 Diagnostic manuals

- Perry, R. N., Hunt, D. J. and Subbotin, S. A. (2020). *Techniques for Work with Plant and Soil Nematodes*. CABI.

11 References

Almeida, N. O., Teixeira, R. A., Carneiro, F. A., de Oliveira, C. M., Ribeiro, V. A., Lobo Júnior, M. and Rocha, M. Rd (2018). Occurrence and correlations of nematodes, *Fusarium oxysporum* and edaphic factors on banana plantations. *Journal of Phytopathology* 166(4): 265–272.

Anderson, J. P. E. (1989). Accelerated microbial degradation of nematicides and other plant-protection chemicals in soil. *Nematropica* 19: 1.

Araya, M. (1999). Effecto del Furadan 4F y vydate 24%L aplicados en el pseudo-tallo de la planta de banano (*Musa* AAA) sobre el rendimiento y las poblaciones de nematodos. CORBANA, 24: 1–10.

Araya, M., De Waele, D. and Vargas, R. (2002). Occurence and population densities of nematode parasites of banana (*Musa* AAA) roots in Costa Rica. *Nematropica* 32: 21–33.

Araya, M. and Vargas, R. (2018). Frecuencia y densidades poblacionales de los nematodos parásitos de las raíces del banano (*Musa* AAA) en Costa Rica del 2000–2008. CORBANA, 44: 71–96.

Araya-Vargas, M. (1998). Comparación de las Poblaciones de Nematodos en Tres Puntos de Muestreo de la Unidad de Producción de Banano *(Musa* AAA*);* Planta Madre, Hijo de Sucesión y en el Intermedio Madre-Hijo. Comparison of Banana *(Musa* AAA*)*

Nematode Population in Three Sampling Points; Mother Plant, Follower Sucker and the Interspace between Both Plants. CORBANA, 23: 69–76.

Araya-Vargas, M. (2004). Los fitonematodos del banano (*Musa* AAA subgrupo Cavendish cultivares Grande Naine, Valery y Williams) su parasitismo y combate. The banana (*Musa* A A A Cavendish subgroup cultivars Grande Naine, Valery and Williams) phytonematodes, their parasistism and control XVI. *Reunión ACORBAT, Oaxaca, MX*.

Araya-Vargas, M., Centeno-Moncada, M. and Carrillo, W. (1995). Densidades Poblacionales y Frecuencias de los Nematodos Parásitos del Banano *(Musa* AAA) en Nueve Cantones de Costa Rica. Frequency and Population Densities of Banana *(Musa* AAA*)* Parasitic Nematodes on Nine Counties from Costa Rica. CORBANA, 20: 6–11.

Araya-Vargas, M. and Cheves-Vargas, A. (1998). Selection of plant type for banana (Musa AAA) nematode sampling. Selección del tipo de planta de banano (*Musa* AAA) para muestreo de nematodos. *Infomusa* 7: 23–26.

Barbosa, R., Monteiro, T., Coutinho, R., Silva, J. G. and Freitas, L. G. (2019). *Pochonia chlamydosporia* no controle do nematoide de galhas em bananeira. *Nematropica* 49: 99–106.

Blake, C. D. (1961). Root rot of bananas caused by *Radopholus similis* (cobb) and its control in New South Wales. *Nematologica* 6(4): 295–310.

Blake, C. D. (1969). Nematode parasites of banana and their control. In: Peachey, J. E. (ed). *Nematodes of Tropical Crops. St.* Albans: Commonwealth Bureau of Helminthology. 108–132.

Blomme, G., Ntamwira, J., Kearsley, E., Bahati, L., Amini, D., Safari, N. and Ocimati, W. (2020). Sensitivity and tolerance of different annual crops to different levels of banana shade and dry season weather. *Frontiers in Sustainable Food Systems* 4: 545926.

Brentu, C. F., Speijer, P. R., Green, K. R., Hemeng, B. M., De Waele, D. and Coyne, D. L. (2004). Micro-plot evaluation of the yield reduction potential of *Pratylenchus coffeae, Helicotylenchus multicinctus* and *Meloidogyne javanica* on plantain cv. Apantu-pa (*Musa* spp., AAB-group) in Ghana. *Nematology* 6: 455–462.

Broadley, R. A. (1974). Nematicide screening in bananas. *Australasian Plant Pathology* 3(1): 23–23.

Broadley, R. A. (1979a). Nematicide treatment of banana planting material. *Australian Journal of Experimental Agriculture* 19(100): 631–633.

Broadley, R. A. (1979b). Non-volatile nematicides for control of burrowing nematode in banana plantations in north Queensland. *Australian Journal of Experimental Agriculture* 19(100): 626–630.

Broadley, R. A. (1979c). A simple method for estimating banana root rot. *Australasian Plant Pathology* 8: 24–25.

Brown, A. M. V. (2018). Endosymbionts of plant-parasitic nematodes. *Annual Review of Phytopathology* 56: 225–242.

Bunt, J. A. (1987). Mode of action of nematicides. In: Veech, J. A. and Dickson, D. W. (Eds.), *Vistas on Nematology*. Hyattsville: Society of Nematologists, 461–468.

Castillo, L. E., Ruepert, C. and Solis, E. (2000). Pesticide residues in the aquatic environment of banana plantation areas in the northern Atlantic zone of Costa Rica. *Environmental Toxicology and Chemistry* 19(8): 1942–1950.

Chabrier, C. and Queneherve, P. (2003). Control of the burrowing nematode (*Radopholus similis* Cobb) on banana: Impact of the banana field destruction method on the efficiency of the following fallow. *Crop Protection* 22(1): 121–127.

Chabrier, C. and Quénéhervé, P. (2008). Preventing nematodes from spreading: A case study with *Radopholus similis* (Cobb) Thorne in a banana field. *Crop Protection* 27(9): 1237–1243.

Chabrier, C. and Quénéhervé, P. (2010). Decline of pesticide use in Martinique: Towards "zero pesticide" in intensive banana cropping systems. In: de Nematologos, O. (Ed.), *de los Tropicos Americanos USA (ed) ONTA Annual Meeting*. Quito, Équateur: s.n., 1 poster.

Chabrier, C., Tixier, P., Duyck, P.-F., Cabidoche, Y. and Quénéhervé, P. (2010a). Survival of the burrowing nematode *Radopholus similis* (Cobb) Thorne without food: Why do males survive so long? *Applied Soil Ecology* 45(2): 85–91.

Chabrier, C., Tixier, P., Duyck, P.-F., Carles, C. and Quénéhervé, P. (2010b). Factors influencing the survivorship of the burrowing nematode, *Radopholus similis* (Cobb.) Thorne in two types of soil from banana plantations in Martinique. *Applied Soil Ecology* 44(2): 116–123.

Charlier, J. B., Cattan, P., Voltz, M. and Moussa, R. (2009). Transport of a nematicide in surface and groundwater in a tropical volcanic catchment. *Journal of Environmental Quality* 38(3): 1031–1041.

Chauvin, C., Dorel, M., Villenave, C., Roger-Estrade, J., Thuries, L. and Risède, J. (2015). Biochemical characteristics of cover crop litter affect the soil food web, organic matter decomposition, and regulation of plant-parasitic nematodes in a banana field soil. *Applied Soil Ecology* 96: 131–140.

Chaves, N. P., Pocasangre, L. E., Elango, F., Rosales, F. E. and Sikora, R. (2009). Combining endophytic fungi and bacteria for the biocontrol of *Radopholus similis* (Cobb) Thorne and for effects in plant growth. *Scientia Horticulturae* 122(3): 472–478.

Chavez, C., Rubio, D., Tobar, M., Salas, E. and Araya, M. (2020). Number of Nematicide Cycles per Year on Banana (*Musa* AAA cv VALERY) Root Nematode Control and Crop Yield. *International Journal of Science and Research* 9(7): 1651–1663.

Chen, J., Li, Q. X. and Song, B. (2020). Chemical nematicides: Recent research progress and outlook. *Journal of Agricultural and Food Chemistry* 68(44): 12175–12188.

Chitamba, J., Manjeru, P., Chinheya, C. C. and Handiseni, M. (2014). Evaluation of legume intercrops on the population dynamics and damage level of burrowing nematode (*Radopholus similis*) in banana (*Musa* spp.). *Archives of Phytopathology and Plant Protection* 47(6): 761–773.

Cobb, N. A. (1915). *Tylenchus similis: The Cause of a Root Disease of Sugar Cane and Banana*. Department of Agriculture.

Colbran, R. C. (1964a). Cover crops for nematode control in old banana land. *Queensland Journal of Agricultural Science* 21: 233-236.

Colbran, R. C. (1964b). Effects of treatments of banana "bits" for nematode control on emergence and yield. *Queensland Journal of Agricultural Science* 21: 237–238.

Colbran, R. C. and Saunders, G. W. (1961). Nematode root-rot of bananas. *Queensland Agricultural Journal* 87: 22–24.

D'Addabbo, T., Laquale, S., Perniola, M. and Candido, V. (2019). Biostimulants for plant growth promotion and sustainable management of phytoparasitic nematodes in vegetable crops. *Agronomy* 9(10): 616.

Declerck, S., Laloux, S., Sarah, J. L. and Delvaux. (1998). Application of a flowing solution culture technique to study the parasitic fitness of the nematodes *Radopholus similis* on banana plantlets under two different nitrogen nutrient regimes. *Plant Pathology* 47(5): 580–585.

Desaeger, J., Wram, C. and Zasada, I. (2020). New reduced-risk agricultural nematicides-rationale and review. *Journal of Nematology* 52: e2020-91.

Devrajan, K., Rajendran, G. and Seenivasan, N. (2003). Nutrient status and phytosynthetic efficiency of banana (*Musa* sp) influenced by *Meloidogyne incognita* infected with *Pasteuria penetrans*. *Nematologie Mediterranea* 2: 197–200.

Dhakshinamoorthy, S., Mariama, K., Elsen, A. and De Waele, D. (2014). Phenols and lignin are involved in the defence response of banana (*Musa*) plants to *Radopholus similis* infection. *Nematology* 16(5): 565–576.

Dita, M., Teixeira, L. A. J., O'Neill, W., Pattison, A. B., Weinert, M. P., Li, C. Y., Zheng, S. J., Staver, C., Thangavelu, R. and Viljoen, A. (2020). *Current State of Fusarium Wilt of Banana in the Subtropics* (1272nd edn.). Leuven, Belgium: International Society for Horticultural Science (ISHS), 45–56.

Djigal, D., Chabrier, C., Duyck, P. F., Achard, R., Quénéhervé, P. and Tixier, P. (2012). Cover crops alter the soil nematode food web in banana agroecosystems. *Soil Biology and Biochemistry* 48: 142–150.

Dochez, C., Dusabe, J., Whyte, J., Tenkouano, A., Ortiz, R. and De Waele, D. (2006). New sources of resistance to *Radopholus similis* in *Musa* germplasm from Asia. *Australasian Plant Pathology* 35(5): 481–485.

Esnard, J., Marban-Mendoza, N. and Zuckerman, B. M. (1998). Effects of three microbial broth cultures and an organic amendment on the growth and populations of free living and plant parasitic nematodes on banana. *European Journal of Plant Pathology* 104(5): 457–463.

Fernandez, E., Mena, J., González, J. and Márquez, M. E. (2005). Biological control of nematodes in banana. In: Turner, D. W. and Rosales, F. E. (Eds.), *Banana Root Systems: Towards a Better Understanding for Its Productive Management*. Montpellier, France: International Network for the Improvement of Banana and Plantain, 193–200.

Fierascu, R. C., Fierascu, I. C., Dinu-Pirvu, C. E., Fierascu, I. and Paunescu, A. (2020). The application of essential oils as a next-generation of pesticides: Recent developments and future perspectives. *Zeitschrift Fur Naturforschung. C, Journal of Biosciences* 75(7–8): 183-204.

Fletcher, S. J., Reeves, P. T., Hoang, B. T. and Mitter, N. (2020). A perspective on RNAi-based biopesticides. *Frontiers in Plant Science* 11: 51.

Freitas, A. S., Pozza, E. A., Alves, M. C., Coelho, G., Rocha, H. S. and Pozza, A. A. A. (2016). Spatial distribution of Yellow Sigatoka Leaf Spot correlated with soil fertility and plant nutrition. *Precision Agriculture* 17(1): 93–107.

Gaidashova, S., Van Asten, P., Dochez, C., Delvaux, B. and De Waele, D. (2010). Impact of the root-lesion nematode *Pratylenchus goodeyi* and mulch on the East African Highland banana crop performance in Kibuye, Wester Rwanda. *Nematology* 12: 349–356.

Gamalero, E. and Glick, B. R. (2020). The use of plant growth-promoting bacteria to prevent nematode damage to plants. *Biology* 9(11): 381.

Garbeva, P., van Veen, J. A. and Van Elsas, J. D. (2004). Microbial diversity in soil: Selection of microbial populations by plant and soil type and implications for disease suppressiveness. *Annual Review of Phytopathology* 42: 243–270.

Gnonhouri, P. G., Zézé, A., Adiko, A. and Kobenan, K. (2020). *Tithonia diversifolia* Crop rotation: An efficient cultural practice for control of burrowing (*Radopholus similis*) and root-lesion (*Pratylenchus coffea*) nematodes in banana orchards in Côte d'Ivoire. *International Journal of Phytopathology* 8(3): 101–109.

Gowen, S. and Queneherve, P. (1990). Nematode parasites of bananas, plantains and abaca. In: Luc, M., Sikora, R. A. and Bridge, J. (Eds.), *Plant Parasitic Nematodes in Subtropical and Tropical Agriculture*. Wallingford: CAB International.

Gowen, S. R. (1995). Pests. In: Gowen, S. R. (Ed.), *Bananas and Plantains*. London, England: Chapman and Hall, 382–402.

Gowen, S. R., Quénéhervé, P. and Fogain, R. (2005). Nematode parasites of bananas and plantains. In: Luc, M., Sikora, L. and Bridge, J. (Eds.), *Plant Parasitic Nematodes in Subtropical and Tropical Agriculture* (2nd edn.). Wallingford, UK: CAB International.

Haegeman, A., Elsen, A., De Waele, D. and Gheysen, G. (2010). Emerging molecular knowledge on *Radopholus similis*, an important nematode pest of banana. *Molecular Plant Pathology* 11(3): 315–323.

Haegeman, A., Jones, J. T. and Danchin, E. G. J. (2011). Horizontal gene transfer in nematodes: A catalyst for plant parasitism? *Molecular Plant-Microbe Interactions* 24(8): 879–887.

Hall, T. J., Walduck, G. D. and Walker, R. W. (1993). Register of Australian herbage plant cultivars. *Australian Journal of Experimental Agricultural* 33: 674–676.

Hamill, S. D., Moisander, J. A. and Smith, M. K. (2009). Micropropagation of vegetatively propagated crops: Accelerating release of new cultivars and providing an important source of clean planting material. *Acta Horticulturae* 829: 213–218.

Hölscher, D., Dhakshinamoorthy, S., Alexandrov, T., Becker, M., Bretschneider, T., Buerkert, A., Crecelius, A. C., De Waele, D., Elsen, A., Heckel, D. G., Heklau, H., Hertweck, C., Kai, M., Knop, K., Krafft, C., Maddula, R. K., Matthäus, C., Popp, J., Schneider, B., Schubert, U. S., Sikora, R. A., Svatoš, A. and Swennen, R. L. (2014). Phenalenone-type phytoalexins mediate resistance of banana plants (*Musa* spp.) to the burrowing nematode *Radopholus similis*. *Proceedings of the National Academy of Sciences of the United States of America* 111(1): 105–110.

Jacob, J., Mitreva, M., Vanholme, B. and Gheysen, G. (2008). Exploring the transcriptome of the burrowing nematode *Radopholus similis*. *Molecular Genetics and Genomics* 280(1): 1–17.

Jaramillo, J., Vintimilla, M., Rubio, D., Soto, G., Tobar, M., Salas, E. and Araya, M. (2019). Efecto de la rotación de nematicida en el control de nematodos y la producción de banano (*Musa*, A. A. A. cv. Williams). Effect of nematicide rotation on banana root nematode control and crop yield. *Agronomía Colombiana* 37.

Jesus, F. N., Damasceno, J. C. A., Barbosa, D. H. S. G., Malheiro, R., Pereira, J. A. and Soares, A. C. F. (2015). Control of the banana burrowing nematode using sisal extract. *Agronomy for Sustainable Development* 35(2): 783–791.

Kavino, M., Harish, S., Kumar, N., Saravanakumar, D. and Samiyappan, R. (2010). Effect of chitinolytic PGPR on growth, yield and physiological attributes of banana (*Musa* spp.) under field conditions. *Applied Soil Ecology* 45(2): 71–77.

Kazuki, S., Yasuhiro, K. and Ken, S. (2019). Plant immune responses to parasitic nematodes. *Frontiers in Plant Science* 10: 1165.

Khan, A., Williams, K. L. and Nevalainen, H. K. M. (2006). Infection of plant-parasitic nematodes by *Paecilomyces lilacinus* and *Macrosporium lysipagum*. *BioControl* 51(5): 659–678.

Kosma, P., Ambang, Z., Begoude, B. A. D., Hoopen, G. M. T., Kuate, J. and Akoa, A. (2011). Assessment of nematicidal properties and phytochemical screening of neem seed formulations using *Radopholus similis*, parasitic nematode of plantain in Cameroon. *Crop Protection* 30(6): 733–738.

Lamour, J., Naud, O., Lechaudel, M., Le Moguédec, G., Taylor, J. and Tisseyre, B. (2020). Spatial analysis and mapping of banana crop properties: Issues of the asynchronicity of the banana production and proposition of a statistical method to take it into account. *Precision Agriculture* 21(4): 897–921.

Leroux, C., Jones, H., Pichon, L., Guillaume, S., Lamour, J., Taylor, J., Naud, O., Crestey, T., Lablee, J. and Tisseyre, B. (2018). GeoFIS: An open source, decision-support tool for precision agriculture data. *Agriculture* 8(6): 73.

Li, J., Zou, C., Xu, J., Ji, X., Niu, X., Yang, J., Huang, X. and Zhang, K. Q. (2015). Molecular mechanisms of nematode-nematophagous microbe interactions: Basis for biological control of plant-parasitic nematodes. *Annual Review of Phytopathology* 53: 67–95.

Mateille, T., Foncelle, B. and Ferrer, H. (1988). Lutte contre les nématodes du bananier par submersion du sol. *Revue de Nématologie* 11: 235–238.

Mendoza, A. R. and Sikora, R. A. (2009). Biological control of *Radopholus similis* in banana by combined application of the mutualistic endophyte *Fusarium oxysporum* strain 162, the egg pathogen *Paecilomyces lilacinus* strain 251 and the antagonistic bacteria *Bacillus firmus*. *BioControl* 54(2): 263–272.

Mendoza, A. R., Sikora, R. A. and Kienwick, S. (2007). Influence of *Paecilomyces lilacinus* strain 251 on the biological control of the burrowing nematode *Radopholus similis* in banana. *Nematropica* 37: 203–213.

Moens, T., Araya, M., Swennen, R. and De Waele, D. (2004). Enhanced biodegradation of nematicides after repetitive applications and its effects on root and yield parameters in commercial banana plantations. *Biology and Fertility of Soils* 39(6): 407–414.

Ocimati, W., Groot, J. C. J., Blomme, G., Timler, C. J., Remans, R., Taulya, G., Ntamwira, J. and Tittonell, P. (2021). A multi-objective model exploration of banana-canopy management and nutrient input scenarios for optimal banana-legume intercrop performance. *Agronomy* 11(2): 311.

Paparu, P., Dubois, T., Coyne, D. L. and Viljoen, A. (2009). Dual inoculation of *Fusarium oxysporum* endophytes in banana: Effect on plant colonization, growth and control of the root burrowing nematode and the banana weevil. *Biocontrol Science and Technology* 19(6): 639–655.

Pattison, A. B. (2021). Rotation crops for banana growers to reduce nematodes. In: *Australian Bananas*. Brisbane: Australian Banana Grower's Council, 1.

Pattison, A. B., Badcock, K. and Sikora, R. A. (2011). Influence of soil organic amendments on suppression of the burrowing nematode, *Radopholus similis*, on the growth of bananas. *Australasian Plant Pathology* 40(4): 385–396.

Pattison, A. B., Moody, P. W., Badcock, K. A., Smith, L. J., Armour, J. A., Rasiah, V., Cobon, J. A., Gulino, L.-M. and Mayer, R. (2008). Development of key soil health indicators for the Australian banana industry. *Applied Soil Ecology* 40(1): 155–164.

Pattison, A. B., Stanton, J. M. and Cobon, J. A. (2000a). Bioassay for enhanced biodegradation of nematicides in soil. *Australasian Plant Pathology* 29(1): 52–58.

Pattison, T., Stanton, J. M., Treverrow, N. L., Lindsay, S. and Campagnolo, D. (2000b). *Managing Banana Nematodes*. Brisbane, Australia: Department of Primary Industries, 49.

Pattison, A. B., Stanton, J. M., Cobon, J. A. and Doogan, V. J. (2002). Population dynamics and economic threshold of the nematodes *Radopholus similis* and *Pratylenchus goodeyi* on banana in Australia. *International Journal of Pest Management* 48(2): 107–111.

Paz Filho, E., Muniz, M., Almeida, A., Soares, N. H. M., Moura Filho, G. and Rocha, F. S. (2019). *Purpureocillium lilacinum* (Lilacel®) in the treatment of banana plantlets infected by plant-parasitic nematodes. *Nematropica* 49: 134–139.

Poeydebat, C., Tixier, P., Chabrier, C., de Bellaire, LdL., Vargas, R., Daribo, M. and Carval, D. (2017). Does plant richness alter multitrophic soil food web and promote plant-parasitic nematode regulation in banana agroecosystems? *Applied Soil Ecology* 117–118: 137–146.

Quénéhérve, P. (1993). Nematode management in intensive banana agrosystems: Comments and outlook from the Cote d'Ivoire experience. *Crop Protection* 12(3): 164–172.

Quénéhervé, P., Chabrier, C., Auwerkerken, A., Topart, P., Martiny, B. and Marie-Luce, S. (2006). Status of weeds as reservoirs of plant parasitic nematodes in banana fields in Martinique. *Crop Protection* 25(8): 860–867.

Quénéhervé, P., Valette, C., Topart, P. and Horry, J. P. (2008). Nematode resistance in bananas: Screening reults on some wild and cultivated accessions of *Musa* spp. *Euphytica* 165: 137–143.

Racke, K. D. and Coats, J. R. (1988). Comparative degradation of organophosphorus insecticides in soil: Specificity of enhanced microbial degradation. *Journal of Agricultural and Food Chemistry* 36(1): 193–199.

Rumbos, C., Mendoza, A., Sikora, R. A. and Kiewnick, S. (2008). Persistence of the nematophagous fungus *Paecilomyces lilacinus* strain 251 in soil under controlled conditions. *Biocontrol Science and Technology* 18(10): 1041–1050.

Salas, E. (2007). Efecto de las cepas nativas de *Paecilomyces lilacinus, Trichoderma* sp., *Metarhizium anisopliae* y *Candelabrella* sp. sobre *Radopholus similis* evaluadas en vitroplantas de banano (*Musa* AAA) cv. Williams. Informe Anual 2006, Costa Rica: CORBANA, 61-62.

Salguero, D., Rudon, G., Blanco, R., Moya, C., Ramclam, W., Medina, L., Azofeifa, D. and Araya, M. (2016). Effect of different nematicide applications per year on banana (*Musa* AAA) root nematode control and crop yield. *Journal of Applied Biological Sciences* 101: 9598–9609.

Seenivasan, N. (2017). Management of *Radopholus similis* and *Helicotylenchus multicinctus* in ratoon banana grown under high density planting systems. *International Journal of Fruit Science* 17(1): 41–62.

Seenivasana, N. (2018). Phytochemical profiling of burrowing nematode (*Radopholus similis*) resistant and susceptible banana (*Musa* spp.) genotypes for detection of marker compounds. *Fruits* 73(1): 48–59.

Seenivasan, N. and Senthilnathan, S. (2018). Effect of humic acid on *Helicotylenchus multicinctus* (Cobb, 1893) Golden, 1956 infesting banana (*Musa* spp.). *Fruits* 73(1): 22–30.

Sikora, R. A., Pocasangre, L., Zum Felde, Az, Niere, B., Vu, T. T. and Dababat, A. A. (2008). Mutualistic endophytic fungi and *in-planta* suppressiveness to plant parasitic nematodes. *Biological Control* 46(1): 15–23.

Silva, J. C. Pd, Medeiros, F. H. Vd and Campos, V. P. (2018). Building soil suppressiveness against plant-parasitic nematodes. *Biocontrol Science and Technology* 28(5): 423–445.

Singh, S. K., Hodda, M. and Ash, G. J. (2013). Plant-parasitic nematodes of potential phytosanitary importance, their main hosts and reported yield losses. *EPPO Bulletin* 43(2): 334–374.

Stanton, J. M., Pattison, A. B. and Kopittke, R. A. (2001). A sampling strategy to assess banana crops for damage by *Radopholus similis* and *Pratylenchus goodeyi*. *Australian Journal of Experimental Agriculture* 41(5): 675–679.

Stirling, G. R. (2014). *Biological Control of Plant-Parasitic Nematodes: Soil Ecosystem Management in Sustainable Agriculture*. Wallingford, UK: CABI Publishing.

Stirling, G. R. and Pattison, A. B. (2008). Beyond chemical dependency for managing plant-parasitic nematodes: Examples from the banana, pineapple and vegetable industries of tropical and subtropical Australia. *Australasian Plant Pathology* 37(3): 254–267.

Stoorvogel, J. J., Bouma, J. and Orlich, R. A. (2004). Participatory research systems analysis: Protyping for a Costa Rican Banana plantation. *Agronomy Journal* 96(2): 323–336.

Tabarant, P., Villenave, C., Risede, J.-M., Roger-Estrade, J., Thuries, L. and Dorel, M. (2011a). Effects of four organic amendments on banana parasitic nematodes and soil nematode communities. *Applied Soil Ecology* 49: 59–67.

Tabarant, P., Villenave, C., Risede, J. M., Roger-Estrade, J. and Dorel, M. (2011b). Effects of organic amendments on plant-parasitic nematode populations, root damage, and banana plant growth. *Biology and Fertility of Soils* 47(3): 341–347.

Tenkouano, A., Hauser, S., Coyne, D. L. and Coulibaly, O. (2006). Clean planting materials and managment practices for sustained production of banana and plantain in Africa. *Chronica Horticulturae* 46: 14–18.

Ternisien, E. and Ganry, J. (1990). Crop rotation in intensive banana cultivation. *Fruits (Paris)*: 98–102.

Thoden, T. C., Korthals, G. W. and Termorshuizen, A. J. (2011). Organic amendments and their influences on plant-parasitic and free-living nematodes: A promising method for nematode management? *Nematology* 13(2): 133–153.

Topalović, O., Hussain, M. and Heuer, H. (2020). Plants and associated soil microbiota cooperatively suppress plant-parasitic nematodes. *Frontiers in Microbiology* 11: 313.

Vargas, R. and Araya, M. (2009). Efecto de enmiendas organicas en el control de *Radopholus similis* y la produccion de banano (*Musa* AAA) cv. Grande Naine. Informe Anual 2008, Cost Rica: CORBANA, 59–61.

Varma, V. and Bebber, D. P. (2019). Climate change impacts on banana yields around the world. *Nature Climate Change* 9(10): 752–757.

Viaene, N., Duran, L., Rivera, J. M., Dueñas, J., Rowe, P. and De Waele, D. (2003). Responses of banana and plantain cultivars, lines and hybrids to the burrowing nematode *Radopholus similis*. *Nematology* 5(1): 85–98.

Wairegi, L. W. I. and van Asten, P. J. A. (2010). The agronomic and economic benefits of fertilizer and mulch use in highland banana systems in Uganda. *Agricultural Systems* 103(8): 543–550.

Wairegi, L. W. I., van Asten, P. J. A., Tenywa, M. M. and Bekunda, M. A. (2010). Abiotic constraints override biotic constraints in East African highland banana systems. *Field Crops Research* 117(1): 146–153.

Wang, K.-H. and Hooks, C. R. R. (2009). Plant-parasitic nematodes and their associated natural enemies with banana (*Musa* spp.) plantings in Hawaii. *Nematropica* 39: 57–73.

Wuyts, N., Lognay, G., Verscheure, M., Marlier, M., De Waele, D. and Swennen, R. (2007). Potential physical and chemical barriers to infection by the burrowing nematode *Radopholus similis* in roots of susceptible and resistant banana (*Musa* spp.). *Plant Pathology* 56(5): 878–890.

Wuyts, N., Swennen, R. and De Waele, D. (2006). Effects of plant phenylpropanoid pathway products and selected terpenoids and alkaloids on the behaviour of the plant-parasitic nematodes *Radopholus similis*, *Pratylenchus penetrans* and *Meloidogyne incognita*. *Nematology* 8(1): 89–101.

Zhong, S., Zeng, H.-C. and Jin, Z.-Q. (2016). Response of soil nematode community composition and diversity to different crop rotations and tillage in the tropics. *Applied Soil Ecology* 107: 134–143.

Published by Burleigh Dodds Science Publishing Limited, 2024.

Part 5

Insect pests

Chapter 18

The banana weevil borer

César Guillén Sanchez, University of Costa Rica, Costa Rica*

1 Introduction

This chapter provides an overview of the epidemiology and impact of the banana weevil borer, *Cosmopolites sordidus* (Germar, 1824) (Coleoptera: Curculionidae) on banana cultivation. An overview of the spread of the pest, its life cycle and the impact on plant health and yields is given before discussing its management. A range of control measures are outlined, including chemical control and its limitations (e.g. due to the development of resistance) as well as cultural/agronomic practices to limit the habitat which allows the pest to become established. Other measures, including the use of pheromones in monitoring and mass trapping, genetic resistance and the use of predator species, nematodes and entomopathogens, are discussed. The chapter also looks at the link between the banana weevil borer and other diseases in banana.

2 Distribution, life cycle and impact

The banana weevil is native to Southeast Asia, probably Malaysia, Indonesia, Java and Borneo (Treverrow, 1985). It occupies the same geographical area as the center of origin of the banana (Stover and Simmonds, 1987). This insect has

* cguillen@corbana.co.cr.

http://dx.doi.org/10.19103/AS.2022.0108.18

moved to all plantations in tropical and subtropical areas where plants of the *Musa* genus (its main host) are grown, such as bananas, plantains, ensetes and manilla hemp (*Musa textilis*).

Its geographic expansion has been gradual, as the insect first occurred in Africa and Latin America between 1836 and 1906, coinciding with the period of major transportation of banana shoots used as planting material from the Dutch East Indies first to Malaysia and Indochina and then to Africa and Latin America to create new areas for banana cultivation (Cuillé and Vilardebo, 1963). It was reported, e.g. in the Pacific islands in 1912 (Cuillé, 1950; Treverrow, 1985), in the French West Indies in 1929 Hustache (1929) and in Cuba in 1944 (Roche, 1975), all of which can be directly related to the movement of vegetative planting material from infected countries to these areas.

2.1 Life cycle of weevil borer

The life cycle of the weevil is characterized by a 'K'-type reproductive strategy with the insect having high longevity (living 1 to 2 years in the wild and up to 3 years in controlled conditions) but low fertility (Gold et al., 2001; Cubillo et al., 2001; Messiaen, 2002). The life cycle involves complete metamorphosis, going through all the stages of development from egg, larva, pre-pupa, pupa to adult (Fig. 1). According to research by Koppenhöfer (1993) and Abera et al. (2000) egg production is low with oviposition estimated from one to four eggs per week, although dissections of field collected weevils revealed an average of ten mature eggs which may suggest that environmental factors may impede weevils from reaching their ovipositional potential.

The selection of a plant by an insect depends on the chemical communication between the insect and the plant, notably volatile compounds

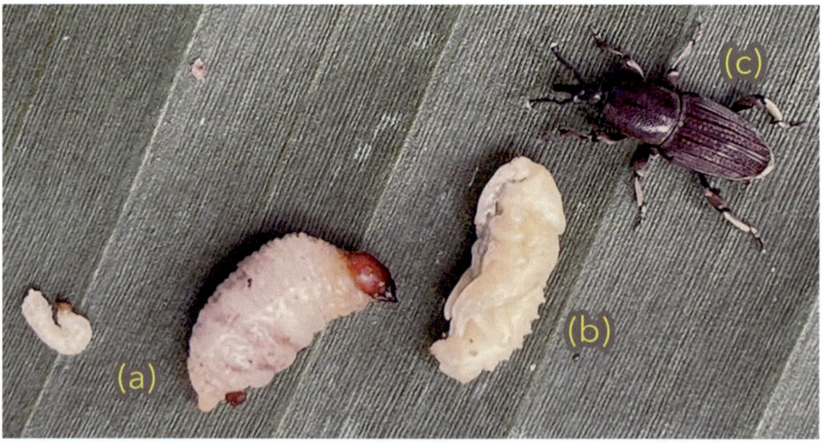

Figure 1 Stage of development of banana weevil borer: (a) larva; (b) pupa; (c) adult.

emanating from fresh or decomposing banana material in the case of *C. sordidus* (Budenberg et al., 1993; Braimah and van Emden, 1999; Sunil and Singh, 2018). *C. sordidus* seems to prefer odors emitted by the corm (where suckers emerge) in plants with fruits as compared to odors from young or adult plants or adult plants with flowers (Cerda et al., 1996). Both male and female weevils are particularly attracted to the pseudostem but oviposition is greater on the corm (Sumani, 1997; Tinzaara, 2005). The female lays individual eggs in holes that she makes in the corm, but she can also lay eggs in leaf sheaths or freshly overturned stems. The eggs are 2.5 mm long, white to slightly yellow, ovoid in shape and take 5–8 days to hatch (Coto and Saunders, 2004; Muñoz, 2006).

The resulting larvae are 10–15 mm long, creamy-white in color, with a reddish-brown head, a thick body, curved in the shape of a 'C' and apodal (without feet). During its development, the larvae go through six (sometimes seven) stages lasting 28–42 days (Coto and Saunders, 2004). The pre-pupa lasts for 1–4 days (Messiaen, 2002). The pupae are yellowish or dark and exarate (i.e. with appendages free to move within the cocoon) (Andrews and Caballero, 1989). Pupae live in galleries found on the periphery of the corm and last for 5–8 days before hatching (Coto and Saunders, 2004) (Fig. 2).

The adult weevils measure between 15 mm and 20 mm long and 4 mm wide. Adults are gregarious but avoid light (negative phototropism). This explains why they are not commonly observed in the field but are found in soil where there is high relative humidity, or in cavities that the adult makes with its beak in depressions in the stem, rhizomes or in harvest residues (Castrillón,

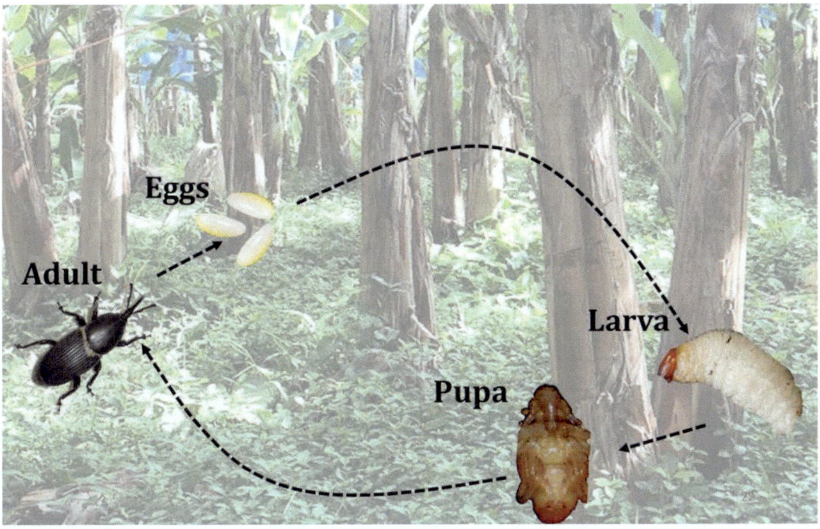

Figure 2 Life cycle of banana weevil borer in tropical humid condition.

2003). Although adult weevils have well-developed wings, flight has only rarely been observed.

The adults of *C. sordidus* live mainly on the soil surface where it is common to find them among crop residues and foliar pods which have dropped on the ground, at the base of the plants or the surrounding vegetation where they take refuge during the day. They are nocturnal and are most active in the evening when the environmental conditions consist of high temperature and high humidity. Their low mobility means that these insects remain very close to the plants where they hatched for long periods of time. When they do move, they typically do so slowly and may take 6 months to cover a distance of 25 m. However, Vinatier et al. (2010) found that some adults of *C. sordidus* can move up to 10 m per day, allowing some insects to reach distant plants quickly. This behavior indicates that colonization and spread of the pest occurs mainly through movement of planting material infested with larvae or eggs within the corm.

2.2 Impact of weevil borer

Damage is mainly caused by the larvae which make galleries in the corms of banana plants at all stages of development (Gold et al., 2001; Carballo, 2001; Muñoz, 2006). At high population numbers, damage to the corm can rupture vascular bundles and destroy roots (Montesdeoca, 1998) (Fig. 3). As well as a deterioration of the root system, weakening of the mother plants can result in the rupture of the stem and/or collapse of the plant, especially when the bunch reaches maturity or as a result of wind currents. The injuries caused by the larvae during feeding and the cavities made by females during oviposition or when feeding can also enable the entry of fungal and bacterial pathogens. Damage to the corm varies based on a range of factors such as variety of banana, the stage of development of the larvae and effectiveness of the pest control strategy used (Fig. 4).

Figure 3 Damage caused by banana weevil borer in banana crop: (a) larva; (b) gallery; (c) plant fall.

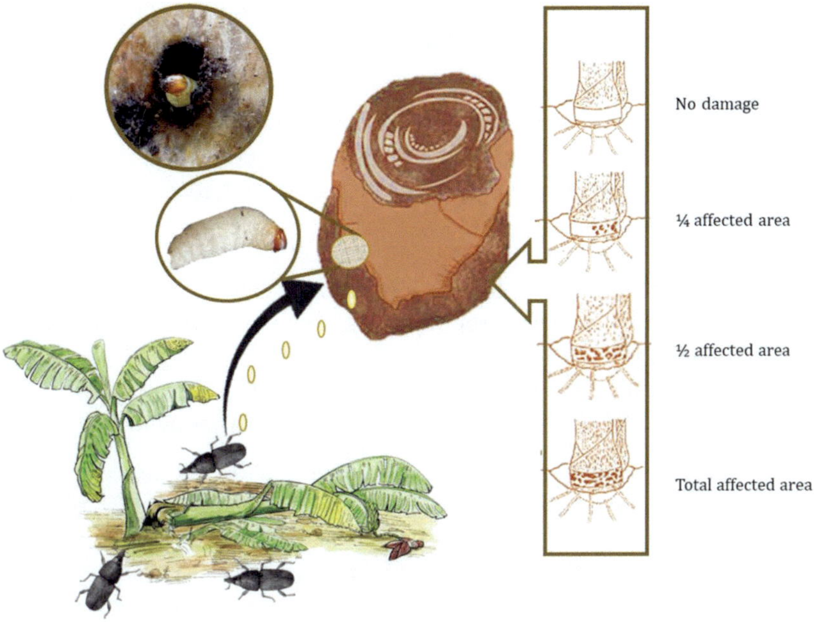

Figure 4 Damage of banana weevil borer in pseudostem and corm of banana plant.

In Uganda, e.g. weevil damage caused a 4% decrease in the first plant crop cycle and a 48% decrease in yield in the fourth cropping cycle (Gold, 1999). Rukazambuga et al. (1998) found that lesions caused by the weevil can reduce plant height and the number of functional leaves, resulting in reduced bunch weight and delayed sucker growth. In Colombia weevil damage in a plantain crop of the Dominico Hartón variety (*Musa* AAB) resulted in a reduction in bunch weight of 4–20% for the first cycle, between 30% and 40% for the second cycle and between 48% and 60% for the third cycle (Castrillón, 2003).

3 Strategies for control of banana weevils

Control of *C. sordidus* in large-scale plantations currently still relies on the application of chemical insecticides. However, to maintain *C. sordidus* populations below the threshold level for economic damage in a more sustainable way, it is necessary to implement integrated pest management (IPM) programs, combining chemical, cultural and biological methods of control (Hilje, 1994; Merchán, 2002; Castrillón, 2003; Muñoz, 2006). Cultural controls remain particularly important in preventing establishment of the pest. Cultural control methods are also the main available means of pest management by smallholder farmers and growers unable to afford synthetic insecticides. Cultural and biological control methods include:

- Farm sanitation (Masanza et al., 2005);
- Planting of host plants with weevil resistance (Kiggundu et al., 2003);
- Use of pseudostem traps (Gold et al., 2002; Masanza et al., 2009);
- Mass trapping with aggregation pheromone lures (Alpizar et al., 1999; Tinzaara, 2005);
- Biological control with entomopathogens (Nankinga and Moore, 2000; Treverrow et al., 1991) and myrmicine ants (Castineiras and Ponce, 1991);
- Biological control with arthropods and entomopathogenic fungi (Braimah and van Emden, 1999); and
- Botanical pesticides such as neem extracts (Musabyimana et al., 2001) and repellents such as the leaves of *Tephrosia* spp. (Walangululu et al., 1993).

A more recent development is Agroecological Crop Protection (ACP), a concept based on ecological practices and principles (Deguine et al., 2017). ACP gives priority to preventive measures rather than curative approaches and emphasizes the protection and promotion of beneficial organisms. It aims to promote ecological functions within agricultural systems by directly or indirectly optimizing interactions between plants, animals and microbial communities, as well as improving soil health. The challenge of ACP is to maintain the populations of pests and diseases within the agroecosystem below thresholds that cause damage that may jeopardize food production. It includes a wide range of levers, such as the conservation of natural enemies that regulate pest insect populations, the use of biopesticides (repellents and microorganisms) and trapping systems with allelochemicals.

4 Chemical and biochemical control

Conventional chemical control involves the use of synthetic nematicides and insecticides and remains the main way to combat the banana weevil borer in intensive commercial production systems. Agrios (2005) describes nematicides as having a dual action effect by controlling both nematodes and insects. Among the synthetic nematicides most used to combat nematodes and weevils are (Araya, 2003):

- organophosphates (terbufos, cadusafos, ethoprophos and fenamiphos); and
- carbamates (carbaryl, oxamyl and carbofuran).

The development and application of some nematicides with no insecticide effect in banana plantations has resulted in dramatic increases in weevil populations and consequent damages to plants (Sponagel et al., 1995).

Challenges facing chemical control include (Muñoz, 2006): the increasing cost of nematicides and the cost of application, the loss of efficacy due to the development of resistant populations, the risk to human health and the environmental impact (e.g. contamination of soil and water) due to high persistence in the environment. Weevil borers have developed resistance to many insecticides. According to Messiaen (2002), resistance has been documented for:

- organochlorines of the cyclodiene group (aldrin, heptachlor, dieldrin) and of the benzene group (hexachlorocyclohexane = HCH);
- organophosphates (chlorpyrifos, ethoprophos, pirimiphos and prothiophos); and
- carbamates (carbofuran).

Resistance has been found to be widespread, e.g. in Australia, Latin America and Africa. Rotation of active ingredients is currently a widespread practice to prevent or delay the acquisition of resistance. According to Gold et al. (2001), organophosphates have the advantage of being less persistent in the environment, but they are costly and are also very toxic, requiring protective measures during application, making them unsuitable for use in small-scale production systems.

Chemical control also includes the use of plant extracts with insecticidal properties such as neem (*Azadirachta indica*) which contains insecticidal compounds such as phytoalexins. Musabyimana et al., (2001) found that neem seed powder applied at 60–100 g/plant at the time of planting, and then at 4-month intervals for a period of 3 years, was effective in combating weevils through a mixture of repelling insects, reduction of oviposition and inhibition of larval growth. Gold and Messiaen (2000) found that immersion of shoots in a 20% solution of *A. indica* seeds during sowing protected young plants from lesions caused by insects and reduced oviposition through a repellent effect on weevils.

5 Cultural control

This strategy is based on the implementation of crop management practices that prevent establishment of the insect and its reproduction in banana plantations. These practices include the following measures to limit habitat which may harbor the insect (Cubillo et al., 2001; Gold et al., 2002):

- destruction of remains of corms and pseudostems;
- good weed control;
- maintenance of well-drained soils; and
- cleaning the base of the plant to remove organic residues (leaf litter or other decomposing material).

In case of new planting areas or plantation renovation work, areas should be free of moss and other plant residues, and the use of pest-free planting material or of plants grown in controlled conditions from tissue culture (Cubillo et al., 2001). If corms and resprouts are used as planting material, they should be subjected to heat treatment with hot water between 52°C and 55°C for 15–27 min (Gold and Messiaen, 2000). This treatment eliminates many of the eggs and larvae that may be in the corm. In research conducted in Uganda, where a rigorous residue management system was implemented in a plantation of the cultivar 'Nabusa' (*Musa* sp., AAA-EA), the weevil population was found to decrease from 52 000 to 13 000/ha, resulting in a decrease in corm injury from 41.0% to 10.0% and a 70.0% increase in bunch weight over a 3-year evaluation period (Masanza, 2003).

6 Ethological control

The aggregation pheromones of *C. sordidus* have the function of attracting individuals of both sexes to a given area to better exploit a food source (Chiri, 1989). The aggregation pheromone of *C. sordidus* has been identified as (IS*,3R*,SR*,7S*) 2,8-dioxa 1-ethyl 3,5,7-trimethyl bicylo [3,2,1] octane 1d, a compound commonly known as sordidin (Beauhaire et al., 1995; Ndiege et al., 1996; Jayaraman et al., 1997) and can be used as a lure in traps. Weevil borer traps can also be made from parts of the corm and fresh or fermented pseudostem which expel volatile compounds that attract the insect (Cerda et al., 1998; Masanza et al., 2009). However, the use of pheromone traps baited specifically with sordidin attract 5–10 times more weevils than traps made from banana tissue, last longer and have an effective attraction range of 10 m, decreasing to 20 m (Tinzaara et al., 1999).

Pheromones can be used in three different ways in the management of weevil borers (Wyatt, 2003):

- monitoring (sampling and detection);
- mass trapping (attract and kill); and
- mating disruption (inhibition of chemical communication).

One or two traps/ha are used for sampling to monitor weevil borer numbers and potential population growth. If trapping is required to reduce the adult population, one to six traps/ha should be used, distributed every 20 m in a transverse line to the cable, moving them 20 m every month in the same direction. In areas where the damage is high but localized, mass trapping using up to 25 traps/ha can be carried out (Cubillo et al., 2001).

Mass trapping aims at the removal of a large proportion of the adult population (Tinzaara et al., 2002). This method assumes that reducing the number of adults in the population will result in a reduction of the number of individuals in subsequent generations (El-Sayed et al., 2006). Between 1997 and 1998, in Dwarf Cavendish banana (*Musa* AAA) plots in Costa Rica, mass trapping of adults using sordidin-baited traps for 12 consecutive months significantly reduced *C. sordidus* populations and damage while increasing the weight of harvested banana bunches by 23% (Alpizar et al., 2012). The potential of this approach can also be seen in the French West Indies where mass trapping also resulted in a significant fall in insect numbers and a substantial reduction in the need for nematicides/insecticides (Rhino et al., 2010).

Traps are often made from reusable plastic containers. There is a range of trap designs which can have a significant effect on performance. Pheromones must be placed in a device which protects them from rain and the sun. Devices must have adequate openings so that the volatiles can diffuse and thus attract insects present in the plantation. They must be located so that the insects can easily reach and walk inside. Once inside, soapy water or some other product that breaks surface tension must be used to prevent insects from escaping and instead falling into a liquid trap and drowning. Cubillo et al. (2008) demonstrated in a 3-year evaluation period that the use of the Cosmolure type of trap at a density of 20 traps/ha captured a greater number of weevils/week/hectares compared to traps of a circular or 'sandwich' type even when the latter were deployed at a density of 60 traps/ha. However, they did not find that the use of weevil traps affected fruit weight, number of hands and fruit thickness.

7 Genetic control

Banana weevil tolerance refers to host plants that can maintain high levels of banana weevil infestation without yield reduction. Characteristics such as degree of sap and latex production, corm diameter and hardness, as well as sprouting ability, make some varieties hardier and thus less susceptible to weevil damage and may also inhibit larval development (Kiggundu and Gold, 2002). For example, Gros Michel (AAA) bananas can tolerate a greater number of lesions due to the size of the corm (Gold et al., 2001).

Other varieties may attract weevils because of higher concentrations of volatile compounds that attract more weevils. According to Fogain and Price (1993), Pavis and Lemaire (1996) and Kiggundu and Gold (2002), the *Musa balbisiana* group, which includes bananas (AAB), is more susceptible than the *Musa acuminata* group (AAA), due to the higher concentration of terpenes in the rhizome, which makes it more olfactorily attractive to weevils.

The preference of the female to oviposit in varieties of the Cavendish subgroup (Williams, Grand Naine, Valery, among others) after flowering is possibly related to a higher concentration of terpenes and other volatile compounds that concentrate in the corm following this phenological stage.

However, authors such as Fogain and Price (1993), Sponagel et al. (1995) and Cubillo and Guzmán (2003) have found that a high percentage of plants maintain the corm without lesions of *C. sordidus* larvae in commercial plantations of the Cavendish type, possibly due to the low susceptibility of this subgroup to the insect. In Colombia, Castrillón et al. (2002) evaluated the susceptibility of 112 varieties of different subgroups of banana to weevil attack based on the infestation coefficient used by Vilardebo (1973) and the capture of adults. They concluded that those of the subgroup (AAB) such as 'Dominico 300, Dominico Maqueño, Dominico Ancuyano, Dominico Común, Dominico Hartón Común, Hartón Maqueño, Hartón Común' and bananas from the Gros Michel (AAA) are more prone to weevil attack.

Pavis and Minost (1993) observed that weevil infestation was heaviest on plantain corms, moderate on those of Cavendish (AAA) and lowest on the corms of AA bananas. According to Gold and Messiaen (2000) the primary sources of resistance to weevil borers appear to be in Yangambi Km5, FHIA-03 (or its progenitors) and some diploid hybrids. These varieties produce secondary compounds toxic to the weevil causing high mortality rates at the larval stage. Research on transgenic plants has focused on protein inhibitors to combat the weevil. According to Kiggundu et al. (2007), transgenic plants incorporating cystatins have been found to inhibit cysteine proteases found in the weevil's intestine, affecting the development and survival of the larvae by up to 60%.

8 Biological control

Research on natural enemies of the weevil has demonstrated the potential of some groups of predators, nematodes and entomopathogens in management of the pest. There are several groups of predators of weevil larvae and eggs, such as:

- *Euborellia annulipes* Lucas (Dermaptera: Carcinophoridae) and *Labia borelli* Burr (Dermaptera: Labiidae) (Koppenhöfer, 1991; Tinzaara et al., 1999);
- *Ontophagus* sp. (Coleoptera: Scarabaeidae) and *Hololepta quadridentata* Fabricius (Coleoptera: Histeridae) (Cubillo et al., 2001; Ríos et al., 2002);
- *Eutochia* sp. (Coleoptera: Tenebrionidae), *Dactylosternum abdominale* Fabricius (Coleoptera: Hydrophilidae) (Koppenhöfer, 1991);
- The ants *Pheidole megacephala* Fabricius and *P. guineense* Fabricius (Hymenoptera: Formicidae) (Castrillón, 2003). *Tetramorium guinieensis, Camponotus* spp. (Neuenschwander, 1988); and
- Larval parasites such as the dipteran *Sarcodexia innota*, and larval and egg predators (*Hololepta* spp. and *Alegoria dilatata*) (Belalcazar, 1991; Roche and Abreu, 1983).

In Kenya, Koppenhöfer (1991) used the predators *Eutochia* sp., *E. annulipes* and *D. abdominale* to achieve reductions in weevil eggs of 19.0%, 28.8% and 43.7%, respectively. In Cuba, Carballo (2001) used nine colonies of *P. megacephala*/ha to achieve a 55.0% reduction in the population of *C. sordidus*, a 65.0% reduction in corm lesions and a 25.0% increase in yield, compared to control without application of ants. Castrillón (2003) also evaluated ants (*P. guineense*) as predators of *C. sordidus* larvae and found a mortality rate of 67.0% to 83.0% under high and low infestation conditions, respectively.

Another biological option to combat the weevil is the use of entomopathogenic nematodes, mainly species within the genera *Heterorhabditis* and *Steinernema*. The entomopathogenic nematodes *Steinernema carpocapsae* and *Heterorhabditis bacteriophora* have been shown to control *C. sordidus* with up to 68% of larvae infected (Treverrow et al., 1991). Sepúlveda and Rubio (2009) found that *Heterorhabditis bacteriophora* caused mortality of between 8.0% and 58.0% in adults and 56.0% and 84.0% in larvae, while *Steinernema carpocapsae* caused mortality between 0.0% and 40.0% in adults and 53.0% and 83.0% in larvae.

Another important group of microorganisms that cause disease in insects are entomopathogenic fungi, which comprise more than 700 species grouped in 90–100 different genera (Monzón, 2001; Wong, 2003). Among the fungi that stand out in the fight against weevils is *Beauveria bassiana* which infects a great variety of insects belonging mainly to the orders of Lepidoptera, Coleoptera and Hemiptera, which include a significant number of pests of economic importance in agriculture (Storey and Gardner, 1986). Biological control of *C. sordidus* with *B. bassiana* has been shown to substantially reduce weevil populations in laboratory experiments (up to 100% larval mortality) (Kaaya et al., 1993; Lopes et al., 2011). Greenhouse trials have shown up to 88.9% larval mortality (Akello et al., 2007) while artificial inoculation resulted in between 42% adult mortality (Godonou et al., 2000) and 72% mortality (Nankinga and Moore, 2000). Although results have been promising under these controlled settings, the efficiency of *B. bassiana* in the field is unclear (Messiaen, 2002; Negrete González et al., 2018; Lopes et al., 2014; Tinzaara et al., 2007). Tinzaara et al. (2007) also explored the possibility of combining dissemination of *B. bassiana* with the use of sordidin. Recent research has also suggested the role of volatile organic compounds from entomopathogens and nematophagous fungi in masking kairomones and repelling weevil borers (Lozano-Soria et al., 2020).

Negrete González et al. (2018) assessed the suitability of the fungus *Metarhizium anisopliae* for microbial control of *C. sordidus* and obtained results that were similar as *B. bassiana* (up to 76.9% adult mortality under laboratory conditions and 48.5% in field trials). To date, however, fungus-based strategies remain little used to control *C. sordidus*. Furthermore, *B. bassiana* has a broad host range (Uma Devi et al., 2008; McKinnon et al., 2017) and may therefore be harmful to non-target species.

9 *Cosmopolites sordidus* and the dissemination of disease

Insects can transmit diseases when they come into contact with plants (Guillén, 2021). Two types of disease transmission mechanism are involved (Willsey et al., 2017):

i) Passive: when the insect disseminates propagules by casual contact, production of feces in the plant environment or by its death in the neighborhood of the plant. Here, the contact with the plant does not involve any exploitation of the plant.
ii) Active: when the insect disseminates propagules while exploiting the plant, e.g. through feeding or oviposition.

When feeding on diseased plants, insects can acquire fungal propagules and then transmit them to healthy plants. The feeding habits of insects like *C. sordidus* also open up a wide range of contamination pathways, particularly for soil-borne diseases (Chakraborty and Roy, 2021). Weevils cause lesions in the epidermis of plants (during feeding or oviposition), creating entry points for fungal infection (Chung et al., 2013). Given their mobility, insects can act as important vectors in fungal and bacteria disease epidemics.

The role of insects in the transmission of phytopathogenic fungi has been shown by Selikhovkin et al. (2018) who demonstrated that certain species of weevils (particularly *Hylastes attenuatus*) can transport pathogenic fungal spores of *Ophiostoma* spp., *Leptographium* spp. and *Fusarium* spp. Romón et al. (2007), Bezos et al. (2017) and Guillén et al. (2021) have shown the potential of *C. sordidus* as a vector for Fusarium wilt, caused by the soil-borne fungus *Fusarium oxysporum* f. sp. *cubense* (Foc). Meldrum et al. (2013) and Guillén et al. (2021) have demonstrated the presence of Foc in the digestive system, excreta and particularly on the exoskeleton of *C. sordidus*.

In addition to the volatiles produced by the plants, some fungi can also produce volatile compounds through their metabolic pathways (Korpi et al., 2009; Müller et al., 2013). These compounds are mainly monoterpenoids, sesquiterpenes, alcohols, aldehydes, aromatic compounds, esters, furans, hydrocarbons, ketones and others that can act like semiochemicals (Morath et al., 2012). Volatile aldehyde production has been used to differentiate between isolates of Foc (Brandes, 1919; Stover, 1962; Moore et al., 1991). Guillén et al. (2022) have found a preference of *C. sordidus* for tissues infected by Foc, suggesting fungal volatiles can play a role in attracting pests which then become a vector in spreading the pathogen.

Were et al. (2015) demonstrated that *C. sordidus* is also an important agent in the dissemination of *Xanthomonas campestris*, and Monge (2015) detected *Ralstonia solanacearum* race 2 on *C. sordidus*. Were et al. (2015) isolated *X. campestris* pv. *musacearum* both in the exoskeleton and digestive system of the weevil borer, as well as from insect excreta. Cubillo and Guzmán (2003) also reported that the presence of the rot caused by bacteria of the genus *Erwinia* in banana corms in Costa Rica was directly proportional to the damage in the corm of the weevil. Castrillón (2000) determined that *C. sordidus* enables the entry of *R. solanaceaurum* in banana and plantain.

10 Conclusion

Weevils are widely distributed in all banana growing areas of the world. They both cause direct damage to the banana plant and can have a significant impact on yield in plantations while they also can potentially be involved in the distribution of diseases. There is a need for further research to understand more holistically the interactions of this insect with plant pathogens. This review documented that the weevil contains infective structures of *Fusarium oxysporum* f. sp. *cubense*, as well as being an important vector of bacterial diseases such as *Xanthomonas campestris* pv. *musacearum*, *Erwinia* sp. and *Ralstonia solanacearum* race 2, which makes it an important pest vector of disease. However, there is a lack of knowledge about the mechanisms involved as well as the precise role of this insect in the epidemiology of banana diseases. Epidemiology has often focused on large-scale patterns of disease distribution and spread but also needs to look at spread at a local level, e.g. by insects within plots, especially if it is to benefit farmers.

Pest management programs should increasingly be based on combining techniques such as mass trapping with pheromone traps and the use of entomopathogenic fungi such as *Beauveria bassiana, Metarhizium anisopliae*, the conservation of predators of *C. sordidus* eggs and larvae and, if required, chemical control with a low impact on beneficial organisms, the wider environment and human health.

11 References

Abera, A. M., Kyamanaywa, S. and Karamura, E. B. 2000. Banana weevil *Cosmopolites sordidus* (Germar) ovipositional preferences, timing of attack and larval survivorship in a mixed cultivar trial in Uganda. In: Craenen, K., Ortiz, R., Karamura, E. B. and Vuylsteke, D. (Eds.), *Proceedings of the First International Conference on Banana and Plantain for Africa*, Kampala, Uganda. *Acta Horticulturae* 540, 487–495.

Agrios, G. N. 2005. *Plant Pathology* (5th edn.). San Diego, CA: Elsevier Academic Press, p. 922.

Akello, J., Dubois, T., Gold, C. S., Coyne, D., Nakavuma, J. and Paparu, P. 2007. *Beauveria bassiana* (Balsamo) Vuillemin as an endophyte in tissue culture banana (Musa spp.). *Journal of Invertebrate Pathology* 96(1), 34–42.

Alpizar, D., Fallas, M., Oehlschlager, A. C. and Gonzalez, L. M. 2012. Management of *Cosmopolites sordidus* and *Metamasius hemipterus* in banana by pheromone-based mass trapping. *Journal of Chemical Ecology* 38(3), 245–252.

Alpizar, D., Fallas, M., Oehlschlager, A. C., Gonzalez, L. M. and Jayaraman, S. 1999. Pheromone-based mass trapping of the banana weevil, *Cosmopolites sordidus* (Germar) and the West Indian sugarcane weevil *Metamasius hemipterus* L. (Coleoptera: Curculionidae) in banana and plantain. In: *Memorias XIII Reunion ACORBAT* (Vols. 23–27). Guayaquil, pp. 515–538.

Andrews, K. L. and Caballero, R. 1989. *Guía para el Estudio de Órdenes y Familias de Insectos de Centroamérica* (4th edn.). Tegucigalpa, Honduras, El Zamorano: Escuela Agrícola Panamericana, p. 179.

Araya, M. 2003. Situación actual del manejo de nematodos en banano (*Musa* AAA) y plátano (*Musa* AAB) en el trópico americano. In: Rivas, G. and Rosales, F. (Eds.), *Actas del Taller Manejo Convencional y Alternativo de la Sigatoka Negra, Nematodos y Otras Plagas Asociadas al Cultivo de Musáceas, Celebrado en Guayaquil, Ecuador* (Vols. 11–13). de agosto, pp. 79–102.

Beauhaire, J., Ducrot, P. H., Malosse, C., Rochat, D., Ndiege, I. O. and Otieno, D. O. 1995. Identification and synthesis of sordidin, a male pheromone emitted by Cosmopolites sordidus. *Tetrahedron Letters* 36(7), 1043–1046.

Belalcazar, S. L. 1991. E1 cultivo del platano (*Musa* A A B Simmonds) en el tropico. Instituto Colombiano agropecuario, Quindio. *Colombia*, p. 376.

Bezos, D., Martínez-Álvarez, P., Fernández, M. and Diez, J. J. 2017. Epidemiology and management of pine pitch canker disease in Europe, a review. *Balt For* 23, 279–293.

Braimah, H. and van Emden, H. F. 1999. Evidence for the presence of chemicals attractive to the banana weevil, *Cosmopolites sordidus* (Coleoptera: Curculionidae) in dead banana leaves. *Bulletin of Entomological Research* 89(6), 485–491.

Brandes, E. W. 1919. Banana wilt. *Phytopathology* 9, 339–389.

Budenberg, W. J., Ndiege, I. O., Karago, F. W. and Hansson, B. S. 1993. Behavioural and electrophysiological responses of the banana weevil Cosmopolites sordidus to host plant volatiles. *Journal of Chemical Ecology* 19(2), 267–277.

Carballo, M. 2001. Opciones para el manejo del picudo negro del plátano. *Manejo Integrado de Plagas* 59, i–iv.

Castineiras, A. and Ponce, E. 1991. Efectividad de la utilizacin de *Pheidole megacephala* (Hymenoptera: Formicidae) en la lucha biológica contra *Cosmopolites sordidus* (Coleoptera: Curculionidae). *Proteccin de Plantas* 1, 15–21.

Castrillón, C. 2000. *Distribución de las Especies de Picudo del Plátano y Evaluación de Sus Entomopatógenos Nativos en el Departamento de Risaralda. Corporación Colombiana de Investigación Agropecuaria (Corpoica).*

Castrillón, C. 2003. Situación actual del picudo negro del banano (*Cosmopolites sordidus* Germar). Coleoptera: Curculionidae en el mundo. In: Rivas, G. and Rosales, F. (Eds.), *Actas del Taller "Manejo Convencional y Alternativo de la Sigatoka Negra, Nematodos y Otras Plagas Asociadas al Cultivo de Musáceas, Celebrado en Guayaquil, Ecuador* (Vols. 11–13). de agosto, pp. 125–138.

Castrillón, C., Valencia, J. A. and Urrea, C. F. 2002. Reacción de diferentes materiales del banco de germoplasma de musáceas al ataque del picudo negro *Cosmopolites sordidus* Germar (Coleoptera: Curculionidae). In: *Acorbat. Memorias XV reunión. Realizada en Cartagena de las Indias Colombia* 27, octubre al 02 noviembre 2002. Medellín, CO, pp. 90–97.

Cerda, H., López, A., Sanoja, O., Sánchez., P. and Jaffé, K. 1996. Atracción olfativa de *Cosmopolites sordidus* Germar (1824) (Coleoptera: Curculionidae) estimulado por volátiles originados en musáceas de distintas edades y variedades genómicas. *Agronomía Tropical* 46, 413–429.

Cerda, H., Mori, K., Nakayama, T. and Jaffé, K. 1998. A synergistic aggregation pheromone component in the banana weevil *Cosmopolites sordidus* Germar 1824 (Coleoptera: Curculionidae). *Acta Cientifica Venezolana* 49(3), 201–203.

Chakraborty, A. and Roy, A. 2021. Microbial influence on plant-insect interaction. In: Singh, I. K. and Singh, A. (Eds.), *Plant-Pest Interactions: From Molecular Mechanisms to Chemical Ecology*. Singapore: Springer, pp. 337–363.

Chiri, A. A. 1989. Utilización del control etológico. In: Andrews, K. L. and Quesada, J. R. (Eds.), *Manejo Integrado de Plagas Insectiles en la Agricultura: Estado Actual y Futuro*. El Zamorano, Honduras: Departamento de Protección Vegetal. Escuela Agrícola Panamericana, pp. 267–282.

Chung, S. H., Rosa, C., Scully, E. D., Peiffer, M., Tooker, J. F., Hoover, K., Luthe, D. S. and Felton, G. W. 2013. Herbivore exploits orally secreted bacteria to suppress plant defenses. *Proceedings of the National Academy of Sciences of the United States of America* 110(39), 15728–15733.

Coto, D. and Saunders, J. L. 2004. *Insectos Plagas de Cultivos Perennes con Énfasis en Frutales en América Latina. CATIE*. Turrialba: Confederación unitaria de trabajadores, p. 400.

Cubillo,D. and Guzmán, M. 2003. Evaluación del daño causado por *Cosmopolites sordidus* en una plantación de banano (Musa AAA) y su relación con la producción, la población de Radophulus similis, el contenido de raíces y la pudrición del Cormo. *Revista CORBANA*, 53–67.

Cubillo, D., Laprade, S. and Obregón, M. 2008. Eficacia biológica de cepas de *Beauveria bassiana* en el control de adultos de *Cosmopolites sordidus* (Coleoptera: Curculionidae), en condiciones de laboratorio y campo. In: Sandoval, J. A. (Ed.), *Dirección de Investigaciones, CORBANA. Informe Anual* (Vol. 2007), pp. 153–156.

Cubillo, D., Laprade, S. and Vargas, R. 2001. *Manual técnico para el manejo integrado de insectos plaga en el cultivo de banano*. Costa Rica, San José: CORBANA, p. 73.

Cuillé, J. 1950. *Recherches sur le Charançon du Bananier Cosmopolites sordidus, Germar: Monographie de l'Insecte et Recherche de Ses Chimiotropismes*. Paris: IFAC Série Technique 4 (Vol. 225).

Cuillé, J. and Vilardebo, A. 1963. Les calandrini nuisibles au bananier. In: Balachowsky, A. S. (Ed.), *Entomologie Appliquée à l'agriculture*. Masson, pp. 1098–1114.

Deguine, J. P., Gloanec, C., Laurent, P., Ratnadass, A. and Aubertot, J. N. 2017. *Agroecological Crop Protection*. Dordrecht: Springer Science, Business Media B.V., p. 280.

El-Sayed, A. M., Suckling, D. M., Wearing, C. H. and Byers, J. A. 2006. Potential of mass trapping for long-term pest management and eradication of invasive species. *Journal of Economic Entomology* 99(5), 1550–1564.

Fogain, R. and Price, N. S. 1993. Varietal screening of some musa cultivars for susceptibility to the banana weevil, *Cosmopolites sordidus* (Coleoptera: Curculionidae). In: Gold,

C. S. and Gemmil, B. (Eds.), *Biological and Integrated Control of Highland Banana and Plantain Pests and Diseases*. Cotonou, Benin, 12–14 November 1991, pp. 49–56.

Godonou, I., Green, K. R., Oduro, K. A., Lomer, C. J. and Afreh-Nuamah, K. 2000. Field evaluation of selected formulations of *Beauveria bassiana* for the management of the banana weevil (*Cosmopolites sordidus*) on plantain (*Musa* spp., AAB group). *Biocontrol Science and Technology* 10(6), 779–788.

Gold, C. S. 1999. Manejo integrado de plagas del gorgojo del banano, con énfasis en África Oriental. In: Rosales, F. E., Tripon, S. C. and Cerna, J. (Eds.), *Producción de Banano Orgánico y/o Ambientalmente Amigable. Memorias del Taller Internacional Realizado en la Escuela de Agricultura de la Región Tropical Húmeda (EARTH) Guácimo, Limón, CR*, pp. 152–172.

Gold, C. S. and Messiaen, S. 2000. El Picudo negro del banano Cosmopolites sordidus. Plagas de Musa - Hoja divulgativa N° 4. En: línea) Consultado 24 October 2021. www.inibap.org.

Gold, C. S., Okech, S. H. and Nokoe, S. 2002. Evaluation of pseudostem trapping as a control measure against banana weevil, *Cosmopolites sordidus* (Coleoptera: Curculionidae) in Uganda. *Bulletin of Entomology Research* 92, 35–44.

Gold, C. S., Pena, J. and Karamura, E. 2001. Biology and integrated pest management for the banana weevil *Cosmopolites sordidus* (Germar) (Coleoptera: Curculionidae). *Integrated Pest Management Reviews* 6(2), 79–155.

Guillén, C. 2021. *The role of banana weevil Cosmopolites sordidus in the epidemiology of the Panama disease Fusarium oxysporum f. sp. cubense race 1*. Ph.D. Thesis. Montpellier, France: University of Montpellier, p. 132.

Guillén, C., de Lapeyre de Bellaire, L., Sandoval Fernández, J. A. and Tixier, P. 2022. Preference of *Cosmopolites sordidus* for fusarium wilt diseased banana plants. *Journal of Applied Entomology* 147, 140–146.

Guillén, C., Tixier, P., Tapia Fernández, A., Conejo Barboza, A. M., Sandoval Fernández, J. A. and de Lapeyre de Bellaire, L. 2021. Can the banana weevil *Cosmopolites sordidus* be a vector of *Fusarium oxysporum* f.sp. *cubense* race 1? Unravelling the internal and external acquisition of effective inoculum. *Pest Management Science* 77(6), 3002–3012.

Hilje, L. 1994. El manejo integrado de plagas como noción y estrategia para enfrentar los problemas de plagas. In: Hilje, L. and Arboleda, O. (Eds.), *Lecturas sobre manejo integrado de plagas. Programa de Agricultura Tropical Sostenible. Área de Fitoprotección*, pp. 1–23.

Hustache, A. 1929. *Les Curculionides de la Guadeloupe*, p. 148.

Jayaraman, S., Ndiege, I. O., Oehlschlager, A. C., Gonzalez, L. M., Alpizar, D., Fallas, M., Budenberg, W. J. and Ahuya, P. 1997. Synthesis, analysis, and field activity of sordidin, a male-produced aggregation pheromone of the banana weevil, *Cosmopolites sordidus*. *Journal of Chemical Ecology* 23(4), 1145–1161.

Kaaya, G. P. and Seshu Reddy, K. V. 1993. Pathogenicity of Beauveria bassiana, Metarhizium anisoplae and Serratia marcescens to the banana weevil Cosmopolites sordidus. *Biocontrol Science and Technology* 3, 177–187.

Kiggundu, A. and Gold, C. 2002. Aspectos de la resistencia al picudo negro del banano en Musa y perspectivas para la ingeniería genética contra el picudo negro. *Infomusa* 11(1), XIII.

Kiggundu, A., Gold, C. S., Labuschagne, M. T., Vuylsteke, D. and Louw, S. 2003. Levels of host plant resistance to banana weevil, *Cosmopolites sordidus* (Germar) (Coleoptera: Curculionidae), in Ugandan Musa germplasm. *Euphytica* 133(3), 267–277.

Kiggundu, A., Kunert, K., Viljoen, A. and Michaud, D. 2007. The control of banana weevil (*Cosmopolites sordidus*) by the utilization of cysteine proteinase inhibitors (cystatins) in transgenic banana plants (Abstracts). In: *Recent Advances in Banana Crop Protection for Sustainable Production and Improved Livelihoods. Greenway Woods Resort, White River, South Africa*, p. 83.

Koppenhöfer, A. M. 1991. Search and evaluation of natural enemies of the banana weevil. In: Gold, C. S. and Gemmill, B. (Eds.), *Biological and Integrated Control of Highland Banana and Plantain Pests and Diseases*. Proceedings of the Research Coordination Meeting. Cotonou, Benin, pp. 87–95.

Koppenhöfer, A. M. 1993. Observations on egg-laying behaviour of the banana weevil, *Cosmopolites sordidus* (Germar.). *Entomologia Experimentalis et Applicata* 68(2), 187–192.

Korpi, A., Järnberg, J. and Pasanen, A. L. 2009. Microbial volatile organic compounds. *Critical Reviews in Toxicology* 39(2), 139–193.

Lopes, R. B., Laumann, R. A., Moore, D., Oliveira, M. W. M. and Faria, M. 2014. Combination of the fungus *Beauveria bassiana* and pheromone in an attract and kill strategy against the banana weevil, *Cosmopolites sordidus*. *Entomologia Experimentalis et Applicata* 151(1), 75–85.

Lopes, R. B., Michereff, M., Tigano, M. S., Neves, P. M. O. J., Lopez, E. L., Fancelli, M. and da Silva, J. P. 2011. Virulence and horizontal transmisión of selected Brazilian strains of *Beauveria bassiana* against *Cosmopolites sordidus* under laboratory conditions. *Bulletin of Insectology* 64, 201–208.

Lozano-Soria, A., Picciotti, U., Lopez-Moya, F., Lopez-Cepero, J., Porcelli, F. and Lopez-Llorca, L. V. 2020. Volatile organic compounds from entomopathogens and nematophagous fungi repel banana black weevil (*Cosmopolites sordidus*). *Insects* 11(8), 509 (DOI: 10.3390/insects11080509).

Masanza, M. 2003. *Effect of crop sanitation on banana weevil Cosmopolites sordidus (Germar) populations and associated damage*. Ph.D. Thesis. University Wageningen, Wageningen, Netherlands, p. 165.

Masanza, M., Gold, C. S., Van Huis, A. and Ragama, P. E. 2009. Influence of plant and residue age on attraction, acceptance and larval survival of the banana weevil *Cosmopolites sordidus* (Coleoptera: Curculionidae). *International Journal of Tropical Insect Science* 29(4), 171–179.

Masanza, M., Gold, C. S., van Huis, A., Ragama, P. E. and Okech, S. H. O. 2005. Effect of crop sanitation on banana weevil *Cosmopolites sordidus* (Germar) (Coleoptera: Curculionidae) populations and crop damage in farmers' fields in Uganda. *Crop Protection* 24(3), 275–283.

McKinnon, A. C., Saari, S., Moran-Diez, M. E., Meyling, N. V., Raad, M. and Glare, T. R. 2017. *Beauveria bassiana* as an endophyte: A critical review on associated methodology and biocontrol potential. *BioControl* 62(1), 1–17.

Meldrum, R. A., Daly, A. M., Tran-Nguyen, L. T. T. and Aitken, E. A. B. 2013. Are banana weevil borers a vector in spreading *Fusarium oxysporum* f. sp. *cubense* tropical race 4 in banana plantations? *Australasian Plant Pathology* 42(5), 543–549.

Merchán, V. M. 2002. Manejo integrado de plagas del plátano y el banano. In: *Acorbat. Memorias XV reunión. Realizada en Cartagena de Indias 27 de oct-02 de nov*. Medellín, Colombia, pp. 353–561.

Messiaen, S. 2002. *Components of strategy for the integrated management of the banana weevil cosmopolites sordidus (Germar) (Coleoptera: Curculionidae).* Ph.D. Thesis. University of Leauven, Leauven, UK, p. 169.

Monge, J. 2015. Detección convencional y molecular de la bacteria *Ralstonia solanacearum* raza 2 agente causal del Moko o marchitez bacteriana en insectos y murciélagos asociados al cultivo del banano (*Musa* AAA) en la zona de Guapiles, Limón, Costa Rica. Tesis de Licenciatura en Ingeniería Agronómica. Universidad Nacional de confederación unitaria de trabajadores, Heredia, p. 90.

Montesdeoca, M. M. 1998. *Empleo de la hormona de agregación sordidin como método de captura y lucha contra Cosmopolites sordidus Germar (Coleoptera: Curculionidae).* Tesis de ingeniero agrónomo. Universidad de La Laguna, Valle Guerra, España, p. 122.

Monzón, A. 2001. Producción, uso y control de calidad de hongos entomopatógenos en Nicaragua. *Manejo Integrado de Plagas (Costa Rica)* 63, 95–103.

Moore, N. Y., Hargreaves, P. A., Pegg, K. and Irwin, J. A. G. 1991. Characterization of strains of *Fusarium oxysporum* f.sp. *cubense* by production of volatiles. *Australian Journal of Botany* 39(2), 161–166.

Morath, S. U., Hung, R. and Bennett, J. W. 2012. Fungal volatile organic compounds: A review with emphasis on their biotechnological potential. *Fungal Biology Reviews* 26(2–3), 73–83.

Müller, A., Faubert, P., Hagen, M., Castell, W., Polle, A., Schnitzler, J. P. and Rosenkranz, M. 2013. Volatile profiles of fungi-chemotyping of species and ecological functions. *Fungal Genetics and Biology* 54, 25–33.

Muñoz, C. M. 2006. Sistemas de producción sostenible de plátano (Musa AAB), fluctuación poblacional y severidad del daño del picudo negro (*Cosmopolitas sordidus* Germar) en San Carlos Costa Rica. Tesis Ph.D. confederación unitaria de trabajadores. Universidad de Costa Rica, San José, p. 116.

Musabyimana, T., Saxena, R. C., Kairu, E. W., Ogol, C. P. K. O. and Khan, Z. R. 2001. Effects of neem seed derivatives on behavioral and physiological responses of the *cosmopolites sordidus* (Coleoptera: Curculionidae). *Journal of Economic Entomology* 94(2), 449–454.

Nankinga, C. M. and Moore, D. 2000. Reduction of banana Weevil populations using different formulations of the entomopathogenic fungus *Beauveria bassiana*. *Biocontrol Science and Technology* 10(5), 645–657.

Ndiege, I. O., Jayaraman, S., Oehlschlager, A. C., Gonzalez, L. M., Alpizar, D. and Fallas, M. 1996. Convenient synthesis and field activity of a male-produced aggregation pheromone for *Cosmopolites sordidus*. *Naturwissenschaften* 83, 280–282.

Negrete González, D., Ávalos Chávez, M. A., Lezama Gutiérrez, R., Chan Cupul, W., Molina Ochoa, J. and Galindo Velasco, E. 2018. Suitability of *Cordyceps bassiana* and *Metarhizium anisopliae* for biological control of *Cosmopolites sordidus* (Germar) (Coleoptera: Curculionidae) in an organic Mexican banana plantation: Laboratory and field trials. *Journal of Plant Diseases and Protection* 125(1), 73–81.

Neuenschwander, P. 1988. Prospects and proposals for biological control of *Cosmopolites sordidus* Germar (*Coleoptera Curculionidae*) in Africa. In: *Nematodes and the Borer Weevil in Bananas*. Proceedings of the a workshop held in Bujumbura, Bumndi, 7–11 December 1987. INIBAP, Montpellier, France, pp. 54–57.

Pavis, C. and Lemaire, L. 1996. Resistencia de los bananos al picudo negro *Cosmopolites sordidus* Germar (Coleoptera: Curculionidae). *Infomusa* 5, 3–9.

Pavis, C. and Minost, C. 1993. Banana resistance to the banana weevil borer Cosmopolites sordidus Germar (Coleoptera: Curculionidae); role of pseudostem attractivity and physical properties of the rhizome. In: Ganry, J. (Ed.), *Breeding Banana and Plantains for Resistance to Diseases and Pests*. Montpellier, France: CIRAD, pp. 129–142.

Rhino, B., Dorel, M., Tixier, P. and Risède, J.-M. 2010. Effect of fallows on population dynamics of *Cosmopolites sordidus*: Toward integrated management of banana fields with pheromone mass trapping. *Agricultural and Forest Entomology* 12(2), 195–202.

Ríos, J. C., Soto, A. and Castrillón, C. 2002. Evaluación de *Beauveria bassiana* (Bals.) Vuill. en formulación comercial y artesanal para el manejo del picudo negro (*Cosmopolites sordidus* GERMAR) en plátano. In: *Acorbat. Memorias XV reunión. Cartagena de Indias,* 27 October–2 November, Medellín, CO, pp. 284–289.

Roche, R. 1975. Comunicacion preliminar sobre la hormiga *Tetramorium guineense* para el control biologico del picudo negro del platano. *Revue de l'Agriculture (Cuba)* 8, 35–37.

Roche, R. and Abreu, S. 1983. Control del picudo negro del plátano (Cosmopolites sordidus) pot la hormiga Tetramorium guinieensis. Ciencias. *Agriculturists (Cuba)* 17, 41–49.

Romón, P., Iturrondobeitia, J. C., Gibson, K., Lindgren, B. S. and Goldarazena, A. 2007. Quantitative association of bark beetles with pitch canker fungus and effects of verbenone on their semiochemical communication in Monterey pine forests in Northern Spain. *Environmental Entomology* 36(4), 743–750.

Rukazambuga, N. D. T. M., Gold, C. S. and Gowen, S. R. 1998. Yield loss in east African highland banana (Musa spp., AAA-EA group) caused by the banana weevil, Cosmopolites sordidus Germar (Abstract). *Crop Protection* 17(7), 581–589.

Selikhovkin, A. V., Markowskaja, S., Vasaitis, R., Martynov, A. N. and Musolin, D. L. 2018. Phytopathogenic fungus *Fusarium circinatum* and potential for its transmission in Russia by insects. *Russian Journal of Biological Invasions* 9(3), 245–252.

Sepúlveda, P. and Rubio, J. 2009. Especies de *Driophthorinae* (Coleoptera: Curculionidae) asociados a plátano y banano (*Musa* sp) en Colombia. *Acta Biológica Colombiana* 14(2), 49–72.

Sponagel, K. W., Díaz, F. J. and Cribas, A. 1995. *El Picudo Negro del Plátano Cosmopolites sordidus. Un Insecto Plaga Importante en Cultivos del Género Musa y Sus Estatus de Peste en Honduras*. La Lima, Cortés, HN: Fundación Hondureña de Investigación Agrícola, p. 35.

Storey, G. K. and Gardner, W. A. 1986. Sensitivity of the entomogenous fungus *Beauveria bassiana* to selected plant growth regulators and spray additives. *Applied and Environmental Microbiology* 52(1), 1–3.

Stover, R. H. 1962. *Fusarial Wilt (Panama Disease) of Bananas and Other Musa Species*. Kew, England: Commonwealth Mycological Institute.

Stover, R. H. and Simmonds, N. W. 1987. *Bananas: Third Edition*, p. 469.

Sumani, A. J. 1997. *Patterns of relationship between banana (Musa spp.) types and the banana weevil, Cosmopolites sordidus Germar (Coleoptera:Curculionidae).* PhD Thesis. University of Zambia, Lusaka, Zambia.

Sunil, K. L. and Singh, J. 2018. Population genetic structure of banana corm weevil cosmopolites sordidus (Germar). *India Journal of Asia-Pacific Entomology* 4, 1222–1232.

Tinzaara, W. 2005. *Chemical ecology and integrated management of the banana weevil Cosmopolites sordidus in Uganda.* PhD Thesis. Wageningen University, p. 184.

Tinzaara, W., Dicke, M., Van Huis, A. and Gold, C. S. 2002. Use of infochemicals in pest management with special reference to the banana weevil, *Cosmopolites sordidus*

(Germar) (Coleoptera: Curculionidae). *International Journal of Tropical Insect Science* 22(4), 241–261.

Tinzaara, W., Gold, C. S., Dicke, M., Huis, A. V., Nankinga, C. M., Kagezi, G. H. and Ragama, P. E. 2007. The use of aggregation pheromone to enhance dissemination of *Beauveria bassiana* for the control of the banana weevil in Uganda. *Biocontrol Science and Technology* 17(2), 111–124.

Tinzaara, W., Karamura, E. and Tushemereirwe, W. 1999. Observaciones preliminares sobre los enemigos naturales asociados con el picudo negro del banano *Cosmopolites sordidus* Germar en Uganda. *Infomusa* 8(1), 28–29.

Treverrow, N. 1985. Banana weevil borer. *Agfacts H6. AE.1*, 1–3.

Treverrow, N. L., Bedding, R. A., Dettmann, E. B. and Maddox, C. 1991. Evaluation of entomopathogenic nematodes for the control of *Cosmopolites sordidus* Germar (Coleoptera: Curculionidae), a pest of bananas in Australia. *Annals of Applied Biology* 119(1), 139–145.

Uma Devi, K., Padmavathi, J., Uma Maheswara Rao, C., Khan, A. A. P. and Mohan, M. C. 2008. A study of host specificity in the entomopathogenic fungus *Beauveria bassiana* (Hypocreales, Clavicipitaceae). *Biocontrol Science and Technology* 18(10), 975–989.

Vilardebo, A. 1973. El coeficiente de infestación. Criterio de evaluación del grado de ataque al bananal por *Cosmopolites sordidus* Germar, el picudo negro del banano. *Fruits* 28(6), 471–426.

Vinatier, F., Chailleux, A., Duyck, P. F., Salmon, F., Lescourret, F. and Tixier, P. 2010. Radiotelemetry unravels movements of a walking insect species in heterogeneous environments. *Animal Behaviour* 80(2), 221–229.

Walangululu, M., Litucha, B. M. and Musasa, M. 1993. *Potential for the Control of Banana Weevil Cosmopolites sordidus Germar With Plants Reported to Have an Insecticidal Effect, Lnfo Musa* (Vol. 2, No. 1, p. 9). Montpellier, France: INIBAP.

Were, E., Nakato, G. V., Ocimati, W., Ramathani, I., Olal, S. and Beed, F. 2015. The banana weevil, *Cosmopolites sordidus* (Germar), is a potential vector of *Xanthomonas campestris* pv. musacearum in bananas. *Canadian Journal of Plant Pathology* 37(4), 427–434.

Willsey, T., Chatterton, S. and Cárcamo, H. 2017. Interactions of root-feeding insects with fungal and oomycete plant pathogens. *Frontiers in Plant Science* 8, 1764.

Wong, W. 2003. *Molecular biology of the entomopathogenic fungus Beauveria bassiana: Insect-cuticle degrading enzymes and development of a new selection marker for fungal transformation.* Ph.D. Thesis. University of Heidelberg, Heidelberg, Germany, p. 82.

Wyatt, T. D. 2003. *Pheromones and Animal Behaviour: Communication by Smell and Taste*. Cambridge: Cambridge University Press.

Chapter 19

Banana Red Rust thrips

Frans Wielemaker Sanderse, Consultant (formerly Director of Research at Dole Fresh Fruit International), Costa Rica

1 Introduction

Red Rust thrips are an important pest in bananas reducing the marketable fruit yield. This chapter presents an overview of the literature complemented by firsthand field experiences and discussions on control strategies and regulatory issues for controlling this pest in both conventional and organic banana plantations. When the author as a young university graduate first started working in banana research with Del Monte in 1979 in Costa Rica, he witnessed very serious fruit rejects due to Red Rust thrips, and this quality issue so intrigued him that he wrote a monograph on the insect that caused it (Wielemaker, 1980). Chlorpyrifos-treated bags to cover banana bunches were not yet widely adopted at that time but soon became the standard, and Red Rust thrips seized to be a problem for conventional banana production.

However, 30 years later when the author was in charge of the Organic Banana Program in the capacity as Director of Research at Dole, this tiny insect

http://dx.doi.org/10.19103/AS.2022.0108.19

again became a huge problem in organic banana production systems and demanded the full attention of the research department to find solutions (Castro and Jair, 2013). Red Rust thrips form a difficult-to-control 'plague' in these sustainable banana systems. Fully understanding the biology, epidemiology, and life cycle of ed rust thrips is important to develop and implement effective integrated pest management options. This chapter gives an in-depth overview of what the pest management practices are in conventional banana production systems before outlining the alternative practices in organic banana production systems. An overview is provided of the challenges and limitations of red rust thrips control, while regulatory issues are also addressed.

2 Pest nomenclature and classification

Common names for the red rust thrip include

- Banana bladderfoots, fringed-winged flies (English);
- *Bananen blasen füsse* (German);
- *Blaaspoten, Franje vleugelen* (Dutch);
- *Thrips de Rouille de banane* (French);
- *Trips de la Roya de Banano, Trips de la Mancha Roja or Oxido Rojo* (Spanish); and
- *Tripes da Ferrugem* (Portuguese).

Other names for thrips are thunder flies, thunder bugs, and storm flies. The causal agent is classified as follows:

Order	Thysanoptera
Family	Thripidae
Genus	*Chaetanophotrips*
Species	*signipennis* (Bagnall) *and* *orchidii* (Moulton)

Red Rust thrips can belong to either *Chaetanophothrips signipennis* (Bagnall) or *Chaetanophothrips orchidii* (Moulton) and are often interchanged or mentioned together. The damage caused to the banana fruit is the same. The two species can be differentiated as the *C. signipennis* females have body hairs that are clearly visible and distinct from *C. orchidii,* using a stereomicroscope. There are 20 species classified in the genus *Chaetanophotrips*, the majority of which occur in Asia which also is the center of origin of the genus Musa to which the banana belongs. The pests' life cycle includes an egg stage, two larval instars, two pupal stages, and an adult stage.

Hara et al. (2002a) call *C. orchidii* the 'Anthurium thrips,' while *C. signipennis* is called the 'Banana thrips,' but they both occur on the same hosts and can

Table 1 Characteristics of Red Rust thrips pests of banana

Causal agent	*Chaetanophothrips orchidii* (Moulton)	*Chaetanophothrips signipennis* (Bagnall)
Common name	Anthurium thrips	Banana thrips
Reproduction	Parthenogenic = Asexual	Nonparthenogenic = Sexual
Feeding on fruit	Yes	Yes
Causing fruit blemish	Yes	Yes
Feed on pseudostem	No	Yes
Feed on followers	No	Yes

only be microscopically differentiated based on the body hairs and glands present on the females of the 'Banana Thrips' (*C. signipennis*) (Sakimura, 1975). Hood (1954) indicated that two structurally different forms of the thrips insect exist, based on its ability to reproduce asexually, called parthenogenic. Roberts (1958) reported that the parthenogenic form is found on the Pacific Coast of Panama and Costa Rica, on the North coast of Honduras, and on the East coast of Guatemala, while on the Caribbean Coast of Panama and Costa Rica both species occur. It has been identified that *C. signipennis* reproduces sexually and *C. orchidii* is the parthenogenic (= asexual) species (Ostmark, 1974). In Hawaii male Anthurium thrips have not been observed such that reproduction of *C. orchidii* occurs without mating in a process whereby virgin females lay eggs that produce female offspring (Hara et al., 2002b). In areas like the Atlantic zone of Costa Rica, where both forms of *C. orchidii* occur the sexual one predominates (Lara, 1970). Besides feeding and ovipositing on the fruit, *C. signipennis* also oviposits on and feeds ('rasps') the pseudostem and suckers which causes characteristic black-colored V-shaped marks, while *C. orchidii* thrips restricts itself to the fruit and does not feed or oviposit on pseudostems. The differences between the species are highlighted in Table 1.

3 Pest distribution and economic impact

Red Rust thrips were first recognized as a pest of bananas in Panama in the early 1930s and were most probably introduced from the Far East with exotic banana planting material imported in the 1920s by the United Fruit Company to Bocas Del Toro island, close to the Almirante Division on the Atlantic Coast of Panama. This planting material was imported as part of the first attempt by the United Fruit Company (UFCo) to breed a Fusarium-resistant Gross Michel type banana variety. As the pest got a foothold on the Gros Michel variety commercially cultivated at that time in that Division of UFCo, it was referred to as the 'Almirante Red Rust Thrip' (Stephens, personal communication; Stephens, 2008, 2012). It is of interest to note that the first major outbreak of fusarium wilt race 1 caused by *Fusarium oxysporum* f. sp. *cubense* also occurred in the Almirante Division in 1905, hence the name Panama disease. It is now generally accepted that *C. signipennis* and *C. orchidii* have a worldwide distribution. Adults are not strong flyers but are easily wind dispersed or by means of infested harvested fruit or planting material. These insects have the ability to hitchhike on planting material, fresh produce, and cut flowers which may explain their global distribution in recent times.

Red Rust thrips belonging to the species *C. signipennis* are of economic importance to banana production in Australia and Central and South America. *C. signipennis* has been of economic importance to bananas for decades in New South Wales and Queensland, Australia (Braithwaite, 1965) and has been reported from Fiji and Papua New Guinea, India, Indonesia, Philippines, and Sri Lanka, from where it was first described by Bagnall. The distribution of *C. orchidii* is worldwide in greenhouses and outdoor landscapes (Hara et al., 2002a,b). In contrast, *C. signipennis* has a somewhat more limited distribution (Schotman, 1989). In the Caribbean *C. orchidii* is widespread and has been postulated to have invaded from Hawaii (Hara et al., 2012). In the Western Hemisphere, it has been recorded in Brazil, Costa Rica, the French West Indies, Honduras, Mexico, Panama, Puerto Rico, and Trinidad and Tobago. In the USA, the pest is present in Florida and Hawaii. Vilardebo (1960) in his 'Mission Report' to Ecuador does not mention either *C. orchidii* or *C. signipennis* as the cause of banana Red Rust which he extensively observed and described in the banana plantations in that country. He identified the causal insect as *Palleucothrips musae* (Hood). Obviously, there have been misidentifications and confusing reports about the species identity of the Red Rust thrips in different banana-growing countries. There are no reports from Africa concerning the occurrence of Red Rust thrips on bananas, and CABI does not report it to be present in Africa. However, a recent report on a 'new banana disease' in Rwanda proved to be caused by thrips with symptoms very similar to those caused by Red Rust thrips (Okonji, 2020).

Due to their excellent natural adaptation, thrips have developed to be one of the most feared and widespread pests in the world and cause serious damage to fruit and vegetable crops and ornamentals. In general, thrips prefer to feed on very young, succulent, immature fruits, flowers, and foliage and are difficult to detect as they are very small and keep themselves concealed throughout most of their life cycle. Banana Red Rust thrips are a serious pest of banana fruit, causing considerable economic losses in banana plantations (Simmonds, 1959; Ostmark, 1976; Simmonds and Stover, 1987). When an outbreak occurs, and this pest is not controlled in a timely manner, an enormous mountain of discarded bananas will soon appear outside the packing house (see Fig. 1). The typical symptoms of Red Rust thrips are often most common on the big top hands of a bunch. These symptoms occur because the adults and larvae punch

Figure 1 Red Rust causing a big mountain of discarded bananas in the packing house. Especially the big top hands are affected.

Figure 2 Red Rust oval-to-lengthwise reddish ugly fruit blemish on bananas caused by *Chaetanophothrips spp*.

and suck plant cells with their mouthparts causing a reddish-colored stain on the peel of the banana fingers (see Fig. 2). Adults also oviposit eggs inside the banana skin (Fig. 3).

Both species, *C. signipennis* and *C. orchidii*, of the genus *Chaetanophotrips*, damage young banana fruit with their feeding activity and by oviposition at finger touch points. When observed closely with a magnifying glass, these creatures produce random reddish squiggles and curlicues on the skin (see Fig. 4). The banana skin is slightly undulated and roughened and discolored from water-soaked dusty to smoky gray in the beginning (between and where

Figure 3 Plantain affected by Red Rust trips.

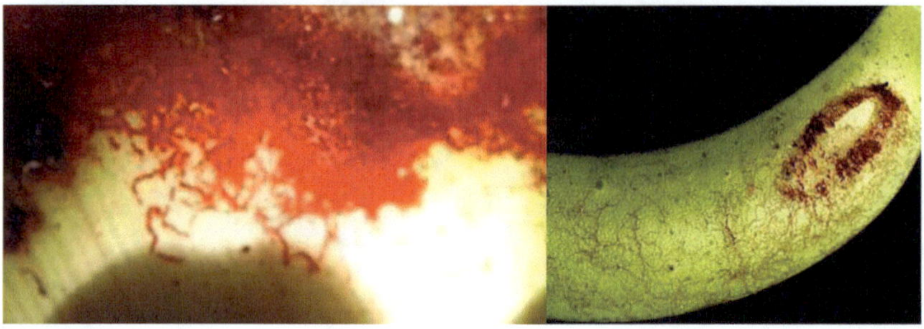

Figure 4 Reddish-brown random squiggles and curlicues develop on the skin of the banana.

the fingers touch) and then turns reddish-brown (which is why it is called Red Rust) and then often shows up as a reddish oval with a lighter center. Later it takes on a corky appearance. In extreme cases, the skin may split because it becomes hard and cannot expand as the finger grows fatter. Damage is mainly restricted to areas surrounding points of contact between the fingers where the thrips shelter. Initially the symptom looks no more than a smoky gray-colored hue which is somewhat water-soaked (see Fig. 5).

By separating the fingers of the still hanging bunch, the circular-shaped reddish damage with a lighter center (where the fingers touch, and where the insects cannot get in between) can be noticed on the - what were - concealed surfaces between adjacent fingers. Some tiny thrips excreta (dust) can be seen on the affected fruit surface. As mentioned earlier, the Red Rust blemish on the banana peel is purely a cosmetic problem as the pulp and taste of the banana is normally not affected unless it splits. Consumers do not like to see a blemish on their bananas and thus supermarkets do not tolerate any Red Rust symptoms at all, not even the minimum, not even on organic bananas. When an 'Ugly Fruit' branded banana line was tried with slight Red Rust symptoms

Figure 5 A water-soaked dusty or smoky gray-colored hue between and where the fingers touch appears as an initial symptom. Later the skin gets a corky appearance and may split.

and other minor blemishes, the market reacted negatively and therefore not even only slightly affected bananas can be processed, packed, and exported; thus, this pest can become a huge loss for growers who did not control the pest (conventional growers) or have no effective organically allowed and affordable control method at hand (organic growers). As far as is known from the trade in Latin America, the presence of thrips on export bananas has not been a cause for quarantine rejection at arrival in any destination country. Legally, probably because *C. signipennis* and *C. orchidii* both have such a worldwide distribution and technically because they seem not to survive the long transit times at low temperatures in ship hauls even though larvae in theory could hatch from the eggs that were deposited in the peel after packing and shipment. No literature was found to confirm this assumptive statement.

Organic banana farms can become seriously affected as not many options for controlling this insect exist. Starting in 2010, organic banana farms in Ecuador and Peru were seriously affected with losses ranging from 35% to 65% and even 100%. Yearly figures of exported organic bananas from Dole's Nueva Esperanza Farm located in Tosagua, Manabí, Ecuador, fell from 48 tons per ha per year in 2009 (which is very high even compared to a conventional farm) to only 8 tons in 2011. Up to 75% of the bananas had to be rejected in the packing house at certain weeks due to Red Rust damage alone and within a matter of half a year the attack totally devastated the production of marketable bananas (Gutierrez, 2015). The next few years had to be bridged by controlling the pest with natural pyrethrin- and rotenone-based insecticides, and later – in 2014 – spinosad was finally approved in Ecuador and production levels of the farm could climb back to more normal levels. The cost of the product and the application is however very expensive and clearly affects the cost per box and thus economic viability of the farm. This episode of high losses in Nueva Esperanza farm and losses due to Red Rust throughout the Ecuadorean and Peruvian organic banana industry led to the investigation of this banana Red Rust pest in detail by Dole's Research department. Many Ecuadorian and Peruvian governmental research institutions investigated this problem and all the findings culminated in an across-industry meeting (Clercx, 2013) with all the stakeholders involved to understand where we stood, what the findings were to date, and what still needed to be done in terms of understanding the problem and how to solve it.

Besides the commercial 'Cavendish' and 'Gross Michel' varieties, many *Musa* species are affected, for example, *Pisang mas* (aka 'Baby'), *P. awak, P. bungay,* and FHIA#2 (aka Monalisa). Especially exotic banana varieties, which have a tight bunch with fingers and hands touching each other, are more prone to damage. Commercial True Horn plantain (aka 'Cuerno Verdadero' in LAT) varieties, which do not have a male bud, and False Horn plantains (aka 'Falso Cuerno' in LAT), which have rudimentary male buds, and what is known as French Plantains (aka 'Dominico'), which do have a male bud, are all affected by Red Rust thrip damage (Fig. 3) but only becomes a problem when these 'cooking bananas' enter high-end commercial local or export markets. The local trade and consumers view bananas and plantains with Red Rust symptoms as 'normal' or 'typical' and is of no concern as it does not negatively affect the quality nor the taste. The symptoms of Red Rust thrips are often thought to be part of exotic banana varieties' distinct characteristics by laymen and tourists traveling to the tropics from temperate countries. In India, the commercial cultivars such as Poovan (aka Mysore), Monthan (aka Bluggoe), Saba, Ney Poovan (aka Pome), and Rasthali (aka Silk) are also susceptible to the Red Rust thrips *C. signipennis* (Padmanaban and Mustaffa, 2010).

The important alternative host for *Chaetanophotrips* species in surrounding areas of commercial banana plantation are feral bananas, plantains, abaca, orchids, ornamental plants, and weeds. Red rust thrips are polyphagous feeders and numerous alternate hosts (crops, ornamentals, and weeds) harbor these thrips as they can be found in the axils of their leaves such as the food crop Elephant Ear (*Xanthosoma* sp., aka 'Tiquisque' in Latin America) and ornamentals like *Strelitzia* (aka 'bird of paradise'), *Dracaena*, *Cordyline* (Hawaiian Ti plant), *Anthurium*, and *Dendrobium*. On most of these plants, thrips can be observed feeding on the whorls (cigars) of immature leaves, causing rust-colored discoloration and silvering. On bananas the cigar leaf is normally, strangely enough, not affected, not even in new plantings where no banana fruit is present. Many of the above-mentioned food plants and ornamentals are often found inside or around banana plantations. But as an overall statement, it can be said that *C. signipennis and C. orchidii* are mainly an economic pest of export bananas and plantain. Nevertheless, a problem in citrus called 'Ring Spot Damage' is caused by the same thrips through feeding injury at fruit touch points.

4 Pest morphology and life cycle

The life cycle of this insect has been studied and reported by Ostmark (1976) and Pinese and Elder (2000, 2004), Hara et al. (2002a), and Arias et al. (2013), a summary of which is shown in Fig. 6. Thrips of the *Chaetanophothrips* genus are tiny slender-bodied insects. The young insect (nymph stage 1 and 2) is whitish in color and still wingless, while the adults are creamy to light brownish in color. The wings of the adults are fringe-winged (with long hairs) and they have protrusible vesicles on the short legs often called bladder foot. They have asymmetrical, piercing, sucking mouth parts with only one mandible with which they puncture ('rasp') the epidermis cells and suck out the contents for food causing damage to the fruit. When the infestation is severe, the injured areas become less elastic resulting in epidermal cracking as the banana fruit enlarges. The females have a big, developed ovipositor at the end of the abdomen, with which eggs, up to 75 in total, are inserted singly into the young

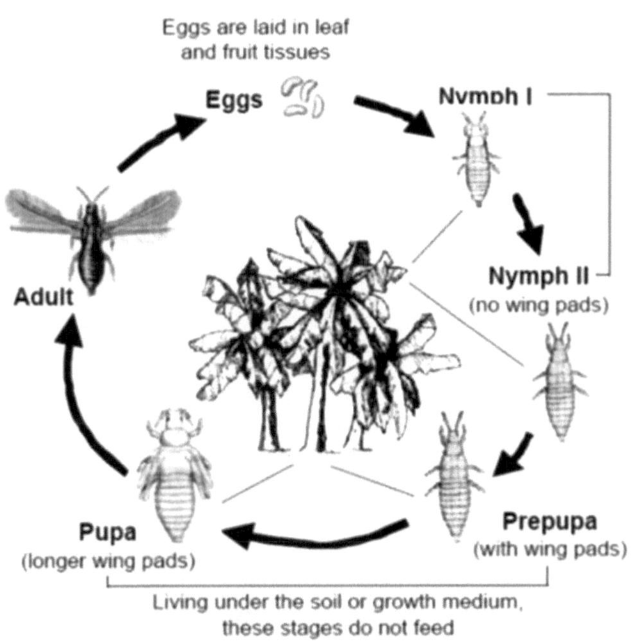

Figure 6 Life cycle diagram of *Chaetanophothrips signipennis*. Source: Drawing by D. Schultz, University of Illinois in Hara et al. (2002a).

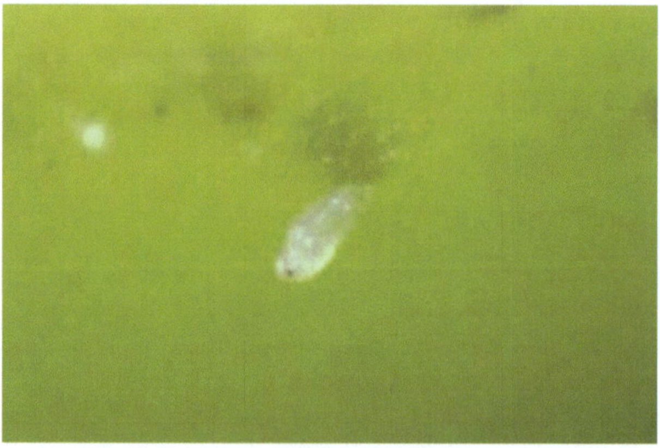

Figure 7 Eggs are oviposited in the banana peel and whitish to transparent nymphs hardly visible to the naked eye hatch days later. Source: Adopted from: Arias de Lopez et al. (2013).

banana peel tissue by *C. orchidii* and also into the stalk or pseudostem tissue by *C. signipennis*.

Oviposition by the females is done on the banana fruit and in the pseudostem under the outer epidermis cell tissue. Eggs last for 7-12 days before they hatch. The kidney-shaped eggs are invisible to the naked eye. The newly hatched first nymphal stage (see Fig. 7) pale in color, also called neonate stage, feeds on the banana peel and changes into the more orange-colored second nymphal (mature nymph) stage and in total the two stages feed for 8–10 days on the banana peel which discolors gray and later reddish in color. The nymphs (as mentioned, sometimes also called larvae) never extend their feeding activity to the pulp. These stages are too small to be seen without a 10× hand lens as they are 1/10th of a millimeter long in the first stage and only slightly larger in the second stage. The second stage nymph insects then migrate to the soil at the base of the banana plant where it molts and passes through two stages (pre-pupae and pupae). During these pupae stages, the insect does not eat. The pre-pupae stage looks similar to the second nymph, but this time shows short wing pads. After 2-5 days, the pre-pupae enter another pupal stage and now show elongated wing pads. This stage lasts for 6–10 days in the soil. The metamorphosis is what is called hemimetabolic (= undergoing a metamorphosis that lacks a real pupal stage). Both semi-pupal stages remain in the top few centimeters of the soil (Pinese and Elder, 2000) beneath the banana plant and are capable of crawling but do not feed. They are very difficult to find as they are only 1 mm long, pupate into adults just below the soil surface near the base of the banana plant and among the followers (aka suckers). Australian scientists found that 79% of the pupation occurs within a 30 cm band around the base of the mat and that 82% of pupation occurs within the top 20 cm of

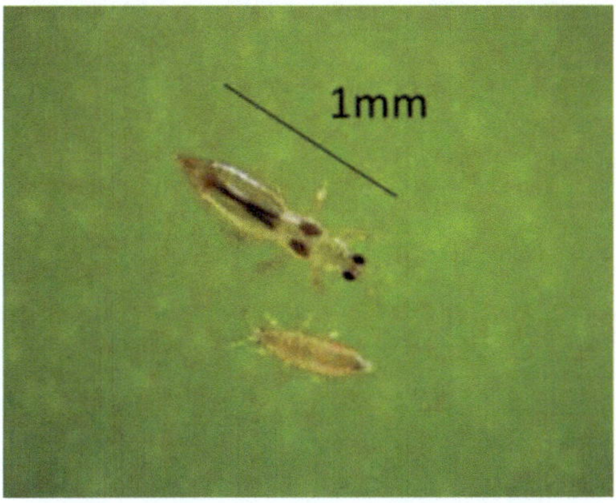

Figure 8 Adult (top) with folded wings and two eye-like dark patches at the base of each wing, and larva (bottom) of *C. signipennis*. Source: Adopted from: Arias de Lopez et al. (2013).

the soil (Pinese and Elder, 2000). The insect emerges as an adult from the soil and lives 27–112 days (average 52 days) and is generally 1.4 to 1.7 mm long (see Fig. 8). Also, from Australia, it is reported that the incidence of Red Rust thrips is more severe in bananas grown in well-drained red clay soils, while wet conditions are thought to cause high mortality of the soil-dwelling pupal stages, either directly from drowning (Pinese and Piper, 1994) or by creating the conditions for the development and spread of entomophagous insects or pathogens.

When the pupae hatch in the soil, the yellow-colored adults crawl out of the soil and up to the pseudostem and will find their way to the fruit. The adults can be distinguished from the first nymphs by the two dark spots at the base of their wings which are fringed (feather-like, although these are a sort of scales). When the wings are folded, the adults appear to have two black converging lines down their back running the length of the abdomen. Two dark patches at the base of the wings are characteristic of Red Rust thrips adults. The eyes are also black.

Adults then go to the banana bunch and rasp and suck the cell juices of the young banana fruit causing superficial damage which first has a water-soaked shade-of-gray appearance and later turns reddish, bronze, brown, and rusty in color – hence the name of the pest – becomes more intense when the fruit reaches harvesting age. They prefer young bunches with 1–4 open hands, although feeding young and adult thrips insects do continue to live on the bunch until harvest. Males and females of *C. signipennis* also feed on the outer epidermis of the leaf-sheets of the pseudostem, which causes characteristic V-shaped black marks where two leaf sheets come together between which they hide from sunlight (and predators) and look for tender tissue to rasp on

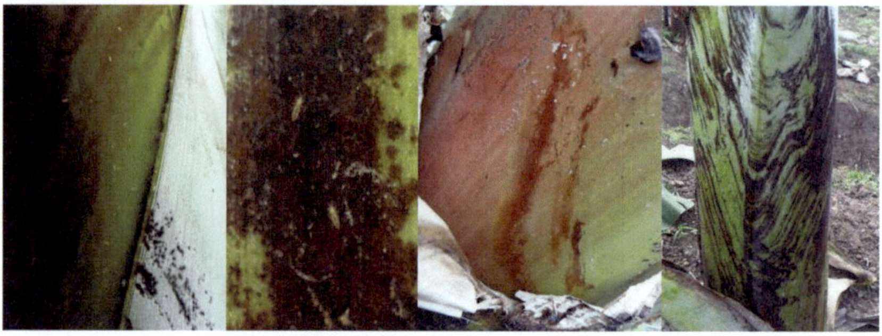

Figure 9 Thrips nymphs and adults cause parallel dark V-shaped black colorations on the pseudostem.

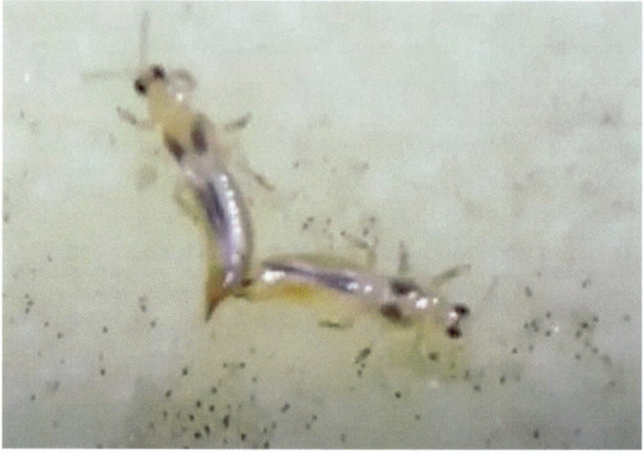

Figure 10 Thrips caught mating under the stereomicroscope in the laboratory. Source: Adopted from: Arias de Lopez et al. (2020).

(see Fig. 9). The nymphs and adults prefer shaded and protected areas and conditions.

Mating of *C. signipennis* occurs continuously (see Fig 10) under the leaf sheath edges of the pseudostem on young plants prior to bunch emergence and later also on the bunch when the bunch has emerged and usually there are many overlapping (aka 'crossed') generations, increasing the difficulty for implementing control strategies. Adult thrips can withstand periods up to 36 h without feeding.

The entire life cycle can be completed in 28 days but may take up to 3 months during the cooler season. Higher temperature and humidity and new growth of banana plants appear to be favorable to thrips' feeding and breeding, leading to heavier infestations and greater damage during the warmer summer months. Damage to the fruit in the packing station at any moment in time more

Figure 11 Adult thrips get into the emerging bud by means of small cracks in the bract. Source: Adopted from: Arias de Lopez et al. (2020).

clearly reflects the level of a previous rather than the present infestation. Adults are weak fliers (Haarer, 1964) and wind plays a key role in the dissemination of the insects over larger distances, in addition to movement of infested plant material and infested bunches or fruit. The infestation of the bunch occurs by flying adults rather than adults and nymphs walking up the pseudostem. The adults fly toward the emerging bud and can even get into the bud (into small cracks of the bract as shown in Fig. 11) which is still mostly inside the pseudostem and emerge with the bud from the throat of the plant in between the male flowers. They can also fly toward emerging stem and walk 'down' the fruit-stalk on to the opening bunch. Months with high humidity and low light intensity are the environmental conditions which thrips prefer. Thrips prefer to get into minute or restricted areas of plants, which is what is called cryptic behavior, which is a behavior pattern that maximizes an organism's ability to conceal itself. Thrips may cause minor irritation when they get on human skin.

5 Detecting pest infestations

The symptoms of Red Rust thrips that develop on the fruit are the result of the nymphs and adults feeding on the young banana skin and ovipositing their eggs under the epidermis. When bunches first show Red Rust thrips symptoms, newly emerging buds and young fruit which are probably infested with thrips should be shaken over a white cloth to detect and better observe and confirm the presence of the insects. Thrips are very small insects that can occur in large numbers and common way to catch thrips used by entomologists is by sucking them into a vial through a Y-shaped device diverting the catch into a glass vial. Thrips can be conserved in a solution of 70% to 80% ethanol. 'Hoyer's' mounting medium is used to slide-mount thrips larvae and adults for preservation and

Figure 12 Inspection of buds and under the leaf sheets for thrip incidence.

later identification with a stereomicroscope. A 10× hand lens will aid in the determination to ascertain that thrips are present. In fact, the thrips get to the bud and underneath the bracts and to the flowers (see Fig. 12) and fingers nearly as soon as the throat of the banana plant shows the tip of the bud. Checks in the field are done by carefully revising the entire bunch or certain predetermined hands, preferably the biggest top hands as they emerged first, and the number of thrips per bunch, hand or finger is counted. Pinese and Piper (1994) recommended to examine at least a total of five bunches per hectare for the presence of Red Rust thrips every two weeks to catch the onset of an infestation, as an infestation can occur very quickly and the threshold in terms of a number of thrips causing severe damage is very low. The method is described as follows: Use a ladder to gain access to the upper part of the bunch and select five central fingers from the top row of the top hand. The number of adult thrips on one side of each finger and the level of gray or red discoloration present on the same side are recorded using a rating scheme and recorded on a field sheet. Thrip counts and damage estimates are well enough correlated to permit their use in estimating the amount of damage associated with various thrips populations. Studies on the relationship between the number of thrips and the incidence of red rust on near-harvest fruit indicated that a population of only one thrip for every two fingers of very young bananas was associated with the presence of symptoms (loss) of Red Rust on 50% of the bananas.

Another method of monitoring the thrips population is using five follower plants per hectare of about 1.5 meters tall and gently pulling away the petioles leaf-sheets from the young pseudostem of leaf number three and counting the number of nymph and adult thrips (using a 10× hand lens) underneath the edges (see Fig. 12) (Castro and Jair, 2013). This method can also be employed on new plantations although usually the thrips populations are low since no canopy is present and thrips do not like bright sunshine. However, heavy weed infestation in the plantation can harbor thrips. Based on the presence and number of Red Rust thrips, it will need to be decided which type of control procedure is to be used. Under heavy infestations, plantations will show the dark V-shaped markings on the pseudostems from the feeding of the nymphs and adults.

The development of economic threshold levels for thrips based on young follower pseudostem monitoring would assist decision-making in relation to expected fruit damage and provide a rational basis for applying control

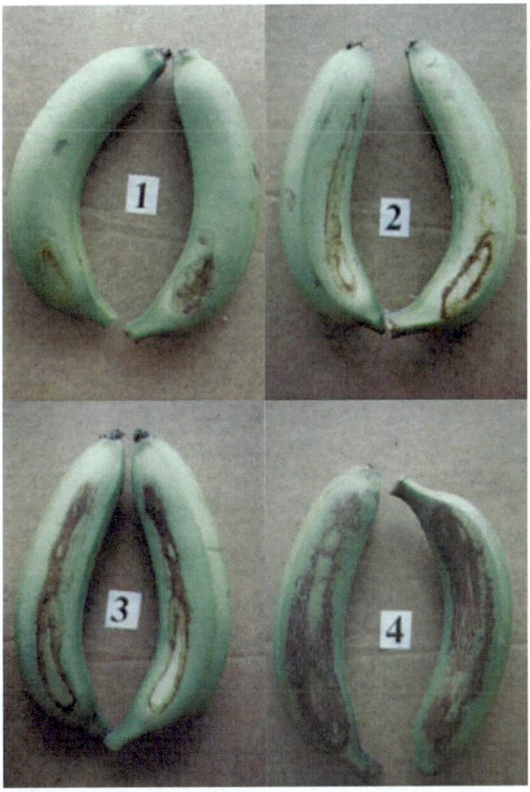

Figure 13 Ratings 1–4 of Red Rust damage for evaluating treatment effects. Source: Adopted from: Pinese and Elder (2000).

treatments. Data concerning the visible V-shaped black markings on the pseudostem related to the resulting fruit damage at harvest would also provide a rapid method to aid decision-making with regard to the need for treatment(s) but it may already be too late to prevent present damage. Assessment of the resulting damage caused by thrips is done in the packing station. To evaluate the effects of treatments in the field, the Quality Control Department checks for instance 10% of the bunches at random that enter a packing station from each section of the farm and records the incidence and severity of Red Rust damage present in the top three hands using the 1 to 4 damage scale as reported by Pinese and Elder (2000), as shown in Fig. 13.

6 Similarities to other pests

The symptoms of Red Rust thrips are sometimes confused with what is called in the banana trade as 'Maturity Stain' (see Fig. 14). Maturity stain affects the entire finger lengthwise in irregular patterns, rather than predominantly the area where the fingers touch(ed) when they emerged. 'Maturity bronzing' symptoms are not confined to where fingers touch and frequently show on the outer curve of the fruit. 'Maturity Stain' or 'Maturity Bronzing' (aka 'Mancha Madurez' in Spanish) is a physiological disorder associated with calcium deficiency at bunch formation (Diaz et al., 2007; Campbell and Williams,1978; Williams et al., 1990). In the case of 'maturity stain', the lack of Calcium leads to rupture of the outer epidermis cells, resulting in oxidation which gives a similar reddish bronze color as seen in areas where the Red Rust thrips feed and oviposit, also giving rise to oxidation and the appearance of a red/bronze color. Detailed comparisons between 'Maturity Stain' and Red Rust symptoms and damage

Figure 14 'Maturity stain' symptoms caused by a physiological imbalance should not be confused with Red Rust blemish.

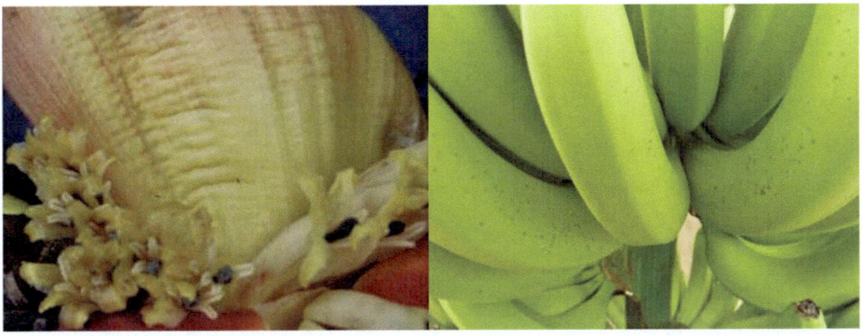

Figure 15 'Flower thrips' blemish fingers by their oviposition leaving braille script 'pimples'.

have been reported in the literature and the symptoms and causes are normally not confused by skilled banana growers.

Another species of thrips, *Frankliniella parvula* Hood, is a very common thrips throughout banana-growing areas and blemishes banana fruit through their oviposition. These thrips are black in color and called 'flower thrips' because they can be found in large numbers in the young banana flowers (see Fig. 15). The oviposition marks on the fruit cause the small 'pimples' found on almost every banana cluster produced anywhere in the world. Fortunately, the damage caused by these 'flower thrips' is not frequently the cause of rejecting bananas at the packing station unless the infestation is very high and produces what is called in the trade 'braille script' which is easily felt by moving one's fingertip over the bananas (see Fig. 15). More 'pimples' can be found on the inner part of the fingers and toward the base of the fruit toward the crown, the same area as where Red Rust blemish occurs as these thrips also like to hide there. The *Frankliniella* thrips enters the bud and crawl underneath the bracts and into the flowers very early when the bud is still mostly inside the pseudostem and only just coming out. They feed on the flowers and numerous dark elongated thrips can be noticed with the naked eye. Just as Red Rust thrips, these flower thrips also pupate in the soil at the base of the banana plant.

Another species of thrips, *Thrips hawaiiensis,* causes a 'Corky Scab' which is a real problem in the commercial banana plantations in Davao, The Philippines. The symptoms are different as a gray brown to reddish roughening (corky-like) of the skin occurs after these thrips feed on the fruit (although they also feed on the bracts and flowers) in the very early stage of bud emergence. A high infestation by these thrips can also cause major losses due to the corky and scabby symptoms on the banana fruit, which is rejected for export.

In Australian commercial banana plantations, 'Silvering Thrips,' *Hercinothrips bicinctus*, causes a silvery blemish on the fruit and the fruit also becomes spotted with the thrips' dark excreta. In severe infestations, the blemish also colors reddish. This same *H. bicinctus* thrip has also been reported from South Africa. This blemish is unsightly on the fruit and this insect is of biosecurity concern to the Latin American banana industry. Another thrip, *H. femoralis*, which causes similar symptoms, is described from the Canary Islands (Perera Gonzalez et al., 2010). CIRAD (Anonymous, 2015) has published the key for banana thrips identification for the Caribbean islands where apparently only *Chaetanaphothrips orchidii* is present, not *C. signipennis*.

7 Integrated pest management: biological and cultural control

It is unknown if there is any genetic resistance to Red Rust thrips and this insect pest has not attracted attention from breeders because it does not cause heavy production losses, does not affect eating qualities, is not a threat to food security in subsistence farming, and can be controlled relatively easily - by chemical means at least - in commercial plantations for the local or export markets. There are no known banana varieties that are inherently resistant to the damage caused by this pest. The physical bunch configuration itself plays a role in the damage caused by this insect. Increasing the space between fingers during the early stages of development of the bunch reduces the damage caused by Red Rust thrips and may be used as a selection criterium in breeding programs. In contrast, resistance to damage by the banana weevil borer (*Cosmopolites sordidus*), a pest which is threatening production of bananas and plantains worldwide, is an important trait that breeding schemes pay attention to. In general, thrips can be vectors of virus diseases, but the *Chaetanophotrips* has not been reported to be a vector of viruses which affect *Musa*.

Losses due to Red Rust thrips can be enormous when a population of this insect increases and the threshold level for infestation to avoid fruit damage is extremely low. Just a few individuals of these thrips feeding on a banana bunch, from the bell's emergence to harvest, can cause severe damage to many fingers and thus huge losses due to the current zero tolerance in export markets for this blemish. Therefore, an integrated pest management system is required which combines several strategies which complement each other.

7.1 Biological control

The banana host and the Red Rust thrips both originated in Asia and subsequently spread around the globe. Thus it is likely that not many biological control agents can be found in Latin America. Natural predators of the thrips nymphs in the form of ants and spiders have been reported. The *Orius spp.* (Fam.

Figure 16 Dorsal and ventral view of *Orius* sp. bug predator of thrips. Source: Adopted from: Castro and Jair (2013).

Anthocoridae) 'Pirate bug' (see Fig. 16) is a general predator of thrips observed in the field although the extent of their effectiveness is not well documented. This insect can be mass reared, and research needs to be initiated to get a high enough population of this natural predator established in a banana plantation. Lacewings, ladybird beetles, and predatory mites may also give some control on nymph and adult thrips. Ants may prey on pre-pupae and pupae in the soil near the base of the banana plant (Pinese and Piper, 1994). Several fungi, including *Paecilomyces* spp. and *Verticillium lecanii*, have been isolated from other thrips species and may well infect banana rust thrips (Buss, 2006).

The Belgian sustainable crop management company Biobest (www .biobest.be) has developed and has commercially available biological control systems for thrips control in greenhouse-grown eggplant, poinsettia, and anthurium based on the predatory mite *Amblyseius cucummeris* and the mold mite *Tyrophagus putrescentiae*. They also mention a biological control system based on *Hypoaspis* to control larval and pupal stages of thrips. It is not known if these are effective on Red Rust thrips in bananas. The Dutch sustainable crop management company Koppert (www.koppert.com) proposed to screen a predatory mite (*Amblyseius swirskii*) and an entomopathogenic fungus (*Verticillium lecanii*) in the laboratory. This research should generate knowledge and undertake validation of the effectiveness of biological control agents in the field. Obviously, in the more controlled environment of a greenhouse, the introduction of sachets with predatory mites is more likely to be successful, as is described from the Canary Islands where bananas are grown inside enclosed shade houses (Nogueroles et al., 2015). In banana plantations, it is harder to bring in a high enough beneficial insect population or to successfully get them to establish and multiply. Furthermore, the quarantine compliance rules and regulations for the importation of insect pathogens and beneficial insects (natural enemies) are not all that clear and easy to comply with in most banana-producing countries. Mating disrupting or other behavior affecting pheromones (aka semio or allelo-chemicals) are being developed as a possible way to attract, catch, and suppress thrips populations early on. Wageningen University has developed a concept called Lurem-TR for greenhouses which helps monitoring populations of thrips used in combination with a sticky trap (Van Tol, 2013). It has been reported that trapping of *C. signipennis* with pheromones is not that effective as these thrips fly very little and only very short distances (Arias et al., 2020). Many sachets with pheromones would then be needed in an open banana plantation. The scientific publications by Clercx et al. (2015) and Arias et al. (2020) are the most up to date and most informative recent articles one can read on the subject of the Red Rust thrips' life cycle and on investigations toward biological control of this pest. More recent publications revealing new insights on the pest and its control have not been published since, but some regulatory issues have substantially changed for certain control compounds. *C. signipennis* flies very little, hence complicating trapping even with pheromones.

7.2 Cultural control

In organic banana plantations where no synthetic chemicals can be used, control of thrips is very challenging. Cultural control methods to aid in lowering thrip insect populations involve removing the banana male bells as early as possible (when the last hand is horizontal) and taking these out of the plantation or putting them in closed plastic bags (see Fig. 17), which will suffocate the thrips. If this is not done, the thrips will continue feeding on the male flowers inside the bell, even when removed from the plant and let to fall to the ground. After a few days, the bells (buds) can be removed from the bags and left to rot in the field or brought to a composting facility. The plastic bags can be reused. Periodic

Figure 17 Early removal of flower buds and enclosing them in plastic bags is used to lower populations. These can later be composted.

de-sheathing of the leaf sheets of the pseudostem is also a common practice to lower the population as the thrips do not like to be exposed to sunlight. Cycles of de-sheathing the pseudostems of mother and follower plants and then immediately applying potassium-based soaps up to runoff lowers the thrips populations in the field.

Good weed control within the plantation and the border areas is recommended. Ornamental plants which can serve as alternate hosts to the pest should be eliminated from the vicinity of the plantation. If corms are used as planting material to establish a new plantation instead of meristems from tissue culture, the corms and the attached soil can harbor thrips pupae, adult, or nymph stages of the insect and be brought to a new farm. In fact, it can be speculated that this pest was disseminated throughout banana growing areas in Central and South America by means of corms used to plant new farms while trying to 'escape' the advancing Fusarium race 1. If the area where the corms were produced had a history of thrips infestation, the corms should be physically cleaned and preferably treated with an insecticide. Heat treatment is also an option (Pinese and Piper, 1994), but it is cumbersome and affects germination success if not executed correctly with regard to temperature and duration of the treatment. When changing banana varieties or renovating plantations, re-sprouting banana plants from the previous planting should be destroyed prior to replanting as they could serve as a source for thrips reinfestation. But even when all these measures are taken, reinfestation is highly likely from whatever neighboring source available and from thrips transported by winds and storms. Good weed control in organic banana farms is difficult as organic herbicides do not exist, and therefore organic banana farms usually have problems with Red Rust thrips infestations.

Trapping thrips using various kinds and colors of plastics (usually white or blue) with a sticky coating applied (see Fig. 18), like Tangle-Trap or Stickem®,

Figure 18 Plastic sheets of different colors with a sticky coating applied can be used for monitoring an infestation. Source: Adopted from: Castro et al. (2013).

have not been reported as very successful in effectively lowering high thrips populations in a banana plantation. This methodology can be used for monitoring the onset of an infestation although the identification of the thrips down to species level is not that easy.

8 Integrated pest management: combined physical and chemical control using stem bags

Covering the banana bunches with a non-insecticide-treated polyethylene plastic bag early provides a good physical barrier and an effective and economically efficient control to various somewhat larger insects but unfortunately does not protect the fruit sufficiently and effectively against Red Rust thrips infestation. The reason is that the thrips are already inside the bud before bagging and also because the bag is perforated, which lets the insects through, and the bag is usually left open at the bottom for sufficient aeration to prevent fungal attacks. Aeration is also important to prevent fruit deformation due to overheating. Because thrips do not like sunlight, transparent bags are less liked by thrips than white opaque or dark blue-colored bags.

Plastic stem bags treated with chlorpyrifos (aka Dursban) coextruded at a 1% (w/w) concentration were developed by Dow Chemical and first introduced in 1965. These were known for years as Poly-D bags. Placing a perforated and chlorpyrifos pesticide impregnated polyethylene stem bag over an emerging stem protected the fruit effectively against thrips. The thrips which are already present under the bracts of the just emerging bud within the pseudostem are killed after placing the impregnated bag. The control is not by contact with the impregnated plastic but by the volatility of the active ingredient into the air enclosed by the bag. Twice weekly cycles of early bagging with this bag are required to 'catch' all the buds in a young stage on time and prevent insect damage. This treatment gave good control of Red Rust in conventional for-export banana cultivation and no resistant thrips population developed over all those decades of use. As far as the author is aware, never has a heavy thrips infestation required a supplemental application of insecticide during the time when chlorpyrifos-treated bags were used. Hence, chlorpyriphos-treated stem bags were a very effective method of thrips and insect control in banana production. However, chlorpyrifos is no longer used for fruit to be sent to the European Union as the MRL for imports were reduced to 0.01 ppm (since October 2020) which is unachievably low. Neither is chlorpyrifos anymore used on fruit to be sent to the US market because it lost its tolerance for that market (effective February 2022). As a result, chlorpyrifos is no longer used in banana plantations producing for the commercial export trade leading to a resurgence of Red Rust thrips problems.

In 2003, the company Chemplast together with FMC developed coextruded perforated bags with the insecticide Bifenthrin at a 0.1% (w/w) concentration (called 'Biflex'). Still later (in 2012) coextruded bags with buprofezin 2.0% (w/w) were developed (called 'Banaflex 2.0'). Later that year a coextruded plastic bag with a combination of buprofezin 2.0% (w/w) and bifentrin 0.1% (w/w) was developed and called 'Banaflex 2.1'. The following year imidacloprid at 0.2% (w/w) coextruded bag was introduced by Chemplast (called 'JadeBag' or 'Gardflex 0.2'). In 2013, a coextruded bag combining imidacloprid at 0.2% (w/w) and bifentrin at 0.1 % (w/w) and called 'Gardflex 0.3' was introduced by Chemplast. All these bags were developed primarily to control mealybugs (Cubillo, 2013). However, they do not control other insect pests as well as chlorpyrifos (Bolaños, 2011), probably because the active ingredients are not that volatile to give good efficacy. Fipronil (aka Regent) impregnated stem bags could be an option (Arias et al., 2000). Since 2020 pyriproxyfen (or Epingle by Valent) is an approved compound and this somewhat more effective synthetic chemical is coextruded at 0.3% (w/w) into the plastic stem bags (called 'Total Flex' by Chemplast). Buprofezin use has been removed from use in bananas since August 2019 because the MRL in the European Union was lowered to 0.01 ppm, which is unachievable and thus an unworkable low level to be able to comply with.

Insecticide-treated bags after perforation should be immediately stored in a tightly wrapped, thick, closed plastic bag and packed in a sealed bag in a cool and dark place to prevent the active ingredient from dissipating and be stored in a shaded closed warehouse. They should be handled on a first in, first out, storage rotation. No bags older that 8 weeks of having been extruded should be used. Bags should be placed on the bunches the same day a sealed package is opened and unused bags properly stored and resealed. Apart from impregnated bags which cover the whole bunch, the use of bow ties (aka 'bufandas') and ties (aka 'corbatas') which are impregnated plastic strips is also common, and these are used inside and under a non-treated bag or as an additional 'booster' several weeks after the initial bagging round done at emergence of the bunch.

Insecticide-impregnated plastic bags mechanically and chemically protect, repel, and/or kill various 'mealybugs' (like *Pseudococcus elisae, Dysmicocus, Ferrisia*, and *Planococcus* spp.), and several members of the 'Armored Scale' family *Diaspididae* and several other important peel-blemishing insects, e.g. 'Carate' caterpillar (*Platynota* sp.), 'Congo' bee (*Trigona* sp.), and 'Colaspis' bug (*Colaspis* sp.), provided perforations that are small enough (pinhole). Plastic bags in general also diminish bird and bat scratches and other animal damages significantly. Fungal infestations of 'Speckling' (*Deightoniella torulosa*) is also diminished if the plastic bag is well enough perforated with at least 5% open area. During the bagging of the fruit, the male bud is removed, and one finger

Figure 19 Stem bag placed early over emerging bunch to protect against red rust thrips and other insects (Cubillo, 2013).

of the last hand is left as the lowest point in order to avoid rotting of the stalk. The bag promotes better growth of the banana bunch reaching a higher calibration faster because of the greenhouse effect of the bag which elevates the temperature around the bunch. A colored ribbon is attached for age identification. Placing plastic bags with or without impregnated insecticide(s), depending on the presence of the different pests, is highly cost effective. The stem bag also protects the banana bunch from receiving fungicide aerial sprays which are frequently employed to control the leaf fungal pathogen Black Sigatoka (*Pseudocercospora fijiensis*) (Fig. 19).

9 Integrated pest management in organic banana production

Without the use of a stem bag with an effective coextruded synthetic insecticide incorporated, which is prohibited in organic agriculture, the control of thrips has proven to be a huge challenge for the organic banana growers in Ecuador and Peru especially and anywhere else where organic bananas are grown. Unfortunately, the development of impregnated plastic stem bags by means of co-extruding an organically approved compound (read botanical) has not been successful as the extrusion temperature is higher than the temperature at which the compounds remain stable. Soaking plastic stem bags in broths of botanicals is an option but is cumbersome and worker safety and environmental safety issues come into play. Only botanicals with a sufficiently volatile component which stays at a high enough concentration within the bag will repel thrips.

Non-woven textile-type mechanical barrier bags (aka Agriban), which are sufficiently densely and evenly woven in order to not let thrips through and tightly closed at the bottom, have also been tried for thrips control. But under heavy infestation, these bags do not work as many thrips are already inside the male bud before the bag can be placed. These cloth-like bags (made from spun polypropylene) can also be soaked in a broth of botanicals. These bags are normally three times more expensive than regular polyethylene-extruded plastic bags. They can however be reused three times if carefully removed before processing. Unfortunately, mealybugs like the enclosed environment within this bag and often populations of these insects explode. Also, under windy conditions, these cloth type bags rub the fruit causing severe damage.

To control Red Rust thrips in organic farms, a combination of procedures needs to be deployed to prevent, delay, and control outbreaks to avoid heavy economic losses. Control procedures should be directed at both the soil-dwelling pupal stage and the adults and larvae on the fruit and pseudostem (in the case of *C. signipennis*). These control procedures should be undertaken all at the same time because the life cycle of thrips is short and continuous, and as a result several overlapping ('crossed') generations occur. Failure to control the pest at any stage will result in continuous reinfestation. *C. signipennis* tends to be more difficult to control than *C. orchidii*, due to the former's ability to reproduce sexually.

9.1 Minerals and botanicals

Elemental sulfur (like Sulfocal) topically applied as a dust to stem-bags (made from waxed paper at that time) was used in the old days in Australia, but this never became a common practice because the putrefying sulfur smell was unbearable. Coextruded sulfur containing polyethylene bags have been successfully produced more recently in Central America but have the same challenge as mentioned previously with the strong smell and worker safety issues. Potassium salts of fatty acids (like Safer Insecticidal Soap, M-Pede, Protec-K) have given some level of control when used as a high-volume spray mixture applied to the pseudostems. These sprays can prove to be phytotoxic to the fruit and thus fruit should not be hit by any spray. Mineral Oil can be used to control pests in general causing death by suffocation but again, these sprays should not be in contact with the banana fruit, thus only be used for pseudostem sprays. There are various 'home remedies' with often questionable control efficacies or at least difficult to repeat every time. Properly designed and replicated independent tests are required to establish the efficacy of such products and procedures.

Reasonably good control has been obtained using pyrethrum extracts called pyrethrins. Pyrethrins are naturally occurring compounds (cinerins, jasmonins, and pyrethrins) with insecticidal properties (Hara et al., 2002b) extracted from certain pyrethrum flowers (*Chrysanthemum cinerariaefolium* or its synonym *Tanacetum cinerariifolium*). Frequent applications - at least at weekly intervals - using the pyrethrum-based natural insecticide formulations (Phyri Plus, Santemo Plus, PyGenic, etc.) is reasonably effective and one of the few treatments used today in commercial organic banana farms in Ecuador and Peru. Pyrethrins only acts on contact and as repellent and should be applied using high total spray volumes. When sprays are directed to the fruit, it can cause fruit burn. Synthetically manufactured chemical insecticides called Pyrethroids (e.g. Decis) are similar in chemical structure to the pyrethrins but are not allowed in organic agriculture. They are synthetic and more stable and toxic to insects as well as to mammals, and last longer in the environment than the natural pyrethrins. Unfortunately, often the natural pyrethrins commercially available are laced with pyrethroids by unscrupulous local formulators which puts exported fruit in jeopardy without the intention of the organic farmer or export company.

Rotenone (as well as its derivative deguelin) is a naturally occurring insecticide also known as Derris (or Barbasco in Spanish). Several rotenone formulations existed in the market (like Atoxin and Rotebiol in Peru, and Root Plus and Rotobiol in Ecuador). They were reasonably effective and were extensively used. However, rotenone is no longer allowed to be used on any crop produced for the US market since 2012 as the exemption of tolerances were revoked (EPA, Docket No. EPA-HQOPP-2012-0171) as a logical procedural sequence after all uses within the USA were revoked in 2010. This stemmed from the fact that all US registrations for agricultural uses of Rotenone (fruits and vegetables) were 'voluntarily' canceled in 2007, as no registrant could afford to support a new registration package concerning this botanical with no patentable active ingredient thus significantly reducing return on investment for manufactures. The organic banana growers objected to no avail. Rotenone has now been added to National List §205.602 as a Prohibited Natural Substance. The European Union legislation on Organic Agriculture still specifically lists rotenone in their regulation (see EC 889-2008 Annex II.1) as an allowed input.

In Europe, rotenone was listed as a pesticide with an MRL at a default level of 0.01 ppm for bananas. Although rotenone is highly toxic to fish, practice had shown that that can be managed and avoided and all commercial labels of rotenone that were used had clear and strong warnings in this regard. The human health toxicity profile of rotenone indicates that it has a high acute toxicity for oral and inhalation exposure and very low toxicity for dermal exposure. However, commercial products of rotenone that were available in agriculture (usually 0.5–1% a.i.) were considered to have very low toxicity to humans for all exposure categories. Nevertheless, the new regulation EU 2021-1165 now also prohibits the use of rotenone in organic agriculture.

Nicotine (from *Nicotiana tabacum*) was a very effective insecticide in the past, including for thrips control. Like what occurred with rotenone recently, nicotine is since many years no longer allowed to be used in organic (or conventional) plantations with export destined to the USA, Canada, or Japan. The European Union legislation on Organic Agriculture still specifically listed nicotine in their Regulation (See EC 889-2008 Annex II.1) as an allowed input. As plantations are normally not producing for the European market only, nicotine as a control compound was not an option unless farms produce exclusively for the European market like is the case mostly in the Dominican Republic. Note should be taken that nicotine is also very toxic to beneficial insects. Since the more recent legislation (EU 2121-1165) and its implementation nicotine is neither allowed anymore in the organic agriculture for export. European and US legislators want to keep each other in check and do not want to lose reciprocal entrance to each other's markets.

The extract of seeds from the Neem tree (*Azadirachta indica*) contains various terpene compounds that are called azadirachtins. These are known to be efficacious for the control of several insects. Nevertheless, the reality is that efficacies of various neem-based extract(s) are often overrated as azadirachtin does not kill insects but acts more as an insect repellent. The resulting neem cake which is left over after extraction still contains some azadirachtins and can be used to suppress thrips (and nematodes at the same time) when applied around each banana mat to control the pupae stage of the thrips in the soil. Cake of the castor bean (*Ricinus communis*), also known as Higuerilla in Latin America, where it is abundantly and cheaply available, can be employed in the same manner.

Other botanicals have been or are still being used or are being experimented with to determine efficacy. To mention a few, extracts derived from the plant sabadilla (*Schoenocaulon officinale*) with veratrine as the active ingredient is a natural botanical insecticide (commercial names e.g. Veratran D from MSG). In the Dominican Republic, an extract from the shrub *Petiveria alliacea* (aka Zorrillo or Anamu) is believed to work for the control of Red Rust thrips (where it is called 'Oxido Rojo') although no evidence has ever been published. Extracts made from garlic, cinnamon, hot chili (containing capsaicin), mustard oil (containing isothiocyanate), cedar oil (from *Juniperus virginiana*), quillaja (*Quillaja saponaria*), quassia (*Quassia amara*, aka 'Hombre Grande', and commercially as Bioquamar marketed in Peru), and epasote (*Dysphania ambrosioides*, aka paico), and 'Madero Negro' (*Gliricidia sepium*) mostly have low efficacy. The legal framework for the use of all the abovementioned botanicals in organic banana plantations needs to be well investigated before using any of them. Even when dealing with botanicals, the worker safety aspects should not be ignored or taken lightly.

As with any control agent, but especially with 'home-brewed' remedies from botanicals, it is important to determine its efficacy first of all, as out of pure desperation, organic banana growers will often buy and apply any 'snake oil' being offered in the market. Officially all commercially sold crop protectant products need to be registered for use in agriculture for its specific pest, crop, and use type. The formulation, concentration, dose range, application volume, and frequency need to be clearly indicated. Phytotoxicity, especially when the sprays are directed directly to the fruit, will depend on these variables. It is also of utmost importance that the active ingredient (even if botanical) is allowed to be used in conventional agriculture first of all and then for organic agriculture following the North American, European, Japanese, and Korean conventional and organic legislation in order to be able to export from the same plantation to all these markets. It should certainly not be assumed that insecticide formulations based on botanicals are allowed or exempt from regulation with regard to agrochemicals and food safety for both conventional or organic agriculture. The additive piperonyl butoxide is often added to pesticide formulations for enhancing efficacy but this compound is not allowed in organic agriculture. Locally manufactured formulations of any of the botanicals may (by scrupulous 'design') contain 'contaminants' of synthetic insecticides which puts the integrity of the organic bananas at risk. Any commercial pesticide should be

Figure 20 Spray runoff toward the lowest part of the hanging fruit may cause phytotoxicity burn damage.

registered at the Ministry of Agriculture of any of the countries of operation. To duly comply with Global Gap certification, an invoice for any pesticide purchase needs to be on file for inspection. Spray runoff toward the lowest part of the developing fruit at the moment of application of several products may cause phytotoxicity (see Fig. 20) and show up several days later as dark circles not to be confused with Red Rust damage.

9.2 Naturals from fermentation

In Queensland, Australia, effective control of Red Rust thrips was obtained and reported as early as the turn of the century (Pinese and Elder, 2000) when using a novel insecticide at the time called Spinosad, which was developed and marketed by Dow AgroSciences. Applications using very low concentrations applied to the pseudostem ensuring coverage up to run-off proved to be very effective in control. Spinosad was also reported in Hawaii as having good to excellent efficacy for 'Banana Thrips' control (Kawate, 2006) and it obtained a label for that specific use in 2006. This insecticide is a mixture of two similar molecules: named spinosyn A and spinosyn D. This insecticide belongs to a new group called the Naturalytes. They are produced by aerobic fermentation of the bacterium *Sacharopolyspora spinosa*. It has high efficacy against various insect groups including thrips but low toxicity to beneficial coleoptera insects (like *Orius*). In general, it is considered to have a low risk to the environment, mammals, applicators, and consumers. Fortunately, for the organic banana farmers, there now is an organically approved formulation of Spinosad called Entrust formulated as a suspension concentrate at 240 g a.i. per liter (also from Dow AgroSciences). Based on the abovementioned references from Australia and Hawaii, it was expected that Entrust would work well for the control of Red Rust thrips in organic banana plantations in Ecuador and Peru where control of this pest was difficult in the organic banana industry as of 2010. This proved to be correct, and the product was registered in Ecuador in 2014 based on the SENASA protocolled study presented by Aycart (2013). Spinosad is active at very low dosages, and it was found to be highly efficacious at 0.08 mL/L of the commercial product Entrust 240 SC. With 48 mL of the commercial product Entrust 240 SC in a total volume of 600 L, all banana plants (spent-stem, mother plant, and followers) in 1 ha could be sufficiently drenched for effective control of the pest. A knapsack (CP-3) sprayer with an extended lance is used for these applications, as especially the throat and any emerging bunch need to be drenched. These applications can be complemented with sprays using a hand-operated pressure pump sprayer directly to the bunch at 2, 4, and 6 weeks after bunch emergence. An army of workers is needed to make all these applications week after week when infestation is high (see Fig. 21).

Figure 21 Knapsack sprayers (left) with an extended lance to spray the banana plants throat and the emerging bunch and the pseudostem and hand-operated pressure pump sprayer ('Star' brand) on the right.

Spinosad's re-entry period of 4 h and pre-harvest period of 24 h are manageable. Because of the fact that insect's resistance to spinosad has been detected in other crops, the number of applications needs to be controlled. Several commercial formulations for conventional agriculture exist in the market, the most common ones being Success, Tracer, and Spintor. These much cheaper formulations of spinosad developed for the conventional market cannot be used in an organic banana plantation because, although spinosad is considered a natural pesticide, the synthetic-based adjuvant(s) are not on the correct EPA exempt list to consider them acceptable for use in organic agriculture according to the USDA-NOP regulations.

For the control of pupae in the growing media of potted flowers affected by western Flower Thrips (*Frankliniella occidentalis*) in greenhouses, the chemical company BASF has developed a thrips control method using a proprietary formulation (aka Nemasys) of the naturally occurring insect-parasitic beneficial nematode called *Steinernema feltiae* reared on a massive scale. These nematodes are released in their infective juvenile stage. Once inside the pupae of the thrips, the nematodes release symbiotic bacteria that quickly kill the thrips in the pupae stage. It is reported that reproduction inside the insect releases new generations of infective juvenile individuals of this entomopathogenic nematode that disperse in search of further prey. Another highly effective biopesticide (aka Grandevo) based on the use of the naturally occurring bacterium *Chromobacterium subtsugae* strain PRAA4-1T, developed by Marrone Bio Innovations, provides long-lasting, broad-spectrum control as it is toxic to the adult and immature thrips when ingested. Both these products are approved for organic agriculture in general according to USDA-NOP rules. If these products also work to control the banana thrips effectively and economically is not known, but the organic banana industry may want to investigate the efficacy in their plantations and consider registration for use in their industry.

10 Integrated pest management: chemical control

In Australia before the Second World War, nicotine, pyrethrum and rotenone (derris), dusts, and sprays were used to lower the populations of Red Rust thrips in the banana industry (Braithwaite, 1965). In the 1960s directed ground sprays and even aerial sprays were used to control thrips using DDT, HCH, and dieldrin (Nickel, 1961; Anonymous, 1970). In Ecuador, dieldrin sprays were recommended at 2-to-3-week intervals for the banana bunch (Vilardebo, 1960). Especially the chlorinated hydrocarbon pesticides are very residual chemicals and they also produced serious biological imbalances and provoked outbreaks of many lepidopterous banana insect pests. These chemicals were phased out over the decades following their use. Later malathion (Champion, 1968), dichlorvos (Robinson and de Villiers, 2007), diazinon (aka Basudin), and carbaryl (aka Sevin) were used (Pardo, 1983).

Because the pupae of these thrips dwell in the soil and pupate near the base of the mat, the control procedures using various chemical nematicides in rotation over the year and directed around the base of the mat against nematodes (like *Rodopholus*, *Meloidogyne* and *Pratylenchus*), and the banana weevil borer (*Cosmopolites sordidus*), control the Red Rust thrips population in conventional plantations as a 'collateral benefit'. When aldicarb (aka Temik) was still allowed in bananas as a nematicide, it indirectly controlled thrips. The same may be true with cadusafos (aka Rugby), terbufos (aka Counter), and ethophos (aka Mocap) applied around the mat and perhaps with oxamil (aka Vydate) injected into the pseudostem. The application of the nematicide Nemagon (1,2-dibromo-3-chloropropane) in the late 1970s and early 1980s had an effect on the Red Rust thrips pupae as well.

Pinese and Piper (1994) reported that 'butt' (the exposed bullhead part of the base of the mat) treatment with insecticides used for weevil control, which is done in Australia using prothiofos (aka Tokuthion) or chlorpyrifos (aka Lorsban), was very effective to control thrip pupae in the soil. They do emphasize though that when infestation levels are high, the bunch and pseudostem need to be treated too and at least every 14 days. Soil applications with fipronil (aka Regent) and bifenthrin and also chlorpyrifos sprayed to the banana bunch were also recommended (Pinese and Elder, 2004). Diazinon was also indicated as an effective treatment for controlling banana thrips but has for phytotoxicity to fruit (Constantinides and McHugh, 2003; Pinese and Piper, 1994). These authors also mentioned that imidacloprid (Provado or Admire), dinotefuran, and flonicamid were effective to control thrips in general while bifenthrin impregnated bags were also used.

A single application (18 g/m of banana row) at planting of a controlled release imidacloprid (Confidor Guard) granular formulation 50 g/kg called suSScon Yellow Intel (meaning Sustained Control) has given extended control up to the second ratoon (so nearly three years) of Red Rust thrips in North Queensland (Anonymus/Crop Care/Bayer, 2015). This treatment also gave good control of the banana weevil borer. In Australia, another method of controlling Red Rust and weevil borers is by injecting the pseudostem with an imidacloprid (Confidor) and spirotetramat (Movento) combo (called Movento Energy).

Again, in Australia the systemic insecticide clothianidin (aka Shield), a nicotinoid developed by Sumitomo Chemical, is used for control of Red Rust thrips and provides long-lasting control. This chemical in a concentrated form is either injected into the base of the mat or sprayed as a solution directed to the bottom part of the soon to shoot but still green or cleaned pseudostem base. This chemical quickly penetrates the plant ready for ingestion by thrips. At the same time, these control procedures control weevil borer larvae.

10.1 Bud Injections

Because the thrips are already in the bud even before it emerges from the throat of the banana plant, a very early injection into the bud can be used to control thrips. This procedure is common in commercial banana plantations focused on export in Asia. Only with very early bud injections (often referred to as SF-101 in Davao, The Philippines) using 60 mL volume of an insecticide application of imidacloprid (aka Confidor), thiamethoxam (aka Actara), or spinosad (aka Success) can this damage be prevented in conventionally grown bananas. Three rounds per week through the plantation are needed to 'catch' all the newly emerging buds in time. These different modes of action insecticides are to be rotated to prevent resistance development in the pest populations. Instead of bud injection which is cumbersome and can cause fruit deformation and phytotoxicity, another method is being developed by the banana industry in Davao by covering the emerging still upright bud with a 'condom-like' polyethylene plastic bag into which a volatile formulation of an insecticide has been sprayed before placement (see Fig. 22). This bag falls to the ground as the bell develops and turns over pointing toward the ground, or it needs to be removed by hand, to let the bunch develop without the physical restriction by this bag. This method may also work against Red Rust thrips with synthetic insecticides or even botanicals in Latin America.

In India, bud injection with Azadirachtin and imidacloprid showed to be equally effective (Bisane, 2018). The bud injection is made at the moment of

Figure 22 Covering the emerging still-upright bud with a polyethylene plastic bag sprayed with an insecticide.

male-bud emergence when the flower is in an upright position and injected at a 30° angle leaving one-fourth length of the bud from the top in order not to injure the developing banana hands inside. In Davao, iPhilippines, a trial was conducted (Celiz and Ubaub, 2018) which showed good effects of a natural extract from Matrine (*Sophora flavescens*), containing oxymatrine (aka Levo), against flower thrips (*Thrips hawaiiensis*) bud-injecting commercial bananas. These two control options may also be effective for the Red Rust thrips.

How different irrigation methods may influence thrips populations has not been properly investigated as far as the author is aware, but it is hard to get to a comparison setup. Flood irrigation could have an impact on the soil dwelling nymphs and pupae, while sprinkler irrigation may have some influence on the nymphs and adults that crawl over the pseudostems. Drip irrigation would have the least of the abovementioned effects.

Red Rust thrips is a difficult 'plague' to control but 'Corky Scab' thrips (caused by *Thrips hawaiiensis*), which is a problem in Davao, and 'Silvering Thrips' (caused by *Hercinothrips bicinctus*), which is problem in Queensland, or the thrips *H. femoralis,* which is a problem in the Canary Islands, are exotic to Latin America and their introduction would mean huge additional complications in producing clean fruit for the mandating consumers. Hence, biosecurity measures and intercontinental movement of plant material needs to carefully be regulated and monitored to prevent this.

11 Summary

Pest problems in many crops, including bananas, are often caused by cultivation of the crop in monoculture in large plantations. In the case of Nueva Esperanza, which was one of the first organic banana farms to be hit hard by the Red Rust thrips infestation, cultivation of bananas in a plantation system surrounded by natural habitat and away from other plantations provided a suitable environment for an epidemic to occur despite being organic and in a high biodiverse area.

As far as the author is aware, never has a heavy thrips infestation required a supplemental application when chlorpyrifos-treated bags were used in conventional farms. But, now that chlorpyrifos is no longer 'allowed' for European or US markets and thus no longer used in the commercial banana export trade to those markets, suddenly, Red Rust thrips are becoming a problem again and soon may pop up in several countries as a quality problem and a cause of rejects.

In organic banana farms, the tremendous problem this pest represented during 2010–2014 is not forgotten but the registration of the very effective natural insecticide called spinosad has provided an effective control option. However, care must be taken in its deployment as the risk of resistance build up in the thrips population to this product is real with few effective alternatives in sight. To date, no control failure due to resistance of the pest to spinosad has been reported from Latin American banana production areas. Some of the suggestions made in this chapter should incentivize further research on other control options.

12 About the author

The author graduated from Tropical Agricultural College in Deventer, the Netherlands. He has worked in the Latin American banana industry since 1978 in research and development positions of which 33 years with Dole Fresh Fruit International, mainly in the field of pest and disease control. Wanting to change the industry, he pioneered Dole's Organic program in 1995 and with his team made Dole today's leader in organic banana (and organic pineapple) production and marketing. He held the position of Dole's director of research at their headquarters in San José Costa Rica. He wrote Chapter 15 of Volume 1 on 'Organic Banana Cultivation' in this same series of books in agricultural science geared toward achieving sustainability. He is now semi-retired and an independent consultant. He is also a co-owner and the director of a materials input company (TRANSASTRA) supplying organic banana farmers.

13 References

Anonymous (1970) U.F.Co. Operation Manual p VII, pp. 12–15.

Anonymous (2015) Banana thrips in the West Indies. Cirad: Institut Technique Tropical, p. 3.

Anonymous and Bayer Crop, C. (2015) New move on the old guard. *Australian Bananas Magazine* 45: 43.

Arias, M., Corozo, R. E. and Vera, E. T. (2013) Manejo integrado del trips de la mancha roja en plantaciones de banano en Ecuador – Avances sobre la biología. *INIAP Ecuador*. Presentacion Reunion TASTE, Sullana, Peru.

Arias, M., Jines, A. and Valle, D. (2000) Evaluación de Fipronyl Impregnado en Fundas Plásticas y Cortadas de Alta Densidad para el control de Thrips de la Flor del Banano. In: *Proceedings of the ACORBAT conferences*, San Juan, PR.

Arias de Lopez, M., Delgado, R., Rengifo, D., Suárez, P., Carlos Rojas, J., Clercx, L., van Tol, R. and Staver, C. (2020a) Fontagro funded project FTG/RF-1332-RG, management of red rust thrips on bananas, strengthening smallholder organic banana growers. *YouTube*. Available at: https://www.youtube.com/watch?v=H9Rf57sWNLQ.

Arias de Lopez, M., Corozo-Ayovi, R. E., Delgado, R., Osorio, B., Moyón, D., Rengifo, D., Suárez, P., Paulino, A., Medrano, S., Sanchez, L., Rojas, J. C., Vegas, U., Alburqueque, D., Staver, C., van Tol, R. and Clercx, L. (2020b) Red rust thrips in smallholder organic export banana in Latin America and the Caribbean: Pathways for control, compatible with organic certification. *Acta Horticulturae* 1272(1272): 153–161. DOI: 10.17660/ActaHortic.2020.1272.19.

Aycart (2013) *Project 03-01-031-0021. Eficacy Trial Protocol for Entrust™ 240 (Spinosad; 240 g i.a./l) for Chaetanaphothrips Signipennis Control in Banana for AGROCALIDAD Registration. DOLE Nueva Esperanza Farm*.

Bisane (2018) Technique for management of banana Red Rust thrips, *Chaetanophothrips signipennis* (Bagnall). *Journal of Entomology and Zoology Studies* 6(5): 1964–1967.

Bolaños, E. (2011). Manual de Buenas Prácticas Agrícolas en el Cultivo de Banano. In: *CORBANA*, pp. 17–23.

Braithwaite, B. M. (1965) Control of banana rust thrips. *Agricultural Gazette* 76: 115–119.

Buss, E. (2006) Thrips on ornamental plants. *Univ. of Florida. IFAS Extension*, pp. 1-6. Document ENY-333.

Campbell, S. J. and Williams, W. T. (1978) Mineral relationships in "maturity bronzing" of banana fruit. *Australian Journal of Experimental Agriculture* 18(95): 603–608.

Castro, C. and Jair, A. (2013) Manejo Integrado del Thrips de la Mancha Roja en Banano Orgánico y Necesidades de Investigacion. Dole´s Research Presentado en Sullana, Perú. *Marzo* 13, durante el Encuentro Internacional sobre el Thrips de la Mancha Roja.

Celiz, J. R. and Ubaub, L. T. (2018) Insecticidal effects of Matrine against flower thrips, Thrips hawaiiensis Morgan on ´Cavendish´ banana. *Journal of South Pacific Agriculture* 21: 9–17.

Champion, J. (1968) El Platano. Tecnicas agricolas y producciones tropicales. *Editorial Blume*, pp. 146–147.

Clercx, L. (2013) *Encuentro Internacional sobre el thrips de la mancha roja (Vols. 12–14) de marzo*. Casa Club, Sullana, Piura, Peru. TASTE communication. 4 páginas.

Clercx, L., Arias Zambrano, M., Dulanto Bejarano, J. and Flores Espinoza, B. (2015) Towards biological control of Red Rust banana thrips in organic and conventional banana. *Acta Horticulturae* 1105(1105): 73–80. DOI: 10.17660/ActaHortic.2015.1105.11.

Constantinides, L. N. and McHugh, J. J. (2003) Pest management strategic plan for banana production in Hawaii. University of Hawaii at Manoa, Honolulu, pp. 15–17. Workshop Summary.

Cubillo, D. (2013) Manejo integrado de insectos plagas en el cultivo de banano-Conceptos y Aplicaciones. In: *Chemplast. confederación unitaria de trabajadores*, San Jose, pp. 36–38.

Diaz, A., Cayón, G. and Jairo, J. (2007) Metabolismo del Calcio y su relación con la "mancha madurez" del fruto de banano. Una revisión. *Agronomía Colombiana* 25(2): 280–287.

Gutierrez, P. (2015) *DOLE Presentation at the VI Cumbre Mundial de Bananos, Guayequil, Ecuador*.

Haarer, A. E. (1964) Modern banana production. *Leonard Hill*, p. 114.

Hara, A. H., Mau, R. F. L., Heu, R., Jacobsen, C. and Niino-DuPonte, R. (2002a) *Banana Rust Thrips Damage to Banana and Ornamentals in Hawaii*. Insect Pests Department of Plant and Environmental Protection Sciences, Hawaii Department of Agriculture, Honolulu, HI, CTAHR, I. P.-10 Publication.

Hara, A. H., Jacobsen, C. and Niino-DuPonte, R. (2002b) *Anthurium Thrips Damage to Ornamentals in Hawaii.* IP-9 - CTAHR Coop. Ext. *Serv*. University Hawaii, pp. 1–4.

Hara, A. H., Jacobsen, C. and Niino-DuPonte, R. (2012) *Chaetanaphothrips Orchidii. Crop Knowledge Master*. University of Hawaii, pp. 1–7.

Hood, J. D. (1954) A new Chaetanophothrips from Formosa, with a not on the banana thrips. *Proceedings of the Biological Society of Washington* 57: 215–218.

Kawate, M. (2006) *Banana Pesticide Update Hawaii Banana Industry Ass. Conference*. Available at: http:/www2.hawaii.edu. 3 pages.

Lara, F. E. (1970) *Problemas y Procedimientos bananeros en la Zona Atlánticade Costa Rica*, p. 62–66.

Nickel, J. L. (1961) Control of Red Rust thrips on bananas in Honduras. *Journal of Economic Entomology* 54(6): 1071–1074.

Nogueroles, C., Lopez-Cepero, J. and Rodriguez, M. (2015) Cultivo Ecológico de la Platanera. *COPLACA – SEAE. Pagina* 75: 76.

Okonji, A. (2020) *New Banana Disease Threatens Uganda and Rwanda Published on ASARECA. 4 Pages*. Available at: https://www.asareca.org.

Ostmark, H. E. (1974) Economic insect pests of banana production. *Annual Review of Entomology* 19: 168.

Ostmark, H. E. (1976) Biology of the banana red rust thrips (*Chaetanaphothrips signipennis* (Bagnall). *Internal Report of the Tropical Research Division of United Brands*.

Padmanaban, B. and Mustaffa, M. M. (2010) Integrated pest management of banana and Plantain. Tech. Bull. No. 09. *Pub*. ICAR-National Research Centre for Banana, Tiruchirappalli (T. N.), pp. 8–9.

Pardo, J. (1983) El cultivo del Banano. *Editorial Universidad Estatal a Distancia Editorial EUNED*, p. 33.

Perera Gonzalez, S., Molina, L. and Jesus, M. (2010) Plagas y Enfermedades de la Platanera Canarias y Control Integrado. *COPLACA*, p. 63.

Pinese, B. and Elder, R. (2000) Biology and IPM control options for banana rust thrips. *Queensland Horticultural Institute, Project HRDC, FR* 96023, p. 31.

Pinese, B. and Elder, R. (2004) Banana rust thrips in bananas. DPI&F. *Horticulture*, p. 3.

Pinese, B. and Piper, R. (1994) Bananas, insect & mite management. *Queensland DPI*, pp. 13–17.

Roberts, F. S. (1958) Insects affecting banana production in Central America. *Proceedings of the International Con*gress 3: 411–415.

Robinson, J. C. and de Villiers, E. A. (2007) *The Cultivation of Banana*. ARC - Institute for Tropical Crops, South Africa, pp. 215–219.

Sakimura, K. (1975) *Danothrips trifasciatus*, new species, and collection notes on Hawaiian species Danothrips (Thysanoptera: Thripidae). *Proceedings of the Hawaiian Entomological Society Volume* 22: 125–132.

Schotman, C. Y. L. (1989) *Plant Pest of Economic Importance in the Region Covered by the Caribbean Plant Protection Commission. Proveg (RLAC/90/03)*, pp. 1–190.

Simmonds, N. W. (1959) *Bananas*. Longmans, Green and Co. Ltd., London, p. 466.

Simmonds, N. W. and Stover, R. H. (1987) *Bananas* (3rd edn.). Longmans, Green and Co. Ltd, London, pp. 330–334.

Stephens, C. (2008) *Outline of History of the Province of Bocas del Toro, Panama. Complete English and Spanish Text. SPS Publications Florida*, p. 12. Available at: www.spsbooks .com.

Stephens, C. (2012) Personal Communication.

Van Tol, R. (2013) Aplicación de semioquímicos para el control de thrips. Plant research International – Wageningen. In: *Presentación reunión TASTE*. Sullana, Perú.

Vilardebo, A. (1960) *IRFA Ecuador Mission Report*.

Wielemaker, F. (1980) Monograph. The banana red rust thrips. *BANDECO*, p. 14.

Williams, M. H., Vesk, M. and Mullins, M. G. (1990) Development of banana fruit and occurrence of the maturity bronzing disorder. *Annals of Botany* 65(1): 9–19.

Chapter 20

Managing banana bunch pests: towards more ecological approaches

Charles Staver, Agricultural Scientist, USA; Luud Clercx, AgroFair, The Netherlands; Myriam Arias, INIAP, Ecuador; César Guillén Sanchez, University of Costa Rica, Costa Rica; Lieselot van der Veken, Pro Terra Agro, Belgium; and Marco Antonio Oviedo Cajas, Banana Specialist, Ecuador

1 Introduction: bananas attractive to consumers but not insect pests

The banana flower and tender new fingers (which will form the final fruit) attract insects and other fauna from the moment the bud emerges from the leaf whorl. Feeding damage to the banana peel causing cosmetic damage, or even the presence of a pest, results in the rejection of fingers, bunches, boxes or even containers for export. The international supply chain requires blemish-free bananas and has low or zero tolerance for the reddish-brown discolouration of banana skin caused by red rust thrips (*Chaetanaphothrips* (also written

http://dx.doi.org/10.19103/AS.2022.0108.20

Chaetanophothrips) *signipennis* and *C. orchidii*), raised pockmarks caused by oviposition by *Frankliniella parvula*, fruit scarring caused by *Colaspis* spp. and sooty mould associated with honeydew generated by diverse species of aphids, whiteflies and mealybugs (see Figs. 5, 7, 9, 13 and 15). Blemished fruit can end up in national and local markets which have typically lower quality requirements. Other insect pests, not necessarily threatening to banana appearance, but considered a quarantine risk for other crops by national plant protection agencies, can also contaminate banana shipments and result in box or container rejection at the moment of detection. Examples include scale insects such as *Diaspis boisduvalii* and mealybugs of the family Pseudococcidae. While demanding blemish-free fruit, supply chain actors and consumers at the same time are increasingly concerned about how their food is produced, in particular the use of toxic pesticides and, more recently, the use of plastics in cultivation. These concerns also affect approved options for grower pest management.

This chapter first reviews trends in the management of banana bunch pests in the context of changing regulatory requirements and market trends. Six major insect pests of the banana bunch are then introduced with a particular focus on understanding their life cycle as a basis for designing more ecologically based pest management. Our objectives in the chapter are to propose management strategies that simultaneously reduce losses from bunch pests, minimize the impact of pest-control practices on human and environment health and are compatible with integrated production systems building on ecological processes.

2 Trends in bunch pest management

Export banana production in its current form developed with the substitution of Gros Michel cultivars by Cavendish cultivars from the 1950s onwards. In the more than 100 years that bananas have been an export commodity, they have become the most traded fruit and, in many countries, the most consumed fruit on a per capita basis. Yellow bananas free of blemishes at favourable prices have become the calling card of supermarkets and often figure prominently in fresh fruit and vegetable sections placed near store entrances. Behind this success are increasingly strict voluntary standards to minimize cosmetic damage to the peel which increase environmental, economic and social costs of production.

In this section, we review the progress made in bunch insect pest management from the first use of insecticides based on generalized spraying techniques through developments in pest management such as more targeted pesticide use, substitution and production system redesign (Bird et al., 1990; Kogan, 1998). This last redesign phase has also been referred to as ecologically based pest management (Altieri et al., 2012). We summarize the growing importance of certification and legal frameworks in determining permitted practices and the challenges linked to the use of plastic bagging in pest control.

The early failure of aerial and generalized broadcast insecticides in banana in the 1950s is documented by Stephens (1984). Such insecticide applications in Costa Rica and Panama reduced damage from the target pest, red rust thrips *Chaetanophothrips orchidii*, but created numerous secondary pest outbreaks, primarily of foliar feeding larvae, caused by the disruption of natural control mechanisms. This experience with banana became a classic example of a progressively worsening pesticide cycle in which initial pesticide use leads both to more pests and to more pesticide use with increasing negative human health and environmental impacts.

As Stephens (1984) describes, the large banana companies deployed entomologists to study the explosion of secondary pests in diverse production zones across Central America. Different insecticides were tested, and application frequency reached up to 12 sprays/year to control secondary pests such as defoliators resulting from the disruption by natural control through excessive pesticide use. By the early 1970s, all insecticide use was stopped and damage by defoliating insects declined. As an alternative, plastic bags were deployed to protect bunches from red rust thrips.

Unlike many other horticultural and field crops, the management of banana bunch pests did not move from regularly scheduled sprays to more selective, threshold-based applications guided by pest monitoring. In practice, this kind of rationalized pesticide use based on thresholds and monitoring the level of pest pressure could not easily be adapted to year-round banana production. This was partly due to zero tolerance to damage or the presence of pests such as scale insects and mealybugs in banana supply chains. Second, unlike most fruits and vegetables which are seasonal, most banana plantations at any moment during the year are characterized by bunches at different stages of growth, from flowering and early bunch formation to final finger filling and bunches ready for harvest. Each bunch age group has different pest management needs, which significantly complicates establishing and monitoring pest thresholds. Third, the zone of flower emergence and bunch formation is 2.5–4 m above ground, which complicates insect pest or damage monitoring, a standard requirement for more targeted pesticide use. A final issue is the time lag of 2–3 months between damage to small tender fingers and the moment when harvested fingers, hands, boxes or containers are inspected for damage at different moments of post-harvest quality control, in the packing shed, before shipment and upon arrival.

Bagging of bunches has become a routine, alternative preventative strategy in pest management (Arias de López et al., 2020). Bagging bunches originated in the 1930s as a way to protect the developing fruit from extremes of weather (Treverrow, 2003). Bags were initially made from burlap or hessian, but, by the early 1950s, plastic bags were being tested in Guatemala (Vezina, 2020). Stephens (1984) documents the routine use of plastic bags in the early 1960s and 1970s to protect bunches from red rust *orchidii* thrips, resulting in the elimination of insecticide use in bananas in the 1980s in some parts of Costa Rica.

While bagging alone has been documented to minimize rejected fingers, hands and bunches (Arias de López et al., 2020; Moscoso-Ramirez and Peña-Peña, 2020) and is common practice in organic banana production, the use of insecticide-impregnated bags has now become standard practice in conventional export banana plantations (Santosh et al., 2017). These bags are typically perforated to ensure adequate air circulation to reduce fungal and bacterial problems (Villalobos et al., 2017). In Australia and other countries in Asia, a small amount of insecticide is injected directly into the erect emerging flower or bell which has reduced the total insecticide use by 95% while offering targeted protection of the bunch (Department of Agriculture and Fisheries/Queensland Government, 2017).

The early attempts to protect the bunch from insect pests were led by entomologists working for banana companies. However, since the 1980s, international and national regulatory frameworks for pesticide use and voluntary certification frameworks have played an increasing role in affecting practices employed in banana plantations. Figure 1 provides a timeline showing how key stages in the development of banana pest management practices have interacted with developments in international and national regulations and voluntary certification frameworks. Future developments in bunch pest management will need to be compatible with these certification and regulatory frameworks.

The Codex Alimentarius was established in 1963 to protect consumers and ensure fair practices in international trade in food products. Since that point in time, restrictions on pesticide use have become increasingly stringent (Pretty and Hine, 2005; FAO and UNEP, 2020; UNEP, 2020a,b). Codex regulations have also changed in response to member country requests for greater regulation of pesticides and residue limits (FAO and WHO, 2021). In 1985, member countries of FAO and WHO adopted an International Code of Conduct on Pesticide Management, updated several times (most recently in 2014). The Code complements legally binding instruments such as the Rotterdam Convention (on the mandatory exchange of reliable information on pesticides), the Stockholm Convention (to protect human health and the environment from persistent pollutants) and the Basel Convention (on transboundary pesticide waste management), also updated several times in Conferences of Parties (COP) (FAO and WHO, 2014; FAO and UNEP, 2020a; UNEP, 2020b).

Theme	Prior to 1950	1950s	1960s	1970s	1980s	1990s	2000s	2010s	Present
Banana bunch technology	Gros Michel Whole bunches	Cavendish Boxed clusters Bagging			Improved plastics for more effective bunch protection				
Programmed pesticide use		General insecticide spraying introduced and abandoned				Generalized spaying organic insecticides			
IPM I: Rational pesticide use				Impregnated bags - chlorpyrifos	Insecticide-impregnated strips inside bags	Insecticide injection into upright bell			
IPM II: Substitution			Early bagging		Substitution of less toxic insecticides in impregnated bags	Bifenthrin Repellents and insecticides for certified organic production Sanitation practices to reduce pest pressure	bifenthrin - buprofezin		
IPM III: Ecosystem redesign						Release and/or spraying of beneficial organisms	Potential for habitat enrichment piloted		
Certification opportunities					Organic certification with regularly updated regulations	Rainforest certification of deforestation and environment Fair trade regulations prices and practices	GlobalGAP regulations for good practices for food safety		
National and international laws and regulations		Codex Alimentarius			Procedures for MRLs set up First International agreement on pesticides			European Green Deal	

Figure 1 Time line of changes in bunch and bunch insect pest management in export banana.

Countries in Europe and North America have implemented restrictions on pesticide use in agriculture in their territories and maximum residue limits (MRLs) allowed on imported food products such as bananas. In the European Union (EU) in 2019, 0.1% of 900 banana samples exceeded the MRL for pesticides, while 7% of pineapple samples were above MRL (Gutierrez, 2021). Although producing countries may have less strict controls on pesticide usage, export banana production practices must meet the standards of importing countries. The first common pesticide policy in 1979 by the EU later became the sustainable pesticide policy in 2009 (Skevas et al., 2013) and, most recently, the Farm to Fork Strategy framework as part of the European Green Deal which aims to create a sustainable and healthy food system. By 2030, the strategy aims to reduce the use of synthetic pesticides by 50%, with 25% of agricultural land under organic production. The ambitions of the Green Deal and the Farm to Fork Strategy go beyond the boundaries of the EU market. The EU aims to support a global transition to sustainable agri-food systems through its international cooperation and trade policies, through the creation of 'Green Alliances'.

More specifically, the EU trade policy is built on a strategy of securing ambitious commitments from third countries in the use of pesticides and the promotion of safer plant protection substances (European Commission, 2020a,b; SAPEA, 2020). Trends are likely to include future prohibition of practices such as the use of insecticide-impregnated bags to ensure blemish-free bananas. Although importing countries cannot legislate practices in banana fields in producing countries, MRLs which are applied to imported foods can be reduced per recent EU actions (EU Pesticide Database, 2022) and USA (EPA, 2022), effectively restricting market access when insecticide-impregnated bags are used.

Regulatory frameworks have also sought to address new developments in pesticides, particularly the development of a wider range of biologically based pest-control products. The U.S. Environmental Protection Agency, e.g. has recognized a distinction between traditional synthetic pesticides and biocontrol products based on naturally occurring materials or organisms with a low risk of harm to humans or the environment. These minimum-risk substances include repellents such as garlic and hot pepper, biopesticides (microbes found in nature, such as *Bacillus thuringiensis*, and substances found in nature, such as insect hormones) and organic pesticide ingredients. Some of these are exempt from the lengthy and expensive approval and registration process required for conventional pesticides and from some regulations to restrict pesticide use.

Consumer concerns about the safety of the food they purchase, the impact of production practices on the environment and fair treatment of growers and workers involved in cultivation are also addressed through voluntary certification programmes. These approaches give different emphases to these three themes. First initiated in the 1940s, organic certification was initially applied to bananas from Mexico and the Dominican Republic in the 1980s.

The sector now represents 11% of global export (Dawson, 2021). Certification prohibits the use of conventional pesticides and chemical fertilizers, thereby reducing risks of environmental contamination and toxicity for workers and for consumers from pesticide residues in fruit. Since the first Fairtrade bananas became available in 1996 with a focus primarily on fair prices for small growers, increasing restrictions have also been placed on allowable pesticides. The number of permitted pesticides will shrink further in the future (Fairtrade International, 2016). The main focus of certification by the Rainforest Alliance, which was first applied to bananas exported from Costa Rica in the 1990s, was initially on deforestation and plastic recycling. Since then, the certification process has expanded to include a list of prohibited pesticides. Rainforest Alliance standards distinguish pass/fail core requirements (fundamentals of sustainable agriculture), mandatory improvement pass/fail requirements (to be achieved in 3 or 6 years) and self-selected improvement requirements. These are monitored by 'Smart Meters' based on self-formulated targets for specific sets of indicators. The GlobalGAP certification framework, founded in 1998 by a group of large multinational supermarkets, focuses on food safety, traceability, meeting regulations governing the fair treatment of agricultural labour and adherence to standards set by producing and consuming countries (GlobalGAP, 2021). Bananas are often exported with multiple certifications – organic, fair trade and GlobalGAP for example – which meet a broad range of market requirements and may ensure more stable demand from exporters and consumers.

These voluntary certification schemes have expanded their focus on pesticide use alongside the development of more stringent international and national regulations. Table 1 illustrates how each certification scheme affects practices for managing bunch insect pests. To meet certification requirements, growers and the associations to which they belong receive regularly scheduled and unscheduled inspections. A core requirement is extensive documentation of field activities, labour use and inputs applied.

Pesticides allowed in organic certification such as spinosad may be overused in frequent broadcast applications, resulting in secondary outbreaks of defoliating insects (Arias de López et al., 2020). In terms of the development stages of integrated pest management (IPM) practices, this represents a return to stage 1 – regular broadcast applications. Although the options for the management of bunch pests have expanded, conventional growers are projecting a large potential increase in losses in response to the increasing pesticide restrictions. The Dole Tropical Fruits Regional Director for Innovation projected losses of 30–75% due to insect bunch pests if insecticide-treated bags are prohibited (Gutierrez, 2021). An upswing in the broadcast application of systemic insecticides such as spirotetramat and imidacloprid in conjunction with fungicide application has also been reported in Central

Table 1 Certification requirements relevant to bunch pest protection

	USDA organic	EU organic	BioSuisse organic	Fairtrade	Rainforest Alliance	GlobalGAP	Demeter
Pesticides permitted	The National List of Allowed and Prohibited Substances[a] – no conventional insecticides permitted	Implementing Regulation 2021/1965 of the Regulation (EU) 2018/848 on organic production, lists authorized active substances[b]	Only substances on 'FiBL Input List' – no synthetic insecticides. Detailed pest-control plan required. Prohibited use of broadcast organic insecticides. No water flow from conventional fields	210 prohibited pesticides – red list FAO/WHO Pesticide Guidelines Highly Hazardous Pesticides (HHP); 39 restricted use pesticides – orange list; 108 hazardous pesticides – yellow list pending decision for orange/red list[c]	Rainforest Alliance manages an extensive list of 156 prohibited pesticides, based on the FAO/WHO Guidelines for HHP[d]	Permitted Plant Protection Products by regulations in each producing country. Detailed application records and knowledge of MRLs in destination markets mandatory[e]	Appendix 5 of the Standard list of permitted natural products – Spinosad and natural pyrethrum are allowed, as exception[f]
Plastics use	Plastics for mulch or row covers cannot be left behind or burned	Not mentioned in the Regulation and Annexes	The use of plastic foil should be kept to a minimum and used foil should be recycled	Waste management plan required for farms with hired labour. No specific mention of plastic	Recycling waste should deploy all available local options	Waste reduction review to promote avoidance, reduction, reuse and recycling	Plastic waste not mentioned
In-field biodiversity	Organic System Plan required for biodiversity use to maintain or improve all natural resources[g]	High soil biodiversity proposed to prevent and control pests and diseases	Measures for self-regulating ecosystem – biodiversity, soil health, pest and disease resistance	Active enhancement natural ecosystem process for natural enemy habitat, soil fertility and water conservation	Required plan to promote habitat for natural enemies	Action plan and tangible actions to enhance habitats and biodiversity on the farm and surroundings	At least 10% farm area or adjacent areas should be biodiversity reserve

Naturally occurring biological control	Organic pest management through prevention and avoidance with monitoring for timely use of suppression.	Use of biodiversity and soil quality as preventive measures for pest and disease control	Control measures must be chosen and carried out in a manner that limits their impact on non-target organisms	Not mentioned in the standards	Integrated pest and resistance management with thresholds, directed use of less toxic and rotating active ingredients	Integrated pest management is expected with at least two preventive measures taken to control key pests	Biological control of pests is encouraged
Impacts on aquatic life	Measures referenced to avoid pollution by nutrients and erosion[h]	Not mentioned	Measures to avoid negative impacts on quality and quantity of surface and groundwater, flora and fauna	Buffer zones around water bodies and watershed recharge – pesticide and fertilizer use prohibited	Riparian buffers of 2–15 m with 20 m exclusion zone around buffers with no pesticide or fertilizer use	Required risk assessment of environmental impacts on water resources of farming activities	Required natural areas may include water bodies and wetlands
Labour safety	National health and safety inspections and reporting included in certification agents[i]	Not mentioned	Health and safety to be guaranteed with training, protective clothing and access to medical care	Health and safety measures (training, facilities and protective clothing) required	Health and safety measures improvement strategy required depending on farm size	Mandatory health and safety risk assessment with detailed control points for training and equipment	Right to health and security of workers identified but not operational
Banana quality standards	Banana quality standards by marketer not certification	EU marketing standards of banana[j]	Banana quality standards by marketer not certification	Quality standards from marketer as part of purchase contracts	Banana quality standards by marketer not certification	Banana quality standards by marketer not certification	Banana quality standards by marketer – not certification

(Continued)

Table 1 (*Continued*)

	USDA organic	EU organic	BioSuisse organic	Fairtrade	Rainforest Alliance	GlobalGAP	Demeter
Banana prices	Price regulations not a part of standards but 50–75% greater	Price regulations not a part of standards, but 50–100% greater	Fair pricing through dialogue for improved future with social auditing is mandatory	Binding Fairtrade conventional and organic Minimum Prices Ex Works and FOB, by banana-exporting country with annual review	A sustainability differential should be paid on top of the market price, and a sustainability investment plan should be financed	No supermarket access in Europe without certification; GlobalGAP became an 'entry ticket'	No reference to price differentials

[a]Details: see https://www.ecfr.gov/current/title-7/part-205/subpart-g.
[b]https://eur-lex.europa.eu/legal-content/EN/TXT/PDF/?uri=CELEX:32021R1165&from=EN.
[c]https://files.fairtrade.net/standards/Hazardous_Materials_List_EN.pdf.
[d]https://www.fao.org/agriculture/crops/thematic-sitemap/theme/pests/code/hhp/en/.
[e]https://www.globalgap.org/uk_en/for-producers/globalg.a.p./integrated-farm-assurance-ifa/.
[f]https://demeter.net/certification/standard/.
[g]http://www.ams.usda.gov/sites/default/files/media/NOP%205020%20Biodiversity%20Guidance%20Rev01%20%28Final%29.pdf.
[h]http://www.ams.usda.gov/sites/default/files/media/NOP%205020%20Biodiversity%20Guidance%20Rev01%20%28Final%29.pdf.
[i]http://www.ams.usda.gov/sites/default/files/media/NOP-PM-11-6-HealthSafetyViolations.pdf.
[j]https://eur-lex.europa.eu/legal-content/EN/TXT/PDF/?uri=CELEX:32011R1333&from=EN.

America, a worrying approach which was abandoned starting in the 1960s shortly after insecticides became available (see Section 9 for more details on insecticides).

These developments suggest that existing bunch pest management practices will need to evolve further if control is to remain both effective and compatible with evolving regulatory and market requirements. The following sections explore how this might be achieved by implementing the fourth stage of IPM in banana bunch pest management involving a redesign of the production ecosystem starting with understanding pest ecology as the foundation for this approach (Nawaz et al., 2019).

3 Major insect pests of the bunch

Insect pests of the banana bunch cause losses to banana growers through cosmetic damage to the fruit or direct physical contamination of fruits as living insects or dead body parts (Figs. 2–15). Some of these insect pests, primarily aphids, are also vectors for viruses such as banana bunchy top virus, cucumber mosaic virus and banana bract mosaic virus. Mealybugs are a vector for banana streak virus. In this section, we introduce six banana bunch insect pests. Understanding the behaviour and ecology of these pests is key to their management, discussed later in the chapter.

3.1 Red rust thrips

Red rust thrips, *Chaetanaphothrips* (also written *Chaetanophothrips*) *signipennis* (Figs. 2 and 3) and *C. orchidii* (Fig. 4), are widespread in Latin America and

Figure 2 Adults on banana pseudostem showing size – red rust thrips *Chaetanaphothrips signipennis* (Bagnall, 1914); 0.7–1.5 mm size of male and female adults with incomplete metamorphosis.

Figure 3 Close-up of male and female adults - red rust thrips *Chaetanaphothrips signipennis* (Bagnall, 1914).

Figure 4 Adult - red rust thrips *Chaetanaphothrips orchidii* (Moulton, 1907); size of 0.7–1.5 mm adults with parthenogenic incomplete metamorphosis.

Asia where they primarily infect banana. They multiply on the banana mat (the mass represented by the fruit-bearing stem, suckers and the rhizome) and are also reported to pupate in the soil (Pinese, 2000). When the flower emerges, the adults lay eggs between the young fingers. The emerging nymphs feed by scraping the skin to generate sap which oxidizes, causing the characteristic red to brown rust pattern (Fig. 5). Severe damage causes skin cracking, although the fruit pulp is not affected. An outbreak in Peru and Ecuador from 2010 to 2013 resulted in reported losses of 30-60% of harvested fruit (Clercx et al., 2015). Follow-up surveys in 2015 showed highly variable losses from grower to grower from 0% to 27% (Staver et al., 2018). Insecticide-impregnated bags are effective control measures in conventional banana production, although this practice is under increasing scrutiny due to the toxicity of insecticides for workers and dealing with used bags which need recycling or disposal. For organic production, bagging before the flower opens ensures minimal access to the young fingers and low levels of bananas discarded due to red rust (Arias de López et al., 2020).

3.2 Banana flower thrips

The banana flower thrips, *F. parvula* (Fig. 6), is common in Latin America and the Caribbean, *Thrips hawaiiensis* is widespread in Asia and the Pacific and *Thrips florum* is found across Africa (Moritz et al., 2013). They can feed on flowers which indirectly reduces finger growth. However, direct damage is caused when adults oviposit on tender fingers. The resulting small, raised pockmarks are tolerated in limited numbers but will result in the rejection of produce for export if numerous (Fig. 7). Losses are less severe than for red rust thrips. *F. parvula* feeds primarily on banana, while *Thrips hawaiiensis* and *Thrips florum* feed on a wide range of cultivated perennial and annual crops, including ornamentals. The use of bags or sleeves effectively minimizes damage. Removal of flowers from fingers as the bunch develops and elimination of the flower bud or bell after the last hand has opened also reduce the presence of banana flower thrips.

Figure 5 Banana peel discoloration caused by both species of red rust thrips.

Figure 6 Adult – flower thrips *Frankliniella parvula* (Hood, 1925); 1.0–2.0 mm size of adult with male and female adults with incomplete metamorphosis.

Figure 7 Raised pockmarks on banana peel caused during egg laying by flower thrips.

3.3 Banana aphids

The specialist banana aphid *Pentalonia nigronervosa* (Fig. 8) is found wherever bananas grow. Other aphid species such as *Aphis gossypii* and *Rhopalosiphum maidis* are also found on banana. Aphids secrete honeydew. When aphid numbers are abundant on the bunch, the resulting honeydew promotes the growth of a black sooty mould. Although not a frequent problem, bananas with black patches are not acceptable for export (Fig. 9). Polyethylene sleeves or

Figure 8 Adult aphid colony - banana aphids *Pentalonia nigronervosa* (Coquerel, 1859); 1.0–2.0 mm size of adult with wingless females in parthenogenic viviparity. Winged adults generated by overcrowding.

Figure 9 Fumagine black soot from aphid presence.

bags impregnated with insecticides are effective in reducing aphid populations in conventional banana production, while organic banana growers depend on bags alone and naturally occurring beneficial organisms to keep aphid populations down. Removing aphid-promoting ants may be helpful if aphid numbers build up.

3.4 Scale insects

The two primary species of scale insects, *Aspidiotus destructor* and *D. boisduvalii*, feed on many economically important plants. Scale insects (Fig. 10), like mealybugs, feed by piercing and sucking plant tissue, which may cause blemishes (Guillén et al., 2010). Scale insects are quite sedentary but can be transported long distances if hidden in infected planting material from ornamental and fruit trees. While scale insects do not often cause serious damage to banana fruits, their presence, either alive or dead, is prohibited on banana fruits for export (Fig. 11). Bagging of the bunch is effective in minimizing scale infestations. Naturally occurring beneficial organisms should be protected by minimizing regular and broadcast application of insecticides, fungicides and nematicides. De-sheathing and accelerated residue decomposition also reduce refuges for scale insects in banana fields (Guillén and Laprade, 2016).

3.5 Mealybugs

Mealybugs are sluggish, crawling insects which pierce and then suck on the tissues of many different food and ornamental crops. As well as causing direct damage, they are also a vector for viruses. Quarantine procedures controlling the movement of plant materials between countries prohibit the import of

Figure 10 Adult - scale insects *Aspidiotus destructor* and *Diaspis boisduvalii;* 2.0–3.0 mm size of an adult female. Immobile wingless females lay eggs which hatch into very small mobile nymphs called crawlers. Smaller males are winged, but adult males do not feed.

plants containing living or dead mealybugs or mealybug parts. Mealybugs also secrete honeydew. The resulting black sooty mould patches (known as fumagine) result in the rejection of bananas for export (Fig. 13). Although

Figure 11 Scale insect outbreak in banana crown representing quarantine risk.

Figure 12 Adult - mealybugs *Pseudococcus elisae*; 2.0–3.0 mm size of adult female. Stages for females are egg, nymph and adult, while smaller winged male pupates before becoming adult; no feeding as adult.

Pseudococcus elisae is known as the banana mealybug (Fig. 12), numerous other species of the family Pseudococcidae are also listed on quarantine lists. Timely bagging of the emerging inflorescence is effective for protecting the bunch. Additional sanitary measures applied to equipment used in the field are also critical. The protective collars used to separate developing hands have been shown to harbour mealybug pupae and should be sterilized or stored long enough to minimize the infection of the forming bunch with viable mealybug pupae. Routine practices such as de-sheathing pseudostems are important to reduce mealybug refuges. Applications of biocides which are toxic for beneficial organisms predating mealybugs should be minimized and preferably avoided.

3.6 Fruit scarring beetle

The fruit scarring beetle, *Colaspis* spp. (Fig. 14), is found in Latin America and the Caribbean. It is the most important of several different species, the adults of which feed on the tender tissue of young bananas, causing persistent scars which increase in size as the fruit develops (Fig. 15). Fruits with scars are rejected for sale. Severe outbreaks produce losses above 50% and are more frequent in the rainy season. *Nodostoma subcostatum* and *Basilepta* sp. in India (Prathapan et al., 2019) and *Myochrous melancholicus* in Mexico (de la Cruz Zapata et al., 2018) are also reported to cause scarring. *Colaspis* beetles lay eggs at the base of the banana stem and the larva feed on young roots of grasses. Adults fly from bunch to bunch. Bunch covers with holes smaller than 4 mm should be placed before bracts begin to open and tied at the bottom to protect the developing bunch. Effective control of grass weeds also reduces feeding options for larva.

Figure 13 Fumagine damage, also quarantine risk due to insect presence.

Figure 14 Adult feeding - fruit scarring beetle *Colaspis* spp.; 5.0–6.0 mm size of adults with complete metamorphosis - egg, larva, pupa and adult.

Figure 15 Fruit scarring by beetle *Colaspis*.

4 Ecologically based pest management

The life cycles of each pest are described in Figs. 2–15 and Table 2. These life cycles and other characteristics are fundamental to identifying potential pest management options. In size, the six-bunch insect pest species range from less than 1.5 mm for thrips, slightly larger for aphids, 2–3 mm for mealybugs and scale insects and 5–6 mm for fruit scarring beetles. These sizes make detection and monitoring challenging. Insect size also affects the perforation size of bags to protect bunches.

The various life stages for red rust thrips, aphids, mealybugs and scale insects are found under leaf sheaths, in leaf whorls and on bunches from flowering onwards. Life stages of flower thrips are concentrated in tender

Table 2 Key characteristics on ecology and behaviour of six insect pests of the banana bunch

Common/scientific name	Location of life stages in banana field	Means of insect to reach bunch	Means of insect to reach new field	Weather favouring outbreak
Red rust thrips *Chaetanaphothrips signipennis*	All stages under leaf sheaths, in leaf whorl and in bunch out of direct light; pupa stage reported in soil	Crawling and short hopping flight from leaf whorl	Short-distance wind dispersal, longer distance on infested plant material	Prolonged weather without rain – winds aid dispersal
Red rust thrips *Chaetanaphothrips orchidii*	All stages on tender leaf and bunch tissues; pupa stage in soil possible	Winged adults flight to inflorescence	Short-distance flight and wind dispersal; longer distance on plant material	Prolonged weather without rain
Banana flower thrips *Frankliniella parvula*	All life stages primarily in flowers and young fruits	Adults fly from flower to flower	Flight by adults	More common in dry season
Aphids *Pentalonia nigronervosa*	In colonies in leaf sheath, leaf whorl and protested sites in bunch – banana primary host	Ant-assisted movement, walking	Winged stage stimulated by overcrowding with wind assistance	Heavy rains unfavourable, drier weather favourable
Scale insects *Aspidiotus destructor* *Diaspis boisduvalii*	Eggs, nymphs, pupae, adults on pseudostem, leaf whorl and bunch	Crawlers move to more favourable sites by crawling. Winged males capable of flight	Infected planting material on protective equipment	More common in dry season and careless removal of dead leaf parts
Mealybug family Pseudoccocidae including *Pseudococcus elisae* *Pseudococcus* jackbeardsleyi	Egg laying in ovisacs in protected sites of pseudostem, leaf whorl and bunch, also at base of corm and on roots	Crawler or nymphs crawl to emerging banana flower. Adult males capable of flight short lived	Pupa relocated to bunch with reused bunch separators	More common in dry season and careless removal of dead leaf parts
Fruit scarring beetles *Colaspis* spp.	Eggs laid at plant base. Larva feed on tender grassroots and pupate in soil. Adults feed on young banana leaves and fingers	Adults are strong fliers	Flight by adults	Associated with periods of greater rain

leaves and flowers. Mealybugs pupate on inanimate surfaces such as styrofoam hand separators. This suggests that insect pest species inhabit protected sites on the banana plant which are more humid and away from light. These sites reduce dehydration risk and protect insects from predators. They also protect pests from pesticide applications and make pest monitoring more challenging.

In the case of the *Colaspis* beetle, the egg, larvae and pupa stages occur in the soil, while the adult inhabits the bunch. The larvae hatch from eggs laid at the base of the stem and feed on plant roots with a preference for grasses. The adults feed on tender fingers and other banana tissues. The emergence of pupae of red rust *signipennis* thrips from soil has also been documented (Pinese, 2000), although experiments in Ecuador failed to reveal red rust thrips emergence from soil samples.

When the bunch is harvested, the leaf whorl and stem end up on the soil surface. This may result in high mortality for those pest species with immobile life stages. However, those with more mobile life stages are able to crawl or fly to adjacent suckers or mats to feed. The six insect pests of the bunch vary greatly in mobility and the mechanism by which they move to the emerging flower (Table 2). *Colaspis* fruit scarring beetles and *F. parvula* flower thrips fly readily from mat to mat and can move readily around the plantation. *Orchidii* red rust thrips also fly from bunch to bunch. *Signipennis* red rust thrips move primarily in short flight-assisted hopping within the stem where they are located. This means that, even at high population densities, *Signipennis* thrips did not appear in sticky or water traps as frequently as *F. parvula* flower thrips. Aphids, mealybugs and scale insects are much less mobile over short distances. Aphids and mealybugs are moved by ants which feed on the sap or honey made available when these insects feed. Immature stages of mealybugs and scale insects known as crawlers disperse through walking. Aphids in the winged stage, primarily in response to adverse conditions or crowding in the colony, move primarily with the wind once airborne.

Clarifying the mechanism for long-distance movement is an important factor in strategies to minimize the introduction of pests into clean fields. Three of the six species (*Colaspis*, *Orchidii* thrips and flower thrips) are capable of independent flight, while *Signipennis* thrips and aphids move longer distances via wind-assisted flight. Mealybugs and scale insects have no mechanism for independent movement. They are spread via the movement of contaminated planting material or as pupae on inanimate objects.

All life stages of red rust thrips, mealybugs, scale insects and aphids feed through rasping and sucking plant cell sap on the surface of stems, tender leaves and banana fingers when tender. The flower thrips feed only on flowers and tender leaves. Fruit damage is created by egg laying rather than feeding.

Colaspis beetles are nocturnal, while red rust thrips and aphids also avoid light. This has implications for the degree of light penetration allowed by bags or sleeves which need to protect the bunch without creating low light conditions favourable for pests. Generally, the temperature range suitable for banana growth is also suitable for all the insect pest species.

To achieve the goal of offering low cost pest management with low toxicity to the environment and human health and low risk of pest damage to fruits, scientists, technicians and growers and their farm teams need ecologically based approaches. Such approaches deploy an understanding of pest characteristics, life cycles and behaviour, their interaction with predators and other elements of the banana ecosystem. We address the following elements of an integrated, holistic approach to pest management:

- Monitoring populations of insect pests and beneficial organisms;
- Creating a less favourable environment for pests to reduce pest damage;
- Creating a more favourable environment for beneficial organisms;
- Strengthening the banana plant to resist insect bunch pest damage; and
- Direct control of insect pests using insecticides and augmentation of beneficials.

5 Monitoring populations of insect pests and beneficial organisms

Monitoring insect pest levels, crop damage and numbers of beneficial insects is the basis for more targeted and environmentally friendly pest management. However, to be of practical use to banana growers and technical field advisors, monitoring must also be relatively simple and low cost. In the short term (days or weeks), early detection of the first appearance of a pest in an otherwise clean plantation needs to trigger effective action to prevent further increase. Scouting pest levels is also commonly used to establish economic thresholds for pest management programmes to be implemented, though this is less applicable in the context of year-round banana production. Scouting may also serve to guide the release of natural enemies or the application of entomopathogenic fungi.

In the medium term (weeks and months), information on damage is useful to monitor the effectiveness of routine practices like bagging, repellents and sanitation. An increase in damage may indicate more lax practices with a changeover in workers or supervision. Observation of damage from secondary pests such as defoliators may also indicate that production practices are reducing the effectiveness of natural control mechanisms. More long-term monitoring is needed to track the success of ecological intensification initiatives such as vegetation strips planted as habitat for beneficials and, in particular, which plant species are appropriate sources of nectar, pollen and alternative prey for beneficials. Long-term monitoring of insect pest levels, damage and numbers of beneficials in response to variations in weather and climate over seasons and years has great potential value to guide improvements to pest management programmes. However, this depends very much on simple and low-cost scouting techniques, data capture and analysis. In Sections 5.1 and 5.2, we provide basic information and an initial proposal to facilitate monitoring.

5.1 Monitoring pests and their damage

Banana exporters exercise strict control on cosmetic and quarantine pest presence just before shipping, when a sample of several boxes per pallet are opened and subject to a very detailed quality control with a scoring system, tolerance levels for certain defects and 'no-go' criteria. Their objective is to ensure that problems of fruit loss or rejection thereafter in the supply chain are minimized but are less concerned about management practices in the field. In this section, we will focus on field monitoring primarily to prevent the rejection of fruit and increase the marketable percentage of fruit reaching the packing shed.

Direct counting and trap methods have been developed to monitor specific insect pests. Crop Science Australia recommends the examination of five immature bunches per hectare every 2 weeks for red rust thrips (https://www.crop.bayer.com.au/pests/pests/banana-rust-thrips). However, it does not offer a comparative scale to establish damage or action thresholds. A brochure on banana thrips management in the French West Indies recommends monitoring thrips in ten young bunches with the appearance of first flowers and on leaves of young suckers but also does not specify a scale for assessing the degree of infestation (Institut Technique Tropical (IT2), 2022). Pinese and Piper (1994) propose biweekly sampling of five bunches for adult thrips and discolouration in the fingers.

Van Tol et al. (2017) reported that sticky traps (of different colours with and without pheromones) and water traps were ineffective in monitoring *signipennis* thrips in spite of high insect levels shown by visual inspection. *F. parvula* were attracted more readily to traps. In other crops, the semiochemical kairomone has been shown to be useful for monitoring and control strategies for flower thrips in closed environments (Cloyd, 2010), and pheromone traps are also available to monitor flying male scale insects in crops like citrus (UC-I. P. M, 2022).

These somewhat incomplete methods to monitor individual insect pests do not offer a basis for building an integrated, simple and low-cost monitoring method to track multiple banana bunch pests. Two different types of monitoring are needed. For scale insects and mealybugs which are not highly mobile and under pressure from natural enemies, banana fields should be relatively free of the pest. The challenge is to detect outbreaks early. A second requirement is medium-term monitoring to ensure that current management practices are keeping losses at relatively low levels across the six pest species.

In monitoring early outbreaks, all field staff should be trained to recognize mealybugs, scale insects and aphids. In the weekly field routine of desheathing, desuckering and harvesting, workers should report suspected cases. Similarly, in the packing shed, workers carrying out debagging, high-pressure water spraying, dehanding, hand and finger selection and boxing should report suspected cases. Marking or georeferencing plants can also be used to facilitate monitoring of infestations and how they spread or contract over time.

For routine loss monitoring, given the challenges in field monitoring of insect pests themselves, Rojas et al. (2016) propose systematic monitoring of the causes of rejected bananas and bunches every 3 months. These authors were working with small growers who normally do not track total losses or their causes, in contrast to large growers for whom this is routine. For small growers, total losses are first estimated by subtracting the weight of bananas in boxes for export from the weight of bananas entering the packing house (average weight of ten bunches × number of bunches processed – the weight of the bunch stem). From the pile of reject bananas, 100 fingers or clusters are classified according to the reason they were rejected. Rojas et al. (2016) proposed nearly 50 separate causes subdivided into biological, physiological, mechanical and physical causes. Biological causes include the damage or presence of the six bunch insects addressed in this chapter among 15 specific biological causes.

5.2 Monitoring natural control agents

A list of natural control organisms, although certainly incomplete due to limited studies (Table 3), provides a starting point for monitoring. Methods used in research studies need to be modified or new methods developed for practical monitoring of natural control agents in commercial plantations, a task which has not yet begun for banana. Some common methods to collect insects may be applicable to monitoring beneficial insects (Table 4). Problems in monitoring include parasitoids such as *Megaphragma* which are extremely small and therefore difficult to observe in sticky or water traps. Parasitized eggs for red rust thrips can be observed with a stereoscope in the lab but not in the field. Regular field monitoring of naturally occurring entomopathogens is also beyond the scope of small and medium growers.

Monitoring of damage from secondary pests can also provide early warning that production practices may be damaging beneficial organisms. For example, damage by leaf-eating worms such as *Ceramidia* indicates overuse of insecticides like spinosad to control red rust thrips in organic banana or overapplication of nematicides in conventional banana production.

Table 3 Naturally occurring organisms affecting population levels of six insect pests of banana bunches

Pest	Control agents: predator	Control agents: parasitoids	Control agents: fungi and bacteria
Red rust *signipennis* thrips	*Chrysoperla carnea* *Blastopthestus pallecens* *Orius insidiosus* *Franklinothrips vespiformis* *Karnyothrips flavipes* *Amblyseius swirskii* Other predatory mites of family Phytoseiidae	*Megaphragma* sp.	*Lecanicillium lecanii*, *Bacillus thuringiensis*, *Metarhizium anisopliae*, *Paecilomyces lilacinus*, *Saccharopolyspora spinosa*
Red rust *orchidii* thrips	*Franklinothrips vespiformis* Other predatory mites *Typhlodromina*, *Amblyseius*, *Lasioseius* y *Neoseiulus* of family Phytoseiidae	No studies	No studies
Banana flower thrips *Frankliniella parvula*	*Chrysoperla* sp. *Blastopthestus pallecens* *Orius* spp., *Anthocoris* spp. *Geocoris* spp., *Nabis* spp. *Macrolophus* spp. and *Aeolothrips* spp. *Diptera*	*Megaphragma* sp., *Ceranisius* spp. and *Thripobius* spp.	*Metarhizium anisopliae*
Scale insects *Diaspis boisduvalii*	*Scymnus* sp. *Cryptolaemus montrouzieri* *Pentilia* spp. *Delphastus* spp. and *Ceraeochrysa* spp.	*Coccobius* sp. *Aphytis* sp. *Plagiomerus* sp. Alberus sp.	No studies
Aphids *Pentalonia nigronervosa*	*Cycloneda sanguinea*; *Crysopa* spp. (Neuroptera: Chrysopidae) *Scymnus* sp. *Acarinus philippinensis* fly larvae Syrphidae *Chrysoperla carnea* *Spiders Araneus* sp., *Argyrodes bonadea* Karsch, *Cheiracanthium* nr. *ligawsolanum* Barrion	*Aphidius* spp. *Lysiphlebus testaceipes* (Hymenoptera: Braconidae) and *Aphelinus* sp.	*Verticillium lecanii* No additional studies
Mealybugs	*Pentilia* sp. *Cryptolaemus montrouzieri*, *Cycloneda* spp. *Chrysoperla carnea*, *Scymnus* sp. and *Ceraeochrysa smithi*	*Hambletonia pseudoccocina* *Leptomastix dactylopii* *Anagyrus* spp.	No studies
Colaspis fruit scarring beetle	*Castollus plagiaticollis*	No studies	*Beauveria bassiana* and *Metarhizium* sp.

Table 4 Sampling methods for use with common beneficial insects of banana bunch insect pests

Method	Possible beneficial insect to be sampled	Applicability in banana
Sweep net	Flying or crawling predators and parasitoids	Not applicable if directed at banana mats, but applicable in associated ground cover vegetation and vegetation strips with small leaves and flexible stems
Pitfall trap	Ground-dwelling crawling fauna	The life cycle of bunch pests and the role of predation when the cut leaf whorl falls to the ground after bunch harvest is poorly documented for red rust thrips, scale insects and mealybugs
Light trap with sheet	No known beneficial insects known to be attracted by night light	General inventory of insect diversity but not for studies of natural biological control organisms
Coloured water trap	Possible use to monitor *Chrysopas*	More potential for insect pests than beneficials. *F. parvula* in blue water traps but no *signipennis thrips*
Beat sheets	Large and small beneficial insects	Not applicable for banana, but applicable in associated shrub-like vegetation strips with small leaves and flexible stems
Trunk intercept	Insects moving on the banana trunk	Placement on suckers or main stem
Coloured sticky traps	Parasitoids with possible coloured traps	Specialized species identification required
Direct observation	Parasitized insect pests	Observation of colonies of mealybugs; Stereoscope observation of parasitized thrips eggs

6 Creating a less favourable environment for pests: bagging and other practices

Practices such as calendarized or rational application of pesticides involve targeting pests directly. However, practices to make the crop environment less favourable for the pest are a very useful approach to reducing crop losses. Techniques currently used include bagging which focuses on excluding pests from the bunch itself, plantation sanitation to reduce refuges or alternate food sources for pests, and the use of repellents combined with bagging to distance pests from the bunch.

6.1 Bagging

Bagging serves multiple functions with relatively few negative trade-offs (Polyban, 2017; Pasberg-Gauhl, 2002; Vezina, 2020; Santosh et al., 2017). The main agronomic benefit is the microenvironment created by the bag which reduces the time from flowering to bunch harvest, increases finger diameter and improves fruit colour. The bag also serves to protect the fingers from mechanical damage from leaves, birds and bats and from sunburn. Pasberg-Gauhl (2002) highlights that bagging also reduces the risk of bunches being infected by fungal spores. As we have already seen, bagging the unopened flower is also an important practice to reduce access to the bunch by the six bunch insect pests.

Bags must be designed to achieve these functions while also ensuring adequate ventilation within the bag to limit humidity and consequent development of diseases (Jackson, 2023, CH21). Adequate airflow is more important in areas with higher relative humidity. Larger perforations in the bag, although fewer in number, increase the airflow. However, larger perforations may allow entry to insects. Dier (2013) contrasted different diameters of perforations and concluded that 3 mm was appropriate to limit red rust damage in a relatively dry zone of Ecuador in bags without insecticide. Larger perforations are more common in conventional banana production systems relying on insecticide-impregnated bags (Polyban, 2017). Smaller perforation sizes also reduce red rust thrip damage (Dier, 2013). However, small perforations of 2 mm require a large number of perforations to achieve adequate ventilation, making bags more fragile.

To be effective, the timing of bagging is critical. In certain areas, bagging may be delayed until after flower remnants are removed from tender fingers and the bud is broken off. However, while this practice reduces problems with latex staining of fruits and flower bracts tangling in the plastic bags, the tender fingers are exposed to thrips and *Colaspis* as well as to mechanical damage. Bagging while the flower bud is still completely closed was found to be highly effective in the reduction of red rust damage (Arias de López et al., 2020). To achieve this in commercial plantations, bagging must be carried out at least twice a week and three times a week when temperatures are higher.

Bag colour has been identified as an important variable with blue and transparent bags commonly used (Santosh et al., 2017). Light penetration should be maximized since *signipennis* thrips and *Colaspis* prefer dark conditions. A bag with little light penetration can become an attractive refuge for these pests, and therefore a translucent rather than opaque bag should be used (Arias de López et al., 2020). Transparent bags had a lower incidence of red rust than other colours. At the same time, certain colours are known attractants of certain pests and should be avoided. For example, blue attracts flower thrips. Blue bags have commonly been used, but experiments have shown that green bags had 20% fewer egg-laying blemishes from flower thrips compared to blue bags (Guillén and Laprade, 2010). Yellow attracts aphids and white flies, another occasional pest of banana bunches. These factors should be taken into account along with perforation size in decisions about the type of bag to use.

Thicker and stronger plastic bags are less susceptible to tearing and puncturing, have a stronger micro-climate effect and, if impregnated with insecticide, can carry a higher dose. The bag should also be long enough to be tied shut at the bottom to reduce insect access. At the same time, the pressure to reduce the use of plastic continues to increase. Nobody knows exactly how much plastic is used in banana production for export. An approximate calculation could assume 2000 bunch bags per hectare per year for 500 000 ha of export bananas (based on the overall area represented by the dozen top banana-exporting countries) (FAO http://www.fao.org/faostat/en/#home). One billion bunch bags per year sized at 85 cm × 150 cm and weighing 20 g per bag convert to 20 million kilos per year and a length long enough to circle the globe nearly 40 times.

As noted earlier, bags have been impregnated with insecticides to improve their effectiveness against pests. Chlorpyrifos, an organophosphate, has been used to impregnate bags for decades but has been subject to increasing strict MRLs. Bifenthrin, a synthetic pyrethrum, and buprofezin, an insect growth regulator, have been used more recently as a substitute (Polyban, 2017; Cubillo Sánchez et al., 2003). Sulphur and organic repellents are also used in response to pressure to use less toxic substitutes. A combination of insecticide-impregnated strips with bagging is also used to reduce the total amount of insecticide, the exposure risk to workers and the burden of recycling bags contaminated with pesticides (Cubillo Sánchez, 2013).

A recent report by the FAO has highlighted the issue of widespread single use of plastic bags in banana cultivation. Plastic waste (particularly if the plastic is impregnated with insecticide) has a high potential to create plastic and microplastic pollution, making improved waste management a priority (FAO, 2021). Global, regional and national policy frameworks to promote good management of agricultural plastic waste are currently fragmented, weak and non-binding. The FAO proposes a mix of potential solutions (FAO, 2021):

- Legal enforcement of responsibility schemes within supply chains to enable appropriate collection and recycling of their products at end of life;
- Training for supply chain actors on the environmental costs of plastics use and management;
- Inclusion of good management practices in stewardship schemes (such as GlobalGAP); and
- The formulation of an international voluntary code of conduct and sets of technical guidelines for all relevant stakeholders, as a step towards a

binding global convention on the disposal of agricultural plastics. In its session of July 2022, the FAO Committee on Agriculture (COAG, the FAO governance body on agricultural matters) endorsed this recommendation (FAO, 2022).

In Costa Rica, the two leading banana producers DOLE and Del Monte have founded RECYPLAST (https://www.recyplast.cr/?lang=en), a company which converts plastic from banana, pineapple and melon production (as well as household recycling) into recycled plastic corner boards for pallets (https://www.youtube.com/watch?v=u4v_RcY81X4). In Peru, AgroFair, together with the Peruvian Banana Cluster and others, has founded the company ECOBAN to recycle bunch bags into corner boards (https://www.youtube.com/watch?v=YFTmRJkRzWU) (Clercx et al, 2023). Similar initiatives are underway in Ecuador, Nicaragua and the Dominican Republic.

Alternatives to plastic bags have also been tested. The Biodynamic Federation Demeter, in trials with paper bags, concluded that the use of paper bags is technically feasible where no sprinkler irrigation is used. However, damage to fingers caused by friction and high material and labour costs (in comparison with plastic bags) significantly reduce grower net income (Brinkmann, 2021). Agribon reusable bags have also been tested. The raw material used is polyethylene which is woven like cloth. However, thorough washing and disinfection are essential to avoid the transmission of scale insects and mealybugs in these bags. During the washing process, bags may break or stretch and lose their effectiveness.

6.2 Plantation sanitation

Another way to make conditions less favourable to pests are targeted agronomic practices to reduce refuges and food sources for pests on the banana mat or other materials used in production. The removal of drying or non-essential leaf sheaths from the pseudostem or the suckers, as well as desuckering, often carried out at the same time, reduce refuges for *signipennis* thrips, scale insects, mealybugs and aphids. The practice also provides the opportunity for early detection of new colonies. Elimination of flowers from tender fingers in the second week after flowering, and removal of the remaining flower bud, reduces food resources for flower thrips and thus risk of infestation. Cutting flowers even earlier with scissors or a knife has been proposed but is very labour intensive. The removed flower bud can also be cut into several pieces to accelerate drying out or decomposition.

Flower removal has been skipped in some production areas due to complications with fruit becoming stained by banana sap caused by bagging. To avoid this problem, the bag can be left up until sap release stops and has dried, but this involves extra labour and increased exposure to other insect pests like red rust thrips. Growers report that flowers should be removed during at least two to three visits to each bunch. When fingers and flowers are tender, flower removal is rapid. The bag can be lowered immediately and the sap of tender flowers is more water like with less potential to stain the fruits. The protectors or separators placed between the banana hands in week 3 can be a source of mealybugs, especially pupae which form in one bunch and are then transferred to a new bunch via separators which are commonly reused several times. Washing, heat disinfection or storage long enough to ensure that all pupae have hatched will eliminate the risk of transmission.

The removal or reduction of alternative food sources for pests may also be helpful. The roots of grass weeds, which serve as a food source for the larvae of *Colaspis* in soil, can be reduced by good residue distribution, the use of leguminous cover crops and timely weeding. Flower thrips and *orchidii* thrips reproduce in other species such as avocado and cocoa, although fortunately these species only flower for a short period. Other vegetation in fence rows may also serve as an alternative host or food source for pests. However, vegetation may also support beneficial insects which contribute to pest control. We return to this subject in Section 7.

6.3 Repellents

Repellents or deterrents are used primarily in organic banana production systems. They are applied to the emerging bell, the bunch stem above the first hand, to the bunch itself and to the impregnated bags. Deletre et al. (2016) propose five modes of action – true repellent, contact irritancy, odour masking, antifeedant and visual masking. However, the product choice in use as repellents in banana is driven more by their availability and cost than by specific mode of action. The products include homemade repellents based on relatively inexpensive ingredients available locally like hot peppers, garlic and onion and commercially available products of mixed composition. Some ingredients may also have insecticidal properties. Repellent mixes have limited toxicity to humans and the environment.

Little formal research has been carried out on the use of repellents. Barrera et al. (2018) sprayed bags with five repellents – salvia *Salvia officinalis* L., anamú *Petiveria alliacea* L., basil *Ocimum basilicum* L., garlic *Allium sativum* L. and hot pepper *Capsicum annuum* in plantain. Garlic provided significantly greater protection than the other products which were not significantly different from bagging alone. Lopez and Vargas (2018) applied for a patent for the use of four repellents – garlic, nutmeg, neem and hot peppers – to impregnate bags. Their data indicate that combined damage to banana fingers with their bags was equivalent to insecticide-treated bags. Impregnated bags are available using hot pepper, mustard, garlic, cinnamon and citronella singly or in combination. Commercial products are also available based on garlic (E-Codallim), garlic and hot peppers (Alitiol and Capsialil) and a non-specified mix of plant oils (Cinna-Mix). Their use in banana is based on research and recommendations from other horticultural crops. Their effects on beneficial insects have not been documented, although parasitoids are disrupted with applications of neem and garlic.

7 Creating a more favourable environment for beneficial organisms

Table 3 summarizes the different predator, parasitoid or entomopathogenic species which can play a role in controlling the six bunch insect pest species, illustrating the wide range of organisms which can reduce pest populations. The differences between pests are, in part, due to the degree to which the relevant beneficial species have been studied. For *signipennis* thrips, multiple insect and mite predators, recently identified parasitoids like *Megaphragma* and numerous entomopathogens have been detected. Arias de López et al. (2017, 2018, 2019a,b) have highlighted the potential for these natural controls to maintain this pest at levels below those causing economic losses (Table 5).

As Table 3 shows, the six insect bench pest species are a food source for numerous predators, parasitoids and entomopathogens. To manage a banana field to create a more favourable habitat for these beneficial organisms, a practice sometimes referred to as conservation biological control is required (Landis et al., 2000). Banana monocultures are usually managed only to optimize banana quality and yields without providing the habitat for beneficial organisms to thrive. Beneficial insects depend on suitable host plants to complete their life cycle. This cycle ranges from laying eggs, larval feeding and adult feeding to breeding sites and refuges for resting. As a starting point, Table 6 lists a range of beneficial organisms, their susceptibility to pesticides and habitat requirements, primarily focused on sources of food such as flowers for pollen and nectar, extrafloral nectaries and honeydew. Alternative prey could also play a role, although limited information is available. Naturally occurring vegetation within the field or in field borders and fences may provide such a habitat but is often removed to reduce the presence of potential host plants for pests. In this section, we address three elements of a more favourable habitat: naturally occurring vegetation, strips of planted vegetation for habitat diversity and altering agronomic practices which are damaging to beneficial organisms.

Table 5 Summary of percentage control by diverse organisms of red rust *signipennis* thrips by life stage

Life cycle stage	Organism and estimated capacity for control
Eggs	50% parasitism (*Megaphragma)*
Nymph I	94% predation (predatory mites)
Nymph II	79% predation (mites)
Nymph III	76% predation (stink bug *Blastopthestus pallescens*)
Pre-pupa	38% predation (predatory mites) 71% predation (stink bug *B. pallescens*)
Pupa	80% predation (stink bug *B. pallescens)*
Adults	91% predation (stink bug *B. pallescens*) 100% mortality (*B. bassiana* and *B. thuringiensis*)

7.1 Naturally occurring vegetation

Naturally occurring vegetation in and around a banana field can be found in the fence rows where shrub species may be planted as a living barrier, with other plant species establishing spontaneously. Spontaneous or planted vegetation may also be found on drainage channel banks. Within fields themselves, low-growing vegetation is usually removed, typically with some weeds escaping control. Rojas (2011) sampled patches of single-species herbaceous vegetation with an insect net in banana fields on the Caribbean coast of Costa Rica. He identified seven species with higher levels of predators and parasitoids, while the remaining 30 species had very low levels of resident populations of beneficial insects. A grass species like *Eleusine indica* which had the highest diversity index of sampled insects was considered inappropriate since it is also a reported host for plant parasitic nematodes. Other species mentioned by Rojas (2011) for their levels of predators and parasitoids included *Clidemia* spp., a shrub with abundant trichomes, *Geophila macropoda*, a shade-tolerant low-growing mat sometimes used as a natural ground cover, and *Cyathula prostrata*, with spiny flowers. The legume *Pueraria phaseoloides*, often planted as a ground cover in banana, did not have notable levels of predators and parasitoids. The possible mechanisms for attractiveness of the high-scoring plants included trichomes with possible secretions and horizontal leaf alignment. Flower abundance was not measured.

7.2 Planted vegetation corridors, living ground covers and banker plants

Banana monoculture, especially with targeted management of weedy low-growing vegetation, lacks plant diversity. This is a common characteristic of

Table 6 Beneficial organisms to guide a strategy of manipulative conservation biological control of banana bunch pests

Beneficial organism	Description	Susceptibility to pesticides	Important habitat features
Ceraeochrysa sp.	Larva of green lace wing predate aphids and thrips	High to repeated broadcast spray	Adults feed on pollen, nectar and honeydew
Blastopthestus pallescens	Stink bug predator of nymph and adult thrips	No data on compatible and non-compatible insecticides	Not known
Amblyseius swirskii	Predatory mite of larval stage of thrips and white flies	High to repeated broadcast sprays, compatible sprays available	Adults feed on pollen, nectar and honeydew. Leaf hairs and domitia for refuge
Orius insidiosus	Nymphs and adults predate thrips, aphids and spider mites	High to repeated broadcast sprays of pyrethroids, neonicotinoids, organophosphates and carbamates	Pollen, sap
Franklinothrips vespiformis	Ant-like predatory thrips nymphs and adults predate thrips, white fly larva and small arthropods	Systemic insecticides particularly damaging	Pollen, nectar
Castollus plagiaticollis	Predatory stink bug	Limited information	No reports
Scymnus sp.	Predatory ladybug larva and adults predate aphids and larva of mealybugs	High for many insecticides, although others are more compatible	Pollen, nectar, honeydew and extrafloral nectaries
Coccobius sp.	Parasitoid laying eggs into larva and adults of scale insects	Compatible and non-compatible biocides need to be identified	Pollen, nectar
Megaphragma sp.	Very small parasitoid-laying eggs into eggs of thrips		Pollen, nectar
Hambletonia pseudoccocina	Parasitoid laying eggs into nymphs and adults of mealybugs		Pollen, nectar

Sources: https://entnemdept.ufl.edu/creatures/BENEFICIAL/swirksi_mite.htm. https://entnemdept.ufl.edu/creatures/beneficial/bugs/Orius_insidiosus.htm. https://entnemdept.ufl.edu/creatures/BENEFICIAL/vespiform_thrips.htm#:~:text=Franklinothrips%20vespiformis%20is%20presumed%20native,New%20Caledonia%20and%2C%20most%20recently%2C. https://entnemdept.ufl.edu/creatures/beneficial/lady_beetles.htm.

crop fields and greenhouse production and has led to a number of strategies to increase plant diversity which may support conservation biological control. Shelter habitats (Griffiths et al., 2008), biological corridors (Nicholls and Altieri, 2004) and banker plants (Huang et al., 2011) are related concepts promoting conservation biological control. This involves the deliberate introduction and management of non-crop vegetation as a habitat for beneficial species. The introduction of trays of young cereal plants infested with cereal aphids into greenhouses for vegetable production along with aphid predators increases and maintains predator populations in advance of the main crop. The generalist aphid predators shift to the main greenhouse crop and keep aphids at very low levels.

For banana fields, numerous sites are available for vegetation strips as habitats. These include field borders and fence rows, drainage and irrigation ditches as well as sites between rows and introduced strips within fields themselves. Vegetation with specific services to beneficial insects could include plants with abundant flowering throughout the year as a source of pollen and nectar or the presence of extrafloral nectaries and domatia (plant structures providing a refuge for arthropods). Following the banker plant model, plants which host prey to promote predator levels might also be identified, although year-round banana production contrasts with many seasonal vegetable and fruit crops. The challenge for banana is to design and maintain a favourable habitat throughout the year based on principles applied in field-grown or greenhouse seasonal crops.

Although no experiments or case studies have been documented on the use of vegetation strips in commercial banana production, the permanent nature of the plantation and the high leaf canopy with densities of 1600–2000 mats in larger plantations suggest a number of features. Strips should be planted between every three and five rows of banana mats. Species selected for prototype strips need the following characteristics:

- Abundant flowering during most months of the year;
- Perennial shrub growth;
- Vigourous growth and flowering in partial shade;
- Features such as nectaries and domatia;
- Relatively easy to establish and to maintain at 1–2 m in height;
- No major pest problems to limit growth;
- Not a host for ants which promote aphids, mealybugs or scale insects; and
- Low demand for water and nutrients.

For continuous flowering, species which appear promising include *Tithonia diversifolia*, *Lantana* spp., perennial basil and *Ricinus communis*. Extrafloral nectaries are found in cotton, *Hibiscus* and *Ipomoea* spp. (morning glory).

Other species for consideration are *Moringa oleifera*, *Hedychium coronarium*, *Malpighia* spp., perennials and sunflower. Leguminous shrubs from the genera *Leucaena*, *Senna* and *Sesbania* might also be useful based on extrafloral nectaries associated with leaves.

Leguminous ground cover plants like *Pueraria* and *Clitoria* are used commercially and have numerous other benefits, including soil erosion protection, soil health improvement through nitrogen fixation and organic matter production, weed suppression and habitat for predators of nematodes and weevils (Huoc et al., 2018). However, they rarely flower in understory conditions and produce limited pollen and nectar. Species with extrafloral nectaries associated with leaf structures such as *Clitoria* have some potential.

A number of potential problems need to be addressed in using vegetation strips. One of the characteristics of the plants used may be invasiveness: their ability to quickly cover the soil. This can lead to obstruction of drainage or entanglement of banana plants. Possible competition with banana plants for nutrients, water and light is also a point of attention (Huoc et al., 2018). Several proposed species for vegetation strips may also be hosts of banana bunch pests like mealybugs, scale insects or aphids. There is a balance between the potential to increase pest levels versus the potential to increase the overall population of beneficials which will reduce pest numbers. Vegetation strips may also be an attractive habitat for ants which move aphids, mealybugs and scale insects from one mat to another or from vegetation strips to banana plants. Monitoring methods like sweep nets to assess relative numbers of pests and beneficials may be more applicable in vegetation strips than in banana mats.

7.3 Reduction in use of substances possibly toxic to beneficial organisms

Insecticides, fungicides, herbicides and nematicides used in conventional banana production have negative impacts on non-target organisms. Although the use of pesticides is substantially lower in organic bananas, the repellents, insecticides and fungicides that are permitted for use can also have negative impacts. Creating a more favourable environment for beneficial organisms needs to include reducing the use of more toxic pesticides and other substances.

There are a few studies on the effects of pesticides specifically on banana-beneficial organisms. Fortunately broader studies substantiate the importance of minimizing pesticide application to protect beneficials. Cheng et al. (2018) looked at the impacts of 13 agrochemicals on three species of *Trichogramma* parasitoids. Only one pesticide was found to be safe for use for the three species, while the other pesticides varied from moderate to extremely toxic. The 13 agrochemicals included three herbicides which also showed slight-to-moderate toxicity. The three *Trichogramma* species had different responses to the different products, suggesting that the search for pesticides compatible with different parasitoids must be done species by species.

Focussing on greenhouse tomato production, Gallego et al. (2019) studied three groups of pesticides related to tomato moth management. One group presented a high compatibility with the release of egg parasitoids, a second group showed some incompatibility, while a third group was incompatible. Spinosad, an insecticide commonly used in organic banana, was included in the last group, suggesting that frequent broadcast application of this product to control red rust thrips is highly problematic for parasitoids like *Megaphragma*. Williams et al. (2003) highlighted the toxicity of spinosad for parasitoids but also found that it was not toxic for predatory insects.

Funderburk (2009) identified pesticides compatible with the release of *Orius insidiosus* in fruiting vegetables. This approach can be seen as a model for banana. The list included spinosad, spinetoram and pyridadyl for thrips and white flies. Biondi et al. (2012) tested lethal and sublethal effects and time of exposure for 14 pesticides, including spinosad, on another Orius predator and concluded that spinosad was toxic up to 7 days after application. Houghgoldstein and Whalen (1993) exposed nymphs of a predatory stink bug to potato foliage after soil-applied insecticide and foliage after foliar-applied insecticides. Stink bug mortality was high with soil-applied insecticides due to the consumption of potato beetle larvae, while nymphs were unaffected by exposure to foliar-applied insecticides. Kumar (2020) documented the negative impacts of three of four insecticides on both adults, nymphs and eggs of the predatory mite *Amblyseius swirskii*. Eggs, for example, were not affected 24 h after application, but at 5, 10 and 15 days after treatment, egg viability was significantly reduced.

These studies, while not conducted in banana, illustrate the destabilizing impact of pesticides on predatory insects, parasitoids and predatory mites, all organisms contributing to conservation biological control. Broadcast aerial applications should be avoided, a lesson learned by banana entomologists in the 1960s. Soil application of systemic insecticides can also be expected to create a highly unfavourable environment for many beneficial organisms. In Section 9, we will discuss the more targeted use of insecticides to reduce overall pesticide use and exposure of beneficial organisms to toxic substances.

8 Improving plant resistance

Three possible approaches to increasing the capacity of the banana plant to resist pest damage are documented. The first is breeding for resistance. Genetically existing banana cultivars have been evaluated for susceptibility to leaf streak, Fusarium, nematodes and bacterial wilts among others (Staver and Capra, 2017) but not for the six insect pests of interest here. Cavendish cultivars, although susceptible to many diseases and pests, have wide acceptance in the market and are very productive, which makes substitution with other potentially more resistant varieties challenging.

Plant nutrition is a second potential area to reduce the attractiveness of banana mats to pests. No specific studies are available for banana, but Altieri et al. (2012) cite the role of plant nitrogen in attracting aphids, whiteflies and thrips. Although insect pests were not studied, Pattison (2006) found that farms which applied less than 400 kg nitrogen/ha, applied organic matter and strategically placed banana residue to feed plant roots had higher performance compared to farms using more than 400 kg/ha of nitrogen. Organic banana producers face greater challenges to source nitrogen and may be less prone to nitrogen overuse. A third, more recent frontier in plant resistance is research into the plant and insect microbiome, but this is unlikely to produce any immediate results for banana producers (Douglas, 2018).

9 Direct insect pest control using augmentation of beneficials and insecticides

In addition to creating an unfavourable environment for target pests, creating a more favourable environment for beneficial organisms and strengthening the resistance of the plant, two direct control strategies can be used as part of an integrated approach: augmentative biological control (including the use of semiochemicals to trap pests) and conventional and organic pesticide use.

Augmentative biological control involves the rearing and release of natural predator species to increase existing populations of beneficials. Three types of biological agents can be applied for the control of insect pests (Hulot and Hiller, 2021):

- Invertebrate organisms (insects, mites and nematodes) are commercially available, although only insects and mites are applicable to bunch insect pests;
- Microbial organisms (bacteria, fungi and viruses) are also commercially available; and
- Semiochemicals can be used to attract insect pests into traps.

Commercially available organisms used in biological augmentative control which have been tested in banana include *Orius insidiosus*, *Amblyseius swirskii* and *Chrysoperla carnea* (Table 7).

Beneficial insects are often released in an inert carrier (e.g. vermiculite or buckwheat husks) to facilitate the spread of natural enemies in the crop. Little heaps of mixed carrier material and beneficial insects are spread throughout the crop or localized in pest hot spots. Broadcasting devices (e.g. a modified

Table 7 Commercially available beneficial organisms for use in banana

Organism	Pests controlled	Strategy for use	Success factors	Practices negative for effective use of biological control
Cryptoleamus montrouzieri	Aphids, mealybugs, scale insects	Rearing and release	High mealybug populations	Spraying with buprofezin
Anagyrus spp.	Mealybugs	Rearing and release	Nymphs – II and III instars	Broadcast use of insecticide
Chrysoperla spp.	Aphids, white flies, thrips, mealybugs, lepidóptera eggs and larva, mites	Rearing and release	Release of 10.000 a 40.000 larvas/ha	Broadcast use of Insecticide
Amblyseius swirskii	Thrips, mites	1 pack 250 mites/bunch	Combined with bagging of closed flower	Broadcast use of Insecticide
Orius insidiosus	Aphids, white flies, thrips, mealybugs, lepidóptera eggs and larva	Inundative release		Broadcast use of Insecticide
Franklinothrips vespiformis	Red rust thrips	Mixed cropping		
Lecanicillium lecanii	Red rust thrips, banana aphid	Apply in early morning or late afternoon		Application with high solar radiation
Bacillus thuringiensis	Red rust thrips			
Metarhizium anisopliae	*Colaspis* sp.			
Paecilomyces lilacinus	Red rust thrips			

leaf blower) can also be used to release beneficials. Another release method is breeding sachets for predatory mites, in which sachets (made from paper or other compostable material) are filled with carrier material, prey and predatory mites and hung in the crop. This technique allows about 4 weeks of continuous release of predatory mites. A study by Moyón (2021) illustrates different options for predator release. They tested the placement of sachets of *Amblyseius* inside and outside the bagged bunch at bagging, deflowering and the placement of separators. All treatments were equally effective in reducing the incidence of red rust thrips compared to unbagged bunches, although no treatment with untreated bags without *Amblyseius* was tested.

Although biological control has been tested at the field scale, strategies for targeted or regular use are still being worked out. Should release be planned according to a regular schedule or timed to coincide with specific pest levels or environmental conditions? What is the optimum number of predators or parasitoids to be released and how can any trigger for release be monitored in a practical and efficient way? How can the effectiveness of release be measured? Predatory mites have been released to re-establish diminished native mite populations with the expectation that released mites will become established which requires favourable conditions for predatory mite survival (as discussed in Section 7). Trapping using pheromones as a lure has been tested for monitoring pest levels but not for the control of bunch pests in banana.

9.1 Conventional and organic insecticides

As discussed earlier, manufactured insecticides to kill target pests have been used to reduce losses to insect pests of the bunch since the 1960s. Broadcast insecticides were found to generate secondary pests due to reductions in the beneficial insect populations. This led to a more targeted approach to protect the bunch through bagging and the impregnation of the bags with insecticides. Chlorpyrifos has been the standard insecticide used for many decades, despite its high toxicity to humans, aquatic organisms, bees and beneficial insects. Altabtabaee et al. (2016) found that the dose from bag to bag was highly variable but was 90% dissipated within 10 days. They calculated that insecticide-impregnated bags are responsible for 22.5 metric tons of chlorpyrifos entering the environment annually in Costa Rica. Toxicity for workers occurs through exposure by contact with the bags or during bag disposal or recycling. Chlorpyrifos is now prohibited in many countries and is being phased out. Bifenthrin and buprofezin are the most common alternative pesticides used to impregnate bags. These products offer diverse benefits in terms of toxicity compared to chlorpyrifos (Table 8).

Diverse substances are applied in organic production to the leaf whorl just prior to shooting, the emerging bud and to tender fingers inside or outside the bag. Organically certified pesticides like spinosad and pyrethrum-based products are widely used. Microbial products and repellents have also been tested. Barreto et al. (2021) compared five organic substances with bagging. Neem and pyrethrum-based Phyriplus were found to have both repellent and insecticidal action. However, they found that lemongrass had the lowest levels of damage to beneficials. Arias de López et al. (2020) found comparable performance with the use of impregnated insecticides, repellents or organic insecticides.

Table 8 Insecticides - toxicity of conventional and organically certified insecticides

Material	Use	Toxicity for beneficials	Toxicity for aquatic life	Toxicity for humans
Chlorpyrifos	Conventional: impregnated bags	Predators: extreme Parasitoids: extreme Bees: extreme	Aquatic: extreme Soil: moderate Birds: extreme	Moderate acute and high chronic toxicity
Bifenthrin synthetic pyrethrum	Conventional: impregnated bags	Predators: moderate Parasitoids: moderate Bees: extreme	Aquatic: extreme acute and chronic Soil: moderate Birds: moderate	II - moderately dangerous or toxic
Buprofezin	Conventional: impregnated bags	Predators: parasitoids: moderate Bees: mild	Aquatic: extreme Earthworms: mild Birds: mild	No acute exposure, danger, chronic exposure not determined
Spinetoram - synthetic spinosad	Conventional: broadcast	Predators: moderate Parasitoids: extreme Bees: extreme	Aquatic: moderate Soil: mild to moderate Birds: mild	IV - no acute risk
Imidacloprid (Kohinor, Plural)	Conventional: broadcast	Predators: extreme Parasitoids: extreme Bees: extreme	Aquatic: extreme chronic Soil: extreme Birds: mild to extreme	II - moderately dangerous or toxic
Spinosad	Organic: application to bell, whorl and peudostem and broadcast	Predators: moderate Parasitoids: extreme Bees: extreme	Aquatic: moderate Soil: mild to moderate Birds: mild	IV - no acute risk
Pyrethrum-based products (Phyriplus, Oleoplus)	Organic: application to bell, whorl, pseudostem and broadcast	Predators: moderate Parasitoids: moderate Bees: extreme	Aquatic: extreme Soil: mild Birds: extreme	II - moderately dangerous or toxic

As noted earlier, there is a need for a reduced, more targeted use of insecticides or repellents to reduce negative effects on beneficials and the environment. However, field reports suggest that broadcast use of insecticides like imidacloprid and spinetoram in certain countries is increasing whilst, in organic production, more frequent and broadcast use of spinosad may also have a damaging effect.

10 Integrating bunch pest management with other management practices

Several practices to protect the banana bunch either from cosmetic damage or from contamination with quarantine insect pests often conflict with environmental protection objectives, labour welfare and consumer preferences or conservation biological control. The management of bunch insect pests has been moving increasingly to less toxic and more targeted management. Alternative methods of control include the use of repellents, release of biological control organisms and habitat management, measures primarily to promote beneficial organisms. These measures should be seen within the broader concept of 'ecological intensification' (Côte et al., 2010) differentiated by Staver et al. (2018) as integrated ecological systems versus simple or single practice-based ecological intensification. This has also been called 'ecostacking' referring to the marshalling of a range of ecosystem service providers (such as beneficials), with each practice contributing synergistically to a more functional biodiverse landscape able to deliver food production and other ecosystem services sustainably (Hokkanen and Menzler-Hokkanen, 2018).

The cornerstone practices for future bunch management are early bagging preferably combined with the use of non-toxic repellents and field sanitation as a stand-alone package. At the same time, conservation biological control can promote beneficials (and thus reduce pest populations needing management) through increased plant diversity. The introduction of increased levels of plant diversity can be visualized in steps with increasing degrees of plantation and management redesign with greater potential for ecostacking. The simplest step is planting of hedges on field borders and drainage ditches as habitat as well as possible contribution of enriched and more varied biomass to the field. Such hedges can also work as barriers to human movement for biosecurity (Staver et al., 2022). Regions where small growers are more frequent have more field borders with hedges than large commercial plantations. Perennial herbaceous cover crops within banana plantations could also be seeded. Both for hedges and for herbaceous cover crops, more studies are needed to evaluate species habitat potential and adaptation to association with bananas. A second step would be the introduction of vegetation strips to further increase available habitat for conservation biological control and diversification of biomass production and root type beyond just banana. A crucial question is whether such strips or diversity corridors can prosper at current population density and understory light availability. A third step to generate a more favourable environment for vegetation corridors and cover crops would be a reduction in mat density or alternation of mat distribution to increase understory solar radiation availability. Increased flowering, greater biomass production and more nitrogen fixation could result in contributing not only to greater background biological control but also to improved soil health, plant nutrition and plant vigour. A challenge might lie in greater water use and possible competition with banana mats. Labour-efficient means to cut back hedges and cover crops may be required to limit competitive effects. A lower density of more vigorous mats may result in bigger bunches, a shorter interval between bunches and lower input of labour and materials for bagging.

11 Conclusion: Scientific and practical frontiers in ecological bunch pest management

The global conscience about the environmental and food web effects of insecticides has continued to expand from the initial alert by Carson (1962) and recent insights by Goulson (2021). Banana entomologists working in the 1960s to reduce cosmetic damage caused by insect pests identified the toxic effects of regular broadcast insecticide sprays within only a few production cycles after ongoing broadcast insecticide spraying (Stephens, 1984) and as a consequence pursued instead a strategy of insecticide elimination to allow naturally occurring biological control to operate more effectively. More than a half century later, the challenge has grown and also has become more specific – how to minimize cosmetic and quarantine damage of insect pests through more effective and practical mobilization of knowledge of banana food web ecology in the context of increasing strict certification and regulatory standards on insecticide use, worker and environmental safety, plastics and climate change. In the following paragraphs, we identify specific research and implementation challenges which require attention in four areas – understanding the banana ecosystem linked to bunch pests, building a more data-driven approach to bunch pest management, plastics reduction and recycling and evolving certifications and regulations which bring together interests of banana-growing communities and countries and consumers.

11.1 Banana ecosystem linked to banana bunch health

The understanding of fundamental ecological processes such as food web dynamics in banana production systems has progressed since the early work of banana entomologists in Central America and Australia. Themes requiring further scientific and financial investment include the following:

- Life cycle of insect pests of the bunch in relation to the banana mat and the cycle of management practices. An important example is the leaf whorl from a recently harvested pseudostem. This biomass falls to the ground along with all the insect inhabitants. Very little is known about insect diversity, population levels and mortality factors linked to the drastic change in microenvironment. What factors reduce survival, and can they be harnessed to reduce insect pest populations? Certain insects are less well studied such as *orchidii* thrips which appears to inhabit largely the bunch (Sections 3 and 4, Table 2).
- Advanced inventory of naturally occurring organisms which affect the life cycle of banana bunch insect pests – what life cycle stage is affected, what is the degree of reduction? Tables 3 and 6 illustrate control organism diversity, but more studies may yield additional parasitoids. Table 5 showing organisms affecting *signipennis* thrips illustrates a framework for such studies.
- The factors which affect the abundance of both bunch insect pests and the biological control organisms are not well understood, especially for the latter group. Food and habitat resources for the different life cycle stages merit targeted studies in banana zones in different surrounding land uses and climatic conditions. Relatively little is known of alternative prey for biological control organisms of bunch insect pests (Section 7). Ants disseminate several bunch insect pests and the habitat factors which operate for and against ants, merit understanding. Beneficial bacteria and fungi represent different study approaches than insects, spiders and mites, especially to understand environmental factors in their abundance.
- Impacts of conventional and organically certified insecticides, fungicides, herbicides, nematicides and repellents on biological control organisms (Section 9). The role of nutrient applications in banana plant attractiveness to insect pest dynamics also merits further clarification (Section 8).
- Integrated perspective taking into account the response of the multiple insect pests and diseases to bag colour, repellents and perforations to identify key combinations maximizing bag contribution to clean bananas appealing to consumers (Section 6).
- The potential to use habitat diversity to increase the effectiveness of conservation biological control. Surrounding land use may also have the potential to make both positive and negative contributions and should be the subject of studies, especially areas in natural vegetation but also other crops.
- Although of less immediate application, the potential for microbiome manipulation to protect banana plants and operate to the detriment of bunch insect pests may be of interest (e.g. Díaz-Castillo, 2016).

11.2 Data-driven approaches to integrated bunch pest management strategies

Export banana production is a data-rich enterprise to ensure fruit calibre and maturity with data on seasonality and bunch production, leaf number at harvest and leaf disease management. We propose the following themes to extend data-driven crop management and production system redesign to the management of clean bananas:

- Improved routine scouting methods of banana bunch pests for field use by banana growers and field technicians building to an integrated approach. Monitoring damage levels due to bunch insects at bunch processing brings an integrated view which can be complemented with targeted field scouting depending on insect behaviour. Combining routine scouting results with local weather data could be a potent tool to fine-tune practices by specific weather events and routine seasonal fluctuations;
- Improved routine scouting methods for beneficial organisms which affect bunch pest population dynamics and damage for use by banana growers and technicians. The lack of such methods is an impediment to the targeted use of habitat diversification. Without structured observation of different species composition and plant density and arrangement, growers cannot easily optimize the use of diversity;
- Strategies for efficient and effective use of predator/parasitoid release based on climatic factors, priority seasons and predictive indicators need to be built on improved scouting methods and results from multi-year databases of pest and damage levels and weather;
- Design and testing of prototypes for the incorporation of diversified vegetation strips with different levels of availability and composition of habitat (within field strips, fence rows and ground cover), interaction with bagging approach and targeted application of compatible repellents and low-toxicity insecticides. Effective, integrated scouting methods are a necessary tool for testing. Flower strips used to increase beneficial fauna in temperate fruit orchards (Pfiffner et al., 2018; Herz et al., 2019) provide a model for banana, although this needs to be expanded to cover beneficial fungi and bacteria;
- Big data approaches which build on the already data-rich routines of export banana need more systematic data collection from routine monitoring of banana losses, levels of insect pest and beneficial organisms in the field using scouting methods proposed earlier and daily weather data; and
- Digital tools, including data platforms, readily available for growers, farm teams and technicians in support of data-driven diversification of banana production to minimize losses to banana bunch pests with reduced environmental impacts from toxic products and plastics.

11.3 Plastics reduction and recycling

The burden of plastics use in export banana production is high, although the multiple benefits in terms of bunch development and reduction in blemishes make the immediate elimination of plastics unlikely. The short-term challenge to the banana sector and the plastics industry is to close the cycle on plastic use and avoid contamination of terrestrial, freshwater and marine ecosystems from dispersed, unrecycled bags (Section 6). These break down into microplastics and microscopic particles, to become part of the soil. The following problems require urgent attention:

- Elimination of insecticides from bags and insecticide-impregnated strips depends on successful advances in the two topics discussed earlier.
- Advanced life cycle analysis of plastics and other materials used in production and export - plastic bags, corner boards, boxes, pallets, straps stickers and mapping of stakeholders involved (manufacturers, distributors, supermarkets and consumers), as a preamble to work out further reduction and recycling approaches together – although beyond the reach of banana agronomists and entomologists. ADI VALOR in France demonstrates what is possible. This initiative which brings together 350 industries, 1300 distributors and over 300 000 farmers collects and recycles almost 80% of all plastic used in agriculture (e.g. containers for pesticides, bags for seeds and fertilizers and foil) (https://www.adivalor.fr/en/).
- Effect of contamination of soils and surface water with microplastics on the environment and nutrition uptake of banana plants.

11.4 Certifications and regulations to balance production and consumption interests

Every container of bananas is loaded for shipment only with complete documentation both for voluntary certification depending on the market targeted and for plant sanitary certification. The requirements established in these two frameworks play an enormous role in how bananas are produced, even though the focus is much more on meeting consumer and importing country demands than on impacts on occupational health and safety of growers and banana workers and the environment in producing countries. The following topics in future approaches to certification and regulations have the potential to bring together more effectively both production and consumption interests around banana bunch pest management and banana cosmetic quality.

- Strengthened certification criteria on conservation biological control which are largely ignored in current certification requirements for organic and GlobalGAP certification.

- Facilitation of data-driven management and big data approaches through requirements for data base establishment and analysis. Software development by certification agencies like GlobalGAP could provide an important basis for such efforts.
- Research coordination and financing for topics on the conservation biological control of global quarantine pests like mealybugs and scale insects from collaboration among regional and global plant protection authorities and certification agencies, as proposed by Baldissera et al. (2020).
- Systematic attention for occupational health and safety related to the use of agrochemicals. The work carried out by the World Banana Forum could be a starting point (https://www.fao.org/world-banana-forum/projects/banana-occupational-health-and-safety-bohesi/en/).

12 Where to look for further information

World Banana Forum – Good Practices in the Banana Industry:

- https://www.fao.org/world-banana-forum/projects/good-practices/en/.

World Banana Forum – Banana Occupational Health and Safety:

- https://www.fao.org/world-banana-forum/projects/bohesi/en/.

YouTube: Fontagro Banano Escalamiento for greater information on banana red rust thrips:

- https://www.youtube.com/@fontagrobananoescalamiento9162/videos.

Australian banana research:

Bunch Pests – Australian Banana Growers' Council.

- https://abgc.org.au/bunch-pests/.
- Bunch pests – Better Bananas https://betterbananas.com.au/2019/08/06/bunch-pest-management/.

The FAO developed a toolkit in Excel for the identification of Highly Hazardous Pesticides (HHP, https://www.fao.org/pesticide-registration-toolkit/special-topics/highly-hazardous-pesticides-hhp/introduction/en/) for pesticide registration authorities in developing countries, to identify, assess and mitigate risks related to the use of certain pesticides.

A list of banned pesticides worldwide can be found in a spreadsheet by the Pesticide Action Network, (https://pan-international.org/pan-international-consolidated-list-of-banned-pesticides/).

A handy overview of links to databases on active ingredients, policies on pesticide use, restrictions and MRL in Australia, the USA and International, FAO, Canada, the EU and more than a dozen other countries can be found at: https://www.agriculture.gov.au/ag-farm-food/food/nrs/databases.

On the website http://www.plaguicidasdecentroamerica.una.ac.cr/index.php/base-de-datos-menu (in Spanish), an overview of the main pesticides used in Central America, their active ingredients, human toxicity, impact on soils and water and ecotocicology. Created by IRET (Instituto Regional de Estudios en Sustancias Tóxicas), part of the UNA (Universidad Nacional de Costa Rica).

13 References

Altabtabaee, R. L., Chaudhary, O., Clement, C. and Polidoro, B. 2016. Loss of chlorpyrifos from plastic bags used in commercial banana production. *International Journal of Environmental Science and Toxicology Research* 4(2), 19–24.

Altieri, M., Nicholls, C. and Fritiz, M. 2012. Manage insects on your farm: A guide to ecological strategies, handbook series Book 7. *Sustainable Agriculture Research and Education (SARE)*. College Park, MD. Sare. https://www.sare.org/wp-content/uploads/Manage-Insects-on-Your-Farm.pdf.

Arias de López, M., Corozo, R., Vera, T., Jines, A., Van Tol, R. and Clercx, L. 2016. Biological control of red rust banana thrips: preliminary results in the laboratory. Available at: https://www.promusa.org/article142-2016-France-symposium. In: Xth International Symposium on Banana/ISHS-ProMusa Symposium Agroecological Approaches to Promote Innovative Banana Production Systems.

Arias de López, M., Corozo-Ayovi,E., Vivas, L., Medrano, S., Sánchez, L., Alburqueque, D., Vera, T., Jines, A., Delgado, R., Rengifo, D., Rojas, J. C. and Staver, C. 2017. Búsqueda de Enemigos Naturales a *Chaetanaphothrips signipennis* y *C. orchidii* en Banano Orgánico en Ecuador. Perú y República Dominicana. Poster IV Congreso Latinoamericano y del Caribe de Plátanos y Bananos. México, COL, México.

Arias de Lopez, M., Corozo, R., Vera, T., Jines, A., Vivas, L., Delgado, R. and Staver, C. 2018. Eficacia de Bioinsecticidas para el Manejo del Trípido de la Mancha Roja *Chaetanaphothrips Signipennis (Thysanoptera: Thrypidae) en banano*.

Arias de Lopez, M., Corozo-Ayovi, E., Moyon, D., Delgado, R., Staver, C. and Polasek, A. 2019a. *First Report of Megaphragma sp. (Hymenoptera: Trichogrammatidae), an Egg Parasitoid of Chaetanaphothrips signipennis and Frankliniella Parvula in Ecuador*. Poster. Annual Meeting of Entomological Society of America, 17-20 November 2019, St Louis, MO.

Arias de López, M., Corozo-Ayovi, E., Delgado, R., Ramos, B.. and Staver, C. 2019b. Predatory efficacy of *Amblyseius swirskii* and effect of *Beauveria bassiana* against the banana red rust thrips, *Chaetanaphothrips signipennis*, under laboratory and screenhouse conditions. *Infogram* D3489, ESA, Missouri, SL.

Arias de López, M., Corozo-Ayovi, R. E., Delgado, R., Osorio, B., Moyón, D., Rengifo, D., Suárez, P., Paulino, A., Medrano, S., Sanchez, L., Rojas, J. C., Vegas, U., Alburqueque, D., Staver, C., van Tol, R. and Clercx, L. 2020. Red rust thrips in smallholder organic export banana in Latin America and the Caribbean: pathways for control, compatible with organic certification. *Acta Horticulturae* 1272, 153–161.

Bananalink 2022. Leading banana governments show unity on shared responsibilities. Available at: https://www.bananalink.org.uk/news/leading-banana-governments-show-unity-on-shared-responsibilities/.

Barrera, J. L., Herrera, C. F. and Pérez, D. J. 2018. Extractos vegetales: alternativa de control de *Colaspis* sp. (Coleoptera: Chrysomelidae) en plátano cv. Harton. *Temas Agrarios* 23(1), 9–17.

Biondi, A., Desneux, N., Siscaro, G. and Zappalà, L. 2012. Using organic-certified rather than synthetic pesticides may not be safer for biological control agents: selectivity and side effects of 14 pesticides on the predator *Orius laevigatus*. *Chemosphere* 87(7), 803–812. https://doi.org/10.1016/j.chemosphere.2011.12.082.

Bird, G., Edens, T., Drummond, F. and Groden, E. 1990. Design of pest management systems for sustainable agriculture. In Francis, C. In: King, L. and Flora, C. (Eds). *Sustainable Agriculture in Temperate Zones*. John Wiley, New York, 55–110.

Brinkmann, C. 2021. *Feasibility Study. Paper versus Plastic Bagging in the Banana Production. Merckens Development Support/Biodynamic Federation Demeter*.

Carson, R. 1962. *Silent Spring*. Houghton Mifflin Company, Boston.

Cheng, S., Lin, R., Wang, L., Qiu, Q., Qu, M., Ren, X., Zong, F., Jiang, H. and Yu, C. 2018. Comparative susceptibility of thirteen selected pesticides to three different insect egg parasitoid Trichogramma species. *Ecotoxicology and Environmental Safety* 166, 86–91.

Clercx, L., Aria Zambrano, M., Dulanto Bejarano, J. and Flores Espinoza, B. 2015. Towards biological control of red rust banana thrips in organic and conventional banana. *Acta Horticulturae* (1105), 73–80. https://doi.org/10.17660/ActaHortic.2015.1105.11.

Clercx, L., Balarezo Camminati, D., Carreño Zapata, M. and Voerman, R. 2023. From bags for bunches to corner boards for pallets: recycling banana plastic in Peru. In: Proceedings of the Banana Symposium, 15-20 August 2022. International Horticultural Congress, Angers, France. Available at: https://musanet.org/wp-content/uploads/2022/09/Luud_Clercx.pdf (poster) and https://www.actahort.org/books/1367/1367_38.htm (article).

Cloyd, R. 2010. Western flower thrips management in greenhouse-grown crops. Kansas State University. Available at: https://bookstore.ksre.ksu.edu/pubs/mf2922.pdf.

Côte, F., Tomekpe, K., Staver, C., Depigny, S., Lescot, T. and Markham, R. 2010. Agro-ecological intensification in banana and plantain: an approach to developing more sustainable cropping systems for both smallholder farmers and large-scale commercial producers. *Acta Horticulturae* 879, 457–463. Available at: https://www.researchgate.net/publication/284062386_Agro-ecological_intensification_in_banana_and_plantain_Musa_spp_An_approach_to_develop_more_sustainable_cropping_systems_for_both_smallholder_farmers_and_large-scale_commercial_producers.

Cubillo Sánchez, D. 2013. *Manejo Integrado de Insectos Plaga en el Cultivo de Banano. Conceptos y Aplicaciones*. confederación unitaria de trabajadores, San José. 105 p.

Cubillo Sánchez, D., Villalta and Guzmán, M. 2003. Evaluación de fundas de polietileno con insecticidas (clorpirifos, bifentrina y retenona) y de polipropileno sin insecticida (Agriban) para insectos plagas. In: Sandoval, J. (Ed.), Informe Anual 2002. CORBANA, SA, San José, CR, 30–40p.

Dawson, C. 2021. Organic World Market Trend 2020: when all the records fell. *Fruitrop* 275, 91–97.

de la Cruz-Zapata, G., Moreno-Jiménez, M., Osorio-Osorio, R., Hernández-Hernández, L. U., Márquez-Quiroz, C., de la Cruz-Lázaro, E. and Valdez-Carrasco, J. M. 2018. Fruit-scarring beetle *Myochrous melancholicus* Jacoby1 as a potential new pest of banana. *Southwestern Entomologist* 43(1), 57–66.

Deletre, E., Schatz, B., Bourguet, D., Chandre, F., Williams, L., Ratnadass, A. and Martin, T. 2016. Prospects for repellent in pest control: current developments and future challenges. *Chemoecology* 26(4), 127–142.

Department of Agriculture and Fisheries/Queensland Department of Agriculture 2017. Banana scab moth (see also Banana Flower Thrips, Banana Red Rust Thrips. Available at: https://www.daf.qld.gov.au/business-priorities/agriculture/plants/fruit-vegetable/insect-pests/banana-scab-moth.

Díaz-Castillo, N. D. 2016. Control biológico de *Chaetanaphothrips signipennis* (Bagnall, 1914) y. *Frankliniella Parvula* (Hood, 1925), mediante el desequilibrio de su microbiota nativa, empleando microorganismos nativos antagonistas obtenidos de la filosfera y rizosfera del banano (Musa acuminata), en Tumbes-Perú, 2014-2015. Universidad Nacional de Tumbes. 96p.

Dier, C. E. 2013. *Fundas de polietileno con orificios de diferentes tamaños para reducir el daño del trips de la mancha roja en banano orgánico*. Tesis Ingeniero Agrónomo. Universidad de Guayaquil. Facultad de Ciencias Agrarias, Guayaquil, Ecuador. 54p.

Douglas, A. E. 2018. Strategies for enhanced crop resistance to insect pests. *Annual Review of Plant Biology* 69, 637–660.

EPA 2022. EPA takes next step to keep chlorpyrifos out of food, protecting farmworkers and children's health. Available at: https://www.epa.gov/newsreleases/epa-takes-next-step-keep-chlorpyrifos-out-food-protecting-farmworkers-and-childrens.

EU Pesticide Database 2022. Pesticide(s) residues and maximum residue levels Chlorpyrifos. Available at: https://ec.europa.eu/food/plant/pesticides/eu-pesticides-database/start/screen/mrls. Consulted 20 November 2022.

European Commission 2020a. Towards a sustainable food system. *Independent Expert Report*. Group of Chief Scientific Advisors. Available at: https://op.europa.eu/en/web/eu-law-and-publications/publication-detail/-/publication/ca8ffeda-99bb-11ea-aac4-01aa75ed71a1.

European Commission 2020b. Farm to fork strategy. For a fair, healthy and environmental-friendly food system. Available at: https://food.ec.europa.eu/document/download/472acca8-7f7b-4171-98b0-ed76720d68d3_en.

Fairtrade International 2016. *Hazardous Materials List*. Available at: https://files.fairtrade.net/standards/Hazardous_Materials_List_EN.pdf.

FAO 2021. *Assessment of Agricultural Plastics and Their Sustainability. A Call for Action*. Rome. Available at: https://www.fao.org/3/cb7856en/cb7856en.pdf.

FAO 2022. Committee on Agriculture. Twenty-eighth session, 18 – 22 July 2022. *Guidance on the use of agricultural plastics*. Rome. Available at: https://www.fao.org/3/nj012en/nj012en.pdf.

FAO and UNEP 2020. Rotterdam Convention on the Prior Informed Consent Procedure of Certain Hazardous Chemicals and Pesticides in International Trade. Text and Annexes, revised in 2019. Available at: www.pic.int/Portals/5/download.aspx?d=UNEP-FAO-RC-CONVTEXT-2019.English.pdf.

FAO and WHO 2014. The international code of conduct on pesticide management. Available at: https://www.fao.org/pest-and-pesticide-management/pesticide-risk-reduction/code-conduct/en/.

FAO and WHO 2021. Codex alimentarius. International Food Standards. Available at: https://www.fao.org/fao-who-codexalimentarius/home/en/.

Funderburk, J. E. 2009. Management of the western flower thrips (Thysanoptera: Thripidae) in fruiting vegetables. *Florida Entomologist* 92(1), 1-6.

Gallego, J. R., Guerrero-Manzano, J., Fernandez-Maldonado, F. J. and Cabello, T. 2019. Susceptibility of the egg parasitoid *Trichogramma achaeae* (Hymenoptera: Trichogrammatidae) to selected insecticides used in tomato greenhouses. *Spanish Journal of Agricultural Research* 17(2), e1009. https://doi.org/10.5424/sjar/2019172-14413.

GlobalGAP, 2021. Módulo base para Todo Tipo de Finca. Módulo base para Cultivos, Frutas y Hortalizas. Available at: https://www.globalgap.org/uk_en/for-producers/globalg.a.p./integrated-farm-assurance-ifa/crops/FV/index.html.

Goulson, D. 2021. *Silent Earth. Averting the Insect Apocalypse*. Vintage Publishing, London.

Griffiths, G. J. K., Holland, J. M., Bailey, A. and Thomas, M. B. 2008. Efficacy and economics of shelter habitats for conservation biological control. *Biological Control* 45(2), 200–209.

Guillén, C. and Laprade, S. 2016. *Manejo y combate de la escama del banano (Diaspis boisduvalii)*. Hoja divulgativa CORBANA, Limón, Costa Rica. N. ° 9-2016.

Guillén, C. and Laprade, S. 2010. Fundas de colores en el control del Tríps de la flor (*Frankliniella parvula*. Thisanoptera: Tripidae) En. Informe Anual 2010. Dirección de Investigaciones. 120. Corbana, Limon, Costa Rica.

Guillén, C., Rodríguez, A., Laprade, S., Valle, H., Segura, R., Uva, V. and Sandoval, J. 2010. *Biología y control de las cochinillas y escamas que atacan al banano*. Hoja divulgativa CORBANA, Limón, Costa Rica. N. ° 5-2010.

Gutierrez, P. 2021. Impact of reduction of MRLs on banana production and export. MRL Harmonization Workshop Webinair 26–27 May 2021 organized by the California Specialty Crop Council. Available at: http://specialtycrops.org/mrlworkshop.html. Available at: http://specialtycrops.org/pdfs/mrl_2021/7.pdf.

Herz, A., Cahenzli, F., Penvern, S., Pfiffner, L., Tasin, M. and Sigsgaard, L. 2019. Managing floral resources in apple orchards for pest control: ideas, experiences and future directions. *Insects* 10(8), 247.

Hokkanen, H. M. T. and Menzler-Hokkanen, I. 2018. Urgent need to develop ecostacking techniques to enhance ecosystem services in cropping systems. *Arthropod-Plant Interactions* 12(4), 477-482.

Houghgoldstein, J. and Whalen, J. 1993. Inundative release of predatory stink bugs for control of Colorado potato beetle. *Biological Control* 3(4), 343-347, https://doi.org/10.1006/bcon.1993.1045.

Huang, N., Enkegaard, A., Osborne, L. S., Ramakers, P. M. J., Messelink, G. J., Pijnakker, J. and Murphy, G. 2011. The banker plant method in biological control. *Critical Reviews in Plant Sciences* 30(3), 259–278.

Hulot, J. F. and Hiller, N. 2021. Exploring the benefits of biocontrol for sustainable agriculture: A literature review on biocontrol in light of the European Green Deal. Institute for European Environmental Policy. Available at: https://ieep.eu/publications/exploring-the-benefits-of-biocontrol-for-sustainable-agriculture/.

Huoc, H. T., Fegeant, N., Chabrier, C., Dorel, M., Le Guen, R., Polanco, R. and Gandini, G. 2018. Plantas de servicio en la Línea Noroeste, República Dominicana. Junta Agroempresarial Dominicana (JAD)/CIRAD. Available at: http://siba.org.do/wp-content/documentacion_general/Guias_tecnicas_CIRAD/Guia_Plantas_de_servicios.pdf.

Institut Technique Tropical 2022. Banana thrips in the West Indies. Available at: https://it2.fr/app/uploads/2022/01/DOC_IT2_2014-Fiche-manuel-BGM-n12-Thrips_ANG_BD.pdf (IT2).

Kogan, M. 1998. Integrated pest management: historical perspectives and contemporary developments. *Annual Review of Entomology* 43(1), 243–270.

Kumar, L. 2020. Banana bunchy top virus (bunchy top of banana). CABI invasive species compendium. Available at: https://www.cabi.org/isc/datasheet/8161#tocontributors.

Landis, D. A., Wratten, S. D. and Gurr, G. M. 2000. Habitat management to conserve natural enemies of arthropod pests in agriculture. *Annual Review of Entomology* 45(1), 175–201.

Lopez, A. and Vargas, O. 2018. Plastic cover consisting of botanical active ingredients for controlling banana pests. Solicitud internacional publicada en virtud del tratado publicado en cooperacion en materia de patentes. Available at: https://patentimages.storage.googleapis.com/49/02/95/614937b160a8ca/WO2018130235A1.pdf.

Moritz, G., Brandt, S., Triapitsyn, S. and Subramanian, S. 2013. Identification and information tools for thrips in East Africa. http://dx.doi.org/10.13140/2.1.5173.8409.

Moscoso-Ramirez, P. A. and Peña-Peña, A. 2020. Effect of treatments with bunch bagging on production, fruit quality and damage by thrips of banana. *Journal of Stored Products and Postharvest Research* 11(2), 15–27.

Moyón, S. D. J. 2021. *Control biológico de Chaetanophothrips signipennis y Frankliniella parvulala en plátano con el ácaro predador Amblyseius swirskii*, Tesis Magister en Sanidad Vegetal. Universidad Agraria del Ecuador. 79p.

Nawaz, A., Sufyan, M., Gogi, M. D. and Javed, M. W. 2019. Sustainable management of insect-pests. In: *Innovations in Sustainable Agriculture*. Springer, Cham, 287–335.

Nicholls, C. I. and Altieri, M. A. 2004. Agroecological bases of ecological engineering for pest management. Eds Gurr, G., Wratten, S. D., & Altieri, M. A. *Ecological Engineering for Pest Management: Advances in Habitat Manipulation for Arthropods*, 33–54. CSIRO Publishing. Collingwood, Australia. https://www.google.com/url?sa=t&rct=j&q=&esrc=s&source=web&cd=&ved=2ahUKEwjxrq_i-quAAxV52QIHHeh3CrQQFnoECCMQAQ&url=https%3A%2F%2Fgtu.ge%2FAgro-Lib%2FIPM-Ecological.pdf&usg=AOvVaw3DjO8xADXKcSav3u4omgcH&opi=89978449.

Pasberg-Gauhl, C. 2002. How can we reduce losses of banana bunches due to 'fruit speckling' in commercial banana plantations? In: *Acorbat 2002. Memorias XV Reunión. Realizada en Cartagena de Indias, Colombia*. 27 de Octubre al 02 Noviembre 2002. Asociación de Bananeros de Colombia AUGURA, Medellín (COL).

Pattison, T. 2006. Banana farm management effects on soil health and plant-parasitic nematodes in Costa Rica. *Final report*. Queensland Department of Primary Industries and Fisheries, University of Southern Queensland, Australia.

Pfiffner, L., Jamar, L., Cahenzli, F., Korsgaard, M., Swiergiel., W. and Sigsgaard, L. 2018. *Perennial Flower Strips: A Tool for Improving Pest Control in Fruit Orchards*. FiBL Technical Guide No 1130.

Pinese, B. 2000. Biology and IPM control options for banana rust thrips. *Final Report Horticulture*. Research and Development Corporation, FR96023.

Pinese, B. and Piper, R. 1994. *Bananas. Insect & Mite Management*. Department of Primary Industries, Queensland. Brisbane. 67p.

Polyban 2017. La contribución del plástico a la productividad bananera. *Congreso Técnico Bananero de Colombia*. AUGURA, Santa Marta (vol. VII). 10 y 11 de Agosto

2017. Available at: https://docplayer.es/80161987-Contribucion-del-plastico-a-la-productividad-bananera-vii-congreso-tecnico-bananero-de-colombia-augura-santa-marta-10-y-11-de-agosto.html.

Prathapan, K. D., Poorani, J., Kumari, S. A., Anuradha, C., Padmanaban, B. and Thanigairaj, R. 2019. Species composition and diagnoses of leaf-and fruit-scarring beetles (Coleoptera, Chrysomelidae) infesting bananas and plantains (Zingiberales, Musaceae) in the Indian subcontinent. *Deutsche Entomologische Zeitschrift* 66(2), 179–202.

Pretty, J. and Hine, R. 2005. Pesticides and the environment. In: Pretty, J. (Ed.). *The Pesticide Detox. Towards a More Sustainable Agriculture*. Earthscan Publications, London.

Rojas, C. 2011. Prospección de la entomofauna benéfica asociada al cultivo del banano (Musa AAA), subgrupo "Cavendish". En: *la Región Atlántica de Costa Rica*. Instituto Tecnológico de Costa Rica, Escuela de Agronomía, San Carlos, Alajuela. confederación unitaria de trabajadores.

Rojas, J., Vegas, U., Rengifo, D., Suarez, P., Corozo, R., Arias, M., Jines, A., Staver, C. and Clercx, L. 2016. Causes of fruit loss pre- and post-harvest in organic export banana systems among organic banana growers in Peru, Ecuador and Dominican Republic. 10–14-October 2016. Available at: https://www.promusa.org/tiki-download_file.php?fileId=2903&display=y. 5th ISHS-Promusa Symposium – Agroecological approaches to promote innovative banana production systems.

Santosh, D. T., Tiwari, K. N. and Reddy, R. G. 2017. Banana bunch covers for quality banana production: a review. *International Journal of Current Microbiology and Applied Sciences* 6(7), 1275–1291.

SAPEA, Science Advice for Policy by European Academies 2020. *A Sustainable Food System for the European Union*. SAPEA, Berlin. Available at: https://sapea.info/topic/food/.

Skevas, T., Lansink, A. O. and Stefanou, S. E. 2013. Designing the emerging EU pesticide policy: a literature review. *NJAS-Wageningen Journal of Life Sciences* 64, 95–103.

Staver, C. and Capra, I. 2017. *Banana Diversity and the Food and Income Threats of Pest and Pathogen Losses: Priority Research Areas to Deploy Diversity to Reduce Pest and Disease Losses*. Available at: https://cgspace.cgiar.org/bitstream/handle/10568/91529/PIM_Bioversity_MusaDiversity_%20FINAL.pdf.

Staver, C., Arias, M., Corozo, R., Delgado, R., Durango, W., Rengifo, D., Pérez, A., Rojas, J. C. and Sánchez, S. 2018. *Informe Final: Fortaleciendo la Base Productiva de Pequeños Productores de Banano Orgánico y Convencional en Latinoamérica y Caribe: Control Biológico de Plagas y Manejo de Suelos Saludables Fontagro FTG/RF-1332-RG*. Available at: https://www.fontagro.org/new/uploads/adjuntos/1332_InformeTecnicoFinal-Tech.pdf.

Staver, C., Arias, M., Guillen, C., Oviedo, M., van der Veken, L. and Clercx, L. 2022. Prospects for biological control and habitat enhancement to minimize quarantine risks and losses from insect bunch pests in organic export banana. International Horticultural Congress, Angers, France. Available at: https://musanet.org/wp-content/uploads/2022/09/Charles_Staver_Prospects.pdf.

Staver, C., Turmel, M., Siles, P., Mpiira, S., Bustamante, O. and Calberto, G. 2018. Step by step tools to identify agroecological intensification alternatives for banana cropping systems. Dita, M., Omondi, A. Alvarez, E., and Zheng, S. *Acta Horticulturae* 1196, 9–18. http://dx.doi.org/10.17660/ActaHortic.2018.1196.02.

Stephens, C. S. 1984. Ecological upset and recuperation of natural control of insect pests in some Costa Rican banana plantations. Perturbación ecológica y recuperación del control natural de plagas de insectos en algunas plantaciones de banano costarricenses. *Turrialba* 34(1), 101–105. Available at: https://repositorio.catie.ac.cr/handle/11554/11889.

Treverrow, N. 2003. Bunch covers for bananas. *Agfact* H6, 2. Available at: www.dpi.nsw.gov.au/__data/assets/pdf_file/0010/119764/banana-bunch-covers.pdf.

UC, I. P. M. 2022. Citrus–California Red scale pheromone trap monitoring. Available at: http://ipm.ucanr.edu/PMG/C107/citrus-redscaletrapmon.pdf. University of California Statewide IPM Program.

UNEP 2020a. Text and Annexes revised //in. Available at: http://chm.pops.int/Portals/0/download.aspx?d=UNEP-POPS-COP-CONVTEXT-2021.English.pdf. Stockholm Convention on Persistent Organic Pollutans (POPS) (vol. 2019).

UNEP 2020b. Protocol on Liability and Compensation for Damage resulting from Transboundary Movements of Hazardous Wastes and their Disposal. Text and Annexes, revised in 2019. Available at: http://www.basel.int/Portals/4/download.aspx?d=UNEP-CHW-IMPL-CONVTEXT.English.pdf. Basel Convention on the Control. of Transboundary Movements of Hazardous Wastes and their Disposal.

Van Tol, R., Arias Zambrano, M., Rojas, J. C., Corozo-Ayovi, E., Albuqueque, D., Jines, A., Delgado, R. and Staver, C. 2017. Monitoreo de trípidos en banano orgánico en Ecuador y Perú mediante el uso de trampas cromáticas y atrayentes. Poster. Iv Congreso Latinoamericano y del Caribe de Plátanos y Bananos. México, COL, México.

Vezina, A. 2020. *Bagging*. Musapedia: PROMUSA. Available at: https://www.promusa.org/Bagging.

Villalobos, R., Villalta, R., Cubillo, D. and Guzmán, M. 2017. Efecto de las características de la funda de polietileno para el racimo de banano (Musa AAA, Cv. Grande Naine) en la producción y la protección contra plagas del fruto. *CORBANA* 37–43(63), 107–123.

Williams, T., Valle, J. and Viñuela, E. 2003. Is the naturally derived insecticide Spinosad® compatible with insect natural enemies? *Biocontrol Science and Technology* 13(5), 459–475.

Part 6

Fruit diseases

Chapter 21

Management of diseases on banana fruit in the field

Marc Jackson, Global Fruit Protection, Australia

1 Introduction

Bananas are the number one fruit sold globally and continue to be one of the most affordable fruits in the market as they are relatively cheap and plentiful all year round. Bananas are a relatively easy crop to grow and growers can reap a return on investment within 9 months, making it an attractive cash crop. These factors along with consumers' expectations of fruit of high-quality drive the management of diseases and pests affecting the banana fruit. Consumers are demanding unblemished fruit of high quality and banana is no exception. Supermarket operators are aware of visual appeal, quality, and shelf life and purchase fruit based on these perceptions. Therefore, it is necessary for growers and marketers to supply a consistently high-quality line of fruit to sustain the demand.

Banana production is a major contributor and employer to the economies of over 130 countries in tropical and subtropical regions. In many regions bananas are a source of export income while other regions supply fruit for domestic consumption. While quality requirements may differ in both cases managing banana diseases is important for long term sustainable supply and reduction of food waste. Waste up to 30% was noted by the author in India (2010–13) due to pre- and post-harvest fruit diseases caused mostly by poor fruit packing, plant hygiene, and lack of pre-cooling.

http://dx.doi.org/10.19103/AS.2022.0108.21

This chapter will focus on Cavendish bananas; however, most diseases are found in many other varieties of bananas throughout the world. Cavendish bananas are the most traded and exported banana variety often involving long supply chains and for this reason, require higher quality standards than locally grown and sold bananas through short supply chains.

There are several diseases that affect fruit quality, some are superficial, on the skin only, while other diseases affect the internal pulp. In this chapter the

Table 1 Bacterial diseases in banana fruit

Common name	Causal agent	Occurrence	Reference
Xanthomonas wilt	*Xanthomonas vasicola* pv. *musacearum*	Africa	GBIF Secretariat, 2021; Blomme et al., 2017
Moko bacterial wilt	*Ralstonia solanacearum* sensu stricto (race 2)	Central and South America, the Caribbean, the Philippines, Peninsular Malaysia	Sequeira, 1958; Yabuuchi et al., 1995
Blood disease	*Ralstonia syzygii* subsp. *celebesensis*	Indonesia, Peninsular Malaysia	Safni et al., 2014; Ray et al., 2021
Mokillo	*Enterobacter cowanii* and *Pantoea agglomerans*		Pathania et al., 2013

Table 2 Fungal diseases affecting banana fruit

Common names of fungal banana diseases	Causal agents	Reference
Anthracnose	*Colletotrichum musae*	Von Arx, 1957
Blotchy mold	*Chaetochyrena musarum*	Singtripop et al., 2016
Cigar-end rot	*Verticillium theobromae; Trachysphaera fructigena*	Mason and Hughes, 1951
Crown rot	*Acremonium sp.; Acremonium strictum; Cephalosporium sp.; Ceratocystis paradoxa; Colletotrichum musae; Fusarium moniliforme; F. pallidoroseum; Fusarium spp.; Lasiodiplodia theobromae; Nigrospora spp.; Verticillium theobromae*	Jones, 2019c
Deightoniella and fruit speckle	*Colletotrichum musae, Fusarium oxysporum and Fusarium semitectum* and *Deightoniella torulosa*	Meredith, 1961; Stover, 1972; Wardlaw, 1972
Freckle	*Phyllosticta cavendishii*	Wong et al., 2012; Drenth and Wong, 2023
Sooty mold	*Chaetothyrina musarum; Cladosporium atriellum; C. cladosporioides; Fumago vagans; Microxyphium* sp.	Jones, 2019b

diseases of the fruit are broken into two sections, bacterial diseases (see Table 1) and fungal diseases (see Table 2). Depending on location, the incidence and severity of some diseases is requiring little or no control, while other diseases require significant efforts to achieve control.

2 Location effect

The impact of diseases depends on the susceptibility of the banana variety, the presence of pathogens and the environmental conditions under which the bananas are grown. It is important to note that many banana diseases have a limited geographic distribution and therefore most governments have strict quarantine guidelines in place related to the movement of planting materials of bananas from one country to another. The current and highest priority disease in bananas is Fusarium wilt Tropical Race 4 caused by *Fusarium oxysporum f. sp. cubense*, recently renamed to *F. odoratissimum* (Maryani et al., 2019), a fungus that can live in the soil for many years. Any disease incursion of this pathogen will impose significant costs and render land for banana cultivation of susceptible varieties such as Cavendish unviable. There are many examples of pathogens moving from one country to another through lack of or non-adherence to strict quarantine protocols (Drenth and Kema, 2021). On a smaller scale, pathogens move from region to region and from farm to farm requiring careful selection of land for banana cultivation. Often the cheapest way to control a disease is not having the pathogen such that prevention is always better than the cure.

The choice of farm location and environmental parameters such as high-water tables, humidity, and weather patterns have a significant effect on the severity of diseases in bananas. Hence, site selection when establishing a new banana plantation is very important to long term sustainable banana production. Correct site selection, farm layout, planting density, and drainage will also reduce overall production costs and lower the costs of chemicals and labor to apply it. To

Figure 1 Establishment of banana plantation in Lakeland, Queensland Australia, with an average rainfall of 900 mm rainfall per year, low relative humidity, and red volcanic soils. Photos by Marc Jackson.

make diseases easier to control plantations are recently more often developed in arid tropical areas, as long as water is available, to reduce the disease pressure and increase the sustainability of banana production (See Fig. 1).

3 Bacterial diseases affecting the fruit and their management

There are several major bacterial diseases affecting banana plants and the fruit, originating in different parts of the world. The three main diseases include Xanthomonas wilt originating from Africa, Moko originating from Central and South America, and Blood disease originating from Indonesia. In addition, there are several minor bacterial diseases such as Mokillo (Table 1) causing damage only to the fruit.

3.1 Banana Xanthomonas wilt

Xanthomonas wilt is caused by *Xanthomonas vasicola* pv. *musacearum* and affects Ensete (aka Ethiopian *banana*, Abyssinian *banana)*, and many species in the genus *Musa*. The disease was first noted in Ethiopia in the 1960s and then spread through Central and Eastern Africa to Rwanda, Uganda, Burundi, Kenya, Congo, and Democratic Republic of the Congo (Mwangi et al., 2007). In some parts of Africa this disease has reduced banana production by 50% to 100%. There are no known cases of Xanthomonas wilt outside Central and Eastern Africa,

Figure 2 Internal discoloration of freshly cut banana fruits affected by Xanthomonas wilt. Photo courtesy of Simon J. Eden-Green/EG Consulting.

therefore strict quarantine guidelines should be adhered to including restricting the movement of banana material and tools from country to country. Strict farm biosecurity should always be maintained to reduce the risk of pathogen invasion.

Common symptoms on the fruit include internal discoloration (Fig. 2) and premature ripening of the fruit. Bananas affected by Xanthomonas wilt are internally characterized by a yellow-orange discoloration of the vascular bundles and dark brown tissue scarring (Biruma et al., 2007).

Xanthomonas wilt is transmitted by insects, soil, and tools over short distances, while the movement of planting material results in long-distance dispersal. Transmission via soil may occur with animal movement in and between plantations, by plantation management practices, or by floods.

3.1.1 Management

There are no chemical or anti-bacterial solutions that effectively control Xanthomonas wilt in the field. Hence, prevention of infection and removal of infected plants are the only currently viable disease control options. It is important to not use any planting material from an infected farm, and only use clean material or tissue-cultured plants when replanting.

Since Xanthomonas wilt doesn't remain in the soil for long periods, about 4 months (Tripathi et al., 2009), replanting in the same field after a fallow period is an option. An effective solution is to remove an affected crop and intercrop with other crops or grasses which are non-hosts of the bacterium. Restricting access of people and animals through fencing and the use of foot baths at entry points of the plantation is advisable. Prevention of flooding through adequate drainage is required, and it is good practice to channel potentially contaminated water away from the plantation. Nematodes are also thought to transmit Xanthomonas wilt via wounds (Denny, 2007; Shehabu et al., 2010), and one may assume that potentially banana weevil borers also play a role in the spread of Xanthomonas wilt. Regular applications of Mocap®, Counter®, Rugby®, or other nematicide will control nematodes and banana weevil borer reducing the risk of spread of Xanthomonas wilt.

Several insects are vectors that transfer the disease to healthy plants. Insects, namely stingless bees (Apidae), fruit flies (Drosophilidae), and grass flies (Chloropidae), transmit the disease from banana to banana after being drawn to the infected ooze exuding from parts of the inflorescence of affected banana plants (Smith et al., 2008). Early bagging will assist by covering the bunches and early removal of the male bell will reduce the risk of insect pests spreading the disease.

Banana tools consisting of de-leafing knives, harvesting equipment, and de-suckering tools require sterilization between each mat using a 10% to 20% bleach solution, examples are Clorox® Concentrated Germicidal Bleach, JIK®, Activate™, or 70% alcohol (Mwangi et al., 2007).

Regular surveillance, detection, and diagnosis of the disease are very important, followed by the destruction of symptomatic plants. Plants need to be removed, including the corm, and destroyed to prevent regrowth. Destruction may involve burning similar to the control of Moko disease in the Philippines using rice husk and burning over a long period of time. More effective is destruction through injection of glyphosate herbicides using a specific applicator (glyphosate: 90 g glyphosate/L water, inject 5–15 mL of prepared solution per banana pseudostem), as this alternative is reported to be 100-fold less labor intensive than removal and burning (Blomme et al., 2010).

3.2 Moko bacterial wilt

Moko is caused by *Ralstonia solanacearum* (race 2) and is found in the Philippines, Central America, and South America and is a major disease, causing serious losses affecting many banana-exporting countries (Buddenhagen, 1994; Rillo, 1979). The current distribution of Moko in Latin America covers the following countries: Belize, Brazil (in the Amazon basin and in Sergipe), Colombia, Costa Rica, Ecuador, El Salvador, Grenada, Grenadines, Guatemala, Guyana, Honduras, Jamaica, Mexico, Nicaragua, Panama, Peru, St. Vincent, Suriname, Trinidad and Tobago, and Venezuela (CABI, 2021).

Figure 3 Internal discoloration of freshly cut banana fruits affected by Moko disease. Photo courtesy of Prof. André Drenth and Dr Lilia Carvalhais, University of Queensland.

Moko expresses disease symptoms in the fruit as deformed young banana fingers which turn black, shrivel up, and ripen prematurely. Bananas infected closer to maturity may show no outward symptoms, but the inner pulp may be discolored and decaying (see Fig. 3) (Denny, 2007; Jones, 2019a).

3.2.1 Transmission

Moko is easily transmitted from one plant to another by insects and by tools used on the plantation. The movement of soil and planting material by humans can spread it over larger distances (French, 1985; Sequeira, 1998).

3.2.2 Management

Moko is a major disease and economic losses will occur if not effectively managed. As preventive strategies it is advisable to limit soil movement using on-farm biosecurity, foot sanitation baths, and controlling movement through the farm. All tools should be sanitized between the use of each plant with a 20% blend of chlorine and water or a 70% mix of alcohol. Moko can be spread via latex transmission by bagging staff therefore it's not recommended that female flowers be removed in the field, flower removal can be performed in the fruit receival patio, just before the de-handing and packing process. A bunch cover/bag must be installed to act as a barrier to limit insect transmission. Flower removal can be performed in the fruit receival patio, just before the de-handing and packing process. The male bud should be broken off as soon as the female flowers are formed as the male flowers are visited by insects that may transmit the disease (Buddenhagen and Kelman, 1964).

Since the symptoms show within two weeks of infection regular surveillance and monitoring are required to detect and destroy infected plants (Lehmann-Danzinger, 1986). Chemical destruction using the injection of herbicides is a highly effective alternative that is 100-fold less labor intensive than manually removing plants including the corm (Blomme et al., 2010). After the destruction of affected plants, a fallow period of a minimum of 12 months, free from weeds and planted with a non-host needs to be implemented to reduce the risk of infection from bacteria present in the soil. Spread of the disease within a plantation is reduced through proper drainage maintenance and flood control (Fucikovsky and Santos, 1993). The use of drip irrigation will assist in limiting the spread of the disease.

3.3 Mokillo

Mokillo is a bacterial disease that affects the banana fruit and is reported to be caused by a range of different bacterial pathogens including *Enterobacter*

Figure 4 Symptoms of Mokillo. Images courtesy of the Queensland Department of Agriculture and Fisheries.

cowanii, Pantoea agglomerans, Burkholderia, cepacia; and Pseudomonas sp. The symptoms are expressed as blackening of the internal center in ripe fruit at time of consumption. Symptoms can also express as single small, mutated banana fruit in a hand which is elongated or deformed at the flower end (see Fig. 4). The disease is mostly diagnosed during the de-handing process, but occasionally fruit is passed on to consumers. The Centre for Tropical Agriculture in Mareeba recovered two bacteria, *Enterobacter cowanii* and *Pantoea agglomerans* from affected fruit. (Pathania et al., 2013).

Mokillo is a sporadic disease that does not cause any economic losses and is a disease that does not require control. The causal bacteria are believed to be opportunistic and enter via a small wound or lesion, damaged flower ends, mechanical damage while bagging, or damage caused by insects and wind. The disease is more prevalent during the wet, hot, and humid season (Pathania et al., 2013). Mokillo does not become systemic in the plant but instead is always confined to a single fruit in a hand. This is in contrast to the other major banana wilts such as Moko, Xanthomonas wilt and Blood disease which are highly systemic and infect every single fruit in a whole bunch.

3.3.1 Management

Careful flower removal at the correct time when the fingers are horizontal is important as at this stage, one wipe will remove the flowers from Cavendish bananas. Removing the flowers too early requires more force which may damage the tip and allow an entry point for Mokillo. Keeping the pseudo stem free from trash and removing down leaves that can damage the bunch at emergence and opening will reduce the incidence of this disease. Early bagging and protecting the bunch will reduce insect transmission and protect against transmission from wind and rain.

4 Fungal diseases affecting the fruit and their management

Key fungal diseases are listed in Table 2.

4.1 Anthracnose Colletotrichum musae

Anthracnose affects bananas both pre-harvest and post-harvest and if left unchecked will cause economic losses. Fruit losses can occur on the farm and throughout the supply chain where fruit will be downgraded and substantial losses can occur during transport, storage, and ripening. The author noted that high economic losses are regularly experienced by banana growers in India due to a lack of hygiene within the plantations, harvesting areas, packaging, and transport made worse by not pre-cooling the fruit prior to packing and transport.

Anthracnose is caused by *Colletotrichum musae,* which commonly occurs in banana plantations and produces spores in large numbers on banana leaves and dead banana flowers. Wind-driven rain disperses these spores which then infect the green fruit. After infection the fungus remains dormant until ripening. Under the right conditions on ripe fruit small round brown to black lesions start to show then the fruit quickly blackens and rots (see Fig. 5). Infield mechanical damage, leaf damage, and sucker damage to the bunch can trigger premature

Figure 5 Anthracnose on banana. Photo courtesy of Plant Village: *https://plantvillage.psu.edu/topics/banana/infos*.

ripening and anthracnose. During the processing and packing of green fruit, take care not to damage the fruit as the pathogen can enter through bruises, scratches, or cuts. Anthracnose is more widespread during wet humid times of the year and special care is required to reduce the inoculum load and prevention of infection during these months.

4.1.1 Management

Crop husbandry and hygiene are beneficial to reduce the inoculum load and includes removal of dead and decaying leaves from the pseudostem, heavy de-leafing, and de-suckering (sucker selection) for airflow within the plantation. Sufficient nutrients and irrigation are required for plant health and a well-maintained drainage system is required. Early bagging also helps to protect the bunch from leaf and insect damage. Control of black leaf streak in the plantation using fungicides also reduces the inoculum load in the field and when high levels of disease are expected the fruit can be treated prior to packing with a registered fungicide.

Regular packing house cleaning with chlorine is required to reduce inoculum. If recycled water is used, then have that water frequently checked for free choline levels and maintain 100 ppm. Cold-rooms, both on farms and at the markets, require frequent cleaning and sterilization.

4.2 Blotchy mold, Black Stain* (Mancha Negra – *Spanish)

Blotchy mold is caused by *Chaetochyrena musarum* and other fungi (Singtripop et al., 2016) which grows on semi-matured fruit, under high humidity, rain, and overcast conditions. Heat and humidity builds up under the covered bunch, providing a perfect environment for disease pressure. The damage to the fruit is superficial only but it affects the visual appearance and quality of the fruit. The fungus forms dark uneven patches over the skin and renders the fruit unsuitable for export markets (Fig. 6). The fungi causing mold are found through the plantations in the soil and on dead and decaying plant matter. The incidence of Blotchy mold increases during periods of high humidity and days with high cloud cover and rain.

Blotchy mold is common throughout the plantation during periods of prolonged rain and high humidity and can affect any part of the plantation. Extreme high disease pressure is mostly found during the first ratoon crop or a nurse suckered area. The plants in these blocks usually produce bunches all at the same time and have a higher number of leaves covering the area with dense foliage. All these factors along with a prolonged wet season will promote a very high incidence of this disease.

Figure 6 Blotchy Mold on banana fruits. Photos by Marc Jackson.

4.2.1 Management

Lowering the disease pressure can be achieved by stripping back the plant of any dead leaf matter and trimming the leaves high, and heavy de-leafing. The opening up of the canopy allows for more airflow which in turn lowers the humidity within the plantation. Removing unwanted suckers also helps improve airflow. The bunch bag also creates a microenvironment that promotes the development of the disease. Therefore, making modifications to bagging during the wet season will also help reduce the disease pressure. Two effective methods which can be used are first, tying the bag with a chimney (see Fig. 7), an escape route for hot humid air to rise and discharge. The second is to use perforated bags, with at least 25% of the bag open to allow for the exchange of air.

Application of fungicides from the ground for the control of Black or Yellow Sigatoka also can reduce the incidence but typically does not give full control. Application of fungicides from the ground covers the undersides of the leaves and dead trash within the plantation and reduces the inoculum load within the plantation.

Blotchy mold can also be removed once the bunch is harvested and in the packing house by spraying a mixture of chlorine (sodium hypochlorite) at a concentration of 100 ppm and leaving it on the fruit for some time before drying and then being washed off with fresh water. Automated pre-spraying with a chlorine mixture before entering a high-pressure bunch wash is used in Australia, additionally, well-chlorinated de-handing and clustering water tanks will aid in the removal of sooty blotch.

Figure 7 The author demonstrating how to leave a chimney for hot humid air to escape. Photo by Marc Jackson. Credit for this method to Naomi Brownrigg from Sellers Bananas, Mission Beach Australia.

4.3 Sooty mold

Sooty mold is caused by several fungi which can grow on the honeydew secreted by aphids, *Pentalonia nigronervosa,* mealy bugs, scale insects, leafhoppers, and white flies. Many saprophytic fungi can grow on this honeydew substance causing sooty mold including *Cladosporium atriellum* and *Penicillium citrinum* (Stover, 1975; Stover, 1972). The growth of these fungi causes a black discoloration over the green banana toward the crown end of the banana to the peduncle. During harvest this fruit would normally be rejected giving rise to considerable losses if left untreated.

Mealy bugs can also cause direct and/or indirect damage to plants. Direct damage occurs through sucking out sap from buds, leaves, or fruits. The indirect damage is by (i) secretion of honeydew, which forms the substrate on which mold fungi grow and produce the sooty (black) mold. The build-up of mold changes the color of the fruits from green to black and affects their aesthetic value (not attractive to consumers) (Fig. 8). Because of this, most fruits are usually rejected, leading to a considerable loss of marketable fruits. In addition, sap-sucking insects may also play a role in the transmission of virus diseases from one plant to another (Okolle et al., 2009).

Sooty mold affects the skin of the fruit only and does not affect the quality of the pulp. But since it affects the visual quality of the fruit it renders it unmarketable. It's also been noted that the use of EPE foam pads can act as a nursery for mealy bugs and without proper washing and treatment of these EPE pads, mealybugs can be spread throughout the farm.

Figure 8 Black sooty mold symptoms on banana fruits. Photo by Marc Jackson.

4.3.1 Management

Sooty mold is not classed as high priority disease and can be culled in the packing house. The incidences are normally low and usually don't require any action. Action is only required when the insect pressure is high within the plantation. Control of sooty mold involves weed control and removing the host plants of aphids. Where grass is required for erosion control or a ground cover crop is recommended, regular mechanical and chemical mowing is recommended, using a wick wiper and glyphosate. A wick applicator is a tool used to apply herbicides directly to weeds without contacting the surrounding crop plants. It consists of a long, flexible, cotton or foam wick that is soaked in the herbicide solution and then dragged along the weed foliage. The herbicide is transferred from the wick to the weed through physical contact.

Integrated Pest Management (IPM) can be used to manage the populations of aphids and mealybugs within the plantation. Many different insects will feed on aphids and mealybugs, such as lacewings and ladybirds. Cleaning and disinfestation of the EPE foam protective pads and correct storage will assist in reducing the spread of mealybugs. The use of insecticide-impregnated bags will assist by removing the mealybug and aphids entering the protected bunch.

4.4 Cigar-end rot

Cigar-end is a fungal disease caused by two different fungi *Verticillium theobromae and Trachysphaera fructigena* which produce similar symptoms but with a difference in the color and appearance of the spores. Both fungi cause the fruit to decay from the flower tip end and continue toward the crown and peduncle along the fingers. The skin blackens and ashy conidia and conidiophores cover the rotted portion giving the banana finger a burned ashy cigar-end appearance. (See Fig. 9).

Both causal agents, *Verticillium theobromae and Trachysphaera fructigena* are commonly present in plantations on dead and rotting plant material, including the flower ends. Wet and humid conditions favor the development of cigar end.

Stover (1972) reported substantial losses in Cameroon in 1968 due to cigar-end rot. However, general field practices assist with keeping this disease under control, and cigar end is not a disease of significant losses for farmers in most parts of the world.

4.4.1 Management

Controlling this disease has many similarities to managing other fungal diseases of banana fruit. Above all it is important to monitor the incidence of cigar-end

Figure 9 Advanced damage of Cigar-end rot showing grayish ash cigar end on banana fingers. Photo by Thomas K. Kwambai, KALRO, Kenya.

rot and ensure inoculum build-up in the plantation is minimized. Good field ventilation is required through de-leafing, de-suckering, and effective weed control. Early bagging and removal of the flower ends and male bud will reduce the disease in field. Inoculum levels can be further reduced by removing all the dead and decaying plant material and cutting the old pseudo stem close to the ground.

Care should be taken when covering the bunch and placing the plastic inserts. Liquid will build-up if wrongly placed and this will create high humidity and enhance fungal development. Ensure the bunch cover is tied high on the peduncle. Remove all trash or bracts. Application of Mancozeb (Dithane M45 at a rate of 60 g/20 L) and Thiophanate-methyl (Topsin M SC at a rate of 15 ml/20 L) (Gekone J) are two chemical controls that effectively control the fungi causing cigar-end rot.

Fruit from the harvested bunches with cigar end should be removed before placing into the de-latex tanks as a cut or bruised fruit that comes into contact with the inoculum could potentially be infected and develop rots associated with cigar end rot. Packing sheds and cooling chambers should be periodically cleaned and sanitized.

4.5 Crown rot

Crown rot is a disease common in Cavendish and other banana varieties in which the fruit is cut from the bunches and placed as hands or clusters in boxes for long-distance transport in a process called de-handing. This process involves cutting the crown which leaves the cut surfaces of the crown exposed to infection by a range of different fungi which can cause the crown to rot. Fungi reported to cause crown rot to evolve include, *Acremonium sp.*; *Acremonium strictum*; *Cephalosporium sp.*; *Ceratocystis paradoxa*; *Colletotrichum musae*; *Fusarium moniliforme*; *F. pallidoroseum*; *Fusarium spp.*; *Lasiodiplodia theobromae*; *Nigrospora spp.*; and *Verticillium theobromae*.

Crown rot is not seen on site in plantations during the packing process. However, the infection of the crown most likely starts here after de-handing and handling of the fruit as part of the normal packing process. The lack of hygiene in the packing house is a major contributing factor. The fungi associated with crown rot are commonly found on leaves and stalks of banana and can transfer from the stalk to the cut crown during the de-handing process in addition to infection in the de-latexing tanks when the water is contaminated with fungal spores (Kamel et al., 2016). The disease can also express itself during prolonged cold storage.

The disease blackens and rots the crown and fungal mycelium may appear to grow from the crown (see Fig. 10), as the disease worsens the rots will move

Figure 10 Crown Rot. Photo by Marc Jackson.

from the crown to the banana stalk and eventually the banana. At this point the bananas will fall away from the stalk. Noted as a significant post-harvest disorder, crown rot is of significant concern to growers and economic losses may occur if the fruit is not treated and control measures are not effectively implemented. Fruit will be either rejected or sold at a lower price, at the risk of losing your retail customers.

4.5.1 Management

4.5.1.1 Chemical control

Chemical control is often required to control crown rot in fruit being exported to distant markets. Two commonly used and effective chemical treatments are prochloraz and thiabendazole. Thiabendazole is currently commonly used in banana-exporting countries.

An unpublished study completed in 2019 written by Solano and Villalobos of Biotech Costa Rica, consisting of 11 different products to control crown rot concluded that Mertect 50 SC (a.i. thiabendazole) inhibited the growth of almost all the pathogens 100% for 36 days. The only exception was *Colletotrichum sp.* which showed some fungal growth in doses at or below 300 ppm after 25 days, although the EC_{50} is reported to be below 50 ppm. Despite this inconsistency, it is generally accepted that thiabendazole is highly effective with the added benefit that it also controls *Penicillium* sp.

Another product, Natap 90% (a.i. Natamycin) was able to successfully inhibit the growth of crown rot-causing fungi *in vitro* with 100% at application doses equal to and higher than 50 ppm. Similar results were observed with Agrobrush Fert (a.i. clove extract).

4.5.1.2 Packing house hygiene

Manual or automated chlorine sprays and then bunch cleaning before de-handing remove spores from the peduncle and fruit, assisting in control.

The author's experience is that the use of recycling de-latex tank water increases the incidence of crown rot such that water care and sanitization are required. It is important to keep the water tank clean of leaf material, damaged bananas, or any form of inoculum-carrying material. An effective solution is to add chlorine and continually measure the free chlorine, not total chlorine, in the water. Installing a probe to actively measure the free chlorine via automated metering linked to a dosing pump will assist to reduce the inoculum load in the water through sanitation. In the author's case Ecolab Australia is assisted with the equipment and the setup (see Fig. 11). It is recommended to use calcium

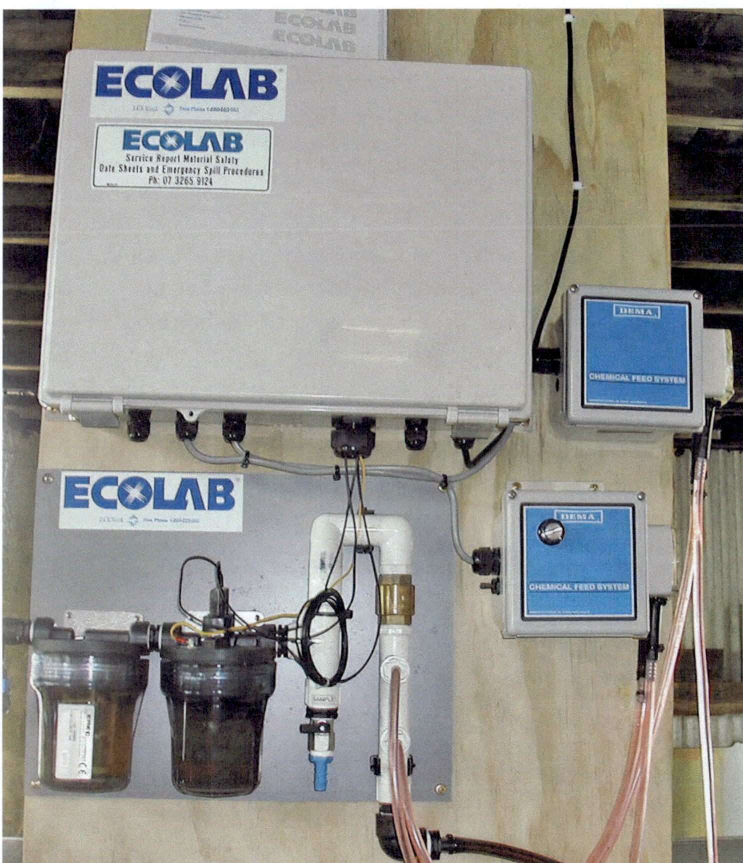

Figure 11 Ecolab monitoring and auto dosing equipment. Jackson Farming Australia. Photo by Marc Jackson.

hypochlorite for the chlorine solution. Sodium hypochlorite reacts with the latex and leaves a black stain on bananas which is further exaggerated when the water contains iron oxide.

Other treatments used in the industry include disinfectants such as quaternary ammonium and hydrogen peroxide.

Weekly deep cleaning of the packing house, consisting of cleaning and removal of all trash, build-up residue stalks, and leaf matter, then washing the floor and equipment with a solution of no less than 100 ppm chlorine. External cleaning and removal of all trash and peduncles (stalks), before sterilization with a chlorine solution of 100 ppm.

4.5.1.3 Pre-cooling

Moving the packed fruit and stacked pallets directly into a pre-cooling room at 14 degrees is an additional preventive measure to reducing crown rots. Regular cleaning and sanitization of these rooms are also required. Other treatments used in the industry include hot water treatments; UV and gamma irradiation treatments (Nelson, 2008; Lassois et al., 2010).

4.6 Diamond fruit spot

Diamond spot is caused by several *Fusarium* species, including *F. solani* and *F. incarnatum*. These species are not able to infect the banana peel but rely on small wounds, insect damage, or lesions caused by other fungi to infect. After infection through a small wound these pathogens colonize the cells surrounding the infection point first producing a raised yellow spot and preventing the cells from expanding with the growing fruit causing the peel to crack in a longitudinal fashion, resulting in a typical diamond shape (see Fig. 12), giving the disease its name. Diamond fruit spot should not be confused with banana diamond leaf

Figure 12 Diamond spot on banana fruits. Photo by Marc Jackson.

spot, which is caused by the fungi *Cercospora hayi*, *Fusarium pallidoroseum*, and/or *F. solani* (Stover, 1972).

4.6.1 Management

In commercial Cavendish plantations diamond fruit spot is managed through regular application of fungicides for the control of Black Sigatoka. Additional disease control measures include plantation hygiene, de-leafing, and removal of dead trash to reduce the inoculum load. Early bagging protects the fruit from mechanical and insect damage.

4.7 Fruit speckle

Fruit speckle is expressed as very small reddish-brown to black spots sometimes surrounded by a water-soaked margin and can be caused by a few different fungi including *Colletotrichum musae*, *Fusarium oxysporum*, *Fusarium semitectum*, and *Deightoniella torulosa*. While there are some variances to the fungi causing speckles as noted in the scientific literature, for the purpose of this chapter which is focused on disease control they are grouped together under a single heading 'Fruit Speckle'. Similar spotting occurs and common names are fruit speckle (see Fig. 13 & 14), pin head black spot, Deightoniella fruit speckle, swamp-spot, Central American speckle, and Costa Rican spot. Fruit speckle is sometimes also associated with early insect lesions, fungi, and oil and solvents from chemical application for Sigatoka control. Losses have been recorded as high as 70% for exportable quality fruit discarded at the

Figure 13 Fruit speckling. Photo by Marc Jackson.

Figure 14 Fruit speckling. Photo by Marc Jackson.

packing houses (Pasberg-Gauhl, 2000). Although fruit speckle becomes almost invisible after ripening (Stover, 1972; Wardlaw, 1972) substantial losses will occur if left untreated.

Speckling is more concentrated around the fingertip and where the banana fingers are attached to the peduncle which may be attributed to two reasons; first, the fungal spores in the flower end is falling and landing in that area on the fruit below. Second, some fruit speckling is attributed to the use of chemicals and oil used in controlling Black and Yellow Sigatoka falling on the formed hands.

Spores are produced in large numbers on dead leaf material during wet conditions and are spread in air currents to the fruit. High humidity, overcast weather, and rainy conditions combined with a lack of airflow increase the disease pressure within the plantation. When the disease pressure is high fruit speckle covers all the fruit and renders the fruit not suitable for export and is sold at a lower grade and to local markets or just discarded.

Symptoms for Deightoniella fruit speckle include sunken black and dark brown spots with a green halo, spotting is greatest at the fingertip end on the banana skin and the lesions are 2–4 mm wide. Deightoniella is associated with wet weather and conditions of water lodging and proper drainage to remove the water and lower the water table is required. Deightoniella fruit speckle is not of particular concern when bananas are grown in well-drained soils. Water-lodged soils will contribute to the development of this disease and economic losses may occur. Farm location and site selection, including avoiding heavy clay low-laying areas, are key factors in producing high-quality fruit suitable for export.

Fruit speckle caused by *Colletotrichum musae* is expressed as small lesions, brown to black in color, 0.5–1 mm in diameter, often with a water-soaked margin. Spots caused by the two *Fusarium* species are brown in color and reach a maximum diameter of 0.5 mm (Pasberg-Gauhl, 2000).

4.7.1 Management

Sap oozing from the fruit after damage or after flower thrip infestations coming into contact with fruit skin can increase the number of spots of speckle on fruit, particularly those caused by Fusarium. Early removal of bracts and bagging bunches to prevent bird and bat damage can result in sap contacting fruit, increasing fruit speckle damage. There is no link between bunch cover type and the incidence of speckle.

A trial conducted on the author's farm Jackson Farming, Queensland, Australia, involving early bud injection (see Fig. 15), gave very good results. Early control of flower thrips by bud injection did lower the incidences of pin head black spot; importantly, bud injection had to take place before the bud emerged from the throat of the plant. Manual removal of the flowers also aided the control. Pinhead blackspot or Costa Rican spot is controlled by spraying the stem and bunch with a solution of either Chlorothalonil – Bravo 720 SC at 5 ml per 1 L water or Mancozeb – Dithiane M45 at 5 gm per 1 L water or Mauzate 75WP at 4 gm per 1 L of water. Topsin® Fungicide, active Thiophanate-methyl, is used in the Philippines and China for the control of fruit speckle and pinhead blackspot.

Bagging with an inner paper bag and or nonwoven cloth bag and tying the knots high on the stalk to prevent the bag from resting on the upper hands also helps to reduce damage. Early bagging at bud emergence also helps, but this is also extra work as the forming bunch will require removal of the bracts and the bell in separate operations afterward.

De-leafing, cutting down old pseudo stems, and cleaning decaying trash will reduce the inoculum load and assist in control. Heavier than normal de-leafing and de-suckering will increase airflow throughout the plantation facilitating drying and reducing inoculum build-up. Using perforated bags or tying the bag with a chimney (Fig. 7) allowing the hot and humid air to escape aids in reducing the disease pressure. Well-maintained drainage is required and will aid in reducing water lodging and drying the farm rapidly. Under-tree misting for control of Black and Yellow Sigatoka is beneficial in reducing the inoculum levels of pin head blackspot. Economic losses will occur from not controlling pinhead blackspot and particular attention should be given to this disease during the wet humid season.

4.8 Banana freckle

Banana Freckle is a disease initially thought to be caused by a single fungal species, *Phyllosticta musarum* and its sexual form, *Guignardia musae* (van der Aa and Vanev, 2002). More recent research revealed that at least three different species cause freckle; *Phyllosticta cavendishii, Phyllosticta musarum*, and *Phyllosticta maculata* (Wong et al., 2012). These fungi have different and

Figure 15 Bell/Bud Injection poster by Richard Piper.

limited geographic distributions causing freckle in different varieties of banana. *P. cavendishii* is the only species able to infect Cavendish AAA.

Figure 16 Banana freckle infected fruit (Image: Jose Liberato, Northern Territory Department of Primary Industry and Fisheries).

Freckle expresses itself on the leaves as numerous small discrete reddish-brown spots on the upper side of the leaf which turn black over time to form patches of protruding fruiting bodies. These give the leaf a sandpaper-like feel to the touch. On the fruit small red-brown to black spots are formed which sometimes are surrounded by a dark green water-soaked halo (Fig. 16). Banana freckle is a leaf disease and spreads to the fruit via rainwater or irrigation water. The spores can also spread by infected leaf material or infected material touching the fruit. The spores land on the fruit and form small dark lesions. The disease favors wet tropical humid climates.

Banana freckle is regarded as an important disease and high economic losses may occur if left untreated. Losses of 78% and downgrading from A class to B class of 59% was measured in the Philippines (Corcolon and Raymundo, 2008). Downgrading or complete destruction of fruit is common due to the market requirements for high-quality fruit. Banana freckle does not affect the eating quality of the fruit but is a cosmetic defect, not tolerated by retail markets in the western world. Banana freckle affecting Cavendish bananas was detected in the Northern Territory Australia in 2013 but an intensive eradication campaign has been successful leading to area freedom being reestablished in 2019 (Drenth and Wong, 2023). Eradication was achieved by destroying all banana plants in certain areas and intensive monitoring (McMaster et al., 2020).

4.8.1 Management

Destroy infected plants and replant them with clean disease-free planting material. Do not use infected sucker material for replanting and stop the movement of leaf and infected fruit. Early bagging to cover the bud on emergence. Fungicides used for Sigatoka control such as Mancozeb, propiconazole, or flusilazole are also effective in controlling freckle. Maintain crop hygiene by cleaning away all dead and decaying material and removing affected leaves over the bunch. Keep the plant's health and nutrition up to date.

5 Conclusion

Infield control of most banana diseases of the fruit involves the application of integrated approaches involving the following simple and cost-effective measures to assist in lowering inoculum build-up and reduce the incidence of disease.

Site selection, insuring well-drained soils, sunshine, and the lack of free water will lower the incidence of fungal diseases that favor high humidity and wet conditions. Seasonal dry tropical arid weather patterns aid in lowering inoculum and these areas are able to produce high-quality fruit free from disease (see Fig. 17). During the rainy season, well-maintained drainage is required to move the surface and sub-surface water away from the plantation. Since some bacterial and fungal diseases can travel via soil movement in turbid water caused by flooding, avoid areas that are prone to flooding.

Tropical arid banana production is common for organic production where inputs are limited, and disease levels remain low. These are classified as sustainable 'climate smart' production areas. The key benefits of farming in these areas are high production with lower inputs which, in turn, increases the profitability of these farmers.

Water is required for banana production and is a requirement in site selection for sustainable tropical arid production. Where water is limited drip irrigation is recommended and where water is more abundantly available under-tree mini sprinklers are recommended as the under-tree mini sprinklers create a microenvironment that enhances banana productivity.

Bacterial and fungal diseases are mostly spread by planting material, leaves and bunches, sap/latex, insects, tools, wind, soil, and water. The best solution in controlling disease is for farmers to have an on-farm biosecurity plan to limit the movement of people, animals, soil, and banana plant material by fencing the farm and having dedicated entry and exit points fitted with foot baths. More information can be found on the Queensland Government website https://www.publications.qld.gov.au/dataset/panama-disease-tropical-race-4-grower-kit and although designed for Fusarium wilt Tropical Race 4 disease, the

Figure 17 Bunches produced in an arid tropical climate, free from fungal diseases. Photo by Marc Jackson.

information is relevant to all disease transmission and can be used as a guide for the implementation of an on-farm quarantine system to stop the movement of planting, leaf and banana plant material off- and on-farm. Best practice also involves the purchase of clean planting material from an accredited provider.

Infield transmission of especially bacterial diseases can be minimized by cleaning and sterilizing tools between each mat. Sterilizing solutions recommended are 5% chlorine, 6% quaternary ammonium, or 70% alcohol. Effective agriculture disinfectants for tools, foot baths, vehicle tires, and equipment wash down are:

- Sporekill® 120 g/L Didecyl Dimathyl-ammonium chloride at 100 mL/10 L of water (1% solution);
- Steri-maX® 120 g/L Didecyl Dimathyl-ammonium chloride at 100 mL/10 L of water (1% solution); and
- Quaternary ammonium at 2 mg/L for tools and 4 mg/L for foot baths.

Good plant health and nutrition will boost the immunity of banana plants. Avoid planting in sandy and heavy clay water-lodged soils and ensure adequate fall for drainage water to escape off-farm. Waterlogged soils, lack of adequate irrigation, poor soil selection (sand ridges), and low nutrition inputs will lower the plant's own defense system (Ferreira et al., 2007). Calcium strengthens cell walls and protects the cells against fungal and bacterial infections (Marschner, 2011). Regular soil and leaf testing will indicate the levels of nutrition required and advise on suitable nutrient applications.

Avoid infield mechanical damage as this can lead to a number of fungal diseases. When investigating care should be taken to isolate the cause. For example, in Australia, health and safety require all tractors to be compliant by having a rollover frame (ROP) mounted on the tractor frame. These ROPs protrude above the head of the driver and easily can damage bunches hanging into the tractor row. The removal of leaf material and accidentally damaging the bunch with de-leafing tools allow diseases to enter. Management of suckers is also important as a sucker not trained can grow up inside the bag and damage the fruit. Train suckers away from the hanging bunch by tying it with string.

Early bagging, bunch protection, and removal of the male bud will protect against insect transmission of fungal and bacterial disease. Bagging will also protect against leaf rub lesions that allow diseases to enter.

Crop husbandry and hygiene by trimming dead and decaying planting material, regular sucker selection, removing weeds, and opening up the

Figure 18 Using IPM for the control of red spider mite. https://bugsforbugs.com.au/. Photo by Marc Jackson.

canopy to allow air to move and travel throughout the plantation will assist in controlling fungi.

Insect control by way of IPM, using naturally or introduced insects to control known pests that are the transmitters of disease is required (Fig. 18). Remove pest host weeds.

Periodic inspections for early detection of disease levels and isolation and destruction of infected plants.

An effective Sigatoka management program involving de-leafing and application of fungicides will assist as most fungal diseases are controlled by the same chemicals used to control Sigatoka.

6 Where to look for further information

- Australian banana best practice factsheet, managing banana speckle. https://www.horticulture.com.au/globalassets/hort-innovation/resource-assets/ba13004-managing-fruit-speckle-banana-industry-pdf.pdf.
- Banana Deightoniella leaf & fruit spot. https://www.pestnet.org/fact_sheets/banana_deightoniella_leaf__fruit_spot_309.htm.
- Banana diseases. https://www.isppweb.org/names_banana_common.asp.
- Better bananas. https://betterbananas.com.au/problem-solver/fruit/.
- Bugs for bugs. https://bugsforbugs.com.au/.
- Cigar-end rot on banana. https://www.plantwise.org/KnowledgeBank/pmdg/20157800297.
- Controlling bacterial wilt of enset using cultural methods. https://www.plantwise.org/KnowledgeBank/factsheetforfarmers/20157800326.
- Controlling bunch pests with Chlorpyrifos dust user guide factsheet. https://abgc.org.au/wp-content/uploads/2018/11/Fact-Sheet-Controlling-bunch-pests-with-Chlorpyrifos-Dust-Nov-18-Update.pdf.
- Ecolab Australia. https://www.ecolab.com/locations/ecolab-australia.
- The cause, distribution and economic importance of fruit speckle of banana in north Queensland. https://era.daf.qld.gov.au/id/eprint/2450/1/The%20cause%2C%20distribution%20and%20economic%20importance%20of%20fruit%20speckle%20of%20banana%20in%20north%20Queensland%20%28BA05001%29.pdf.
- The cause and management of crown rot of banana. https://www.horticulture.com.au/globalassets/laserfiche/assets/project-reports/ba13011/ba13011-final-report-386.pdf.
- Pacific pests and pathogens - Fact sheets. https://www.pestnet.org/fact_sheets/banana_freckle_124.pdf.

7 References

Bagging technique improves fruit quality for Sellars Bananas. https://betterbananas.com.au/2022/10/07/grower-case-study-bagging-technique-improves-fruit-quality-for-sellars-bananas/.

Biruma, M., Pillay, M., Tripathi, L., Blomme, G., Abele, S., Mwangi, M., Bandyopadhyay, R., Muchunguzi, P., Kassim, S. and Nyine, M. (2007). Banana Xanthomonas wilt: A review of the disease, management strategies and future research directions. *African Journal of Biotechnology* 6(8), 953–962.

Blomme, G., Dita, M., Jacobsen, K. S., Pérez Vicente, L., Molina, A., Ocimati, W., Poussier, S. and Prior, P. (2017). Bacterial diseases of bananas and enset: Current state of knowledge and integrated approaches toward sustainable management. *Frontiers in Plant Science* 8, 1290.

Blomme, G., Turyagyenda, L., Mukasa, H. and Eden-Green, S. (2010). The effectiveness of different herbicides in the destruction of banana Xanthomonas wilt infected plants. *African Crop Science Journal* 16(1), 103–110.

Buddenhagen, I. (1994). Banana diseases caused by bacteria. In: Ploetz, R. C., Zentmyer, G. A., Nishijima, W. T., Rohrbach, K. G. and Ohr, H. D. (Eds.), *Compendium of Tropical Fruit Diseases*. St Paul, MN: APS Press, pp. 15–17.

Buddenhagen, I. and Kelman, A. (1964). Biological and physiological aspects of bacterial wilt caused by *Pseudomonas solanacearum*. *Annual Review of Phytopathology* 2(1), 203–230.

CABI. (2021). *Ralstonia solanacearum Race 2 (Moko Disease)*. Wallingford, UK: Invasive Species Compendium. Available at: https://www.cabi.org/isc/datasheet/44999.

Corcolon, B. M. and Raymundo, A. D. (2008). Estimating yield losses in banana due to freckle disease caused by *Phyllosticta musarum* (Cke.) Van der Aa. *Philippine Journal of Crop Science* 33(2), 75–85.

Denny, T. (2007). Plant pathogenic *Ralstonia* species. In: Gnanamanickam, S. S. (Ed.), *Plant-Associated Bacteria*. Dordrecht, South Holland: Springer, pp. 573–644.

Drenth, A. and Kema, G. (2021). The vulnerability of bananas to globally emerging disease threats. *Phytopathology* 111(12), 2146–2161.

Ferreira, R. B., Monteiro, S., Freitas, R., Santos, C. N., Chen, Z., Batista, L. M., Duarte, J., Borges, A. and Teixeira, A. R. (2007). The role of plant defence proteins in fungal pathogenesis. *Molecular Plant Pathology* 8(5), 677–700.

French, E. (1985). Interaction between strains of *Pseudomonas solanacearum*, its hosts and the environment. In: *Proceedings of an International Workshop, Philippine Council for Agriculture and Resources Research and Development*, Philippines, 8–10 October 1985, pp. 99–104.

Fucikovsky, L. and Santos, M. (1993). Advance of bacterial wilt in bananas in Mexico. In: *Proceedings of the-ACIAR, Australian Centre for International Agricultural Research, Australia*, pp. 341–342.

Global Biodiversity Information Facility (GBIF Secretariat). (2021). *Xanthomonas campestris (Pammel, 1895) Dowson 1939, the Integrated Taxonomic Information System*. Denmark: Checklist Dataset. Available at: https://www.gbif.org/species/5427664. DOI: 10.15468/39omei.

Jones, D. R. (2019). Diseases caused by bacteria and phytoplasmas – Bacterial wilt diseases. In: Jones, D. R. (Ed.), *Handbook of Diseases of Banana, Abacá and Enset*. Wallingford, UK: CAB International, pp. 296–329.

Jones, D. R. and Stover, R. H. (2019a). Fungal diseases of banana fruit – Preharvest diseases. In: Jones, D. R. (Ed.), *Handbook of Diseases of Banana, Abaca and Enset*. Wallingford, UK: CAB International, pp. 255–270.

Jones, D. R. and Muirhead, I. F. (2019b). Fungal diseases of banana fruit – Postharvest diseases. In: Jones, D. R. (Ed.), *Handbook of Diseases of Banana, Abaca and Enset*. Wallingford, UK: CAB International, pp. 271–287.

Kamel, M. A. M., Cortesi, P. and Saracchi, M. (2016). Etiological agents of crown rot of organic bananas in Dominican Republic. *Postharvest Biology and Technology* 120, 112–120.

Lassois, L., Jijakli, M. H., Chillet, M. and De Lapeyre De Bellaire, L. (2010). Crown rot of bananas: Preharvest factors involved in postharvest disease development and integrated control methods. *Plant Disease* 94(6), 648–658.

Lehmann-Danzinger, H. (1986). The distribution of Moko disease in Central and South America and its control on plantains and bananas. In: *Proceedings of the for Improving Citrus and Banana Production in the Caribbean Through Phytosanitation*, St. Lucia, Australia, 2–5 December 1986, pp. 130–152.

Marschner, H. (2011). *Marschner's Mineral Nutrition of Higher Plants*. London: Academic Press.

Maryani, N., Lombard, L., Poerba, Y. S., Subandiyah, S., Crous, P. W. and Kema, G. H. J. (2019). Phylogeny and genetic diversity of the banana Fusarium wilt pathogen *Fusarium oxysporum* f. sp. *cubense* in the Indonesian centre of origin. *Studies in Mycology* 92(1), 155–194.

Mason, E. W. and Hughes, S. J. (1951). Phragmocephala gen. nov. Hyphomycetarum. *The Naturalist* 76(838), 97–105.

McMaster, C. A., Tran-Nguyen, L. T. T., Voutsinos, M. Y., Cook, S. E., Condé, B. D., West, S. J., Nguyen, T. V. and Liberato, J. R. (2020). Outbreak of freckle disease (*Phyllosticta cavendishii*) on Cavendish bananas in Australia and the initial biosecurity response. *Plant Pathology* 69(1), 38–49.

Meredith, D. S. (1961). Fruit-spot ('speckle') of Jamaican bananas caused by *Deightoniella torulosa* (Syd.) Ellis: I. *Transactions of the British Mycological Society* 44(1), 95–IN10.

Mwangi, M., Bandyopadhyay, R., Ragama, P. and Tushemereirwe, W. K. (2007). Assessment of banana planting practices and cultivar tolerance in relation to management of soilborne *Xanthomonas campestris* pv *musacearum*. *Crop Protection* 26(8), 1203–1208.

Nelson, S. (2008). Postharvest rots of banana. *Plant Disease* 54, 1–4.

Okolle, J. N., Fansi, G. H., Lombi, F. M., Sama Lang, P. and Loubana, P. M. (2009). Banana entomological research in Cameroon: How far and what next. *The African Journal of Plant Science and Biotechnology* 3(1), 1–19.

Pasberg-Gauhl, C. (2000). *Fruit Speckling' on Bananas in the Atlantic Zone of Costa Rica*. Costa Rica: BASF.

Pathania, N., Vawdrey, L. and Grice, K. (2013). Wet-season watch for Mokillo. *Australian Bananas*. Available at: https://abgc.org.au/wp-content/uploads/2013/03/abgc_banana_mag_iss38_autumn2013_DRAFT-04.pdf.

Ray, J. D., Subandiyah, S., Rincon-Florez, V. A., Prakoso, A. B., Mudita, I. W., Carvalhais, L. C., Markus, J. E. R., O'Dwyer, C. A. and Drenth, A. (2021). Geographic expansion of banana blood disease in Southeast Asia. *Plant Disease* 105(10), 2792–2800.

Rillo, A. (1979). Bacterial wilt of banana in the Philippines. *Bulletin Phytosanitaire de la FAO* 27(4), 105–108.

Safni, I., Cleenwerck, I., De Vos, P., Fegan, M., Sly, L. and Kappler, U. (2014). Polyphasic taxonomic revision of the *Ralstonia solanacearum* species complex: Proposal to emend the descriptions of *Ralstonia solanacearum* and *Ralstonia syzygii* and reclassify current *R. syzygii* strains as *Ralstonia syzygii* subsp. *syzygii* subsp. nov., *R. solanacearum* phylotype IV strains as *Ralstonia syzygii* subsp. *indonesiensis* subsp. nov., banana blood disease bacterium strains as *Ralstonia syzygii* subsp. *celebesensis* subsp. nov. and *R. solanacearum* phylotype I and III strains as *Ralstoniapseudosolanacearum* sp. nov. *International Journal of Systematic and Evolutionary Microbiology* 64(9), 3087–3103.

Sequeira, L. (1958). Bacterial wilt of bananas: Dissemination of the pathogen and control of the disease. *Phytopathology* 48, 64–69.

Sequeira, L. (1998). Bacterial wilt: The missing element in international banana improvement programs. In: Prior, P. H., Allen, C. and Elphinstone, J. (Eds.), *Bacterial Wilt Disease*. Berlin, Heidelberg: Springer, pp. 6–14.

Shehabu, M., Addis, T., Mekonen, S., De Waele, D. and Blomme, G. (2010). Nematode infection predisposes banana to soil-borne *Xanthomonas campestris* pv. *musacearum* transmission. *Tree and Forestry Science and Biotechnology* 4(2), 63–64.

Singtripop, C., Hongsanun, S., Li, J., De Silva, N. I., Phillips, A. J. L., Jones, G. E. B., Bahkali, A. H. and Hyde, K. D. (2016). *Chaetothyrina mangiferae* sp. nov., a new species of *Chaetothyrina*. *Phytotaxa* 255(1), 21–33.

Smith, J., Jones, D., Karamura, E., Blomme, G. and Turyagyenda, F. (2008). *An Analysis of the Risk from Xanthomonas campestris pv. Musacearum to Banana Cultivation in Eastern, Central and Southern Africa*. Montpellier, France: Bioversity International.

Stover, R. H. (1972). *Banana, Plantain and Abaca Diseases*. Kew, UK: Commonwealth Mycological Institute.

Stover, R. H. (1975). Sooty moulds of bananas. *Transactions of the British Mycological Society* 65(2), 328.

Tripathi, L., Mwangi, M., Abele, S., Aritua, V., Tushemereirwe, W. K. and Bandyopadhyay, R. (2009). Xanthomonas wilt: A threat to banana production in East and Central Africa. *Plant Disease* 93(5), 440–451.

Van Der Aa, H. A. and Vanev, S. (2002). *A Revision of the Species Described in Phyllosticta*. Utrecht, Netherlands: Centraalbureau voor Schimmelcultures (CBS).

Von Arx, J. A. (1957). Die arten der Gattung *Colletotrichum* Corda. *Phytopathologische Zeitschrift* 29, 413–468.

Wardlaw, C. W. (1972). *Banana Diseases Including Plantains and Abaca*. London, UK: Longmans, Green & Co., Ltd.

Wong, M. H., Crous, P. W., Henderson, J., Groenewald, J. Z. and Drenth, A. (2012). *Phyllosticta* species associated with freckle disease of banana. *Fungal Diversity* 56(1), 173–187.

Yabuuchi, E., Kosako, Y., Yano, I., Hotta, H. and Nishiuchi, Y. (1995). Transfer of two *Burkholderia* and an *Alcaligenes* species to *Ralstonia* gen. nov.: Proposal of *Ralstonia pickettii* (Ralston, Palleroni and Doudoroff 1973) comb. nov., *Ralstonia solanacearum* (Smith 1896) comb. nov. and *Ralstonia eutropha* (Davis 1969) comb. nov. *Microbiology and Immunology* 39(11), 897–904.

Chapter 22

Postharvest diseases of bananas and their sustainable management

M. Gloria Lobo and Goretti Díaz-Delgado, Instituto Canario de Investigaciones Agrarias (ICIA), Spain; and Francisco Javier Fernández-Rojas, Cooperativa Platanera de Canarias (COPLACA), Spain

1 Introduction

Bananas (*Musa acuminata*) are a climacteric perishable fruit. They are susceptible to a variety of postharvest diseases during storage, transportation and commercialization causing severe fruit rot which affects quality and results in major economic losses. The most common and serious diseases include:

- crown rot;
- anthracnose;
- mokillo fruit speckle;
- squirter; and
- ceratocystis fruit stem-end rot.

Most of these diseases can be controlled by hygiene and the use of fungicides, refrigeration, controlled atmospheres or physical treatments such as hot water (HW), ultraviolet C (UV-C) irradiation, etc. Consumers are increasingly demanding fruit that is non-chemically treated and sustainably produced

http://dx.doi.org/10.19103/AS.2022.0108.22

to secure food safety and environmental sustainability. Good management practices during cultivation, harvest, transportation to the packaging facilities, packing, refrigeration, ripening and commercialization can avoid pathogen infestation and proliferation. More recently, physical and biological control strategies have emerged as promising alternatives to the use of synthetic fungicides to reduce postharvest diseases, especially when combined with current good management practices. Given their impact, this chapter discusses crown rot and anthracnose in detail, with a briefer coverage of minor diseases such as mokillo fruit speckle, squirter and ceratocystis fruit stem-end rot.

2 Crown rot

2.1 Symptoms, pathogen characteristics and diagnosis

Crown rot is a postharvest disease and is a major cause of economic loss in banana-producing countries when control strategies/treatments are not effectively implemented. Lassois and de Lapeyre de Bellaire (2014) reported losses of 10–86%, Alvindia et al. (2000) described losses of 86% in untreated fruits from the Philippines, Kuyu and Tola (2018) reported 55% losses in southwest Ethiopia and Anthony et al. (2004) reported 20% losses in Sri Lanka. Krauss and Johanson (2000) recorded a 10% loss in the United Kingdom of bananas from the Windward Islands and indicated that this incidence rises periodically during the rainy season. Crown rot pathogens are found wherever bananas are grown but are more prevalent in areas of high rainfall, especially where growers do not follow good field and packinghouse sanitation practices, and in the fruit of poor quality.

The disease was not a problem prior to the 1960s when the Gros Michel subgroup cultivars were shipped as complete bunches (Fig. 1a). After these cultivars were replaced by cultivars in the Cavendish subgroup due to their resistance to Fusarium wilt disease caused by *Fusarium oxysporum* f. sp. *cubense* Race 1 postharvest problems started to occur. Cavendish fruit is not as robust as Gros Michel and bunch shipping was discontinued and replaced by 'de-handing' the bunches and cutting the fruit into clusters, consisting of several banana fruits joined by the crown tissue and packed into cartons (Fig. 1b) (Lassois et al., 2010a). The cut surfaces at the pedicel created in the packaging-houses, when the bunch is de-handed using a knife, is a prime site for infection by different opportunistic plant pathogens. Fungal pathogens may infect the cut surface and colonise the fruit tissues, from water contaminated with fungal spores used for washing bananas (Hailu et al., 2013). Although infection mainly occurs during harvest and processing time, the visible symptoms appear after transportation and commercialization.

Figure 1 (a) Gros Michel subgroup cultivars commercialized in bunches and (b) Cavendish subgroup commercialized in hands or clusters. Images from Francisco Javier Fernández-Rojas.

2.2 Crown rot disease symptoms

Crown rot is caused by a consortium of different fungi and symptom expression and severity depends on the quantity and proportion of the fungal species involved in the infection; however, the progression of disease symptoms is very similar. The first symptoms appear at the cut crown surface where the hands are separated from the bunch stem. Fungi colonise the crown, giving it a whitish, grey or pink colour (Lukezic et al., 1967) (Fig. 2). The mycelium and the rot itself spoil the fresh, clean appearance of the ripening fruit and the fingers eventually fall from the weakened crown if the rot penetrates deeply resulting in detaching of the fruit from the peduncle/crown at an advanced stage of rot (Fig. 2). Incorrect postharvest ripening and storage conditions accelerate disease development. Moreover, the disease can also trigger early banana ripening causing wastage during shipping and marketing due to the ethylene released by stressed and necrotic tissue (Simmonds, 1963; Dominguez and Vendrell, 1993), and by mycelia of fungi such as *Colletotrichum musae* (Peacock and Muirhead, 1974; Daundasekera et al., 2003, 2008; Lassois et al., 2010a).

2.3 Pathogen characteristics and diagnosis

The disease is caused by several different fungi that grow and sporulate on leaves, flowers, bracts and transitional leaves that are dead, or decomposing, in plantations. The number and composition of the fungal species complex involved usually vary and depend upon the crop location and season. Fungal

Figure 2 Crown rot disease evolution. (a) Percentage of crown covered by mycelium; (b) Disease progression from the crown to the pedicel and banana fingers. Images from M. Gloria Lobo and Francisco Javier Fernández-Rojas.

populations causing crown rot may differ between banana plants and even between clusters originating from the same plant (Lukezic and Kaiser, 1966).

The most frequently isolated and identified species involved in causing crown rot are *Musicillium theobromae* (formerly *Verticillium theobromae*), *Colletotrichum musae, Ceratocystis paradoxa* (*Thielaviopsis paradoxa; Chalara paradoxa* (synonym)) being *Thielaviopsis musarum* specific for the *Musa* genus, *Lasiodiplodia theobromae, Nigrospora sphaerica, Cladosporium* spp., *Acremonium* spp., *Penicillium* spp., *Aspergillus* spp., different *Fusarium* such as *F. oxysporum, F. proliferatum, F. sporotrichoides, F. semitectum, F. solani* and *F. verticillioides* (Lassois and de Lapeyre de Bellaire, 2014).

Colletotrichum musae is considered to be the most prevalent in the Windward Islands (Finlay and Brown, 1993) and Ethiopia (Alemu, 2014). Indrakeerthi and Adikaram (2011) identified *F. semitectum, Colletotrichum musae* and *Colletotrichum gloeosporioides* as the major fungi causing crown rot in Sri Lanka. In Mauritius and Malaysia, the dominant fungi isolated were *Colletotrichum musae* together with *F. incarnatum* and/or *F. verticillioides; Fusarium* spp. feature prominently in China and Central and South America (Jones and Muirhead, 2019). *Curvularia* spp. and *Gliomastrix musicola* have also been found in Malaysia (Sepiah and Nik Mohd, 1987). *L. theobromae* (formerly *Botryodiplodia theobromae*) was isolated and identified by Kumah et al. (2013) as the main causative pathogen involved in crown rot disease of 'Medium Cavendish' bananas from different districts in the eastern region of Ghana. In diverse regions of India, the fungus *L. theobromae* was the major pathogen responsible for crown rot disease, with *Colletotrichum musae* and *Fusarium*. spp. identified as minor pathogens (Thangavelu et al., 2007).

- *F. pallidoroseum* (41.3%), *Colletotrichum musae* (24.3%), *V. theobromae* (7.2%) and *Fusarium* complex (*F. oxysporum, F. solani, F. moniliforme, F. equisetti, F. ventricosum* and *Fusarium* sp.) (19.3%) were identified in Cuba (Perez Vincente et al., 2001), while in 'Prata Anã' bananas from Brazil, the more frequent observed fungi were *Colletotrichum musae* (the primary agent of examined rotting fruit), *Trichoderma harzianum, F. equisetii, Penicillium* sp., *Aspergillus parasiticus, Trichothecium roseum, Colletotrichum acucatum, Alternaria* sp., *Cladosporium musae* and *Curvularia lunata* (Moraes et al., 2006). According to the findings of Marin et al. (1996) in Grande Naine bananas from Latin America (Mexico, Guatemala and Costa Rica), *F. verticillioides* and *F. semitectum* were the most pathogenic species, while Griffee and Burden (1976) considered *Colletotrichum musae, L. theobromae, Ceratocystis paradoxa, F. semitectum* and *F. graminearum* as major pathogens involved in this disease in the Caribbean.

Du et al. (2017) first isolated *F. chlamydosporum* from symptomatic fruit in Fujian Province (China), and Wailiullah et al. (2022) first isolated *F. proliferatum* in Georgia in mature banana fruits (cv. 'Pisang Awak') with crown rot symptoms. Kamel et al. (2016) found that the complex was dominated by *Fusarium* (47.0%), followed by *Lasiodiplodia* (7.4%) and *Colletotrichum* (1.1%) in organic bananas from the Dominican Republic. Nevertheless, they concluded that the aetiological agents of crown rot disease ranked, based on their presence and pathogenicity, as *F. incarnatum*, *Colletotrichum musae*, *F. verticillioides*, *F. sacchari* and *L. theobromae*. Kouadio et al. (2017) identified 13 fungal species involved in banana postharvest diseases in Côte d'Ivoire during rainy seasons. In Australia, *F. pallidoraseum* and *V. theobromae* were more common in the dry period with *Colletotrichum musae* being more prominent in the wet periods. *N. sphaerica*, although rare, is also known to induce crown rot, especially in fruits produced during low winter temperatures (Gowen, 1995). In the Canary Islands, Spain, *Acremonium* sp., *Alternaria alternata*, *Cladosporium cladosporoides*, *Colletotrichum* sp., *Colletotrichum musae*, *Fusarium* spp., *F. moniliforme*, *F. moniliforme* var. subglutinans, *F. oxysporum*, *F. proliferatum*, *F. roseum*, *F. musae*, *Geotrichum* sp., *Gliocladium* sp., *Nigrospora* sp., *N. oryzae*, *Penicillium* spp., *Penicillium steckii*, *Stemphylium* sp., *M. theobromae*, *Simplicillium* sp. and *Thielaviopsis musarum* were isolated in the crowns and the water tanks at the packing facilities (Hernández et al., 1984; Perera et al., 2012; Díaz-Delgado et al., 2021) (Figs. 3 and 4).

Many authors have reported on the strong pathogenicity of *Colletotrichum musae*, which can trigger an infection from a minimal amount of inoculum

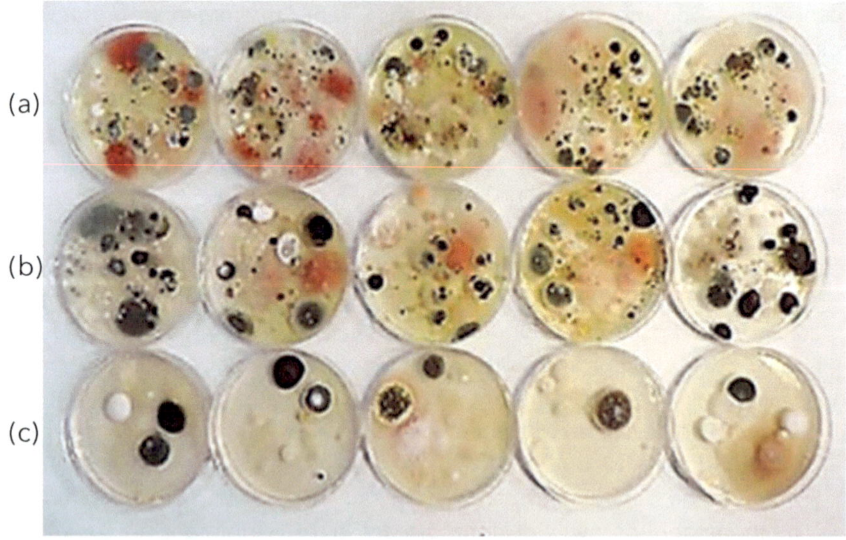

Figure 3 Fungi colonies from washing water tanks of packing facilities in Tenerife, Canary Islands (Spain). (a) Water, (b) 10^{-1} and (c) 10^{-2} dilutions on potato dextrose agar (PDA) after 7 days at 25°C. Image from M. Gloria Lobo.

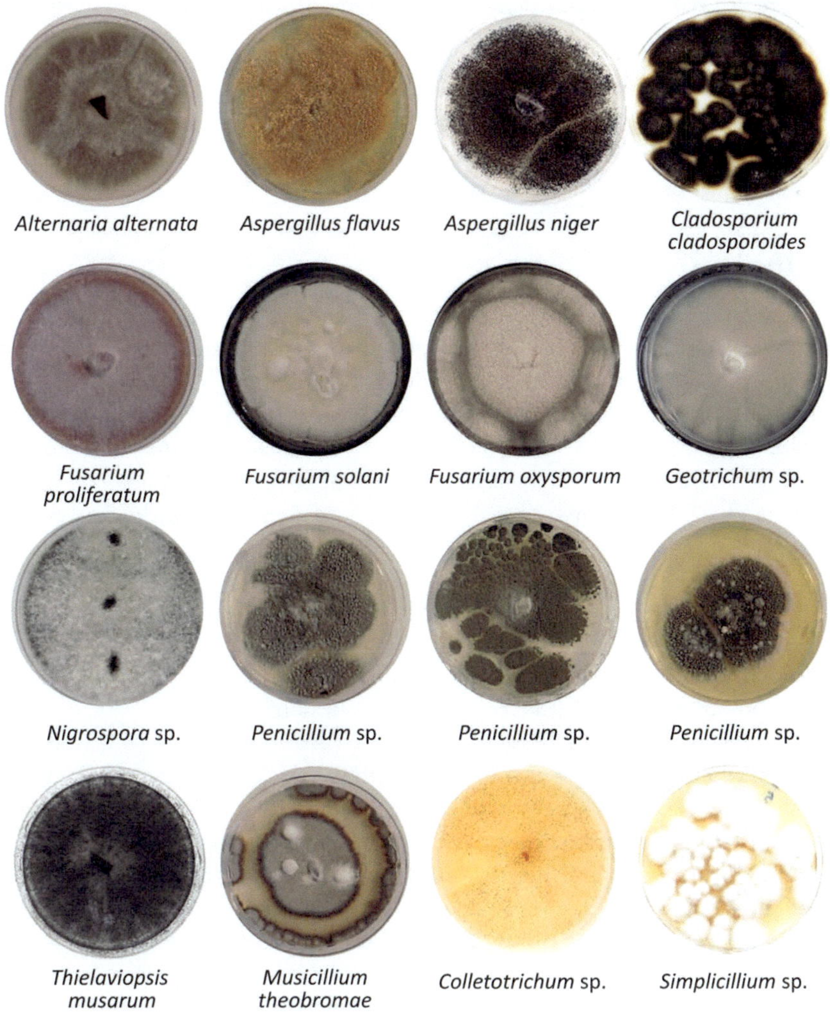

Figure 4 Fungi complex causing crown rot in banana isolated in washing water from packing facilities in Tenerife (Canary Islands, Tenerife, Spain). Images from Francisco Javier Fernández-Rojas, Goretti L. Díaz-Delgado and M. Gloria Lobo.

(Lukezic et al., 1967; Finlay and Brown, 1993), although the damage is greater when co-inoculated with other pathogens such as *L. theobromae* and *F. verticillioides* (Anthony et al., 2004). The specific pathogenicity of the different microorganisms involved as well as their interaction requires a better understanding of the role of each fungus in the complex to better characterise their pathogenicity and select an effective treatment.

2.4 Crown rot: disease cycle and epidemiology

The disease cycle itself is very complicated as the disease is caused by different fungal species in different ratios, but there are some general features and disease progression depends on the susceptibility of each banana variety, amount of inoculum, the physiological quality of the fruit and conditions under which the fruit processed and held.

2.5 The host

Bananas and the closely related plantain (cooking banana) are the hosts, with commercial dessert banana cultivars (AAA genome) more susceptible to crown rot than cultivars of the plantain subgroup (AAB) (Slabaugh and Grove, 1982). In the Cavendish subgroup, 'Robusta' and 'Valery' are more susceptible than 'Lacatan' (Shillinford and Sinclair, 1977). 'Kluai Hom Thong' (AAA) and 'Pisang Berangen' (AAA) are highly susceptible (Jitareerat and Uthairatanakij, 2013; Nuratika et al., 2017) while 'Karpuravalli' (ABB, 'Pisang Awak' subgroup) is less susceptible (Thangavelu et al., 2007).

2.6 Crown rot pathogens

Fungi responsible for crown rot live saprophytically on dead banana leaves, flower bracts, discarded fruits and the bunch stem (Meredith, 1962a,b; de Lapeyre de Bellaire and Mourichon, 1997). Airborne spores settle on the bunches while waterborne spores such as those of *Colletotrichum musae* are splashed onto fruit by rain or irrigation (Jones and Muirhead, 2019). Moreover, banana contamination by spores of *Colletotrichum musae* and various *Fusarium* spp. mainly occurs during the first 40 days following bunch emergence, and these are gradually replaced by other fungi in the complex, without causing any symptoms of decay on the green fruits (de Lapeyre de Bellaire and Mourichon, 1997). Spores are carried from the plantations to the packinghouse on bunches, plastic sleeves or blankets used to protect the bunch during transport. De-handing the bunch, cutting the clusters and immersion of hands and clusters in de-latexing and washing water baths enable fungal spores to be drawn deeply into the wounded tissues (Shillingford, 1977; Slabaugh and Grove, 1982). In addition, the knife used may become contaminated during the de-handing operation, transferring the inoculum from one hand to another. Jones (1991) indicated that wounding activated previously established latent/quiescent infections of *Colletotrichum musae*, which is also the causal agent of anthracnose. The spores adhere tightly to the fruit surface and can penetrate 5–7 mm into the crown tissues within only 3 min (Barkai-Golan, 2001). The crowns may also be contaminated by airborne spores before the fruit is packed in boxes, especially when the facilities are dirty or if debris (pistils, stalks and fruits) are stacked nearby. Moreover, conidia can survive for several months under many temperature and humidity conditions before germination (Meredith, 1971).

The potential quality of the fruit consists of two different components influenced by management and environmental factors during banana growth. One is a pathogenic component which reflects the level of crown contamination and pathogenicity of the fungi present. The other is physiological and determines the susceptibility of the fruit to the disease (Lassois et al., 2010a; Ewané et al., 2012).

Several pre-harvest factors influence the susceptibility of fruit to crown rot (Fig. 5):

- Geographical and seasonal variations: Ewané et al. (2013) showed that crown rot developed to a greater degree at lower (80 m) compared with higher altitudes (500 m) in Cameroon over a year. Umana-Rojas et al. (2011) found a positive correlation between crown rot incidence and both precipitation and high relative humidity in Costa Rica. It was also observed that the microbial load increased in the washing tanks during the rainy season due to an increased airborne spore load of some species of the pathogen complex.
- Production system (integrated/organic): Umana-Rojas et al. (2011) reported a lower presence of crown rot in bananas produced in an organic system than in the integrated system.
- Foliar diseases such as *Pseudocercospora fijiensis* (previously known as *Mycosphaerella fijiensis*), responsible for black leaf streak disease, and *Pseudocercospora musicola*, which causes Yellow Sigatoka disease, negatively affect fruit quality and increase susceptibility to crown rot (Ewané et al., 2013).
- Hand removal in the field has advantages in the context of integrated pest management, as fruit susceptibility to crown rot was reduced while fruit size increased (Lassois et al., 2010b).
- Bunch age: there is an intra-bunch banana crown rot susceptibility gradient. Clusters growing on the first hands of a bunch, which are on average, 70 degree days (dd) more advanced than the hands initiated last (Jullien et al., 2001), are more susceptible (Forret, 2008; Lassois et al., 2010b). The later the harvest (beyond 900 dd), the more susceptible the fruit will be to postharvest diseases.
- Biotic stress factors: phenolic compounds have been proposed as potential participants in banana tree defence mechanisms against various pathogens (Ewané et al., 2012, 2020).
- Harvesting, handling, transport, storage and commercialization (postharvest stages) have to be well managed in order to avoid fruit

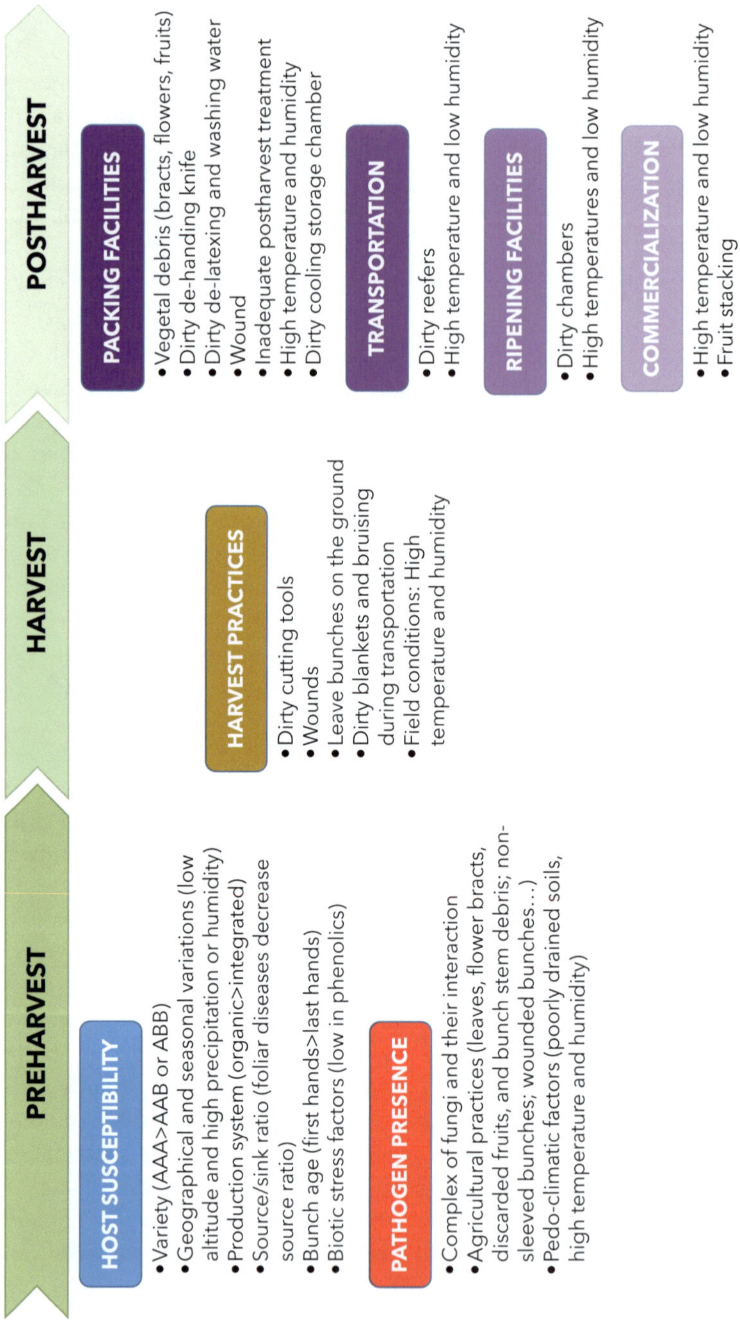

Figure 5 Conditions and factors increasing crown rot and other postharvest diseases.

injuries that enhance ethylene production by the fruit and wounds that favour infection by pathogens.

2.7 Assessing incidence and crown rot disease severity

The incidence (%) of crown rot disease on banana fruit is calculated as follows:

Disease incidence (DI%) = number of infected fruits/total number of fruits × 100

Disease severity can be ranked by observing the percentage of rotting symptoms based on a linear scale (0–4) (Zoeir et al., 2017a) as follows:

0 = healthy fruit free of rot and discolouration.
1 = 1–25% rotten and discolouration area.
2 = 26–50% rotten and discolouration area.
3 = 51–75% rotten and discolouration area.
4 = 76–100% rotten and discolouration area.

Disease severity (%) is calculated as follows:

$$\text{Disease severity (DS\%)} = \sum(n \times r) \times 100/N$$

where n = the number of fruits in each numerical disease grade, r = the disease grade number and N = the total number of infected fruits multiplied by the maximum numerical disease grade.

Fruit susceptibility to crown rot is generally assessed using the International Necrotic Surface method. Briefly, crowns are inoculated with a conidial suspension of *Colletotrichum musae* and after simulating transport, artificial ripening and storage, the progression of rot is determined by cutting the cluster crown longitudinally in two and measuring the area of rot spread into the crown, from the original inoculation point.

3 Management of crown rot

3.1 Pre-harvest

Control measures for crown rot include sanitation in the field and at the packinghouse, postharvest treatment and adequate environmental conditions before and after ripening (Muirhead and Jones, 2000; Slabaugh, 1994). Crown rot control starts in the field with the regular removal of banana debris (senescent leaves, pistils, stalks and rotten fruit) to reduce the number of fungal spores that can be disseminated via wind and water splash from irrigation or rain (Ploetz, et al., 2003; Lassois et al., 2010a). Odame (2010) found that weed control, three times a year, reduced the breeding locations and survival of vectors that facilitate crown rot inoculum transfer from plant to plant and from one location to another. In the same way, the practice of mulching serves as a barrier to field infection as spores in the soil are prevented from being carried by rain or wind onto healthy banana plants. Rodents, birds and insects (banana weevils, white flies and ants) play an important role in the transmission of diseases as vectors for primary inoculum (de Lapeyre de Bellaire and Mourichon, 1997). Finlay et al. (1992) highlighted the potential role of bunch stalks as a source of fungal inoculum which causes crown rot. The early elimination of flower parts from the newly formed fingers in the field is also essential to reduce bunch contamination by reducing the amount of pathogenic inoculum present in the field (Odame, 2010; Lobo and Fernández-Rojas, 2020). Other cultural practices, such as bunch sleeving, reduce contamination by *Colletotrichum musae* by over 80% (de Lapeyre de Bellaire et al., 2000) and limit the development of other diseases such as pitting (Guyon, 1970) and speckling (Jones and Muirhead, 2019).

Breeding programmes, most notably at FHIA (Fundación Hondureña de Investigación Agrícola) in Honduras, have produced crown rot-resistant hybrids, but these have different organoleptic characteristics than Cavendish clones and are generally less accepted by mainstream customers (Piña et al., 2006).

Plant waste and rotting fruits and bunch stalks should be removed from packinghouses and good cultural packing practices reduce crown rot incidence. Finlay et al. (1992) reduced disease incidence by more than 50% by de-handing (without washing) in a relatively clean packing shed instead of the field. This observation provides a good argument for keeping the packing area free of leaf trash, floral remnants and discarded fruit (Jones and Muirhead, 2019). Some packing facilities wash the banana bunches with water jets prior to being de-handed, which diminished the amount of spores in the water tanks (Lobo and Montero-Calderón, 2020). To maintain water quality in closed systems, the water is treated with active chlorine or quaternary ammonium disinfectants (Shillingford, 1977; Slabaugh and Grove, 1982) and regularly adjusted due to redox reactions with the latex and organic matter. Some packing facilities prefer to change the water frequently. Adding aluminium sulphate (alum) to the de-latexing baths is highly recommended as it helps remove the latex, controls pathogens in the wash water and promotes satisfactory healing of the cut wounds (Yañez et al., 2004; Lobo and Montero-Calderón, 2020).

De-handing should be done carefully with a sharp hygienic knife to avoid crown contamination and a ragged cut. It is also important to cut wide crown sections containing as much crown tissue as possible, a technique that seems to enhance crown resistance to rot and seldom leads to the spread of rot into the fruit pedicels (Muirhead and Jones, 2000). Clusters should not be immersed for too long in the contaminated water to avoid being infected by fungal spores.

3.2 Application of synthetic fungicides

Crown rot has been principally controlled by postharvest fungicides such as benomyl, thiabendazole, imazalil, myclobutanil, azoxystrobin, thiophanate-methyl, carbendazím, tebuconazole alone or in combination, as some pathogens develop resistance (Aked et al., 2001; Salazar et al., 2012; Scribano and Garcete, 2016; Fernández-Rojas et al., 2021). Before exporting, it is important to know which fungicides are permitted in the importing country and their maximum residue limits to avoid legal infractions. The application methods vary markedly, including dipping, spraying and cascade treatments, but in every case, the crowns of the bananas must be thoroughly wetted to ensure the efficacy of the fungicide treatment.

Although crown rot is mainly controlled by postharvest fungicide treatments, alternative control methods have been developed due to: (1) fungal resistance to some commonly used fungicides (De Lapeyre de Bellaire and Maurichon, 1998; Salazar et al., 2012); (2) environmental problems linked to the disposal of fungicide mixtures used at packinghouses (Lassois and de Lapeyre de Bellaire, 2014); (3) consumer aversion to chemical residues in food (De Costa and Erabadupitiya, 2005; Lassois and de Lapeyre de Bellaire, 2014) and (4) the prohibition of its use in pesticide- or fungicide-free bananas.

A combination of fungicide and physical treatments, salts, coatings, etc. can significantly minimize the requirement for fungicide use on harvested commodities (Fallik, 2011). Consequently, physical treatments, eco-friendly antifungal agents such as plant-based essential oils, salts, organic acids or coatings, and biocontrol agents, alone or in combination, have been extensively investigated as alternative strategies for the control of banana crown rot.

3.3 Physical treatments

Several studies have proved that HW (45–50°C) is an effective non-chemical method for controlling postharvest diseases (López-Cabrera and Marrero-Domínguez, 1996; Reyes et al., 1998; Alvindia, 2012) if suitable combinations of temperatures and exposure times are selected to prevent loss of fruit quality (Lurie, 1998). López-Cabrera and Marrero-Domínguez (1996) indicated that HW treatment (cluster immersion for 15–30 min at 45–47.5°C) provides an excellent way of controlling the growth of *F. proliferatum* and *Colletotrichum musae*. In Brazil, a 20-min treatment at 45°C effectively decreased crown rot from 100% to less than 15% in cv 'Santa Catarina Prata' (AAB) and 'Williams' (Cavendish, AAA) (Reyes et al., 1998). Alvindia (2012) reported that 50°C for 20 min was the optimum exposure treatment for cv. Bungulan (AAA), finding fruit injury (scalding, shrivelling and failure to soften) with longer or hotter treatments; crown rot severity decreased by 50% after 7 days and 33% after 14 days. The HW treatment was ineffective in killing dormant spores but caused a delay in mycelium growth and conidial germination. Alvindia (2013a) pointed out that the combination of sodium carbonate (Na_2CO_3; 1%, 30 min) followed by an HW dip (50°C, 20 min) for cultures of *L. theobromae*, *Thielaviopsis paradoxa, Colletotrichum musae* and *Fusarium verticillioides* resulted in fruit with excellent appearance, reduced loss in weight, firm texture and low disease incidence (12%, comparable 10% disease incidence of fruit treated with fungicide). HW alone showed an efficacy of 52% in controlling fruit decay.

- UV-C light (between 200 nm and 280 nm) is an effective non-thermal method to disinfect water, sterilise surfaces and improve food preservation (Alothman et al., 2009) and may have the potential to control postharvest diseases. Moreover, in appropriate doses, it increases the nutraceutical properties of foods, promotes the synthesis of compounds with antifungal properties and slows the processes of maturation and senescence (Cisneros-Cevallos, 2003; Rivera-Pastrana et al., 2007). A UV-C dose effectively controlled mycelial growth, sporulation and conidial germination of the isolated fungal pathogens (*L. theobromae, Colletotrichum musae* and *Fusarium* spp.) at a dose of 0.015 kJ m^{-2}. Moreover, UV-C (0.01 kJ m^{-2}) was able to reduce severity by 46% in Berangan bananas without peel browning or negative effects on any of the physiological variables evaluated (Nuratika et al., 2017). Uribe-Palacio et al. (2022) observed that the fungicidal effect of UV-C decreased when the inoculum or the period between inoculation and treatment increased.

Kanapathipillai et al. (1987) pointed out that gamma-ray treatment (38 min at 4 kGy) inhibited spore germination, the formation of *Colletotrichum musae* appressoria and all fungal development on the banana surface. Nevertheless, Marriott and Palmer (1980) noted that irradiation can alter the banana peel; hence, Wills et al. (1998) proposed 0.5 kGy as the maximum dose. The development of postharvest fruit irradiation has been hindered due to the high cost of its application and consumer dislike of irradiated products.

In recent years, other physical barriers have been developed to keep the fungal complex that causes crown rot at bay. The idea is to deprive possible rot and mould spores of oxygen, thus creating a miniature modified atmosphere (MA) environment around each crown, protecting the crown and maintaining the turgor of the tissue. Thus, the development of postharvest diseases can be controlled by wrapping ParaSeal (a paraffin-based semi-transparent stretchable plastic moisture-proof self-sealing wrap), a biodegradable film or sealing wax (Fig. 6). Initial data reveal that wrapping a biodegradable film around the crowns and the use of sealing wax gives protection against crown rot development in a study comparing different eco-friendly postharvest treatments (Lobo et al., in preparation).

Figure 6 Physical barriers against banana crown rot. (a) Para-Seal, (b) biodegradable film and (c,d) sealing wax. Images from M. Gloria Lobo and Francisco Javier Fernández-Rojas.

3.4 Managing environmental conditions before and after ripening

Disease development of crown rot depends on the environmental conditions (temperature, relative humidity and atmospheric composition) as it directly affects pathogen biology and has an indirect effect on fruit metabolism. As fruit resistance decreases at the onset of the ripening process, it is important to carefully manage these factors (Krauss and Johanson, 2000). Hence, one way to slow down banana metabolism to delay crown rot development is to refrigerate the fruit without inducing chilling injury. During shipping, container temperatures are set at 13–14°C because lower temperatures provoke peel browning due to oxidation. At this temperature, fungi grow and germinate slower than at higher temperatures (Slabaugh and Grove, 1982). Moreover, the senescence of banana crown tissues, which is conducive to crown rot development, is slowed down by maintaining their turgidity (Badger, 1965), by maintaining a high relative humidity that limits the loss of water via fruit transpiration to ensure a long green life. Crown rot can be partially controlled by packing bananas in MA (Ranasinghe et al., 2005; Bastiaanse et al., 2010) achieved through the use of sealed plastic bags. This reduces the fruit respiration rate, hampers endogenous ethylene synthesis and inhibits the metabolism of some pathogenic agents. Bastiaanse et al. (2010) observed that the content of O_2 and CO_2 ranged between 1–2% and 5–7%, respectively, when bananas were packed in 50-mL thick non-perforated polyethylene bags called banavacs. MAs are often used during shipping but it is important that bananas avoid O_2 levels below 1–2% and CO_2 levels above 7–12% to maintain fruit quality (Daun et al., 1973).

3.5 Eco-friendly antifungal agents for crown rot control

The effects of salts, plant extracts, organic acids, coatings, etc. have been evaluated on the development of crown rot. Na_2CO_3, sodium bicarbonate ($NaHCO_3$), calcium chloride ($CaCl_2$), sodium chloride and sodium hypochlorite (NaClO) are common food additives that can be used to control food spoilage by fungi, bacteria and yeast. They are classified as 'generally recognised as safe' (GRAS) by the U.S. Food and Drug Administration and are exempt from residue tolerances on all agricultural commodities (Palou et al., 2016) and their use is approved in 'organic' products. The addition of a surfactant (APSA 80) increased the fungistatic action of the salts against pathogens causing banana crown rot (Alvindia and Natsuaki, 2007). The effective treatments to control crown rot of bananas and avoid phytotoxicity were 5 g/L of NaClO, $NaHCO_3$ and 5 g/L $CaCl_2$ ameliorated with a surfactant. Chirawut and Chinaphuti (2013) showed that potassium sorbate at 500 mg/L inhibited disease incidence by 47% and disease severity by 82%.

Bananas dipped in 1500 μg/mL soluble silicon (potassium silicate) for 20 min exhibited longer shelf life, lower rates of ethylene production and respiration and delayed time taken for fungal rot development compared with non-Si-treated (control) fruit (Nikagolla et al., 2019). The use of potassium aluminium sulphate (alum) at a rate of 1–2% is usually added to de-latexing tanks to remove the latex, to control pathogens in the wash water and to promote satisfactory healing of the wound at the crown (Anyasi et al., 2013). Alum treatment, combined with vacuum packaging, controlled crown rot disease of Cavendish bananas completely for up to 2 weeks in cold storage (Siriwardana et al., 2016, 2017). Treatment with a potassium sorbate heated solution delayed crown rot severity, as did the thiabendazole treatment, up to 21 days at 13°C (Jitareerat and Uthairatanakij, 2013).

Essential oils are important compounds among the wide range of natural substances exhibiting antifungal properties extracted from plants. They comprise a combination of volatile secondary metabolites that are bioactive in the vapour phase, show direct activity against phytopathogens and can also enhance plant defence mechanisms against these microorganisms (Bautista-Baños et al., 2013; Palou et al., 2016). Cinnamaldehyde, citral, eugenol, limonene and thymol are some of the key active ingredients that are safe for human consumption and have GRAS status. Other aqueous or organic solvent extracts of plants or herbs have also been reported to reduce decay on harvested horticultural products. In general, they are biodegradable substances composed of compounds that exhibit direct fungicidal or fungistatic activity, along with some ability to delay ripening and extend the shelf life of produce. Typically, these compounds are products of secondary metabolism produced by the plant for its own protection against pests and pathogens. *Cymbopogon nardus* and *Ocimum basilicum* oils (0.2–0.6% (v/v)) displayed fungicidal activity against *Colletotrichum musae* and *F. proliferatum,* demonstrating synergistic action when combined (Anthony et al., 2003). *O. basilicum* oil (0.16% (v/v)) controlled crown rot and anthracnose, enabling bananas to be stored for up to 21 days at 13.5 ± 1°C with no detrimental effect on their organoleptic properties (Anthony et al., 2004). *Allium sativum* extracts and essential oils of *Cinnamomum zeylanicum* and *Syzygium aromaticum* also have fungicidal activity (Lassois et al., 2010a). Win et al. (2007) showed that cinnamon extracts reduced crown rot, increased green life and had no negative effects on postharvest banana quality. These plant extracts have fungistatic and fungitoxic activity and inhibit conidial germination and mycelial growth of *Colletotrichum musae, Fusarium* spp. and *L. theobromae*.

Ranasinghe et al. (2002) performed a liquid bioassay using cinnamon leaf, bark and clove oils against anthracnose and crown rot pathogens. The test oils were fungistatic and fungicidal within a range of 0.03–0.11% (v/v). Moreover, Ranasinghe et al. (2003) used wastewater samples collected from a cinnamon bark oil distillery and found that cinnamaldehyde (66.2%) was the major constituent of the extracted oil solution. The oil was fungistatic and fungicidal within the range of 0.64–1.00 mg/mL. Spraying Embul bananas with emulsions of the oil extract prior to being stored, controlled crown rot, enabling bananas to be stored for up to 14 days at ambient temperature (28 ± 2°C) and 21 days at 14°C; 90% r.h. in MA.

Dipping banana fruits in Zimmu leaf extract at 25% conc. resulted in 100% inhibition of crown rot disease in cold storage (14°C) for up to 35 days and increased the shelf life to 64 days. However, when stored at 28°C, the same treatment exhibited 86% inhibition of crown rot disease for up to 12 days.

Indrakeerthi and Adikaram (2011) showed that conidia of major crown rot pathogens were digested completely when exposed to a water-soluble fraction (WSF) of *Carica papaya* (papaya) latex. One minute was enough for a WSF preparation (50%) to digest a dense suspension of *Colletotrichum gloeosporioides* conidia. Moreover, the application of the WSF of papaya latex directly onto the cut surface of the banana crown pre-inoculated with *F. semitectum*, *Colletotrichum musae* and *Colletotrichum gloeosporioides* prevented the development of crown rot. The activity of enzymes in the WSF of latex, particularly chitinase, is probably implicated in the conidia digestion process. The authors indicated that the WSF latex is more effective when applied at least 60 min after de-handing, to allow crown exudates to dry out. This appears to be due to the coagulation of proteins in papaya latex in the presence of watery exudate of the freshly exposed banana crown. The latex treatment also hardened and caused a black colouration in the exposed crown surface forming a physical barrier to invading pathogens.

Oxalic acid (100 mg/L) and salicylic acid (250 mg/L) are effective GRAS compounds for the control of crown rot disease in bananas (Chirawut and Chinaphuti, 2013). It has been suggested that *F. proliferatum* depolymerises the cell wall barrier, impairs the defence system in banana fruit and activates the non-defensive Jasmonic acid (JA) signaling pathway accelerating the senescence of banana fruit.

The occurrence of crown rot disease appears to correlate with the physiological state of the fruits at harvest. Plant polyphenols are secondary metabolites involved in the defence mechanisms of plants against fungal pathogens. Therefore, phenolics have a possible role in the biochemical defence of banana crowns and could be used by producers as a chemical criterion for estimating the level of susceptibility of the banana to crown rot (Ewané et al., 2020).

Khan et al. (2001) showed that some antioxidants such as butylated hydroxyanisole (BHA), benzoic acid, propyl gallate and propyl paraben had antifungal activity against *Colletotrichum musae,* particularly in the *in vitro* assays. They suggested that BHA, a food-grade chemical, enhances the activity of fungicides currently used to control this pathogen on bananas, allowing lower concentrations of fungicide to be used. Acetic and propionic acids (Krauss, 1995) and citric seed extracts (Krauss, 1996; Dorta et al., 2010) are useful against crown rot at low levels of incidence and severity but not when the infection is extensive, for example, during the rainy season.

Antimicrobial emulsion-based coating materials have been used extensively to inhibit the growth of fungi and extend the shelf life of bananas. Targeted emulsion-based film functionalities offer excellent mechanical, physicochemical and barrier properties and limit fungal growth kinetics. Antifungal compounds, such as essential oils or phenolic compounds, are added to emulsion-based (wax/biopolymer) coatings to enhance their antimicrobial properties. Agyemang et al. (2022) formulated a beeswax-based emulsion for banana coatings using high-pressure homogenisation (HPH) and hydrophilic food-grade/organic coating materials. Sucrose emulsions were tested as coatings on banana fruits to assess their efficacy against moisture and firmness loss during storage. The addition of thymol (0.5% w/w) to the sucrose emulsions further enhanced the inhibition of *Colletotrichum musae* growth to a minimum of <20%. Emulsion stability was clearly enhanced using HPH while the most effective inhibition was obtained using a 20% diluted sucrose emulsion containing 0.5% thymol, with or without HPH. Demerutis et al. (2008) combined Biocto 6 (seed extract from citrus) (Productos Biogenicos S.A., San Jose, Costa Rica) with a wax-based adjuvant (Verdiol, Productos Biogenicos S.A.) and the organic product provided the same level of crown rot control as the fungicide treatments used to export bananas. Dudoit et al. (2020) reported the antifungal activity of Brazilian red propolis extract as a potential natural alternative for controlling banana crown rot.

3.6 Biological control agents for crown rot

Different fungi, bacteria and yeasts have been studied to control crown rot alone or in combination with other postharvest methods. The efficacy of a biocontrol agent depends upon the approaches used, which include environment manipulation, use of mixtures of antagonists, physiological and genetic manipulation of antagonists, combinations of field and postharvest applications, manipulation of formulations and combination with other non-biological methods, in the framework of an integrated control system (Janisiewicz and Korsten, 2002). Antagonists can be used to significantly reduce lesions induced by species in the fungal complex that causes crown rot, but the control efficacy of biological control agents by themselves is limited and variable, so to enhance control, it has to be combined with other control tactics such as inorganic salts, HW treatments, UVC and MA (De Costa and Erabadupitiya, 2005; Bastiaanse et al., 2010).

Alvindia and Acda (2012) showed that HW treatment (50°C, 20 min) before dipping the fruit in a conidial suspension of *Trichoderma harzianum* strain DGA01 (106 conidia/mL) enhanced the control of crown rot. Likewise, Alvindia (2013d) pointed out that the fungus *Trichoderma harzianum* strain DGA01 (10^6 conidia/mL) combined with $NaHCO_3$ (1%) was capable of reducing the incidence of crown rot by 92% when fruits were dipped for 30 min in the solution.

Alvindia (2013b) isolated two bacteria from the surface of banana fruits and screened for *in vitro* antagonism toward *L. theobromae*. The isolates, identified as *Bacillus amyloliquefaciens*, were further tested for antibiosis against other crown rot-causing pathogens (*Thielaviopsis paradoxa*, *Colletotrichum musae* and *F. verticillioides*). Postharvest application of *B. amyloliquefaciens* in the packinghouse reduced the incidence of crown rot to a level significantly lower than in fungicide-treated or control fruits. The effect of *B. amyloliquefaciens* was enhanced when combined with $NaHCO_3$ (Alvindia, 2013b), or with HW (Alvindia, 2013c), both controlling 93% of fruit decay, comparable to synthetic fungicide control.

Lobo et al. (2022) performed an *in vivo* trial using commercial biocontrol agents in a banana packinghouse. De-handed bananas were washed in the water tank prior to the application of Serenade (*Bacillus subtilis* strain QST 713, 1.34%) and Nexy (*Candida oleophila* strain 0, 8.1%) from DECCO, combined with other postharvest treatments. Serenade combined with macro perforated bags was as effective as those treated with the fungicide Azoxystrobin.

3.7 Crown rot: conclusions and future trends

At present, severely affected banana fruits are still found in consumer markets containing *Fusarium musae*, *F. verticillioides*, *F. oxysporum*, *F. proliferatum*, etc. These fungi cause crown rot disease in bananas and lead to economic loss and may cause keratitis, skin infections and systemic infections in immunocompromised patients (Triest and Hendrickx, 2016). The improvement of postharvest fungicidal treatments to prevent banana crown rot postharvest disease, better control measurements and better hygiene criteria for the processing of banana fruits are recommended. Considering the results presented in this chapter, an integrated strategy is required to control banana crown rot under commercial conditions. In addition, resistant banana cultivars may need to be explored.

4 Anthracnose

Anthracnose is a postharvest disease of bananas caused by the fungus *Colletotrichum musae* (Berk. & M.A. Curtis) Arx, which results in major economic losses during transportation and storage with potential losses up to 30–40% loss of marketable fruit (Ranasinghe et al., 2003).

4.1 Disease symptoms

Immature green bananas are infected by spores of *Colletotrichum musae* in the field causing a latent infection. Symptoms occur during storage or after ripening as peel blemishes, that is, black or brown sunken spots of various sizes on fruit that may bear masses of salmon-coloured acervuli with their associated conidia on the fruit peel (Ranasinghe et al., 2005) (Fig. 7). Control measures that either prevent infection or can effectively delay symptom expression of anthracnose have an important role in extending the shelf life of banana fruit during storage and marketing.

4.2 Pathogen characteristics and diagnosis

Colletotrichum musae is a specialised pathogen of bananas and also causes other diseases of banana such as crown rot, main stalk rot, stem-end rot and anthracnose rot of pre-harvest fruit. Although *Colletotrichum musae* is the most common species associated with anthracnose in bananas, other *Colletotrichum* species cause similar symptoms. Among others, *Colletotrichum gloeosporioides*, *Colletotrichum siamense*, *Colletotrichum theobromicola*, *Colletotrichum tropicale* and *Colletotrichum scovillei* have also been reported in different countries (Table 1), while Huang et al. (2021) reported *Colletotrichum*

Figure 7 Anthracnose caused by *Colletotrichum* sp. Images from Javier Fernández-Rojas and Goretti L. Díaz-Delgado.

fructicola and *Colletotrichum cliviicola* are associated with banana anthracnose worldwide.

Traditionally, identification of *Colletotrichum musae* was based on morphological characteristics such as abundant sporulation, straight and cylindrical conidia and irregularly shaped appressoria (Sutton, 1980, 1992); however, they often overlap and are ambiguous among the *Colletotrichum gloeosporioides* species complex. Thus, molecular analysis is required to identify the *Colletotrichum* species that causes banana anthracnose.

4.3 Anthracnose: disease cycle and epidemiology

Fungi causing anthracnose infect bananas during the spring, just as the first new leaves begin to expand, and then continue through the summer while environmental conditions are suitable. Spores are released from last season's diseased and senescing tissue, including leaves, bracts, discarded fruit and fruit stems. The fungus, *Colletotrichum musae*, infects the fruit in the first month after flowering (de Lapeyre de Bellaire and Mourichon, 1997). The conidia reach the surface of the fruit in rainwater trickling over the bunch or irrigation water and in the packinghouse in recirculated washing water, which accumulates

Table 1 References of different *Colletotrichum* species causing anthracnose disease.

Reported *Colletotrichum* spp.	Country	References
C. musae	Jamaica	Meredith (1960)
	Brazil	Peres et al. (2002)
	China	Huang et al. (2021)
C. paxtonii	St Lucia	Damm et al. (2012)
C. gloeosporioides	Malaysia	Intan Sakinah et al. (2013)
	Ecuador	Riera et al. (2019)
	Pakistan	Alam et al. (2021)
C. scovillei	China	Sakinah et al. (2014), Zhou et al. (2017)
C. *siamense*	India	Kumar et al. (2017)
	Turkey	Uysal and Kurt (2020)
	Brazil	Vieira et al. (2017)
	China	Huang et al. (2021)
C. tropicale	Brazil	Vieira et al. (2017)
C. chrysophilum	Brazil	Vieira et al. (2017)
	Mexico	Fuentes-Aragón et al. (2021)
C. theobromicola	Brazil	Vieira et al. (2017)
C. karstii	China	Huang et al. (2021)
C. fructicola	China	Huang et al. (2021)
C. cliviicola	China	Huang et al. (2021)

spores from plant debris. The optimum growth and sporulation temperature of the fungus is 27-30°C while there is little growth below 15°C. Conidia adhere to the fruit surface, germinate, penetrate the surface of the fruit and form a melanic appressorium, which is a dormant structure of the pathogen, within 24–48 h. The melanic appressoria remains latent and only germinates during fruit ripening to form an infection hypha that colonizes the peel. The hypha penetrates and colonizes the peel, and the mycelium enters the skin and later the fruit pulp, forming brown lesions (Muirhead and Deverall, 1981) mainly located on the angular corners of the fruit, often scraped, scratched or bruised during packing and general handling in commercial trade. Infections can also occur directly through wounds, which stimulate conidial germination and mycelial growth but inhibit appressorial formation (Yang and Chuang, 1996). These lesions, also called anthracnose wounds, evolve rapidly during transportation and storage and seriously affect product quality. If the fruit is not injured after harvest, latent infections resume activity during ripening and similar lesions form later in the ripening process and are less damaging and of less concern commercially.

The potential quality of the fruit depends upon its susceptibility to anthracnose, which is affected by the soil and climatic conditions of the production zone, and the cultural practices used by growers. The physiological component of fruit quality is also influenced by mineral nutrition and water supply to the banana plants. Chillet et al. (2000, 2006a,b) found that stressful conditions, especially soil flooding, had no direct effect on fruit susceptibility to *Colletotrichum musae*, while fruit of the same physiological age developed in hotter locations, or plants with high foliar manganese and low calcium concentrations grown in acid soils were more susceptible. High humidity, high temperatures and the presence of bruises enable infection, which probably occurs mainly at the remains of the perianth (Wardlaw, 1961). Fruit of a great calliper grade is also more likely to develop wound anthracnose as it is more susceptible to wounding (Stover, 1972).

Of the dessert cultivars, 'Gros Michel' (AAA) is less susceptible to wounding the green fruit and developing anthracnose than different cultivars in the Cavendish subgroup (AAA). Shillingford and Sinclair (1977) found that 'Robusta' and 'Valery' developed more wound anthracnose than 'Lacatan'. Sakinah et al. (2014) reported susceptibility of 'Pisang Mas' (AA, Sucrier subgroup), 'Pisang Berangan' (AAA, Lakatan subgroup), 'Pisang Nangka' (AAA), 'Pisang Awak' (ABB) and 'Pisang Rastali' (AAB, Silk subgroup) grown in Malaysia. In Brazil, Prata Aña (AAB, Pome subgroup) was susceptible (Silva et al., 2016), and from the banana genotypes with resistance to black leaf streak, 'Thap Maeo' (AAB, Mysore subgroup) showed the lowest susceptibility while 'Ambrosia' (AAAA, Jamaican bred hybrid), 'PV 42-53' (AAAB, EMBRAPA bred hybrid), 'FHIA-18' (AAAB, FHIA bred hybrid) and 'FHIA-01' (AAAB, FHIA bred hybrid) were the most susceptible (Pinho et al., 2010). Ootani et al. (2017) reported that the genotypes 'FHIA-01, 'Maca' and 'Prata' were more susceptible to *Colletotrichum musae* than 'Caipira', 'FHIA-18' and 'Thap maeo'. The Jamaican-bred tetraploid 'Calypso' ('65–3405–1') was resistant to wound anthracnose, 'Lacatan' was intermediate, and both 'Valery' and 'Giant Cavendish' were equally susceptible.

5 Management of anthracnose

New control strategies have been developed as the fungicidal treatments against anthracnose are not effective in all producing countries and consumers demand pesticide-free fruit. Fungal strains with resistance to fungicides have also emerged. Intensive monoculture production systems in industrial plantations may give rise to the development of new strains of *Colletotrichum musae* resistant to some fungicides used post-harvest when these same products are applied regularly in the field to control leaf spot diseases (de Lapeyre de Bellaire and Dubois, 1997).

The control of anthracnose depends on the same principles as discussed for crown rot. Minimizing fruit damage both in the field and in the packing facilities, reducing inoculum levels by protecting the bunches from rainwater or by using fresh or clean water to wash fruit, using fungicides or eco-friendly treatments to protect fruit and maintaining it in a hard, green state until ripening, will control anthracnose development. Thus, sanitation in the field and at the packinghouse is extremely important to reduce the amount of inoculum that can cause anthracnose. In organic bananas, the incidence of this pathogen is higher than in traditional crops due to the lack of use of synthetic fungicides during the pre-harvest period. Placing a plastic sleeve around the banana bunches markedly reduces the levels of fruit contamination, by acting directly on the occurrence and amount of rainwater contaminated with anthracnose spores (de Lapeyre de Bellaire et al., 2000). This practice alone appears to be more effective than the removal of the floral parts and the bunch bracts, which are labour-intensive operations. Nevertheless, the combination of the two practices leads to a satisfactory reduction in disease incidence and may enable avoidance of fungicide use. Bruising and wounds during harvest and at the packinghouse need to be avoided to prevent the development of wound anthracnose. Prompt cooling to 14°C after packing is also recommended to extend green fruit life and diminish anthracnose development.

Synthetic fungicides such as benomyl, imazalil and thiabendazole have been commonly used in the past to control postharvest diseases (Khan et al., 2001). However, persistent use of these fungicides has resulted in the emergence of resistant strains of *Colletotrichum musae* (de Lapeyre de Bellaire and Chilin-Charles, 2008). Sponholz et al. (2004) evaluated different fungicides showing that procloraz (100 mg/L) and propiconazole (250 mg/L) controlled the disease after 15 days of storage while the non-treated bananas developed lesions on 60% of their fruit surface area. Guzmán and Villalta (2008) mixed azoxystrobin (100–150 mg/L) and thiabendazole (250 mg/L) and the result was similar to the thiabendazole and imazalil mixture (400 + 400 mg/L) used commercially. This mixture was able to reduce the fungicide dose by 50% without diminishing control of anthracnose and other postharvest diseases, in addition to reducing the economic cost by approx. 20%. Variability in fungicidal sensitivity among the *Colletotrichum* species has been observed (Bele et al., 2018). Greater sensitivity of *Colletotrichum gloeosporioides* to imazalil was observed compared with *Colletotrichum musae,* which showed moderate sensitivity. The inhibition of ergosterol biosynthesis induced by imazalil was more pronounced for *Colletotrichum gloeosporioides*. The indiscriminate use of these fungicides without rotation or alternating with other fungicides to control pre-harvest and postharvest banana diseases has led to resistance to azoxystrobin, boscalid, benomyl and benzomidazole fungicides by fungal pathogens of banana fruits in many countries (FRAC, 2010).

Khan et al. (2001) reported that the combination of the synthetic antioxidant BHA (0.5 mM) with imazalil (1.8 μM) completely prevented mycelial growth and germination of *Colletotrichum musae in vitro*. BHA (5 mM) inhibited fungal growth at 25°C by 35–41%. Moreover, when it was combined with imazalil (0.45 mM) better control (63–80% inhibition) was achieved, even greater than with higher concentrations of imazalil (1.78 mM) or thiabendazole (2.46 mM) applied alone (53–58% and 43–54% inhibition, respectively). The results suggest that BHA, which is a food-grade chemical, might have the potential to enhance the activity of fungicides currently used to control *Colletotrichum musae* on bananas, allowing lower concentrations of fungicide to be used.

Considering the increased consumer awareness of environmental and human health issues, there is an urgent need to develop safe, efficient and eco-friendly procedures and products to control postharvest anthracnose of bananas. HW treatment (20 min at 45°C) controlled disease incidence after 12 days of cold storage (Sponholz et al., 2004). Organic bananas dipped in HW (20 min at 40°C) caused severe inhibition of *Colletotrichum musae* (49.5%), which was significantly higher than with other HW dips without showing a negative effect on the physicochemical and sensory fruit qualities (Vilaplana et al., 2018). Nolasco et al. (2008) were able to control anthracnose in 'Prata' bananas for 12 storage days after HW treatment (6 or 12 min at 50°C). Fernandes et al. (2018) showed that an HW dip (8 min at 52°C) reduced anthracnose in 'Prata Anã' bananas in up to 81.6% of the injured area and delayed their maturation. This fact confirms that thermal immersion treatments at high temperatures produce changes in maturation as was reported by Lobo et al. (2000) working with Cavendish bananas treated for 15 min at 50°C. The association of hydrothermal treatment with the fungicide imazalil resulted in better control of anthracnose in the 'Prata Anã' banana (Fernandes et al., 2018). Postharvest treatment involving fruit dipped for 30 min in Na_2CO_3 (1%) following HW treatment (3 min at 50°C) reduced the severity of anthracnose, which was statistically similar to that of the fungicide treatment (Sinthuja and Prasannath, 2020).

Anthracnose was completely inhibited on bananas artificially inoculated with *Colletotrichum musae* and then irradiated with UV-C light (wavelength 254 nm) for 2 h (Bokhari et al., 2013). Salts, essential oils, plant extracts, coatings and antagonistic organisms are emerging as alternatives to conventional fungicides as they are generally safe for humans and environmentally friendly.

Rizwana (2018) screened some oils both *in vitro* and *in vivo* for their antifungal activity against *Colletotrichum musae*. Clove oil (0.1 μL/mL) completely stopped the conidial germination and mycelial growth of *Colletotrichum musae*. Fenugreek and almond oil, at a concentration of 2 μL/mL exhibited significant inhibition of mycelial growth, 61% and 57%, respectively, and demonstrated fungicidal and fungistatic activities.

Extracts from *Solanum torvum*, *Jatropha curcas* and *Emblica officinalis* inhibited the mycelial growth of *Colletotrichum musae* and *S. torvum* extract at 25%, and at 50% completely inhibited the growth of *Colletotrichum musae*. Moreover, the *S. torvum* extract was found to be highly effective in reducing disease incidence, better than the standard treatment with the fungicide benomyl (0.1%) *in vivo* at room temperature and in cool storage on the three banana varieties 'Robusta' (AAA), 'Rasthali' (AAB) and 'Ney Poovan' (AB) tested. The extracts also significantly increased the shelf life of bananas, particularly their green life (Thangavelu et al., 2004). Anthony et al. (2003) pointed out that spraying Embul bananas with emulsions of *Ocimum basilicum* (0.16%) prior to cool storage can be recommended as a safe, cost-effective method, with commercial potential for controlling postharvest diseases and extending storage life. Hinokitiol, a natural product extracted by steam distillation from the woody tissues of *Chymacyparis,* exhibits antifungal activity against several agricultural pathogens such as *Botrytis cinerea*, *Alternaria alternata* and *Fusarium* sp. Qiao et al. (2022) reported that hinokitiol exhibited promising activity against *Colletotrichum musae* both *in vitro* and *in vivo* because it destroyed the morphology and the cell membrane integrity of the fungus. Moreover, it was able to delay banana ripening and enhance antioxidant enzyme activities.

Zahid et al. (2012) showed the potential of chitosan-loaded nano-emulsions to control *Colletotrichum musae*. Chitosan at 1.0% concentration with a droplet size of 200 nm exhibited the best results in delaying the onset of anthracnose and maintaining the quality of all the fruits for up to 28 days of cold storage.

5.1 Biological control agents for anthracnose

A suspension of the bacteria *Burkholderia spinose* (10^5 CFU/mL) containing Tween 20 (0.02%), isolated from the fructosphere of a range of dessert and cooking-type banana varieties grown in Sri Lanka effectively controlled anthracnose and blossom end (De Costa et al., 2008).

Damasceno et al. (2019) reported that the following bacteria were efficient in controlling anthracnose: *Bacillus velezensis, Enterobacter cloacae, Serratia marcescens* and *Stenotrophomonas maltophilia*. *B. velezensis* showed similar performance to Thiabendazole (Tecto SC®). Moreover, *B. velezensis* inhibited mycelial growth and pathogen sporulation and was positive for cellulase and xylanase activity, which are mechanisms of action relevant to a biocontrol agent. Rivarez et al. (2021) isolated seven bacteria from wild honey produced by *Apis mellifera*. Three of the genus *Lactobacillus* spp. (BC2, BC3, BC7) and one of the genera *Bacillus* spp. (BC6) were found to be effective antagonists against *Colletotrichum musae in vitro* and *in vivo*.

A combination of treatments is recommended to achieve better control of anthracnose. Dipping banana fruit in a 300 mM $NaHCO_3$ solution for 10 min followed by dipping in a *B. spinosa* suspension in nutrient broth (10^8 CFU/mL) effectively controlled anthracnose, crown rot and blossom end rot of bananas (var. Kolikuttu). $NaHCO_3$ increased pH, total soluble solids and peel thickness which may have an indirect or cumulative effect on the reduction of postharvest disease development in bananas (De Costa and Gunawardhana, 2012).

6 Other postharvest diseases: mokillo, fruit speckle, squirter and ceratocystis fruit stem-end rot

There are several other plant pathogens that can affect the banana post-harvest but they are in general less common. Nevertheless, it is important to control them to avoid fruit quality problems and the spread of these pathogens.

6.1 Mokillo (finger-tip rot)

Mokillo or finger-tip rot is a minor disease caused by an aerobic bacterium (*Pseudomonas* sp.) that affects bananas, including the heavily traded Cavendish banana cultivars. Losses from the disease are usually not severe (0.5–3%) and only one or a few fingers at the most from the youngest hands are affected.

Diseased fingers are recognised externally by their deformed shape. They are smaller, straighter and often taper towards the tip, which is a condition known as 'bottle-neck'. As a consequence, they protrude out of the line of the other fingers (Stover, 1972; Lee et al., 2003). At the tip of the flower end, a brown discolouration may be evident. Initially, parts of the pulp within the finger appear slightly gelatinous and yellow. Later, sections of the pulp become rusty red in colour and degenerate to a gummy, sap-like constituency (Fig. 8).

The bacterial invasion is through the flower end of the finger. Stover (1972) reported that mechanical inoculation of *Pseudomonas* species into the banana pulp reproduced the symptoms of finger-tip rot within 14 days. Lee et al. (2003) first noted *Burkholderia cenocepacia* as a pathogen of banana finger-tip rot in Taiwan. In 2022, Zang et al. also stated *B. cepacia* produced finger-tip rot disease in Guangxi province, China. Pathania et al. (2017) observed, in far north Queensland during the processing of ripe banana into pulp, reddish brown to blackish centres and pulp breakdown, symptoms very

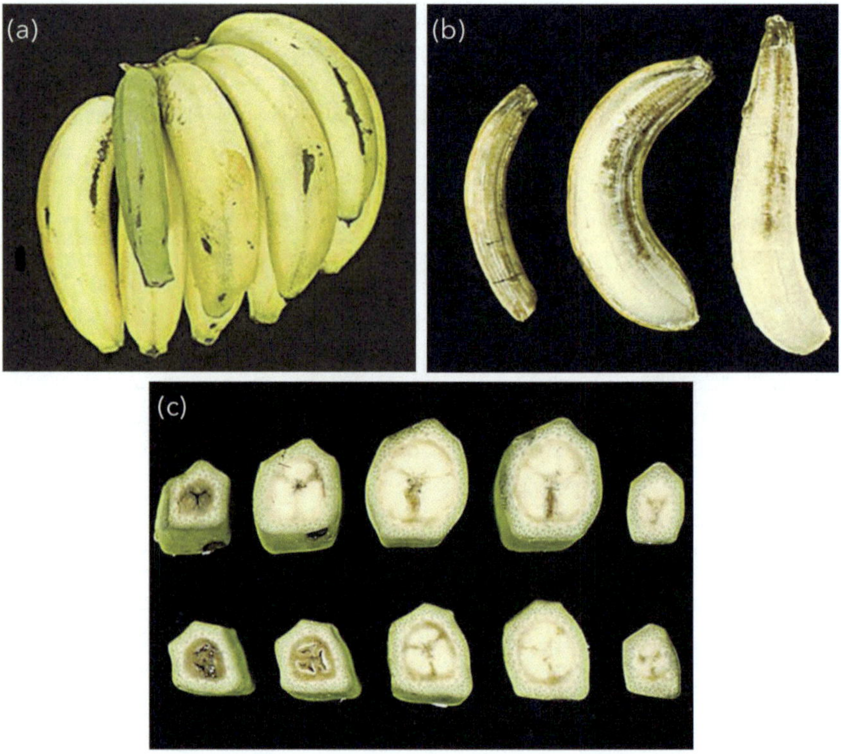

Figure 8 Mokillo (finger-tip rot) produced by *Burkholderia cenocepacia (*Syn. *Pseudomonas cenocepacia)*. (a) Diseased banana hand with typical finger shape and external discolouration, (b) longitudinal sections of fruit with internal discolouration and rot and (c) cross sections of fruit with internal discolouration and rot. Source: https://www.plantdiseases.org/search?query=finger-tip+rot.

similar to those of finger-tip rot. Two bacteria from the Enterobacteraceae family, *P. agglomerans* and *Enterobacter cowanii,* were frequently recovered and isolated from the infected fruits. Artificial inoculations of banana fruit with the bacteria were conducted to determine their pathogenic nature. Fingers on bunches at the bract-lift stage were injected with suspensions (10^8 CFU/mL) of each of the bacteria. One month after inoculation, fingers were cut to check for disease development. Both bacteria produced similar symptoms to the fruit samples taken from the pulp processing facility. Inoculated fingers were easily distinguished from others in the bunch by their deformed shape (Fig. 8a), being out of line with other fingers, and their premature yellowing and reduced fruit size. Internally, inoculated fingers exhibited a rusty red discolouration of the pulp with a deep gelatinous cavity. This study provided the first report of *P. agglomerans* and *E. cowanii* causing internal finger rot or finger-tip rot in bananas (Pathania et al., 2017).

No treatment has been identified to manage the bacteria that cause 'Mokillo' or finger-tip rot. However, the risk of infection can be reduced through the management of insects, maintaining proper drainage and field sanitation during warm and humid weather conditions, decontaminating cutting tools and farm equipment and discarding fruit with obvious symptoms of deformity (narrow at the flower end or straight and out of alignment) at the packing shed.

6.2 Fruit speckle

Recent studies have provided more insights into this disease, which may cause in some cases losses of up to 20%, and how to manage it. It was previously known as 'Deightoniella spot' caused by *Deightoniella torulosa* (Merdedith, 1961). More recent studies have shown that it is caused by *Colletotrichum musae, F. oxysporum* and *F. semitectum* (Vawdrey et al., 2008), pathogens usually found on banana bracts, old leaves and flowers. Other common names of the disease include 'swamp spot' and 'salt and pepper spot'.

Fruit speckle is more aggressive in subtropical production areas during warm, humid weather conditions. High-density planting, inadequate weed control and standing water on the ground create a high-humidity atmosphere, which encourages speckle (Jones and Stover, 2019). All banana cultivars are susceptible and become less susceptible as the fruit matures. At bract lift, *Colletotrichum musae* caused significantly more speckle lesions per square centimetre than the *Fusarium* spp., suggesting it was more pathogenic at this stage of fruit maturity (Vawdrey et al., 2008).

Fruit speckle typically appears as a small reddish-brown to black spot (0.5–1 mm in diameter) with a water-soaked margin. All surfaces of the fruit can be affected, but symptoms are often more pronounced at the neck and flower end of fingers. Spots caused by *Colletotrichum musae* are brown to black in colour, 0.5–1 mm in diameter, often surrounded by a water-soaked halo. Spots caused by the two *Fusarium* species are brown in colour and reach a maximum diameter of 0.5 mm. As the fruit ripens, the spots caused by *Colletotrichum musae* can enlarge to 3–4 mm in diameter and become dark, sunken and circular. Spots caused by *Fusarium* almost disappear as the fruit ripens. This disease should not be confused with the normal brown sugar spots that develop on over-ripe bananas (Fig. 9).

As with many postharvest diseases, hygiene is the key to successful management. Growers should de-leaf and de-sucker, particularly prior to the wet season, to help reduce the number of fungal spores in a plantation. Meredith (1963) pointed out that de-trashing two or three times a year reduced disease severity to a level at which it ceased to be of commercial importance. Bagging bunches shortly after emergence also provide good control of fruit speckles. It has to be noted that sap coming into contact with fruit skin, along with flower thrips infestations, can result in a 10-fold increase in the incidence

Figure 9 Fruit speckle previously known as 'Deightoniella spot' caused by *Colletotrichum musae*, *Fusarium oxysporum* and *Fusarium semitectum*. Images from Francisco Javier Fernández-Rojas.

of speckle spots on fruit caused by *Fusarium*. Thus, it is of great importance to control thrips. It is also recommended to spray with a registered fungicide such as mancozeb under the banana plant before and during the wet season to reduce the fungal spore load in the plantation. Another management option is to treat emerging bunches with a fungicide such as mancozeb before the bracts are fully open and again at bunch covering. Finally, avoid physical damage to the developing bunch as that may cause sap release onto fruit (Vawdrey et al., 2008, 2010).

Oil sprays used with some fungicides to control leaf spot diseases such as yellow and black Sigatoka can damage the fruit skin, particularly in hot dry conditions; this damage can be confused with fruit speckles. Bagging bunches prior to the application of oil is the only way to stop this damage.

6.3 Squirter

Squirter is a disease that affects fingers when packed singly; hands and clusters are usually not affected. Squirter at present is a disease of little economic importance. The pathogen penetrates through broken stem ends (cut pedicel) and advances into the fruit as it ripens. It discolours and converts the inner pulp into a dark watery fluid and may squirt from the end of fruit if squeezed

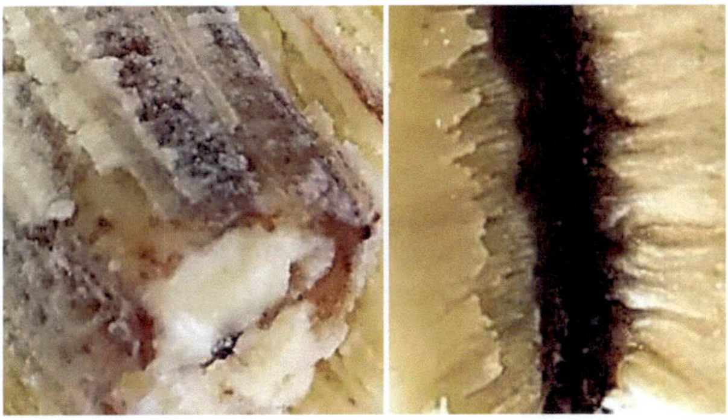

Figure 10 Squirter produced by *Nigrospora* sp. Images from Francisco Javier Fernández-Rojas and M. Gloria Lobo.

(Fig. 10). External symptoms become visible as a bluish-tan discolouration of the peel as the fruit ripens (Stover, 1972). This disease is generally attributed to *N. sphaerica*. Other *Nigrospora* species can cause the same symptoms as *N. sphaerica* such as *N. musae* (McLennan and Hoëtte, 1933).

The fungus forms conidiospores on dead vegetation, including banana trash. Airborne conidia are common in banana plantations (Meredith, 1961b) and in wash water in packing stations. Like stem-end rot, dipping fruit in a solution containing a fungicide controls the problem (Stover, 1972).

The application of fungicides containing benzimidazole compounds such as thiabendazole, and in the past, benomyl has shown good results in controlling *N. sphaerica* (Allen, 1970). Using molecular methods, Wang et al. (2017) studied the phylogenetic relationships between isolates of *Nigrospora* and recognised *N. musae* and *N. sphaerica* as separate species. It is important to highlight that *N. sphaerica* can act as an opportunistic pathogen causing onychomycosis in humans (De Hoog et al., 2000; Fan et al., 2009) and corneal ulcers (Kindo et al., 2014).

6.4 *Ceratocystis* fruit stem-end rot

This disease is only occasionally serious where infection levels are high and favourable conditions occur. It is caused by *Thielaviopsis paradoxa; Chalara paradoxa* (syn. *Ceratocystis paradoxa*), which is a polyphagous fungus responsible for the development of the disease not only in bananas but also causing black rot in crops such as pineapple, sugarcane and coconut. *Ceratocystis paradoxa* has been associated with the genus *Musa* in several countries in the Americas, Africa and Oceania (Farr and Rosmman, 2015). Pereira de Melo et al. (2016) reported that *Thielaviopsis musarum* causes postharvest crown and fruit rot of bananas in northeastern Brazil. They inoculated the fungus in healthy bananas which caused black rot followed by a browning and softening of the pulp.

Symptoms include rotting and blackening of the cushion, stalk, and crown-end of the banana finger which can cause fruit to ripen prematurely and detach from the pedicel (Fig. 11). The invaded flesh becomes soft and water soaked.

The conidia (Fig. 12) remain viable in the soil and plant debris, being freely found in banana-growing areas. This pathogen grows more actively at temperatures between 21°C and 22°C and is almost completely inactivated below 10°C.

The level of humidity in the soil and in the plantations is an important factor for the germination of conidia of the fruit and penetration of the host via a germ tube. The infective asexual conidia disseminate by splashing on the fruit during rain. Infection occurs mainly before harvest, via insect punctures, growth cracks or directly via the natural crevices between individual fruitlets or peduncles. Being a wound pathogen, the fungus gains entry through the cut stem of the peduncle or through wounds sustained during the handling of the fruit.

Ceratocystis stem end rot caused by *Thielaviopsis paradoxa* can be avoided if the bunches do not become soiled during harvest if the hands are cut with clean knives and if the packing facilities are kept free from banana, pineapple and sugarcane waste products. Minimising bruising, prompt cooling to 14°C, proper sanitation of handling facilities and HW treatment of hands (such as 5 min in 50°C water) prior to packing helps to control the disease in cold storage.

Figure 11 Ceratocystis fruit (stem end rot) produced by *Thielaviopsis musarum* (syn. *Ceratocystis paradoxa*). Images from Francisco Javier Fernández Rojas.

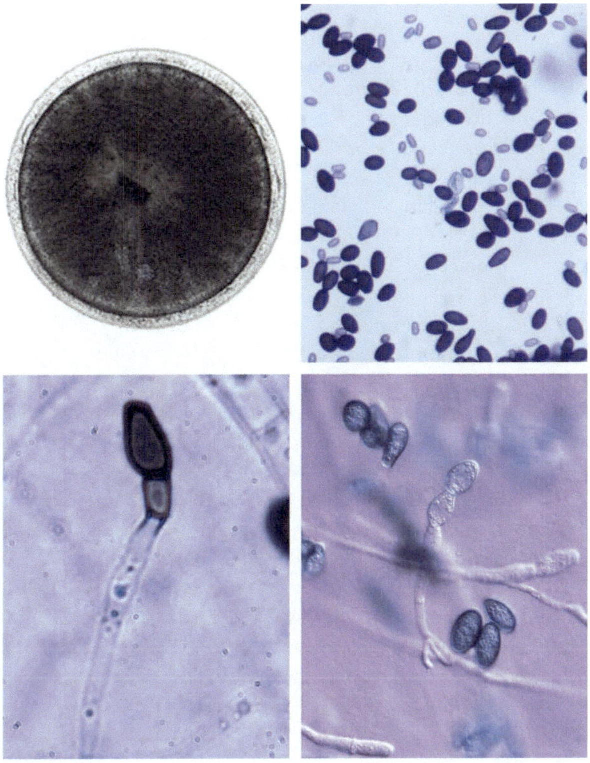

Figure 12 *Thielaviopsis musarum* conidia and conidiophores. Images from Francisco Javier Fernández-Rojas.

7 Conclusion

Prevention, through Good Agricultural Practices, to reduce the risk of microbial contamination of banana bunches is highly recommended as it can prevent the development of postharvest diseases. It is of great importance to correctly identify the causal agent responsible for the disease in order to choose an effective and sustainable management practice.

8 Acknowledgements

The authors gratefully acknowledge the financial support from CAIA 2023-001-04-3 project from the Government of the Canary Islands, Spain and to Noreima Barroso for her participation in the development of the sealing wax.

9 Where to look for further information

https://betterbananas.com.au/2020/06/08/postharvest-research/.

10 References

Agyemang, B., Grabulos, J., Hubert, O., Bourlieu, C., Nigen, M., Lebrun, M., Coffigniez, F., Guillard, V. and Brat, P. (2022). Properties of beeswax antifungal coatings obtained by high-pressure homogenisation and their application for preserving bananas during storage, *International Journal of Food Science and Technology* 57(8): 5349–5356 DOI:10.1111/ijfs.15865.

Alam, M. W., Malik, A., Rehman, A., Hameed, A., Sarwar, M., Hussain, M., Hussain, D. and Shafeeq, T. (2021). First record of *Colletotrichum gloeosporioides* causing anthracnose of banana in Pakistan, *Plant Disease* 105(7): 2013. DOI:10.1094/PDIS-01-21-0215-PDN.

Alemu, K. (2014). Importance and pathogen spectrum of crown rot of banana in Jimma town, Southwestern Ethiopia, *Journal of Biology, Agriculture and Healthcare* 4(23): 106–111.

Allen, R. N. (1970). Control of black-end and squirter diseases in bananas with benzunidazole and salicylanilide compounds, *Australian Journal of Experimental Agriculture and Animal Husbandry* 10: 490–492.

Alothman, M., Bhat, R. and Karim, A. A. (2009). UV radiation induced changes of antioxidant capacity of fresh-cut tropical fruits, *Innovative Food Science and Emerging Technologies* 10(4): 512–516. DOI:10.1016/j.ifset.2009.03.004.

Alvindia, D. G. (2012). Revisiting hot water treatments in controlling crown rot of banana cv. Bungulan, *Crop Protection* 33: 59–64. DOI:10.1016/j.cropro.2011.09.023.

Alvindia, D. G. (2013a). An integrated approach with hot water treatment and salt in the control of crown rot disease and preservation of quality in banana, *International Journal of Pest Management* 59(4): 271–278. DOI:10.1080/09670874.2013.845927.

Alvindia, D. G. (2013b). Enhancing the bioefficacy of *Bacillus amyloliquefaciens* DGA14 with inorganic salts for the control of banana crown rot, *Crop Protection* 51: 1–6. DOI:10.1016/j.cropro.2013.03.013.

Alvindia, D. G. (2013c). Improving control of crown rot disease and quality of pesticide-free banana fruit by combining *Bacillus amyloliquefaciens* DGA14 and hot water treatment, *European Journal of Plant Pathology* 136(1): 183–191. DOI:10.1007/s10658-012-0154-7.

Alvindia, D. G. (2013d). Sodium bicarbonate enhances efficacy of *Trichoderma harzianum* DGA01 in controlling crown rot of banana, *Journal of General Plant Pathology* 79(2): 136–144. DOI:10.1007/s10327-013-0432-z.

Alvindia, D. G. and Acda, M. A. (2012). An integrated approach with *Trichoderma harzianum* DGA01 and hot water treatment on control of crown rot disease and retention of overall quality in banana, *Biocontrol Science and Technology* 22(9): 1021–1033. DOI:10.1080/09583157.2012.707640.

Alvindia, D. G. and Natsuaki, K. T. (2007). Control of crown rot-causing fungal pathogens of banana by inorganic salts and a surfactant, *Crop Protection* 26(11): 1667–1673. DOI:10.1016/j.cropro.2007.02.008.

Alvindia, D. G. and Natsuaki, K. T. (2008). Evaluation of fungal epiphytes isolated from banana fruit surfaces for biocontrol of banana crown rot disease, *Crop Protection* 27(8): 1200–1207. DOI:10.1016/j.cropro.2008.02.007.

Alvindia, D. G., Kobayashi, T., Yaguchi, Y. and Natsuaki, K. T. (2000). Symptoms and the associated fungi of postharvest diseases on non-chemical bananas imported from the Philippines, *Japanese Journal of Tropical Agriculture* 44: 87–93.

Anthony, S., Abeywickrama, K. and Wijeratnam, S. W. (2003). The effect of spraying essential oils of *Cymbopogon nardus*, *Cymbopogon flexuosus* and Ocimum *basilicum* on postharvest diseases and storage life of Embul banana, *Journal of Horticultural Science and Biotechnology* 78(6): 780–785. DOI:10.1080/14620316.2003.11511699.

Anthony, S., Abeywickrama, K., Dayananda, R., Wijeratnam, S. W. and Arambewela, L. (2004). Fungal pathogens associated with banana fruit in Sri Lanka, and their treatment with essential oils, *Mycopathologia* 157(1): 91–97.

Anyasi, T. A., Jideani, A. I. O. and Mchau, G. R. A. (2013). Functional properties and postharvest utilization of commercial and noncommercial banana cultivars, *Comprehensive Reviews in Food Science and Food Safety* 12(5): 509–522. DOI:10.1111/1541-4337.12025.

Badger, A. M. (1965). Influence of relative humidity on fungi causing crown rot of boxed bananas, *Phytopathology* 55: 688–692.

Barkai-Golan, R. (2001). *Postharvest Diseases of Fruits and Vegetables: Development and Control*. Elsevier, Amsterdam. 418pp.

Bastiaanse, H., de Lapeyre de Bellaire, L., Lassois, L., Misson, C. and Jijakli, M. H. (2010). Integrated control of crown rot of banana with *Candida oleophila* strain O, calcium chloride and modified atmosphere packaging, *Biological Control* 53(1): 100–107.

Bautista-Baños, S., Sivakumar, D., Bello-Pérez, A., Villanueva-Arce, R. and Hernández-López, M. (2013). A review of the management alternatives for controlling fungi on papaya fruit during the postharvest supply chain, *Crop Protection* 49: 8–20.

Bele, L., Kouamé, D. K. and Atta, H. D. (2018). Sensitivity of *Colletotrichum* species responsible for banana anthracnose disease to some fungicides used in postharvest treatments in Côte d'Ivoire, *International Journal of Environment, Agriculture and Biotechnology* 3(2): 537–542. DOI:10.22161/ijeab/3.2.30.

Bokhari, N. A., Siddiqui, I., Siddique, K. P. I., Rizwana, H. and Soliman, D. A. W. (2013). Management of anthracnose of banana by UV irradiation, *Journal of Animal and Plant Sciences* 23: 1211–1214.

Chillet, M., de Lapeyre de Bellaire, L., Dorel, M., Joas, J., Dubois, C., Marchal, J. and Perrier, X. (2000). Evidence for the variation in susceptibility of bananas to wound anthracnose due to Colletotrichum musae and the influence of edaphic conditions, *Scientia Horticulturae* 86(1): 33–47.

Chillet, M., Hubert, O. and de Lapeyre de Bellaire, L. (2006a). Relationship between ripening and the development of banana anthracnose caused by *Colletotrichum musae* (Berck. and Curt.) Arx, *Journal of Phytopathology* 154(3): 143–147.

Chillet, M., Hubert, O., Rives, M. J. and de Lapeyre de Bellaire, L. L. (2006b). Effects of the physiological age of bananas on their susceptibility to wound anthracnose due to *Colletotrichum musae*, *Plant Disease* 90(9): 1181–1185.

Chirawut, B. and Chinaphuti, A. (2013). Evaluation of GRAS compounds for the control of crown rot disease of banana 'Hom Thong'. In: Acta Horticulturae conference on postharvest pest and disease management in exporting horticultural crops international, I. (Ed.) (973): 125–130, Hale, C. and Jitareerat, P. (Eds.), *Acta Horticulturae* 973: 125–130.

Cisneros-Cevallos, L. (2003). The use of controlled postharvest abiotic stresses as a tool for enhancing the nutraceutical content and add-in value of fresh fruits and vegetables, *Journal of Food Science* 68(5): 1560–1565. DOI:10.1111/j.1365-2621.2003.tb12291.x.

Damasceno, C. L., Duarte, E. A. A., dos Santos, L. B. P. R., de Oliveira, T. A. S., de Jesus, F. N., de Oliveira, L. M., Goes-Neto, A. and Soares, A. C. F. (2019). Postharvest biocontrol of anthracnose in bananas by endophytic and soil rhizosphere bacteria associated with sisal (*Agave sisalana*) in Brazil, *Biological Control* 137: 104016. DOI:10.1016/j.biocontrol.2019.104016.

Daun, H., Gilbert, S. G., Ashkenazi, Y. and Henig, Y. (1973). Storage quality of bananas packaged in selected permeability films, *Journal of Food Science* 38(7): 1247–1250.

Daundasekera, W. A. M., Joyce, D. C., Adikaram, N. K. B. and Terry, L. A. (2008). Pathogen-produced ethylene and the Colletotrichum musae-banana fruit pathosystem, *Australasian Plant Pathology* 37(5): 448–453. DOI:10.1071/AP08038.

Daundasekera, W. A. M., Joyce, D. C., Aked, J. and Adikaram, N. K. B. (2003). Ethylene production by *Colletotrichum musae* in vitro, *Physiological and Molecular Plant Pathology* 62(1): 21–28.

De Costa, D. M. and Erabadupitiya, H. R. U. T. (2005). An integrated method to control postharvest diseases of banana using a member of the *Burkholderia cepacia* complex, *Postharvest Biology and Technology* 36(1): 31–39. DOI:10.1016/j.postharvbio.2004.11.007.

De Costa, D. M. and Gunawardhana, H. M. D. M. (2012). Effects of sodium bicarbonate on pathogenicity of *Colletotrichum musae* and potential for controlling postharvest diseases of banana, *Postharvest Biology and Technology* 68: 54–63. DOI:10.1016/j.postharvbio.2012.02.002.

De Costa, D. M., Zahra, A. R. F., Kalpage, M. D. and Rajapakse, E. M. G. (2008). Effectiveness and molecular characterization of *Burkholderia spinosa*, a prospective biocontrol agent for controlling postharvest diseases of banana, *Biological Control* 47(3): 257–267. DOI:10.1016/j.biocontrol.2008.08.010.

De Hoog, G. S., Guarro, J., Gene, J. and Figueras, M. J. (2000). *Atlas of Clinical Fungi* (vols. 708-711). Centraalbureau voor Schimmelcultures, Utrecht, The Netherlands: and Universitat Rovira i Virgili, Reus, Spain.

de Lapeyre de Bellaire, L. and Mourichon, X. (1997). The pattern of fungal contamination of the banana bunch during its development and potential influence on incidence of crown rot and anthracnose diseases, *Plant Pathology* 46(4): 481–489.

de Lapeyre de Bellaire, L., Chillet, M., Dubois, C. and Mourichon, X. (2000). Importance of different sources of inoculum and dispersal methods of conidia of *Colletotrichum musae*, the causal agent of banana anthracnose, for fruit contamination, *Plant Pathology* 49(6): 782–790.

de Melo, M. P., da Silva Matos, K., Júnior, J. E. A. B., Moreira, S. I., Silva, F. F., de Alcântara Neto, F., Alves, E. and Pereira, O. L. (2016). Thielaviopsis musarum causes postharvest crown and fruit rot of banana in Northeastern Brazil, *Tropical Plant Pathology* 41(4): 258–263.

Demerutis, C., Quirós, L., Martinuz, A., Alvarado, E., Williams, R. N. and Ellis, M. A. (2008). Evaluation of an organic treatment for post-harvest control of crown rot of banana, *Ecological Engineering* 34(4): 324–327.

Díaz-Delgado, G., Fernández-Rojas, F. J., Espino de Paz, A., Laich, F., Dorta, E. and Lobo, M. G. (2021). *Aguas de Lavado: Fuente de Inóculo para el Desarrollo de la Pudrición de Corona del Plátano Durante la Postcosecha. Encuentro Internacional Phytoma-España: "La Sanidad Vegetal en Cultivos Mediterráneos y Subtropicales. Retos Ante una Transición Agroecológica*, 27 a 29 octubre 2021. Adeje, Tenerife (España).

Dominguez, M. and Vendrell, M. (1993). Wound ethylene biosynthesis in preclimacteric banana slices, *Acta Horticulturae* (343): 270–274.

Dorta, E., Marrero, A., Lobo, G. and Hernández, J. (2010). *Evaluación de Cuatro Extractos de Plantas, una sal Inorgánica y un Fungicida para el Control de la Podredumbre de Corona del Plátano. Trabajo Fin de Carrera*. Escuela Técnica Superior de Ingeniería Agraria. Universidad de La Laguna.

Du, Y. X., Chen, F. R., Shi, N. N. and Ruan, H. C. (2017). First Report of Fusarium chlamydosporum Causing Banana crown Rot in Fujian Province, China, *Plant Disease* 101(6). DOI:10.1094/PDIS-11-16-1674-PDN.

Dudoit, A., Mertz, C., Chillet, M., Cardinault, N. and Brat, P. (2020). Antifungal activity of Brazilian red propolis extract and isolation of bioactive fractions by thin-layer chromatography-bioautography, *Food Chem.* 15(327): 127060. Doi: 10.1016/j.foodchem.2020.127060.

Ewané, C. A., de Lapeyre de Bellaire, L., Lepoivre, P. and Lassois, L. (2012). Involvement of phenolic compounds in the susceptibility of bananas to crown rot. A review, *Biotechnology, Agronomy and Society and Environment* 16(3): 393–404.

Ewané, C. A., Chillet, M., Castelan, F., Brostaux, Y., Lassois, L., Essoh Ngando, J., Hubert, O., Chilin-Charles, Y., Lepoivre, P. and de Lapeyre de Bellaire, L. (2013). Impact of the extension of black leaf streak disease on banana susceptibility to post-harvest diseases, *Fruits* 68(5): 351–365. DOI:10.1051/fruits/2013081.

Ewané, C. A., Nott, K., Lassois, L., Lepoivre, P. and de Lapeyre de Bellaire, L. (2020). Severe modifications in source-sink ratio influence the susceptibility of bananas to crown rot and its phenolics content, *Plant Pathology* 69(9): 1740–1753. DOI:10.1111/ppa.13243.

Fallik, E. (2011). Hot water treatments of fruits and vegetables for postharvest storage, *Horticulture Research* 38: 191–212.

Fan, Y. M., Huang, W. M., Li, W. and Zhang, G. X. (2009). Onychomycosis caused by *Nigrospora sphaerica* in an immunocompetent man, *Archives of Dermatology* 145(5): 611–612.

Fernandes, M. B., Ribeiro, R. C. F., Silva, M. L. M., Fernandes, M. B. and Mizobutsi, E. H. (2018). Hydrothermal treatment in the management of anthracnose in 'Prata-Ana' banana produced in the semiarid region of Minas Gerais, Brazil, *Revista Brasileira de Fruticultura* 40(2): e-871. DOI:10.1590/0100-29452018871.

Fernández-Rojas, F. J., Duque Yanes, M., Lobo, M. G., Álvarez Acosta, C. and Domínguez Palarea, M. E. (2021). *Evaluación de la Eficacia de Productos Autorizados en Postcosecha para el Control de la Pudrición de Corona -Crown Rot- del Plátano*.

Fernández-Rojas, F. J. (2016). Técnicas no convencionales en el manejo postcosecha de la pudrición de corona del plátano (crown rot). Trabajo Final de Carrera de Ingeniero Agrónomo. Universidad de La Laguna, Tenerife, España.

Finlay, A. R. and Brown, A. E. (1993). The relative importance of *Colletotrichum musae* as a crown rot pathogen on Windward Island bananas, *Plant Pathology* 42(1): 67–74.

Finlay, A. R., Lubin, C. and Brown, A. E. (1992). The banana stalk as a source of inoculum of fungal pathogens which cause crown rot, *Tropical Science* 32: 343–352.

Forret, M. (2008). *Etude de la Variation de Sensibilité des Bananes d'Exportation aux Pourritures de Couronne en Fonction du Stade de Récolte*. Travail de Fin d'Étude, Faculté des Sciences Agronomiques de Gembloux.

FRAC (2010). Fungicide Resistance Action Committee; Banana working Group. Available at: https://www.frac.info/docs/default-source/working-groups/banana-group/group/2010-meeting-minutes---english.pdf?sfvrsn=cc97419a_4.

Fuentes-Aragón, D., Rebollar-Alviter, A., Osnaya-González, M., Enciso-Maldonado, G. A., González-Reyes, H. and Silva-Rojas, H. V. (2021). Multilocus phylogenetic analyses suggest the presence of Colletotrichum chrysophilum causing banana anthracnose in Mexico, *Journal of Plant Diseases and Protection* 128(2): 589–595. DOI:10.1007/s41348-020-00396-w.

Gowen, S. (1995). *Banana and Plantain*. Chapman and Hall, London, UK. pp. 337–417.

Griffee, P. J. and Burden, O. J. (1976). Fungi associated with crown rot of boxed bananas in the Windward Islands, *Journal of Phytopathology* 85(2): 149–158.

Guyon, M. (1970). Essais de lutte chimique contre la "Johnson fruit spot" au Nicaragua, *Fruits* 25: 685–691.

Guzmán, M. and Villalta, R. (2008). Efficacy of azoxystrobin and thiabendazole mixtures in the control of post-harvest diseases in bananas (Musa AAA), *CORBANA* 34(61): 1–10.

Hailu, M., Workneh, T. S. and Belew, D. (2013). Review on postharvest technology of banana fruit, *African Journal of Biotechnology* 12(7): 635–647. DOI:10.5897/AJBX12.020.

Hernández, J., Sala, L. and Gallo, L. (1984). Crown rot del plátano. I. Patogenicidad de cepas y resistencia a productos. In: Sociedad Española de Fitopatología (ed.). Comunicaciones III Congreso de Fitopatología. 29 octubre-2 noviembre, Tenerife. pp. 256–261.

Huang, R., Sun, W., Wang, L., Li, Q., Huang, S., Tang, L., Guo, T., Mo, J. and Hsiang, T. (2021). Identification and characterization of Colletotrichum species associated with anthracnose disease of banana, *Plant Pathology* 70(8): 1827–1837. DOI:10.1111/ppa.13426.

Indrakeerthi, S. R. P. and Adikaram, N. K. B. (2011). Control of crown rot of banana using *Carica papaya* latex, *Journal of the National Science Foundation of Sri Lanka* 39(2): 155–162.

Intan Sakinah, M. A., Suzianti, I. V. and Latiffah, Z. (2013). First report of *Colletotrichum gloeosporioides* causing anthracnose of banana (Musa spp.) in Malaysia, *Plant Disease* 97(7): 991. DOI:10.1094/PDIS-10-12-0985-PDN.

Janisiewicz, W. J. and Korsten, L. (2002). Biological control of postharvest diseases of fruits, *Annual Review of Phytopathology* 40: 411–441.

Jitareerat, P. and Uthairatanakij, A. (2013). Integrated control of crown rot disease in 'Kluai Hom Thong' banana. In: *Acta Horticulturae* International Postharvest Symposium Abdullah, H., M. N. and Latifah (Eds) 1012: 719–725. DOI: 10.17660/ActaHortic.2013.1012.96.

Jitareerat, P., Srilaong, V., Srepong, K. and Aoki, S. (2010). Effects of the Combined Treatments of chitinase Producing Bacterium, 1-MCP and Active Packaging on crown Rot Disease and qualities of Banana 'Kluai Hom Thong'. VI, *Acta Horticulturae* International Postharvest Symposium Erkan, M., Aksoy, U. (Eds) 877: 1457–1464. DOI: 10.17660/ActaHortic.2010.877.199

Jones, D. R. (1991). Chemical control of crown rot in Queensland bananas, *Australian Journal of Experimental Agriculture* 31(5): 693–698.

Jones, D. R. and Muirhead, I. F. (2019). Fungal disease of banana fruit: postharvest Diseases. In: *Handbook of Diseases of Banana, Abaca and Enset* Jones, D. R. (Ed.). Available at: https://lccn.loc.gov/2018034301. CABI Publishing, Boston. pp. 271-294.

Jones, D. R. and Stover, R. H. (2019). Fungal disease of banana fruit: preharvest Diseases. In: *Handbook of Diseases of Banana, Abaca and Enset* Jones, D. R. (Ed.). CABI Publishing, Boston. pp. 255–271. Available at: https://lccn.loc.gov/2018034301

Jullien, A., Malezieux, E., Michaux-Ferriere, N., Chillet, M. and Ney, B. (2001). Within-bunch variability in banana fruit weight: importance of developmental lag between fruits, *Annals of Botany* 87(1): 101–108. DOI:10.1006/anbo.2000.1309.

Kanapathipillai, V. S., Ahmad, R. and Mahamad, M. I. (1987). The effect of sterile filtrates of *Trichoderna* spp. and *Penicillium* spp. and 4 KGy irradiation on the spore germination of *Colletotrichum* musae. In: *Movements of Pests and Control Strategies* Singh, K. G., Manalo, P. L., Sastrontomo, S. S., Chan, K. C., Lim, L. G., Ganapathi, A. N., Rahim, M. A. A., Durai, P. S. S. and Doss, M. C. (Eds). Association of Southeast Asian Nations Plant and Quarantine Centre and Training Institute, Kuala Lumpur, Malaysia. pp. 283–292.

Kamel, M. A. M., Cortesi, P. and Saracchi, M. (2016). Etiological agents of crown rot of organic bananas in Dominican Republic, *Postharvest Biology and Technology* 120: 112–120. DOI:10.1016/j.postharvbio.2016.06.002.

Khan, S. H., Aked, J. and Magan, N. (2001). Control of the anthracnose pathogen of banana (*Colletotrichum musae*) using antioxidants alone and in combination with thiabendazole or imazalil, *Plant Pathology* 50(5): 601–608. DOI:10.1046/j.1365-3059.2001.00599.x.

Kindo, A. J., Subramanian, A. J., Subramanian, A. and Suresh, K. (2014). *Nigrospora sphaerica* causing corneal ulcer in an immunocompetent woman: a case report, *International Journal of Case Reports and Images* 5(10): 675–679.

Krauss, U. (1995). Evaluation of natural substances as alternatives for crown rot control, *NRI Crown Rot Project F0048: R5697*, WINBAN R&D Division, St. Lucia.

Krauss, U. (1996). *Evaluation of LonlifeTM and Propolis as Organic Alternatives to Fungicidal Control of Crown Rot, NRI Crown Rot Project F0048, R5697*, WIBDECO Technical Services Division, St. Lucial.

Krauss, U. and Johanson, A. (2000). Recent advances in the control of crown rot of banana in the Windward Islands, *Crop Protection* 19(3): 151–159.

Kulkarni, S. A., Sellamuthu, P. S., Nagarajan, S. K., Madhavan, T. and Sadiku, E. R. (2022). Antifungal activity of wild bergamot (*Monarda fistulosa*) essential oil against postharvest fungal pathogens of banana fruits, *South African Journal of Botany* 144: 166–174. DOI:10.1016/j.sajb.2021.08.019.

Kumah, P., Ampomah, E. O., Olympio, N. S. and Moses, E. (2013). Efficacy of four botanicals and two chemical fungicides in the control of crown rot disease of banana (Musa spp. AAA) 'medium Cavendish', *Acta Horticulturae* (1007): 385–392.

Kumar, V. S., Nair, B. A., Nair, P. V. R., Annamalai, A., Jaishanker, R., Umamaheswaran, K., Sooraj, N. P. and Peethambaran, C. K. (2017). First Report of Colletotrichum siamense Causing Anthracnose of Cliff Banana in India, *Plant Disease* 101(2): 390. DOI:10.1094/PDIS-07-16-0961-PDN.

Kuyu, C. G. and Tola, Y. B. (2018). Assessment of banana fruit handling practices and associated fungal pathogens in Jimma town market, southwest Ethiopia, *Food Science and Nutrition* 6(3): 609–616. DOI:10.1002/fsn3.591.

Lassois, L. and de Lapeyre de Bellaire, L. (2014). Crown rot disease of bananas. In: *Postharvest Decay. Control Strategies* Bautista-Baños, S. (Ed.). Academic Press, CA. pp. 103–130. DOI:10.1016/B978-0-12-411552-1.00003-X.

Lassois, L., Bastiaanse, H., Chillet, M., Jullien, A., Jijakli, M. H. and de Lapeyre de Bellaire, L. (2010). Hand position on the bunch and source-sink ratio influence the banana susceptibility to crown rot disease, *Annals of Applied Biology* 156(2): 221-229.

Lassois, L., de Clerck, C., Frettinger, P., de Lapeyre De Bellaire, L., Lepoivre, P. and Haissam Jijakli, M. (2011a). Catecholamine biosynthesis pathway potentially involved in banana defense mechanisms to crown rot disease, *Communications in Agricultural and Applied Biological Sciences* 76(4): 591-601.

Lassois, L., Frettinger, P., de Lapeyre de Bellaire, L., Lepoivre, P. and Jijakli, H. (2011b). Identification of genes involved in the response of banana to crown rot disease, *Molecular Plant-Microbe Interactions* 24(1): 143-153.

Lassois, L., de Lapeyre de Bellaire, L. and Jijakli, M. H. (2008). Biological control of crown rot of bananas with *Pichia anomala* strain K and *Candida oleophila* strain O, *Biological Control* 45(3): 410-418. DOI:10.1016/j.biocontrol.2008.01.013.

Lassois, L., Jijakli, M. H., Chillet, M. and de Lapeyre de Bellaire, L. (2010). Crown rot of bananas: pre-harvest factors involved in post-harvest disease development and integrated control methods, *Plant Disease* 94(6): 648-658.

Lee, Y. A., Shiao, Y. Y. and Chao, C. P. (2003). First report of Burkholderia cepacia as a pathogen of banana finger-tip rot in Taiwan, *Plant Disease* 87(5): 601. DOI:10.1094/PDIS.2003.87.5.601A:0.1094.

Lobo, M. G. and Fernández-Rojas, F. J. (2020). *Biology and Postharvest Physiology of Banana, In: Handbook of Banana Production, Postharvest Science, Processing Technology, and Nutrition* Siddiq, M., Ahmed, J. and Lobo, M. G. (Eds). John Wiley & Sons Ltd, Hoboken. pp. 19-44. DOI:10.1002/9781119528265.ch2.

Lobo, M. G. and Montero-Calderón, M. (2020). Harvesting and postharvest technology of banana. In: *Handbook of Banana Production, Postharvest Science, Processing Technology, and Nutrition* Siddiq, M., Ahmed, J. and Lobo, M. G. (Eds). John Wiley & Sons Ltd, Hoboken, NJ. pp. 61-80. DOI:10.1002/9781119528265.ch4.

Lobo, M. G., Díaz-Delgado, G. L., Álvarez-Acosta, C., García Luque, M., Pérez-Lozano, R. and Domínguez-Palarea, M. E. (2022). Evaluación de la eficacia de productos postcosecha alternativos a los fungicidas de síntesis para el control de la pudrición de corona (crown rot) del plátano. Informe técnico del proyecto MAC2/1.1B/310 FRUTTMAC. 44p. Available at: https://fruttmac.com/publicaciones/.

Lobo, M. G., Pomar, M. and Marrero, A. (2000). Simulación de la cadena comercial del plátano de Canarias, In: *V Simposio Nacional y. Ibérico de Post Recolección de Frutos y Hortalizas. Tenerife* Lobo, M. G. y Marrero, A. (Eds) (vol. II). pp.215-218.

López-Cabrera, J. and Marrero-Domínguez, A. (1998). Use of hot water dips to control the incidence of banana crown rot, *Acta Horticulturae* (490): 563-570. DOI:10.17660/ActaHortic.1998.490.60.

Lukezic, F. L. and Kaiser, W. J. (1966). Aerobiology of Fusarium roseum 'Gibbosum' associated with crown rot of boxed bananas, *Phytopathology* 56: 545-548.

Lukezic, F. L., Kaiser, W. J. and Martinez, M. M. (1967). The incidence of crown rot of boxed bananas in relation to microbial populations of the crown tissues, *Canadian Journal of Botany* 45(4): 413-421.

Lurie, S. (1998). Postharvest heat treatments, *Postharvest Biology and Technology* 14(3): 257-269.

Marin, D. H., Sutton, T. B., Blankenship, S. M. and Swallow, W. H. (1996). Pathogenicity of fungi associated with crown rot of bananas in Latin America on Grande Naine and disease resistant hybrid bananas, *Plant Disease* 80(5): 525–528.

Marriott, J. and Palmer, J. K. (1980). Bananas - Physiology and biochemistry of storage and ripening for optimum quality, *Critical Reviews in Food Science and Nutrition* 13(1): 41–88.

McLennan, E. I. and Hoëtte, S. (1933). *Nigrospora musae* n. sp. and its connexion with 'squirter' disease in bananas, *Bulletin of the Australian Institute of Scientific and Industrial Research* 75: 1–36.

Meredith, D. S. (1960). Studies on *Gloeosporium musarum* Cke. and Massee causing storage rots of Jamaican bananas. I. Anthracnose and its chemical control, *Annals of Applied Biology* 48(2): 279–290. DOI:10.1111/j.1744-7348.1960.tb03524.x.

Meredith, D. S. (1961a). Fruit-spot ('speckle') of Jamaican bananas caused by *Deightoniella torulosa* (Syd.) Ellis: I, *Transactions of the British Mycological Society* 44(1): 95–IN10. DOI:10.1016/S0007-1536(61)80012-0.

Meredith, D. S. (1961b). Atmospheric content of *nigrospora* spores in Jamaican banana plantations, *Journal of General Microbiology* 26: 343–349.

Meredith, D. S. (1962a). Some components of the air-spora in Jamaican banana plantations, *Annals of Applied Biology* 50(3): 577–594.

Meredith, D. S. (1962b). Some fungi on decaying banana leaves in Jamaica, *Transactions of the British Mycological Society* 45(3): 335–347.

Meredith, D. S. (1963). Fruit-spot ('speckle') of Jamaican bananas caused by *Deightoniella torulosa* (Syd.) Ellis, *Annals of Applied Biology* 52(1): 55–62. DOI:10.1111/j.1744-7348.1963.tb03727.x.

Meredith, D. S. (1971). Transport and storage diseases of bananas, biology and control, *Tropical Agriculture (Trinidad)* 48: 35–50.

Moraes, W. S., Zambolim, L. and Lima, J. D. (2006). Incidência de fungos em pós-colheira de banana (Musa spp.) 'Prata anã' (AAB), *Summa Phytopathologica* 32(1): 67–70. DOI:10.1590/S0100-54052006000100010.

Muirhead, I. F. and Deverall, B. J. (1981). Role of appressoria in latent infection of banana fruits by *Colletotrichum musae*, *Physiological Plant Pathology* 19(1): 77–IN28.

Muirhead, I. F. and Jones, D. R. (2000). Fungal diseases of banana fruit. Postharvest diseases. In: *Diseases of Banana* Jones, D. R. (Ed.). Abaca and Enset. CABI Publishing, Wallingford. pp. 190–206.

Nikagolla, N. G. D. N., Udugala-Ganehenege, M. Y. and Daundasekera, W. A. M. (2019). Postharvest application of potassium silicate improves keeping quality of banana, *Journal of Horticultural Science and Biotechnology* 94(6): 735–743. DOI:10.1080/14620316.2019.1614486.

Nolasco, C. A. and Salomão, L. C. C., Cecon, P. R., Bruckner, C. H. and Rocha, A. (2008). Eficiência da Hidrotermia no Controle de Podridões Pós-Colheita da Banana 'Prata', *Revista Brasileira de Armazenamento* 33(2): 134–139.

Nuratika, T. S. M., Ding, P., Jugah, K. and Hasanah, M. G. (2017). Potential of UVC germicidal irradiation in suppressing crown rot disease, improving postharvest quality and antioxidant capacity of Musa AAA 'Berangan' during fruit ripening, *Food Science and Nutrition* 5(5): 967–980. DOI:10.1002/fsn3.482.

Odame, A. E. (2010). Occurrence and pathogenicity of crown rot disease organisms in major banana producing Areas in Ashanti Region, Thesis (MSc. Postharvest physiology degree) Kwame Nkrumah University of Science and Technology. p. 112. Available at: http://hdl.handle.net/123456789/405.

Ootani, M. A., dos Santos, G. R., Maciel, G. Pd. S., de Souza, C. M., Aguiar, R. W. D. and Coutinho, I. B. D. (2017). Agronomic characterization and resistance of banana genotypes to *Colletotrichum musae*, *Bioscience Journal* 33(6): 1506–1512.

Palou, L., Ali, A., Fallik, E. and Romanazzi, G. (2016). GRAS, plant-and animal-derived compounds as alternatives to conventional fungicides for the control of postharvest diseases of fresh horticultural produce, *Postharvest Biology and Technology* 122: 41–52.

Pathania, N., Vawdrey, L. and Grice, C. (2017). Cause of internal finger tip rot of banana in far north Queensland. In: Proceedings of the Australasian Plant Pathology Society Biennial Conference, Science Protecting Plant Health, 23–29 September 2017, Brisbane, Australia. Poster 227.

Peacock, B. C.and Muirhead, I. F. (1974). Ethylene production by *Colletotrichum musae*, *Queensland Journal of Agricultural and Animal Sciences* 31: 249–252.

Perera, S. D., Hernández, J. M., Piedra, A. and Duque, M. (2012). Eficacia in vitro del sorbato potásico sobre el crecimiento miceliar de diez especies fúngicas componentes del complejo de la pudrición de corona (crown rot) del plátano. Available at: http://www.agrocabildo.org/publicaciones.asp.

Peres, N. A. R., Kurumae, E. E., Dias, M. S. C. and De Souza, N. L. (2002). Identification and characterization of Colletotrichum spp., affecting fruit after harvest in Brazil, *Journal of Phytopathology* 150(3): 128–134.

Perez Vicente, L. and Hernandez, A. (2002). Reaction of black sigatoka resistant FHIA hybrids to *Fusarium pallidoroseum* and *Colletotrichum musae* causal agents of crown rot and anthracnose diseases of bananas, *Phytopathology* 92: S136.

Perez Vicente, L., Sáenz, M., Milanés, M., López, M. O. and Mauri, F. (2001). Pudrición de la corona de los bananos en Cuba. Etiología y dinámica de las especies de hongos asociadas, *Fitosanidad* 5(4): 15–20.

Pinho, D. B., Mizobutsi, E. H., Silva, E., S. O., Mizobutsi, G. P., Xavier, A. A., Ribeiro, R. C. F. and Maia, V. M. (2010). Avaliação de genótipos de bananeira à *Colletotrichum musae* em póscolheita, *Revista Brasileira de Fruticultura* 32(3): 786–790.

Piña, G., Laborem Escalona, G., Surga, J., Marín, C., Rangel, L., Espinoza, M. and Delgado, A. (2006). Atributos de calidad en frutos de híbridos FHIA (Musa) para tres ciclos de cosecha, *Revista de la Facultad de Agronomía* 23(4): 429–447. Available at: http://ve.scielo.org/scielo.php?script=sci_arttext&pid=S0378-78182006000400006&lng=es&tlng=es.

Ploetz, R. C., Thomas, J. E. and Slabaugh, W. (2003). Disease of banana and plantain. In: *Diseases of Tropical Fruit Crops* Ploetz, R. C. (Ed.). CABI Publishing International, London. pp. 73–134.

Qiao, Y., Xu, L., Xu, G., Cao, Y., Gao, Y., Wang, Y. and Feng, J. (2022). Efficacy and potential mechanism of hinokitiol against postharvest anthracnose of banana caused by Colletotrichum musae. *LWT* 161: 113334. https://doi.org/10.1016/j.lwt.2022.113334.

Ranasinghe, L. S., Jayawardena, B. and Abeywickrama, K. (2002). Fungicidal activity of essential oils of Cinnamomum zeylanicum (L.) and Syzygium aromaticum (L.) Merr et LMPerry against crown rot and anthracnose pathogens isolated from banana, *Letters in Applied Microbiology* 35(3): 208–211.

Ranasinghe, L. S., Jayawardena, B. and Abeywickrama, K. (2003). Use of waste generated from cinnamon bark oil (*Cinnamomum zeylanicum* Blume) extraction as a postharvest treatment for Embul banana, *Journal of Food, Agriculture and Environment* 1(2): 340–344.

Ranasinghe, L. S., Jayawardena, B. and Abeywickrama, K. (2005). An integrated strategy to control post-harvest decay of Embul banana by combining essential oils with modified atmosphere packaging, *International Journal of Food Science and Technology* 40(1): 97–103. DOI:10.1111/j.1365-2621.2004.00913.x.

Riera, N., Ramirez-Villacis, D., Barriga-Medina, N., Alvarez-Santana, J., Herrera, K., Ruales, C. and Leon-Reyes, A. (2019). First Report of Banana Anthracnose Caused by Colletotrichum gloeosporioides in Ecuador, *Plant Disease* 103(4): 763–763. DOI:10.1094/PDIS-01-18-0069-PDN.

Rivarez, M. P. S., Parac, E. P., Laurel, N. R., Cunanan, B. V., Magarro, A. B. and Dimasingkil, S. F. M. (2021). Antagonistic activity of bacteria from wild honey against *Colletotrichum musae*, and testing of wild honey as biopesticide spray to control banana anthracnose, *Bioscience Journal* 37: e37089. DOI:10.14393/BJ-v37n0a2021-49921.

Rivera-Pastrana, D. M., Gardea-Bejar, A. A., Martinez-Tellez, M. A., Rivera-Dominguez, M. and Gonzales-Aguilar, G. A. (2007). Efectos bioquímicos postcosecha de la irradiación UV-C en frutas y hortalizas, *Revista Fitotecnia Mexicana* 30(4): 361–372.

Rizwana, H. (2018). Postharvest control of anthracnose lesions and its causative agent, Colletotrichum musae by some oils, *Cellular and Molecular Biology* 64(4): 52–58. DOI:10.14715/cmb/2018.64.4.9.

Reyes, M. E. Q., Nishijima, W. and Paull, R. E. (1998). Control of crown rot in 'Santa Catarina Prata' and 'Williams' banana with hot water treatments, *Postharvest Biology and Technology* 14(1): 71–75.

Sakinah, M. A. I., Suzianti, I. V. and Latiffah, Z. (2014). Phenotypic and molecular characterization of *Colletotrichum* species associated with anthracnose of banana (Musa spp) in Malaysia, *Genetics and Molecular Research* 13(2): 3627–3637.

Salazar, E., Hernández, R., Tapia, A. and Gómez-Alpízar, L. (2012). Identificación molecular del hongo *Colletotrichum* spp., aislado de banano (Musa spp) de altura en la zona de Turrialba y determinación de su sensibilidad a fungicidas poscosecha, *Agronomía Costarricense* 36(1): 53–68. Available at: http://www.redalyc.org/articulo.oa?id=43623289006.

Scribano, F. R. and Garcete, V. (2016). Eficiencia de fungicidas de síntesis y orgánicos sobre la pudrición de corona del fruto de banano Musa acuminataColla en la provincia de Formosa, Argentina, *Revista de Investigaciones Agropecuarias* 42(2): 201–206.

Sepiah, M. and Nik Mohd, N. A. M. (1987). Effects of benomyl and prochloraz on post-harvest diseases of banana, *Asean Food Journal* 3: 101–104.

Shillingford, C. A. (1977). Control of banana fruit rots and of fungi that contaminate washing water, *Tropical Science* 19(4): 197–203.

Shillingford, C. A. and Sinclair, J. B. (1977). Susceptibility of five banana cultivars to anthracnose and crown rotting fungi, *Plant Disease* 61: 797–801.

Silva, L., Maia, E., Barbosa, M. G., Fernandes, M. B., Ribeiro, R. C. F. and Mizobutsi, E. H. (2016). Temporal progress and control of anthracnose on banana in semiarid in the north of Minas Gerais, *Revista Brasileira de Fruticultura* 38: 81–91.

Simmonds, J. H. (1963). Studies in the latent phase of *Colletotrichum* species causing ripe rots of tropical fruits, *Queensland Journal of Agricultural and Animal Sciences* 20: 373–424.

Sinthuja, S. and Prasannath, K. (2020). Management of postharvest anthracnose of banana using inorganic salts alone and in combination with hot water, *Plant Pathology and Quarantine* 10(1): 44–52. DOI:10.5943/ppq/10/1/5.

Siriwardana, H., Abeywickrama, K. and Kannangara, S. (2016). Efficacy of Basil oil and alum in controlling Crown rot disease of Cavendish banana (Musa acuminata, AAA) during cold storage, *International Journal of Applied Engineering Research* 2(5): 1224–1239.

Siriwardana, H., Abeywickrama, K., Kannangara, S., Jayawardena, B. and Attanayake, S. (2017). Basil oil plus aluminium sulfate and modified atmosphere packaging controls Crown rot disease in Embul banana (*Musa acuminata*, AAB) during cold storage, *Scientia Horticulturae* 217: 84–91. DOI:10.1016/j.scienta.2017.01.032.

Slabaugh, W. R. (1994). Crown mold, crown rot and pedicel rot. In: *Compendium of Tropical Fruit Diseases* Ploetz, R. C., G. A., Zentmyer, W. T., Nishijima, K. G., Rohrbach, H. D. and Ohr (Eds). APS Press, Minnesota. pp. 8–9.

Slabaugh, W. R. and Grove, M. D. (1982). Post-harvest diseases of bananas and their control, *Plant Disease* 66(1): 746–750. DOI:10.1094/PD-66-746.

Sponholz, C., Batista, U. G., Zambolim, L., Salomão, L. C. C. and Cardoso, A. A. (2004). Thermotherapy of banana 'Prata' to control post-harvest anthracnose, *Fitopatologia Brasileira* 29(5): 480–485.

Stover, R. H. (1972). *Banana, Plantain and Abaca Diseases*. Commonwealth Mycological Institute, Kew, UK, 316 pp.

Sutton, B. C. (1980). *The Coelomycetes*. Commonwealth Mycological Institute, Kew.

Sutton, B. C. (1992). The genus *Glomerella* and its anamorph *Colletotrichum*. In: *Colletotrichum: Biology, Pathology and Control* Bailey, J. A., M. J. and Jeger (Eds). CAB International, Wallingford. pp. 1-26.

Thangavelu, R., Sangeetha, G. and Mustaffa, M. M. (2007). Cross-infection potential of crown rot pathogen (*Lasiodiplodia theobromae*) isolates and their management using potential native bioagents in banana, *Australasian Plant Pathology* 36(6): 595–605. DOI:10.1071/AP07072.

Thangavelu, R., Sundararaju, P. and Sathiamoorthy, S. (2004). Management of anthracnose disease of banana caused by *Colletotrichum musae* using plant extracts, *Journal of Horticultural Science and Biotechnology* 79(4): 664–668. DOI:10.1080/14620316.2004.11511823.

Triest, D. and Hendrickx, M. (2016). Postharvest disease of banana caused by *Fusarium musae*: A public health concern?, *PLOS Pathogens* 12(11): e1005940. DOI:10.1371/journal.ppat.1005940.

Umana-Rojas, G., Garcia, J. and Arauz, L. F. (2011). Influence of production system and weather variables in the incidence and severity of crown rot in bananas, *Acta Horticulturae* 906: 197–204.

Uribe-Palacio, S., Ramírez-Sáncheza, M., Umaña-Rojas, G. and Sáenz-Murillo, M. V. (2022). Studies on the potential for treatment with short wave ultraviolet light (UV-C) to reduce postharvest diseases in banana fruit crown (*Musa* sp., group AAA, subgroup Cavendish), *Fruits* 77(1): 1–10. DOI:10.17660/th2022/005.

Uysal, A. and Kurt, Ş. (2020). First report of *Colletotrichum siamense* causing anthracnose on banana fruits in Turkey, *Journal of Plant Pathology* 102(3): 967. DOI:10.1007/s42161-020-00534-1.

Vawdrey, L. L., Langdon, P. and Westerhuis, D. (2010). Aetiology, effect of chemicals, and influence of fruit exudates, insects and fruit maturity on the incidence of fruit speckle of banana, *Australasian Plant Pathology* 39(6): 524–529. DOI:10.1071/AP09098.

Vawdrey, L. L., Langdon, P., Westerhuis, D. and Astridge, D. (2008). *The Cause Distribution and Economic Importance of Fruit Speckle of Banana in North Queensland. Final Report Project Number: BA05001*. Horticulture Australia Limited, Sydney. p. 34.

Vieira, W. A. S., Lima, W. G., Nascimento, E. S., Michereff, S. J., Câmara, M. P. S. and Doyle, V. P. (2017). The impact of phenotypic and molecular data on the inference of *Colletotrichum* diversity associated with Musa, *Mycologia* 109(6): 912–934.

Vilaplana, R., Hurtado, G. and Valencia-Chamorro, S. (2018). Hot water dips elicit disease resistance against anthracnose caused by *Colletotrichum* musae in organic bananas (*Musa acuminata*), *LWT* 95: 247–254. DOI:10.1016/j.lwt.2018.04.085.

Waliullah, S., Fonsah, E. G., Brock, J., Li, Y. and Ali, M. E. (2022). First report of crown rot of banana caused by *Fusarium proliferatum* in Georgia, U.S.A., *Plant Disease* 106(5): 1526. DOI:10.1094/PDIS-08-21-1809-PDN.

Wang, M., Liu, F., Crous, P. W. and Cai, L. (2017). Phylogenetic reassessment of *nigrospora: ubiquitous* endophytes, plant and human pathogens, *Persoonia* 39: 118–142. DOI:10.3767/persoonia.2017.39.06.

Wardlaw, C. W. (1961). *Banana Diseases: Including Plantains and Abaca*, Longmans, London.

Wills, R., McGlasson, W. B., Graham, D. and Joyce, D. C. (1998). *Postharvest. An Introduction to the Physiology and Handling of Fruit, Vegetables and Ornamentals*. CABI Publishing, Wallingford.

Win, N. K. K., Jitareerat, P., Kanlayanarat, S. and Sangchote, S. (2007). Effects of cinnamon extract, chitosan coating, hot water treatment and their combinations on crown rot disease and quality of banana fruit, *Postharvest Biology and Technology* 45(3): 333–340.

Yang, H. R. and Chuang, T. Y. (1996). The nutrient effect on the appressorial behaviour and latent infection of *Colletotrichum musae*, *Plant Protection Bulletin (Taiwan)* 3: 247–259.

Yañez, L., Armenta, M., Mercado, E., Yahia, E. M. and Gutierrez, P. (2004). Integral handling of banana. In: *Production Practices and Quality Assessment of Food Crops (vol. 3) 'Quality Handling and Evaluation'* Dris, R. and S. M. Jain (Eds). Kluwer Publishers, The Netherlands. pp. 129–168.

Zahid, N., Ali, A., Manickam, S., Siddiqui, Y. and Maqbool, M. (2012). Potential of chitosan-loaded nanoemulsions to control different *Colletotrichum* spp. and maintain quality of tropical fruits during cold storage, *Journal of Applied Microbiology* 113(4): 925–939. DOI:10.1111/j.1365-2672.2012.05398.x.

Zakaria, L., Sahak, S., Zakaria, M. and Salleh, B. (2009). Characterisation of *Colletotrichum* species associated with anthracnose of banana, *Tropical Life Sciences Research* 20(2): 119–125.

Zhang, Y., Liu, F., Wang, B., Qiu, D., Liu, J., Wu, H., Cheng, C., Bei, X. and Lü, P. (2022). First report of *Burkholderia cepacia* causing finger-tip rot on banana fruit in the Guangxi Province of China, *Plant Disease* 1979: PDIS05211083PDN. DOI:10.1094/PDIS-05-21-1083-PDN.

Zhimo, V. Y., Dilip, D., Sten, J., Ravat, V. K., Bhutia, D. D., Panja, B. and Saha, J. (2017). Antagonistic yeasts for biocontrol of the banana postharvest anthracnose pathogen *Colletotrichum musae*, *Journal of Phytopathology* 165(1): 35–43. DOI:10.1111/jph.12533.

Zhou, Y., Huang, J. S., Yang, L. Y., Wang, G. F. and Li, J. Q. (2017). First report of banana anthracnose caused by *Colletotrichum scovillei* in China, *Plant Disease* 101(2): 381. DOI:10.1094/PDIS-08-16-1135-PDN.

Zoeir, H. A., El Zahaby, Z. H. M., Ziedan, E. H. and Maswada, H. F. (2017). Efficacy of antifungal activity of essential oils, salts and antioxidants acids on pathogenic fungi and their application methods for controlling postharvest diseases on banana fruits, *African Journal of Microbiology Research* 11(29): 1162–1170.

Index